MW00655652

HEAT CONDUCTION

HEAT CONDUCTION

Second Edition

M. NECATI ÖZIŞIK
Department of Mechanical and Aerospace Engineering
North Carolina State University
Raleigh, North Carolina

A Wiley-Interscience Publication
JOHN WILEY & SONS, INC.
New York · Chichester · Brisbane · Toronto · Singapore

This text is printed on acid-free paper.

Library of Congress Cataloging in Publication Data:

Özişik, M. Necati.
Heat conduction / M. Necati Özisik. — 2nd ed.
p. cm.
Includes bibliographical references and index.
ISBN 0-471-53256-8 (cloth : alk. paper)
1. Heat—Conduction. I. Title.
QC321.034 1993
621.402′2—dc20 92-26905

Printed in the United States of America

10 9 8 7 6 5 4 3 2 1

To Gül

CONTENTS

PREFACE

In preparing the second edition of this book, the changes have been motivated by the desire to make this edition a more application-oriented book than the first one in order to better address the needs of the readers seeking solutions to heat conduction problems without going through the details of various mathematical proofs. Therefore, emphasis is placed on the understanding and use of various mathematical techniques needed to develop exact, approximate, and numerical solutions for a broad class of heat conduction problems. Every effort has been made to present the material in a clear, systematic, and readily understandable fashion. The book is intended as a graduate-level textbook for use in engineering schools and a reference book for practicing engineers, scientists and researchers. To achieve such objectives, lengthy mathematical proofs and developments have been omitted, instead examples are used to illustrate the applications of various solution methodologies.

During the twelve years since the publication of the first edition of this book, changes have occurred in the relative importance of some of the application areas and the solution methodologies of heat conduction problems. For example, in recent years, the area of inverse heat conduction problems (IHCP) associated with the estimation of unknown thermophysical properties of solids, surface heat transfer rates, or energy sources within the medium has gained significant importance in many engineering applications. To answer the needs in such emerging application areas, two new chapters are added, one on the theory and application of IHCP and the other on the formulation and solution of moving heat source problems. In addition, the use of enthalpy method in the solution of phase-change problems has been expanded by broadening its scope of applications. Also, the chapters on the use of Duhamel's method, Green's function, and finite-difference methods have been revised in order to make them application-oriented. Green's function formalism provides an efficient, straightforward approach for developing exact analytic solutions to a broad class of heat conduction problems in the rectangular, cylindrical, and spherical coordinate systems, provided that appropriate Green's functions are available. Green's functions needed for use in such formal solutions are constructed by utilizing the tabulated eigenfunctions, eigenvalues and the normalization integrals presented in the tables in Chapters 2 and 3.

Chapter 1 reviews the pertinent background material related to the heat conduction equation, boundary conditions, and important system parameters. Chapters 2, 3, and 4 are devoted to the solution of time-dependent homogeneous heat conduction problems in the rectangular, cylindrical, and spherical coordinates, respectively, by the application of the classical method of separation of variables and orthogonal expansion technique. The resulting eigenfunctions, eigenconditions, and the normalization integrals are systematically tabulated for various combinations of the boundary conditions in Tables 2-2, 2-3, 3-1, 3-2, and 3-3. The results from such tables are used to construct the Green functions needed in solutions utilizing Green's function formalism.

Chapters 5 and 6 are devoted to the use of Duhamel's method and Green's function, respectively. Chapter 7 presents the use of Laplace transform technique in the solution of one-dimensional transient heat conduction problems.

Chapter 8 is devoted to the solution of one-dimensional, time-dependent heat conduction problems in parallel layers of slabs and concentric cylinders and spheres. A generalized orthogonal expansion technique is used to solve the homogeneous problems, and Green's function approach is used to generalize the analysis to the solution of problems involving energy generation.

Chapter 9 presents approximate analytical methods of solving heat conduction problems by the integral and Galerkin methods. The accuracy of approximate results are illustrated by comparing with the exact solutions. Chapter 10 is devoted to the formulation and the solution of moving heat source problems, while Chapter 11 is concerned with the exact, approximate, and numerical methods of solution of phase-change problems.

Chapter 12 presents the use of finite difference methods for solving the steady-state and time-dependent heat conduction problems. Chapter 13 introduces the use of integral transform technique in the solution of general time-dependent heat conduction equations. The application of this technique for the solution of heat conduction problems in rectangular, cylindrical, and spherical coordinates requires no additional background, since all basic relationships needed for constructing the integral transform pairs have already been developed and systematically tabulated in Chapters 2 to 4. Chapter 14 presents the formulation and methods of solution of inverse heat conduction problems and some background information on statistical material needed in the inverse analysis. Finally, Chapter 15 presents the analysis of heat conduction in anisotropic solids. A host of useful information, such as the roots of transcendental equations, some properties of Bessel functions, and the numerical values of Bessel functions and Legendre polynomials are included in Appendixes IV and V for ready reference.

I would like to express my thanks to Professors J. P. Bardon and Y. Jarny of University of Nantes, France, J. V. Beck of Michigan State University, and Woo Seung Kim of Hanyang University, Korea, for valuable discussions and suggestions in the preparation of the second edition.

Raleigh, North Carolina
December 1992

M. NECATI ÖZIŞIK

HEAT CONDUCTION

1

HEAT CONDUCTION FUNDAMENTALS

The energy given up by the constituent particles such as atoms, molecules, or free electrons of the hotter regions of a body to those in cooler regions is called *heat*. Conduction is the mode of heat transfer in which energy exchange takes place in solids or in fluids in rest (i.e., no convective motion resulting from the displacement of the macroscopic portion of the medium) from the region of high temperature to the region of low temperature due to the presence of temperature gradient in the body. The heat flow cannot be measured directly, but the concept has physical meaning because it is related to the measurable scalar quantity called *temperature*. Therefore, once the temperature distribution $T(\mathbf{r}, t)$ within a body is determined as a function of position and time, then the heat flow in the body is readily computed from the laws relating heat flow to the temperature gradient. The science of heat conduction is principally concerned with the determination of temperature distribution within solids. In this chapter we present the basic laws relating the heat flow to the temperature gradient in the medium, the differential equation of heat conduction governing the temperature distribution in solids, the boundary conditions appropriate for the analysis of heat conduction problems, the rules of coordinate transformation needed to write the heat conduction equation in different orthogonal coordinate systems, and a general discussion of various methods of solution of the heat conduction equation.

1-1 THE HEAT FLUX

The basic law that gives the relationship between the heat flow and the temperature gradient, based on experimental observations, is generally named after the

French mathematical physicist Joseph Fourier [1], who used it in his analytic theory of heat. For a homogeneous, isotropic solid (i.e., material in which thermal conductivity is independent of direction) the *Fourier law* is given in the form

$$\mathbf{q}(\mathbf{r},t) = -k\nabla T(\mathbf{r},t) \quad \mathrm{W/m^2} \tag{1-1}$$

where the temperature gradient is a vector normal to the isothermal surface, the *heat flux vector* $\mathbf{q}(\mathbf{r},t)$ represents heat flow per unit time, per unit area of the isothermal surface in the direction of the decreasing temperature, and k is called the *thermal conductivity* of the material which is a positive, scalar quantity. Since the heat flux vector $\mathbf{q}(\mathbf{r},t)$ points in the direction of decreasing temperature, the minus sign is included in equation (1-1) to make the heat flow a positive quantity. When the heat flux is in $\mathrm{W/m^2}$ and the temperature gradient in °C/m, the thermal conductivity k has units W/(m·°C). In the rectangular coordinate system, for example, equation (1-1) is written as

$$\mathbf{q}(x,y,z,t) = -\hat{\mathbf{i}}k\frac{\partial T}{\partial x} - \hat{\mathbf{j}}k\frac{\partial T}{\partial y} - \hat{\mathbf{k}}k\frac{\partial T}{\partial z} \tag{1-2}$$

where $\hat{\mathbf{i}}, \hat{\mathbf{j}}$, and $\hat{\mathbf{k}}$ are the unit direction vectors along the x, y, and z directions, respectively. Thus, the three components of the heat flux vector in the x, y, and z directions are given, respectively, by

$$q_x = -k\frac{\partial T}{\partial x}, \qquad q_y = -k\frac{\partial T}{\partial y}, \qquad \text{and} \qquad q_z = -k\frac{\partial T}{\partial z} \tag{1-3a,b,c}$$

Clearly, the heat flow rate for a given temperature gradient is directly proportional to the thermal conductivity k of the material. Therefore, in the analysis of heat conduction, the thermal conductivity of the material is an important property, which controls the rate of heat flow in the medium. There is a wide difference in the thermal conductivities of various engineering materials. The highest value is given by pure metals and the lowest value by gases and vapors; the amorphous insulating materials and inorganic liquids have thermal conductivities that lie in between. To give some idea of the order of magnitude of thermal conductivity for various materials, Fig. 1-1 illustrates the typical ranges. Thermal conductivity also varies with temperature. For most pure metals it decreases with temperature, whereas for gases it increases with increasing temperature. For most insulating materials it increases with increasing temperatures. Figure 1-2 illustrates the effect of temperature on thermal conductivity of materials. At very low temperature approaching absolute zero, thermal conductivity first increases rapidly and then exhibits a sharp descent as shown in Fig. 1-3. A comprehensive compilation of thermal conductivities of materials may be found in references 2–4.

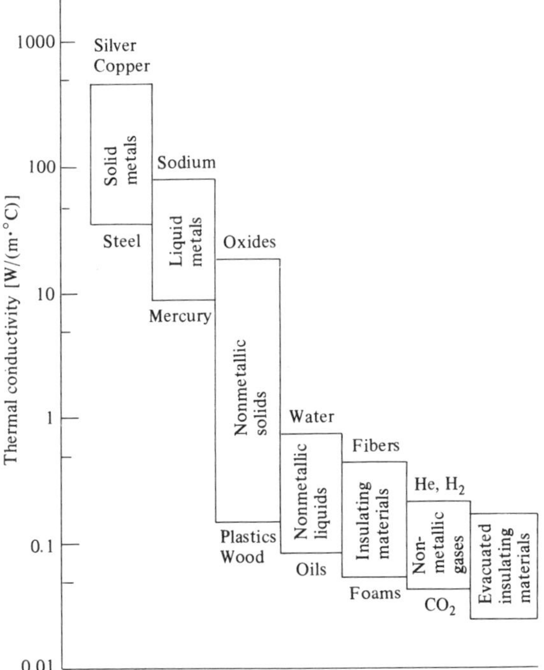

Fig. 1-1 Typical range of thermal conductivity of various materials.

We present in Appendix I the thermal conductivity of typical engineering materials together with the specific heat C_p, density ρ, and the thermal diffusivity α.

1-2 THE DIFFERENTIAL EQUATION OF HEAT CONDUCTION

We now derive the differential equation of heat conduction for a stationary, homogeneous, isotropic solid with heat generation within the body. Heat generation may be due to nuclear, electrical, chemical, γ-ray, or other sources that may be a function of time and/or position. The heat generation rate in the medium, generally specified as heat generation per unit time, per unit volume, is denoted by the symbol $g(\mathbf{r}, t)$, and if SI units are used, is given in the units W/m^3.

We consider the energy-balance equation for a small control volume V, illustrated in Fig. 1-4, stated as

$$\begin{bmatrix} \text{Rate of heat entering through} \\ \text{the bounding surfaces of } V \end{bmatrix} + \begin{bmatrix} \text{rate of energy} \\ \text{generation in } V \end{bmatrix} = \begin{bmatrix} \text{rate of storage} \\ \text{of energy in } V \end{bmatrix} \tag{1-4}$$

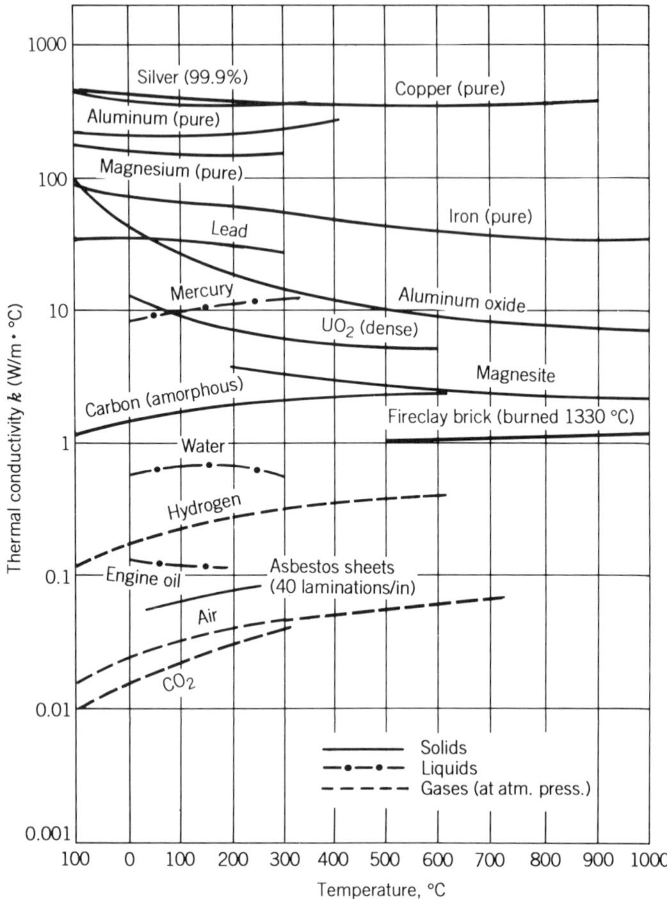

Fig. 1-2 Effect of temperature on thermal conductivity of materials.

Various terms in this equation are evaluated as

$$\left[\begin{array}{c}\text{Rate of heat entering through}\\ \text{the bounding surfaces of } V\end{array}\right] = -\int_A \mathbf{q}\cdot\hat{\mathbf{n}}\,dA = -\int_V \nabla\cdot\mathbf{q}\,dv \qquad \text{(1-5a)}$$

where A is the surface area of the volume element V, $\hat{\mathbf{n}}$ is the outward-drawn normal unit vector to the surface element dA, $\mathbf{q}$ is the heat flux vector at dA; here, the minus sign is included to ensure that the heat flow is into the volume element V, and the divergence theorem is used to convert the surface integral to volume integral. The remaining two terms are evaluated as

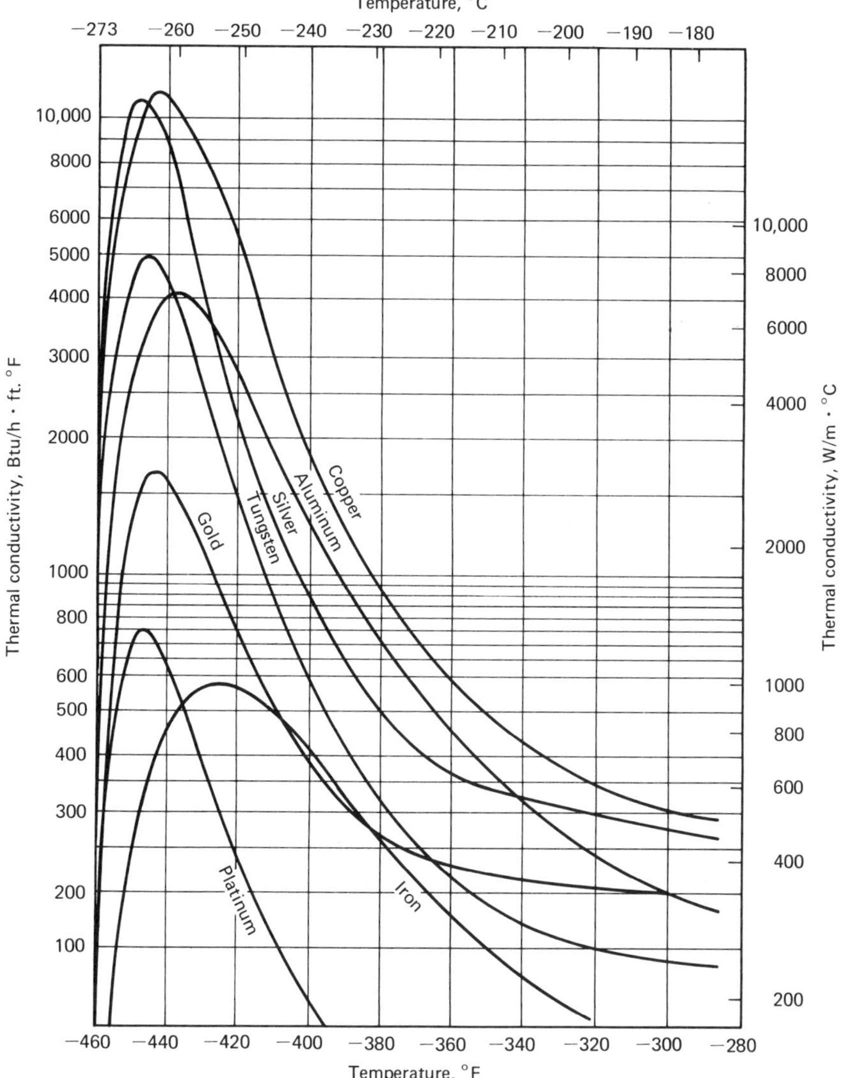

Fig. 1-3 Thermal conductivity of metals at low temperatures. (From Powell et al. [2])

$$(\text{Rate of energy generation in } V) = \int_V g(\mathbf{r}, t)\, dv \tag{1-5b}$$

$$(\text{Rate of energy storage in } V) = \int_V \rho C_p \frac{\partial T(\mathbf{r}, t)}{\partial t}\, dv \tag{1-5c}$$

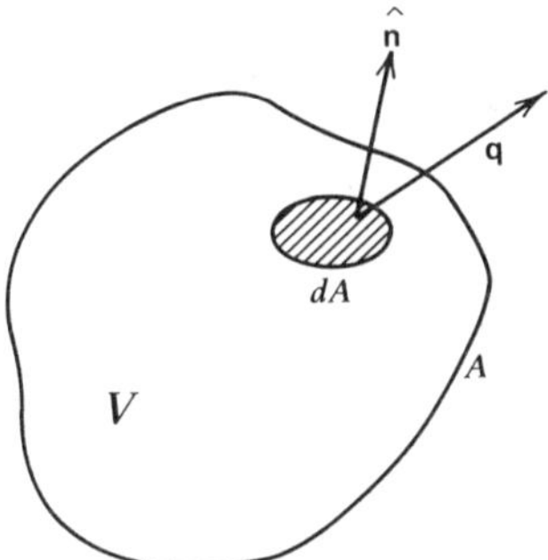

Fig. 1-4 Nomenclature for the derivation of heat conduction equation.

The substitution of equations (1-5) into equation (1-4) yields

$$\int_V \left[-\nabla \cdot \mathbf{q}(\mathbf{r}, t) + g(\mathbf{r}, t) - \rho C_p \frac{\partial T(\mathbf{r}, t)}{\partial t} \right] dv = 0 \tag{1-6}$$

Equation (1-6) is derived for an arbitrary small-volume element V within the solid; hence the volume V may be chosen so small as to remove the integral. We obtain

$$-\nabla \cdot \mathbf{q}(\mathbf{r}, t) + g(\mathbf{r}, t) = \rho C_p \frac{\partial T(\mathbf{r}, t)}{\partial t} \tag{1-7}$$

Substituting $\mathbf{q}(\mathbf{r}, t)$ from equation (1-1) into equation (1-7), we obtain the *differential equation of heat conduction* for a stationary, homogeneous, isotropic solid with heat generation within the body as

$$\nabla \cdot [k \nabla T(\mathbf{r}, t)] + g(\mathbf{r}, t) = \rho C_p \frac{\partial T(\mathbf{r}, t)}{\partial t} \tag{1-8}$$

This equation is intended for temperature or space dependent k as well as temperature dependent C_p. When the thermal conductivity is assumed to be constant (i.e., independent of position and temperature), equation (1-8) simplifies to

$$\nabla^2 T(\mathbf{r}, t) + \frac{1}{k} g(\mathbf{r}, t) = \frac{1}{\alpha} \frac{\partial T(\mathbf{r}, t)}{\partial t} \tag{1-9a}$$

where

$$\alpha = \frac{k}{\rho C_p} = \text{thermal diffusivity} \tag{1-9b}$$

TABLE 1-1 Effect of Thermal Diffusivity on the Rate of Heat Propagation

Material	Silver	Copper	Steel	Glass	Cork
$\alpha \times 10^6\ \mathrm{m^2/s}$	170	103	12.9	0.59	0.155
Time	9.5 min	16.5 min	2.2 h	2.00 days	7.7 days

For a medium with constant thermal conductivity and no heat generation, equations (1-9) become the diffusion or the Fourier equation

$$\nabla^2 T(\mathbf{r}, t) = \frac{1}{\alpha} \frac{\partial T(\mathbf{r}, t)}{\partial t} \tag{1-10}$$

Here, the thermal diffusivity α is the property of the medium and has a dimension of length^2/time, which may be given in the units $\mathrm{m^2/h}$ or $\mathrm{m^2/s}$. The physical significance of thermal diffusivity is associated with the speed of propagation of heat into the solid during changes of temperature with time. The higher the thermal diffusivity, the faster is the propagation of heat in the medium. This statement is better understood by referring to the following specific heat conduction problem: Consider a semiinfinite medium, $x \geqslant 0$, initially at a uniform temperature T_0. For times $t > 0$, the boundary surface at $x = 0$ is kept at zero temperature. Clearly, the temperature in the body will vary with position and time. Suppose we are interested in the time required for the temperature to decrease from its initial value T_0 to half of this value, $\frac{1}{2}T_0$, at a position, say, 30 cm from the boundary surface. Table 1-1 gives the time required for several different materials. It is apparent from these results that the larger the thermal diffusivity, the shorter is the time required for the applied heat to penetrate into the depth of the solid.

1-3 HEAT CONDUCTION EQUATION IN CARTESIAN, CYLINDRICAL, AND SPHERICAL COORDINATE SYSTEMS

The first step in the analytic solution of a heat conduction problem for a given region is to choose an orthogonal coordinate system such that its coordinate surfaces coincide with the boundary surfaces of the region. For example, the rectangular coordinate system is used for rectangular bodies, the cylindrical and the spherical coordinate systems are used for bodies having shapes such as cylinder and sphere, respectively, and so on. Here we present the heat conduction equation for an homogeneous, isotropic solid in the rectangular, cylindrical, and spherical coordinate systems.

Equations (1-8) and (1-9) in the *rectangular coordinate system* (x, y, z), respectively, become

$$\frac{\partial}{\partial x}\left(k\frac{\partial T}{\partial x}\right) + \frac{\partial}{\partial y}\left(k\frac{\partial T}{\partial y}\right) + \frac{\partial}{\partial z}\left(k\frac{\partial T}{\partial z}\right) + g = \rho C_p \frac{\partial T}{\partial t} \tag{1-11a}$$

$$\frac{\partial^2 T}{\partial x^2} + \frac{\partial^2 T}{\partial y^2} + \frac{\partial^2 T}{\partial z^2} + \frac{1}{k} g = \frac{1}{\alpha} \frac{\partial T}{\partial t} \tag{1-11b}$$

Figure 1-5*a,b* show the coordinate axes for the cylindrical (r, ϕ, z) and the spherical (r, ϕ, θ) coordinate systems. In the cylindrical coordinate system equations (1-8) and (1-9), respectively, become

$$\frac{1}{r} \frac{\partial}{\partial r}\left(kr \frac{\partial T}{\partial r} \right) + \frac{1}{r^2} \frac{\partial}{\partial \phi}\left(k \frac{\partial T}{\partial \phi} \right) + \frac{\partial}{\partial z}\left(k \frac{\partial T}{\partial z} \right) + g = \rho C_p \frac{\partial T}{\partial t} \tag{1-12a}$$

$$\frac{1}{r} \frac{\partial}{\partial r}\left(r \frac{\partial T}{\partial r} \right) + \frac{1}{r^2} \frac{\partial^2 T}{\partial \phi^2} + \frac{\partial^2 T}{\partial z^2} + \frac{1}{k} g = \frac{1}{\alpha} \frac{\partial T}{\partial t} \tag{1-12b}$$

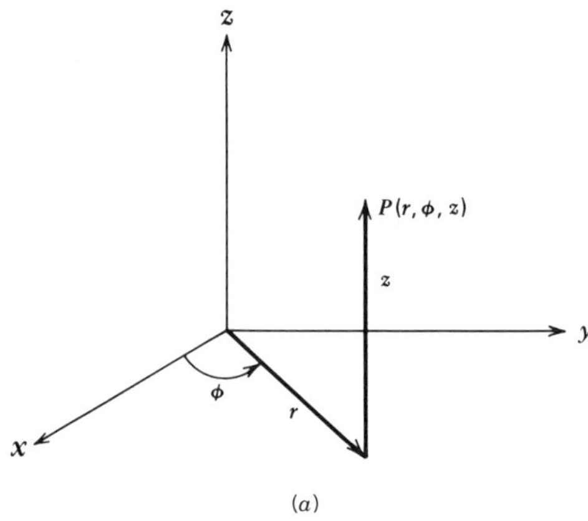

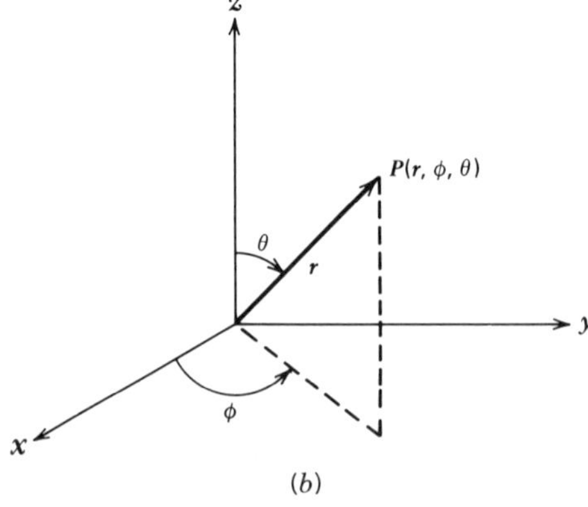

Fig. 1-5 (*a*) Cylindrical coordinate system (r, ϕ, z); (*b*) Spherical coordinate system (r, ϕ, θ).

and in the spherical coordinate system they take the form

$$\frac{1}{r^2}\frac{\partial}{\partial r}\left(kr^2\frac{\partial T}{\partial r}\right)+\frac{1}{r^2\sin\theta}\frac{\partial}{\partial\theta}\left(k\sin\theta\frac{\partial T}{\partial\theta}\right)+\frac{1}{r^2\sin^2\theta}\frac{\partial}{\partial\phi}\left(k\frac{\partial T}{\partial\phi}\right)+g=\rho C_p\frac{\partial T}{\partial t} \tag{1-13a}$$

$$\frac{1}{r^2}\frac{\partial}{\partial r}\left(r^2\frac{\partial T}{\partial r}\right)+\frac{1}{r^2\sin\theta}\frac{\partial}{\partial\theta}\left(\sin\theta\frac{\partial T}{\partial\theta}\right)+\frac{1}{r^2\sin^2\theta}\frac{\partial^2 T}{\partial\phi^2}+\frac{1}{k}g=\frac{1}{\alpha}\frac{\partial T}{\partial t} \tag{1-13b}$$

1-4 HEAT CONDUCTION EQUATION IN OTHER ORTHOGONAL COORDINATE SYSTEMS

In this book we shall be concerned particularly with the solution of heat conduction problems in the rectangular, cylindrical, and spherical coordinate systems; therefore, equations needed for such purposes are immediately obtained from equations (1-11)–(1-13) given above. The heat conduction equations in other *orthogonal curvilinear coordinate systems* (i.e., a coordinate system in which the coordinate lines intersect each other at right angles) are readily obtained by the coordinate transformation. Here we present a brief discussion of the transformation of the heat conduction equation into a general *orthogonal curvilinear coordinate system.* The reader is referred to references 5–7 for further details.

Let u_1, u_2, and u_3 be the three space coordinates, and $\hat{\mathbf{u}}_1, \hat{\mathbf{u}}_2$, and $\hat{\mathbf{u}}_3$ be the unit direction vectors in the u_1, u_2, and u_3 directions in a general orthogonal curvilinear coordinate system shown in Fig. 1-6. A differential length dS in the rectangular coordinate system (x, y, z) is given by

$$(dS)^2=(dx)^2+(dy)^2+(dz)^2 \tag{1-14}$$

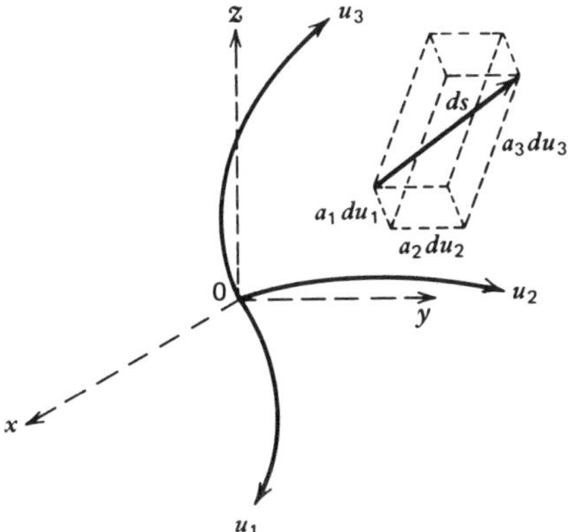

Fig. 1-6 A differential length ds in a curvilinear coordinate system (u_1, u_2, u_3).

Let the functional relationship between orthogonal curvilinear coordinates (u_1, u_2, u_3) and the rectangular coordinates (x, y, z) be given as

$$x = X(u_1, u_2, y_3), \qquad y = Y(u_1, u_2, u_3) \qquad \text{and} \qquad z = Z(u_1, u_2, u_3) \tag{1-15}$$

Then, the differential lengths dx, dy, and dz are obtained from equations (1-15) by differentiation

$$dx = \sum_{i=1}^{3} \frac{\partial X}{\partial u_i} du_i \tag{1-16a}$$

$$dy = \sum_{i=1}^{3} \frac{\partial Y}{\partial u_i} du_i \tag{1-16b}$$

$$dz = \sum_{i=1}^{3} \frac{\partial Z}{\partial u_i} du_i \tag{1-16c}$$

Substituting equations (1-16) into equation (1-14), and noting that the dot products must be zero when u_1, u_2, and u_3 are mutually orthogonal yields the following expression for the differential length dS in the orthogonal curvilinear coordinate system u_1, u_2, u_3

$$(dS)^2 = a_1^2(du_1)^2 + a_2^2(du_2)^2 + a_3^2(du_3)^2 \tag{1-17}$$

where

$$a_i^2 = \left(\frac{\partial x}{\partial u_i}\right)^2 + \left(\frac{\partial y}{\partial u_i}\right)^2 + \left(\frac{\partial z}{\partial u_i}\right)^2, \qquad i = 1, 2, 3 \tag{1-18}$$

Here, the coefficients a_1, a_2, and a_3 are called the *scale factors*, which may be constants or functions of the coordinates. Thus, when the functional relationship between the rectangular and the orthogonal curvilinear system is available [i.e., as in equation (1-15)], then the scale factors a_i are evaluated by equation (1-18).

Once the scale factors are known, the gradient of temperature in the orthogonal curvilinear coordinate system (u_1, u_2, u_3) is given by

$$\nabla T = \hat{\mathbf{u}}_1 \frac{1}{a_1} \frac{\partial T}{\partial u_1} + \hat{\mathbf{u}}_2 \frac{1}{a_2} \frac{\partial T}{\partial u_2} + \hat{\mathbf{u}}_3 \frac{1}{a_3} \frac{\partial T}{\partial u_3} \tag{1-19}$$

The expression defining the heat flux vector $\mathbf{q}$ becomes

$$\mathbf{q} = -k\nabla T = -k \sum_{i=1}^{3} \hat{\mathbf{u}}_i \frac{1}{a_i} \frac{\partial T}{\partial u_i} \tag{1-20}$$

and the three components of the heat flux vector along the u_1, u_2, and u_3

coordinates are given by

$$q_i = -k\frac{1}{a_i}\frac{\partial T}{\partial u_i}, \qquad i = 1, 2, 3 \tag{1-21}$$

The divergence of the heat flux vector $\mathbf{q}$ in the orthogonal curvilinear coordinate system (u_1, u_2, u_3) is given by

$$\nabla \cdot \mathbf{q} = \frac{1}{a}\left[\frac{\partial}{\partial u_1}\left(\frac{a}{a_1}q_1\right) + \frac{\partial}{\partial u_2}\left(\frac{a}{a_2}q_2\right) + \frac{\partial}{\partial u_3}\left(\frac{a}{a_3}q_3\right)\right] \tag{1-22a}$$

where

$$a = a_1 a_2 a_3 \tag{1-22b}$$

The differential equation of heat conduction in a general orthogonal curvilinear coordinate system is now obtained by substituting the results given by equations (1-21) and (1-22) into equation (1-7)

$$\frac{1}{a}\left[\frac{\partial}{\partial u_1}\left(k\frac{a}{a_1^2}\frac{\partial T}{\partial u_1}\right) + \frac{\partial}{\partial u_2}\left(k\frac{a}{a_2^2}\frac{\partial T}{\partial u_2}\right) + \frac{\partial}{\partial u_3}\left(k\frac{a}{a_3^2}\frac{\partial T}{\partial u_3}\right)\right] + g = \rho C_p\frac{\partial T}{\partial t} \tag{1-23}$$

The heat conduction equations in the cylindrical and spherical coordinates given previously by equations (1-12) and (1-13) are readily obtainable as special cases from the general equation (1-23) if the appropriate values of the scale factors are introduced.

Length, Area, and Volume Relations

In the analysis of heat conduction problems integrations are generally required over a length, an area, or a volume. If such an operation is to be performed in an orthogonal curvilinear coordinate system, expressions are needed for a differential length dl, a differential area dA, and a differential volume dV. These relations are determined as now described.

In the case of rectangular coordinate system, a differential volume element dV is given by

$$dV = dx\,dy\,dz \tag{1-24a}$$

and the differential areas dA_x, dA_y, and dA_z cut from the planes $x = \text{constant}$, $y = \text{constant}$, and $z = \text{constant}$ are given, respectively, by

$$dA_x = dy\,dz, \qquad dA_y = dx\,dz, \qquad \text{and} \qquad dA_z = dx\,dy \tag{1-24b}$$

In the case of an orthogonal curvilinear coordinate system, the elementary lengths dl_1, dl_2, and dl_3 along the three coordinate axes u_1, u_2, and u_3 are given, respectively, by

$$dl_1 = a_1\, du_1, \qquad dl_2 = a_2\, du_2, \qquad \text{and} \qquad dl_3 = a_3\, du_3 \tag{1-25a}$$

Then, an elementary volume element dV is expressed as

$$dV = a_1 a_2 a_3\, du_1\, du_2\, du_3 = a\, du_1\, du_2\, du_3, \qquad \text{where} \qquad a \equiv a_1 a_2 a_3 \tag{1-25b}$$

The differential areas dA_1, dA_2, and dA_3 cut from the planes $u_1 = \text{constant}$, $u_2 = \text{constant}$, and $u_3 = \text{constant}$ are given, respectively, by

$$dA_1 = dl_2\, dl_3 = a_2 a_3\, du_2\, du_3, \qquad dA_2 = dl_1\, dl_3 = a_1 a_3\, du_1\, du_3 \qquad \text{and}$$
$$dA_3 = dl_1\, dl_2 = a_1 a_2\, du_1\, du_2 \tag{1-25c}$$

Example 1-1

Determine the scale factors for the cylindrical coordinate system (r, ϕ, z) and write the expressions for the heat flux components.

Solution. The functional relationships between the coordinates (r, ϕ, z) and the rectangular coordinates (x, y, z) are given by

$$x = r\cos\phi, \qquad y = r\sin\phi, \qquad z = z$$

Let

$$u_1 \equiv r, \qquad u_2 = \phi, \qquad \text{and} \qquad u_3 \equiv z \tag{1-26}$$

The scale factors $a_1 \equiv a_r, a_2 \equiv a_\phi$, and $a_3 \equiv a_z$ for the (r, ϕ, z) coordinate system are determined by equation (1-18) as

$$a_1^2 \equiv a_r^2 = \left(\frac{\partial x}{\partial r}\right)^2 + \left(\frac{\partial y}{\partial r}\right)^2 + \left(\frac{\partial z}{\partial r}\right)^2 = \cos^2\phi + \sin^2\phi + 0 = 1$$

$$a_2^2 \equiv a_\phi^2 = \left(\frac{\partial x}{\partial \phi}\right)^2 + \left(\frac{\partial y}{\partial \phi}\right)^2 + \left(\frac{\partial z}{\partial \phi}\right)^2 = (-r\sin\phi)^2 + (r\cos\phi)^2 + 0 = r^2$$

$$a_3^2 \equiv a_z^2 = \left(\frac{\partial x}{\partial z}\right)^2 + \left(\frac{\partial y}{\partial z}\right)^2 + \left(\frac{\partial z}{\partial z}\right)^2 = 0 + 0 + 1 = 1$$

Hence the scale factors for the cylindrical coordinate system become

$$a_r = 1, \qquad a_\phi = r, \qquad a_z = 1, \qquad \text{and} \qquad a = r \tag{1-27a}$$

and the three components of the heat flux are given as

$$q_r = -k\frac{\partial T}{\partial r}, \qquad q_\phi = -\frac{k}{r}\frac{\partial T}{\partial \phi}, \qquad \text{and} \qquad q_z = -k\frac{\partial T}{\partial z} \tag{1-27b}$$

Example 1-2

Determine the scale factors for the spherical coordinate system (r, ϕ, θ).

Solution. The functional relationships between the coordinates (r, ϕ, θ) and the rectangular coordinates (x, y, z) are given by

$$x = r\sin\theta\cos\phi, \qquad y = r\sin\theta\sin\phi, \qquad z = r\cos\theta$$

Let

$$u_1 \equiv r, \qquad u_2 \equiv \phi, \qquad \text{and} \qquad u_3 = \theta \tag{1-28}$$

Then, by utilizing equation (1-18), the scale factors $a_1 \equiv a_r, a_2 \equiv a_\phi$, and $a_3 \equiv a_\theta$ are determined as

$$a_1^2 \equiv a_r^2 = (\sin\theta\cos\phi)^2 + (\sin\theta\sin\phi)^2 + (\cos\theta)^2 = 1$$

$$a_2^2 \equiv a_\phi^2 = r^2\sin^2\theta\sin^2\phi + r^2\sin^2\theta\cos^2\phi + 0 = r^2\sin^2\theta$$

$$a_3^2 \equiv a_\theta^2 = r^2\cos^2\theta\cos^2\phi + r^2\cos^2\theta\sin^2\phi + r^2\sin^2\phi = r^2$$

Hence the scale factors become

$$a_r = 1, \qquad a_\phi = r\sin\theta, \qquad a_\theta = r, \qquad \text{and} \qquad a = r^2\sin\theta \tag{1-29}$$

1-5 GENERAL BOUNDARY CONDITIONS

The differential equation of heat conduction will have numerous solutions unless a set of boundary conditions and an initial condition (for the time-dependent problem) are prescribed. The initial condition specifies the temperature distribution in the medium at the origin of the time coordinate (that is, $t = 0$), and the boundary conditions specify the temperature or the heat flow at the boundaries of the region. For example, at a given boundary surface, the temperature distribution may be prescribed, or the heat flux distribution may be prescribed, or there may be heat exchange by convection and/or radiation with an environment at a prescribed temperature. The boundary condition can be derived by writing an energy balance equation at the surface of the solid.

We consider a surface element having an outward-drawn unit normal vector $\hat{\mathbf{n}}$, subjected to convection, radiation, and external heat supply as illustrated in Fig. 1-7. The physical significance of various heat fluxes shown in this figure is as follows.

The quantity q_{sup} represents energy supplied to the surface, in W/m^2, from an external source.

The quantity q_{conv} represents heat loss from the surface at temperature T by convection with a heat transfer coefficient h into an external ambient at a temperature T_∞, and is given by

$$q_{\text{conv}} = h(T - T_\infty), \qquad \text{W/m}^2 \tag{1-30a}$$

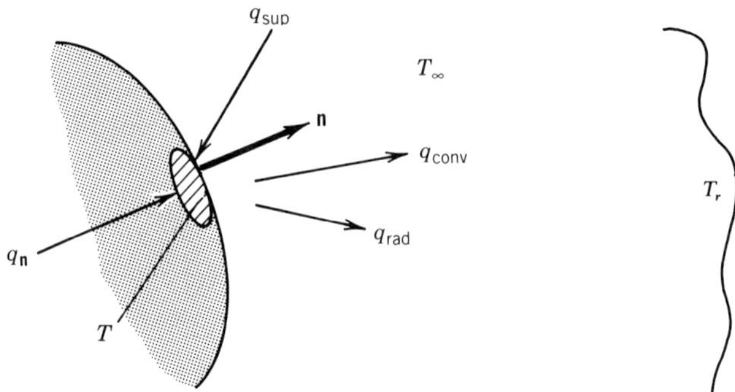

Fig. 1-7 Energy balance at the surface of a solid.

TABLE 1-2 Typical Values of the Convective Heat Transfer Coefficient *h*

Type of Flow	h, W/(m^2·°C)
Free Convection, $\Delta T = 25$°C	
0.25-m vertical plate in	
Atmospheric air	5
Engine oil	37
Water	440
0.02-m-OD horizontal cylinder in	
Atmospheric air	8
Engine oil	62
Water	741
Forced Convection	
Atmospheric air at 25°C with $U_\infty = 10$ m/s over	
$L = 0.1$-m flat plate	40
Flow at 5 m/s across 1-cm-OD cylinder of	
Atmospheric air	85
Engine oil	1,800
Water flow at 1 kg/s inside 2.5-cm-ID tube	10,500
Boiling of Water at 1 atm	
Pool boiling in a container	3,000
Pool boiling at peak heat flux	35,000
Film boiling	300
Condensation of Steam at 1 atm	
Film condensation on horizontal tubes	9,000–25,000
Film condensation on vertical surfaces	4,000–11,000
Dropwise condensation	60,000–120,000

Here the heat transfer coefficient h varies with the type of flow (laminar, turbulent, etc.), the geometry of the body and flow passage area, the physical properties of the fluid, the average temperature, and many others. There is a wide difference in the range of values of the heat transfer coefficient for various applications. Table 1-2 lists the typical values of h, in W/m^2°C, encountered in some applications.

The quantity q_{rad} represents heat loss from the surface by radiation into an ambient at an effective temperature T_r, and is given by

$$q_{\text{rad}} = \epsilon\sigma(T^4 - T_r^4) \quad \text{W/m}^2 \tag{1-30b}$$

where ϵ is the emissivity of the surface and σ is the Stefan–Boltzmann constant, that is, $\sigma \equiv 5.6697 \times 10^{-8}$ W/(m$^2 \cdot$K^4).

The quantity q_n represents the component of the conduction heat flux vector normal to the surface element and is

$$q_n = \mathbf{q} \cdot \hat{\mathbf{n}} = -k\nabla T \cdot \hat{\mathbf{n}} \tag{1-31a}$$

For the Cartesian coordinates we have

$$\nabla T = \hat{\mathbf{i}}\frac{\partial T}{\partial x} + \hat{\mathbf{j}}\frac{\partial T}{\partial y} + \hat{\mathbf{k}}\frac{\partial T}{\partial z} \tag{1-31b}$$

$$\hat{\mathbf{n}} = \hat{\mathbf{i}}l_x + \hat{\mathbf{j}}l_y + \hat{\mathbf{k}}l_z \tag{1-31c}$$

Introducing equations (1-31b,c) into (1-31a), the normal component of the heat flux vector at the surface becomes

$$q_n = -k\left(l_x\frac{\partial T}{\partial x} + l_y\frac{\partial T}{\partial y} + l_z\frac{\partial T}{\partial z}\right) = -k\frac{\partial T}{\partial n} \tag{1-32}$$

where l_x, l_y, and l_z are the direction cosines (i.e., cosine of the angles) of the unit normal vector $\hat{\mathbf{n}}$ with the x, y, and z coordinate axes, respectively. Similar expressions can be developed for the cylindrical and spherical coordinate systems.

To develop the boundary condition, we consider the energy balance at the surface as

$$\text{Heat supply} = \text{heat loss}$$

or

$$q_n + q_{\text{sup}} = q_{\text{conv}} + q_{\text{rad}} \tag{1-33}$$

Introducing the expressions (1-30a,b) and (1-32) into (1-33), the boundary condition becomes

$$-k\frac{\partial T}{\partial n} + q_{\text{sup}} = h(T - T_\infty) + \epsilon\sigma(T^4 - T_r^4) \tag{1-34a}$$

which can be rearranged as

$$k\frac{\partial T}{\partial n}+hT+\epsilon\sigma T^4=hT_\infty+q_{\text{sup}}+\epsilon\sigma T_r^4 \tag{1-34b}$$

where all the quantities on the right-hand side of equation (1-34b) are known and the surface temperature T is unknown.

The general boundary condition given by equations (1-34) is nonlinear because it contains the fourth power of the unknown surface temperature T^4. In addition, the absolute temperatures need to be considered when radiation is involved. If $(|T-T_r|)/T_r \ll 1$, the radiation term can be linearized and equation (1-34a) takes the form

$$-k\frac{\partial T}{\partial n}+q_{\text{sup}}=h(T-T_\infty)+h_r(T-T_r) \tag{1-35a}$$

where the heat transfer coefficient for radiation is defined as

$$h_r\equiv 4\epsilon\sigma T_r^3 \tag{1-35b}$$

1-6 LINEAR BOUNDARY CONDITIONS

In this book, for the analytic solution of linear heat conduction problems, we shall consider the following three different types of linear boundary conditions.

1. *Boundary Condition of the First Kind.* This is the situation when the temperature distribution is prescribed at the boundary surface, that is

$$T=f(\mathbf{r},t) \qquad \text{on} \qquad S \tag{1-36a}$$

where the prescribed surface temperature $f(\mathbf{r},t)$ is, in general, a function of position and time. The special case

$$T=0 \qquad \text{on} \qquad S \tag{1-36b}$$

is called the *homogeneous boundary condition of the first kind.*

2. *Boundary Condition of the Second Kind.* This is the situation in which the heat flux is prescribed at the surface, that is

$$k\frac{\partial T}{\partial n}=f(\mathbf{r},t) \qquad \text{on} \qquad S \tag{1-37a}$$

where $\partial T/\partial n$ is the derivative along the outward drawn normal to the surface.

Here $f(\mathbf{r}, t)$ is the prescribed heat flux, W/m^2. The special case

$$\frac{\partial T}{\partial n} = 0 \qquad \text{on} \qquad S \tag{1-37b}$$

is called the *homogeneous boundary condition of the second kind.*

3. *Boundary Condition of the Third Kind.* This is the convection boundary condition which is readily obtained from equation (1-35a) by setting the radiation term and the heat supply equal to zero, that is

$$k\frac{\partial T}{\partial n} + hT = hT_\infty(\mathbf{r}, t) \qquad \text{on} \qquad S \tag{1-38a}$$

where, for generality, the ambient temperature $T_\infty(\mathbf{r}, t)$ is assumed to be a function of position and time. The special case

$$k\frac{\partial T}{\partial n} + hT = 0 \qquad \text{on} \qquad S \tag{1-38b}$$

is called the *homogeneous boundary condition of the third kind.* It represents convection into a medium at zero temperature. Clearly, the boundary conditions of the first and second kind are obtainable from the boundary condition of the third as special cases if k and h are treated as coefficients. For example, by setting $k = 0$ and $T_\infty(\mathbf{r}, t) \equiv f(\mathbf{r}, t)$, equation (1-38a) reduces to equation (1-36a). Similarly, by setting $hT_\infty(\mathbf{r}, t) \equiv f(\mathbf{r}, t)$ and then letting $h = 0$ on the left-hand side, equation (1-38a) reduces to equation (1-37a).

4. *Interface Boundary Condition.* When two materials having different thermal conductivities k_1 and k_2 are in imperfect contact and have a common boundary as illustrated in Fig. 1-8, the temperature profile through the solids experiences a sudden drop across the interface between the two materials. The physical significance of this temperature drop is envisioned better if we consider an enlarged view of the interface as shown in this figure and note that actual metal-to-metal contact takes place at a limited number of spots and the void between them is filled with air, which is the surrounding fluid. As thermal conductivity of air is much smaller than that of metal, a steep temperature drop occurs across the gap. To develop the boundary condition for such an interface, we write the energy balance as

$$\begin{pmatrix}\text{Heat conduction}\\ \text{thru. solid 1}\end{pmatrix} = \begin{pmatrix}\text{heat transfer}\\ \text{across the gap}\end{pmatrix} = \begin{pmatrix}\text{heat conduction}\\ \text{thru. solid 2}\end{pmatrix} \tag{1-39a}$$

$$-k_1 \left.\frac{\partial T_1}{\partial x}\right|_i = h_c(T_1 - T_2)_i = -k_2 \left.\frac{\partial T_2}{\partial x}\right|_i \tag{1-39b}$$

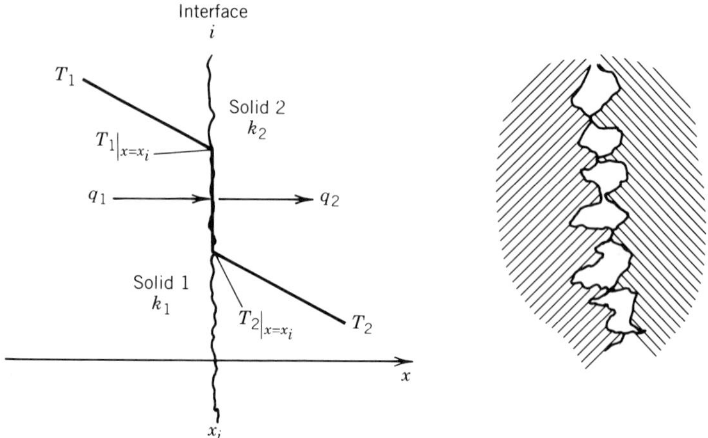

Fig. 1-8 Boundary condition at the interface of two contacting surfaces.

where subscript i denotes the inferface and h_c, in W/(m$^2 \cdot$°C), is called the *contact conductance* for the interface. Equation (1-39b) provides two expressions for the boundary condition at the interface of two contacting solids, and it is generally called the *interface boundary conditions.*

For the special case of *perfect thermal contact* between the surfaces, we have $h_c \to \infty$, and equation (1-39b) reduces to

$$T_1 = T_2 \qquad \text{at} \quad S_i \tag{1-40a}$$

$$-k_1 \frac{\partial T_1}{\partial x} = -k_2 \frac{\partial T_2}{\partial x} \qquad \text{at} \quad S_i \tag{1-40b}$$

where equation (1-40a) is the continuity of temperature, and equation (1-40b) is the continuity of heat flux at the interface.

The experimentally determined values of contact conductance for typical materials in contact can be found in references 8–10. The surface roughness, the interface pressure and temperature, thermal conductivities of the contacting metal and the type of fluid in the gap are the principal factors that affect contact conductance.

To illustrate the effects of various parameters such as the surface roughness, the interface temperature, the interface pressure, and the type of material, we present in Fig. 1-9*a,b* the *interface thermal contact conductance h* for stainless steel-to-stainless steel and aluminum-to-aluminum joints. The results on these figures show that interface conductance increases with increasing interface pressure, increasing interface temperature, and decreasing surface roughness. The interface conductance is higher with a softer material (aluminum) than with a harder material (stainless steel).

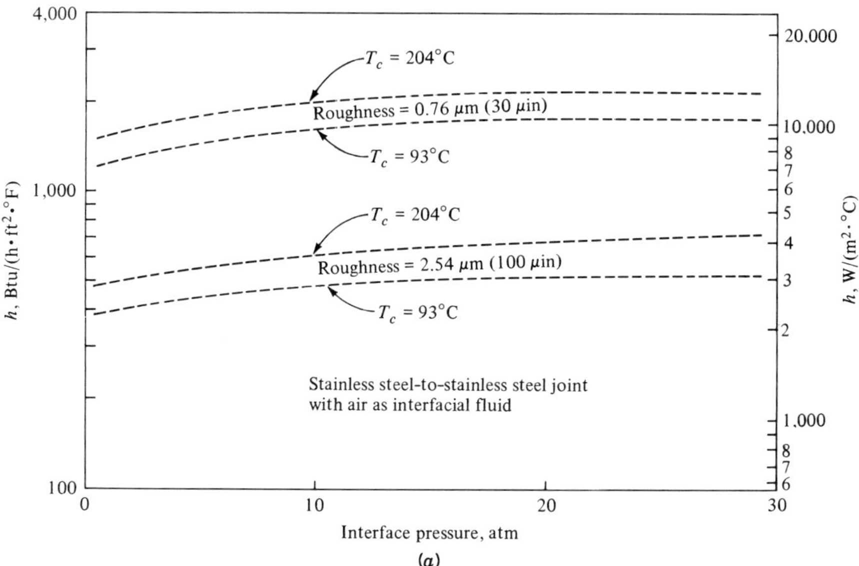

(a)

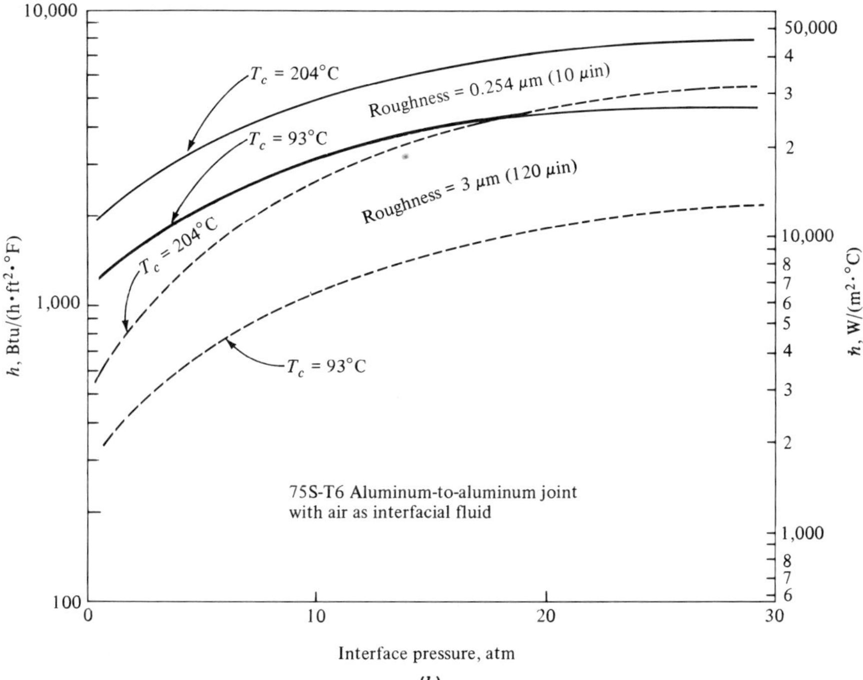

(b)

Fig. 1-9 Effects of interface pressure, contact temperature, and roughness on interface conductance h. (Based on data from reference 8).

The smoothness of the surface is another factor that affects contact conductance; a joint with a superior surface finish may exhibit lower contact conductance owing to waviness. The adverse effect of waviness can be overcome by introducing between the surfaces an interface shim from a soft material such as lead.

Contact conductance also is reduced with a decrease in the ambient-air pressure, because the effective thermal conductance of the gas entrapped in the interface is lowered.

Example 1-3

Consider a plate subjected to heating at the rates of f_1 and f_2, in W/m^2, at the boundary surfaces $x=0$ and $x=L$, respectively. Write the boundary conditions.

Solution. The prescribed heat flux boundary condition is given by equation (1-37a) as

$$k\frac{\partial T}{\partial n}=f \qquad \text{on} \qquad S \tag{1-41}$$

The outward-drawn normal vectors at the boundary surfaces $x=0$ and $x=L$ are in the negative x and positive x directions, respectively. Hence the boundary conditions become

$$-k\frac{\partial T}{\partial x}=f_1 \qquad \text{at} \qquad x=0 \tag{1-42a}$$

$$k\frac{\partial T}{\partial x}=f_2 \qquad \text{at} \qquad x=L \tag{1-42b}$$

Example 1-4

Consider a hollow cylinder subjected to convection boundary conditions at the inner $r=a$ and outer $r=b$ surfaces into ambients at temperatures $T_{\infty 1}$

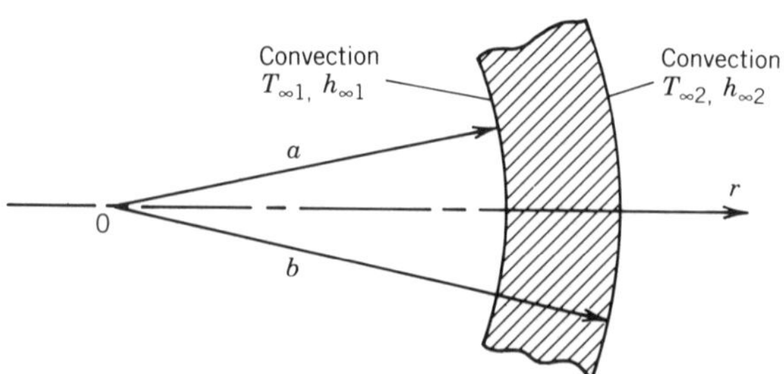

Fig. 1-10 Boundary conditions for Example 1-4.

and $T_{\infty 2}$, with heat transfer coefficients $h_{\infty 1}$ and $h_{\infty 2}$ respectively, as illustrated in Fig. 1-10. Write the boundary conditions.

Solution. The convection boundary condition is given by equation (1-38a) in the form

$$k\frac{\partial T}{\partial n}+hT=hT_{\infty} \qquad \text{at} \qquad S \tag{1-43}$$

The outward-drawn normal at the boundary surfaces $r=a$ and $r=b$ are in the negative r and positive r directions. Hence the boundary condition (1-43) gives

$$-k\frac{\partial T}{\partial r}+h_{\infty 1}T=h_{\infty 1}T_{\infty 1} \qquad \text{at} \qquad r=a \tag{1-44a}$$

$$k\frac{\partial T}{\partial r}+h_{\infty 2}T=h_{\infty 2}T_{\infty 2} \qquad \text{at} \qquad r=b \tag{1-44b}$$

1-7 TRANSFORMATION OF NONHOMOGENEOUS BOUNDARY CONDITIONS INTO HOMOGENEOUS ONES

In the solution of transient heat conduction problems with the orthogonal expansion technique, the contribution of nonhomogeneous terms of the boundary conditions in the solution generally gives rise to convergence difficulties when the solution is evaluated near the boundary. Therefore, whenever possible, it is desirable to transform the nonhomogeneous boundary conditions into homogeneous ones. Here we present a methodology for performing such transformations for some special cases.

We consider one-dimensional transient heat conduction with energy generation and nonhomogeneous convection boundary conditions for a slab, hollow cylinder and sphere given by

$$\frac{1}{x^p}\frac{\partial}{\partial x}\left(x^p\frac{\partial T}{\partial x}\right)+\frac{1}{k}g(x,t)=\frac{1}{\alpha}\frac{\partial T}{\partial t}, \qquad \text{in} \qquad x_0<x<x_L, \qquad t>0 \tag{1-45a}$$

$$-k\frac{\partial T}{\partial x}+h_1T=h_1f_1(t) \qquad \text{at} \qquad x=x_0 \qquad t>0 \tag{1-45b}$$

$$k\frac{\partial T}{\partial x}+h_2T=h_2f_2(t) \qquad \text{at} \qquad x=x_L \qquad t>0 \tag{1-45c}$$

$$T=F(x) \qquad \text{for} \qquad t=0 \qquad x_0\leqslant x\leqslant L \tag{1-45d}$$

where

$$p = \begin{cases} 0 & \text{slab} \\ 1 & \text{cylinder} \\ 2 & \text{sphere} \end{cases} \tag{1-45e}$$

Here, $f_1(t)$ and $f_2(t)$ are the ambient temperatures.

We assume that the temperature $T(x, t)$ can be split up into three components as

$$T(x,t) = \theta(x,t) + \phi_1(x) f_1(t) + \phi_2(x) f_2(t) \tag{1-46}$$

where the dimensionless functions $\phi_1(x)$ and $\phi_2(x)$ are the solutions of the following two steady-state problems

$$\frac{d}{dx}\left(x^p \frac{d\phi_1}{dx}\right) = 0 \qquad \text{in} \qquad x_0 < x < x_L \tag{1-47a}$$

$$-k\frac{d\phi_1}{dx} + h_1\phi_1 = h_1 \qquad \text{at} \qquad x = x_0 \tag{1-47b}$$

$$k\frac{d\phi_1}{dx} + h_2\phi_1 = 0 \qquad \text{at} \qquad x = x_L \tag{1-47c}$$

and

$$\frac{d}{dx}\left(x^p \frac{d\phi_2}{dx}\right) = 0 \qquad \text{in} \qquad x_0 < x < x_L \tag{1-48a}$$

$$-k\frac{d\phi_2}{dx} + h_1\phi_2 = 0 \qquad \text{at} \qquad x = x_0 \tag{1-48b}$$

$$k\frac{d\phi_2}{dx} + h_2\phi_2 = h_2 \qquad \text{at} \qquad x = x_L \tag{1-48c}$$

Then, it can be shown that the function $\theta(x, t)$ is the solution of the following one-dimensional transient heat conduction with homogeneous convection boundary conditions, a modified energy generation term $g^*(x, t)$ and a modified initial condition function $F^*(x)$, given in the form

$$\frac{1}{x^p}\frac{\partial}{\partial x}\left(x^p \frac{\partial\theta}{\partial x}\right) + g^*(x,t) = \frac{1}{\alpha}\frac{\partial\theta}{\partial t} \qquad \text{in} \qquad x_0 < x < x_L, \quad t > 0 \tag{1-49a}$$

$$-k\frac{\partial\theta}{\partial x} + h_1\theta = 0 \qquad \text{at} \qquad x = x_0, \qquad t > 0 \tag{1-49b}$$

$$k\frac{\partial\theta}{\partial x}+h_2\theta=0 \qquad \text{at} \quad x=x_L, \quad t>0 \tag{1-49c}$$

$$\theta=F^*(x) \qquad \text{for} \quad t=0, \quad x_0\leqslant x\leqslant x_L \tag{1-49d}$$

where $g^*(x,t)$ and $F^*(x)$ are defined by

$$g^*(x,t)=\frac{1}{k}g(x,t)-\frac{1}{\alpha}\left(\phi_1(x)\frac{df_1(t)}{dt}+\phi_2(x)\frac{df_2(t)}{dt}\right) \tag{1-50a}$$

$$F^*(x)=F(x)-\{\phi_1(x)f_1(0)+\phi_2(x)f_2(0)\} \tag{1-50b}$$

The validity of the above splitting-up procedure can be verified by introducing equation (1-46) into equations (1-45) and utilizing equations (1-47), (1-48) and (1-49).

The above splitting-up procedure can be extended to the multidimensional problems provided that the nonhomogeneous terms in the boundary conditions do not vary with the position, but may depend on time.

1-8 HOMOGENEOUS AND NONHOMOGENEOUS PROBLEMS

For convenience in the anaysis, the time-dependent heat conduction problems will be considered in two groups: *homogeneous problems* and *nonhomogeneous problems.*

The problem will be referred to *homogeneous* when both the differential equation and the boundary conditions are homogeneous. Thus the problem

$$\nabla^2T=\frac{1}{\alpha}\frac{\partial T}{\partial t} \qquad \text{in region } R, \quad t>0 \tag{1-51a}$$

$$k_i\frac{\partial T}{\partial n_i}+h_iT=0 \qquad \text{on boundary } S_i, \quad t>0 \tag{1-51b}$$

$$T=F(\mathbf{r}) \qquad \text{in region } R, \quad t=0 \tag{1-51c}$$

will be referred to homogeneous because both the differential equation and the boundary condition are homogeneous.

The problem will be referred to *nonhomogeneous* if the differential equation, or the boundary conditions, or both are nonhomogeneous. For example, the problem

$$\nabla^2T+\frac{g(\mathbf{r},t)}{k}=\frac{1}{\alpha}\frac{\partial T}{\partial t} \qquad \text{in region } R, \quad t>0 \tag{1-52a}$$

$$k_i \frac{\partial T}{\partial n_i} + h_i T = f_i(\mathbf{r}, t) \qquad \text{on boundary } S_i, \quad t > 0 \tag{1-52b}$$

$$T = F(\mathbf{r}) \qquad \text{in region } R, \quad t = 0 \tag{1-52c}$$

is nonhomogeneous because the differential equation and the boundary condition are nonhomogeneous.

The problem

$$\nabla^2 T + \frac{g(\mathbf{r}, t)}{k} = \frac{1}{\alpha}\frac{\partial T}{\partial t} \qquad \text{in region } R, \quad t > 0 \tag{1-53a}$$

$$k_i \frac{\partial T}{\partial n_i} + h_i T = 0 \qquad \text{on boundary } S_i, \quad t > 0 \tag{1-53b}$$

$$T = F(\mathbf{r}) \qquad \text{in region } R, \quad t = 0 \tag{1-53c}$$

is also nonhomogeneous because the differential equation is nonhomogeneous.

1-9 HEAT CONDUCTION EQUATION FOR MOVING SOLIDS

So far we considered stationary solids. Suppose the solid is moving with a velocity $\mathbf{u}$ and we have chosen the rectangular coordinate system. Let u_x, u_y, and u_z be the three components of the velocity in the x, y and z direction, respectively. For solids, assuming that ρC_p is *constant*, the motion of the solid is regarded to give rise to convective or enthalpy fluxes

$$\rho C_p T u_x, \qquad \rho C_p T u_y, \qquad \rho C_p T u_z$$

in the x, y, and z directions, respectively, in addition to the conduction fluxes in those directions. With these considerations the components of the heat flux vector $\mathbf{q}$ are taken as

$$q_x = -k\frac{\partial T}{\partial x} + \rho C_p T u_x \tag{1-54a}$$

$$q_y = -k\frac{\partial T}{\partial x} + \rho C_p T u_y \tag{1-54b}$$

$$q_z = -k\frac{\partial T}{\partial z} + \rho C_p T u_z \tag{1-54c}$$

Clearly, on the right-hand sides of these equations, the first term is the conduction flux and the second term is the convection flux due to the motion of the solid. For the case of no motion, equations (1-54) reduces to equations (1-3).

The heat conduction equation for the moving solid is obtained by introducing equations (1-54) into the energy equation (1-7):

$$k\nabla^2 T + g(\mathbf{r}, t) = \rho C_p \left(\frac{\partial T}{\partial t} + u_x \frac{\partial T}{\partial x} + u_y \frac{\partial T}{\partial y} + u_z \frac{\partial T}{\partial z} \right) \tag{1-55}$$

This equation is written more compactly as

$$\alpha \nabla^2 T + \frac{1}{\rho C_p} g(\mathbf{r}, t) = \frac{DT}{Dt} \tag{1-56}$$

which are strictly applicable for constant ρC_p. Here, $\alpha = (k/\rho C_p)$ is the thermal diffusivity and D/Dt is the *substantial* (*or total*) *derivative* defined by

$$\frac{D}{Dt} \equiv \frac{\partial}{\partial t} + u_x \frac{\partial}{\partial x} + u_y \frac{\partial}{\partial y} + u_z \frac{\partial}{\partial z} \tag{1-57}$$

For the case of no motion, equation (1-56) reduces to equations (1-9).

1-10 HEAT CONDUCTION EQUATION FOR ANISOTROPIC MEDIUM

So far we considered the heat flux law for isotropic media, that is, thermal conductivity k is independent of direction, and developed the heat conduction equation accordingly. However, there are natural as well as synthetic materials in which thermal conductivity varies with direction. For example, in a tree trunk the thermal conductivity may vary with direction; that is, the thermal conductivities along the grain and across the grain are different. In laminated sheets the thermal conductivity along and across the laminations are not the same. Other examples include sedimentary rocks, fibrous reinforced structures, cables, heat shielding for space vehicles, and many others.

Orthotropic Medium

First we consider a situation in the rectangular coordinates in which the thermal conductivities k_x, k_y, and k_z in the x, y, and z directions, respectively, are different. Then the heat flux vector $\mathbf{q}(x, y, z, t)$ given by equation (1-2) is modified as

$$\mathbf{q}(x, y, z, t) = -\left(\hat{\mathbf{i}} k_x \frac{\partial T}{\partial x} + \hat{\mathbf{j}} k_y \frac{\partial T}{\partial y} + \hat{\mathbf{k}} k_z \frac{\partial T}{\partial z} \right) \tag{1-58}$$

and the three components of the heat flux vector in the x, y, and z directions, respectively, become

$$q_x = -k_x \frac{\partial T}{\partial x}, \qquad q_y = -k_y \frac{\partial T}{\partial y} \qquad \text{and} \qquad q_z = -k_z \frac{\partial T}{\partial z} \tag{1-59}$$

Similar relations can be written for the heat flux components in the cylindrical and spherical coordinates. The materials in which thermal conductivity vary in the (x, y, z) or (r, θ, z) or (r, θ, ϕ) directions are called *orthotropic* materials. The heat conduction equation for an orthotropic medium in the rectangular coordinate system is obtained by introducing the heat flux vector given by equation (1-58) into equation (1-7). We find

$$\frac{\partial}{\partial x}\left(k_x \frac{\partial T}{\partial x}\right) + \frac{\partial}{\partial y}\left(k_y \frac{\partial T}{\partial y}\right) + \frac{\partial}{\partial z}\left(k_z \frac{\partial T}{\partial z}\right) + g = \rho C_p \frac{\partial T}{\partial t} \tag{1-60}$$

Thus thermal conductivity has three distinct components.

Anisotropic Medium

In a more general situation encountered in heat flow through *crystals*, at any point in the medium, each component q_x, q_y, and q_z of the heat flux vector is considered a linear combination of the temperature gradients $\partial T/dx, \partial T/dy$, and $\partial T/dz$, that is

$$q_x = -\left(k_{11} \frac{\partial T}{\partial x} + k_{12} \frac{\partial T}{\partial y} + k_{13} \frac{\partial T}{\partial z}\right) \tag{1-61a}$$

$$q_y = -\left(k_{21} \frac{\partial T}{\partial x} + k_{22} \frac{\partial T}{\partial y} + k_{23} \frac{\partial T}{\partial z}\right) \tag{1-61b}$$

$$q_z = -\left(k_{31} \frac{\partial T}{\partial x} + k_{32} \frac{\partial T}{\partial y} + k_{33} \frac{\partial T}{\partial z}\right) \tag{1-61c}$$

Such a medium is called an *anisotropic* medium and the thermal conductivity for such a medium has nine components, k_{ij}, called the *conductivity coefficients* that are considered to be the components of a second-order tensor $\bar{\bar{k}}$:

$$\bar{\bar{k}} \equiv \begin{vmatrix} k_{11} & k_{12} & k_{13} \\ k_{21} & k_{22} & k_{23} \\ k_{31} & k_{32} & k_{33} \end{vmatrix} \tag{1-62}$$

Crystals are typical example of anisotropic material involving nine conductivity coefficients [11,12]. The heat conduction equation for anisotropic solids in the rectangular coordinate system is obtained by introducing the expressions for the three components of heat flux given by equations (1-61) into the energy equation (1-7). We find

$$k_{11}\frac{\partial^2 T}{\partial x^2}+k_{22}\frac{\partial^2 T}{\partial y^2}+k_{33}\frac{\partial^2 T}{\partial z^2}+(k_{12}+k_{21})\frac{\partial^2 T}{\partial x\partial y}+(k_{13}+k_{31})\frac{\partial^2 T}{\partial x\partial z}$$

$$+(k_{23}+k_{32})\frac{\partial^2 T}{\partial y\partial z}+g(x,y,z,t)=\rho C_p\frac{\partial T(x,y,z,t)}{\partial t} \tag{1-63}$$

where $k_{12}=k_{21}, k_{13}=k_{31}$, and $k_{23}=k_{32}$ by the reciprocity relation. This matter will be discussed further in Chapter 15.

1-11 LUMPED SYSTEM FORMULATION

The transient heat conduction formulations considered previously assume temperature varying both with time and position. There are many engineering applications in which the variation of temperature within the medium can be neglected and temperature is considered to be a function of time only. Such formulations, called *lumped system formulation*, provide great simplification in the analysis of transient heat conduction; but their range of applicability is very restricted. Here we illustrate the concept of lumped formulation approach and examine its range of validity.

Consider a small, high-conductivity material, such as a metal, initially at a uniform temperature T_0, suddenly immersed into a well-stirred hot bath maintained at a uniform temperature T_∞. Let V be the volume, A the surface area, ρ density, C_p specific heat of the solid, and h the heat transfer coefficient between the solid surface and the fluid. We assume that the temperature distribution within the solid remains sufficiently uniform for all times due to its small size and high thermal conductivity. Then the temperature $T(t)$ of the solid can be considered to be a function of time only. The energy-balance equation on the solid is stated as

$$\begin{pmatrix}\text{Rate of heat flow into the}\\ \text{solid through its boundaries}\end{pmatrix}=\begin{pmatrix}\text{rate of increase of the}\\ \text{internal energy of the solid}\end{pmatrix} \tag{1-64}$$

When the appropriate mathematical expressions are written, the energy equation (1-64) takes the form

$$hA[T_\infty - T(t)]=\rho C_p V\frac{dT(t)}{dt} \tag{1-65}$$

which is rearranged as

$$\frac{dT(t)}{dt} + \frac{hA}{\rho C_p V}[T(t) - T_\infty] = 0 \qquad \text{for} \qquad t > 0 \tag{1-66a}$$

$$T(t) = T_0 \qquad \text{for} \qquad t = 0 \tag{1-66b}$$

A temperature excess $\theta(t)$ is defined as

$$\theta(t) = T(t) - T_\infty \tag{1-67}$$

Then, the lumped formulation becomes

$$\frac{d\theta(t)}{dt} + m\theta(t) = 0 \qquad \text{for} \qquad t > 0 \tag{1-68a}$$

$$\theta(t) = T_0 - T_\infty = \theta_0 \qquad \text{for} \qquad t = 0 \tag{1-68b}$$

where

$$m = \frac{hA}{\rho C_p V} \tag{1-68c}$$

and the solution is given by

$$\frac{\theta(t)}{\theta_0} = e^{-mt} \tag{1-69}$$

This is a very simple expression for temperature varying with time and the parameter m has the unit of $(\text{time})^{-1}$.

The physical significance of the parameter m is better envisioned if its definition is rearranged in the form

$$\begin{aligned} \frac{1}{m} &= (\rho C_p V)\left(\frac{1}{hA}\right) \\ &= (\text{thermal capacitance})\begin{pmatrix}\text{external thermal} \\ \text{resistance}\end{pmatrix} \end{aligned} \tag{1-70}$$

Then, the smaller is the thermal capacitance or the external thermal resistance, the larger is the value of m, and hence the faster is the rate of change of temperature $\theta(t)$ of the solid according to equation (1-69).

In order to establish some criteria for the range of validity of such a simple method for the analysis of transient heat conduction, we consider the definition of the Biot number Bi, and rearrange it in the form

$$\mathrm{Bi} = \frac{hL}{k_s} = \frac{(L/k_sA)}{(1/hA)} = \frac{\begin{pmatrix}\text{internal thermal}\\ \text{resistance}\end{pmatrix}}{\begin{pmatrix}\text{external thermal}\\ \text{resistance}\end{pmatrix}} \tag{1-71}$$

where k_s = thermal conductivity of the solid and $L = V/A$ = *characteristic length* of the solid.

We recall that the lumped system analysis is applicable if the temperature distribution within the solid remains sufficiently uniform during the transients, whereas the temperature distribution in a solid becomes uniform if the internal resistance of the solid to heat flow is negligible. Now we refer to the above definition of the Biot number and note that the internal thermal resistance of solid is small in comparison to the external thermal resistance if the Biot number is small. Therefore, we conclude that the lumped system analysis is valid only for small values of the Biot number. For example, exact analytic solutions of transient heat conduction for solids in the form of a slab, cylinder or sphere, subjected to convective cooling show that for $\mathrm{Bi} < 0.1$, the variation of temperature within the solid during transients is less than 5%. Hence it may be concluded that the lumped system analysis may be applicable for most engineering applications if the Biot number is less than about 0.1.

Example 1-5

The temperature of a gas stream is to be measured with a thermocouple. The junction may be approximated as a sphere of diameter $D = \frac{3}{4}$ mm, $k = 30$ W/(m·°C), $\rho = 8400$ kg/m^3 and $C_p = 0.4$ kJ/(kg·°C). If the heat transfer coefficient between the junction and the gas stream is $h = 600$ W/(m^2·°C), how long does it take for the thermocouple to record 99% of the temperature difference between the gas temperature and the initial temperature of the thermocouple?

Solution. The characteristic length L is

$$L = \frac{V}{A} = \frac{(4/3)\pi r^3}{4\pi r^2} = \frac{r}{3} = \frac{D}{6} = \frac{3/4}{6} = \frac{1}{8}\,\text{mm} = \frac{10^{-3}}{8}\,\text{m}$$

The Biot number becomes

$$\mathrm{Bi} = \frac{hL}{k} = \frac{600}{30}\,\frac{10^{-3}}{8} = 2.5 \times 10^{-3}$$

hence the lumped system analysis is applicable since $\mathrm{Bi} < 0.1$. From equation (1-69) we have

$$\frac{T(t) - T_\infty}{T_0 - T_\infty} = \frac{1}{100} = e^{-mt}$$

or

$$e^{mt} = 100, \qquad mt = 4.6$$

The value of m is determined from its definition

$$m = \frac{hA}{\rho C_p V} = \frac{h}{\rho C_p L} = \frac{600}{8400 \times 400} \frac{8}{10^{-3}} = 1.428 \text{ s}^{-1}$$

Then

$$t = \frac{4.6}{m} = \frac{4.6}{1.428} \cong 3.22 \text{ s}$$

That is, about 3.22 s is needed for the thermocouple to record 99% of the applied temperature difference.

Partial Lumping

In the lumped system analysis described above, we considered a total lumping in all the space variables; as a result, the temperature for the lumped system became a function of the time variable.

It is also possible to perform a *partial* lumping, such that the temperature variation is retained in one of the space variables but lumped in the others. For example, if temperature gradient in a solid is very steep, say, in the x direction and very small in the y and z directions, then it is possible to lump the system in the y and z variables. To illustrate this matter we consider a solid as shown in Fig. 1-11, in which temperature gradients are assumed to be large along the x direction, but small over the y–z plane perpendicular to the x axis. Let the solid dissipate heat by convection from its lateral surfaces into an ambient at a constant temperature T_∞ with a heat transfer coefficient h.

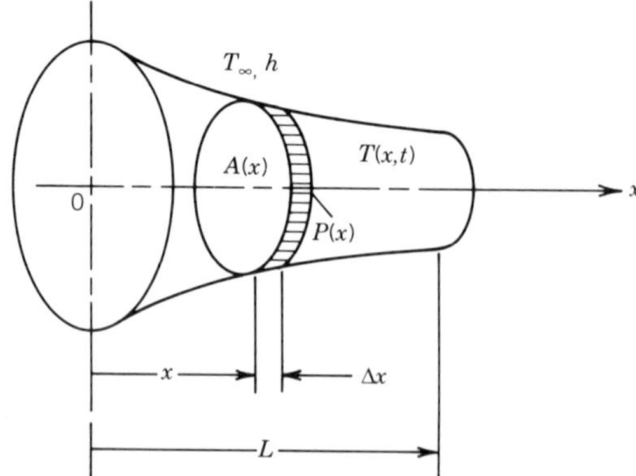

Fig. 1-11 Nomenclature for the derivation of the partially lumped heat conduction equation.

To develop the heat conduction equation with lumping over the plane perpendicular to the x axis, we consider an energy balance for a disk of thickness Δx about the axial location x given by

$$\begin{pmatrix}\text{Net rate of heat}\\ \text{gain by conduction}\\ \text{in the } x \text{ direction}\end{pmatrix} + \begin{pmatrix}\text{rate of heat gain}\\ \text{by convection from}\\ \text{the lateral surfaces}\end{pmatrix} = \begin{pmatrix}\text{rate of increase}\\ \text{of internal energy}\\ \text{of the disk}\end{pmatrix} \tag{1-72}$$

When the appropriate mathematical expressions are introduced for each of these three terms, we obtain

$$-\frac{\partial}{\partial x}(Aq)\Delta x + hp(x)\Delta x[T_\infty - T(x,t)] = \rho C_p \Delta x A(x)\frac{\partial T(x,t)}{\partial t} \tag{1-73a}$$

where the heat flux q is given by

$$q = -k\frac{\partial T(x,t)}{\partial x} \tag{1-73b}$$

and other quantities are defined as

$A(x)$ = cross-sectional area of the disk
$p(x)$ = perimeter of the disk
h = heat transfer coefficient
k = thermal conductivity of the solid
T_∞ = ambient temperature

We introduce a new temperature $\theta(x,t)$ as

$$\theta(x,t) = T(x,t) - T_\infty \tag{1-74}$$

and substitute the expression for q into the energy equation (1-73a). Then equation (1-73a) takes the form

$$\frac{1}{A(x)}\frac{\partial}{\partial x}\left[A(x)\frac{\partial \theta}{\partial x}\right] - \frac{hp(x)}{kA(x)}\theta(x,t) = \frac{1}{\alpha}\frac{\partial \theta(x,t)}{\partial t} \tag{1-75}$$

For the steady state, equation (1-75) simplifies to

$$\frac{d}{dx}\left[A(x)\frac{d\theta(x)}{dx}\right] - \frac{hp(x)}{k}\theta(x) = 0 \tag{1-76}$$

If we further assume that the cross-sectional area $A(x) = A_0$ = constant, equation

(1-76) reduces to

$$\frac{d^2\theta(x)}{\partial x^2} - \frac{hp}{kA_0}\theta(x) = 0 \tag{1-77}$$

which is the fin equation for fins of uniform cross-section.

The solution to the fin equation (1-77) can be constructed in the form

$$\theta(x) = c_1 \cosh mx + c_2 \sinh mx \tag{1-78a}$$

or

$$\theta(x) = c_1^* e^{-mx} + c_2^* e^{mx} \tag{1-78b}$$

The two unknown coefficients are determined by the application of boundary conditions at $x = 0$ and $x = L$, and the solutions can be found in any one of the standard books on heat transfer [13].

The solution of equation (1-76) for fins of variable cross section is more involved. Analytic solutions of fins of various cross sections can be found in the references 14 and 15.

REFERENCES

1. J. B. Fourier, *Theorie Analytique de la Chaleur*, Paris, 1822 (English trans. by A. Freeman, Dover Publications, New York, 1955).
2. R. W. Powell, C. Y. Ho, and P. E. Liley, *Thermal Conductivity of Selected Materials*, NSRDS-NBS 8, U.S. Department of Commerce, National Bureau of Standards, 1966.
3. *Thermophysical Properties of Matter*, Vols. 1–3, 1F1/Plenum Data Corp., New York, 1969.
4. C. Y. Ho, R. W. Powell, and P. E. Liley, *Thermal Conductivity of Elements*, Vol. 1, first supplement to *J. Phys. Chem. Ref. Data* (1972).
5. P. Moon and D. E. Spencer, *Field Theory for Engineers*, Van Nostrand, Princeton, N.J., 1961.
6. M. P. Morse and H. Feshbach, *Methods of Theoretical Physics*, Part I, McGraw-Hill, New York, 1953.
7. G. Arfken, *Mathematical Methods for Physicists*, Academic Press, New York, 1966.
8. M. E. Barzelay, K. N. Tong, and G. F. Holloway, NACA Tech. Note, 3295, May 1955.
9. E. Fried and F. A. Castello, *ARS J.* **32**, 237–243, 1962.
10. H. L. Atkins and E. Fried, AIAA Paper No. 64–253, 1964.
11. W. A. Wooster, *A Textbook in Crystal Physics*, Cambridge University Press, London.
12. J. F. Nye, *Physical Properties of Crystals*, Clarendon Press, London, 1957.
13. M. N. Ozisik, *Heat Transfer*, McGraw-Hill, New York, 1985.
14. D. A. Kern and A. D. Kraus, *Extended Surface Heat Transfer*, McGraw-Hill, New York, 1972.
15. M. D. Mikhailov and M. N. Ozisik, *Unified Analysis and Solutions of Heat and Mass Diffusion*, Wiley, New York, 1984.

PROBLEMS

1-1 Verify that ∇T and $\nabla \cdot \mathbf{q}$ in the cylindrical coordinate system (r, ϕ, z) are given as

$$\nabla T = \hat{\mathbf{u}}_r \frac{\partial T}{\partial r} + \hat{\mathbf{u}}_\phi \frac{1}{r}\frac{\partial T}{\partial \phi} + \hat{\mathbf{u}}_z \frac{\partial T}{\partial z}$$

$$\nabla \cdot \mathbf{q} = \frac{1}{r}\frac{\partial}{\partial r}(r q_r) + \frac{1}{r}\frac{\partial q_\phi}{\partial \phi} + \frac{\partial q_z}{\partial z}$$

1-2 Verify that ∇T and $\nabla \cdot \mathbf{q}$ in the spherical coordinate system (r, ϕ, θ) are given as

$$\nabla T = \hat{\mathbf{u}}_r \frac{\partial T}{\partial r} + \hat{\mathbf{u}}_\phi \frac{1}{r \sin\theta}\frac{\partial T}{\partial \phi} + \hat{\mathbf{u}}_\theta \frac{1}{r}\frac{\partial T}{\partial \theta}$$

$$\nabla \cdot \mathbf{q} = \frac{1}{r^2}\frac{\partial}{\partial r}(r^2 q_r) + \frac{1}{r \sin\theta}\frac{\partial q_\phi}{\partial \phi} + \frac{1}{r \sin\theta}\frac{\partial}{\partial \theta}(q_\theta \sin\theta)$$

1-3 By using the appropriate scale factors in equation (1-23) show that the heat conduction equation in the cylindrical and spherical coordinate systems are given by equations (1-12) and (1-13).

1-4 Obtain expressions for elemental areas dA cut from the surfaces r = constant, θ = constant, and z = constant, also for an elemental volume dV in the cylindrical coordinate system (r, θ, z).

1-5 Repeat Problem 1-4 for the spherical coordinate system (r, ϕ, θ).

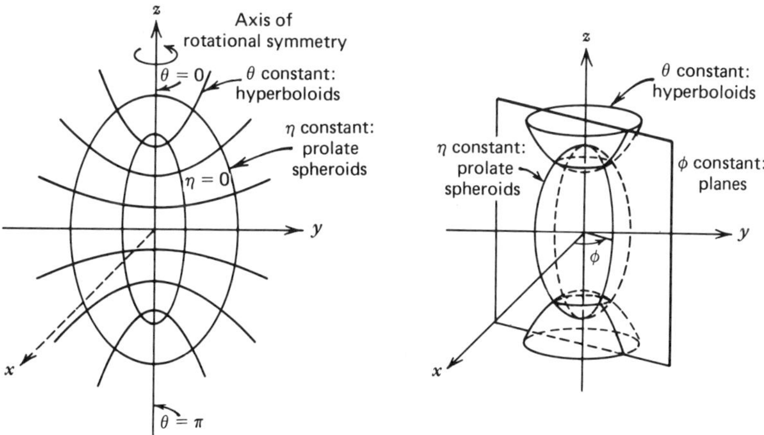

Fig. 1-12 Prolate spheroidal coordinates (η, θ, ϕ).

1-6 The *prolate spheroidal* coordinate system (η, θ, ϕ) as illustrated in Fig. 1-12 consists of prolate spheroids $\eta =$ constant, hyperboloids $\theta =$ constant, and planes $\phi =$ constant. Note that as $\eta \to 0$ spheroids become straight lines of length $2A$ on the z axis and as $\eta \to \infty$ spheroids become nearly spherical. For $\theta = 0$, hyperboloids degenerate into z axis from A to $+\infty$, and for $\theta = \pi$ hyperboloids degenerate into z axis from $-A$ to $-\infty$, and for $\theta = \pi/2$ hyperboloids become the x–y plane. If the coordinates (η, θ, ϕ) of the prolate spheroidal system are related to the rectangular coordinates by

$$x = A \sinh \eta \sin \theta \cos \phi$$

$$y = A \sinh \eta \sin \theta \sin \phi$$

$$z = A \cosh \eta \cos \theta$$

show that the scale factors are given by

$$a_1 \equiv a_\eta = A(\sin^2 \theta + \sinh^2 \eta)^{1/2}$$

$$a_2 \equiv a_\theta = A(\sin^2 \theta + \sinh^2 \eta)^{1/2}$$

$$a_3 \equiv a_\phi = A \sinh \eta \sin \theta$$

1-7 Using the scale factors determined in Problem 1-6, show that the expression for $\nabla^2 T$ in the prolate spheroidal coordinates (η, θ, ϕ) is given as

$$\nabla^2 T = \frac{1}{A^2(\sinh^2 \eta + \sin^2 \theta)} \left[\frac{\partial^2 T}{\partial \eta^2} + \coth \eta \frac{\partial T}{\partial \eta} + \frac{\partial^2 T}{\partial \theta^2} + \cot \theta \frac{\partial T}{\partial \theta} \right] + \frac{1}{A^2 \sinh^2 \eta \sin^2 \theta} \frac{\partial^2 T}{\partial \phi^2}$$

1-8 Obtain expressions for elemental areas dA cut from the surfaces $\eta =$ constant, $\theta =$ constant, and $\phi =$ constant, and also for an elemental volume element dV in the prolate spheroidal coordinate system (η, θ, ϕ) discussed above.

1-9 The coordinates (η, θ, ϕ) of an *oblate spheroidal* coordinate system are related to the rectangular coordinates by

$$x = A \cosh \eta \sin \theta \cos \phi$$

$$y = A \cosh \eta \sin \theta \sin \phi$$

$$z = A \sinh \eta \cos \theta$$

Show that the scale factors are given by

$$a_1^2 \equiv a_\eta^2 = A^2(\cosh^2 \eta - \sin^2 \theta)$$

$$a_2^2 \equiv a_\theta^2 = A^2(\cosh^2 \eta - \sin^2 \theta)$$

$$a_3^2 \equiv a_\phi^2 = A^2 \cosh^2 \eta \sin^2 \theta$$

1-10 Using the scale factors in Problem 1-9, show that the expression for $\nabla^2 T$ in the oblate spheroidal coordinate system (η, θ, ϕ) is given by

$$\nabla^2 T = \frac{1}{A^2(\cosh^2\eta - \sin^2\theta)}\left[\frac{\partial^2 T}{\partial \eta^2} + \tanh\eta\frac{\partial T}{\partial \eta} + \frac{\partial^2 T}{\partial \theta^2} + \cot\theta\frac{\partial T}{\partial \theta}\right] + \frac{1}{A^2\cosh^2\eta\sin^2\theta}\frac{\partial^2 T}{\partial \phi^2}$$

1-11 Show that the following three different forms of the differential operator in the spherical coordinate system are equivalent.

$$\frac{1}{r^2}\frac{d}{dr}\left(r^2\frac{dT}{dr}\right) = \frac{1}{r}\frac{d^2}{dr^2}(rT) = \frac{d^2T}{dr^2} + \frac{2}{r}\frac{dT}{dr}$$

1-12 Set up the mathematical formulation of the following heat conduction problems:

1. A slab in $0 \leqslant x \leqslant L$ is initially at a temperature $F(x)$. For times $t > 0$, the boundary at $x = 0$ is kept insulated and the boundary at $x = L$ dissipates heat by convection into a medium at zero temperature.
2. A semiinifinite region $0 \leqslant x < \infty$ is initially at a temperature $F(x)$. For times $t > 0$, heat is generated in the medium at a constant rate of g_0 W/m^3, while the boundary at $x = 0$ is kept at zero temperature.
3. A solid cylinder $0 \leqslant r \leqslant b$ is initially at a temperature $F(r)$. For times $t > 0$, heat is generated in the medium at a rate of $g(r)$, W/m^3, while the boundary at $r = b$ dissipates heat by convection into a medium at zero temperature.
4. A solid sphere $0 \leqslant r \leqslant b$ is initially at temperature $F(r)$. For times $t > 0$, heat is generated in the medium at a rate of $g(r)$, W/m^3, while the boundary at $r = b$ is kept at a uniform temperature T_0.

1-13 For an anisotropic solid, the three components of the heat conduction vector q_x, q_y and q_z are given by equations (1-61). Write the similar expressions in the cylindrical coordinates for q_r, q_ϕ, q_z and in the spherical coordinates for q_r, q_ϕ, q_θ.

1-14 Prove the validity of the transformation of the heat conduction problem [equation (1-45)] into the three simpler problems given by equations (1-47), (1-48) and (1-49) by using the splitting-up procedure defined by equation (1-46).

1-15 A long cylindrical iron bar of diameter $D = 5$ cm, initially at temperature $T_0 = 650$°C, is exposed to an air stream at $T_\infty = 50$°C. The heat transfer coefficient between the air stream and the surface of the bar is $h =$ 80 W/(m$^2\cdot$°C). Thermophysical properties may be taken as $\rho =$

7800 kg/m^3, $C_p = 460$ J/(kg·°C), and $k = 60$ W/(m·°C). Determine the time required for the temperature of the bar to reach 250°C by using the lumped system analysis.

1-16 A thermocouple is to be used to measure the temperature in a gas stream. The junction may be approximated as a sphere having thermal conductivity $k = 25$ W/(m·°C), $\rho = 8400$ kg/m^3, and $C_p = 0.4$ kJ/(kg·°C). The heat transfer coefficient between the junction and the gas stream is $h = 560$ W/(m^2·°C). Calculate the diameter of the junction if the thermocouple should measure 95% of the applied temperature difference in 3s.

2

THE SEPARATION OF VARIABLES IN THE RECTANGULAR COORDINATE SYSTEM

The method of separation of variables has been widely used in the solution of heat conduction problems. The homogeneous problems are readily handled with this method. The multidimensional steady-state heat conduction problems with no generation can also be solved with this method if only one of the boundary conditions is nonhomogeneous; problems involving more than one nonhomogeneous boundary conditions can be split up into simpler problems each containing only one nonhomogeneous boundary condition. In this chapter we discuss the general problem of the separability of the heat-conduction equation; examine the separation in the rectangular coordinate system; determine the elementary solutions, the norms, and the eigenvalues of the resulting separated equations for different combinations of boundary conditions and present these results systematically in a tabulated form for ready reference; examine the solution of one and multidimensional homogeneous problems by the method of separation of variables; examine the solution of multidimensional steady-state heat conduction problems with and without heat generation; and describe the splitting up of a nonhomogeneous problem into a set of simpler problems that can be solved by the separation of variable technique. The reader should consult references 1–4 for a discussion of the mathematical aspects of the method of separation of variables and references 5–8 for additional applications on the solution of heat conduction problems.

2-1 BASIC CONCEPTS IN THE SEPARATION OF VARIABLES

To illustrate the basic concepts associated with the method of separation of variables we consider a homogeneous boundary-value problem of heat conduction for a slab in $0 \leqslant x \leqslant L$. Initially the slab is at a temperature $T = F(x)$, and for times $t > 0$ the boundary surface at $x = 0$ is kept insulated while the boundary at $x = L$ dissipates heat by convection with a heat-transfer coefficient h into a medium at zero temperature. There is no heat generation in the medium. The mathematical formulation of this problem is given as (see Fig. 2-1)

$$\frac{\partial^2 T(x,t)}{\partial x^2} = \frac{1}{\alpha}\frac{\partial T(x,t)}{\partial t} \quad \text{in} \quad 0 < x < L, \quad t > 0 \tag{2-1a}$$

$$\frac{\partial T}{\partial x} = 0 \quad \text{at} \quad x = 0, \quad t > 0 \tag{2-1b}$$

$$k\frac{\partial T}{\partial x} + hT = 0 \quad \text{at} \quad x = L, \quad t > 0 \tag{2-1c}$$

$$T = F(x) \quad \text{for} \quad t = 0, \quad 0 \leqslant x \leqslant L \tag{2-1d}$$

To solve this problem we assume the separation of function $T(x,t)$ into a space- and time-dependent functions in the form

$$T(x,t) = X(x)\Gamma(t) \tag{2-2}$$

The substituting of equation (2-2) into equation (2-1a) yields

$$\frac{1}{X(x)}\frac{d^2X(x)}{dx^2} = \frac{1}{\alpha\Gamma(t)}\frac{d\Gamma(t)}{dt} \tag{2-3}$$

In this equation, the left-hand side is a function of the space variable x, alone, and the right-hand side of the time variable t, alone; the only way this equality holds if both sides are equal to the same constant, say $-\beta^2$; thus, we have

$$\frac{1}{X(x)}\frac{d^2X(x)}{dx^2} = \frac{1}{\alpha\Gamma(t)}\frac{d\Gamma(t)}{dt} = -\beta^2 \tag{2-4}$$

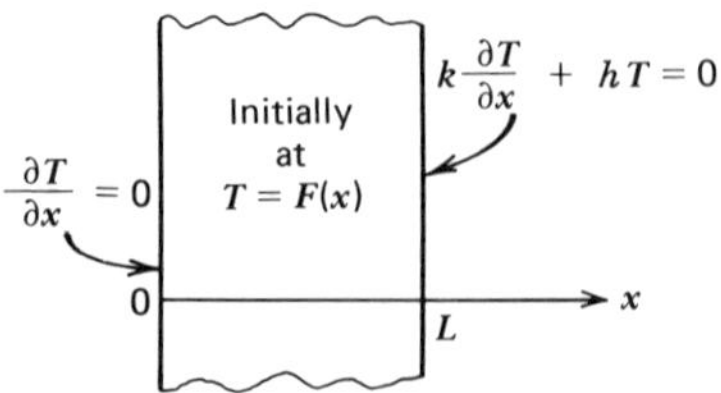

Fig. 2-1 Heat conduction in a slab.

Then, the function $\Gamma(t)$ satisfies the differential equation

$$\frac{d\Gamma(t)}{dt} + \alpha\beta^2\Gamma(t) = 0 \tag{2-5}$$

which has a solution in the form

$$\Gamma(t) = e^{-\alpha\beta^2 t} \tag{2-6}$$

Here, we note that the negative sign chosen above for β^2 now ensures that the solution $\Gamma(t)$ approaches zero as time increases indefinitely because both α and t are positive quantities. This is consistent with the physical reality for the problem (2-1) in that the temperature tends to zero as $t \to \infty$.

The space-variable function $X(x)$ satisfies the differential equation

$$\frac{d^2X(x)}{dx^2} + \beta^2 X(x) = 0 \qquad \text{in} \qquad 0 < x < L \tag{2-7a}$$

The boundary conditions for this equation are obtained by introducing the separated solution (2-2) into the boundary conditions (2-1b) and (2-1c); we find

$$\frac{dX}{dx} = 0 \qquad \text{at} \qquad x = 0 \tag{2-7b}$$

$$k\frac{dX}{dx} + hX = 0 \qquad \text{at} \qquad x = L \tag{2-7c}$$

The auxiliary problem defined by equations (2-7) is called an *eigenvalue problem*, because it has solutions only for certain values of the separation parameter $\beta = \beta_m$, $m = 1, 2, 3, \ldots$, which are called the *eigenvalues*; the corresponding solutions $X(\beta_m, x)$ are called the *eigenfunctions* of the problem. When β is not an eigenvalue, that is, when $\beta \neq \beta_m$, the problem has trivial solutions (i.e., $X = 0$ if $\beta \neq \beta_m$). We now assume that these eigenfunctions $X(\beta_m, x)$ and the eigenvalues β_m are available and proceed to the solution of the above heat conduction problem. The complete solution for the temperature $T(x, t)$ is constructed by a linear superposition of the above separated elementary solutions in the form

$$T(x, t) = \sum_{m=1}^{\infty} c_m X(\beta_m, x) e^{-\alpha\beta_m^2 t} \tag{2-8}$$

This solution satisfies both the differential equation (2-1a) and the boundary conditions (2-1b) and (2-1c) of the heat conduction problem, but it does not necessarily satisfy the initial condition (2-1d). Therefore, the application of the initial condition to equation (2-8) yields

$$F(x)=\sum_{m=1}^{\infty} c_m X(\beta_m, x) \qquad \text{in} \qquad 0<x<L \tag{2-9}$$

This result is a representation of an arbitrary function $F(x)$ defined in the interval $0<x<L$ in terms of the eigenfunctions $X(\beta_m, x)$ of the eigenvalue problem (2-7). The unknown coefficients c_m's can be determined by making use of the orthogonality of the eigenfunctions given as

$$\int_0^L X(\beta_m, x) X(\beta_n, x)\, dx = \begin{cases} 0 & \text{for } m \neq n \\ N(\beta_m) & \text{for } m = n \end{cases} \tag{2-10}$$

where, the *normalization integral* (or the *norm*), $N(\beta_m)$, is defined as

$$N(\beta_m)=\int_0^L [X(\beta_m, x)]^2\, dx \tag{2-11}$$

The eigenvalue problem given by equations (2-7) is a special case of a more general eigenvalue problem called the *Sturm–Liouville* problem. A discussion of the orthogonality property of the Sturm–Liouville problem can be found in the references 4, 5, 7, and 8.

To determine the coefficients c_m we operate on both sides of equation (2-9) by the operator $\int_0^L X(\beta_n, x)\, dx$ and utilize the orthogonality property given by equations (2-10); we find

$$c_m=\frac{1}{N(\beta_m)} \int_0^L X(\beta_m, x) F(x)\, dx \tag{2-12}$$

The substitution of equation (2-12) into equation (2-8) yields the solution for the temperature as

$$T(x, t)=\sum_{m=1}^{\infty} e^{-\alpha \beta_m^2 t} \frac{1}{N(\beta_m)} X(\beta_m, x) \int_0^L X(\beta_m, x') F(x')\, dx' \tag{2-13}$$

Thus the temperature distribution in the medium can be determined as a function of position and time from equation (2-13) once the explicit expressions are available for the eigenfunctions $X(\beta_m, x)$, the eigenvalues β_m and the norm $N(\beta_m)$. This matter will be discussed later in this chapter.

2-2 GENERALIZATION TO THREE-DIMENSIONAL PROBLEMS

The method of separation of variables illustrated above for the solution of the one-dimensional homogeneous heat conduction problem is now formally generalized to the solution of the following three-dimensional homogeneous problem

$$\nabla^2 T(\mathbf{r},t) = \frac{1}{\alpha}\frac{\partial T(\mathbf{r},t)}{\partial t} \qquad \text{in region } R, \qquad t>0 \tag{2-14}$$

$$k_i \frac{\partial T}{\partial n_i} + h_i T = 0 \qquad \text{on boundary } S_i, \quad t>0 \tag{2-15a}$$

$$T(\mathbf{r},t) = F(\mathbf{r}) \qquad \text{for } t=0, \text{ in region } R \tag{2-15b}$$

where $\partial/\partial n_i$ denotes differentiation along the outward-drawn normal to the boundary surface S_i and $\mathbf{r}$ denotes the general space coordinate. It is assumed that the region R has a number of continuous boundary surfaces $S_i, i = 1, 2, .., N$ in number, such that each boundary surface S_i fits the coordinate surface of the chosen orthogonal coordinate system. Clearly the slab problem considered above is obtainable as a special case from this more general problem; that is, the slab has two continuous boundary surfaces one at $x = 0$ and the other at $x = L$. The boundary conditions for the slab problem are readily obtainable from the general boundary condition (2-15a) by choosing the coefficients h_i and k_i, accordingly.

To solve the above general problem we assume a separation in the form

$$T(\mathbf{r},t) = \psi(\mathbf{r})\Gamma(t) \tag{2-16}$$

where function $\psi(\mathbf{r})$, in general, depends on three space variables. We substitute equation (2-16) into equation (2-14) and carry out the analysis with a similar argument as discussed above to obtain

$$\frac{1}{\psi(\mathbf{r})}\nabla^2\psi(\mathbf{r}) = \frac{1}{\alpha\Gamma(t)}\frac{d\Gamma(t)}{dt} = -\lambda^2 \tag{2-17}$$

where λ is the separation variable. Clearly, the function $\Gamma(t)$ satisfies an ordinary differential equation of the same form as equation (2-5) and its solution is taken as $\exp(-\alpha\lambda^2 t)$. The space-variable function $\psi(\mathbf{r})$ satisfies the following auxiliary problem

$$\nabla^2\psi(\mathbf{r}) + \lambda^2\psi(\mathbf{r}) = 0 \qquad \text{in region } R \tag{2-18a}$$

$$k_i\frac{\partial\psi}{\partial n_i} + h_i\psi = 0 \qquad \text{on boundary } S_i \tag{2-18b}$$

where $i = 1, 2, .., N$. The differential equation (2-18a) is called the *Helmholtz*

TABLE 2-1 Orthogonal Coordinate Systems Allowing Simple Separation of the Helmholtz and Laplace Equations[a]

Coordinate System	Functions That Appear in Solution
1 Rectangular	Exponential, circular, hyperbolic
2 Circular–cylinder	Bessel, exponential, circular
3 Elliptic–cylinder	Mathieu, circular
4 Parabolic-cylinder	Weber, circular
5 Spherical	Legendre, power, circular
6 Prolate spheroidal	Legendre, circular
7 Oblate spheroidal	Legendre, circular
8 Parabolic	Bessel, circular
9 Conical	Lamé, power
10 Ellipsoidal	Lamé
11 Paraboloidal	Baer

[a] From references 1, 3, and 10.

equation, and it is a partial-differential equation, in general, in the three space variables. The solution of this partial-differential equation is essential for the solution of the above heat conduction problem. The *Helmholtz* equation (2-18a) can be solved by the method of separation of variables provided that its separation into a set or ordinary differential equation is possible. The separability of the Helmholtz equation has been studied and it has been shown that a simple separation of the Helmholtz equation (also of the Laplace equation) into ordinary differential equations is possible in eleven orthogonal coordinate system. We list in Table 2-1 these 11 orthogonal coordinate systems and also indicate the type of functions that may appear as solutions of the separated functions [1, 3, 10]. A discussion of the separation of the Helmholtz equation will be presented in this chapter for the rectangular coordinate system and in the following two chapters for the cylindrical and spherical coordinate systems. The reader should consult references 10 and 11 for the definition of various functions listed in Table 2-1.

2-3 SEPARATION OF THE HEAT CONDUCTION EQUATION IN THE RECTANGULAR COORDINATE SYSTEM

Consider the three-dimensional, homogeneous heat conduction equation in the rectangular coordinate system

$$\frac{\partial^2 T}{\partial x^2} + \frac{\partial^2 T}{\partial y^2} + \frac{\partial^2 T}{\partial z^2} = \frac{1}{\alpha}\frac{\partial T}{\partial t} \qquad \text{where} \quad T \equiv T(x, y, z, t) \tag{2-19}$$

Assume a separation of variables in the form

$$T(x, y, z, t) = \psi(x, y, z)\Gamma(t) \tag{2-20}$$

Equation (2-19) becomes

$$\frac{1}{\psi}\left(\frac{\partial^2\psi}{\partial x^2}+\frac{\partial^2\psi}{\partial y^2}+\frac{\partial^2\psi}{\partial z^2}\right)=\frac{1}{\alpha\Gamma(t)}\frac{d\Gamma(t)}{dt}=-\lambda^2 \tag{2-21}$$

Then, the separated functions $\Gamma(t)$ and ψ satisfy the equations

$$\frac{d\Gamma(t)}{dt}+\alpha\lambda^2\Gamma(t)=0 \tag{2-22}$$

$$\frac{\partial^2\psi}{\partial x^2}+\frac{\partial^2\psi}{\partial y^2}+\frac{\partial^2\psi}{\partial z^2}+\lambda^2\psi=0 \tag{2-23}$$

Equation (2-23) is the Helmholtz equation; we assume a separation in the form

$$\psi(x,y,z)=X(x)Y(y)Z(z) \tag{2-24}$$

The substitution of equation (2-24) into equation (2-23) yields

$$\frac{1}{X}\frac{d^2X}{dx^2}+\frac{1}{Y}\frac{d^2Y}{dy^2}+\frac{1}{Z}\frac{d^2Z}{dz^2}+\lambda^2=0 \tag{2-25}$$

Here, since each term is a function of a single independent variable, the only way this equality is satisfied is if each term is equated to an arbitrary separation constant, say, in the form

$$\frac{1}{X}\frac{d^2X}{dx^2}=-\beta^2,\qquad \frac{1}{Y}\frac{d^2Y}{dy^2}=-\gamma^2,\qquad \text{and}\qquad \frac{1}{Z}\frac{d^2Z}{dz^2}=-\eta^2 \tag{2-26}$$

Then the separated equations become

$$\frac{d^2X}{dx^2}+\beta^2X=0 \tag{2-27a}$$

$$\frac{d^2Y}{dy^2}+\gamma^2Y=0 \tag{2-27b}$$

$$\frac{d^2Z}{dz^2}+\eta^2Z=0 \tag{2-27c}$$

where

$$\beta^2+\gamma^2+\eta^2=\lambda^2 \tag{2-27d}$$

Clearly, the solutions of the separated equations for the functions X, Y, and Z are sines and cosines, and the solution of equation (2-22) for the function $\Gamma(t)$ is given as

$$\Gamma(t) = e^{-\alpha(\beta^2 + \gamma^2 + \eta^2)t} \tag{2-28}$$

The complete solution for the temperature $T(x, y, z, t)$ is constructed by a linear superposition of the separated solutions X, Y, Z, and Γ. When the region is finite, say, in the x direction, the separation constant β associated with it takes discrete values and the superposition of the separated solutions for the x variable is performed by summation over all permissible values of β_m. On the other hand, when the region is infinite or semiinfinite, the separation constant assumes all values from zero to infinity continuously and superposition is done by integration over all values of β. In the following sections we examine the explicit functional forms of the separated solutions for finite, semiinfinite, and infinite regions. The elementary solutions obtained in this manner are tabulated systematically for ready reference in the solution of heat conduction problems by the method of separation of variables.

2-4 ONE-DIMENSIONAL HOMOGENEOUS PROBLEMS IN A FINITE MEDIUM ($0 \leqslant x \leqslant L$)

Here we consider the application of the method of separation of variables to the solution of the homogeneous boundary-value problem of heat conduction for a slab. That is, a slab, $0 \leqslant x \leqslant L$, initially at a temperature $F(x)$, dissipates heat by convection for times $t > 0$ from its boundary surfaces into an environment at zero temperature. For generality we assumed that the heat transfer coefficients at the two boundaries are not the same. The mathematical formulation of this problem is given as

$$\frac{\partial^2 T(x,t)}{\partial x^2} = \frac{1}{\alpha}\frac{\partial T(x,t)}{\partial t} \quad \text{in} \quad 0 < x < L, \quad t > 0 \tag{2-29a}$$

$$-k_1 \frac{\partial T}{\partial x} + h_1 T = 0 \quad \text{at} \quad x = 0, \quad t > 0 \tag{2-29b}$$

$$k_2 \frac{\partial T}{\partial x} + h_2 T = 0 \quad \text{at} \quad x = L, \quad t > 0 \tag{2-29c}$$

$$T = F(x) \quad \text{for} \quad t = 0, \quad \text{in} \quad 0 \leqslant x \leqslant L \tag{2-29d}$$

Clearly, the heat conduction problems for a slab for other combinations of boundary conditions are readily obtainable as special cases from the problem considered here by setting any one of the coefficients k_1, k_2, h_1, and h_2 equal to zero. Nine different combinations of these boundary conditions are possible.

We assume a separation in the form

$$T(x,t) = X(x)\Gamma(t) \tag{2-30}$$

and separate the equation in a manner described above. The solution for the function $\Gamma(t)$ is given as

$$\Gamma(t) = e^{-\alpha\beta^2 t} \tag{2-31}$$

and the space-variable function $X(\beta, x)$ satisfies the following eigenvalue problem:

$$\frac{d^2X(x)}{dx^2} + \beta^2 X(x) = 0 \quad \text{in} \quad 0 < x < L \tag{2-32a}$$

$$-k_1 \frac{dX}{dx} + h_1 X = 0 \quad \text{at} \quad x = 0 \tag{2-32b}$$

$$k_2 \frac{dX}{dx} + h_2 X = 0 \quad \text{at} \quad x = L \tag{2-32c}$$

This problem is a special case of the Sturm–Liouville problem discussed in Note 1, with $p(x) = 1, w(x) = 1, q(x) = 0$, and $\lambda = \beta^2$. Then, the eigenfunctions $X(\beta_m, x)$ are orthogonal, that is

$$\int_0^L X(\beta_m, x) X(\beta_n, x)\, dx = \begin{cases} 0 & \text{for } m \neq n \\ N(\beta_m) & \text{for } m = n \end{cases} \tag{2-33}$$

The solution of the problem (2-29) is now constructed as

$$T(x,t) = \sum_{m=1}^{\infty} c_m X(\beta_m, x) e^{-\alpha\beta_m^2 t} \tag{2-34}$$

The application of the initial condition (2-29d) gives

$$F(x) = \sum_{m=1}^{\infty} c_m X(\beta_m, x) \quad \text{in} \quad 0 < x < L \tag{2-35a}$$

This is a representation of an arbitrary function $F(x)$ defined in the interval $0 < x < L$ in terms of the eigenfunction $X(\beta_m, x)$ of the eigenvalue problem (2-32). Suppose such a representation is permissible, the coefficients c_m can be determined by operating on both sides of equation (2-35a) by the operator $\int_0^L X(\beta_m, x)\,dx$ and utilizing the orthogonality property of the eigenfunctions. We find

$$c_m = \frac{1}{N(\beta_m)} \int_0^L X(\beta_m, x')F(x')\,dx' \tag{2.35b}$$

where the *norm* N is defined as

$$N(\beta_m) = \int_0^L [X(\beta_m, x)]^2\,dx \tag{2.35c}$$

The substitution of equation (2-35b) into equation (2-34) yields the solution for the temperature $T(x, t)$ as

$$T(x,t) = \sum_{m=1}^{\infty} e^{-\alpha\beta_m^2 t} \frac{1}{N(\beta_m)} X(\beta_m, x) \int_0^L X(\beta_m, x')F(x')\,dx' \tag{2-36a}$$

This solution is valid for times $t > 0$; as $t \to 0$, it approaches to the initial value of the temperature in the medium. Therefore, by substituting $t = 0$ in equation (2-36), we obtain

$$F(x) = \sum_{m=1}^{\infty} \frac{1}{N(\beta_m)} X(\beta_m, x) \int_0^L X(\beta_m, x')F(x')\,dx' \quad \text{in} \quad 0 < x < L \tag{2-36b}$$

This equation is a representation of an arbitrary function $F(x)$ defined in the interval $0 < x < L$ in terms of the eigenfunctions $X(\beta_m, x)$ of the eigenvalue problem given by equations (2-32).

The eigenfunctions $X(\beta_m, x)$ of the eigenvalue problem (2-32) are given as

$$X(\beta_m, x) = \beta_m \cos \beta_m x + H_1 \sin \beta_m x \tag{2-37a}$$

where the eigenvalues β_m are the roots of the following transcendental equation

$$\tan \beta_m L = \frac{\beta_m(H_1 + H_2)}{\beta_m^2 - H_1 H_2} \tag{2-37b}$$

and the normalization integral $N(\beta_m)$ is given by

$$N(\beta_m) = \frac{1}{2}\left[(\beta_m^2 + H_1^2)\left(L + \frac{H_2}{\beta_m^2 + H_2^2}\right) + H_1\right] \tag{2-37c}$$

where

$$H_1 \equiv \frac{h_1}{k_1} \quad \text{and} \quad H_2 \equiv \frac{h_2}{k_2} \tag{2-37d}$$

The reader should consult reference 8 [p. 80] for the derivation of these results.

The eigenfunctions, eigencondition and the normalization integral given by equations (2-37) are for the general case of boundary condition of the third kind at both boundaries, $x = 0$ and $x = L$. The results for other combinations of boundary conditions are obtainable by setting in equations (2-37) H_1 or H_2 equal to zero or infinity.

We list in Table 2-2 the eigenfunctions $X(\beta_m, x)$, the eigenvalues β_m, and the normalization integral $N(\beta_m)$ of the eigenvalue problem (2-32) for nine different combinations of boundary conditions at $x = 0$ and $x = L$. We note that, for the boundary condition of the second kind at both boundaries (i.e., case 5), $\beta_0 = 0$ is also an eigenvalue corresponding to the eigenfunction $X(\beta_m) = 1$ as shown in note 2 at the end of this chapter.

2-5 COMPUTATION OF EIGENVALUES

Once the eigenvalues β_m are computed from the solution of the transcendental equation, the eigenfunctions $X(\beta_m)$ and the normalization integral $N(\beta_m)$ become known, and the temperature distribution $T(x, t)$ in the medium is determined from the solution given by equation (2-36a). Some of the transcendental equations, such as $\sin \beta_m L = 0$ or $\cos \beta_m L = 0$, are simple expressions; hence the β_m are readily evaluated. Consider, for example, the transcendental equation for the case 1 in Table 2-2 with $H_1 = H_2 \equiv H$. The resulting expression is written as

$$\tan \beta L = 2\frac{(\beta L)(HL)}{(\beta L)^2 - (HL)^2} \tag{2-38a}$$

and for convenience this result is rearranged as

$$\cot \xi = \frac{1}{2}\left(\frac{\xi}{B} - \frac{B}{\xi}\right) \equiv Z \tag{2-38b}$$

where $\xi = \beta L$ and $B = HL$. Clearly, the solution of this transcendental equation is not so easy. First we present a graphical interpretation of the roots of this transcendental equation before discussing its computer solution.

Graphical Representation

The result given by equation (2-38b) represents the following two curves:

$$Z = \frac{1}{2}\left(\frac{\xi}{B} - \frac{B}{\xi}\right) \qquad \text{and} \qquad Z = \cot \xi \tag{2-38c, d}$$

The first of these curves represents a hyperbola whose center is at the origin and its asymptotes are $\xi = 0$ and $Z = \xi/2B$, while the second represents a set of cotangent curves as illustrated in Fig. 2-2. The ξ values corresponding to the

TABLE 2-2 The Solution $X(\beta_m, x)$, the Norm $N(\beta_m)$ and the Eigenvalues β_m of the Differential Equation

$$\frac{d^2X(x)}{dx^2} + \beta^2 X(x) = 0 \qquad \text{in} \qquad 0 < x < L$$

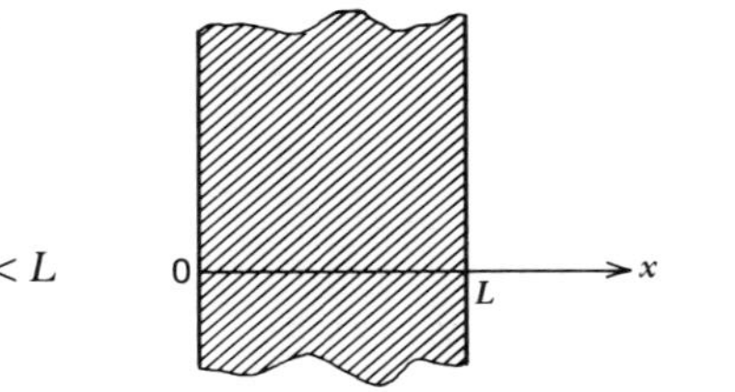

Subject to the Boundary Conditions Shown in the Table Below

No.	Boundary Condition at $x=0$	Boundary Condition at $x=L$	$X(\beta_m, x)$	$1/N(\beta_m)$	Eigenvalues β_m's are Positive Roots of
1	$-\frac{dX}{dx} + H_1 X = 0$	$\frac{dX}{dx} + H_2 X = 0$	$\beta_m \cos \beta_m x + H_1 \sin \beta_m x$	$2\left[(\beta_m^2 + H_1^2)\left(L + \frac{H_2}{\beta_m^2 + H_2^2}\right) + H_1\right]^{-1}$	$\tan \beta_m L = \frac{\beta_m (H_1 + H_2)}{\beta_m^2 - H_1 H_2}$
2	$-\frac{dX}{dx} + H_1 X = 0$	$\frac{dX}{dx} = 0$	$\cos \beta_m (L - x)$	$2\frac{\beta_m^2 + H_1^2}{L(\beta_m^2 + H_1^2) + H_1}$	$\beta_m \tan \beta_m L = H_1$
3	$-\frac{dX}{dx} + H_1 X = 0$	$X = 0$	$\sin \beta_m (L - x)$	$2\frac{\beta_m^2 + H_1^2}{L(\beta_m^2 + H_1^2) + H_1}$	$\beta_m \cot \beta_m L = -H_1$
4	$\frac{dX}{dx} = 0$	$\frac{dX}{dx} + H_2 X = 0$	$\cos \beta_m x$	$2\frac{\beta_m^2 + H_2^2}{L(\beta_m^2 + H_2^2) + H_2}$	$\beta_m \tan \beta_m L = H_2$

5	$\frac{dX}{dx}=0$	$\frac{dX}{dx}=0$	$*\cos\beta_m x$	$\frac{2}{L}$ for $\beta_m \neq 0$; $\frac{1}{L}$ for $\beta_0 = 0^a$	$\sin\beta_m L = 0^a$
6	$\frac{dX}{dx}=0$	$X=0$	$\cos\beta_m x$	$\frac{2}{L}$	$\cos\beta_m L = 0$
7	$X=0$	$\frac{dX}{dx}+H_2X=0$	$\sin\beta_m x$	$2\frac{\beta_m^2+H_2^2}{L(\beta_m^2+H_2^2)+H_2}$	$\beta_m \cot\beta_m L = -H_2$
8	$X=0$	$\frac{dX}{dx}=0$	$\sin\beta_m x$	$\frac{2}{L}$	$\cos\beta_m L = 0$
9	$X=0$	$X=0$	$\sin\beta_m x$	$\frac{2}{L}$	$\sin\beta_m L = 0$

[a] For this particular case $\beta_0 = 0$ is also an eigenvalue corresponding to $X = 1$.

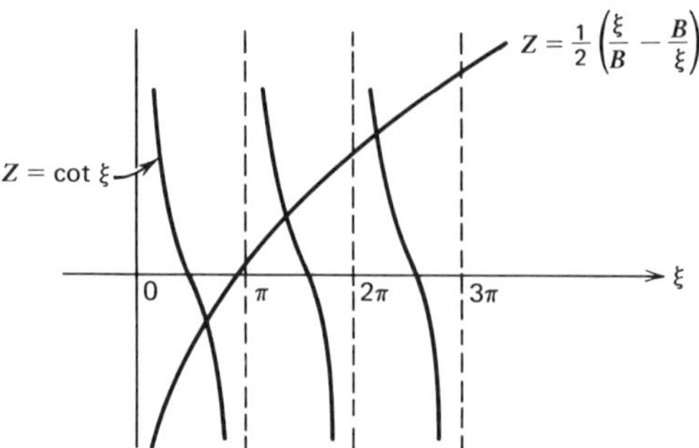

Fig. 2-2 Geometrical representation of the roots of $\cot \xi = (\frac{1}{2})[(\xi/B) - (B/\xi)]$.

intersections of the hyperbola with the cotangent curves are the roots of the above transcendental equation. Clearly there are an infinite number of such points, each successively located in intervals $(0 - \pi)$, $(\pi - 2\pi)$, $(2\pi - 3\pi)$, etc. Because of symmetry, the negative roots are equal in absolute value to the positive ones; therefore, only the positive roots need to be considered in the solution since the solution remains unaffected by the sign of the root. The graphical representation of the roots shown in Fig. 2-2 is useful to establish the regions where the roots lie; but accurate values of the roots are determined by numerical solution of the transcendental equations as described next.

Numerical Solutions

Various methods are available for solving transcendental equations numerically [14, 15]. Here we consider the bisection, Newton–Raphson and Secant methods for the determination of the roots of transcendental equations.

Bisection Method Consider a transcendental equation written compactly in the form

$$F(\xi) = 0 \tag{2-39}$$

and suppose it has only one root in the region $\xi_i \leqslant \xi \leqslant \xi_{i+1}$ as illustrated in Fig. 2-3. We wish to determine this root by the bisetion method. The interval $\xi_i \leqslant \xi \leqslant \xi_{i+1}$ is divided into two subintervals by a point $\xi_{i+(1/2)}$ defined by

$$\xi_{i+(1/2)} = \tfrac{1}{2}(\xi_i + \xi_{i+1}) \tag{2-40a}$$

and the sign of the product $F(\xi_i)$. $F(\xi_{i+(1/2)})$ is examined. If the product

$$F(\xi_i)F(\xi_{i+(1/2)}) < 0 \tag{2-40b}$$

then the root lies in the first subinterval $\xi_i \leqslant \xi \leqslant \xi_{i+(1/2)}$, since the sign change

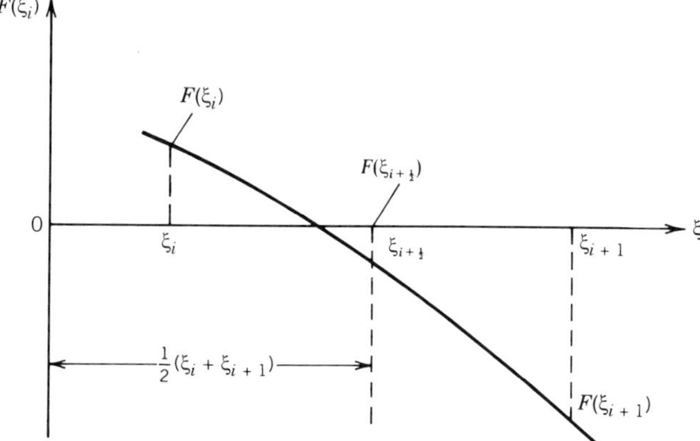

Fig. 2-3 The bisection method.

occurs in this interval. If the product is positive, the root must lie in the second subinterval. If the product is exactly zero, $\xi_{i+(1/2)}$ is the exact root. The procedure For determining the root is now apparent. The subinterval containing the root is bisected and the bisection procedure is continued until the change in the value of root from one bisection to the next becomes less than a specified tolerance ϵ.

The bisection procedure always yields a root if a region is found over which $F(\xi)$ changes sign and has only one root. Therefore, the graphical interpretation of roots as illustrated in Fig. 2-2 is useful to locate the regions where the roots lie. In the absence of graphical representation, one starts with $\xi = 0$ and evaluates $F(\xi)$ for each small increment of ξ until $F(\xi)$ changes sign. Then, a root must lie in that interval and the bisection procedure is applied for its determination.

In each bisection, the interval is reduced by half; therefore, after n bisections the original interval is reduced by a factor 2^n. For example, 10 bisections reduce the original interval by a factor more than 1000, and 20 bisections reduce more than one million.

Newton–Raphson Method Consider a function $F(\xi) = 0$ plotted against ξ as illustrated in Fig. 2-4. Let the tangent drawn to this curve at $\xi = \xi_i$ intersect the ξ axis at $\xi = \xi_{i+1}$. The slope of this tangent is given by

$$F'(\xi_i) = \frac{F(\xi_i)}{\xi_i - \xi_{i+1}} \tag{2-41a}$$

where prime denotes derivative with respect to ξ. Solving this equation for ξ_{i+1} we obtain

$$\xi_{i+1} = \xi_i - \frac{F(\xi_i)}{F'(\xi_i)} \tag{2-41b}$$

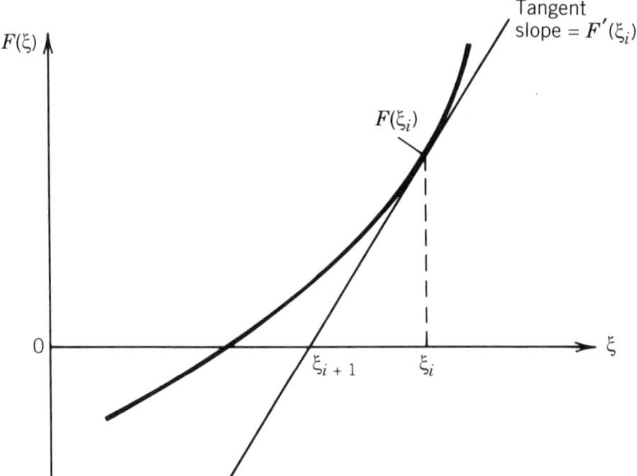

Fig. 2-4 Newton–Raphson method.

Equation (2-41b) provides an expression for calculating ξ_{i+1} from the knowledge of $F(\xi_i)$ and $F'(\xi_i)$ by iteration, until the change in the value of ξ_{i+1} from one iteration to the next is less than a specified convergence criteria ε. The method is widely used in practice because of its rapid convergence; however, there are situations that may give rise to convergence difficulties. For example, if the initial approximation to the root is not sufficiently close to the exact value of the root or the second derivative $F''(\xi)$ changes sign near the root convergence difficulties arise.

Secant Method The Newton–Raphson method requires the derivative of the function for each iteration. However, if the function is difficult to differentiate, the derivative is approximated by a difference approximation, hence equation (2-41b) takes the form

$$\xi_{i+1} = \xi_i - \frac{F(\xi_i)}{[F(\xi_i) - F(\xi_{i-1})]/(\xi_i - \xi_{i-1})} \tag{2-42}$$

The secant method may not be as rapidly convergent as the Newton–Raphson method; but if the evaluation of $F'(\xi)$ is time-consuming, then the secant method may require less computer time than Newton's method.

Tabulated Eigenvalues In Appendix II we tabulated first six roots of the transcendental equations

$$F(\beta) \equiv \beta \tan \beta - C = 0 \tag{2-43a}$$

$$F(\beta) \equiv \beta \cot \beta + C = 0 \tag{2-43b}$$

for several different values of C. These transcendental equations are associated with the cases 2 and 3 in Table 2-2, respectively.

When using the secant or Newton–Raphson method for solving such transcendental equations, it is preferable to establish the region where a root lies by a bisection method or a graphical approach, and then apply the secant or Newton–Raphson method. An examination of the roots listed in the Appendix II reveals that the roots lie in the intervals which are multiples of π. Consider, for example, the transcendental equation (2-43a). For large values of C, the roots lie in the regions where the slope of the tangent curve is very steep; hence difficulty is experienced in the determination of roots from Eq. (2-43b) when the roots lie in the regions where the slope of the cotangent curve is very steep. In such situations, the convergence difficulty is alleviated if equations (2-43a) and (2-43b) are rearranged, respectively, in the forms

$$F(\beta) \equiv \beta \sin \beta - C \cos \beta = 0 \tag{2-44a}$$

$$F(\beta) = \beta \cos \beta + C \sin \beta = 0 \tag{2-44b}$$

Example 2-1

A slab in $0 \leqslant x \leqslant L$ is initially at a temperature $F(x)$; for times $t > 0$, the boundary at $x = 0$, is kept insulated and the boundary at $x = L$ dissipates heat by convection into a medium at zero temperature, that is

$$\frac{\partial T}{\partial x} = 0 \quad \text{at} \quad x = 0 \quad \text{and} \quad \frac{\partial T}{\partial x} + H_2 T = 0 \quad \text{at} \quad x = L$$

Obtain an expression for the temperature distribution $T(x, t)$ in the slab. Also consider the case when $F(x) = T_0 = \text{constant}$.

Solution. The boundary conditions for this problem correspond to case 4 in Table 2-2. Therefore, when the eigenfunctions $X(\beta_m, x)$ and the norm $N(\beta_m)$ are obtained from this table and introduced into equation (2-36a), the solution becomes

$$T(x,t) = 2 \sum_{m=1}^{\infty} e^{-\alpha \beta_m^2 t} \frac{\beta_m^2 + H_2^2}{L(\beta_m^2 + H_2^2) + H_2} \cdot \cos \beta_m x \int_{x'=0}^{L} F(x') \cos \beta_m x' \, dx' \tag{2-45a}$$

where β_m values are the positive roots of

$$\beta_m \tan \beta_m L = H_2 \tag{2-45b}$$

For the special case of $F(x) = T_0 = \text{constant}$, the integration in equation

(2-45a) can be performed and the solution reduces to

$$T(x,t) = 2T_0 \sum_{m=1}^{\infty} e^{-\alpha\beta_m^2 t} \frac{\beta_m^2 + H_2^2}{L(\beta_m^2 + H_2^2) + H_2} \frac{\sin \beta_m L}{\beta_m} \cos \beta_m x \tag{2-45c}$$

and by making use of the transcendental equation (2-45b) this result is written as

$$T(x,t) = 2T_0 \sum_{m=1}^{\infty} e^{-\alpha\beta_m^2 t} \frac{H_2}{L(\beta_m^2 + H_2^2) + H_2} \frac{\cos \beta_m x}{\cos \beta_m L} \tag{2-45d}$$

Example 2-2

A slab, $0 \leqslant x \leqslant L$, is initially at a temperature $F(x)$, for times $t > 0$ the boundaries at $x = 0$ and $x = L$ are kept insulated, that is, $\partial T/\partial x = 0$ at $x = 0$ and $x = L$. Obtain an expression for the temperature distribution $T(x,t)$ in the slab.

Solution. The boundary conditions for this problem correspond to case 5 in Table 2-2. Obtaining $X(\beta_m, x)$ and $N(\beta_m)$ from this table and introducing them into equation (2-36) and noting that for this special case $\beta_0 = 0$ is also an eigenvalue, the solution of the problem becomes

$$T(x,t) = \frac{1}{L}\int_0^L F(x')\,dx' + \frac{2}{L}\sum_{m=1}^{\infty} e^{-\alpha\beta_m^2 t} \cdot \cos \beta_m x \int_{x'=0}^{L} F(x') \cos \beta_m x'\,dx' \tag{2-46}$$

where β_m values are the roots of $\sin \beta_m L = 0$ or given as $\beta_m = m\pi/L, m = 1, 2, 3, \ldots$. Here, the first term on the right-hand side of the equation results from the fact that $\beta_0 = 0$ is also an eigenvalue. The physical significance of this term is as follows: It represents the temperature in the solid as $t \to \infty$ (i.e., after the transients have passed); it is an arithmetic mean of the initial temperature over the region $0 \leqslant x \leqslant L$. This is to be expected by physical considerations, since heat cannot escape from the insulated boundaries, eventually the temperature equalizes over the region.

2-6 ONE-DIMENSIONAL HOMOGENEOUS PROBLEMS IN A SEMIINFINITE MEDIUM

We now consider the solution of a homogeneous heat conduction problem for a semiinfinite region. That is, a semiinfinite region, $0 \leqslant x < \infty$, is initially at a temperature $F(x)$ and for times $t > 0$ the boundary surface at $x = 0$ dissipates heat by convection into a medium at zero temperature as illustrated in Fig. 2-5.

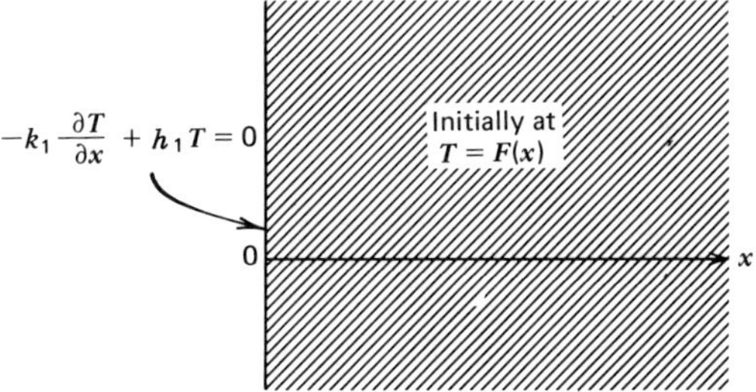

Fig. 2-5 Heat conduction in a semiinfinite region.

The mathematical formulation of this problem is given as

$$\frac{\partial^2 T(x,t)}{\partial x^2} = \frac{1}{\alpha}\frac{\partial T(x,t)}{\partial t} \quad \text{in} \quad 0 < x < \infty, \quad t > 0 \tag{2-47a}$$

$$-k_1 \frac{\partial T}{\partial x} + h_1 T = 0 \quad \text{at} \quad x = 0, \quad t > 0 \tag{2-47b}$$

$$T = F(x) \quad \text{for} \quad t = 0, \quad \text{in} \quad 0 \leqslant x < \infty \tag{2-47c}$$

We assume a separation in the form $T(x,t) = X(x)\Gamma(t)$; then, the solution for the function $\Gamma(t)$ is as given previously by equation (2-31); that is:

$$\Gamma(t) = e^{-\alpha\beta^2 t} \tag{2-31}$$

where β is the separation constant, and the space-variable function $X(\beta, x)$ satisfies the following problem:

$$\frac{d^2 X(x)}{dx^2} + \beta^2 X(x) = 0 \quad \text{in} \quad 0 < x < \infty \tag{2-48a}$$

$$-k_1 \frac{dX(x)}{dx} + h_1 X(x) = 0 \quad \text{at} \quad x = 0 \tag{2-48b}$$

The solution of equations (2-48) may be taken in the form

$$X(\beta, x) = \beta \cos \beta x + H_1 \sin \beta x \tag{2-49a}$$

where

$$H_1 = \frac{h_1}{k_1} \tag{2-49b}$$

and the separation variable β assumes all values from zero to infinity continuously. The general solution for $T(x, t)$ is constructed by the superposition of all these elementary solutions by integrating over the value of β from zero to infinity

$$T(x,t) = \int_{\beta=0}^{\infty} c(\beta) e^{-\alpha\beta^2 t} (\beta \cos \beta x + H_1 \sin \beta x)\, d\beta \tag{2-50}$$

The application of the initial condition to equation (2-50) yields

$$F(x) = \int_{\beta=0}^{\infty} c(\beta) X(\beta, x)\, d\beta \qquad \text{in} \qquad 0 < x < \infty \tag{2-51}$$

where

$$X(\beta, x) = \beta \cos \beta x + H_1 \sin \beta x$$

This result is a representation of an arbitrary function $F(x)$ defined in the semi-infinite interval $0 < x < \infty$ in terms of the solution of the auxiliary problem defined by equations (2-48). A similar representation has been developed [11, p. 228] when solving the heat-conduction problem (2-47) by the Laplace transform technique, and that result can be expressed in the form

$$F(x) = \int_{\beta=0}^{\infty} X(\beta, x) \left[\frac{1}{N(\beta)} \int_{x'=0}^{\infty} X(\beta, x') F(x')\, dx' \right] d\beta \tag{2-52}$$

where

$$X(\beta, x) = \beta \cos \beta x + H_1 \sin \beta x \qquad \text{and} \qquad \frac{1}{N(\beta)} = \frac{2}{\pi} \frac{1}{\beta^2 + H_1^2}$$

The representation given by equation (2-52) is valid when $F(x)$ and dF/dx are sectionally continuous on each finite interval in the range $0 < x < \infty$, provided the integral $\int_0^\infty |F(x)|\, dx$ exists, if $F(x)$ is defined as its mean value at each point of discontinuity.

By comparing equations (2-51) and (2-52) we obtain the unknown coefficient $c(\beta)$ as

$$c(\beta) = \frac{1}{N(\beta)} \int_0^{\infty} X(\beta, x') F(x')\, dx' \tag{2-53}$$

where $N(\beta)$ and $X(\beta, x)$ are as defined previously. The substitution of equation (2-53) into equation (2-50) yields the solution for the heat conduction problem (2-47) as

$$T(x,t) = \int_{\beta=0}^{\infty} e^{-\alpha\beta^2 t} \frac{1}{N(\beta)} X(\beta, x) \int_{x'=0}^{\infty} X(\beta, x') F(x')\, dx'\, d\beta \tag{2-54}$$

and the Norm $N(\beta)$ of the Differential Equation

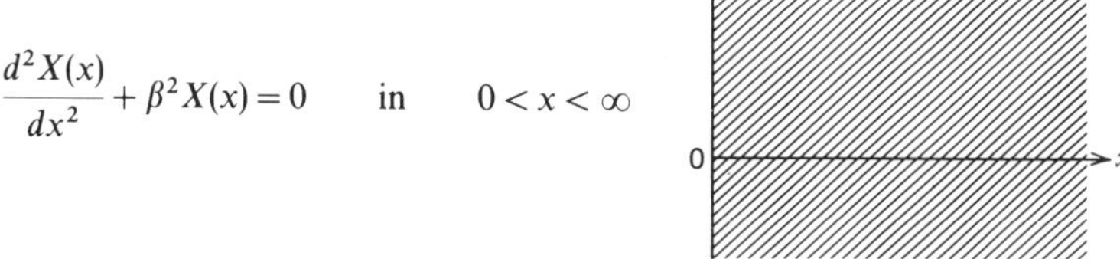

$$\frac{d^2X(x)}{dx^2} + \beta^2 X(x) = 0 \qquad \text{in} \qquad 0 < x < \infty$$

Subject to the Boundary Conditions Shown in the Table Below

No.	Boundary Condition at $x = 0$	$X(\beta, x)$	$1/N(\beta)$
1	$-\frac{dX}{dx} + H_1 X = 0$	$\beta \cos \beta x + H_1 \sin \beta x$	$\frac{2}{\pi} \frac{1}{\beta^2 + H_1^2}$
2	$\frac{dX}{dx} = 0$	$\cos \beta x$	$\frac{2}{\pi}$
3	$X = 0$	$\sin \beta x$	$\frac{2}{\pi}$

where

$$X(\beta, x) = \beta \cos \beta x + H_1 \sin \beta x \tag{2-55a}$$

$$\frac{1}{N(\beta)} = \frac{2}{\pi} \frac{1}{\beta^2 + H_1^2} \qquad \text{and} \qquad H_1 \equiv \frac{h_1}{k_1} \tag{2-55b}$$

The functions $X(\beta, x)$ and $N(\beta)$ given by equations (2-55) are for a boundary condition of the third kind at $x = 0$. The boundary condition at $x = 0$ may also be of the second or the first kind. We list in Table 2-3 the functions $X(\beta, x)$ and $N(\beta)$ for these three different boundary conditions at $x = 0$. Thus, the solution of the homogeneous heat conduction problem for a semiinfinite medium $0 \leqslant x < \infty$ given by equations (2-47) is obtainable from equation (2-54) for the three different boundary conditions at $x = 0$ if $X(\beta, x)$ and $N(\beta)$ are taken from Table 2-3, accordingly.

Example 2-3

A semiinfinite region $0 \leqslant x < \infty$ is initially at temperature $F(x)$. For time $t > 0$ the boundary at $x = 0$ is kept at zero temperature. Obtain an expression for the temperature distribution $T(x, t)$ in the medium. Also, examine the case when $F(x) = T_0 = \text{constant}$.

Solution. The boundary condition for this problem corresponds to case 3 in Table 2-3. Obtaining the functions $X(\beta, x)$ and $N(\beta)$ from this table and substituting in equation (2-54) the solution becomes

$$T(x,t) = \frac{2}{\pi} \int_{x'=0}^{\infty} F(x') \int_{\beta=0}^{\infty} e^{-\alpha\beta^2 t} \sin \beta x \sin \beta x' \, d\beta \, dx' \tag{2-56}$$

The integration with respect to β is evaluated by making use of the following relations

$$2 \sin \beta x \sin \beta x' = \cos \beta(x - x') - \cos \beta(x + x') \tag{2-57a}$$

and from Dwight [17, #861.20] we have

$$\int_{\beta=0}^{\infty} e^{-\alpha\beta^2 t} \cos \beta(x - x') \, d\beta = \sqrt{\frac{\pi}{4\alpha t}} \cdot \exp\left[-\frac{(x - x')^2}{4\alpha t} \right] \tag{2-57b}$$

$$\int_{\beta=0}^{\infty} e^{-\alpha\beta^2 t} \cos \beta(x + x') \, d\beta = \sqrt{\frac{\pi}{4\alpha t}} \cdot \exp\left[-\frac{(x + x')^2}{4\alpha t} \right] \tag{2-57c}$$

Then

$$\frac{2}{\pi} \int_{\beta=0}^{\infty} e^{-\alpha\beta^2 t} \sin \beta x \sin \beta x' \, d\beta$$

$$= \frac{1}{(4\pi\alpha t)^{1/2}} \left[\exp\left(-\frac{(x - x')^2}{4\alpha t} \right) - \exp\left(-\frac{(x + x')^2}{4\alpha t} \right) \right] \tag{2-57d}$$

and the solution (2-56) becomes

$$T(x,t) = \frac{1}{(4\pi\alpha t)^{1/2}} \int_{x'=0}^{\infty} F(x') \left[\exp\left(-\frac{(x - x')^2}{4\alpha t} \right) - \exp\left(-\frac{(x + x')^2}{4\alpha t} \right) \right] dx' \tag{2-58a}$$

For a constant initial temperature in the solid, $F(x) = T_0 = \text{constant}$, equation (2-58a) becomes

$$\frac{T(x,t)}{T_0} = \frac{1}{(4\pi\alpha t)^{1/2}} \left[\int_{x'=0}^{\infty} \exp\left(-\frac{(x - x')^2}{4\alpha t} \right) dx' - \int_{x'=0}^{\infty} \exp\left(-\frac{(x + x')^2}{4\alpha t} \right) dx' \right] \tag{2-58b}$$

Introducing the following new variables,

$$-\eta = \frac{x - x'}{\sqrt{4\alpha t}}, \qquad dx' = \sqrt{4\alpha t} \, d\eta \qquad \text{for the first integral}$$

$$\eta = \frac{x + x'}{\sqrt{4\alpha t}}, \qquad dx' = \sqrt{4\alpha t} \, d\eta \qquad \text{for the second integral}$$

equation (2-58b) becomes

$$\frac{T(x,t)}{T_0} = \frac{1}{\sqrt{\pi}} \left[\int_{-x/\sqrt{4\alpha t}}^{\infty} e^{-\eta^2} \, d\eta - \int_{x/\sqrt{4\alpha t}}^{\infty} e^{-\eta^2} \, d\eta \right] \tag{2-58c}$$

Since $e^{-\eta^2}$ is symmetrical about $\eta = 0$, equation (2-58c) is written in the form

$$\frac{T(x,t)}{T_0} = \frac{2}{\sqrt{\pi}} \int_0^{x/\sqrt{4\alpha t}} e^{-\eta^2} \, d\eta \tag{2-58d}$$

The right-hand side of this equation is called the *error function* of argument $x/\sqrt{4\alpha t}$ and the solution is expressed in the form

$$\frac{T(x,t)}{T_0} = \operatorname{erf}\left(\frac{x}{\sqrt{4\alpha t}} \right) \tag{2-58e}$$

The values of the error functions are tabulated in Appendix III. Also included in this appendix is a brief discussion of the properties of the error function.

2-7 FLUX FORMULATION

The one-dimensional transient heat conduction equation, customarily given in terms of temperature $T(x,t)$, can be expressed in terms of heat flux, $q(x,t)$. Such a formulation is useful for solving heat conduction in a semiinfinite medium with prescribed heat flux boundary condition. Consider the heat conduction equation

$$\frac{\partial^2 T}{\partial x^2} = \frac{1}{\alpha} \frac{\partial T(x,t)}{\partial t} \tag{2-59a}$$

and the definition of the heat flux

$$q(x,t) = -k \frac{\partial T(x,t)}{\partial x} \tag{2-59b}$$

Equation (2-59a) is differentiated with respect to space variable and the result is manipulated by utilizing equations (2-59a) and (2-59b). We obtain

$$\frac{\partial^2 q}{\partial x^2} = \frac{1}{\alpha} \frac{\partial q(x,t)}{\partial t} \tag{2-59c}$$

which is a differential equation in flux and is of the same form as equation (2-59a).

To illustrate its application, we consider heat conduction in a semiinfinite medium $0 < x < \infty$, initially at zero temperature and for times $\tau > 0$, a constant heat flux f_0 is applied at the boundary surface $x = 0$. The mathematical formulation of this problem, in the flux formulation, is given by

$$\frac{\partial^2 q}{\partial x^2} = \frac{1}{\alpha}\frac{\partial q(x,t)}{\partial t} \quad \text{in} \quad 0 < x < \infty, \quad t > 0 \tag{2-59d}$$

$$q(x,t) = f_0 = \text{constant} \quad \text{at} \quad x = 0, \quad t > 0 \tag{2-59e}$$

$$q(x,t) = 0 \quad \text{for} \quad t = 0 \tag{2-59f}$$

A new dependent variable $Q(x,t)$ is defined as

$$Q(x,t) = q(x,t) - f_0 \tag{2-59g}$$

Then the problem takes the form

$$\frac{\partial^2 Q}{\partial x^2} = \frac{1}{\alpha}\frac{\partial Q(x,t)}{\partial t} \quad \text{in} \quad 0 < x < \infty, \quad t > 0 \tag{2-60a}$$

$$Q(x,t) = 0 \quad \text{at} \quad x = 0, \quad t > 0 \tag{2-60b}$$

$$Q(x,t) = -f_0 \quad \text{for} \quad t = 0 \tag{2-60c}$$

The solution of this problem is immediately obtained from equation (2-58e) as

$$Q(x,t) = -f_0 \operatorname{erf}(x/\sqrt{4\alpha t}) \tag{2-61a}$$

or

$$q(x,t) = f_0 + Q(x,t) = f_0[1 - \operatorname{erf}(x/\sqrt{4\alpha t})] \tag{2-61b}$$

$$q(x,t) = f_0 \operatorname{erfc}(x/\sqrt{4\alpha t}) \tag{2-61c}$$

Once $q(x,t)$ is known, the temperature distribution $T(x,t)$ is determined by the integration of equation (2-59b). We obtain

$$T(x,t) = \frac{f_0}{k}\int_x^\infty \operatorname{erfc}(x'/\sqrt{4\alpha t})\, dx' \tag{2-62a}$$

The integration is performed by utilizing the relationship given by equation (6) in Appendix III.

$$T(x,t) = \frac{2f_0}{k}\left[\left(\frac{\alpha t}{\pi}\right)^{1/2} e^{-x^2/4\alpha t} - \frac{x}{2}\operatorname{erfc}(x/\sqrt{4\alpha t})\right] \tag{2-62b}$$

and the temperature at the surface $x=0$ becomes

$$T(0,t)=\frac{2f_0}{k}\left(\frac{\alpha t}{\pi}\right)^{1/2} \tag{2-62c}$$

Example 2-4

A semiinfinite region $0<x<\infty$ is initially at temperature $F(x)$. For times $t>0$ the boundary at $x=0$ is kept insulated. Obtain an expression for the temperature distribution in the medium.

Solution. The boundary condition for this problem corresponds to case 2 in Table 2-3. Therefore the formal solution given by equation (2-54) becomes

$$T(x,t)=\frac{2}{\pi}\int_{x'=0}^{\infty} F(x')\int_{\beta=0}^{\infty} e^{-\alpha\beta^2 t}\cos\beta x\cos\beta x'\,d\beta\,dx' \tag{2-63a}$$

From the trigonometric relations we have

$$2\cos\beta x\cos\beta x'=\cos\beta(x-x')+\cos\beta(x+x') \tag{2-63b}$$

Therefore, the integration with respect to β in the above solution is performed by utilizing equations (2-57b) and (2-57c); then the solution becomes

$$T(x,t)=\frac{1}{(4\pi\alpha t)^{1/2}}\int_{x'=0}^{\infty} F(x')\left[\exp\left(-\frac{(x-x')^2}{4\alpha t}\right)\right.$$
$$\left.+\exp\left(-\frac{(x+x')^2}{4\alpha t}\right)\right]dx' \tag{2-63c}$$

A comparison of the solutions (2-58a) and (2-63c) reveals that the two exponential terms are subtracted in the former and added in the latter.

2-8 ONE-DIMENSIONAL HOMOGENEOUS PROBLEMS IN AN INFINITE MEDIUM

We now consider the homogeneous heat conduction problem for a one-dimensional infinite medium, $-\infty<x<\infty$, which is initially at a temperature $F(x)$. We are interested in the determination of the temperature $T(x,t)$ of the medium for time $t>0$. No boundary conditions are specified for the problem since the medium extends to infinity in both directions; but the problem consists of a boundedness condition on $T(x,t)$. The mathematical formulation is given as

$$\frac{\partial^2 T(x,t)}{\partial x^2}=\frac{1}{\alpha}\frac{\partial T(x,t)}{\partial t} \quad \text{in} \quad -\infty<x<\infty,\quad t>0 \tag{2-64a}$$

$$T = F(x) \qquad \text{for} \quad t = 0, \quad \text{in} \quad -\infty < x < \infty \tag{2-64b}$$

By separating the variables in the form $T(x,t) = X(x)\Gamma(t)$, the solution for the function $\Gamma(t)$ is given as

$$\Gamma(t) = e^{-\alpha\beta^2 t} \tag{2-64c}$$

and the function $X(x)$ satisfies the equation

$$\frac{d^2X(x)}{dx^2} + \beta^2 X(x) = 0 \qquad \text{in} \qquad -\infty < x < \infty \tag{2-64d}$$

Two linearly independent solutions of this equation are $\cos \beta x$ and $\sin \beta x$, corresponding to each value of β. As negative values of β generates no additional solutions, we consider only $\beta \geqslant 0$. The general solution of the heat conduction problem is constructed by the superposition of $X(\beta, x)\Gamma(t)$ in the form

$$T(x,t) = \int_{\beta=0}^{\infty} e^{-\alpha\beta^2 t}[a(\beta)\cos \beta x + b(\beta)\sin \beta x]\, d\beta \tag{2-65}$$

The unknown coefficients $a(\beta)$ and $b(\beta)$ are to be determined so that for $t = 0$ this solution represents the initial temperature distribution $F(x)$ in the medium $-\infty < x < \infty$. The application of the initial condition to equation (2-65) yields

$$F(x) = \int_{\beta=0}^{\infty} [a(\beta)\cos \beta x + b(\beta)\sin \beta x]\, d\beta, \qquad -\infty < x < \infty \tag{2-66a}$$

This equation is the Fourier formula for the integral representation of an arbitrary function $F(x)$ defined in the interval $-\infty < x < \infty$; the coefficients $a(\beta)$ and $b(\beta)$ are given as [4, p. 114; 14, p. 1]

$$a(\beta) = \frac{1}{\pi}\int_{x'=-\infty}^{\infty} F(x')\cos \beta x'\, dx' \tag{2-66b}$$

$$b(\beta) = \frac{1}{\pi}\int_{x'=-\infty}^{\infty} F(x')\sin \beta x'\, dx' \tag{2-66c}$$

Equations (2-66b,c) are substituted into equation (2-66a), the trigonometric terms are combined and the order of integration is changed. We obtain

$$F(x) = \int_{\beta=0}^{\infty} \left[\frac{1}{\pi}\int_{x'=-\infty}^{\infty} F(x')\cos \beta(x - x')\, dx'\right] d\beta \tag{2-66d}$$

The representation given by equation (2-66d) is valid if function $F(x)$ and dF/dx

are sectionally continuous on every finite interval on the x axis, $F(x)$ is defined as its mean value at each point of discontinuity, and the integral $\int_{-\infty}^{\infty} |F(x)|\,dx$ exists [4, p. 115].

A comparison of the results in equations (2-66a) and (2-66d) implies that the coefficients are given by

$$[a(\beta)\cos\beta x + b(\beta)\sin\beta x] \equiv \frac{1}{\pi}\int_{x'=-\infty}^{\infty} F(x')\cos\beta(x-x')\,dx' \tag{2-67}$$

Then the solution given by equation (2-65) becomes

$$T(x,t) = \frac{1}{\pi}\int_{\beta=0}^{\infty} e^{-\alpha\beta^2 t}\int_{x'=-\infty}^{\infty} F(x')\cos\beta(x-x')\,dx'\,d\beta \tag{2-68}$$

In view of the integral, Dwight [17, #861.20]

$$\int_{\beta=0}^{\infty} e^{-\alpha\beta^2 t}\cos\beta(x-x')\,d\beta = \sqrt{\frac{\pi}{4\alpha t}}\exp\left[-\frac{(x-x')^2}{4\alpha t}\right] \tag{2-69}$$

The solution (2-68) takes the form

$$T(x,t) = \frac{1}{[4\pi\alpha t]^{1/2}}\int_{x'=-\infty}^{\infty} F(x')\exp\left[-\frac{(x-x')^2}{4\alpha t}\right]dx' \tag{2-70}$$

Example 2-5

In a one-dimensional infinite medium $-\infty < x < \infty$, the region $-L < x < L$ is initially at a constant temperature T_0, and everywhere outside this region is at zero temperature. Obtain an expression for the temperature distribution $T(x,t)$ in the medium for times $t > 0$.

Solution. For this particular case the initial condition function is of the form

$$Fx) = \begin{cases} T_0 & \text{in} \quad -L < x < L \\ 0 & \text{everywhere outside this region} \end{cases}$$

and the solution (2-70) becomes

$$T(x,t) = \frac{T_0}{(4\pi\alpha t)^{1/2}}\int_{-L}^{L} \exp\left[-\frac{(x-x')^2}{4\alpha t}\right]dx' \tag{2-71}$$

A new variable is defined as

$$\eta = \frac{x-x'}{\sqrt{4\alpha t}} \qquad \therefore dx' = -\sqrt{4\alpha t}\,d\eta \tag{2-72}$$

Then equation (2-71) becomes

$$T(x,t) = \frac{T_0}{2}\left[\frac{2}{\sqrt{\pi}}\int_0^{(L+x)/\sqrt{4\alpha t}} e^{-\eta^2}\,d\eta + \frac{2}{\sqrt{\pi}}\int_0^{(L-x)/\sqrt{4\alpha t}} e^{-\eta^2}\,d\eta\right] \tag{2-73}$$

which is written in the form

$$\frac{T(x,t)}{T_0} = \frac{1}{2}\left[\operatorname{erf}\left(\frac{L+x}{\sqrt{4\alpha t}}\right) + \operatorname{erf}\left(\frac{L-x}{\sqrt{4\alpha t}}\right)\right] \quad \text{in} \quad -\infty < x < \infty \tag{2-74}$$

2-9 MULTIDIMENSIONAL HOMOGENEOUS PROBLEMS

Having established the eigenfunctions, eigenconditions, and the normalization integrals for one-dimensional problems of finite, semiinfinite, and infinite regions, we are now in a position to apply the method of separation of variables to the solution of multidimensional homogeneous heat conduction problems as illustrated below with representative examples.

Example 2-6

A rectangular region $0 \leqslant x \leqslant a, 0 \leqslant y \leqslant b$ is initially at temperature $F(x, y)$. For times $t > 0$ the boundary at $x = 0$ is kept insulated, the boundary at $y = 0$ is kept at zero temperature, and boundaries at $x = a$ and $y = b$ dissipate heat by convection into an environment at zero temperature as illustrated in Fig. 2-6. Obtain an expression for the temperature distribution $T(x, y, t)$ for times $t > 0$.

Solution. The mathematical formulation of the problem is given as

$$\frac{\partial^2 T}{\partial x^2} + \frac{\partial^2 T}{\partial y^2} = \frac{1}{\alpha}\frac{\partial T}{\partial t} \quad \text{in} \quad 0 < x < a, \quad 0 < y < b, \quad t > 0 \tag{2-75a}$$

$$\frac{\partial T}{\partial x} = 0 \quad \text{at} \quad x = 0; \qquad \frac{\partial T}{\partial x} + H_2 T = 0 \quad \text{at} \quad x = a \quad \text{for} \quad t > 0 \tag{2-75b}$$

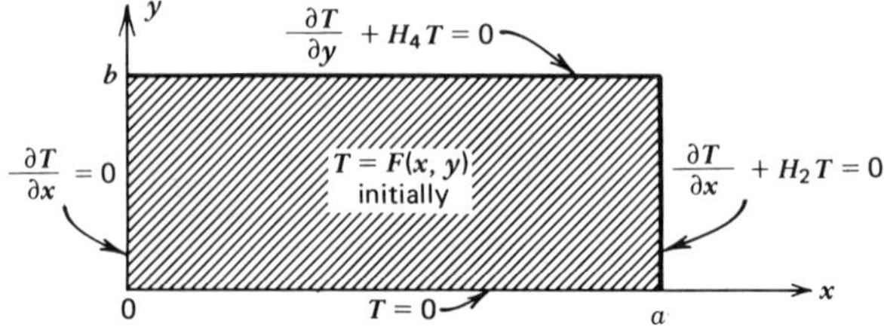

Fig. 2-6 Boundary and initial conditions for a rectangular region considered in Example 2-6.

$$T=0 \quad \text{at} \quad y=0; \qquad \frac{\partial T}{\partial y}+H_4 T=0 \quad \text{at} \quad y=b \quad \text{for} \quad t>0 \tag{2-75c}$$

$$T=F(x,y) \qquad \text{for} \quad t=0, \text{in the region} \tag{2-75d}$$

Assuming a separation in the form

$$T(x,y,t)=\Gamma(t)X(x)Y(y) \tag{2-76}$$

the problems defining the $X(x)$ and $Y(y)$ functions become

$$\frac{d^2X(x)}{dx^2}+\beta^2X(x)=0 \qquad \text{in} \qquad 0<x<a \tag{2-77a}$$

$$\frac{dX}{dx}=0 \quad \text{at } x=0; \qquad \frac{dX}{dx}+H_2X=0 \qquad \text{at } x=a \tag{2-77b}$$

and

$$\frac{d^2Y(y)}{dy^2}+\gamma^2Y(y)=0 \qquad \text{in} \qquad 0<y<b \tag{2-78a}$$

$$Y=0 \quad \text{at} \quad y=0; \qquad \frac{dY}{dy}+H_4Y=0 \qquad \text{at } y=b \tag{2-78b}$$

and the solution for $\Gamma(t)$ is given by

$$\Gamma(t)=e^{-\alpha(\beta^2+\gamma^2)t} \tag{2-79}$$

The complete solution for the problem is constructed as

$$T(x,y,t)=\sum_{m=1}^{\infty}\sum_{n=1}^{\infty}c_{mn}e^{-\alpha(\beta_m^2+\gamma_n^2)t}X(\beta_m,x)Y(\gamma_n,y) \tag{2-80}$$

For $t=0$ equation (2-80) becomes

$$F(x,y)=\sum_{m=1}^{\infty}\sum_{n=1}^{\infty}c_{mn}X(\beta_m,x)Y(\gamma_n,y) \qquad \text{in} \qquad 0<x<a, \quad 0<y<b \tag{2-81}$$

The unknown coefficient c_{mn} is determined by operating on both sides of equation (2-81) successively by the operators

$$\int_0^a X(\beta_m,x)\,dx \qquad \text{and} \qquad \int_0^b Y(\gamma_n,y)dy$$

and utilizing the orthogonality of these eigenfunctions. We obtain

$$c_{mn} = \frac{1}{N(\beta_m)N(\gamma_n)} \int_{x'=0}^{a} \int_{y'=0}^{b} X(\beta_m, x')Y(\gamma_n, y')F(x', y')\,dx'\,dy' \tag{2-82}$$

where

$$N(\beta_m) \equiv \int_0^a X^2(\beta_m, x)dx \qquad \text{and} \qquad N(\gamma_n) \equiv \int_0^b Y^2(\gamma_n, y)\,dy$$

The substitution of equation (2-82) into equation (2-80) gives the solution of this problem as

$$T(x, y, t) = \sum_{m=1}^{\infty} \sum_{n=1}^{\infty} e^{-\alpha(\beta_m^2 + \gamma_n^2)t} \frac{1}{N(\beta_m)N(\gamma_n)} X(\beta_m, x)Y(\gamma_n, y) \int_0^a \int_0^b X(\beta_m, x') \cdot Y(\gamma_n, y')F(x', y')\,dx'\,dy' \tag{2-83}$$

The eigenfunctions, the eigenvalues, and the norms appearing in equation (2-83) are immediately obtainable from Table 2-2; that is, $X(\beta_m, x)$ satisfying the eigenvalue problem (2-77) corresponds to case 4 and is given by

$$X(\beta_m, x) = \cos \beta_m x \tag{2-84a}$$

$$\frac{1}{N(\beta_m)} = 2\frac{\beta_m^2 + H_2^2}{a(\beta_m^2 + H_2^2) + H_2} \tag{2-84b}$$

and the β_m values are the positive roots of

$$\beta_m \tan \beta_m a = H_2 \tag{2-84c}$$

The function $Y(\gamma_n, y)$ satisfying the eigenvalue problem (2-78) corresponds to case 7 in Table 2-2; after replacing L by b, β by γ, and H_2 by H_4, we find

$$Y(\gamma_n, y) = \sin \gamma_n y \tag{2-85a}$$

$$\frac{1}{N(\gamma_n)} = 2\frac{\gamma_n^2 + H_4^2}{b(\gamma_n^2 + H_4^2) + H_4} \tag{2-85b}$$

and the γ_n values are the positive roots of

$$\gamma_n \cot \gamma_n b = -H_4 \tag{2-85c}$$

Introducing equations (2-84) and (2-85) into equation (2-83), the solution

becomes

$$T(x,y,t) = 4 \sum_{m=1}^{\infty} \sum_{n=1}^{\infty} e^{-\alpha(\beta_m^2+\gamma_n^2)t} \cdot \frac{\beta_m^2 + H_2^2}{a(\beta_m^2 + H_2^2) + H_2} \frac{\gamma_n^2 + H_4^2}{b(\gamma_n^2 + H_4^2) + H_4}$$

$$\cdot \cos\beta_m x \sin\gamma_n y \int_{x'=0}^{a} \int_{y'=0}^{b} \cos\beta_m x' \sin\gamma_n y' F(x',y')\, dx'\, dy' \quad (2\text{-}86)$$

where β_m and γ_n are the positive roots of the equations (2-84c) and (2-85c), respectively.

Example 2-7

A rectangular parallelepiped $0 \leqslant x \leqslant a, 0 \leqslant y \leqslant b, 0 \leqslant z \leqslant c$ is initially at temperature $F(x,y,z)$. For times $t > 0$ all boundary surfaces are kept at zero temperature. Obtain an expression for $T(x,y,z,t)$ for times $t > 0$.

Solution. The mathematical formulation of this problem is given as

$$\frac{\partial^2 T}{\partial x^2} + \frac{\partial^2 T}{\partial y^2} + \frac{\partial^2 T}{\partial z^2} = \frac{1}{\alpha}\frac{\partial T}{\partial t} \quad \text{in} \quad 0<x<a, \quad 0<y<b, \quad 0<z<c,$$

$$\text{for} \quad t>0 \quad (2\text{-}87a)$$

$$T = 0 \quad \text{on all boundaries, for } t>0 \quad (2\text{-}87b)$$

$$T = F(x,y,z) \quad \text{for } t=0\text{, in the region} \quad (2\text{-}87c)$$

Assuming a separation in the form $T(x,y,z,t) = \Gamma(t)X(x)Y(y)Z(z)$, the complete solution for $T(x,y,z,t)$ in terms of these separated functions is written as

$$T(x,y,z,t) = \sum_{m=1}^{\infty} \sum_{n=1}^{\infty} \sum_{p=1}^{\infty} c_{mnp} e^{-\alpha(\beta_m^2+\gamma_n^2+\eta_p^2)t} X(\beta_m,x)Y(\gamma_n,y)Z(\eta_p,z) \quad (2\text{-}88)$$

The application of the initial condition gives

$$F(x,y,z) = \sum_{m=1}^{\infty} \sum_{n=1}^{\infty} \sum_{p=1}^{\infty} c_{mnp} X(\beta_m,x)Y(\gamma_n,y)Z(\eta_p,z) \quad (2\text{-}89)$$

The unknown coefficient c_{mnp} is determined by operating on both sides of equation (2-89) successively by the operators

$$\int_0^a X(\beta_m,x)\,dx, \qquad \int_0^b Y(\gamma_n,y)\,dy, \qquad \text{and} \qquad \int_0^c Z(\eta_p,z)\,dz$$

and utilizing the orthogonality of these eigenfunctions. We obtain

$$c_{mnp} = \frac{1}{N(\beta_m)N(\gamma_n)N(\eta_p)} \int_{x'=0}^{a} \int_{y'=0}^{b} \int_{z'=0}^{c} X(\beta_m, x')Y(\gamma_n, y')Z(\eta_p, z') \cdot F(x', y', z')\, dx'\, dy'\, dz' \tag{2-90}$$

where

$$N(\beta_m) \equiv \int_0^a X^2(\beta_m, x)\, dx, \qquad N(\gamma_n) \equiv \int_0^b Y^2(\gamma_n, y)\, dy,$$

and

$$N(\eta_p) \equiv \int_0^c Z^2(\eta_p, z)\, dz$$

The substitution of equation (2-90) into equation (2-88) gives the solution as

$$T(x, y, z, t) = \sum_{m=1}^{\infty} \sum_{n=1}^{\infty} \sum_{p=1}^{\infty} e^{-\alpha(\beta_m^2 + \gamma_n^2 + \eta_p^2)t} \frac{1}{N(\beta_m)N(\gamma_n)N(\eta_p)} X(\beta_m, x)Y(\gamma_n, y)Z(\eta_p, z) \cdot \int_{x=0}^{a} \int_{y=0}^{b} \int_{z=0}^{c} X(\beta_m, x')Y(\gamma_n, y')Z(\eta_p, z')F(x', y', z')\, dx'\, dy'\, dz' \tag{2-91}$$

Here, the functions X, Y, Z satisfy the eigenvalue problems whose solutions corresponds to those given by case 9 in Table 2-2. Therefore, from Table 2-2 we immediately obtain

$$X(\beta_m, x) = \sin \beta_m x, \qquad \frac{1}{N(\beta_m)} = \frac{2}{a} \qquad \text{and} \qquad \beta_m\text{'s are roots of } \sin \beta_m a = 0$$

$$Y(\gamma_n, y) = \sin \gamma_n y, \qquad \frac{1}{N(\gamma_n)} = \frac{2}{b} \qquad \text{and} \qquad \gamma_n\text{'s are roots of } \sin \gamma_n b = 0$$

$$Z(\eta_p, z) = \sin \eta_p z, \qquad \frac{1}{N(\eta_p)} = \frac{2}{c} \qquad \text{and} \qquad \eta_p\text{'s are roots of } \sin \eta_p c = 0$$

Hence, the solution (2-91) becomes

$$T(x, y, z, t) = \frac{8}{abc} \sum_{m=1}^{\infty} \sum_{n=1}^{\infty} \sum_{p=1}^{\infty} e^{-\alpha(\beta_m^2 + \gamma_n^2 + \eta_p^2)t} \cdot \sin \beta_m x \sin \gamma_n y \sin \eta_p z \cdot \int_{x'=0}^{a} \int_{y'=0}^{b} \int_{z'=0}^{c} \sin \beta_m x' \sin \gamma_n y' \sin \eta_p z' F(x', y', z')\, dx'\, dy'\, dz' \tag{2-92}$$

where

$$\beta_m = \frac{m\pi}{a}, \qquad m = 1, 2, 3, \ldots$$

$$\gamma_n = \frac{n\pi}{b}, \qquad n = 1, 2, 3, \ldots$$

$$\eta_p = \frac{p\pi}{c}, \qquad p = 1, 2, 3, \ldots$$

Example 2-8

A semiinfinite rectangular strip $0 \leqslant y \leqslant b, 0 \leqslant x < \infty$ is initially at temperature $F(x, y)$. For times $t > 0$ the boundaries at $x = 0$ and $y = b$ are kept at zero temperature and the boundary at $y = 0$ dissipates heat by convection into an environment at zero temperature as illustrated in Fig. 2-7. Obtain an expression for the temperature $T(x, y, t)$ for times $t > 0$.

Solution. The mathematical formulation of this problem is given as

$$\frac{\partial^2 T}{\partial x^2} + \frac{\partial^2 T}{\partial y^2} = \frac{1}{\alpha}\frac{\partial T}{\partial t} \quad \text{in} \quad 0 < x < \infty, \quad 0 < y < b, \quad t > 0 \tag{2-93a}$$

$$T = 0 \quad \text{at} \quad x = 0, \quad t > 0 \tag{2-93b}$$

$$-\frac{\partial T}{\partial y} + H_1 T = 0 \quad \text{at} \quad y = 0, \quad t > 0 \tag{2-93c}$$

$$T = 0 \quad \text{at} \quad y = b, \quad t > 0 \tag{2-93d}$$

$$T = F(x, y) \quad \text{for} \quad t = 0, \text{ in the region} \tag{2-93e}$$

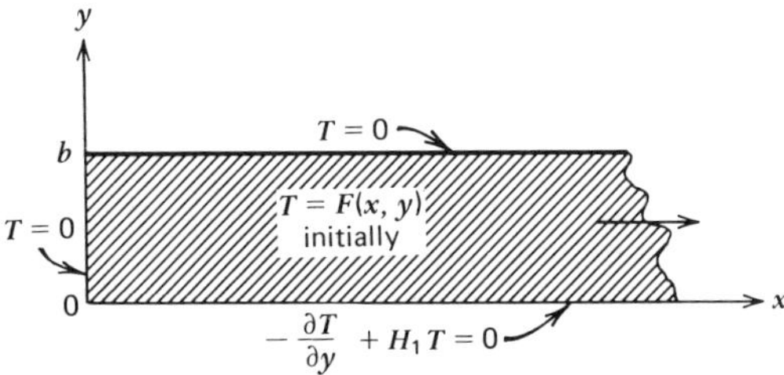

Fig. 2-7 Boundary and initial conditions for a semiinfinite strip considered in Example 2-8.

The general solution for $T(x, y, t)$ is written in the form

$$T(x, y, t) = \sum_{n=1}^{\infty} \int_{\beta=0}^{\infty} c_n(\beta) e^{-\alpha(\beta^2 + \gamma_n^2)t} X(\beta, x) Y(\gamma_n, y)\, d\beta \tag{2-94}$$

where $X(\beta, x)$ satisfies the auxiliary problem

$$\frac{d^2 X(x)}{dx^2} + \beta^2 X(x) = 0 \quad \text{in} \quad 0 < x < \infty \tag{2-95a}$$

$$X = 0 \quad \text{at} \quad x = 0 \tag{2-95b}$$

and $Y(\gamma_n, y)$ satisfies the eigenvalue problem

$$\frac{d^2 Y(y)}{dy^2} + \gamma_n^2 Y(y) = 0 \quad \text{in} \quad 0 < y < b \tag{2-96a}$$

$$-\frac{dY}{dy} + H_1 Y = 0 \quad \text{at} \quad y = 0 \tag{2-96b}$$

$$Y = 0 \quad \text{at} \quad y = b \tag{2-96c}$$

The application of the initial condition to equation (2-94) yields

$$F(x, y) = \sum_{n=1}^{\infty} \int_{\beta=0}^{\infty} c_n(\beta) X(\beta, x) Y(\gamma_n, y)\, d\beta \tag{2-97}$$

To determine the unknown coefficients $c_n(\beta)$ we first operate on both sides of equation (2-97) by the operator $\int_0^b Y(\gamma_n, y)\, dy$ and utilize the othogonality of the eigenfunctions $Y(\gamma_n, y)$. We obtain

$$f^*(x) = \int_{\beta=0}^{\infty} c_n(\beta) X(\beta, x)\, d\beta \quad \text{in} \quad 0 < x < \infty \tag{2-98a}$$

where

$$f^*(x) \equiv \frac{1}{N(\gamma_n)} \int_{y=0}^{b} Y(\gamma_n, y) F(x, y)\, dy, \qquad N(\gamma_n) \equiv \int_0^b Y^2(\gamma_n, y)\, dy \tag{2-98b}$$

Equation (2-98a) is a representation of an arbitrary function $f^*(x)$, defined in the interval $0 < x < \infty$ in terms of the functions $X(\beta, x)$, which are the solution of the auxiliary problem (2-95). This representation is of exactly the same form as that given by equation (2-51); the coefficient of equation (2-51) is given by equation (2-53). Therefore, the unknown coefficient $c(\beta)$ in equation (2-98a) is determined according to equation (2-53); we obtain

$$c_n(\beta) = \frac{1}{N(\beta)} \int_{x=0}^{\infty} X(\beta, x) \left[\frac{1}{N(\gamma_n)} \int_{y=0}^{b} Y(\gamma_n, y) F(x, y)\, dy \right] dx \tag{2-99}$$

The substitution of equation (2-99) into equation (2-94) gives the solution for $T(x, y, t)$ as

$$T(x, y, t) = \sum_{n=1}^{\infty} \int_{\beta=0}^{\infty} e^{-\alpha(\beta^2 + \gamma_n^2)t} \frac{1}{N(\beta) N(\gamma_n)} X(\beta, x) Y(\gamma_n, y) \cdot$$

$$\cdot \left[\int_{x'=0}^{\infty} \int_{y'=0}^{b} X(\beta, x') Y(\gamma_n, y') F(x', y')\, dx'\, dy' \right] d\beta \tag{2-100}$$

The eigenfunctions $Y(\gamma_n, y)$, the norm $N(\gamma_n)$ and the eigenvalues γ_n for the y separation are immediately obtainable from case 3 of Table 2-2 by appropriate changes in the symbols. We find

$$Y(\gamma_n, y) = \sin \gamma_n (b - y) \tag{2-101a}$$

$$\frac{1}{N(\gamma_n)} = 2 \frac{\gamma_n^2 + H_1^2}{b(\gamma_n^2 + H_1^2) + H_1} \tag{2-101b}$$

and the γ_n values are the positive roots of

$$\gamma_n \cot \gamma_n b = -H_1 \tag{2-101c}$$

The function $X(\beta, x)$ and the norm $N(\beta)$ are obtained from case 3 of Table 2-3, as

$$X(\beta, x) = \sin \beta x \tag{2-102a}$$

$$\frac{1}{N(\beta)} = \frac{2}{\pi} \tag{2-102b}$$

Substituting equations (2-101) and (2-102) into equation (2-100) and after changing the orders of integration, we obtain

$$T(x, y, t) = \frac{4}{\pi} \sum_{n=1}^{\infty} e^{-\alpha \gamma_n^2 t} \frac{\gamma_n^2 + H_1^2}{b(\gamma_n^2 + H_1^2) + H_1} \sin \gamma_n (b - y)$$

$$\int_{x'=0}^{\infty} \int_{y'=0}^{b} F(x', y') \sin \gamma_n (b - y')\, dx'\, dy' \int_{\beta=0}^{\infty} e^{-\alpha \beta^2 t} \sin \beta x' \sin \beta x\, d\beta \tag{2-103}$$

The last integral with respect to β was evaluated previously and the result given by equation (2-57d); then equation (2-103) becomes

$$T(x,y,t)=\frac{1}{(\pi\alpha t)^{1/2}}\sum_{n=1}^{\infty}e^{-\alpha\gamma_n^2 t}\frac{\gamma_n^2+H_1^2}{b(\gamma_n^2+H_1^2)+H_1}\sin\gamma_n(b-y)$$
$$\cdot\int_{x'=0}^{\infty}\int_{y'=0}^{b}F(x',y')\sin\gamma_n(b-y')\left[\exp\left(-\frac{(x-x')^2}{4\alpha t}\right)\right.$$
$$\left.-\exp\left(-\frac{(x+x')^2}{4\alpha t}\right)\right]dx'\,dy' \tag{2-104}$$

2-10 PRODUCT SOLUTION

In the rectangular coordinate system, the solution of multidimensional homogeneous heat conduction problems can be written down very simply as the product of the solutions of one-dimensional problems if the initial temperature distribution in the medium is expressible as a product of single space variable functions. For example, for a two-dimensional problem it may be in the form $F(x,y)=F_1(x)\,F_2(y)$, or for a three-dimensional problem in the form $F(x,y,z)=F_1(x)F_2(y)F_3(z)$. Clearly, the case of uniform temperature initial condition also is expressible in the product form.

To illustrate this method we consider the following two-dimensional homogeneous heat conduction problem for a rectangular region $0\leqslant x\leqslant a$, $0\leqslant y\leqslant b$:

$$\frac{\partial^2 T}{\partial x^2}+\frac{\partial^2 T}{\partial y^2}=\frac{1}{\alpha}\frac{\partial T}{\partial t}\quad\text{in}\quad 0<x<a,\quad 0<y<b,\quad t>0 \tag{2-105a}$$

$$-k_1\frac{\partial T}{\partial x}+h_1T=0\quad\text{at}\quad x=0,\quad t>0 \tag{2-105b}$$

$$k_2\frac{\partial T}{\partial x}+h_2T=0\quad\text{at}\quad x=a,\quad t>0 \tag{2-105c}$$

$$-k_3\frac{\partial T}{\partial y}+h_3T=0\quad\text{at}\quad y=0,\quad t>0 \tag{2-105d}$$

$$k_4\frac{\partial T}{\partial y}+h_4T=0\quad\text{at}\quad y=b,\quad t>0 \tag{2-105e}$$

$$T=F_1(x)F_2(y)\quad\text{for}\quad t=0,\text{ in the region} \tag{2-105f}$$

where

$$T=T(x,y,t)$$

To solve this problem we consider the following two one-dimensional homogeneous heat conduction problems for slabs $0 \leqslant x \leqslant a$ and $0 \leqslant y \leqslant b$, given as

$$\frac{\partial^2 T_1}{\partial x^2} = \frac{1}{\alpha}\frac{\partial T_1}{\partial t} \quad \text{in} \quad 0 < x < a, \quad t > 0 \tag{2-106a}$$

$$-k_1 \frac{\partial T_1}{\partial x} + h_1 T_1 = 0 \quad \text{at} \quad x = 0, \quad t > 0 \tag{2-106b}$$

$$k_2 \frac{\partial T_1}{\partial x} + h_2 T_1 = 0 \quad \text{at} \quad x = a, \quad t > 0 \tag{2-106c}$$

$$T_1 = F_1(x) \quad \text{for} \quad t = 0, \quad \text{in} \quad 0 \leqslant x \leqslant a \tag{2-106d}$$

and

$$\frac{\partial^2 T_2}{\partial y^2} = \frac{1}{\alpha}\frac{\partial T_2}{\partial t} \quad \text{in} \quad 0 < y < b, \quad t > 0 \tag{2-107a}$$

$$-k_3 \frac{\partial T_2}{\partial y} + h_3 T_2 = 0 \quad \text{at} \quad y = 0, \quad t > 0 \tag{2-107b}$$

$$k_4 \frac{\partial T_2}{\partial y} + h_4 T_2 = 0 \quad \text{at} \quad y = b, \quad t > 0 \tag{2-107c}$$

$$T_2 = F_2(y) \quad \text{for} \quad t = 0, \quad \text{in} \quad 0 \leqslant y \leqslant b \tag{2-107d}$$

Here we note that the boundary conditions for the problem (2-106) are the same as those given by equations (2-105b,c) and those for the problem (2-107) are the same as those given by equations (2-105d,e). Then the solution of the two-dimensional problem (2-105) is given as the product solution of the above one-dimensional problems as

$$T(x, y, t) = T_1(x, t) \cdot T_2(y, t) \tag{2-108}$$

To prove the validity of this result we substitute equation (2-108) into equations (2-105) and utilize the equations (2-106) and (2-107). For example, the substitution of equation (2-108) into the differential equation (2-105a) gives

$$T_2 \frac{\partial^2 T_1}{\partial x^2} + T_1 \frac{\partial^2 T_2}{\partial y^2} = \frac{1}{\alpha} T_2 \frac{\partial T_1}{\partial t} + \frac{1}{\alpha} T_1 \frac{\partial T_2}{\partial t}$$

or

$$T_2\left(\frac{\partial^2 T_1}{\partial x^2} - \frac{1}{\alpha}\frac{\partial T_1}{\partial t}\right) + T_1\left(\frac{\partial^2 T_2}{\partial y^2} - \frac{1}{\alpha}\frac{\partial T_2}{\partial t}\right) = 0 \tag{2-109}$$

Thus, the differential equation is satisfied in view of equations (2-106a) and (2-107a).

Similarly, the substitution of equation (2-107) into the boundary conditions (2-105) shows that they are also satisfied. Hence, equation (2-108) is the solution of the problem (2-105).

Example 2-9

A semiinfinite corner, $0 \leqslant x < \infty$ and $0 \leqslant y < \infty$, is initially at a constant temperature T_0; for times $t > 0$ the boundaries at $x = 0$ and $y = 0$ are kept at zero temperatures. Obtain an expression for the temperature $T(x, y, t)$ in the region for times $t > 0$.

Solution. The solution of this problem can be expressed as the product of the solutions of the following two one-dimensional problems: (1) $T_1(x, t)$, the solution for a semifinite region $0 \leqslant x < \infty$ initially at a temperature $F_1(x) = 1$ and for times $t > 0$ the boundary surface at $x = 0$ is kept at zero temperature, and (2) $T_2(y, t)$, the solution for a semiinfinite region $0 \leqslant y < \infty$ initially at a temperature $F_2(y) = T_0$ and for times $t > 0$ the boundary at $y = 0$ is kept at zero temperature. Clearly, the initial condition for the two-dimensional problem is expressible as a product, $T_0 = 1 \cdot T_0$. The solution of such one-dimensional problem was considered previously in Example 2-3; thus obtaining these solutions from equation (2-58e), we write

$$T_1(x, t) = \operatorname{erf}\left(\frac{x}{\sqrt{4\alpha t}}\right) \quad \text{and} \quad T_2(y, t) = T_0 \operatorname{erf}\left(\frac{y}{\sqrt{4\alpha t}}\right)$$

Then, the solution for the above two-dimensional problem becomes

$$T(x, y, t) = T_1(x, t) T_2(y, t) = T_0 \operatorname{erf}\left(\frac{x}{\sqrt{4\alpha t}}\right) \operatorname{erf}\left(\frac{y}{\sqrt{4\alpha t}}\right) \tag{2-110}$$

Example 2-10

A rectangular region $0 \leqslant x \leqslant a, 0 \leqslant y \leqslant b$ is initially at a uniform temperature $F(x, y) = T_0$. For times $t > 0$, the boundaries at $x = 0$ and $y = 0$ are insulated and the boundaries at $x = a$ and $y = b$ dissipate heat by convection into an environment at zero temperature with a heat transfer coefficient h (or $H = h/k$). Figure 2-8 illustrates the boundary conditions for this problem. Obtain an expression for the temperature distribution $T(x, y, t)$ for times $t > 0$.

Solution. The solution of this problem can be expressed as the product of the solutions of the following two slab problems: (1) $T_1(x, t)$, for a slab, $0 \leqslant x \leqslant a$, initially at a temperature $F(x) = 1$ and for times $t > 0$ the boundary at $x = 0$ is insulated and the boundary at $x = a$ dissipates heat by convection into an environment at zero temperature with a heat transfer coefficient h (or $H = h/k$);

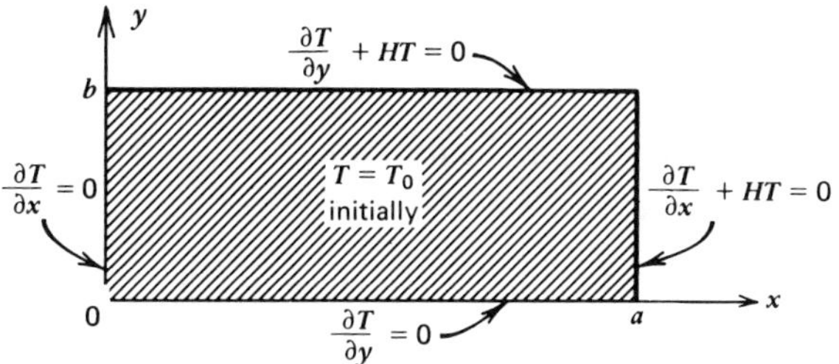

Fig. 2-8 Boundary and initial conditions for a rectangular region considered in Example 2-10.

and (2) $T_2(y,t)$, for a slab, $0 \leqslant y \leqslant b$, initially at a temperature $F(y) = T_0$ and for times $t > 0$ the boundary at $y = 0$ is insulated and the boundary at $y = b$ dissipates heat by convection into an environment at zero temperature with a heat transfer coefficient h (or $H = h/k$). These slab problems were solved previously in Example 2-1 and the solutions for $T_1(x,t)$ and $T_2(y,t)$ are readily obtainable from equation (2-45d) by appropriate changes in the parameters. We set $T_0 = 1$, $L = a$, and $H_2 = H$ to obtain

$$T_1(x,t) = 2 \sum_{m=1}^{\infty} e^{-\alpha\beta_m^2 t} \frac{H}{a(\beta_m^2 + H^2) + H} \frac{\cos \beta_m x}{\cos \beta_m a} \tag{2-111a}$$

where the β_m values are the positive roots of $\beta_m \tan \beta_m a = H$.

We set $L = b$, $H_2 = H$, $x = y$, and $\beta_m = \gamma_n$ to find

$$T_2(y,t) = 2T_0 \sum_{n=1}^{\infty} e^{-\alpha\gamma_n^2 t} \frac{H}{b(\gamma_n^2 + H^2) + H} \frac{\cos \gamma_n y}{\cos \gamma_n b} \tag{2-111b}$$

where the γ_n values are the positive roots of $\gamma_n \tan \gamma_n b = H$.

Then the solution of the above problem for the rectangular region becomes

$$T(x,y,t) = T_1(x,t) T_2(y,t) \tag{2-111c}$$

$$T(x,y,t) = 4T_0 \sum_{m=1}^{\infty} \sum_{n=1}^{\infty} \frac{H^2 e^{-\alpha(\beta_m^2 + \gamma_n^2)t}}{[a(\beta_m^2 + H^2) + H][b(\gamma_n^2 + H^2) + H]} \frac{\cos \beta_m x \cos \gamma_n y}{\cos \beta_m a \cos \gamma_n b} \tag{2-111d}$$

2-11 MULTIDIMENSIONAL STEADY-STATE PROBLEMS WITH NO HEAT GENERATION

The multidimensional steady-state heat conduction problem with no heat generation can be solved by the separation of variables technique when only one of the boundary conditions is nonhomogeneous. If the problem involves more than one nonhomogeneous boundary condition, it can be split up into a set of simpler problems each containing only one nonhomogeneous boundary condition; the method of separation of variables can then be used to solve the resulting simpler problems. Consider, for example the following steady-state problem subject to more than one nonhomogeneous boundary condition

$$\nabla^2 T(\mathbf{r}) = 0 \qquad \text{in region } R \tag{2-112a}$$

$$k_i \frac{\partial T}{\partial n_i} + h_i T = f_i \qquad \text{on boundary } S_i \tag{2-112b}$$

where $\partial/\partial n_i$ is the derivative along the outward-drawn normal to the boundary surface S_i, $i = 1, 2, \ldots, N$ and N is the number of continuous boundary surfaces of the region, and f_i is the nonhomogeneous part of the boundary condition at the surface S_i. This problem can be split up into a set of simpler problems for the temperatures $T_j(\mathbf{r})$ in the form

$$\nabla^2 T_j(\mathbf{r}) = 0 \qquad \text{in region } R \tag{2-113a}$$

$$k_i \frac{\partial T_j}{\partial n_i} + h_i T_j = \delta_{ij} f_i \qquad \text{on boundary } S_i \tag{2-113b}$$

where

$$i = 1, 2, \ldots, N$$

$$j = 1, 2, \ldots, N$$

$$\delta_{ij} = \text{Kronecker delta} = \begin{cases} 0 & \text{for } i \neq j \\ 1 & \text{for } i = j \end{cases}$$

Clearly, each of the steady-state problems given by equations (2-113) has only one nonhomogeneous boundary condition. Then, the solution of the heat conduction problem (2-112) is obtained by the superposition of these simpler problems in the form

$$T(\mathbf{r}) = \sum_{j=1}^{N} T_j(\mathbf{r}) \tag{2-114}$$

The validity of this result is readily verified by substituting equation (2-114) into equations (2-112) and utilizing equations (2-113).

Example 2-11

Obtain an expression for the steady-state temperature distribution $T(x, y)$ in a rectangular region $0 \leqslant x \leqslant a, 0 \leqslant y \leqslant b$ for the boundary conditions shown in Fig. 2-9.

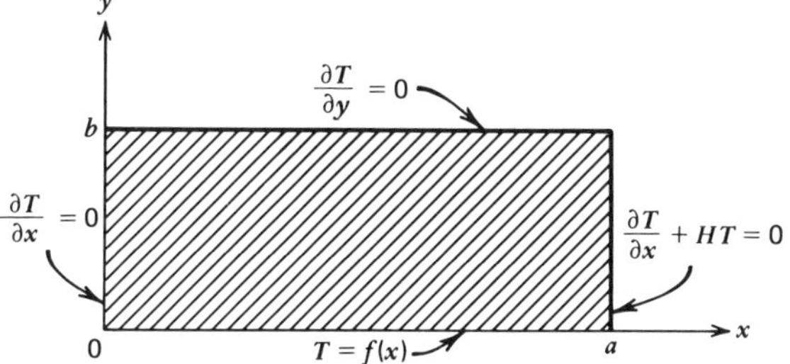

Fig. 2-9 Boundary conditions for a rectangular region considered in Example 2-11.

Solution. The mathematical formulation of the problem is given as

$$\frac{\partial^2 T(x, y)}{\partial x^2} + \frac{\partial^2 T(x, y)}{\partial y^2} = 0 \qquad \text{in} \qquad 0 < x < a, \quad 0 < y < b \tag{2-115a}$$

$$\frac{\partial T}{\partial x} = 0 \qquad \text{at } x = 0, \qquad \frac{\partial T}{\partial x} + HT = 0 \qquad \text{at } x = a \tag{2-115b}$$

$$T = f(x) \qquad \text{at } y = 0, \qquad \frac{\partial T}{\partial y} = 0 \qquad \text{at } y = b \tag{2-115c}$$

In this problem the boundary condition at $y = 0$ is nonhomogeneous; looking ahead in the analysis, we conclude that the nonhomogeneous part $f(x)$ of this boundary condition should be represented in terms of the separated solution $X(x)$ for the problem. Therefore, when separating the temperature in the form $T(x, y) = X(x)Y(y)$, the sign of the separation constant should be so chosen as to produce an eigenvalue problem for the function $X(x)$. With this consideration the separated problems become

$$\frac{d^2 X(x)}{dx^2} + \beta^2 X(x) = 0 \qquad \text{in} \qquad 0 < x < a \tag{2-116a}$$

$$\frac{dX}{dx} = 0 \qquad \text{at} \qquad x = 0 \tag{2-116b}$$

$$\frac{dX}{dx} + HX = 0 \qquad \text{at} \qquad x = a \tag{2-116c}$$

and

$$\frac{d^2 Y(y)}{dy^2} - \beta^2 Y(y) = 0 \qquad \text{in} \qquad 0 < y < b \tag{2-117a}$$

$$\frac{dY}{dy} = 0 \qquad \text{at} \qquad y = b \tag{2-117b}$$

The solution of the eigenvalue problem (2-116) is immediately obtainable from Table 2-2 as case 4; by replacing L by a and H_2 by H we find

$$X(\beta_m, x) = \cos \beta_m x, \qquad \frac{1}{N(\beta_m)} = 2\frac{\beta_m^2 + H^2}{a(\beta_m^2 + H^2) + H} \tag{2-118a}$$

and the β_m values are the positive roots of

$$\beta_m \tan \beta_m a = H \tag{2-118b}$$

The solution of equations (2-117) is taken as

$$Y(\beta_m, y) = \cosh \, \beta_m(b - y) \tag{2-118c}$$

The complete solution for $T(x, y)$ is constructed as

$$T(x, y) = \sum_{m=1}^{\infty} c_m \cosh \beta_m(b - y) \cos \beta_m x \tag{2-119}$$

which satisfies the heat-conduction equation (2-115a) and its three homogeneous boundary conditions; the coefficients c_m should be so determined that this solution also satisfies the nonhomogeneous boundary condition. The application of the boundary condition at $y = 0$ yields

$$f(x) = \sum_{m=1}^{\infty} c_m \cosh \beta_m b \cos \beta_m x \qquad \text{in} \qquad 0 < x < a \tag{2-120}$$

The coefficients c_m are determined by utilizing the orthogonality of the functions $\cos \beta_m x$; we find

$$c_m = \frac{1}{N(\beta_m) \cosh \beta_m b} \int_0^a \cos \beta_m x' f(x') \, dx' \tag{2-121}$$

The substitution of this expression into equation (2-119) together with the value of $N(\beta_m)$ as given above, results in the solution

$$T(x, y) = 2 \sum_{m=1}^{\infty} \frac{\beta_m^2 + H^2}{a(\beta_m^2 + H^2) + H} \, \frac{\cosh \beta_m(b - y)}{\cosh \beta_m b} \cos \beta_m x \int_0^a \cos \beta_m x' f(x') \, dx' \tag{2-122}$$

where the β_m values are the roots of equation (2-118b).

Example 2-12

Obtain an expression for the steady-state temperature $T(x, y)$ in a semiinfinite strip $0 \leqslant y \leqslant b, 0 \leqslant x < \infty$ for the boundary conditions shown in Fig. 2-10.

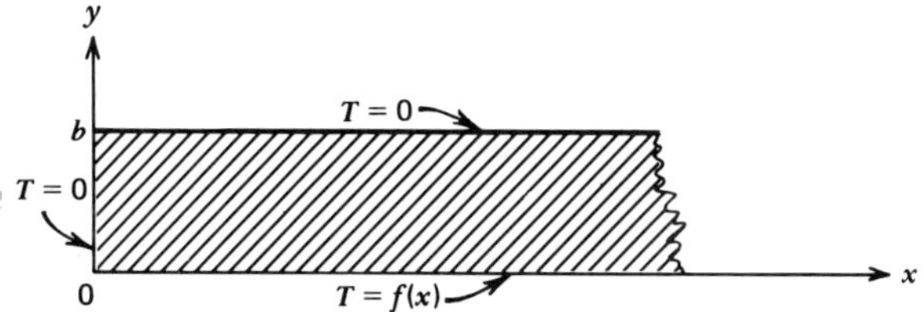

Fig. 2-10 Boundary conditions for a semiinifinite strip considered in Example 2-12.

Solution. The mathematical formulation of this problem is given as

$$\frac{\partial^2 T(x,y)}{\partial x^2} + \frac{\partial^2 T(x,y)}{\partial y^2} = 0 \qquad \text{in} \qquad 0 < y < b, \quad 0 < x < \infty \tag{2-123a}$$

$$T = 0 \qquad \text{at } x = 0, \tag{2-123b}$$

$$T = f(x) \qquad \text{at } y = 0, \quad T = 0 \quad \text{at} \quad y = \text{b} \tag{2-123c}$$

The separated equations for the functions $X(x)$ and $Y(y)$ are now constructed by considering the fact that the nonhomogeneous boundary condition function $f(x)$ defined in the interval $0 < x < \infty$ should be represented by the function $X(x)$. Then the separated problems become

$$\frac{d^2 X(x)}{dx^2} + \beta^2 X(x) = 0 \qquad \text{in} \qquad 0 < x < \infty \tag{2-124a}$$

$$X = 0 \qquad \text{at} \qquad x = 0 \tag{2-124b}$$

and

$$\frac{d^2 Y(y)}{dy^2} - \beta^2 Y(y) = 0 \qquad 0 < y < b \tag{2-125a}$$

$$Y = 0 \qquad \text{at} \qquad y = b \tag{2-125b}$$

The solution of the problem (2-124) is obtainable from Table 2-3, case 3, as

$$X(\beta, x) = \sin \beta x \qquad \text{and} \qquad \frac{1}{N(\beta)} = \frac{2}{\pi} \tag{2-126a}$$

and the solution of (2-125) is given as

$$Y(\beta, y) = \sinh \beta(b - y) \tag{2-126b}$$

Then the complete solution for $T(x, y)$ is constructed as

$$T(x, y) = \int_{\beta=0}^{\infty} A(\beta) \sinh \beta(b - y) \sin \beta x \, d\beta \tag{2-127}$$

If this solution should also satisfy the nonhomogeneous boundary condition at $y = 0$ for the above heat conduction problem, we obtain

$$f(x) = \int_{\beta=0}^{\infty} A(\beta) \sinh \beta b \sin \beta x \, d\beta \qquad \text{in} \qquad 0 < x < \infty \tag{2-128a}$$

This is a representation of function $f(x)$ defined in the interval $0 < x < \infty$; but it is a special case of the representation given by equation (2-51). The coefficient for equation (2-51) is given by equation (2-53). Therefore, the coefficient of equation (2-128a) is determined from the result in equation (2-53) as

$$A(\beta) \sinh \beta b = \frac{1}{N(\beta)} \int_0^{\infty} \sin \beta x' f(x') \, dx' \tag{2-128b}$$

where

$$\frac{1}{N(\beta)} = \frac{2}{\pi}$$

as given previously. The substitution of $A(\beta)$ into equation (2-127) gives

$$T(x, y) = \frac{2}{\pi} \int_{\beta=0}^{\infty} \frac{\sinh \beta(b - y)}{\sinh \beta b} \sin \beta x \, d\beta \int_{x'=0}^{\infty} \sin \beta x' f(x') \, dx' \tag{2-129a}$$

or changing the order of integration we obtain

$$T(x, y) = \frac{2}{\pi} \int_{x'=0}^{\infty} f(x') \, dx' \int_{\beta=0}^{\infty} \frac{\sinh \beta(b - y)}{\sinh \beta b} \sin \beta x \sin \beta x' \, d\beta \tag{2-129b}$$

The integral with respect to β has been evaluated [16, Section 10.11]; then the solution for the temperature becomes

$$T(x, y) = \frac{1}{2b} \sin \frac{\pi y}{b} \int_{x'=0}^{\infty} f(x') \left[\frac{1}{\cos [\pi(b - y)/b] + \cosh [\pi(x - x')/b]} - \frac{1}{\cos [\pi(b - y)/b] + \cosh [\pi(x + x')/b]} \right] dx' \tag{2-130}$$

Example 2-13

Obtain an expression for the steady-state temperature $T(x, y)$ in an infinite strip $0 \leqslant y \leqslant b$, $-\infty < x < \infty$ for the boundary conditions shown in Fig. 2-11.

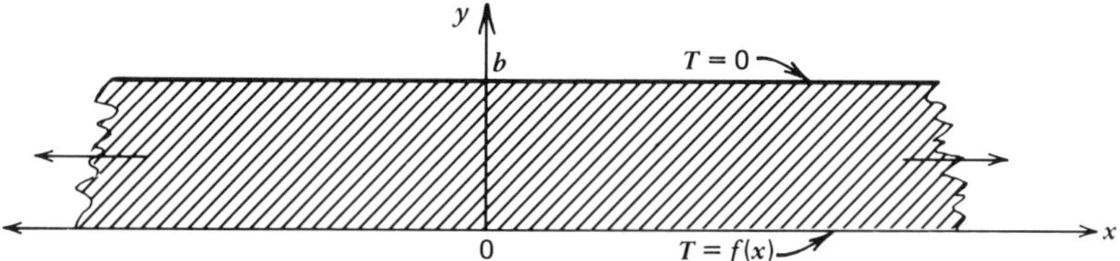

Fig. 2-11 Boundary conditions for an infinite strip considered in Example 2-13.

Solution. The mathematical formulation of this problem is given as

$$\frac{\partial^2 T(x,y)}{\partial x^2}+\frac{\partial^2 T(x,y)}{\partial y^2}=0 \qquad \text{in} \qquad -\infty<x<\infty, \quad 0<y<b \tag{2-131a}$$

$$T=f(x) \qquad \text{at } y=0, \qquad T=0 \qquad \text{at} \qquad y=b \tag{2-131b}$$

The separated problems are taken as

$$\frac{d^2X(x)}{dx^2}+\beta^2 X(x)=0 \qquad \text{in} \qquad -\infty<x<\infty \tag{2-132a}$$

and

$$\frac{d^2Y(y)}{dy^2}-\beta^2 Y(y)=0 \qquad \text{in} \qquad 0<y<b \tag{2-132b}$$

$$Y(y)=0 \qquad \text{at} \qquad y=b \tag{2-132c}$$

Then the general solution for $T(x, y)$ is constructed as

$$T(x,y)=\int_{\beta=0}^{\infty}\sinh\beta(b-y)[A(\beta)\cos\beta x+B(\beta)\sin\beta x]\,d\beta \tag{2-133}$$

If this solution should satisfy the boundary condition at $y=0$, we obtain

$$f(x)=\int_{\beta=0}^{\infty}\sinh\beta b[A(\beta)\cos\beta x+B(\beta)\sin\beta x]\,d\beta \qquad \text{in} \qquad -\infty<x<\infty \tag{2-134a}$$

This is a representation of function $f(x)$ defined in the interval $-\infty<x<\infty$ in a form similar to that given by equation (2-66a); the coefficients of equation (2-66a) is given by equation (2-67). Then the coefficients of (2-134a) are obtained according to the relation given by equation (2-67). We find

$$\sinh\beta b[A(\beta)\cos\beta x+B(\beta)\sin\beta x]=\frac{1}{\pi}\int_{x'=-\infty}^{\infty}f(x')\cos\beta(x-x')\,dx' \tag{2-134b}$$

The substitution of these coefficients into equation (2-133) yields

$$T(x, y) = \frac{1}{\pi} \int_{\beta=0}^{\infty} \frac{\sinh \beta(b-y)}{\sinh \beta b} \int_{x'=-0}^{\infty} f(x') \cos \beta(x-x')\, dx'\, d\beta \tag{2-135a}$$

or changing the order of integration we obtain

$$T(x, y) = \frac{1}{\pi} \int_{x'=-0}^{\infty} f(x') \int_{\beta=0}^{\infty} \frac{\sinh \beta(b-y)}{\sinh \beta b} \cos \beta(x-x')\, d\beta\, dx' \tag{2-135b}$$

The integral with respect to β is available in the integral tables of Dwight [17, #862.41]. Then the solution becomes

$$T(x, y) = \frac{1}{2b} \sin \frac{y\pi}{b} \int_{x'=-\infty}^{\infty} \frac{f(x')}{\cos[\pi(b-y)/b] + \cosh[\pi(x-x')/b]}\, dx' \tag{2-136}$$

Example 2-14

Obtain an expression for the steady-state temperature $T(x, y)$ in a semiinfinite strip $0 \leqslant y \leqslant b, 0 \leqslant x < \infty$ for the boundary conditions shown in Fig. 2-12.

Solution. The mathematical formulation is given as

$$\frac{\partial^2 T(x, y)}{\partial x^2} + \frac{\partial^2 T(x, y)}{\partial y^2} = 0 \quad \text{in} \quad 0 \leqslant y < b, \quad 0 < x < \infty \tag{2-137a}$$

$$T = f(y) \qquad \text{at } x = 0 \tag{2-137b}$$

$$T = 0 \qquad \text{at } y = 0, \qquad T = 0 \quad \text{at} \quad y = b \tag{2-137c}$$

The sign of the separation constant must be so chosen that the separation function $T(y)$ results in an eigenvalue problem. Then the separated problems are taken as

$$\frac{d^2 Y(y)}{dy^2} + \gamma^2 Y(y) = 0 \quad \text{in} \quad 0 < y < b \tag{2-138a}$$

$$Y = 0 \quad \text{at } y = 0, \qquad Y = 0 \quad \text{at} \quad y = b \tag{2-138b}$$

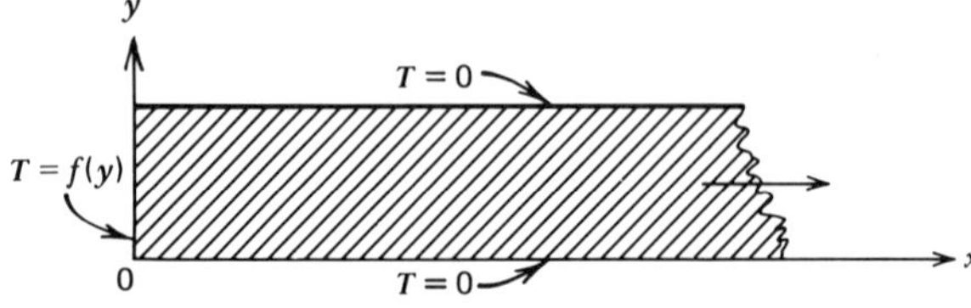

Fig. 2-12 Boundary conditions for a semiinfinite strip considered in Example 2-14.

and

$$\frac{d^2X(x)}{dx^2} - \gamma^2 X(x) = 0 \qquad \text{in} \qquad 0 < x < \infty \tag{2-139}$$

The solution of equations (2-138) is obtainable from Table 2-2, case 9, as

$$Y(\gamma_n, y) = \sin \gamma_n y, \qquad N(\gamma_n) = \frac{b}{2} \tag{2-140a}$$

where the γ_n values are the positive roots of $\sin \gamma_n b = 0$ or $\gamma_n = n\pi/b, n = 1, 2, 3, \ldots$. The solution of (2-139) that does not diverge at infinity is

$$X(\gamma_n x) = e^{-\gamma_n x} \tag{2-140b}$$

Then the complete solution for $T(x, y)$ is constructed as

$$T(x, y) = \sum_{n=1}^{\infty} c_n e^{-\gamma_n x} \sin \gamma_n y \tag{2-141}$$

The application of the boundary condition at $x = 0$ gives

$$f(y) = \sum_{n=1}^{\infty} c_n \sin \gamma_n y \qquad \text{in} \qquad 0 < y < b \tag{2-142a}$$

The coefficients c_n are determined by utilizing the orthogonality of the eigenfunctions $\sin \gamma_n y$; we find

$$c_n = \frac{1}{N(\gamma_n)} \int_0^b \sin \gamma_n y' f(y')\, dy' \tag{2-142b}$$

The substitution of c_n into equation (2-141) together with the value of $N(\gamma_n)$ as given above results in

$$T(x, y) = \frac{2}{b} \sum_{n=1}^{\infty} e^{-\gamma_n x} \sin \gamma_n y \int_0^b \sin \gamma_n y' f(y')\, dy' \tag{2-143}$$

where

$$\gamma_n = \frac{n\pi}{b}, \qquad n = 1, 2, 3 \ldots$$

For the special case of $f(y) = T_0 =$ constant, the integral is performed and the solution becomes

$$\frac{T(x, y)}{T_0} = \frac{4}{\pi} \sum_{n=\text{odd}}^{\infty} \frac{1}{n} e^{-\gamma_n x} \sin \gamma_n y \tag{2-144}$$

where only the odd values of n are considered in the summation because the terms for the even values of n vanish.

2-12 SPLITTING UP OF NONHOMOGENEOUS PROBLEMS INTO SIMPLER PROBLEMS

When the heat conduction problem is nonhomogeneous due to the nonhomogeneity of the differential equation and/or the boundary conditions, it can be split up into a set of simpler problems that may be solved by the method of the separation of variables. Here we consider a nonhomogeneous problem in which the generation term and the nonhomogeneous parts of the boundary-condition functions *do not depend on time*:

$$\nabla^2 T(\mathbf{r},t) + \frac{1}{k} g(\mathbf{r}) = \frac{1}{\alpha}\frac{\partial T(\mathbf{r},t)}{\partial t} \qquad \text{in region } R, \qquad t>0 \tag{2-145a}$$

$$k_i \frac{\partial T}{\partial n_i} + h_i T = f_i(\mathbf{r}) \qquad \text{on boundary } S_i, \quad t>0 \tag{2-145b}$$

$$T = F(\mathbf{r}) \qquad \text{for } t=0, \text{ in region } R \tag{2-145c}$$

where $(\partial/\partial n_i)$ = derivative along the outward-drawn normal to the boundary surface $S_i (i = 1, 2, \ldots, N)$ and N = number of continuous boundary surfaces of region R. Here we note that $g(\mathbf{r})$ and $f_i(\mathbf{r})$ *do not depend on time*. Clearly, many special cases are obtainable from the general problem given above. We shall now split up this problem into a number of simpler problems in the following manner:

1. A set of steady-state problems defined by the temperatures $T_{0j}(\mathbf{r})$, $j = 0, 1, 2, \ldots, N$.
2. A homogeneous time-dependent problem defined by the temperature $T_h(\mathbf{r}, t)$.

The temperatures $T_{0j}(\mathbf{r})$ are taken as the solutions of the following set of steady-state problems

$$\nabla^2 T_{0j}(\mathbf{r}) + \delta_{0j}\frac{1}{k} g(\mathbf{r}) = 0 \qquad \text{in region } R \tag{2-146a}$$

$$k_i \frac{\partial T_{0j}}{\partial n_i} + h_i T_{0j} = \delta_{ij} f_i(\mathbf{r}) \qquad \text{on boundary } S_i \tag{2-146b}$$

where

$$i = 1, 2, \ldots, N$$
$$j - 0, 1, 2, , \ldots, N$$

N = number of continuous boundary surfaces of region R

$$\delta_{ij} = \text{Kronecker delta} = \begin{cases} 0 & \text{for } i \neq j \\ 1 & \text{for } i = j \end{cases}$$

The temperature $T_h(\mathbf{r}, t)$ is taken as the solution of the following homogeneous problem:

$$\nabla^2 T_h(\mathbf{r}, t) = \frac{1}{\alpha} \frac{\partial T_h(\mathbf{r}, t)}{\partial t} \qquad \text{in region } R, \quad t > 0 \tag{2-147a}$$

$$k_i \frac{\partial T_h}{\partial n_i} + h_i T_h = 0 \qquad \text{on boundary } S_i \tag{2-147b}$$

$$T_h = F(\mathbf{r}) - \sum_{j=0}^{N} T_{0j}(\mathbf{r}) \qquad \text{for } t = 0, \quad \text{in region } R \tag{2-147c}$$

Then, the solution $T(\mathbf{r}, t)$ of the problem (2-145) is given in terms of the solutions of the above problems as

$$T(\mathbf{r}, t) = T_h(\mathbf{r}, t) + \sum_{j=0}^{N} T_{0j}(\mathbf{r}) \tag{2-148}$$

The validity of equation (2-148) can be verified by substituting this equation into equation (2-145) and by utilizing equations (2-146) and (2-147).

We note that equations (2-146) corresponds to a set of steady-state heat conduction problems. The function $T_{00}(\mathbf{r})$ for $j = 0$ corresponds to a steady-state heat conduction problem with heat generation in the medium, but subject to all homogeneous boundary conditions. The functions $T_{01}(\mathbf{r})$, $T_{02}(\mathbf{r})$, $T_{03}(\mathbf{r}), \ldots$ for $j = 1, 2, 3, \ldots$, respectively, corresponds to heat conduction problems with no heat generation, but only one of the boundary conditions, $i = j$, is nonhomogeneous.

The homogeneous problem given by equations (2-147) is the homogeneous version of the original problem (2-145), except the initial condition is modified by subtracting from it the sum of the solutions of the steady-state problems (2-146).

Clearly, the problems defined by equations (2-146) and (2-147), when given in the rectangular coordinate system, are soluble with the techniques discussed in this chapter. The more general case will be discussed in Chapter 13 in connection with the general method of solution of heat-conduction problems by the integral transform technique.

Example 2-15

A slab, $0 \leqslant x \leqslant L$, is initially at temperature $F(x)$. For times $t > 0$ the boundaries at $x = 0$ and $x = L$ are kept at constant temperatures T_1 and T_2, respectively. Obtain an expression for the temperature distribution $T(x, t)$ in the slab.

Solution. The mathematical formulation of this problem is given as

$$\frac{\partial^2 T}{\partial x^2} = \frac{1}{\alpha}\frac{\partial T}{\partial t} \qquad \text{in} \qquad 0 < x < L, \quad t > 0 \tag{2-149a}$$

$$T = T_1 \qquad \text{at} \qquad x = 0, \qquad t > 0 \tag{2-149b}$$

$$T = T_2 \qquad \text{at} \qquad x = L, \qquad t > 0 \tag{2-149c}$$

$$T = F(x) \qquad \text{for} \qquad t = 0, \qquad \text{in the region} \tag{2-149d}$$

Since the problem is one-dimensional, we split it into a steady-state problem for $T_s(x)$ given as

$$\frac{d^2 T_s}{dx^2} = 0 \qquad \text{in} \qquad 0 < x < L \tag{2-150a}$$

$$T_s = T_1 \qquad \text{at} \qquad x = 0 \tag{2-150b}$$

$$T_s = T_2 \qquad \text{at} \qquad x = L \tag{2-150c}$$

and to a homogeneous problem for $T_h(x, t)$ given by

$$\frac{\partial^2 T_h}{\partial x^2} = \frac{1}{\alpha}\frac{\partial T_h}{\partial t} \qquad \text{in} \qquad 0 < x < L, \quad t > 0 \tag{2-151a}$$

$$T_h = 0 \qquad \text{at} \qquad x = 0 \quad \text{and} \quad x = L, \quad t > 0 \tag{2-151b}$$

$$T_h = F(x) - T_s(x) \equiv f^*(x) \qquad \text{for} \qquad t = 0, \quad \text{in the region} \tag{2-151c}$$

Then, the solution for the original problem (2-149) is determined from

$$T(x, t) = T_s(x) + T_h(x, t) \tag{2-152}$$

The solution of the steady-state problem (2-150) is given as

$$T_s(x) = T_1 + (T_1 - T_1)\frac{x}{L} \tag{2-153}$$

The solution of the problem (2-151) is immediately written from equation (2-36a) as

$$T_h(x, t) = \sum_{m=1}^{\infty} e^{-\alpha\beta_m^2 t} \frac{1}{N(\beta_m)} X(\beta_m x) \int_0^L X(\beta_m x') f^*(x')\, dx' \tag{2-154}$$

where the eigenfunctions $X(\beta_m x)$, the norm $N(\beta_m)$ and the eigenvalues β_m are obtained from Table 2-2, case 9, as

$$X(\beta_m, x) = \sin \beta_m x, \qquad \frac{1}{N(\beta_m)} = \frac{2}{L} \tag{2-155a}$$

and the β_m values are the roots of

$$\sin \beta_m L = 0 \tag{2-155b}$$

and the initial condition function $f^*(x)$ is defined as

$$f^*(x) \equiv F(x) - T_s(x) = F(x) - T_1 - (T_2 - T_1)\frac{x}{L} \tag{2-155c}$$

The solution $T(x, t)$ of the problem (2-149) is obtained by introducing equations (2-153) and (2-154) into equation (2-152). We find

$$\begin{aligned} T(x,t) = T_1 + (T_2 - T_1)\frac{x}{L} \sum_{m=1}^{\infty} e^{-\alpha\beta_m^2 t} \sin \beta_m x \\ \cdot \int_0^L \left[F(x') - T_1 - (T_2 - T_1)\frac{x'}{L} \right] \sin \beta_m x' \, dx' \end{aligned} \tag{2-156a}$$

Performing the integrations we obtain

$$\begin{aligned} T(x,t) = T_1 + (T_2 - T_1)\frac{x}{L} + \frac{2}{L} \sum_{m=1}^{\infty} e^{-\alpha\beta_m^2 t} \sin \beta_m x \int_0^L F(x') \sin \beta_m x' \, dx' \\ + \frac{2}{L} \sum_{m=1}^{\infty} e^{-\alpha\beta_m^2 t} \frac{1}{\beta_m} \sin \beta_m x [T_2 \cos m\pi - T_1] \end{aligned} \tag{2-156b}$$

where $\cos m\pi = (-1)^m$ and $\beta_m = m\pi/L$.

Example 2-16

A slab, $0 \leqslant x \leqslant L$, is initially at temperature $F(x)$. For times $t > 0$, heat is generated in the solid at a constant rate of g_0 per unit volume, the boundary at $x = 0$ is kept insulated and the boundary at $x = L$ is kept at zero temperature. Obtain an expression for the temperature distribution $T(x, t)$ in the slab.

Solution. The mathematical formulation of this problem is given as

$$\frac{\partial^2 T}{\partial x^2} + \frac{1}{k} g_0 = \frac{1}{\alpha} \frac{\partial T}{\partial t} \qquad \text{in} \qquad 0 < x < L, \quad t > 0 \tag{2-157a}$$

$$\frac{\partial T}{\partial x}=0 \qquad \text{at} \qquad x=0, \qquad t>0 \tag{2-157b}$$

$$T=0 \qquad \text{at} \qquad x=L, \qquad t>0 \tag{2-157c}$$

$$T=F(x) \qquad \text{for} \qquad t=0, \qquad \text{in } 0 \leqslant x \leqslant L \tag{2-157d}$$

This problem is split up into a steady-state problem for $T_s(x)$ as

$$\frac{d^2T}{dx^2}+\frac{1}{k}g_0=0 \qquad \text{in} \qquad 0<x<L \tag{2-158a}$$

$$\frac{dT_s}{dx}=0 \quad \text{at} \quad x=0 \qquad \text{and} \qquad T_s=0 \qquad \text{at} \qquad x=L \tag{2-158b}$$

and a homogeneous problem for $T_h(x,t)$ as

$$\frac{\partial^2 T_h}{\partial x^2}=\frac{1}{\alpha}\frac{\partial T_h}{\partial t} \qquad \text{in} \qquad 0<x<L, \quad t>0 \tag{2-159a}$$

$$\frac{\partial T_h}{\partial x}=0 \qquad \text{at} \qquad x=0, \quad T_h=0 \qquad \text{at} \qquad x=L, \quad \text{for } t>0 \tag{2-159b}$$

$$T_h=F(x)-T_s(x)\equiv f^*(x) \qquad \text{for} \qquad t=0, \text{ in } 0 \leqslant x \leqslant L \tag{2-159c}$$

Then, the solution of the original problem (2-157) is determined from

$$T(x,t)=T_s(x)+T_h(x,t) \tag{2-160}$$

The solution of the steady-state problem (2-158) is

$$T_s(x)=\frac{1}{2k}g_0L^2\left(1-\frac{x^2}{L^2}\right) \tag{2-161}$$

and the solution of the homogeneous problem (2-159) is obtained as

$$T_h(x,t)=\frac{2}{L}\sum_{m=0}^{\infty}e^{-\alpha\beta_m^2 t}\cos\beta_m x\int_0^L f^*(x')\cos\beta_m x'\,dx' \tag{2-162a}$$

where

$$f^*(x)\equiv F(x)-\frac{1}{2k}g_0L^2\left(1-\frac{x^2}{L^2}\right) \tag{2-162b}$$

and the β_m values are the positive roots of

$$\cos \beta_m L = 0 \qquad \text{or} \qquad \beta_m = \frac{(2m+1)\pi}{2L}, \qquad m = 0, 1, 2 \ldots \tag{2-162c}$$

Introducing equations (2-161) and (2-162) into (2-160) and performing the integrations we obtain

$$T(x,t) = \frac{g_0 L^2}{2k}\left(1 - \frac{x^2}{L^2}\right) + \frac{2}{L}\sum_{m=0}^{\infty} e^{-\alpha\beta_m^2 t} \cos \beta_m x \int_0^L F(x') \cos \beta_m x' \, dx'$$
$$- \frac{2g_0}{kL}\sum_{m=0}^{\infty} (-1)^m e^{-\alpha\beta_m^2 t} \frac{1}{\beta_m^3} \cos \beta_m x \tag{2-163}$$

2-13 USEFUL TRANSFORMATIONS

In this section we present some transformations that are useful in reducing the differential equation into a more convenient form.

1. We consider an equation containing convective and generation terms in the form

$$\frac{\partial T}{\partial t} = \alpha \frac{\partial^2 T}{\partial x^2} - \beta \frac{\partial T}{\partial x} + \gamma T + g \tag{2-164}$$

where α, β, and γ are constants, $\beta(\partial T/\partial x)$ represents a convective term and γT represents generation proportional to the local temperature. We define a new dependent variable $W(x,t)$ as

$$T(x,t) = W(x,t) \exp\left[\frac{\beta}{2\alpha}x - \left(\frac{\beta^2}{4\alpha} - \gamma\right)t\right] \tag{2-165}$$

Then, under this transformations, equation (2-164) reduces to

$$\frac{\partial W}{\partial t} = \alpha \frac{\partial^2 W}{\partial x^2} + g \cdot \exp\left\{-\left[\frac{\beta}{2\alpha}x - \left(\frac{\beta^2}{4\alpha} - \gamma\right)t\right]\right\} \tag{2-166}$$

which is easier to solve than equation (2-164). The boundary and the initial conditions for the problem should be transformed with the same transformation.

2. We now generalize the above procedure to three-dimensional equation given as

$$\frac{\partial T}{\partial t} = \alpha\left(\frac{\partial^2 T}{\partial x^2} + \frac{\partial^2 T}{\partial y^2} + \frac{\partial^2 T}{\partial z^2}\right) - \beta_1 \frac{\partial T}{\partial x} - \beta_2 \frac{\partial T}{\partial y} - \beta_3 \frac{\partial T}{\partial z} + \gamma T + g \tag{2-167}$$

where $\alpha, \beta_1, \beta_2, \beta_3$, and γ are constants. We define a new dependent variable $W(x, y, z, t)$ as

$$T(x, y, z) = W(x, y, z, t) \exp\left[\frac{\beta_1}{2\alpha}x - \left(\frac{\beta_1^2}{4\alpha} - \gamma\right)t\right] \cdot \exp\left[\frac{\beta_2}{2\alpha}y - \frac{\beta_2^2}{4\alpha}t\right] \cdot \exp\left[\frac{\beta_3}{2\alpha}z - \frac{\beta_3^2}{4\alpha}t\right] \tag{2-168}$$

Under this transformation equation (2-167) reduces to

$$\frac{\partial W}{\partial t} = \alpha\left(\frac{\partial^2 W}{\partial x^2} + \frac{\partial^2 W}{\partial y^2} + \frac{\partial^2 W}{\partial z^2}\right) + G \tag{2-169a}$$

where

$$G \equiv g \cdot \exp\left[-\frac{\beta_1}{2\alpha}x + \left(\frac{\beta_1^2}{4\alpha} - \gamma\right)t\right] \cdot \exp\left[-\frac{\beta_2}{2\alpha}y + \frac{\beta_2^2}{4\alpha}t\right] \cdot \exp\left[-\frac{\beta_3}{2\alpha}z + \frac{\beta_3^2}{4\alpha}t\right] \tag{2-169b}$$

which is easier to solve than equation (2-167).

2-14 TRANSIENT-TEMPERATURE CHARTS

Temperature–time charts are useful for rapid estimation of temperature history in solids, and for some specific situations such charts can be prepared. Here we consider a plate of thickness $2L$, initially at a uniform temperature T_i and for times $t > 0$ it is subjected to convection from both its surfaces into an ambient at a constant temperature T_∞ with a heat transfer coefficient h. Because of symmetry, we choose the origin of the x coordinate at the center of the plate and consider only half of the plate. The mathematical formulation of this transient heat conduction problem is given in the dimensionless form as

$$\frac{\partial^2 \theta}{\partial X^2} = \frac{\partial \theta}{\partial \tau} \quad \text{in} \quad 0 < X < 1, \quad \text{for} \quad \tau = 0 \tag{2-170a}$$

$$\frac{\partial \theta}{\partial X} = 0 \quad \text{at} \quad X = 0, \quad \text{for} \quad \tau > 0 \tag{2-170b}$$

$$\frac{\partial \theta}{\partial X} + \text{Bi}\,\theta = 0 \quad \text{at} \quad X = 1, \quad \text{for} \quad \tau > 0 \tag{2-170c}$$

$$\theta = 1 \quad \text{in} \quad 0 \leqslant X \leqslant 1, \quad \text{for} \quad \tau = 0 \tag{2-170d}$$

where the following dimensionless parameters are introduced

$$\theta = \frac{T(x,t) - T_\infty}{T_i - T_\infty} = \text{dimensionless temperature} \tag{2-171a}$$

$$X = \frac{x}{L} = \text{dimensionless coordinate} \tag{2-171b}$$

$$\text{Bi} = \frac{hL}{k} = \text{Biot number} \tag{2-171c}$$

$$\tau = \frac{\alpha t}{L^2} = \text{dimensionless time or Fourier number} \tag{2-171d}$$

It is instructive to examine the physical significance of the dimensionless parameters τ and Bi.

The dimensionless time τ is rearranged in the form

$$\tau = \frac{\alpha t}{L^2} = \frac{k(1/L)L^2}{\rho c_p L^3/t} = \frac{\begin{pmatrix}\text{rate of heat conduction}\\ \text{across } L \text{ in volume } L^3, \text{W}/^\circ\text{C}\end{pmatrix}}{\begin{pmatrix}\text{rate of heat storage}\\ \text{in volume } L^3, \text{W}/^\circ\text{C}\end{pmatrix}}$$

Thus, the Fourier number is a measure of the rate of heat conduction compared with the rate of heat storage in a given volume element. Therefore, the larger the Fourier number, the deeper the penetration of heat into a solid over a given time.

The physical significance of the Biot number has already been discussed in the previous chapter in connection with the lumped analysis. It represents the ratio of the "internal thermal resistance" to the "external thermal resistance."

The solution of the transient heat conduction problem (2-170) is presented in the graphical form in Fig. 2-13 *a,b*. Here, Fig. 2-13 *a* gives the midplane temperature T_0 or $\theta(0, \tau)$ at $X = 0$ as a function of the dimensionless time τ for several different values of the parameter 1/Bi. The curve for $1/\text{Bi} = 0$ corresponds to the case in which $h \to \infty$, or the surfaces of the plate are maintained at the ambient temperature T_∞. For large values of 1/Bi, the Biot number is small, or the internal conductance of the solid is large in comparison with the heat transfer coefficient at the surface. This, in turn, implies that the temperature distribution within the solid is sufficiently uniform, and hence lumped system analysis becomes applicable.

Figure 2-13 *b* relates the temperature at six different locations within the slab to the midplane temperature T_0 [i.e., $\theta(0, \tau)$]. Thus, given T_0, temperature at these locations can be determined. An examination of Fig. 2-13*b* reveals that for values of 1/Bi larger than 10, or $\text{Bi} < 0.1$, the temperature distribution within the slab may be considered uniform with an error of less than about 5%; hence for such

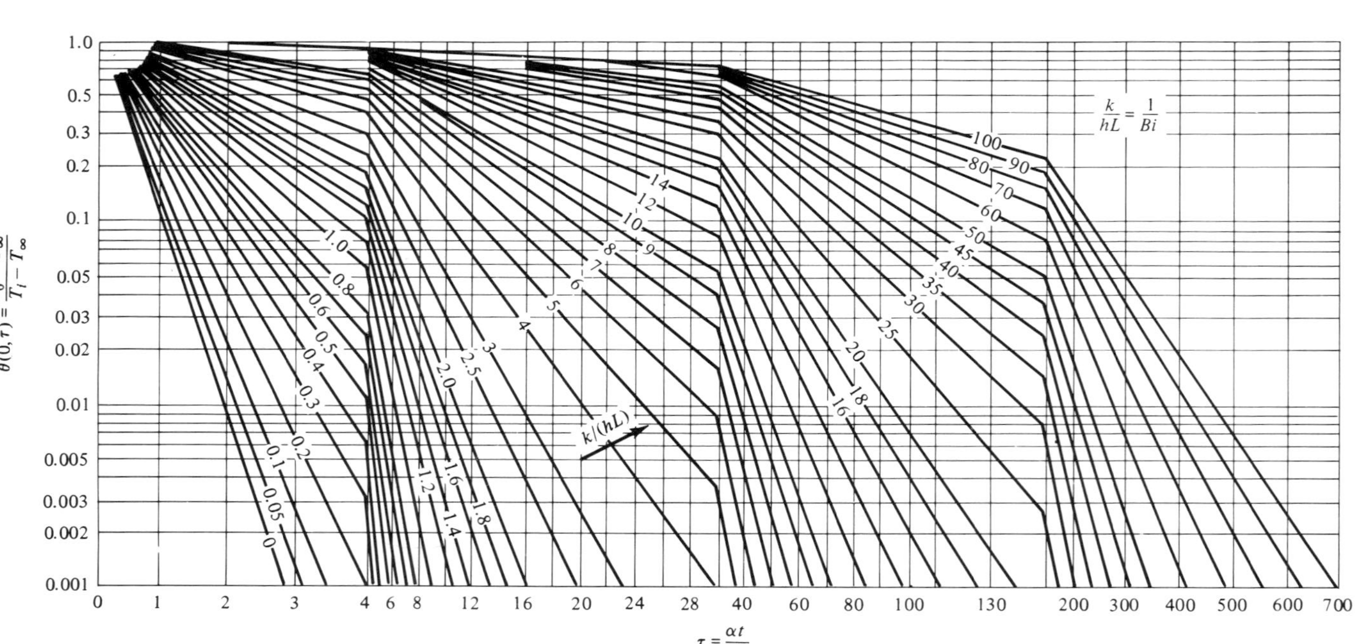

Fig. 2-13 Transient temperature chart for a slab of thickness $2L$ subjected to convection at both boundary surfaces. (From Heisler [18].) (*a*) Dimensionless temperature $\theta(0, \tau)$ at the center plane $x = 0$; (*b*) position correction chart for use with part *a*.

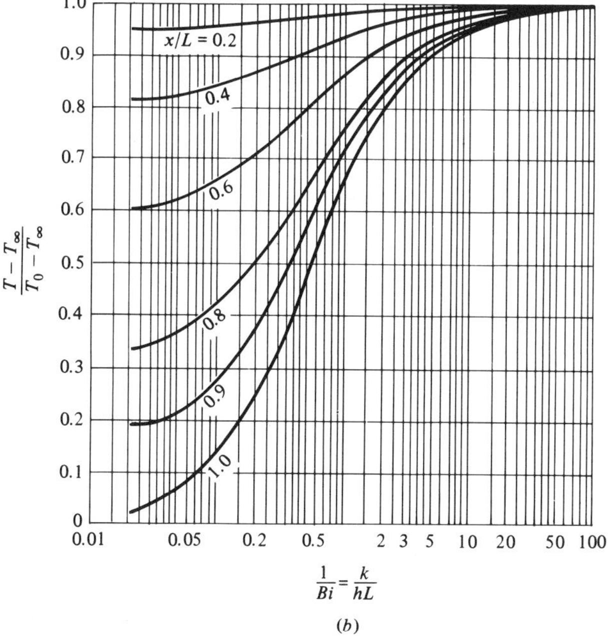

Fig. 2-13 *(Continued)*

situations the spatial variation of temperature within the medium can be neglected and the *lumped system analysis* can be applicable.

Example 2-17

A 10 cm thick brick wall is initially at a uniform temperature $T_i = 240°C$. At time $t = 0$, both surfaces of the wall are subjected to convective cooling into an ambient at temperature $T_\infty = 40°C$ with a heat transfer coefficient $h = 60\,W/(m^2 \cdot °C)$. Using the transient temperature chart, calculate the midplane temperature at 2 h after exposure to the cool environment. Take the physical properties as

$$\alpha = 0.5 \times 10^{-6}\,m^2/s; \quad k = 0.69\,W/(m \cdot °C); \quad \rho = 2300\,kg/m^3$$

Solution. We determine L, τ, and $1/\text{Bi}$.

$$L = \frac{0.1}{2} = 0.05\,m, \quad \tau = \frac{\alpha t}{L^2} = \frac{0.5 \times 10^{-6}}{(0.05)^2}(2 \times 3600) = 1.44$$

$$\frac{1}{\text{Bi}} = \frac{k}{hL} = \frac{0.69}{60 \times 0.05} = 0.23$$

From Fig. 2-13a, for $\tau = 1.44$ and $1/\text{Bi} = 0.23$ we have

$$\theta(0, \tau) = \frac{T_0 - T_\infty}{T_i - T_\infty} = \frac{T_0 - 40}{240 - 40} = 0.12$$

Thus

$$T_0 = 40 + 24 = 64°\text{C}$$

Thus the midplane temperature is approximately 64°C.

REFERENCES

1. M. Philip Morse and H. Feshbach, *Methods of Theoretical Physics*, Part I, McGraw-Hill, New York, 1953.
2. John W. Dettman, *Mathematical Methods in Physics and Engineering*, McGraw-Hill, New York, 1962.
3. Parry Moon and Domino Eberle Spencer, *Field Theory for Engineers*, Van Nostrand, Princeton, N.J., 1961.
4. R. V. Churchill, *Fourier Series and Boundary Value Problems*, McGraw-Hill, New York, 1963.
5. H. S. Carslaw and J. C. Jaeger, *Conduction of Heat in Solids*, Oxford at the Clarendon Press, London, 1959.
6. V. S. Arpaci, *Conduction Heat Transfer*, Addison-Wesley, Reading, Mass., 1966.
7. M. N. Özişik, *Boundary Value Problems of Heat Conduction*, International Textbook Company, Scranton, Pa., 1968. Also Dover Publication, New York, 1989.
8. M. N. Özişik, *Heat Conduction*, 1st ed., Wiley, New York, 1980.
9. M. D. Mikhailov and M. N. Özişik, *Unified Analysis and Solutions of Heat and Mass Diffusion*, Wiley, New York, 1984.
10. P. Moon and D. E. Spencer, *Q. Appl. Math.* **16**, 1–10, 1956.
11. A. Erdelyi, W. Magnus, F. Oberhettinger, and F. G. Tricomi, *Higher Transcendental Functions*, McGraw-Hill., New York, 1953.
12. R. V. Churchill, *Operational Mathematics*, McGraw-Hill, New York, 1958.
13. E. C. Titchmarsh, *Eigenfunction Expansions*, Clarendon Press, London, 1962.
14. M. L. James, G. M. Smith, and J. C. Wolford, *Applied Numerical Methods for Digital Computations with Fortran and CSMP*, 2nd ed., IEP, New York, 1977.
15. Y. Jaluria, *Computer Methods for Engineering*, Allyn and Bacon, London, 1988.
16. E. C. Titchmarsh, *Fourier Integrals*, 2nd ed., Clarendon Press, London, 1962.
17. H. B. Dwight, *Tables of Integrals and Other Mathematical Data*, 4th ed., Macmillan, New York, 1957.
18. M. P. Heisler, Temperature Charts for Induction and Constant Heating, *Trans. ASME*, **69**, 227–236, 1947.

PROBLEMS

2-1 A slab, $0 \leqslant x \leqslant L$, is initially at a temperature $F(x)$. For times $t > 0$ the boundaries at $x = 0$ and $x = L$ are kept at zero temperature. Derive an expression for the temperature $T(x, t)$ in the slab for times $t > 0$. Determine the temperature $T(x, t)$ for the special case $F(x) = T_0 =$ constant.

2-2 A slab, $0 \leqslant x \leqslant L$, is initially at a temperature $F(x)$. For times $t > 0$ the boundary surface at $x = 0$ is kept insulated and that at $x = L$ dissipates heat by convection into a medium at zero temperature with a heat transfer coefficient h. Obtain an expression for the temperature distribution $T(x, t)$ in the slab for times $t > 0$ and for the heat flux at the boundary surface $x = L$.

2-3 A slab, $0 \leqslant x \leqslant L$, is initially at a temperature $F(x)$. For times $t > 0$ the boundary surface at $x = 0$ is kept at zero temperature, whereas the boundary at $x = L$ dissipates heat by convection into a medium at zero temperature with a heat transfer coefficient h. Obtain an expression for the temperature $T(x, t)$ in the slab and the heat flux at the boundary surface $x = L$ for times $t > 0$. Also consider the case when $F(x) = T_0 =$ constant.

2-4 A semiinfinite medium, $0 \leqslant x < \infty$, is initially at zero temperature. For times $t > 0$ the boundary surface at $x = 0$ is kept at a constant temperature T_0. Obtain an expression for the temperature distribution $T(x, t)$ in the slab for times $t > 0$.

2-5 A semiinfinite medium, $0 \leqslant x < \infty$, is initially at a uniform temperature T_0 and for times $t > 0$ it dissipates heat by convection from the boundary surface $x = 0$ into an environment at zero temperature. Obtain an expression for the temperature distribution $T(x, t)$ in the medium for times $t > 0$. Determine an expression for the heat flux at the surface $x = 0$.

2-6 In a one-dimensional infinite medium, $-\infty < x < \infty$, initially, the region $a < x < b$ is at a constant temperature T_0, and everywhere outside this region is at zero temperature. Obtain an expression for the temperature distribution $T(x, t)$ in the medium for times $t > 0$.

2-7 A rectangular region $0 \leqslant x \leqslant a, 0 \leqslant y \leqslant b$ is initially at a temperature $F(x, y)$. For times $t > 0$ it dissipates heat by convection from all its boundary surfaces into an environment at zero temperature. The heat transfer coefficient is the same for all the boundaries. Obtain an expression for the temperature distribution $T(x, y, t)$ in the region for times $t > 0$.

2-8 A region $x > 0, y > 0, z > 0$ is initially at a uniform temperature T_0. For times $t > 0$ all the boundaries are kept at zero temperature. Using the product solution, obtain an expression for the temperature distribution $T(x, y, z, t)$ in the medium.

2-9 A region $x > 0, y > 0$ is initially at a uniform temperature T_0. For times $t > 0$, both boundaries dissipate heat by convection into an environment at zero temperature. The heat transfer coefficients are the same for both boundaries. Using the product solution, obtain an expression for the temperature distribution $T(x, y, t)$ in the medium.

2-10 A rectangular region $0 \leqslant x \leqslant a, 0 \leqslant y \leqslant b$ is initially at a uniform temperature T_0. For times $t > 0$ the boundaries at $x = 0$ and $y = 0$ are kept at zero temperature and the boundaries at $x = a$ and $y = b$ dissipate heat by convection into an environment at zero temperature. The heat transfer coefficients are the same for both of these boundaries. Using the product solution, obtain an expression for the temperature distribution $T(x, y, t)$ n the medium for times $t > 0$.

2-11 A rectangular parallelepiped $0 \leqslant x \leqslant a, 0 \leqslant y \leqslant b, 0 \leqslant z \leqslant c$ is initially at a uniform temperature T_0. For times $t > 0$ the boundaries at $x = 0$, $y = 0$, and $z = 0$ are insulated and the boundaries at $x = a$, $y = b$, and $z = c$ are kept at zero temperature. Using the product solution, obtain an expression for the temperature distribution $T(x, y, z, t)$ in the region.

2-12 Repeat problem (2-11) for the case when the boundaries at $x = a$, $y = b$, and $z = c$ dissipate heat by convection into an environment at zero temperature. Assume the heat transfer coefficients to be the same at all these boundaries.

2-13 Obtain an expression for the steady-state temperature distribution $T(x, y)$ in a semiinfinite strip $0 \leqslant x \leqslant a, 0 \leqslant y < \infty$, for the case when the boundary at $x = 0$ is kept at a temperature $f(y)$ and the boundaries at $y = 0$ and $x = a$ are kept at zero temperature.

2-14 Obtain an expression for the steady-state temperature distribution $T(x, y)$ in an infinite strip $0 \leqslant x \leqslant a, -\infty < y < \infty$, for the case when the boundary surface at $x = 0$ is kept at a temperature $f(y)$ and the boundary surface at $x = a$ is kept at zero temperature.

2-15 Obtain an expression for the steady-state temperature distribution $T(x, y)$ in a rectangular region $0 \leqslant x \leqslant a, 0 \leqslant y \leqslant b$ for the following boundary conditions: the boundary at $x = 0$ is kept insulated, the boundary at $y = 0$ is kept at a temperature $f(x)$ and the boundaries at $x = a$ and $y = b$ dissipate heat by convection into an environment at zero temperature. Assume the heat transfer coefficient to be the same for both boundaries.

2-16 Obtain an expression for the steady-state temperature distribution $T(x, y, z)$ in a rectangular parallelepiped $0 \leqslant x \leqslant a, 0 \leqslant y \leqslant b$, $0 \leqslant z \leqslant c$ for the following boundary conditions: the boundary surfaces at $x = 0$ is kept at temperature T_0, the boundaries at $y = 0$ and $z = 0$ are kept insulated, the boundary at $x = a$ is kept at zero temperature, and the boundaries at $y = b$ and $z = c$ dissipate heat by convection into an environment at zero temperature. The heat transfer coefficient are the same for all these surfaces.

2-17 Obtain an expression for the steady-state temperature distribution $T(x, y)$ in a rectangular region $0 \leqslant x \leqslant a, 0 \leqslant y \leqslant b$ in which heat is generated at a constant rate $g(x, y) = g_0 =$ ollowing boundary conditions: boundaı ısulated, whereas the boundaries at $x =$ ›erature.

2-18 A slab, $0 \leqslant x \leqslant L$, is initially at zero temperature. For times $t > 0$ the boundary at $x = 0$ is kept insulated, the boundary at $x = L$ is kept at zero temperature, and there is heat generation within the solid at a constant rate of g_0. Obtain an expression for the temperature distribution $T(x, t)$ in the slab for times $t > 0$.

2-19 Obtain an expression for the steady-state temperature distribution $T(x, y)$ in an infinite strip $0 \leqslant y \leqslant b$, $0 \leqslant x < \infty$, for the case where the boundary at $x = 0$ is kept at zero temperature, the boundary at $y = b$ is insulated and the boundary at $y = 0$ is subjected to a heat supply at a rate of $f(x)$, $\mathrm{W/m^2}$.

NOTES

1. The properties of the following homogeneous boundary value problem, called a Sturm–Liouville problem, were first studied by J. C. F. Sturm and J. Liouville in *Journal de Mathématique*, 1836–1838. Here we present the orthogonality of the eigenfunctions

$$\frac{d}{dx}\left[p(x)\frac{d\psi(\lambda, x)}{dx}\right] + [q(x) + \lambda w(x)]\psi(\lambda, x) = 0 \quad \text{in} \quad a < x < b \tag{1a}$$

$$A_1\frac{d\psi(\lambda, x)}{dx} + A_2\psi(\lambda, x) = 0 \quad \text{at} \quad x = a \tag{1b}$$

$$B_1\frac{d\psi(\lambda, x)}{dx} + B_2\psi(\lambda, x) = 0 \quad \text{at} \quad x = b \tag{1c}$$

where the functions $p(x), q(x), w(x)$ and $dp(x)/dx$ are assumed to be real valued, and continuous, and $p(x) > 0$ and $w(x) > 0$ over the interval (a, b). The constants A_1, A_2, B_1, B_2 are real and independent of the parameter λ. Let

$$L[\psi(\lambda, x)] \equiv \frac{d}{dx}\left[p(x)\frac{d\psi(\lambda, x)}{dx}\right] + q(x)\psi(\lambda, x) \tag{2}$$

We then write equation (1a) for any two eigenfunctions $\psi(\lambda_m, x)$ and $\psi(\lambda_n, x)$ as

$$L[\psi_m(x)] + \lambda_m w(x)\psi_m(x) = 0 \tag{3a}$$

$$L(\psi_n(x)] + \lambda_n w(x)\psi_n(x) = 0 \tag{3b}$$

where

$$\psi_n(x) \equiv \psi(\lambda_n x)$$

We multiply equation (3a) by $\psi_n(x)$ and equation (3b) by $\psi_m(x)$, then subtract the results

$$\frac{d}{dx}[p(\psi_n\psi'_m - \psi_m\psi'_n)] = (\lambda_n - \lambda_m)w\psi_m\psi_n \tag{4}$$

Both sides of equation (4) is integrated from $x = a$ to $x = b$ and the result rearranged

$$\int_{x=a}^{b} w\psi_m\psi_n\,dx = \frac{1}{\lambda_n - \lambda_m}\int_{x=a}^{b}\frac{d}{dx}[p(\psi_n\psi'_m - \psi_m\psi'_n)]\,dx \tag{5}$$

For $m \neq n$, the argument of the integral on the right-hand side of equation (5) vanishes because of the homogeneous boundary conditions (1b) and (1c) for the problem. For $m = n$ (i.e., $\lambda_m \to \lambda_n$), the left-hand side of equation (5) is the *norm* N, but the right-hand side is indefinite because both the numerator and the denominator vanish. However, for $\lambda_m \to \lambda_n$, the right-hand side is evaluated by L'Hospital's rule. Thus equation (5) is written more compactly as

$$\int_{x=a}^{b} w\psi_m\psi_n\,dx = \begin{cases} 0 & \text{for} \quad m \neq n \\ N(\lambda_n) & \text{for} \quad m = n \end{cases} \tag{6}$$

where

$$N(\lambda_n) \equiv \int_a^b w\psi_n^2\,dx = \int_a^b p\left(\frac{\partial\psi_n}{\partial\lambda_n}\cdot\frac{\partial\psi_n}{\partial x} - \psi_n\frac{\partial^2\psi_n}{\partial\lambda_n\partial x}\right)dx \tag{7}$$

which proves that eigenfunctions of the Sturm–Liouville system are orthogonal with respect to the weighting function $w(x)$ in the interval (a, b).

2. For a boundary condition of the second kind at both boundaries, the eigenvalue problem is given as

$$\frac{d^2X(x)}{dx^2} + \beta_m^2 X(x) = 0 \qquad \text{in} \qquad 0 < x < L \tag{1a}$$

$$\frac{dX}{dx} = 0 \qquad \text{at} \qquad x = 0 \quad \text{and} \quad x = L \tag{1b}$$

From equation (1a) we have

$$\beta^2\int_0^L X^2(x)\,dx = -\left[X\frac{dX}{dx}\right]_0^L + \int_0^L\left(\frac{dX}{dx}\right)^2 dx \tag{2}$$

The first term on the right vanishes in view of the boundary conditions. Then $\beta_0 = 0$ is also an eigenvalue corresponding to the eigenfunction $X_0(x) = \text{constant} \neq 0$. For $X_0 = 1$, the norm N becomes

$$N = \int_0^L X_0^2\,dx = \int_0^L dx = L \tag{3}$$

3

THE SEPARATION OF VARIABLES IN THE CYLINDRICAL COORDINATE SYSTEM

In this chapter we examine the separation of the homogeneous heat conduction equation in the cylindrical coordinate system; determine the elementary solutions, the norms, and the eigenvalues of the separated problems for different combinations of boundary conditions and systematically tabulate the resulting expressions for ready reference; discuss the solution of the one- and multidimensional homogeneous problems by the method of separation of variables; examine the solutions of steady-state multidimensional problems with and without the heat generation in the medium; and illustrate the splitting up of nonhomogeneous problems into a set of simpler problems. The reader should consult references 1–4 for additional applications on the solution of heat conduction problems in the cylindrical coordinate system.

3-1 SEPARATION OF HEAT CONDUCTION EQUATION IN THE CYLINDRICAL COORDINATE SYSTEM

Consider the three-dimensional, homogeneous differential equation of heat conduction in the cylindrical coordinate system,

$$\frac{\partial T}{\partial r^2}+\frac{1}{r}\frac{\partial T}{\partial r}+\frac{1}{r^2}\frac{\partial^2 T}{\partial \phi^2}+\frac{\partial^2 T}{\partial z^2}=\frac{1}{\alpha}\frac{\partial T}{\partial t} \tag{3-1}$$

where $T \equiv T(r, \phi, z, t)$. Assume a separation of variables in the form

$$T(r, \phi, z, t) = \phi(r, \phi, z)\Gamma(t) \tag{3-2}$$

Equation (3-1) becomes

$$\frac{1}{\psi}\left(\frac{\partial^2\psi}{\partial r^2}+\frac{1}{r}\frac{\partial\psi}{\partial r}+\frac{1}{r^2}\frac{\partial^2\psi}{\partial\phi^2}+\frac{\partial^2\psi}{\partial z^2}\right)=\frac{1}{\alpha\Gamma(t)}\frac{d\Gamma(t)}{\partial t}=-\lambda^2 \tag{3-3}$$

Then, the separated equations for $\Gamma(t)$ and ψ are taken as

$$\frac{d\Gamma(t)}{dt}+\alpha\lambda^2\Gamma(t)=0 \tag{3-4}$$

$$\frac{\partial^2\psi}{\partial r^2}+\frac{1}{r}\frac{\partial\psi}{\partial r}+\frac{1}{r^2}\frac{\partial^2\psi}{\partial\phi^2}+\frac{\partial^2\psi}{\partial z^2}+\lambda^2\psi=0 \tag{3-5}$$

Equation (3-5) is the *Helmholtz equation*; we assume a separation in the form

$$\psi(r,\phi,z)=R(r)\Phi(\phi)Z(z) \tag{3-6}$$

Then equation (3-5) becomes

$$\frac{1}{R}\left(\frac{d^2R}{dr^2}+\frac{1}{r}\frac{dR}{dr}\right)+\frac{1}{r^2}\frac{1}{\Phi}\frac{d^2\Phi}{d\phi^2}+\frac{1}{Z}\frac{d^2Z}{dz^2}+\lambda^2=0 \tag{3-7}$$

The only way this equality is satisfied is if each group of functions is equated to an arbitrary separation constant in the form

$$\frac{1}{Z}\frac{d^2Z}{dz^2}=-\eta^2,\quad \frac{1}{\Phi}\frac{d^2\Phi}{d\phi^2}=-v^2,\quad \text{and}\quad \frac{1}{R_v}\left(\frac{d^2R_v}{dr^2}+\frac{1}{r}\frac{dR_v}{dr}\right)-\frac{v^2}{r^2}=-\beta^2 \tag{3-8}$$

Then, the separated equations and their elementary solutions become

$$\frac{d^2Z}{dz^2}+\eta^2Z=0 \qquad Z(\eta,z)\text{: }\sin\eta z\text{ and }\cos\eta z \tag{3-9a}$$

$$\frac{d^2\Phi}{d\phi^2}+v^2\Phi=0 \qquad \Phi(v,\phi)\text{: }\sin v\phi\text{ and }\cos v\phi \tag{3-9b}$$

$$\frac{d^2R_v}{dr^2}+\frac{1}{r}\frac{dR_v}{dr}+\left(\beta^2-\frac{v^2}{r^2}\right)R_v=0 \qquad R_v(\beta,r)\text{: }J_v(\beta r)\text{ and }Y_v(\beta r) \tag{3-9c}$$

and the function $\Gamma(t)$ satisfies equation (3-4), that is,

$$\frac{d\Gamma}{dt}+\alpha\lambda^2\Gamma=0 \qquad \Gamma(t)\text{: }e^{-\alpha\lambda^2t} \tag{3-9d}$$

where

$$\lambda^2 = \beta^2 + \eta^2 \tag{3-9e}$$

Here we note that the separation constant λ^2 does not include ν^2 because of the nature of the separation. Equation (3-9c) is called *Bessel's differential equation* of order ν, and its solutions, $J_\nu(\beta r)$ and $Y_\nu(\beta r)$, are the *Bessel functions* of order ν of the first and second kind, respectively. Clearly, the order ν of the Bessel functions is due to the presence of the separation equation (3-9b) resulting from the azimuthal dependence of temperature.

A discussion of the properties of Bessel functions is given in Appendix IV; the reader should consult references 5–8 for further information on Bessel functions.

Figure 3-1 shows $J_0(x)$, $J_1(x)$, $Y_0(x)$, and $Y_1(x)$ functions. Both $J_\nu(x)$ and $Y_\nu(x)$ functions have oscillatory behavior like trigonometric functions, but $Y_\nu(x)$ functions become infinite at $x = 0$.

Having established the separation equations associated with the r, ϕ, z, and t variables of the transient heat conduction equation (3-1), we now examine the separation equations associated with some special cases equation (3-1).

1. *Temperature has no ϕ dependence.* Equation (3-1) becomes

$$\frac{\partial^2 T}{\partial r^2} + \frac{1}{r}\frac{\partial T}{\partial r} + \frac{\partial^2 T}{\partial z^2} = \frac{1}{\alpha}\frac{\partial T}{\partial t} \tag{3-10}$$

The separated equations and their elementary solutions become

$$\frac{d^2 Z}{dz^2} + \eta^2 Z = 0 \qquad Z(\eta, z)\text{: } \sin \eta z \text{ and } \cos \eta z \tag{3-11a}$$

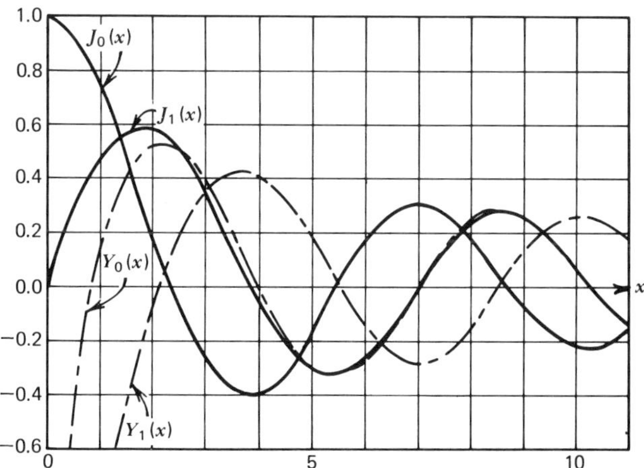

Fig. 3-1 $J_0(x)$, $Y_0(x)$ and $J_1(x)$, $Y_1(x)$ functions.

$$\frac{d^2R_0}{dr^2}+\frac{1}{r}\frac{dR_0}{dr}+\beta^2R_0=0 \qquad R_0(\beta,r)\colon J_0(\beta r) \text{ and } Y_0(\beta r) \tag{3-11b}$$

$$\frac{d\Gamma}{dt}+\alpha\lambda^2\Gamma=0 \qquad \Gamma(t)\colon e^{-\alpha\lambda^2 t} \tag{3-11c}$$

where

$$\lambda^2=\beta^2+\eta^2 \tag{3-11d}$$

For this particular case temperature has no ϕ dependence, hence there is no separation equation for the ϕ variable for $v=0$. It is for this reason, the solutions for the separation are zero-order Bessel functions.

2. *Temperature has no z dependence.* Equation (3-1) becomes

$$\frac{\partial^2 T}{\partial r^2}+\frac{1}{r}\frac{\partial T}{\partial r}+\frac{1}{r^2}\frac{\partial^2 T}{\partial \phi^2}=\frac{1}{\alpha}\frac{\partial T}{\partial t} \tag{3-12}$$

The separated equations and their elementary solutions are

$$\frac{d^2\Phi}{d\phi^2}+v^2\Phi=0 \qquad \Phi(v,\phi)\colon \sin v\phi \text{ and } \cos v\phi \tag{3-13a}$$

$$\frac{d^2R_v}{dr^2}+\frac{1}{r}\frac{dR_v}{dr}+\left(\beta^2-\frac{v^2}{r^2}\right)=0 \qquad R_v(\beta,r)\colon J_v(\beta r) \text{ and } Y_v(\beta r) \tag{3-13b}$$

$$\frac{d\Gamma}{dt}+\alpha\lambda^2\Gamma=0 \qquad \Gamma(t)\colon e^{-\alpha\lambda^2 t} \tag{3-13c}$$

where

$$\lambda^2=\beta^2 \tag{3-13d}$$

3. *Temperature has no time dependence.* Equation (3-1) reduces to

$$\frac{\partial^2 T}{\partial r^2}+\frac{1}{r}\frac{\partial T}{\partial r}+\frac{1}{r^2}\frac{\partial^2 T}{\partial \phi^2}+\frac{\partial^2 T}{\partial z^2}=0 \tag{3-14}$$

The separated equations and their elementary solution become

$$\frac{d^2\Phi}{d\phi^2}+v^2\Phi=0 \qquad \Phi(v,\phi)\colon \sin v\phi \text{ and } \cos v\phi \tag{3-15a}$$

$$\frac{d^2Z}{dz^2}+\eta^2 Z=0 \qquad Z(\eta,z)\colon \sin \eta z \text{ and } \eta z \tag{3-15b}$$

$$\frac{d^2R_v}{dr^2}+\frac{1}{r}\frac{dR_v}{dr}-\left(\eta^2+\frac{v^2}{r^2}\right)R_v=0 \qquad R_v(\eta,r)\colon I_v(\eta r) \text{ and } K_v(\eta r) \tag{3-15c}$$

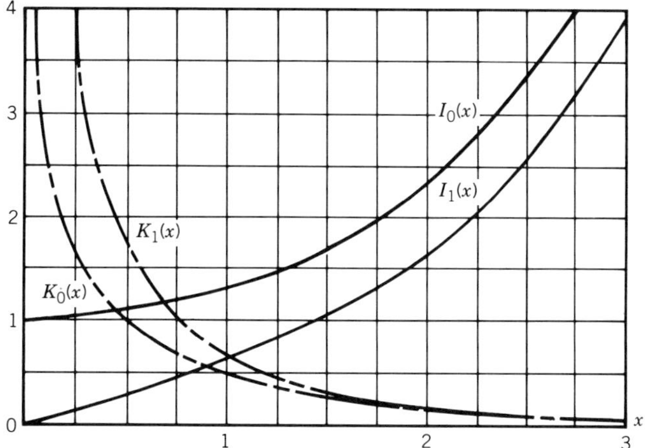

Fig. 3-2 $I_0(x), K_0(x)$ and $I_1(x), K_1(x)$ functions.

For this particular case the separation equation (3-15c) for the r variable is obtainable from that given by equation (3-13b) by setting $\beta = i\eta$ where $i = \sqrt{-1}$. Then the solutions for equation (3-15c) can be written as $J_v(i\eta)$ and $Y_v(i\eta)$; however, to alleviate the complex argument notation, these functions are denoted by $I_v(\eta)$ and $K_v(\eta)$, which are called the *modified Bessel functions of order* v of the first kind and of the second kind, respectively. Figure 3-2 shows a plot of $I_0(x)$, $I_1(x)$, $K_0(x)$, and $K_1(x)$ functions. We note that $I_v(x)$ functions become infinite as $x \to \infty$ and $K_v(x)$ functions become infinite as $x \to 0$. A discussion of the properties of modified Bessel functions and their numerical values are given in Appendix IV. There is another possibility for the separation of equation (3-14), obtainable by replacing η^2 by $-\eta^2$ in equations (3-15b) and (3-15c). In this case, the elementary solutions for the Z separation are taken as $e^{-\eta z}, e^{\eta z}$ or $\sinh \eta z$, $\cosh \eta z$; the equation for the R separation becomes *Bessel's differential equation* of order v and its solutions are taken as: $J_v(\eta r)$, $Y_v(\eta r)$.

4. *The temperature has no t and z dependence.* Then equation (3-1) simplifies to

$$\frac{\partial^2 T}{\partial r^2} + \frac{1}{r}\frac{\partial T}{\partial r} + \frac{1}{r^2}\frac{\partial^2 T}{\partial \phi^2} = 0 \tag{3-16}$$

The separated equations and their elementary solutions become

$$\frac{d^2\Phi}{d\phi^2} + v^2\Phi = 0 \qquad \Phi(v, \phi)\colon \sin v\phi \text{ and } \cos v\phi \tag{3-17a}$$

$$\frac{d^2R}{dr^2} + \frac{1}{r}\frac{dR}{dr} - \frac{v^2}{r^2}R = 0 \qquad R(r)\colon \begin{cases} r^v \text{ and } r^{-v} & \text{for } \quad v \neq 0 \\ c_1 + c_2 \ln r & \text{for } \quad v = 0 \end{cases} \tag{3-17b}$$

We note that for this particular case the equation for the function $R(r)$ is *Euler's homogeneous* differential equation.

3-2 REPRESENTATION OF AN ARBITRARY FUNCTION IN THE CYLINDRICAL COORDINATE SYSTEM

Basic to the solutions with the orthogonal expansion technique is the representation of an arbitrary function in terms of the eigenfunctions of the resulting eigenvalue problems. In the cylindrical coordinate system we have three distinct eigenvalue problems associated with the separated differential equations (3-9a), (3-9b), and (3-9c).

The eigenvalue problem associated with the differential equation (3-9a) is exactly the same as that considered in Chapter 2. Therefore, all the results presented in Chapter 2 for the rectangular coordinate system are applicable for this eigenvalue problem.

The differential equation (3-9b) appears to be similar to equations (3-9a); but, the eigenvalue problem associated with it, for the case of full circular cylinder, is cyclic with a period of 2π. Therefore, we need to examine the representation of an arbitrary function $F(\phi)$ in terms of the eigenfunctions of such an eigenvalue problem.

Finally, the differential equation (3-9c) is Bessel's differential equation which is different from those considered previously. Therefore, we need to examine the representation of an arbitrary function $F(r)$ in terms of the eigenfunctions of the eigenvalue problem associated with Bessel's equation (3-9c).

Such representations are now developed for arbitrary functions $F(r)$ and $F(\phi)$ for use as ready reference in the solution of heat conduction equation with the separation of variables in the cylindrical coordinates. As we have done in the rectangular coordinates, we develop such representations for each specific spatial domain separately. The representation of $F(r)$ is considered over the regions $0 \leqslant r \leqslant b$, $0 \leqslant r < \infty$, $a < r < \infty$, and $a < r < b$, while the representation of $F(\phi)$ is considered over the region $0 \leqslant \phi \leqslant 2\pi$ with the condition of periodicity of solution with a period of 2π.

Representation of $F(r)$ over $0 \leqslant r < b$

We consider the representation of an arbitrary function $F(r)$ defined in a finite interval $0 \leqslant r < b$ in terms of the eigenfunctions $R_\nu(\beta_m, r)$ of the eigenvalue problem

$$\frac{d^2R_\nu(r)}{dr^2} + \frac{1}{r}\frac{dR_\nu(r)}{dr} + \left(\beta^2 - \frac{\nu^2}{r^2}\right)R_\nu(r) = 0 \qquad \text{in} \qquad 0 \leqslant r < b \tag{3-18a}$$

$$\frac{dR_\nu}{dr} + HR_\nu = 0 \qquad \text{at} \qquad r = b \tag{3-18b}$$

Such an eigenvalue problem is encountered in the solution of heat conduction problem for a full solid cylinder with temperature varying with the azimuthal angle ϕ. For generality, the boundary condition at $r = b$ is chosen of the third kind. The results for the boundary conditions of the second and first kinds are obtainable from those for the third kind as special cases by setting $H = 0$ and $H \to \infty$, respectively.

The system (3-18) is a special case of the general Sturm–Liouville problem considered in Chapter 2. Therefore, the eigenfunctions $R_\nu(\beta_m, r)$ have the following orthogonality property:

$$\int_0^b rR_\nu(\beta_m, r)R_\nu(\beta_n, r)\,dr = \begin{cases} 0 & \text{for} \quad m \neq n \\ N(\beta_m) & \text{for} \quad m = n \end{cases} \tag{3-19}$$

We now consider the representation of an arbitrary function $F(r)$ defined in the finite interval $0 \leqslant r < b$ in terms of the eigenfunctions $R_\nu(\beta_m, r)$ in the form

$$F(r) = \sum_{m=1}^{\infty} c_m R_\nu(\beta_m, r) \qquad \text{in} \qquad 0 \leqslant r < b \tag{3-20}$$

The unknown coefficients c_m are determined by operating on both sides of equation (3-20) by the operator $\int_0^b rR_\nu(\beta_n, r)\,dr$ and utilizing the orthogonality relation (3-19). We find

$$c_m = \frac{1}{N(\beta_m)} \int_0^b rR_\nu(\beta_m, r)F(r)\,dr \tag{3-21}$$

where the norm, $N(\beta_m)$ is

$$N(\beta_m) = \int_0^b rR_\nu^2(\beta_m, r)\,dr \tag{3-22}$$

The substitution of equation (3-21) into (3-20) gives

$$F(r) = \sum_{m=1}^{\infty} \frac{1}{N(\beta_m)} R_\nu(\beta_m, r) \int_0^b r' R_\nu(\beta_m, r')F(r')\,dr' \qquad \text{in} \qquad 0 \leqslant r < b \tag{3-23}$$

where the function $R_\nu(\beta_m, r)$ is given by

$$R_\nu(\beta_m, r) = J_\nu(\beta_m r) \tag{3-24}$$

Here, the function $Y_\nu(\beta_m, r)$ is excluded from the solution, because the region includes the origin $r = 0$ where $Y_\nu(\beta_m r)$ becomes infinite.

The eigencondition for determining β_m is obtained by introducing equation (3-24) into the boundary condition (3-18b). We obtain

$$\beta_m J_\nu'(\beta_m b) + HJ_\nu(\beta_m b) = 0 \tag{3-25}$$

where we defined

$$J'_\nu(\beta_m b) \equiv \left[\frac{d}{dr} J_\nu(r) \right]_{r=\beta_m b} \tag{3-26}$$

with H and ν are real constants and [5, p. 597]

$$\nu + \tfrac{1}{2} \geqslant 0$$

The eigenvalues β_m are the positive roots of the transcendental equation (3-25). In this equation, the prime over the Bessel function denotes the derivative in the sense defined by equation (3-26). From equations (3-22) and (3-24) the norm becomes

$$N(\beta_m) = \int_0^b r J_\nu^2(\beta_m r)\, dr \tag{3-27}$$

When the integration is performed by utilizing the integration formula given by equation (25b) of the Appendix IV, the norm is determined as

$$N(\beta_m) = \frac{b^2}{2}\left[J_\nu'^2(\beta_m b) + \left(1 - \frac{\nu^2}{\beta_M^2 b^2}\right) J_\nu^2(\beta_m b) \right] \tag{3-28a}$$

In view of the transcendental equation (3-25), this relation may be written in the alternative form as

$$N(\beta_m) = \frac{b^2}{2}\left[\frac{H^2}{\beta_m^2} + \left(1 - \frac{\nu^2}{\beta_m^2 b^2}\right)\right] J_\nu^2(\beta_m b) \tag{3-28b}$$

The above expressions for the eigencondition and the norm are developed for boundary condition of the third kind as given by equation (3-18b). Expressions for the case of boundary conditions of the second and first kinds are obtained from these general expressions as special cases as described below.

Boundary Conditon at $r = b$ is of the Second Kind. For this special case we have

$$R_\nu(\beta_m, r) = J_\nu(\beta_m r) \tag{3-29}$$

the eigenvalues β_m are the positive roots of

$$J'_\nu(\beta_m b) = 0 \tag{3-30}$$

The norm is obtained from equation (3-28b) by noting that for the boundary condition of the second kind we have $H = 0$:

$$N(\beta_m) = \frac{b^2}{2}\left(1 - \frac{\nu^2}{\beta_m^2 b^2}\right) J_\nu^2(\beta_m b) \qquad \text{for} \qquad \beta_m \neq 0 \tag{3-31a}$$

Note that, for the boundary condition of the second kind, $\beta_0 = 0$ is also an eigenvalue for $\nu = 0$; then the corresponding eigenfunction and the norm for this special case are

$$R_0(\beta_0 r) = 1 \qquad \text{and} \qquad N(\beta_0) = \int_0^b r\,dr = \frac{b^2}{2} \qquad \text{for} \qquad \beta_0 = 0 \quad \text{(3-31b)}$$

See note 1 at end of this chapter for a discussion of $\beta_0 = 0$.

Boundary Condition at $r = b$ is of the First Kind. For this case we have

$$R_\nu(\beta_m, r) = J_\nu(\beta_m r) \tag{3-32}$$

and the eigenvalues β_m are the positive roots of

$$J_\nu(\beta_m b) = 0 \tag{3-33}$$

The norm $N(\beta_m)$ is obtained from equation (3-28a) by utilizing equation (3-33).

$$N(\beta_m) = \frac{b^2}{2} J_\nu'^2(\beta_m b) \tag{3-34}$$

In Table 3-1 we summarize the eigenfunctions $R_\nu(\beta_m, r)$, the norms $N(\beta_m)$, and the eigenconditions. They will be needed in the solution of heat conduction problems for a solid cylinder $0 \leqslant r < b$ when temperature varies with azimuth angle ϕ.

For problems with aximuthal symmetry, the eigenvalue problem (3-18) is applicable with $\nu = 0$. Therefore, the results in Table 3-1 are also applicable for heat conduction problems in a solid cylinder with azimuthally symmetric temperature if we set $\nu = 0$ in these results.

Representation of $F(r)$ over $0 \leqslant r < \infty$

We now consider the representation of an arbitrary function $F(r)$ defined in the infinite interval $0 \leqslant r < \infty$ in terms of the solutions of the following differential equation

$$\frac{d^2 R_\nu(r)}{dr^2} + \frac{1}{r}\frac{dR_\nu(r)}{dr} + \left(\beta^2 - \frac{\nu^2}{r^2}\right) R_\nu(r) = 0 \qquad \text{in} \qquad 0 \leqslant r < \infty \tag{3-35}$$

subject to the condition that $R_\nu(r)$ remains finite at $r = 0$. Expansions of this type will be needed in the solution of heat-conduction problems in a region $0 \leqslant r < \infty$, $0 \leqslant \phi \leqslant 2\pi$ in the cylindrical coordinate system with temperature varying radially and azimuthally.

TABLE 3-1 The Eigenfunctions $R_\nu(\beta_m, r)$, the Norm $N(\beta_m)$, and the Eigenvalues β_m of the Differential Equation

$$\frac{d^2R_\nu}{dr^2} + \frac{1}{r}\frac{dR_\nu}{dr} + \left(\beta^2 - \frac{\nu^2}{r^2}\right)R_\nu = 0 \quad \text{in} \quad 0 \leqslant r < b$$

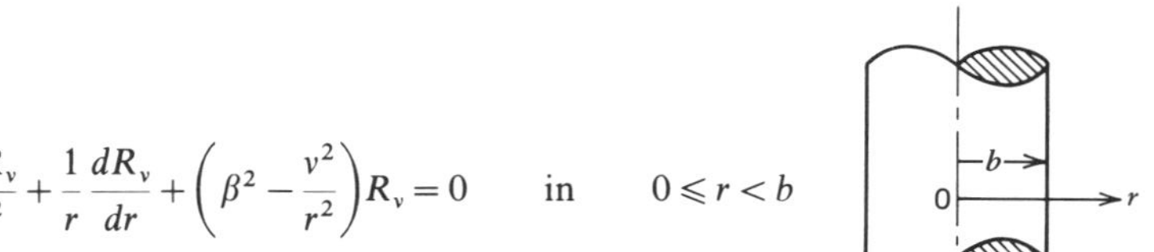

Subject to the Boundary Conditions Shown in the Table Below

No.	Boundary Condition at $r = b$	$R_\nu(\beta_m, r)$	$\dfrac{1}{N(\beta_m)}$	Eigenvalues β_m are the Positive Roots of
1	$\dfrac{dR_\nu}{dr} + HR_\nu = 0$	$J_\nu(\beta_m r)$	$\dfrac{2}{J_\nu^2(\beta_m b)} \cdot \dfrac{\beta_m^2}{b^2(H^2 + \beta_m^2) - \nu^2}$	$\beta_m J'_\nu(\beta_m b) + HJ_\nu(\beta_m b) = 0$
2	$\dfrac{dR_\nu}{dr} = 0$	$J_\nu(\beta_m r)^a$	$\dfrac{2}{J_\nu^2(\beta_m b)} \cdot \dfrac{\beta_m^2}{b^2\beta_m^2 - \nu^2}{}^a$	$J'_\nu(\beta_m b) = 0^a$
3	$R_\nu = 0$	$J_\nu(\beta_m r)$	$\dfrac{2}{b^2 J'^2_\nu(\beta_m b)}$	$J_\nu(\beta_m b) = 0$

[a]For this particular case $\beta_0 = 0$ is also an eigenvalue with $\nu = 0$; then the corresponding eigenfunction is $R_0 = 1$ and the norm $1/N(\beta_0) = 2/b^2$.

The solution of equation (3-35) that remains finite at $r=0$ is

$$R_\nu(\beta, r) = J_\nu(\beta r) \tag{3-36}$$

An arbitrary function $F(r)$ defined in the interval $0 \leqslant r < \infty$ can be represented in terms of $J_\nu(\beta r)$ functions for $\nu \geqslant -\frac{1}{2}$ in the form [9, p. 88; 10, p. 52; 5, p. 453]

$$F(r) = \int_{\beta=0}^{\infty} r^{1/2} \beta J_\nu(\beta r)\, d\beta \int_{r'=0}^{\infty} r'^{1/2} J_\nu(\beta r') F(r')\, dr' \qquad 0 \leqslant r < \infty \tag{3-37}$$

if the integral $\int_0^\infty F(r')\,dr'$ is absolutely convergent, and if the function $F(r)$ is of bounded variation in the neighborhood of the point r.

If we now replace $F(r)$ by $r^{1/2}F(r)$ in the equation (3-37), we obtain

$$F(r) = \int_{\beta=0}^{\infty} \beta J_\nu(\beta r)\, d\beta \int_{r'=0}^{\infty} r' J_\nu(\beta r') F(r')\, dr' \qquad 0 \leqslant r < \infty \tag{3-38}$$

which is the representation of a function $F(r)$ in the interval $0 \leqslant r < \infty$ that will be needed for the solution of heat conduction problems in an infinite region $0 \leqslant r < \infty$.

For problems with azimuthal symmetry, the eigenvalue problem (3-35) is applicable with $\nu = 0$. For such a case the representation (3-38) is applicable by setting $\nu = 0$.

Representation of $F(r)$ over $a < r < \infty$

We now examine the representation of an arbitrary function $F(r)$ defined in the interval $a < r < \infty$ in terms of the solutions of the following problem

$$\frac{d^2 R_0(r)}{dr^2} + \frac{1}{r}\frac{dR_0(r)}{dr} + \beta^2 R_0(r) = 0 \qquad \text{in} \qquad a < r < \infty \tag{3-39a}$$

$$-\frac{dR_0}{dr} + HR_0 = 0 \qquad \text{at} \qquad r = a \tag{3-39b}$$

such a representation is needed in the solution of heat conduction problems in the region $a < r < \infty$ in the cylindrical coordinate system for an azimuthally symmetric temperature i.e., temperature does not depend on ϕ. The representation of an arbitrary function $F(r)$ in the region $a < r < \infty$ in terms of the solutions $R_0(\beta, r)$ of the problem (3-39) is considered in reference [12] and the result can be written in the form

$$F(r) = \int_{\beta=0}^{\infty} \frac{1}{N(\beta)} \beta R_0(\beta, r)\, d\beta \int_{r'=a}^{\infty} r' R_0(\beta, r') F(r')\, dr' \qquad \text{in} \qquad a < r < \infty \tag{3-40}$$

Here, the norm $N(\beta)$, the function $R_0(\beta, r)$ depend on the type of the boundary condition at $r = a$; that is, whether it is of the first, second, or the third kind. We present the expressions for $R_0(\beta, r)$ and $N(\beta)$ for these three different types of boundary conditions at $r = a$.

The Boundary Condition at $r = a$ of the Third Kind. The solution of equation (3-39a) satisfying the boundary condition (3-39b) is taken as

$$R_0(\beta, r) = J_0(\beta r)[\beta Y_1(\beta a) + H Y_0(\beta a)] - Y_0(\beta r)[\beta J_1(\beta a) + H J_0(\beta a)] \quad (3\text{-}41)$$

and the norm $N(\beta)$ is given by

$$N(\beta) = [\beta J_1(\beta a) + H J_0(\beta a)]^2 + [\beta Y_1(\beta a) + H Y_0(\beta a)]^2 \quad (3\text{-}42)$$

The Boundary Condition at $r = a$ is of the Second Kind. For this special case we have $H = 0$. The solution of equation (3-39a) satisfying this boundary condition is taken as

$$R_0(\beta, r) = J_0(\beta r) Y_1(\beta a) - Y_0(\beta r) J_1(\beta a) \quad (3\text{-}43)$$

and the norm becomes

$$N(\beta) = J_1^2(\beta a) + Y_1^2(\beta a) \quad (3\text{-}44)$$

The Boundary Condition at $r = a$ is of the First Kind. For this special case we have $H \to \infty$. The solution of equation (3-39a) satisfying this boundary condition is taken as

$$R_0(\beta, r) = J_0(\beta r) Y_0(\beta a) - Y_0(\beta r) J_0(\beta a) \quad (3\text{-}45)$$

and the corresponding norm becomes

$$N(\beta) = J_0^2(\beta a) + Y_0^2(\beta a) \quad (3\text{-}46)$$

We summarize in Table 3-2 the above results for $R_0(\beta, r)$ and $N(\beta)$ for the boundary conditions of the first, second, and third kinds at $r = a$.

Representation of $F(r)$ over $a < r < b$

We now consider the representation of an arbitrary function $F(r)$ defined in a finite interval $a \leqslant r \leqslant b$ in terms of the eigenfunctions of the following eigenvalue problem:

$$\frac{d^2 R_\nu(R)}{dr^2} + \frac{1}{r}\frac{dR_\nu(r)}{dr} + \left(\beta^2 - \frac{\nu^2}{r^2}\right) R_\nu(r) = 0 \qquad \text{in} \qquad a < r < b \quad (3\text{-}47a)$$

TABLE 3-2 The Solution $R_0(\beta, r)$, and the Norm $N(\beta)$ of the Differential Equation

$$\frac{dR_0^2(r)}{dr^2} + \frac{1}{r}\frac{dR_0(r)}{dr} + \beta^2 R_0(r) = 0 \qquad \text{in} \qquad a < r < \infty$$

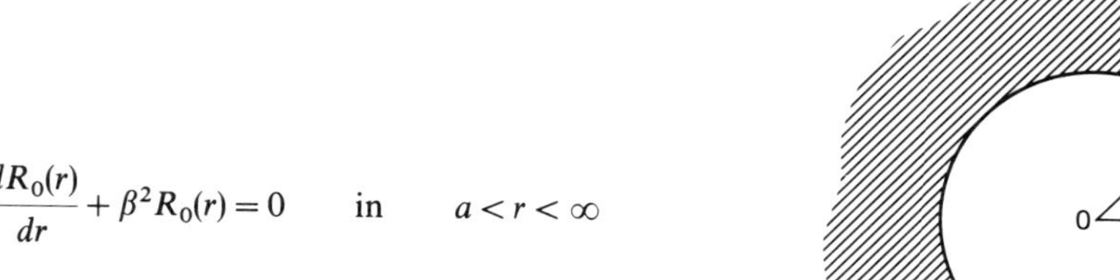

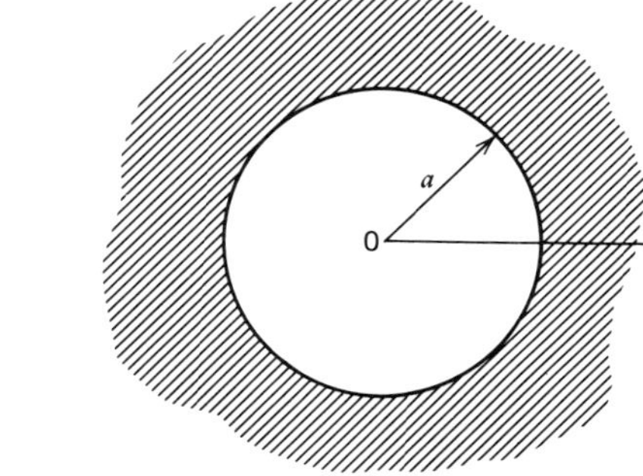

Subject to the Boundary Conditions Shown in the Table Below

No.	Boundary Condition at $r = a$	$R_0(\beta, r)$ and $1/N(\beta)$
1	$-\frac{dR_0}{dr} + HR = 0$	$R_0(\beta, r) = J_0(\beta r)[\beta Y_1(\beta a) + HY_0(\beta a)] - Y_0(\beta r)[\beta J_1(\beta a) + HJ_0(\beta a)]$ $\frac{1}{N(\beta)} = \{[\beta J_1(\beta a) + HJ_0(\beta a)]^2 + [\beta Y_1(\beta a) + HY_0(\beta a)]^2\}^{-1}$
2	$\frac{dR_0}{dr} = 0$	$R_0(\beta, r) = J_0(\beta r)Y_1(\beta a) - Y_0(\beta r)J_1(\beta a)$ $\frac{1}{N(\beta)} = [J_1^2(\beta a) + Y_1^2(\beta a)]^{-1}$
3	$R_0 = 0$	$R_0(\beta, r) = J_0(\beta r)Y_0(\beta a) - Y_0(\beta r)J_0(\beta a)$ $\frac{1}{N(\beta)} = [J_0^2(\beta a) + Y_0^2(\beta a)]^{-1}$

TABLE 3-3 The Solution $R_\nu(\beta_m, r)$, the Norm $N(\beta_m)$, and the Eigenvalues β_m of the Differential Equation

$$\frac{d^2R_\nu(r)}{dr^2} + \frac{1}{r}\frac{dR_\nu(r)}{dr} + \left(\beta^2 - \frac{\nu^2}{r^2}\right)R_\nu(r) = 0 \quad \text{in} \quad a < r < b$$

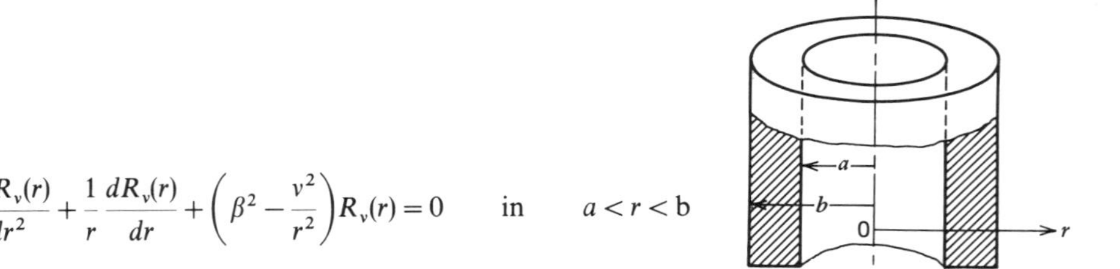

Subject to the Boundary Conditions Shown in the Table Below

No.	Boundary Condition at $r = a$	Boundary Condition at $r = b$	$R_\nu(\beta_m, r)$ and $1/N(\beta_m)$	β_m Values are the Positive roots of
1	$\frac{dR_\nu}{dr} = 0$	$\frac{dR_\nu}{dr} = 0$	$R_\nu(\beta_m, r) = J_\nu(\beta_m r)Y'_\nu(\beta_m b) - J'_\nu(\beta_m b)Y_\nu(\beta_m r)$ $\frac{1}{N(\beta_m)} = \frac{\pi^2}{2}\frac{\beta_m^2 J'^2_\nu(\beta_m a)}{\left[1-\left(\frac{\nu}{\beta_m b}\right)^2\right]J'^2_\nu(\beta_m a) - \left[1-\left(\frac{\nu}{\beta_m a}\right)^2\right]J'^2_\nu(\beta_m b)}$	$J'_\nu(\beta_m a)Y'_\nu(\beta_m b)$ $- J'_\nu(\beta_m b)Y'_\nu(\beta_m a) = 0$[a]

2	$\frac{dR_\nu}{dr}=0$	$R_\nu=0$	$R_\nu(\beta_m,r)=J_\nu(\beta_m r)Y_\nu(\beta_m b)-J_\nu(\beta_m b)Y_\nu(\beta_m r)$ $\frac{1}{N(\beta_m)}=\frac{\pi^2}{2}\frac{\beta_m^2 J_\nu'^2(\beta_m a)}{J_\nu'^2(\beta_m a)-\left[1-\left(\frac{\nu}{\beta_m a}\right)^2\right]J_\nu^2(\beta_m b)}$	$J_\nu'(\beta_m a)Y_\nu(\beta_m b)$ $-J_\nu(\beta_m b)Y_\nu'(\beta_m a)=0$
3	$R_\nu=0$	$\frac{dR_\nu}{dr}=0$	$R_\nu(\beta_m,r)=J_\nu(\beta_m r)Y_\nu'(\beta_m b)-J_\nu'(\beta_m b)Y_\nu(\beta_m r)$ $\frac{1}{N(\beta_m)}=\frac{\pi^2}{2}\frac{\beta_m^2 J_\nu^2(\beta_m a)}{\left[1-\left(\frac{\nu}{\beta_m b}\right)^2\right]J_\nu^2(\beta_m a)-J_\nu'^2(\beta_m b)}$	$J_\nu(\beta_m a)Y_\nu'(\beta_m b)$ $-J_\nu'(\beta_m b)Y_\nu(\beta_m a)=0$
4	$R_\nu=0$	$R_\nu=0$	$R_\nu(\beta_m,r)=J_\nu(\beta_m r)Y_\nu(\beta_m b)-J_\nu(\beta_m b)Y_\nu(\beta_m r)$ $\frac{1}{N(\beta_m)}=\frac{\pi^2}{2}\frac{\beta_m^2 J_\nu^2(\beta_m a)}{J_\nu^2(\beta_m a)-J_\nu^2(\beta_m b)}$	$J_\nu(\beta_m a)Y_\nu(\beta_m b)$ $-J_\nu(\beta_m b)Y_\nu(\beta_m a)=0$

[a]For this particular case $\beta_0=0$ is also an eigenvalue with $\nu=0$; the corresponding eigenfunction is $R_0(\beta_0,r)=1$ and the norm $1/N(\beta_0)=2/(b^2-a^2)$.

$$-\frac{dR_\nu}{dr} + H_1 R_\nu = 0 \qquad \text{at} \qquad r = a \tag{3-47b}$$

$$\frac{dR_\nu}{dr} + H_2 R_\nu = 0 \qquad \text{at} \qquad r = b \tag{3-47c}$$

The eigenvalue problem of this type is encountered in the solution of heat conduction for a hollow cylinder with azimuthally varying temperature distribution. For generality, boundary condition of the third kind is chosen for both boundaries. Other combinations of boundary conditions are obtainable by setting the coefficients H_1, H_2 equal to zero or infinity; thus, nine different combinations are possible.

The system (3-47) is a special case of the Sturm–Liouville problem, hence the eigenfunctions $R_\nu(\beta_m, r)$ have the following orthogonality property

$$\int_a^b r R_\nu(\beta_m, r) R_\nu(\beta_n, r)\, dr = \begin{cases} 0 & \text{for} \quad m \neq n \\ N(\beta_m) & \text{for} \quad m = n \end{cases} \tag{3-48}$$

where

$$N(\beta_m) = \int_a^b r R_\nu^2(\beta_m, r)\, dr \tag{3-49}$$

Now we consider the representation of an arbitrary function $F(r)$ defined in the interval $a < r < b$ in terms of the eigenfunctions $R_\nu(\beta_m, r)$ of the above eigenvalue problem (3-55) in the form

$$F(r) = \sum_{m=1}^{\infty} c_m R_\nu(\beta_m, r) \qquad \text{in} \qquad a < r < b \tag{3-50}$$

The unknown coefficients c_m are determined by following a procedure described previously; then the representation (3-50) becomes

$$F(r) = \sum_{m=1}^{\infty} \frac{1}{N(\beta_m)} R_\nu(\beta_m, r) \int_a^b r' R_\nu(\beta_m, r') F(r')\, dr' \qquad \text{in} \qquad a < r < b \tag{3-51}$$

We present in Table 3-3 the eigenfunctions $R_\nu(\beta_m, r)$, the norm $N(\beta_m)$ and the eigenconditions of the eigenvalue problem (3-47) for four different combinations of the boundary conditions of the first and second kind at the boundaries. The boundary conditions of the third kind are not included in this table, because the resulting expressions are too complex to be practical for computational purposes.

Representation of $F(\phi)$ over $0 \leqslant \phi \leqslant 2\pi$

We now consider the representation of an arbitrary function $F(\phi)$ defined in the interval $0 \leqslant \phi \leqslant 2\pi$ in terms of the eigenfunctions of the eigenvalue problem associated with the separation equation (3-9b). We have the following eigenvalue problem

$$\frac{d^2\Phi}{d\phi^2} + v^2\Phi = 0 \quad \text{in} \quad 0 \leqslant \phi \leqslant 2\pi \tag{3-52a}$$

The solution may be taken as

$$\Phi(v, \phi) = A_v \sin v\phi + B_v \cos v\phi \tag{3-52b}$$

We now examine the representation of a function $F(\phi)$ that is periodic in ϕ with period 2π in terms of $\Phi(v, \phi)$ functions in the form

$$F(\phi) = \sum_v (A_v \sin v\phi + B_v \cos v\phi) \quad \text{in} \quad 0 \leqslant \phi \leqslant 2\pi \tag{3-53}$$

The condition that $F(\phi)$ is periodic in ϕ with period 2π requires that the separation constants v should be taken as integers, that is

$$v = 0, 1, 2, 3 \ldots$$

To determine the coefficients A_v, we operate on both sides of equation (3-53) by the operator $\int_0^{2\pi} \sin v'\phi \, d\phi$ and utilize the orthogonality of functions $\sin v\phi$. We obtain

$$A_v = \frac{1}{\pi}\int_0^{2\pi} F(\phi) \sin v\phi \, d\phi \quad \text{for} \quad v = 0, 1, 2, 3 \ldots \tag{3-54a}$$

since $\int_0^{2\pi} \sin^2 v\phi \, d\phi = \pi$ and the integrals of the product of $\sin v\phi$, $\cos v\phi$ vanish. To determine the coefficients B_v we operate on both sides of equation (3-53) by the operator $\int_0^{2\pi} \cos v'\phi \, d\phi$ and utilize the orthogonality of functions $\cos v\phi$. We find

$$B_v = \begin{cases} \dfrac{1}{\pi}\displaystyle\int_0^{2\pi} F(\phi) \cos v\phi \, d\phi & \text{for} \quad v = 1, 2, 3 \ldots \quad (3\text{-}54\text{b}) \\ \dfrac{1}{2\pi}\displaystyle\int_0^{2\pi} F(\phi) \, d\phi & \text{for} \quad v = 0 \quad (3\text{-}54\text{c}) \end{cases}$$

since $\int_0^{2\pi} \cos^2 v\phi \, d\phi$ is equal to π for $v = 1, 2, 3, \ldots$ and equal to 2π for $v = 0$.

The substitution of the above expressions for A_v and B_v into equation (3-53) yields the representation in the form

$$\begin{aligned} F(\phi) &= \frac{1}{2\pi}\int_0^{2\pi} F(\phi') \, d\phi' + \frac{1}{\pi}\sum_{v=1}^{\infty}\int_0^{2\pi} F(\phi')(\sin v\phi \sin v\phi' + \cos v\phi \cos v\phi') \, d\phi' \\ &= \frac{1}{2\pi}\int_0^{2\pi} F(\phi') \, d\phi' + \frac{1}{\pi}\sum_{v=1}^{\infty}\int_0^{2\pi} F(\phi') \cos v(\phi - \phi') \, d\phi' \end{aligned} \tag{3-55a}$$

This representation may be writte

$$F(\phi)=\frac{1}{\pi}\sum_{\nu}\int_{0}^{2\pi}F(\phi')\cos \ldots \ldots 2\pi \qquad \text{(3-55b)}$$

where

$$\nu=0,1,2,3\ldots$$

and replace π *by* 2π *for* $\nu=0$. If we compare the representations given by equations (3-53) and (3-55b) we conclude that

$$[A_{\nu}\sin\nu\phi+B_{\nu}\cos\nu\phi]\equiv\frac{1}{\pi}\int_{0}^{2\pi}F(\phi')\cos\nu(\phi-\phi')\,d\phi' \qquad \text{(3-56)}$$

where

$$\nu=0,1,2,3\ldots$$

and replace π *by* 2π *for* $\nu=0$.

The representation of $F(\phi)$ as given above will be needed in the solution of heat conduction in a full cylinder (i.e., $0\leqslant\phi\leqslant 2\pi$) with azimuthally varying temperature.

Representation of $F(\phi)$ over $0<\phi<\phi_0(<2\pi)$

In the case of a portion of a cylinder the range of ϕ variable is $0<\phi<\phi_0(<2\pi)$. For such a case, equation (3-52a) should be solved over the range $0<\phi<\phi_0(<2\pi)$ with prescribed boundary conditions at the boundary surfaces $\phi=0$ and $\phi=\phi_0$. For such a case, the eigenvalue problem for the function $\Phi(\phi,\nu)$ is similar to that of a slab in the region $0<\phi<\phi_0$ and the results presented in Table 2-1 may be utilized to determine the eigenfunctions, the norm, and the eigenconditions.

3-3 HOMOGENEOUS PROBLEMS IN (r,t) VARIABLES

Having established the representation of an arbitrary function $F(r)$ in terms of the solutions of Bessel's differential equation as discussed previously, the solution of the one-dimensional homogeneous heat conduction problems in the (r,t) variables becomes a straightforward matter as now illustrated with examples.

Example 3-1

A solid cylinder, $0\leqslant r\leqslant b$, is initially at a temperature $F(r)$; for times $t>0$ the boundary surface at $r=b$ dissipates heat by convection into a medium at zero temperature as illustrated in Fig. 3-3*a*. Obtain an expression for the temperature distribution $T(r,t)$ for times $t>0$.

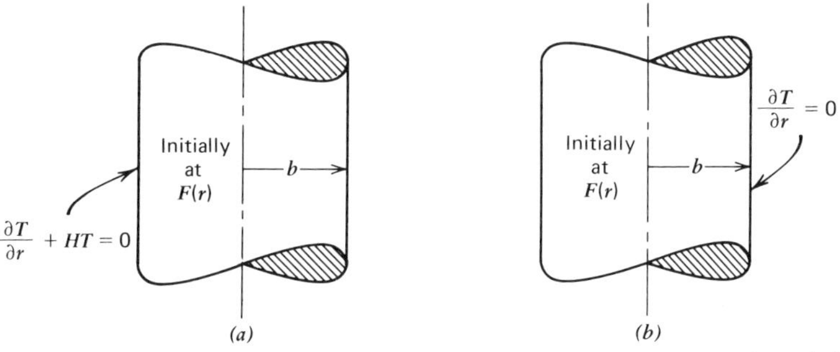

Fig. 3-3 Boundary and initial conditions for a solid cylinder considered in (*a*) Example 3-1 and (*b*) Example 3-2.

Solution. The mathematical formulation of this problem is taken as

$$\frac{\partial^2 T}{\partial r^2} + \frac{1}{r}\frac{\partial T}{\partial r} = \frac{1}{\alpha}\frac{\partial T(r,t)}{\partial t} \qquad \text{in} \qquad 0 \leqslant r < b, \quad t > 0 \tag{3-57a}$$

$$\frac{\partial T}{\partial r} + HT = 0 \qquad \text{at} \qquad r = b, \qquad t > 0 \tag{3-57b}$$

$$T = F(r) \qquad \text{for} \qquad t = 0, \qquad \text{in} \qquad 0 \leqslant r \leqslant b \tag{3-57c}$$

Separating the variables, it can be shown that the solution for the time separation is given by

$$\Gamma(t) = r^{-\alpha\beta_m^2 t} \tag{3-58}$$

and the space-variable function $R(\beta_m, r)$ satisfies the following eigenvalue problem:

$$\frac{d^2 R_0(r)}{dr^2} + \frac{1}{r}\frac{dR_0(r)}{dr} + \beta^2 R_0(r) = 0 \qquad \text{in} \qquad 0 \leqslant r < b \tag{3-59a}$$

$$\frac{dR_0}{dr} + HR_0 = 0 \qquad \text{at} \qquad r = b \tag{3-59b}$$

The complete solution for $T(r, t)$ is constructed as

$$T(r,t) = \sum_{m=1}^{\infty} c_m e^{-\alpha\beta_m^2 t} R_0(\beta_m, r) \tag{3-60}$$

The application of the initial condition gives

$$F(r) = \sum_{m=1}^{\infty} c_m R_0(\beta_m, r) \qquad \text{in} \qquad 0 \leqslant r < b \tag{3-61}$$

This is an expansion of an arbitrary function $F(r)$ defined in the interval $0 \leqslant r < b$ in the same form as given by equation (3-20). The unknown coefficients c_m are determined by utilizing the orthogonality of the eigenfunction as given by equation (3-19). Then, the solution for the problem becomes

$$T(r,t) = \sum_{m=1}^{\infty} \frac{1}{N(\beta_m)} e^{-\alpha\beta_m^2 t} R_0(\beta_m, r) \int_0^b r' R_0(\beta_m, r') F(r')\, dr' \tag{3-62}$$

where $R_0(\beta_m, r)$, $N(\beta_m)$ and the eigenvalues β_m are immediately obtainable from Table 3-1, case 1, by setting $v = 0$, because the eigenvalue problem (3-59) is a special case of the general problem (3-18) with $v = 0$. Then the solution (3-62) takes the form

$$T(r,t) = \frac{2}{b^2} \sum_{m=1}^{\infty} e^{-\alpha\beta_m^2 t} \frac{\beta_m^2 J_0(\beta_m r)}{(\beta_m^2 + H^2) J_0^2(\beta_m b)} \int_0^b r' J_0(\beta_m r') F(r')\, dr' \tag{3-63}$$

where the β_m values are the positive roots of

$$\beta_m J_0'(\beta_m b) + H J_0(\beta_m b) = 0 \qquad \text{or} \qquad \beta_m J_1(\beta_m b) = H J_0(\beta_m b) \tag{3-64}$$

For the special case of $F(r) = T_0 = \text{constant}$, the solution (3-63) reduces to

$$T(r,t) = \frac{2T_0}{b} \sum_{m=1}^{\infty} e^{-\alpha\beta_m^2 t} \frac{\beta_m J_0(\beta_m r) J_1(\beta_m b)}{(\beta_m^2 + H^2) J_0^2(\beta_m b)} \tag{3-65a}$$

$$= \frac{2HT_0}{b} \sum_{m=1}^{\infty} e^{-\alpha\beta_m^2 t} \frac{J_0(\beta_m r)}{(\beta_m^2 + H^2) J_0(\beta_m b)} \tag{3-65b}$$

where we utilized equation (3-64) to obtain the alternative form given by equation (3-65b)

Example 3-2

A solid cylinder, $0 \leqslant r \leqslant b$, is initially at a temperature $F(r)$; for times $t > 0$ the boundary surface at $r = b$ is kept insulated as illustrated in Fig. 3-3b. Obtain an expression for the temperature distribution $T(r, t)$ for times $t > 0$.

Solution. The solution for this problem is written formally exactly in the same form as that given by equation (3-62); but, $R_0(\beta_m, r)$, $N(\beta_m)$ and eigenvalues β_m are taken from Table 3-1, case 2, by setting $v = 0$. We obtain

$$T(r,t) = \frac{2}{b^2} \int_{r'=0}^{b} r' F(r')\, dr' + \frac{2}{b^2} \sum_{m=1}^{\infty} e^{-\alpha\beta_m^2 t} \cdot \frac{J_0(\beta_m r)}{J_0^2(\beta_m b)} \int_{r'=0}^{b} r' J_0(\beta_m r') F(r')\, dr' \tag{3-66a}$$

where the β_m values are the positive roots of

$$J'_0(\beta_m b) = 0 \qquad \text{or} \qquad J_1(\beta_m b) = 0 \tag{3-66b}$$

The first term on the right-hand side of equation (3-66a) is due to the fact that $\beta_0 = 0$ is also an eigenvalue for this special case. The region being insulated, heat cannot escape from the boundaries, hence the temperature, after the transients have passed, becomes the average of the initial temperature distribution over the cross section of the cylinder as given by the first term on the right-hand side of equation (3-66a).

Example 3-3

A solid cylinder, $0 \leqslant r \leqslant b$, is initially at a temperature $F(r)$; for times $t > 0$ the boundary surface at $r = b$ is kept at zero temperature. Obtain an expression for the temperature distribution $T(r,t)$ for times $t > 0$.

Solution. The solution is written formally exactly in the same form as that given by equation (3-62); but, $R_0(\beta_m, r), N(\beta_m)$ and eigenvalues β_m are taken from Table 3-1, case 3, by setting $v = 0$. We obtain

$$T(r,t) = \frac{2}{b^2} \sum_{m=1}^{\infty} e^{-\alpha\beta_m^2 t} \frac{J_0(\beta_m r)}{J_1^2(\beta_m b)} \int_{r'=0}^{b} r' J_0(\beta_m r') F(r')\, dr' \tag{3-67a}$$

where the β_m values are the positive roots of

$$J_0(\beta_m b) = 0 \tag{3-67b}$$

For the case of constant initial temperature $F(r) = T_0$, equation (3-67a) becomes

$$T(r,t) = \frac{2T_0}{b} \sum_{m=1}^{\infty} e^{-\alpha\beta_m^2 t} \frac{J_0(\beta_m r)}{\beta_m J_1(\beta_m b)} \tag{3-68}$$

Example 3-4

A hollow cylinder, $a < r < b$, is initially at a temperature $F(r)$ (Fig. 3-4). For times $t > 0$, the boundary surfaces at $r = a$ and $r = b$ are kept insulated. Develop an expression for the temperature distribution $T(r,t)$ for times $t > 0$.

Solution. The mathematical formulation of the problem is given by

$$\frac{\partial^2 T}{dr^2} + \frac{1}{r}\frac{\partial T}{\partial r} = \frac{1}{\alpha}\frac{\partial T}{\partial t} \qquad \text{in} \qquad a < r < b, \quad t > 0 \tag{3-69a}$$

$$\frac{\partial T}{\partial r} = 0 \qquad \text{at} \qquad r = a, \qquad t > 0 \tag{3-69b}$$

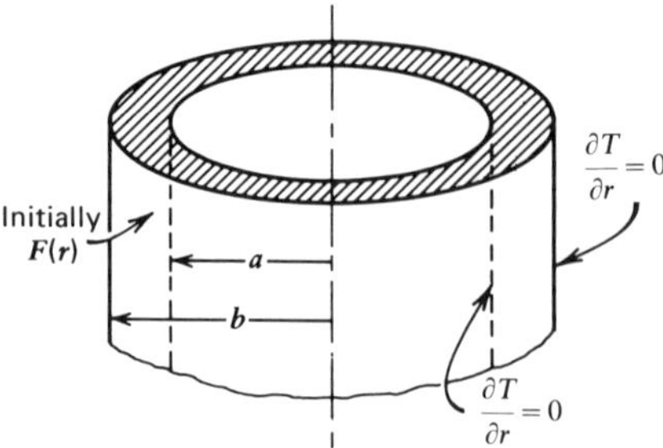

Fig. 3-4 Boundary and initial conditions for a hollow cylinder considered in Example 3-4.

$$\frac{\partial T}{\partial r} = 0 \qquad \text{at} \qquad r = b, \qquad t > 0 \tag{3-69c}$$

$$T = F(r) \qquad \text{for} \qquad t = 0, \qquad \text{in the region} \tag{3-69d}$$

Separating the variables, it can be shown that the solution for the time variable function is given by $\exp(-\alpha\beta_m^2 t)$, and the space-variable function $R_0(\beta_m, r)$ is the solution of the following eigenvalue problem:

$$\frac{d^2 R_0}{dr^2} + \frac{1}{r}\frac{dR_0}{dr} + \beta_m^2 R_0 = 0 \qquad \text{in} \qquad a < r < b \tag{3-70a}$$

$$\frac{dR_0}{dr} = 0 \qquad \text{at} \qquad r = a \tag{3-70b}$$

$$\frac{dR_0}{dr} = 0 \qquad \text{at} \qquad r = b \tag{3-70c}$$

Then, the complete solution for $T(r, t)$ is written as

$$T(r, t) = \sum_{m=1}^{\infty} c_m e^{-\alpha\beta_m^2 t} R_0(\beta_m, r) \tag{3-71}$$

The application of the initial condition (3-69d) yields

$$F(r) = \sum_{m=1}^{\infty} c_m R_0(\beta_m, r) \qquad \text{in} \qquad a < r < b \tag{3-72}$$

This is an expansion of an arbitrary function $F(r)$ defined in the interval $a < r < b$ in terms of the eigenfunctions $R_0(\beta_m, r)$ of the eigenvalue problem

(3-70). The unknown coefficients c_m are readily determined by utilizing the orthogonality of the eigenfunctions as given by equations 3-48 and 3-49. Then, the solution for the temperature $T(r, t)$ is written as

$$T(r,t) = \sum_{m=1}^{\infty} \frac{1}{N(\beta_m)} e^{-\alpha\beta_m^2 t} R_0(\beta_m, r) \int_a^b r' R_0(\beta_m, r') F(r')\, dr' \tag{3-73}$$

where the eigenfunctions $R_0(\beta_m, r)$, eigenvalues β_m, and the norm $N(\beta_m)$ are obtained from Table 3-3, case 1 by setting $v = 0$.

$$R_0(\beta_m, r) = J_0(\beta_m r) Y_0'(\beta_m b) - J_0'(\beta_m b) Y_0(\beta_m r) \tag{3-74a}$$

$$\frac{1}{N(\beta_m)} = \frac{\pi^2}{2} \frac{\beta_m^2 J_0'^2(\beta_m a)}{J_0'^2(\beta_m a) - J_0'^2(\beta_m b)} \tag{3-74b}$$

and the eigenvalues β_m are the positive roots of the following transcendental equation:

$$J_0'(\beta_m a) Y_0'(\beta_m b) - J_0'(\beta_m b) Y_0'(\beta_m a) = 0 \tag{3-74c}$$

In addition, for this particular case $\beta_0 = 0$ is also an eigenvalue; then the corresponding eigenfunction and the norm are taken as

$$R_0(\beta_0, r) = 1, \qquad \frac{1}{N(\beta_0)} = \frac{2}{b^2 - a^2} \tag{3-75}$$

Therefore, we start the summation in equation (3-73) from $m = 0$ and the solution becomes

$$T(r,t) = \frac{2}{b^2 - a^2} \int_a^b r' F(r')\, dr' + \sum_{m=1}^{\infty} \frac{1}{N(\beta_m)} e^{-\alpha\beta_m^2 t} R_0(\beta_m, r) \int_a^b r' R_0(\beta_m, r') F(r')\, dr' \tag{3-76}$$

where $R_0(\beta_m, r)$ and $N(\beta_m)$ are as defined by equations (3-74a, b). Note that, the first term on the right-hand side of the solution (3-76) represents the steady-state temperature in the cylinder after the temperature transients have passed. It is the average of the initial temperature distribution over the region.

Example 3-5

A hollow cylinder, $a \leqslant r \leqslant b$, is initially at a temperature $F(r)$; for times $t > 0$ the boundary surfaces at $r = a$ and $r = b$ are kept at zero temperature. Obtain an expression for the temperature distribution $T(r, t)$ for times $t > 0$.

Solution. This problem is similar to the one considered above, except the boundary conditions at $r=a$ and $r=b$ are both of the first kind. Therefore, the solution for $T(r,t)$ is of the same form as given by equation (3-73); that is

$$T(r,t)=\sum_{m=1}^{\infty}\frac{1}{N(\beta_m)}e^{-\alpha\beta_m^2 t}R_0(\beta_m,r)\int_a^b r'R_0(\beta_m,r')F(r')\,dr' \tag{3-77}$$

except $R_0(\beta_m,r)$, $N(\beta_m)$ and the eigencondition should be obtained from case 4 of Table 3-3 by setting $v=0$. Then the solution becomes

$$T(r,t)=\frac{\pi^2}{2}\sum_{m=1}^{\infty}\frac{\beta_m^2 J_0^2(\beta_m a)}{J_0^2(\beta_m a)-J_0^2(\beta_m b)}e^{-\alpha\beta_m^2 t}R_0(\beta_m,r)\int_a^b r'R_0(\beta_m,r')F(r')\,dr' \tag{3-78a}$$

where

$$R_0(\beta_m,r)=J_0(\beta_m r)Y_0(\beta_m b)-J_0(\beta_m b)Y_0(\beta_m r) \tag{3-78b}$$

and the β_m values are the positive roots of

$$J_0(\beta_m a)Y_0(\beta_m b)-J_0(\beta_m b)Y_0(\beta_m a)=0 \tag{3-79}$$

For the special case of $F(r')=T_0=\text{constant}$, the integral in equation (3-78a) is evaluated, the resulting expression is simplified by utilizing the Wronskian relationship of Bessel functions given in the Appendix IV, equation (27). Then the temperature distribution for this special case becomes

$$T(r,t)=T_0\pi\sum_{m=1}^{\infty}e^{-\alpha\beta_m^2 t}\frac{J_0(\beta_m a)}{J_0(\beta_m a)+J_0(\beta_m b)}R_0(\beta_m,r) \tag{3-80}$$

where $R_0(\beta_m,r)$ as given by equation (3-78b).

Example 3-6

An infinite region $0\leqslant r<\infty$ is initially at a temperature $F(r)$. Obtain an expression for the temperature distribution $T(r,t)$ for times $t>0$

Solution. The heat conduction problem is given by

$$\frac{\partial^2 T}{\partial r^2}+\frac{1}{r}\frac{\partial T}{\partial r}=\frac{1}{\alpha}\frac{\partial T}{\partial t} \quad \text{in} \quad 0\leqslant r<\infty, \quad t>0 \tag{3-81a}$$

$$T=F(r) \quad \text{for} \quad t=0, \quad \text{in the region} \tag{3-81b}$$

and subject to the condition that temperature remains finite at $r=0$. Separating the variables, it can be shown that the solution for the time-variable function is given by $\exp(-\alpha\beta^2 t)$, where β is the separation variable. The space-variable function $R_0(\beta, r)$ is the solution of the following equation

$$\frac{d^2 R_0}{dr^2} + \frac{1}{r}\frac{dR_0}{dr} + \beta^2 R_0 = 0 \qquad \text{in} \qquad 0 \leqslant r < \infty \tag{3-82}$$

subject to the condition that $R_0(\beta, r)$ remains finite at $r = 0$. The solution of equation (3-82), which is finite at $r = 0$, is

$$R_0(\beta, r) = J_0(\beta r) \tag{3-83}$$

Then, the complete solution for $T(r, t)$ is constructed as

$$T(r, t) = \int_{\beta=0}^{\infty} c(\beta) e^{-\alpha\beta^2 t} J_0(\beta r)\, d\beta \tag{3-84}$$

The application of the initial condition (3-81b) yields

$$F(r) = \int_{\beta=0}^{\infty} c(\beta) J_0(\beta r)\, d\beta \qquad \text{in} \qquad 0 \leqslant r < \infty \tag{3-85}$$

This is an expansion of an arbitrary function $F(r)$ defined in the interval $0 \leqslant r < \infty$ in terms of $J_0(\beta_r)$ functions. Such a representation was given in the previous section by equation (3-38) in terms of $J_\nu(\beta r)$ functions. Therefore, by setting $\nu = 0$ in equation (3-38) we obtain

$$F(r) = \int_{\beta=0}^{\infty} \beta J_0(\beta r)\, d\beta \int_{r'=0}^{\infty} r' J_0(\beta r') F(r')\, dr' \qquad \text{in} \qquad 0 \leqslant r < \infty \tag{3-86}$$

By comparing equations (3-85) and (3-86) we find the coefficient $c(\beta)$ as

$$c(\beta) = \beta \int_{r'=0}^{\infty} r' J_0(\beta r') F(r')\, dr' \tag{3-87}$$

The substitution of equation (3-87) into equation (3-84) yields

$$T(r, t) = \int_{\beta=0}^{\infty} r^{-\alpha\beta^2 t} \beta J_0(\beta r)\, d\beta \int_{r'=0}^{\infty} r' J_0(\beta r') F(r')\, dr' \tag{3-88}$$

By changing the order of integration and making use of the following integral (Appendix IV, equation (24)]

$$\int_{\beta=0}^{\infty} e^{-\alpha\beta^2 t} \beta J_0(\beta r) J_0(\beta r')\, d\beta = \frac{1}{2\alpha t} \exp\left(-\frac{r^2 + r'^2}{4\alpha t}\right) I_0\left(\frac{rr'}{2\alpha t}\right) \tag{3-89}$$

the solution (3-88) becomes

$$T(r,t)=\frac{1}{2\alpha t}\int_{r'=0}^{\infty} r'\exp\left(-\frac{r^2+r'^2}{4\alpha t}\right)F(r')I_0\left(\frac{rr'}{2\alpha t}\right)dr' \tag{3-90}$$

For the special case of

$$F(r)=\begin{cases} T_0, \text{constant} & \text{for} \quad 0<r<b \\ 0 & \text{for} \quad r>b \end{cases} \tag{3-91}$$

the solution (3-90) takes the form

$$\frac{T(r,t)}{T_0}=\frac{1}{2\alpha t}\exp\left(-\frac{r^2}{4\alpha t}\right)\int_{r'=0}^{b} r'\exp\left(-\frac{r'^2}{4\alpha t}\right)I_0\left(\frac{rr'}{2\alpha t}\right)dr' \equiv P \tag{3-92}$$

This result is called a *P function*, which has been numerically evaluated and the results are tabulated [14].

Example 3-7

A region $a \leqslant r < \infty$ in the cylindrical coordinate system is initially at a temperature $F(r)$; for times $t>0$ the boundary surface at $r=a$ is kept at zero temperature. Obtain an expression for the temperature distribution $T(r,t)$ in the region for times $t>0$.

Solution. The heat-conduction problem is given by

$$\frac{\partial^2 T}{\partial r^2}+\frac{1}{r}\frac{\partial T}{\partial r}=\frac{1}{\alpha}\frac{\partial T}{\partial t} \quad \text{in} \quad a<r<\infty, \quad t>0 \tag{3-93a}$$

$$T=0 \quad \text{at} \quad r=a, \quad t>0 \tag{3-93b}$$

$$T=F(r) \quad \text{for} \quad t=0, \quad \text{in the region} \tag{3-93c}$$

By separating the variables it can be shown that the time-variable function is given by $\exp(-\alpha\beta^2 t)$ and the space-variable function $R_0(\beta,r)$ is the solution of the following problem

$$\frac{d^2R_0}{dr^2}+\frac{1}{r}\frac{dR_0}{dr}+\beta^2R_0=0 \quad \text{in} \quad a<r<\infty \tag{3-94a}$$

$$R_0=0 \quad \text{at} \quad r=a \tag{3-94b}$$

Then, the complete solution for $T(r,t)$ is constructed as

$$T(r,t)=\int_{\beta=0}^{\infty} c(\beta)e^{-\alpha\beta^2 t}R_0(\beta,r)\,d\beta \tag{3-95}$$

The application of the initial condition (3-93c) yields

$$F(r) = \int_{\beta=0}^{\infty} c(\beta) R_0(\beta, r)\, d\beta \qquad \text{in} \qquad a < r < \infty \tag{3-96}$$

This is an expansion of an arbitrary function $F(r)$ defined in the interval $a < r < \infty$ in terms of the solutions of the eigenvalue problem (3-94). Such an expansion was given previously as given by equation (3-40) for a more general case. By comparing equation (3-96) with equation (3-40), we obtain the expansion coefficient $c(\beta)$ as

$$c(\beta) \equiv \frac{1}{N(\beta)} \beta \int_{r'=a}^{\infty} r' R_0(\beta, r') F(r')\, dr' \tag{3-97}$$

The substitution of equation (3-97) into equation (3-95) gives

$$T(r, t) = \int_{\beta=0}^{\infty} \frac{\beta}{N(\beta)} e^{-\alpha\beta^2 t} R_0(\beta, r)\, d\beta \int_{r'=a}^{\infty} r' R_0(\beta, r') F(r')\, dr' \tag{3-98}$$

The functions $R_0(\beta, r)$ and $N(\beta)$ are obtained from Table 3-2 case 3; the solution (3-98) becomes

$$T(r, t) = \int_{\beta=0}^{\infty} \frac{\beta}{J_0^2(\beta a) + Y_0^2(\beta a)} e^{-\alpha\beta^2 t} [J_0(\beta r) Y_0(\beta a) - Y_0(\beta r) J_0(\beta a)]\, d\beta$$

$$\cdot \int_{r'=a}^{\infty} r' [J_0(\beta r') Y_0(\beta a) - Y_0(\beta r') J_0(\beta a)] F(r')\, dr' \tag{3-99}$$

Example 3-8

A region $a \leqslant r < \infty$ in the cylindrical coordinate system is initially at a temperature $F(r)$; for times $t > 0$ the boundary at $r = a$ dissipates heat by convection into a medium at zero temperature. Obtain an expression for the temperature distribution $T(r, t)$ for times $t > 0$.

Solution. The heat conduction problem is given by

$$\frac{\partial^2 T}{\partial r^2} + \frac{1}{r}\frac{\partial T}{\partial r} = \frac{1}{\alpha}\frac{\partial T}{\partial t} \qquad \text{in} \qquad a < r < \infty, \quad t > 0 \tag{3-100a}$$

$$-\frac{\partial T}{\partial r} + HT = 0 \qquad \text{at} \qquad r = a, \qquad t > 0 \tag{3-100b}$$

$$T = F(r) \qquad \text{for} \qquad t = 0 \qquad \text{in } a \leqslant r < \infty \tag{3-100c}$$

By carrying out the analysis as described in the previous example, the solution is written in the form as given by equation (3-98), that is

$$T(r,t)=\int_{\beta=0}^{\infty}\frac{\beta}{N(\beta)}e^{-\alpha\beta^2 t}R_0(\beta,r)\,d\beta\int_{r'=a}^{\infty}r'R_0(\beta,r')F(r')\,dr' \tag{3-101}$$

where the functions $R_0(\beta, r)$ and $N(\beta)$ are obtained from Table 3-2, case 1, as

$$R_0(\beta,r)=J_0(\beta r)[\beta Y_1(\beta a)+HY_0(\beta a)]-Y_0(\beta r)[\beta J_1(\beta a)+HJ_0(\beta a)] \tag{3-102a}$$

$$N(\beta)=[\beta J_1(\beta a)+HJ_0(\beta a)]^2+[\beta Y_1(\beta a)+HY_0(\beta a)]^2 \tag{3-102b}$$

3-4 HOMOGENEOUS PROBLEMS IN (r, z, t) VARIABLES

The general solution of the homogeneous problems of heat conduction in (r, z, t) variables is constructed by the superposition of the separated solutions $\Gamma(t)$, $R_0(\beta, r)$ and $Z(\eta, z)$ for the t, r, and z variables, respectively. The analysis is straightforward because explicit expressions for the separated solutions are available in tabulated form for various combinations of boundary conditions. That is, the functions $R_0(\beta_m, r)$, the norm $N(\beta_m)$, and the eigenvalues β_m for finite regions (i.e., $0 \leqslant r \leqslant a$ and $a \leqslant r \leqslant b$) are obtainable from Tables 3-1 and 3-3 by setting $v = 0$; and the corresponding expressions for a semiinfinite region $a \leqslant r < \infty$ are obtainable from Table 3-2. Similarly, the expressions defining the functions $Z(\eta_p, z)$, the norm $N(\eta_p)$ and the eigenvalues η_p for a finite region $0 \leqslant z \leqslant c$ are available in Table 2-2 and the corresponding expressions for a semiinfinite region

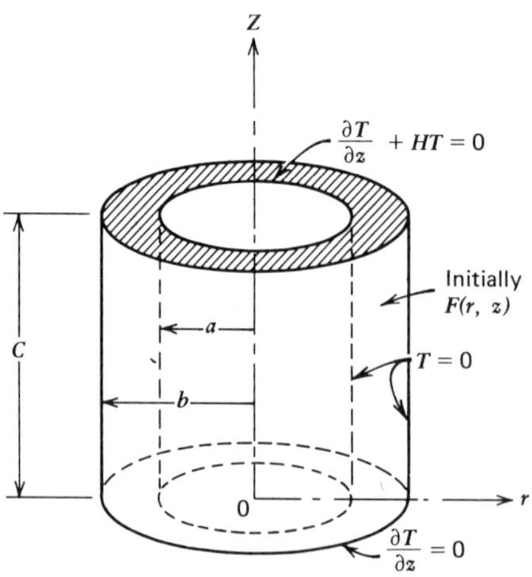

Fig. 3-5 Boundary and initial conditions for a hollow cylinder considered in Example 3-9.

$0 \leqslant z < \infty$ are obtainable from Table 2-3. We illustrate below the application with several representative examples.

Example 3-9

A hollow cylinder of finite length, in the region $a \leqslant r \leqslant b$, $0 \leqslant z \leqslant c$, is initially at a temperature $F(r, z)$. For times $t > 0$, the boundaries at $r = a$ and $r = b$ are kept at zero temperatures, the boundary at $z = 0$ is insulated, and the boundary at $z = c$ is dissipating heat by convection into a medium at zero temperature as illustrated in Fig. 3-5. Obtain an expression for the temperature distribution $T(r, z, t)$ for times $t > 0$.

Solution. The mathematical formulation of the problem is given as

$$\frac{\partial^2 T}{\partial r^2} + \frac{1}{r}\frac{\partial T}{\partial r} + \frac{\partial^2 T}{\partial z^2} = \frac{1}{\alpha}\frac{\partial T}{\partial t} \quad \text{in} \quad a < r < b, \quad 0 < z < c, \quad t > 0 \tag{3-103}$$

$$T = 0 \quad \text{at} \quad r = a, \quad r = b, \quad t > 0 \tag{3-104a}$$

$$\frac{\partial T}{\partial z} = 0 \quad \text{at} \quad z = 0, \quad t > 0 \tag{3-104b}$$

$$\frac{\partial T}{\partial z} + HT = 0 \quad \text{at} \quad z = c, \quad t > 0 \tag{3-104c}$$

$$T = F(r, z) \quad \text{for} \quad t = 0, \quad \text{in the region} \tag{3-104d}$$

The separation of variables lead to a set of equations as given by equations (3-11); the separated solutions are taken as

$$e^{-\alpha(\beta_m^2 + \eta_p^2)t}, \qquad R_0(\beta_m, r), \qquad \text{and} \qquad Z(\eta_p, z) \tag{3-105}$$

Here, the eigenvalues β_m and η_p are discrete because the regions in the r and z directions are both finite. The complete solution for $T(r, z, t)$ is constructed as

$$T(r, z, t) = \sum_{m=1}^{\infty} \sum_{p=1}^{\infty} c_{mp} R_0(\beta_m, r) Z(\eta_p, z) e^{-\alpha(\beta_m^2 + \eta_p^2)t} \tag{3-106}$$

The application of the initial condition (3-104d) yields

$$F(r, z) = \sum_{m=1}^{\infty} \sum_{p=1}^{\infty} c_{mp} R_0(\beta_m, r) Z(\eta_p, z) \quad \text{in} \quad a < r < b, \quad 0 < z < c \tag{3-107}$$

The coefficients c_{mp} are determined by operating on both sides of equation

(3-107) successively by the operators

$$\int_a^b rR_0(\beta_{m'}, r)dr \qquad \text{and} \qquad \int_0^c Z(\eta_{p'}, z)dz \tag{3-108}$$

and utilizing the orthogonality of these eigenfunctions. We obtain

$$c_{mp} = \frac{1}{N(\beta_m)N(\eta_p)} \int_{r=a}^{b} \int_{z=0}^{c} rR_0(\beta_m, r)Z(\eta_p, z)F(r, z)dr\, dz \tag{3-109}$$

Then the solution (3-106) becomes

$$T(r, z, t) = \sum_{m=1}^{\infty} \sum_{p=1}^{\infty} \frac{e^{-\alpha(\beta_m^2 + \eta_p^2)t}}{N(\beta_m)N(\eta_p)} R_0(\beta_m, r)Z(\eta_p, z) \int_{r'=a}^{b} \int_{z'=0}^{c} r'R_0(\beta_m, r') \cdot Z(\eta_p, z')F(r', z')dz'\, dr' \tag{3-110}$$

where the eigenfunctions $R_0(\beta_m, r)$, the norm $N(\beta_m)$ and the eigenvalues β_m are obtained from Table 3-3, case 4 by setting $\nu = 0$. We obtain

$$R_0(\beta_m, r) = J_0(\beta_m r)Y_0(\beta_m b) - J_0(\beta_m b)Y_0(\beta_m r) \tag{3-111a}$$

$$\frac{1}{N(\beta_m)} = \frac{\pi^2}{2} \frac{\beta_m^2 J_0^2(\beta_m a)}{J_0^2(\beta_m a) - J_0^2(\beta_m b)} \tag{3-111b}$$

and the β_m values are the positive roots of

$$J_0(\beta_m a)Y_0(\beta_m b) - J_0(\beta_m b)Y_0(\beta_m a) = 0 \tag{3-111c}$$

The eigenfunctions $Z(\eta_p, z)$, the norm $N(\eta_p)$ and the eigenvalues η_p are obtained from Table 2-2, case 4, by making appropriate changes in the symbols. We find

$$Z(\eta_p, z) = \cos \eta_p z \tag{3-112a}$$

$$\frac{1}{N(\eta_p)} = 2 \frac{\eta_p^2 + H^2}{c(\eta_p^2 + H^2) + H} \tag{3-112b}$$

and the η_p values are the positive roots of

$$\eta_p \tan \eta_p c = H \tag{3-112c}$$

Example 3-10

A solid cylinder, $0 \leqslant r \leqslant b, 0 \leqslant z < \infty$, is initially at temperature $F(r, z)$. For times $t > 0$, the boundaries are kept at zero temperature as illustrated in

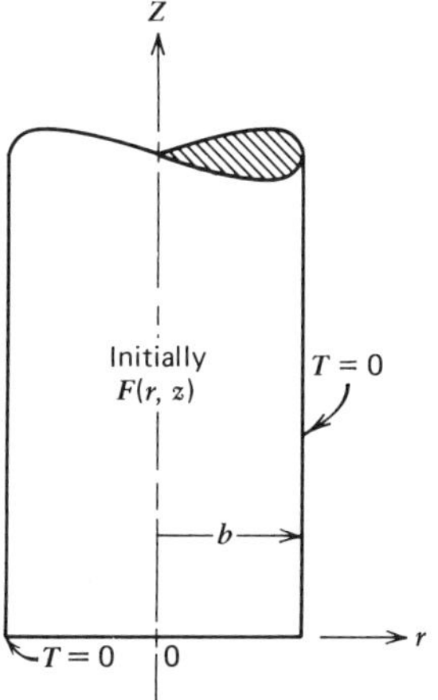

Fig. 3-6 Boundary and initial conditions for a solid cylinder considered in Example 3-10.

Fig. 3-6. Obtain an expression for the temperature distribution $T(r, z, t)$ in the cylinder for times $t > 0$.

Solution. The mathematical formulation of the problem is given as

$$\frac{\partial^2 T}{\partial r^2} + \frac{1}{r}\frac{\partial T}{\partial r} + \frac{\partial^2 T}{\partial z^2} = \frac{1}{\alpha}\frac{\partial T}{\partial t} \qquad \text{in} \quad 0 \leqslant r < b, \quad 0 < z < \infty, \quad t > 0 \tag{3-113}$$

$$T = 0 \qquad \text{at} \quad r = b, \quad z = 0 \text{ for } t > 0 \tag{3-114a}$$

$$T = F(r, z) \qquad \text{for} \quad t = 0, \quad \text{in the region} \tag{3-114b}$$

The separated solutions are taken as

$$e^{-\alpha(\beta_m^2 + \eta^2)t}, \qquad R_0(\beta_m, r), \qquad \text{and} \qquad Z(\eta, z) \tag{3-115}$$

Here we note that the eigenvalues β_m are discrete because the region in the r direction is finite, but the separation constant η takes all values from zero to infinity because the region in the z direction is semiinfinite.
The complete solution for $T(r, z, t)$ is constructed as

$$T(r, z, t) = \sum_{m=1}^{\infty} \int_{\eta=0}^{\infty} c_m(\eta) R_0(\beta_m, r) Z(\eta, z) e^{-\alpha(\beta_m^2 + \eta^2)t} d\eta \tag{3-116}$$

The application of the initial condition (3-114b) yields

$$F(r, z) = \sum_{m=1}^{\infty} \int_{\eta=0}^{\infty} c_m(\eta) R_0(\beta_m, r) Z(\eta, z) d\eta \qquad \text{in} \quad 0 \leqslant r < b, \quad 0 < z < \infty \tag{3-117}$$

Both sides of equation (3-117) are operated on by the operator

$$\int_{r=0}^{b} rR_0(\beta_{m'}, r)dr \tag{3-118}$$

and the orthogonality of $R_0(\beta_m, r)$ functions is utilized. We obtain

$$f^*(z) = \int_{\eta=0}^{\infty} c_m(\eta)Z(\eta, z)d\eta \qquad \text{in} \qquad 0 < z < \infty \tag{3-119a}$$

where we defined

$$f^*(z) \equiv \frac{1}{N(\beta_m)} \int_{r=0}^{b} rR_0(\beta_m, r)F(r, z)dr \tag{3-119b}$$

The representation given by equation (3-119a) is exactly the same as that given by equation (2-51) for a semiinfinite region. Therefore, the unknown coefficient $c_m(\eta)$ is determined according to the result in equation (2-53); we find

$$c_m(\eta) \equiv \frac{1}{N(\eta)} \int_{z=0}^{\infty} Z(\eta, z)f^*(z)dz \tag{3-120}$$

The substitution of equation (3-120) together with equation (3-119b) into equation (3-116) gives the solution for $T(r, z, t)$ in the form

$$T(r, z, t) = \sum_{m=1}^{\infty} \int_{\eta=0}^{\infty} \frac{e^{-\alpha(\beta_m^2 + \eta^2)t}}{N(\beta_m)N(\eta)} R_0(\beta_m, r)Z(\eta, z)\, dn \int_{r'=0}^{b} \int_{z'=0}^{\infty} r'R_0(\beta_m, r') \cdot Z(\eta, z')F(r', z')dz'\, dr' \tag{3-121}$$

The eigenfunctions $R_0(\beta_m, r)$, the norm $N(\beta_m)$, and the eigenvalues β_m are obtained from Table 3-1, case 3, by setting $v = 0$; we find

$$R_0(\beta_m, r) = J_0(\beta_m r), \qquad \frac{1}{N(\beta_m)} = \frac{2}{b^2 J_0'^2(\beta_m b)} = \frac{2}{b^2 J_1^2(\beta_m b)} \tag{3-122a}$$

and the β_m values are the positive roots of

$$J_0(\beta_m b) = 0 \tag{3-122b}$$

The functions $Z(\eta, z)$ and $N(\eta)$ are obtained from Table 2-3, case 3, as

$$Z(\eta, z) = \sin \eta z \qquad \text{and} \qquad \frac{1}{N(\eta)} = \frac{2}{\pi} \tag{3-123}$$

When the results in equations (3-122) and (3-123) are introduced into equation (3-121) and the order of integration is changed, we obtain

$$T(r,z,t) = \frac{4}{\pi b^2} \sum_{m=1}^{\infty} \frac{J_0(\beta_m r)}{J_1^2(\beta_m b)} e^{-\alpha\beta_m^2 t} \int_{r'=0}^{b} \int_{z'=0}^{\infty} r' J_0(\beta_m r') F(r', z') dz'\, dr'$$

$$\cdot \int_{\eta=0}^{\infty} e^{-\alpha\eta^2 t} \sin \eta z \sin \eta z'\, d\eta \tag{3-124}$$

The last integral with respect to η is similar to the one given by equation (2-57d); then this integral is evaluated as

$$\frac{2}{\pi} \int_{\eta=0}^{\infty} e^{-\alpha\eta^2 t} \sin \eta z \sin \eta z'\, d\eta = \frac{1}{(4\pi\alpha t)^{1/2}}$$

$$\cdot \left[\exp\left(-\frac{(z-z')^2}{4\alpha t} \right) - \exp\left(-\frac{(z+z')^2}{4\alpha t} \right) \right] \tag{3-125}$$

and this result is introduced into equation (3-124).

3-5 HOMOGENEOUS PROBLEMS IN (r, ϕ, t) VARIABLES

In the analysis of heat conduction problems involving (r, ϕ, t) variables, the following two situations require different considerations: (1) the range of ϕ variable is $0 \leqslant \phi \leqslant 2\pi$ as in the case of a *full cylinder*—in this case no boundary conditions are prescribed in ϕ except the requirement that the temperature should be periodic in ϕ with period 2π; and (2) the range of ϕ variable is $0 \leqslant \phi \leqslant \phi_0 < 2\pi$ as in the case of a *portion of a cylinder*—in this case boundary conditions should be prescribed at $\phi = 0$ and $\phi = \phi_0$.

Example 3-11

A solid cylinder, $0 \leqslant r \leqslant b, 0 \leqslant \phi \leqslant 2\pi$ is initially at temperature $F(r, \phi)$. For times $t > 0$, heat is dissipated by convection from the boundary surface at $r = b$ into an environment at zero temperature. Obtain an expression for the temperature distribution $T(r, \phi, t)$ in the cylinder.

Solution. The mathematical formulation of this problem is given as

$$\frac{\partial^2 T}{\partial r^2} + \frac{1}{r}\frac{\partial T}{\partial r} + \frac{1}{r^2}\frac{\partial^2 T}{\partial \phi^2} = \frac{1}{\alpha}\frac{\partial T}{\partial t} \quad \text{in} \quad 0 \leqslant r < b,\ 0 \leqslant \phi \leqslant 2\pi,\ t > 0 \tag{3-126}$$

$$\frac{\partial T}{\partial r} + HT = 0 \quad \text{at} \quad r = b, \quad t > 0 \tag{3-127}$$

$$T = F(r, \phi) \qquad \text{for} \qquad t = 0, \qquad \text{in the region} \tag{3-128}$$

The separated solutions are taken as

$$e^{-\alpha\beta_m^2 t}, \qquad \Phi(v, \phi) = A \sin v\phi + B \cos v\phi, \qquad R_v(\beta_m, r) \tag{3-129}$$

The complete solution of $T(r, \phi, t)$ is constructed by the superposition of these elementary solutions as

$$T(r, \phi, t) = \sum_{m=1}^{\infty} \sum_{v=0}^{\infty} e^{-\alpha\beta_m^2 t}(A_{mv} \sin v\phi + B_{mv} \cos v\phi)R_v(\beta_m, r) \tag{3-130}$$

The application of the initial condition (3-128) gives

$$F(r, \phi) = \sum_{m=1}^{\infty} \sum_{v} (A_{mv} \sin v\phi + B_{mv} \cos v\phi)R_v(\beta_m, r) \qquad \text{in } 0 \leqslant r < b, 0 \leqslant \phi \leqslant 2\pi \tag{3-131}$$

We now operate on both sides of this expression by the operator

$$\int_0^b rR_v(\beta_{m'}, r)dr \tag{3-132}$$

and utilize the orthogonality property of the functions $R_v(\beta_m, r)$. We obtain

$$f(\phi) = \sum_{v} (A_{mv} \sin v\phi + B_{mv} \cos v\phi)N(\beta_m) \qquad \text{in} \qquad 0 \leqslant \phi \leqslant 2\pi \tag{3-133}$$

where we defined

$$f(\phi) \equiv \int_0^b rR_v(\beta_m, r)F(r, \phi)dr \tag{3-134}$$

Equation (3-133) is representation of a function $f(\phi)$ periodic in ϕ with period 2π similar to the representation considered by equation (3-53). We recall that the coefficients of equation (3-53) are given by equation (3-56). Therefore, the coefficients of equation (3-133) are immediately obtainable from the result given by equation (3-56) as

$$[A_{mv} \sin v\phi + B_{mv} \cos v\phi]N(\beta_m) \equiv \frac{1}{\pi} \int_{\phi=0}^{2\pi} f(\phi') \cos v(\phi - \phi')d\phi' \tag{3-135}$$

where

$$v = 0, 1, 2, 3 \ldots \tag{3-136}$$

and replace π *by* 2π *for* $v = 0$. The substitution of equation (3-135) together with equation (3-134) into equation (3-130) gives the temperature distribution as

$$T(r, \phi, t) = \frac{1}{\pi} \sum_{m=1}^{\infty} \sum_{v=0}^{\infty} \frac{e^{-\alpha\beta_m^2 t}}{N(\beta_m)} R_v(\beta_m, r) \cdot \int_{\phi'=0}^{2\pi} \int_{r'=0}^{b} r' R_v(\beta_m, r') \cos v(\phi - \phi') F(r', \phi') dr' d\phi' \tag{3-137}$$

where

$$v = 0, 1, 2, 3 \ldots$$

and replace π *by* 2π *for* $v = 0$. The eigenfunctions $R_v(\beta_m, r)$, the norm $N(\beta_m)$, and the eigenvalues β_m are obtained from Table 3-1, case 1, as

$$R_v(\beta_m, r) = J_v(\beta_m r), \qquad \frac{1}{N(\beta_m)} = \frac{2}{J_v^2(\beta_m b)} \frac{\beta_m^2}{b^2(H^2 + \beta_m^2) - v^2} \tag{3-138}$$

and the β_m values are the positive roots of

$$\beta_m J_v'(\beta_m b) + H J_v(\beta_m b) = 0 \tag{3-139}$$

Example 3-12

Repeat Example 3-11 for the case when the boundary surface at $r = b$ is kept at zero temperature.

Solution. The mathematical formulation of this problem is similar to the one given above except the boundary condition (3-127) should be replaced by the boundary condition $T = 0$ at $r = b$. Therefore, the general solution given above by equation (3-137) is also applicable for this case provided that the functions defining $R_v(\beta_m, r)$, $N(\beta_m)$, and β_m are obtained from Table 3-1, case 3, as

$$R(\beta_m, r) = J_v(\beta_m r), \qquad \frac{1}{N(\beta_m)} = \frac{2}{b^2 J_v'^2(\beta_m b)} \tag{3-140}$$

and the β_m values are the roots of

$$J_v(\beta_m b) = 0 \tag{3-141}$$

The substitution of equations (3-140) into equation (3-137) gives the solution as

$$T(r,\phi,t)=\frac{2}{\pi b^2}\sum_{m=1}^{\infty}\sum_{\nu}\frac{e^{-\alpha\beta_m^2 t}}{J_\nu'^2(\beta_m b)}J_\nu(\beta_m r)$$

$$\cdot\int_{\phi'=0}^{2\pi}\int_{r'=0}^{b} r' J_\nu(\beta_m r')\cos\nu(\phi-\phi')F(r',\phi')dr'\,d\phi' \tag{3-142}$$

where

$$\nu=0,1,2,3\ldots$$

and replace π *by* 2π *for* $\nu=0$, β_m's are the positive roots of $J_\nu(\beta_m b)=0$.

Example 3-13

The portion of a solid cylinder, $0\leqslant r\leqslant b, 0\leqslant\phi\leqslant\phi_0<2\pi$ is initially at temperature $F(r,\phi)$. For times $t>0$ the boundaries at $r=b, \phi=0$ and $\phi=\phi_0$ are kept at zero temperature as illustrated in Fig. 3-7. Obtain an expression for the temperature distribution $T(r,\phi,t)$ for times $t>0$.

Solution. The mathematical formulation of this problem is given as

$$\frac{\partial^2 T}{\partial r^2}+\frac{1}{r}\frac{\partial T}{\partial r}+\frac{1}{r^2}\frac{\partial^2 T}{\partial\phi^2}=\frac{1}{\alpha}\frac{\partial T}{\partial t}$$

$$\text{in}\qquad 0\leqslant r<b,\quad 0<\phi<\phi_0,\quad t>0 \tag{3-143a}$$

$$T=0\qquad \text{at}\qquad r=b,\qquad \phi=0,\quad \phi=\phi_0,\quad t>0 \tag{3-143b}$$

$$T=F(r,\phi)\qquad \text{for}\qquad t=0,\qquad \text{in the region} \tag{3-143c}$$

The separated solutions are taken as

$$e^{-\alpha\beta_m^2 t},\qquad \Phi(\nu,\phi),\qquad \text{and}\qquad R_\nu(\beta_m,r) \tag{3-144}$$

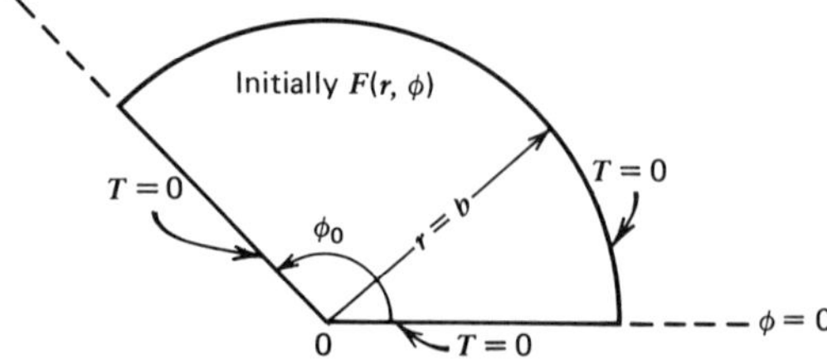

Fig. 3-7 Boundary and initial conditions for a portion of a cylinder considered in Example 3-13.

where the function $\Phi(\nu, \phi)$ is the solution of the eigenvalue problem

$$\frac{d^2\Phi}{d\phi^2} + \nu^2\Phi = 0 \quad \text{in} \quad 0 < \phi < \phi_0(\phi_0 < 2\pi) \tag{3-145a}$$

$$\Phi(\nu, \phi) = 0 \quad \text{at} \quad \phi = 0 \quad \text{and} \quad \phi = \phi_0 \tag{3-145b,c}$$

and the function $R_\nu(\beta_m, r)$ is the solution of the eigenvalue problem

$$\frac{d^2R_\nu(r)}{dr^2} + \frac{1}{r}\frac{dR_\nu(r)}{dr} + \left(\beta^2 - \frac{\nu^2}{r^2}\right)R_\nu(r) = 0 \quad \text{in} \quad 0 \leqslant r < b \tag{3-146a}$$

$$R_\nu = \text{finite} \quad \text{at} \quad r = 0 \tag{3-146b}$$

$$R_\nu = 0 \quad \text{at} \quad r = b \tag{3-146c}$$

The complete solution for $T(r, \phi, t)$ is constructed by the superposition of these separated solutions as

$$T(r, \phi, t) = \sum_{m=1}^{\infty} \sum_{\nu} c_{m\nu} R_\nu(\beta_m, r)\Phi(\nu, \phi)e^{-\alpha\beta_m^2 t} \tag{3-147}$$

The application of the initial condition (3-143c) gives

$$F(r, \phi) = \sum_{m=1}^{\infty} \sum_{\nu} c_{m\nu} R_\nu(\beta_m, r)\Phi(\nu, \phi) \quad \text{in} \quad 0 \leqslant r < b, \quad 0 < \phi < \phi_0 \tag{3-148}$$

To determine the coefficients $c_{m\nu}$, both sides of equation (3-148) are operated on successively by the operators

$$\int_{\phi=0}^{\phi_0} \Phi(\nu', \phi)d\phi \quad \text{and} \quad \int_{r=0}^{b} rR_\nu(\beta_{m'}, r)dr$$

and the orthogonality property of these eigenfunctions are utilized. We find

$$c_{m\nu} = \frac{1}{N(\beta_m)N(\nu)} \int_{r=0}^{b} \int_{\phi=0}^{\phi_0} rR_\nu(\beta_m, r)\Phi(\nu, \phi)F(r, \phi)d\phi\, dr \tag{3-149}$$

This result is now introduced into equation (3-147) to obtain the solution for $T(r, \phi, t)$ in the form

$$T(r, \phi, t) = \sum_{m=1}^{\infty} \sum_{\nu} \frac{e^{-\alpha\beta_m^2 t}}{N(\beta_m)N(\nu)} R_\nu(\beta_m, r)\Phi(\nu, \phi) \cdot \int_{r'=0}^{b} \int_{\phi'=0}^{\phi_0} r'R_\nu(\beta_m, r')\Phi(\nu, \phi')F(r', \phi')d\phi'\, dr' \tag{3-150}$$

where $R_\nu(\beta_m, r)$, $N(\beta_m)$, and β_m's are obtained from Table 3-1, case 3, as

$$R_\nu(\beta_m, r) = J_\nu(\beta_m r), \qquad \frac{1}{N(\beta_m)} = \frac{2}{b^2 J_\nu'^2(\beta_m b)} \tag{3-151a}$$

and the β_m values are the positive roots of

$$J_\nu(\beta_m b) = 0 \tag{3-151b}$$

The expressions defining $\Phi(\nu, \phi)$, $N(\nu)$, and ν are obtained from Table 2-2, case 9, by appropriate change of the notation. We find

$$\Phi(\nu, \phi) = \sin \nu\phi, \qquad \frac{1}{N(\nu)} = \frac{2}{\phi_0} \tag{3-152a}$$

and the ν values are the positive roots of

$$\sin \nu\phi_0 = 0 \tag{3-152b}$$

When the results given by equations (3-151) and (3-152) are introduced into equation (3-150) the solution becomes

$$T(r, \phi, t) = \frac{4}{b^2\phi_0} \sum_{m=1}^{\infty} \sum_{\nu} e^{-\alpha\beta_m^2 t} \frac{J_\nu(\beta_m r)}{J_\nu'^2(\beta_m b)} \sin \nu\phi$$

$$\cdot \int_{r'=0}^{b} \int_{\phi'=0}^{\phi_0} r' J_\nu(\beta_m r') \sin \nu\phi' F(r', \phi') d\phi' dr' \tag{3-153}$$

where the β_m values are positive roots of $J_\nu(\beta_m b) = 0$, and ν values are given by

$$\nu = \frac{n\pi}{\phi_0}, \qquad n = 1, 2, 3 \ldots$$

For the special case of $F(r, \phi) = T_0 =$ constant, the solution (3-153) becomes

$$T(r, \phi, t) = \frac{8T_0}{b^2\phi_0} \sum_{m=1}^{\infty} \sum_{\nu} e^{-\alpha\beta_m^2 t} \frac{J_\nu(\beta_m r)}{J_\nu'^2(\beta_m b)} \frac{\sin \nu\phi}{\nu} \int_{r'=0}^{b} r' J_\nu(\beta_m r') dr' \tag{3-154}$$

where the β_m values are the positive roots of $J_\nu(\beta b) = 0$, and the ν values are given by

$$\nu = \frac{(2n-1)\pi}{\phi_0}, \qquad n = 1, 2, 3 \ldots$$

3-6 HOMOGENEOUS PROBLEMS IN (r, ϕ, z, t) VARIABLES

The general solution of the homogeneous heat-conduction problem in (r, ϕ, z, t) variables is constructed by the superposition of all permissible elementary solutions; the resulting expansion coefficients are then determined by a procedure described previously. The analysis is straightforward because all the elementary solutions are now available and systematically tabulated for all combinations of boundary conditions. The application is illustrated with the following examples.

Example 3-14

A solid cylinder, $0 \leqslant r \leqslant b, 0 \leqslant \phi \leqslant 2\pi, 0 \leqslant z \leqslant c$, is initially at a temperature $F(r, \phi, z)$. For times $t > 0$ the boundary at $z = 0$ is insulated, the boundary at $z = c$ is kept at zero temperature, and the boundary at $r = b$ dissipates heat by convection into a medium at zero temperature as illustrated in Fig. 3-8. Obtain an expression for the temperature distribution $T(r, \phi, z, t)$ for times $t > 0$.

Solution. The mathematical formulation of this problem is given as

$$\frac{\partial^2 T}{\partial r^2} + \frac{1}{r}\frac{\partial T}{\partial r} + \frac{1}{r^2}\frac{\partial^2 T}{\partial \phi^2} + \frac{\partial^2 T}{\partial z^2} = \frac{1}{\alpha}\frac{\partial T}{\partial t} \quad \text{in} \quad 0 \leqslant r < b, \quad 0 \leqslant \phi \leqslant 2\pi, \quad 0 < z < c, \quad t > 0 \tag{3-155}$$

$$\frac{\partial T}{\partial r} + HT = 0 \quad \text{at} \quad r = b, \quad t > 0 \tag{3-156a}$$

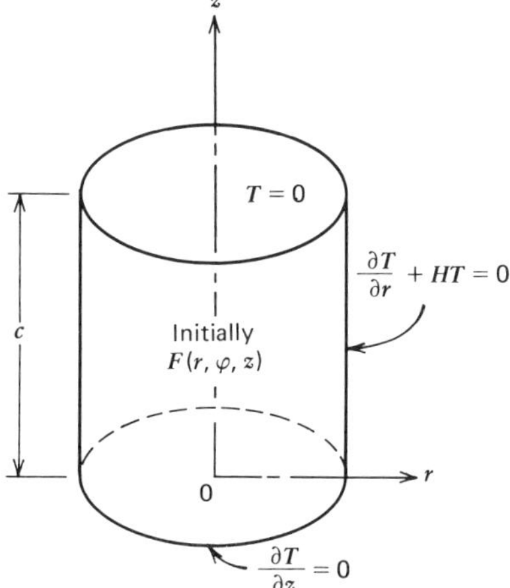

Fig. 3-8 Boundary and initial conditions for a solid cylinder considered in Example 3-14.

$$\frac{\partial T}{\partial z}=0 \qquad \text{at} \qquad z=0, \qquad t>0 \tag{3-156b}$$

$$T=0 \qquad \text{at} \qquad z=c, \qquad t>0 \tag{3-156c}$$

$$T=F(r,\phi,z) \qquad \text{for} \qquad t=0, \quad \text{in the region} \tag{3-156d}$$

The elementary solutions are taken as

$$e^{-\alpha(\beta_m^2+\eta_p^2)t}, \qquad R_\nu(\beta_m,r), \qquad Z(\eta_p,z), \qquad (A\sin\nu\phi+B\cos\nu\phi)$$

The complete solution for $T(r,\phi,z,t)$ is constructed by the linear superposition of these elementary solutions as

$$T(r,\phi,z,t)=\sum_{m=1}^{\infty}\sum_{p=1}^{\infty}\sum_{\nu}R_\nu(\beta_m,r)Z(\eta_p,z) \cdot[A_{mp\nu}\sin\nu\phi+B_{mp\nu}\cos\nu\phi]e^{-\alpha(\beta_m^2+\eta_p^2)t} \tag{3-157}$$

The application of the initial condition yields

$$F(r,\phi,z,t)=\sum_{m=1}^{\infty}\sum_{p=1}^{\infty}\sum_{\nu}R_\nu(\beta_m,r)Z(\eta_p,z)[A_{mp\nu}\sin\nu\phi+B_{mp\nu}\cos\nu\phi] \tag{3-158}$$

To determine the coefficients, we operate on both sides of this equation successively by the operators

$$\int_0^b rR_\nu(\beta_{m'},r)\,dr \qquad \text{and} \qquad \int_0^c Z(\eta_{p'},z)\,dz$$

and utilize the orthogonality of the eigenfunctions $R_\nu(\beta_m,r)$ and $Z(\eta_p,r)$. We find

$$f(\phi)=\sum_{\nu}N(\beta_m)N(\eta_p)[A_{mp\nu}\sin\nu\phi+B_{mp\nu}\cos\nu\phi] \qquad \text{in} \qquad 0\leqslant\phi\leqslant 2\pi \tag{3-159a}$$

where we defined

$$f(\phi)\equiv\int_{z=0}^{c}\int_{r=0}^{b}rR_\nu(\beta_m,r)Z(\eta_p,z)F(r,\phi,z)\,dr\,dz \tag{3-159b}$$

Equation (3-159a) is a representation of function $f(\phi)$ periodic in ϕ with period 2π similar to the representation considered in equation (3-53); the coefficients of equation (3-53) are given by equation (3-56). Therefore, the coefficients of equation (3-159a) are obtained from the result in equation (3-56) as

$$N(\beta_m)N(\eta_p)[A_{mp\nu}\sin\nu\phi + B_{mp\nu}\cos\nu\phi] \equiv \frac{1}{\pi}\int_{\phi'=0}^{2\pi} f(\phi')\cos\nu(\phi - \phi')\,d\phi' \tag{3-160}$$

where

$$\nu = 0, 1, 2, 3 \ldots$$

and replace π by 2π for $\nu = 0$. The substitution of equation (3-160) into equation (3-157) together with equation (3-159b) gives the temperature distribution in the form

$$\begin{aligned} T(r, \phi, z, t) = & \sum_{m=1}^{\infty}\sum_{p=1}^{\infty}\sum_{\nu=0}^{\infty} \frac{e^{-\alpha(\beta_m^2+\eta_p^2)t}}{N(\beta_m)N(\eta_p)} R_\nu(\beta_m, r)Z(\eta_p, z) \\ & \cdot \int_{\phi'=0}^{2\pi}\int_{z'=0}^{c}\int_{r'=0}^{b} r' R_\nu(\beta_m, r')Z(\eta_p, z')\cos\nu(\phi - \phi') \\ & \cdot F(r', \phi', z')\,dr'\,dz'\,d\phi' \end{aligned} \tag{3-161}$$

where

$$\nu = 0, 1, 2, 3 \ldots$$

and replace π by 2π for $\nu = 0$. The expressions defining $R_\nu(\beta_m, r)$, $N(\beta_m)$, and β_m are obtained from Table 3-1, case 1, as

$$R_\nu(\beta_m, r) = J_\nu(\beta_m r), \qquad \frac{1}{N(\beta_m)} = \frac{2}{J_\nu^2(\beta_m b)}\frac{\beta_m^2}{b^2(H^2 + \beta_m^2) - \nu^2} \tag{3-162a}$$

and the β_m values are the positive roots of

$$\beta_m J_\nu'(\beta_m b) + HJ_\nu(\beta_m b) = 0 \tag{3-162b}$$

and the expressions defining $Z(\eta_p, z)$, $N(\eta_p)$ and η_p are obtained from Table 2-2, case 6, by making appropriate changes in the symbols. We find

$$Z(\eta_p, z) = \cos\eta_p z, \qquad \frac{1}{N(\eta_p)} = \frac{2}{c} \tag{3-163a}$$

and the η_p values are the positive roots of

$$\cos\eta_p c = 0 \qquad \left(\text{or } \eta_p = \frac{(2p-1)\pi}{2c}, p = 1, 2, 3 \ldots\right) \tag{3-163b}$$

3-7 MULTIDIMENSIONAL STEADY-STATE PROBLEM WITH NO HEAT GENERATION

The multidimensional steady-state heat conduction problem with no heat generation can be solved by the separation of variables if only one of the boundary conditions is nonhomogeneous. If the problem involves more than one nonhomogeneous boundary condition, it can be split up into a set of simpler problems each containing only one nonhomogeneous boundary condition as discussed in Section 2-10. To illustrate the application we consider the following examples.

Example 3-15

Obtain an expression for the steady-state temperature distribution $T(r, z)$ in a solid cylinder $0 \leqslant r \leqslant b, 0 \leqslant z \leqslant c$, when the boundary surface at $z = 0$ is kept at a temperature $f(r)$, boundary at $z = c$ is kept at zero temperature, and that at $r = b$ dissipates heat by convection into a medium at zero temperature.

Solution. The mathematical formulation of the problem is given as

$$\frac{\partial^2 T}{\partial r^2} + \frac{1}{r}\frac{\partial T}{\partial r} + \frac{\partial^2 T}{\partial z^2} = 0 \qquad \text{in} \qquad 0 \leqslant r < b, 0 < z < c \tag{3-164a}$$

$$\frac{\partial T}{\partial r} + HT = 0 \qquad \text{at} \qquad r = b \tag{3-164b}$$

$$T = f(r) \qquad \text{at} \qquad z = 0 \tag{3-164c}$$

$$T = 0 \qquad \text{at} \qquad z = c \tag{3-164d}$$

In this problem the boundary condition at $z = 0$ is nonhomogeneous; looking ahead in the analysis we conclude that the nonhomogeneous part $f(r)$ of the boundary condition should be represented in terms of the separated solutions $R_0(\beta_m, r)$. Therefore, in separating the variables the sign of the separation constant should be so chosen as to produce an eigenvalue problem for the functions $R_0(\beta_m, r)$. With this consideration the separated equations are taken as

$$\frac{d^2 R_0}{dr^2} + \frac{1}{r}\frac{dR_0}{dr} + \beta^2 R_0 = 0 \qquad \text{in} \qquad 0 \leqslant r < b \tag{3-165a}$$

$$\frac{dR_0}{dr} + HR_0 = 0 \qquad \text{at} \qquad r = b \tag{3-165b}$$

and

$$\frac{d^2 Z}{dz^2} - \beta^2 Z = 0 \qquad \text{in} \qquad 0 < z < c \tag{3-166a}$$

$$Z = 0 \qquad \text{at} \qquad z = c \tag{3-166b}$$

Then, the solution for $T(r, z)$ is constructed as

$$T(r, z) = \sum_{m=1}^{\infty} A_m \sinh \beta_m (c - z) R_0(\beta_m, r) \tag{3-167}$$

The application of the boundary condition at $z = 0$ gives

$$f(r) = \sum_{m=1}^{\infty} A_m \sinh \beta_m c R_0(\beta_m, r) \qquad \text{in} \qquad 0 \leqslant r < b \tag{3-168}$$

where, the coefficients A_m are determined as

$$A_m = \frac{1}{N(\beta_m) \sinh \beta_m c} \int_0^b r R_0(\beta_m, r) f(r)\, dr \tag{3-169}$$

Introducing equation (3-169) into equation (3-167) the solution becomes

$$T(r, z) = \sum_{m=1}^{\infty} \frac{1}{N(\beta_m)} \frac{\sinh \beta_m (c - z)}{\sinh \beta_m c} R_0(\beta_m, r) \int_0^b r' R_0(\beta_m, r') f(r')\, dr' \tag{3-170}$$

where the expressions defining the functions $R_0(\beta_m, r), N(\beta_m)$, and β_m are obtained from Table 3-1, case 1, by setting $v = 0$. We find

$$R_0(\beta_m, r) = J_0(\beta_m, r), \qquad \frac{1}{N(\beta_m)} = \frac{2}{J_0^2(\beta_m b)} \frac{\beta_m^2}{b^2(H^2 + \beta_m^2)} \tag{3-171a}$$

and the β_m values are the positive roots of

$$\beta_m J_0'(\beta_m b) + H J_0(\beta_m b) = 0 \qquad \text{or} \qquad \beta_m J_1(\beta_m b) = H J_0(\beta_m b) \tag{3-171b}$$

For the special case of $f(r) = T_0 = \text{constant}$, the integral in equation (3-170) is performed and the solution becomes

$$\frac{T(r, z)}{T_0} = \frac{2}{b} \sum_{m=1}^{\infty} \frac{\beta_m J_1(\beta_m b)}{J_0^2(\beta_m b)(H^2 + \beta_m^2)} \frac{\sinh \beta_m (c - z)}{\sinh \beta_m c} J_0(\beta_m r) \tag{3-172a}$$

or, by utilizing equation (3-171b), we find

$$\frac{T(r,z)}{T_0}=\frac{2}{b}\sum_{m=1}^{\infty}\frac{H}{H^2+\beta_m^2}\,\frac{\sinh\beta_m(c-z)}{\sinh\beta_m c}\,\frac{J_0(\beta_m r)}{J_0(\beta_m b)} \tag{3-172b}$$

Example 3-16

Obtain an expression for the steady-state temperature distribution $T(r,\phi)$ in a solid cylinder $0\leqslant r\leqslant b$, $0\leqslant\phi\leqslant 2\pi$, which is subjected to convective heat transfer at the boundary surface $r=b$ with an environment whose temperature varies around the circumference.

Solution. The mathematical formulation of this problem is given as

$$\frac{\partial^2 T}{\partial r^2}+\frac{1}{r}\frac{\partial T}{\partial r}+\frac{1}{r^2}\frac{\partial^2 T}{\partial\phi^2}=0 \qquad \text{in} \qquad 0\leqslant r<b, \quad 0\leqslant\phi\leqslant 2\pi \tag{3-173a}$$

$$\frac{\partial T}{\partial r}+HT=f(\phi) \qquad \text{at} \qquad r=b \tag{3-173b}$$

The separated equations and their elementary solutions are as given by equations (3-17). The general solution for $T(r,\phi)$ is constructed in terms of these solutions as

$$T(r,\phi)=\sum_{\nu}r^{\nu}(C_\nu\sin\nu\phi+D_\nu\cos\nu\phi) \tag{3-174}$$

where we excluded the elementary solutions $r^{-\nu}$ and $\ln r$ because they diverge at $r=0$. This solution is introduced into the boundary condition (3-173b); we find

$$\sum_{\nu}b^{\nu-1}(\nu+Hb)(C_\nu\sin\nu\phi+D_\nu\cos\nu\phi)=f(\phi) \qquad \text{in} \qquad 0\leqslant\phi\leqslant 2\pi \tag{3-175}$$

This equation is a representation of function $f(\phi)$ periodic in ϕ with period 2π similar to the representation considered in equation (3-53); the coefficient of equation (3-53) are given by equation (3-56). Therefore, by comparing equation (3-175) with equations (3-53) and (3-56) we conclude that the coefficients are given by

$$b^{\nu-1}(\nu+Hb)(C_\nu\sin\nu\phi+D_\nu\cos\nu\phi)\equiv\frac{1}{\pi}\int_{\phi'=0}^{2\pi}f(\phi')\cos\nu(\phi-\phi')\,d\phi' \tag{3-176}$$

where

$$\nu=0,1,2,3,\ldots$$

and replace π by 2π for $\nu = 0$. When these coefficients are introduced into equation (3-174), the solution for the temperature becomes

$$T(r,\phi) = \frac{b}{\pi}\sum_{\nu}\left(\frac{r}{b}\right)^{\nu}\frac{1}{\nu + Hb}\int_{\phi'=0}^{2\pi} f(\phi')\cos\nu(\phi-\phi')\,d\phi' \tag{3-177}$$

where

$$\nu = 0, 1, 2, 3, \ldots$$

and replace π by 2π for $\nu = 0$.

Example 3-17

Obtain an expression for the steady-state temperature $T(r,z)$ in a solid cylinder $0 \leqslant r \leqslant b, 0 \leqslant z \leqslant c$, when the boundary at $r = b$ is at temperature $f(z)$ and the boundaries at $z = 0$ and $z = c$ are at zero temperature.

Solution. The mathematical formulation of this problem is given as

$$\frac{\partial^2 T}{\partial r^2} + \frac{1}{r}\frac{\partial T}{\partial r} + \frac{\partial^2 T}{\partial z^2} = 0 \quad \text{in} \quad 0 \leqslant r < b, \quad 0 < z < c \tag{3-178a}$$

$$T = f(z) \quad \text{at} \quad r = b \tag{3-178b}$$

$$T = 0 \quad \text{at} \quad z = 0 \quad \text{and} \quad z = c \tag{3-178c}$$

The separated equations are taken as

$$\frac{\partial^2 Z}{\partial z^2} + \eta^2 Z = 0 \quad \text{in} \quad 0 < z < c \tag{3-179a}$$

$$Z = 0 \quad \text{at} \quad z = 0 \quad \text{and} \quad z = c \tag{3-179b}$$

and

$$\frac{d^2 R_0}{dr^2} + \frac{1}{r}\frac{dR_0}{dr} - \eta^2 R_0 = 0 \quad \text{in} \quad 0 \leqslant r < b \tag{3-180}$$

We note that the sign of the separation constant is so chosen as to produce an eigenvalue problem for $Z(\eta, z)$, because the boundary condition function $f(z)$ should be represented in terms of $Z(\eta, z)$. The general solution for $T(r,z)$ is constructed as

$$T(r,z) = \sum_{m=1}^{\infty} A_m I_0(\eta_m r) Z(\eta_m, z) \tag{3-181}$$

The application of the boundary condition (3-178b) yields

$$f(z) = \sum_{m=1}^{\infty} A_m I_0(\eta_m b) Z(\eta_m, z) \qquad \text{in} \qquad 0 < z < c \tag{3-182}$$

The coefficients A_m are determined as

$$A_m = \frac{1}{I_0(\eta_m b) N(\eta_m)} \int_0^c Z(\eta_m, z) f(z)\, dz \tag{3-183}$$

Introducing equation (3-183) into equation (3-181) the solution becomes

$$T(r, z) = \sum_{m=1}^{\infty} = \frac{1}{N(\eta_m)} \frac{I_0(\eta_m r)}{I_0(\eta_m b)} Z(\eta_m, z) \int_0^c Z(\eta_m, z') f(z')\, dz' \tag{3-184}$$

where $Z(\eta_m, z)$, $N(\eta_m)$, and η_m are obtained from Table 2-2, case 9, as

$$Z(\eta_m, z) = \sin \eta_m z, \qquad \frac{1}{N(\eta_m)} = \frac{2}{c} \tag{3-185a}$$

and the η_m values are the roots of

$$\sin \eta_m c = 0 \tag{3-185b}$$

Substituting equations (3-185) into equation (3-184) we find

$$T(r, z) = \frac{2}{c} \sum_{m=1}^{\infty} \frac{I_0(\eta_m r)}{I_0(\eta_m b)} \sin \eta_m z \int_0^c \sin \eta_m z' f(z')\, dz' \tag{3-186}$$

where

$$\eta_m = \frac{m\pi}{c}$$

3-8 SPLITTING UP OF NONHOMOGENEOUS PROBLEMS INTO SIMPLER PROBLEMS

When the heat conduction problem is nonhomogeneous because of the nonhomogeneity of the differential equation and/or the boundary conditions, it can be split into a set of simpler problems, as discussed in the Section 2-12, if the generation term and the nonhomogeneous part of the boundary conditions *do not depend on time.*

Example 3-18

A solid cylinder, $0 \leqslant r \leqslant b$, is initially at temperature $F(r)$. For times $t > 0$, heat is generated within the solid at a constant rate of g_0 and the boundary surface at $r = b$ is kept at zero temperature. Obtain an expression for the temperature distribution $T(r, t)$ in the cylinder for times $t > 0$.

Solution. The mathematical formulation of this problem is given as

$$\frac{\partial^2 T}{\partial r^2} + \frac{1}{r}\frac{\partial T}{\partial r} + \frac{1}{k} g_0 = \frac{1}{\alpha}\frac{\partial T}{\partial t} \quad \text{in} \quad 0 \leqslant r < b, \quad t > 0 \tag{3-187a}$$

$$T = 0 \quad \text{at} \quad r = b, \quad t > 0 \tag{3-187b}$$

$$T = F(r) \quad \text{for} \quad t = 0, \quad \text{in the region} \tag{3-187c}$$

This problem is split into a steady-state problem for $T_s(r)$ as

$$\frac{d^2 T_s}{dr^2} + \frac{1}{r}\frac{dT_s}{dr} + \frac{1}{k} g_0 = 0 \quad \text{in} \quad 0 \leqslant r < b \tag{3-188a}$$

$$T = 0 \quad \text{at} \quad r = b \tag{3-188b}$$

and into a homogeneous problem for $T_h(r, t)$ as

$$\frac{\partial^2 T_h}{\partial r^2} + \frac{1}{r}\frac{\partial T_h}{\partial r} = \frac{1}{\alpha}\frac{\partial T_h}{\partial t} \quad \text{in} \quad 0 \leqslant r < b, \quad t > 0 \tag{3-189a}$$

$$T = 0 \quad \text{at} \quad r = b, \quad t > 0 \qquad \text{(3-188b)}$$

$$T = F(r) - T_s(r) \quad \text{for} \quad t = 0, \quad \text{in the region} \tag{3-189c}$$

Then, the solution $T(r, t)$ of the original problem (3-187) is obtained as

$$T(r, t) = T_s(r) + T_h(r, t) \tag{3-190}$$

The steady-state problem is readily solved

$$T_s(r) = \frac{g_0}{4k}(b^2 - r^2) \tag{3-191}$$

The homogeneous problem (3-189) is exactly the same as considered in Example 3-3; hence its solution is immediately obtained from equation (3-67) as

$$T_h(r, t) = \frac{2}{b^2}\sum_{m=1}^{\infty} e^{-\alpha\beta_m^2 t}\frac{J_0(\beta_m r)}{J_1^2(\beta_m b)}\int_0^b r' J_0(\beta_m r')[F(r') - T_s(r')]\,dr' \tag{3-192a}$$

where the β_m values are the roots of

$$J_0(\beta_m b) = 0 \tag{3-192b}$$

When the results in equations (3-191) and (3-192) are introduced into equation (3-190) and some of the integrals are performed, we obtain

$$T(r,t) = \frac{g_0(b^2 - r^2)}{4k} - \frac{2g_0}{bk} \sum_{m=1}^{\infty} e^{-\alpha\beta_m^2 t} \frac{J_0(\beta_m r)}{\beta_m^3 J_1(\beta_m b)}$$

$$+ \frac{2}{b^2} \sum_{m=1}^{\infty} e^{-\alpha\beta_m^2 t} \frac{J_0(\beta_m r)}{J_1^2(\beta_m b)} \int_0^b r' J_0(\beta_m r') F(r') dr' \tag{3-193}$$

Example 3-19

A solid cylinder is initially at temperature $F^*(r)$. For times $t > 0$ heat is generated in the region at a constant rate of g_0 per unit volume and the boundary surface at $r = b$ is subjected to convection with an environment at temperature T_∞. Obtain an expression for the temperature distribution $T(r,t)$ in the solid for times $t > 0$.

Solution. The mathematical formulation of this problem is given as

$$\frac{\partial^2 T}{\partial r^2} + \frac{1}{r}\frac{\partial T}{\partial r} + \frac{g_0}{k} = \frac{1}{\alpha}\frac{\partial T}{\partial t} \quad \text{in} \quad 0 \leqslant r < b, \quad t < 0 \tag{3-194a}$$

$$\frac{\partial T}{\partial r} + HT = HT_\infty \quad \text{at} \quad r = b, \quad t > 0 \tag{3-194b}$$

$$T = F^*(r) \quad \text{for} \quad t = 0, \quad \text{in} \quad 0 \leqslant r \leqslant b \tag{3-194c}$$

and the temperature should remain finite at $r = 0$. This problem is split into a steady-state problem for $T_s(r)$ as

$$\frac{d^2 T_s}{dr^2} + \frac{1}{r}\frac{dT_s}{dr} + \frac{g_0}{k} = 0 \quad \text{in} \quad 0 \leqslant r < b \tag{3-195a}$$

$$\frac{dT_s}{dr} + HT_s = HT_\infty \quad \text{at} \quad r = b \tag{3-195b}$$

and into a homogeneous problem for $T_h(r,t)$ as

$$\frac{\partial^2 T_h}{\partial r^2} + \frac{1}{r}\frac{\partial T_h}{\partial r} = \frac{1}{\alpha}\frac{\partial T_h}{\partial t} \quad \text{in} \quad 0 \leqslant r < b, \quad t > 0 \tag{3-196a}$$

$$\frac{\partial T_h}{\partial r} + HT_h = 0 \qquad \text{at} \qquad r = b, \qquad t > 0 \tag{3-196b}$$

$$T_h = F^*(r) - T_s(r) \equiv F(r) \qquad \text{for} \qquad t = 0, \qquad \text{in } 0 \leqslant r \leqslant b \tag{3-196c}$$

Then, the solution of the problem (3-194) is given by

$$T(r, t) = T_s(r) + T_h(r, t) \tag{3-197}$$

The solution $T_s(r)$ of the steady-state problem (3-195) is a straightforward matter. The homogeneous problem (3-196) is exactly the same as the problem (3-57) considered in Example (3-1); therefore, the solution of $T_h(r, t)$ is obtainable from equation (3-63) by setting in that equation $F(r) = F^*(r) - T_s(r)$.

3-9 TRANSIENT-TEMPERATURE CHARTS

In the previous chapter we presented transient temperature charts for a slab of thickness $2L$ subjected to convection at both surfaces. We now consider one-dimensional, transient heat conduction in a long cylinder of radius b, which is initially at a uniform temperature T_i. Suddenly, at time $t = 0$, the boundary surface at $r = b$ is subjected to convection with a heat transfer coefficient h into an ambient at temperature T_∞ and maintained so for $t > 0$. The mathematical formulation of this heat conduction problem is given in the dimensionless form as

$$\frac{1}{R}\frac{\partial}{\partial R}\left(R\frac{\partial \theta}{\partial R}\right) = \frac{\partial \theta}{\partial \tau} \qquad \text{in} \qquad 0 < R < 1, \quad \text{for } \tau > 0 \tag{3-198a}$$

$$\frac{\partial \theta}{\partial R} = 0 \qquad \text{at} \qquad R = 0, \qquad \text{for } \tau > 0 \tag{3-198b}$$

$$\frac{\partial \theta}{\partial R} + \text{Bi}\,\theta = 0 \qquad \text{at} \qquad R = 1, \qquad \text{for } \tau > 0 \tag{3-198c}$$

$$\theta = 1 \qquad \text{in} \qquad 0 \leqslant R \leqslant 1, \text{ for } \tau = 0 \tag{3-198d}$$

where various dimensionless quantities are defined as follows:

$$\text{Bi} = \frac{hb}{k} = \text{Biot number} \tag{3-199a}$$

$$\tau = \frac{\alpha t}{b^2} = \text{dimensionless time, or Fourier number} \tag{3-199b}$$

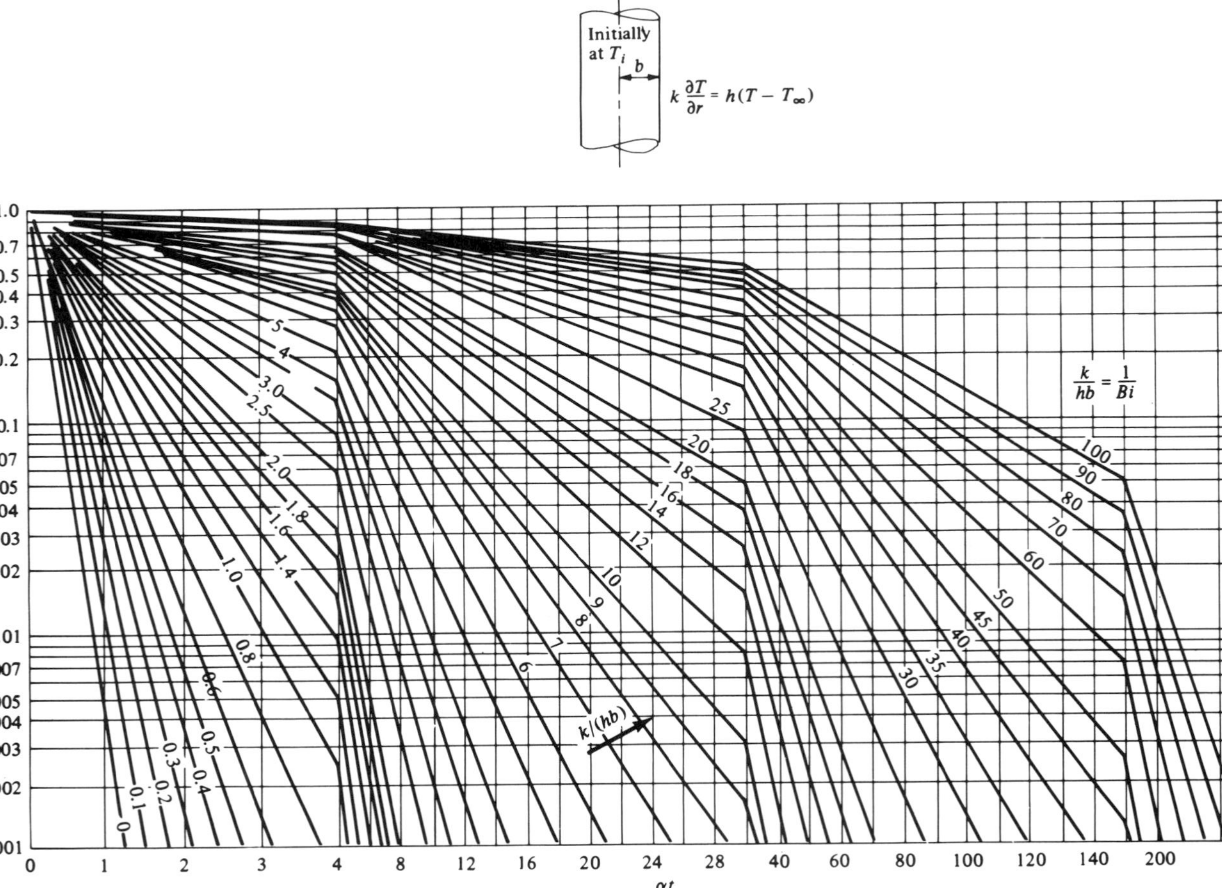

Fig. 3-9 Transient-temperature chart for a long solid cylinder subjected to convection. (From Heisler [15]). (*a*) Dimensionless temperature $\theta(0, \tau)$ at the center; (*b*) position correction chart for use with part *a*.

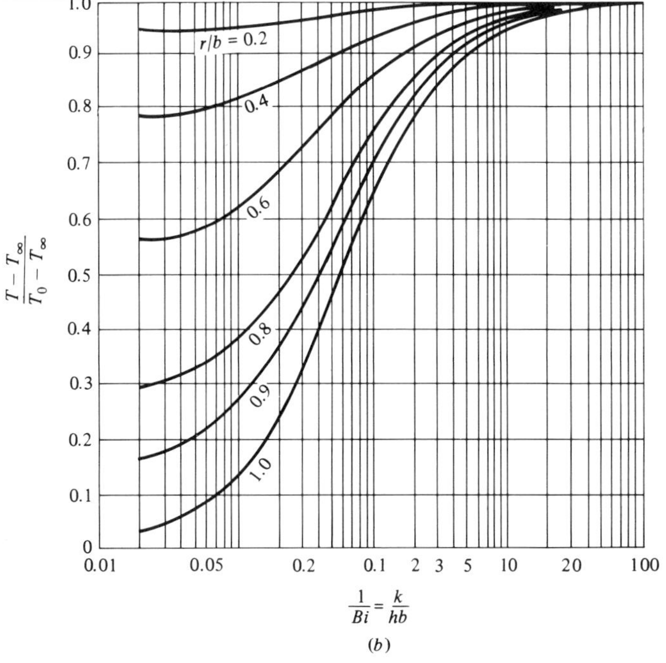

(*b*)

Fig. 3-9 (*Continued*)

$$\theta = \frac{T(r,t) - T_\infty}{T_i - T_\infty} = \text{dimensionless temperature} \tag{3-199c}$$

$$R = \frac{r}{b} = \text{dimensionless radial coordinate} \tag{3-199d}$$

The solution of this transient heat conduction problem is presented in graphical form in Fig. 3-9*a*,*b*. Here, Fig. 3-9*a* gives the centerline temperature $\theta(0, \tau)$ as a function of the dimensionless time τ for several different values of the parameter 1/Bi. The curve for $1/\text{Bi} = 0$ corresponds to the case $h \rightarrow \infty$, or the surface of the cylinder maintained at the ambient temperature T_∞. Figure 3-9*b* relates the temperature at six different locations within the cylinder to the cylinder centerline temperature $\theta(0, \tau)$. An examination of Fig. 3-9*b* reveals that for values of 1/Bi larger than 10 or $\text{Bi} < 0.1$, the temperature distribution within the cylinder is considered uniform with an error of less than about 5%. For such cases, the lumped system is applicable. The use of the charts 3–9 is similar to that described in Example 2-17 for the case of a slab.

REFERENCES

1. H. S. Carslaw and J. C. Jaeger, *Conduction of Heat in Solids*, Clarendon Press, London, 1959.
2. M. N. Özişik, *Heat Conduction*, 1st. ed., Wiley, New York, 1980.
3. M. D. Mikhailov and M. N. Özişik, *Unified Analysis and Solutions of Heat and Mass Diffusion*, Wiley, New York, 1984.
4. P. Moon and D. E. Spencer, *Field Theory for Engineers*, Van Nostrand, Princeton, N.J., 1961.
5. G. N. Watson, *A Treatise on the Theory of Bessel Functions*, 2nd ed., Cambridge University Press, London, 1966.
6. N. W. McLachlan, *Bessel Functions for Engineers*, 2nd ed., Clarendon Press, London, 1961.
7. E. T. Whittaker and G. N. Watson, *A Course of Modern Analysis*, Cambridge University Press, London, 1965.
8. M. Abramowitz and I. A. Stegun, *Handbook of Mathematical Functions*, National Bureau of Standards, Applied Mathematic Series 55, U.S. Government Printing Office, Washington, D.C., 20402, 1964.
9. E. C. Titchmarsh, *Eigenfunction Expansions*, Part I, Clarendon Press, London, 1962.
10. I. N. Sneddon, *Fourier Transforms*, McGraw-Hill, New York, 1951.
11. G. Cinelli, *Int. J. Eng. Sci.*, **3**, 539–559, 1965.
12. S. Goldstein, *Proc. Lond. Math. Soc. (2)* **34**, 51–88, 1932.
13. E. C. Titchmarsh, *Proc. Lond. Math. Soc. (2)* **22**, 15–28, 1923.
14. J. I. Masters, *J. Chem. Phys.*, **23**, 1865–1874, 1955.
15. M. P. Heisler, *Trans. ASME*, **69**, 227–236, 1947

PROBLEMS

3-1 A hollow cylinder, $a \leqslant r \leqslant b$, is initially at a temperature $F(r)$. For times $t > 0$ the boundaries at $r = a$ and $r = b$ are kept insulated. Obtain an expression for temperature distribution $T(r, t)$ in the solid for times $t > 0$.

3-2 A region $a \leqslant r < \infty$ in the cylindrical coordinate system is initially at a temperature $F(r)$. For times $t > 0$ the boundary at $r = a$ is kept insulated. Obtain an expression for the temperature distribution $T(r, t)$ in the region for times $t > 0$.

3-3 A solid cylinder, $0 \leqslant r \leqslant b, 0 \leqslant z \leqslant c$, is initially at temperature $F(r, z)$. For times $t > 0$, the boundary at $z = 0$ is insulated, the boundary at $z = c$ is dissipating heat by convection into a medium at zero temperature, and the boundary at $r = b$ is kept at zero temperature. Obtain an expression for the temperature distribution $T(r, z, t)$ in the solid for times $t > 0$.

3-4 A semiinfinite solid cylinder, $0 \leqslant r \leqslant b, 0 \leqslant z < \infty$, is initially at temperature $F(r, z)$. For times $t > 0$, the boundary at $z = 0$ is kept insulated and the boundary at $r = b$ is dissipating heat by convection into a medium at zero temperature. Obtain an expression for the temperature distribution $T(r, z, t)$ in the solid for times $t > 0$.

3-5 A semiinfinite hollow cylinder, $a \leqslant r \leqslant b, 0 \leqslant z < \infty$, is initially at temperature $F(r, z)$. For times $t > 0$, the boundaries at $z = 0, r = a$, and $r = b$ are all kept at zero temperature. Obtain an expression for the temperature distribution $T(r, z, t)$ in the solid for times $t > 0$.

3-6 A solid cylinder, $0 \leqslant r \leqslant b, 0 \leqslant \phi \leqslant 2\pi$, is initially at temperature $F(r, \phi)$. For times $t > 0$, the boundary at $r = b$ is kept insulated. Obtain an expression for the temperature distribution $T(r, \phi, t)$ in the solid for times $t > 0$.

3-7 A hollow cylinder, $a \leqslant r \leqslant b, 0 \leqslant \phi \leqslant 2\pi$, is initially at temperature $F(r, \phi)$. For times $t > 0$, the boundaries at $r = a$ and $r = b$ are kept insulated. Obtain an expression for the temperature distribution $T(r, \phi, t)$ in the region for times $t > 0$.

3-8 A portion of a solid cylinder $0 \leqslant r \leqslant b, 0 \leqslant \phi \leqslant \phi_0 (< 2\pi)$ is initially at temperature $F(r, \phi)$. For times $t > 0$, the boundary at $r = b$ dissipates heat by convection into a medium at zero temperature, the boundaries at $\phi = 0$ and $\phi = \phi_0$ are kept at zero temperature. Obtain an expression for the temperature distribution $T(r, \phi, t)$ in the solid for times $t > 0$.

3-9 A portion of a hollow cylinder $a \leqslant r \leqslant b, 0 \leqslant \phi \leqslant \phi_0 < 2\pi$ is initially at temperature $F(r, \phi)$. For times $t > 0$, the boundaries at $r = a, r = b, \phi = 0$, and $\phi = \phi_0$ are all kept at zero temperature. Obtain an expression for the temperature distribution $T(r, \phi, t)$ in the solid for times $t > 0$.

3-10 Repeat problem 3-6 for the case when the boundary at $r = b$ is kept at constant temperature T_0.

3-11 A solid cylinder $0 \leqslant r \leqslant b, 0 \leqslant z \leqslant c, 0 \leqslant \phi \leqslant 2\pi$, is initially at temperature $F(r, \phi, z)$. For times $t > 0$, the boundary at $z = 0$ is kept insulated, the boundaries at $z = c$ and $r = b$ are kept at zero temperature. Obtain an expression for the temperature distribution $T(r, z, \phi, t)$ in the solid for times $t > 0$.

3-12 A portion of a solid cylinder, $0 \leqslant r \leqslant b, 0 \leqslant \phi \leqslant \phi_0 < 2\pi, 0 \leqslant z \leqslant c$, as illustrated in Fig. 3-8, is initially at temperature $F(r, \phi, z)$. For times $t > 0$, the boundary surface at $z = 0$ is kept insulated, the boundary at $z = c$ dissipates heat by convection into an environment at zero temperature, and the remaining boundaries are kept at zero temperature. Obtain an expression for temperature distribution $T(r, \phi, z, t)$ in the solid for times $t > 0$.

3-13 Solve problem 3-3 by using product solution for the case solid is initially at a uniform temperature T_0.

3-14 Obtain an expression for the steady-state temperature distribution $T(r, z)$ in a solid cylinder, $0 \leqslant r \leqslant b, 0 \leqslant z \leqslant c$, when the boundary at $z = 0$ is kept at temperature $F(r)$, and there is convection into a medium at zero temperature from the surfaces $r = b$ and $z = c$. Assume heat transfer coefficients to be the same for both of these surfaces.

3-15 Obtain an expression for the steady-state temperature distribution $T(r, z)$ in a hollow cylinder $a \leqslant r \leqslant b, 0 \leqslant z \leqslant c$, when the boundary at $r = a$ is kept at temperature $F(z)$ and other boundaries at $r = b, z = 0$, and $z = c$ are kept at zero temperature.

3-16 Obtain an expression for the steady-state temperature $T(r, z)$ in a hollow cylinder $a \leqslant r \leqslant b, 0 \leqslant z \leqslant c$, when the heat flux into the surface at $r = a$ is $f(z)$ [i.e., $-k(\partial T/\partial r) = f(z)$ at $r = a$] and the other boundaries at $r = b$, $z = 0$ and $z = c$ are kept at zero temperature.

3-17 Obtain an expression for the steady-state temperature distribution $T(r, \phi)$ in a solid cylinder $0 \leqslant r \leqslant b, 0 \leqslant \phi \leqslant 2\pi$, when the boundary at $r = b$ is subjected to a prescribed temperature distribution $f(\phi)$.

3-18 Obtain an expression for the steady-state temperature distribution $T(r, z)$ in a solid, semiinfinite cylinder $0 \leqslant r \leqslant b, 0 \leqslant z < \infty$, when the boundary at $r = b$ is kept at prescribed temperature $f(z)$ and the boundary at $z = 0$ is kept at zero temperature.

3-19 Obtain an expression for the steady-state temperature, distribution $T(r, z)$ in a solid, semiinfinite cylinder $0 \leqslant r \leqslant b, 0 \leqslant z < \infty$, when the boundary at $z = 0$ is kept at temperature $f(r)$ and the boundary at $r = b$ dissipates heat by convection into a medium at zero temperature.

NOTE

1. Consider the eigenvalue problem

$$\frac{1}{r}\frac{d}{dr}\left(r\frac{dR(r)}{dr}\right) + \left(\beta^2 - \frac{v^2}{r^2}\right)R(r) = 0 \qquad \text{in} \qquad 0 \leqslant r < b \tag{1a}$$

$$\frac{dR}{dr} = 0 \qquad \text{at} \qquad r = b \tag{1b}$$

For equation (1a) for $v = 0$, we find

$$\beta^2 \int_0^b rR^2(r)dr = -\int_0^b R\frac{d}{dr}\left(r\frac{dR}{dr}\right)dr \tag{2}$$

Integrating the right-hand side by parts, we obtain

$$\beta^2 \int_0^b rR^2(r)dr = -\left[rR\frac{dR}{dr}\right]_0^b + \int_0^b r\left(\frac{dR}{dr}\right)^2 dr \tag{3}$$

The first term on the right vanishes in view of the boundary condition (1b); then

$$\beta^2 = \frac{1}{N}\int_0^b r\left(\frac{dR}{dr}\right)^2 dr \qquad \text{where} \qquad N \equiv \int_0^b rR^2\,dr \tag{4}$$

Clearly, $\beta_0 = 0$ is also an eigenvalue corresponding to $R_0(\beta_0, r) = \text{constant} \neq 0$. Then, for $R_0(\beta_0, r) = 1$ the corresponding norm becomes

$$N(\beta_0) = \frac{b^2}{2} \tag{5}$$

4

THE SEPARATION OF VARIABLES IN THE SPHERICAL COORDINATE SYSTEM

In this chapter we present the separation of the homogeneous heat conduction equation in the spherical coordinate system and examine the solution of homogeneous problems of spheres involving (r, t), (r, μ, t), and (r, μ, ϕ, t) variables by the method of separation of variables. The solution of multidimensional steady-state problems on sphere is also presented. The reader should consult references 1–7 for further application of the method of separation of variables to the solution of homogeneous heat-conduction problems in the spherical coordinate system.

4-1 SEPARATION OF THE HEAT CONDUCTION EQUATION IN THE SPHERICAL COORDINATE SYSTEM

Consider the three-dimensional, homogeneous heat conduction equation in the spherical coordinate system (r, θ, ϕ) given as

$$\frac{\partial^2 T}{\partial r^2}+\frac{2}{r}\frac{\partial T}{\partial r}+\frac{1}{r^2 \sin\theta}\frac{\partial}{\partial \theta}\left(\sin\theta\frac{\partial T}{\partial \theta}\right)+\frac{1}{r^2\sin^2\theta}\frac{\partial^2 T}{\partial \phi^2}=\frac{1}{\alpha}\frac{\partial T}{\partial t} \tag{4-1}$$

where $T \equiv T(r, \theta, \phi, t)$. This equation is put into a more convenient form by defining a new independent variable μ as

$$\mu = \cos\theta \tag{4-2}$$

Equation (4.1) becomes

$$\frac{\partial^2 T}{\partial r^2}+\frac{2}{r}\frac{\partial T}{\partial r}+\frac{1}{r^2}\frac{\partial}{\partial \mu}\left[(1-\mu^2)\frac{\partial T}{\partial \mu}\right]+\frac{1}{r^2(1-\mu^2)}\frac{\partial^2 T}{\partial \phi^2}=\frac{1}{\alpha}\frac{\partial T}{\partial t} \tag{4-3}$$

where $T \equiv T(r, \mu, \phi, t)$. If we define a new dependent variable V as

$$V = r^{1/2} T \tag{4-4}$$

equation (4-3) takes the form

$$\frac{\partial^2 V}{\partial r^2} + \frac{1}{r}\frac{\partial V}{\partial r} - \frac{1}{4}\frac{V}{r^2} + \frac{1}{r^2}\frac{\partial}{\partial \mu}\left[(1-\mu^2)\frac{\partial V}{\partial \mu}\right] + \frac{1}{r^2(1-\mu^2)}\frac{\partial^2 V}{\partial \phi^2} = \frac{1}{\alpha}\frac{\partial V}{\partial t} \tag{4-5}$$

Equations (4-3) and (4-5) are the two different forms of the heat conduction equation for sphere that will be considered in this chapter. Equation (4-5) will be used only when temperature depends on the (r, μ, ϕ, t) or (r, μ, t) variables. The reason for this is that, when equation (4-5) is used in such situations, the elementary solutions of the differential equation for $R(r)$ become *Bessel functions* which have already been discussed in the previous chapter. However, if equation (4-3) is used, the elementary solutions are *spherical Bessel functions.* For all other cases, including the time-dependent problems involving (r, μ, ϕ), (r, μ), (r, t), (r) variables as well as all steady-state problems, the governing equation will be obtained from equation (4-3).

We now examine the separation of equations (4-3) and (4-5) for typical cases.

1. *Temperature depends on* (r, μ, ϕ, t). The heat conduction equation (4-3) for $T(r, \mu, \phi, t)$ is transformed into equation (4-5) for $V(r, \mu, \phi, t)$ by the transformation (4-4). Thus we consider the separation of the following equation:

$$\frac{\partial^2 V}{\partial r^2} + \frac{1}{r}\frac{\partial V}{\partial r} - \frac{1}{4}\frac{V}{r^2} + \frac{1}{r^2}\frac{\partial}{\partial \mu}\left[(1-\mu^2)\frac{\partial V}{\partial \mu}\right] + \frac{1}{r^2(1-\mu^2)}\frac{\partial^2 V}{\partial \phi^2} = \frac{1}{\alpha}\frac{\partial V}{\partial t} \tag{4-6a}$$

If we assume a separation of variables in the form

$$V(r, \mu, \phi, t) = \Gamma(t) R(r) M(\mu) \Phi(\phi) \tag{4-6b}$$

the separated equations become

$$\frac{d\Gamma(t)}{dt} + \alpha\lambda^2 \Gamma(t) = 0 \tag{4-7a}$$

$$\frac{d^2\Phi(\phi)}{d\phi^2} + m^2 \Phi(\phi) = 0 \tag{4-7b}$$

$$\frac{d^2 R}{dr^2} + \frac{1}{r}\frac{dR}{dr} + \left[\lambda^2 - \left(n + \frac{1}{2}\right)^2 \frac{1}{r^2}\right] R = 0 \tag{4-7c}$$

$$\frac{d}{d\mu}\left[(1-\mu^2)\frac{dM}{d\mu}\right] + \left[n(n+1) - \frac{m^2}{1-\mu^2}\right] M = 0 \tag{4-7d}$$

Equation (4-7c) is the Bessel differential equation of order $(n+\frac{1}{2})$, which has solutions $J_{n+1/2}(\lambda r)$ and $Y_{n+1/2}(\lambda r)$. When the order of the Bessel function is not zero or positive integer, the solution $Y_{n+1/2}(\lambda r)$ can be replaced by $J_{-n-1/2}(\lambda r)$ as discussed in Appendix IV. The differential equation (4-7d) is called *Legendre's associated differential equation*, its solutions $P_n^m(\mu)$ and $Q_n^m(\mu)$ are called *associated Legendre functions of degree n and order m, of the first and second kind*, respectively.

The elementary solutions of the separated equations (4-7) can be summarized as

$$\Gamma(t): \quad e^{-\alpha\lambda^2 t} \tag{4-8a}$$

$$\Phi(\phi): \quad \sin m\phi \quad \text{and} \quad \cos m\phi \tag{4-8b}$$

$$R(r): \quad J_{n+1/2}(\lambda r) \quad \text{and} \quad Y_{n+1/2}(\lambda r) \tag{4-8c}$$

$$M(\mu): \quad P_n^m(\mu) \quad \text{and} \quad Q_n^m(\mu) \tag{4-8d}$$

A brief discussion of Legendre functions is given in the next paragraph.

2. *Temperature depends on* (r, μ, t). We consider the transformed heat conduction equation (4-5). For the case of no dependence on the azimuth angle ϕ, this equation simplifies to

$$\frac{\partial^2 V}{\partial r^2}+\frac{1}{r}\frac{\partial V}{\partial r}-\frac{1}{4}\frac{V}{r^2}+\frac{1}{r^2}\frac{\partial}{\partial \mu}\left[(1-\mu^2)\frac{\partial V}{\partial \mu}\right]=\frac{1}{\alpha}\frac{\partial V}{\partial t} \tag{4-9}$$

where $V \equiv V(r, \mu, t)$.

The separation of equation (4-9) results in the following equations

$$\frac{d\Gamma(t)}{dt}+\alpha\lambda^2\Gamma(t)=0 \tag{4-10a}$$

$$\frac{d^2 R}{dr^2}+\frac{1}{r}\frac{dR}{dr}+\left[\lambda^2-\left(n+\frac{1}{2}\right)^2\frac{1}{r^2}\right]R=0 \tag{4-10b}$$

$$\frac{d}{d\mu}\left[(1-\mu^2)\frac{dM}{d\mu}\right]+n(n+1)M=0 \tag{4-10c}$$

The elementary solutions of these equations are taken as

$$\Gamma(t): \quad e^{-\alpha\lambda^2 t} \tag{4-11a}$$

$$R(r): \quad J_{n+1/2}(\lambda r) \quad \text{and} \quad Y_{n+1/2}(\lambda r) \tag{4-11b}$$

$$M(\mu): \quad P_n(\mu) \quad \text{and} \quad Q_n(\mu) \tag{4-11c}$$

We note that when the temperature is independent of the azimuth angle ϕ, the separated equation (4-10c) for the function $M(\mu)$ becomes the Legendre's differential equation. The solutions $P_n(\mu)$ and $Q_n(\mu)$ are called the *Legendre functions of degree n*, of the first and second kinds, respectively.

3. *Temperature depends on* (r, μ, ϕ). The governing heat conduction equation for this case is obtained from equation (4-3) by omitting the time derivative term

$$\frac{\partial^2 T}{\partial r^2} + \frac{2}{r}\frac{\partial T}{\partial r} + \frac{1}{r^2}\frac{\partial}{\partial \mu}\left[(1-\mu^2)\frac{\partial T}{\partial \mu}\right] + \frac{1}{r^2(1-\mu^2)}\frac{\partial^2 T}{\partial \phi^2} = 0 \tag{4-12}$$

where $T \equiv T(r, \mu, \phi)$.

Assuming a separation in the form

$$T(r, \mu, \phi) = R(r)M(\mu)\Phi(\phi)$$

The resulting separated equations becomes

$$\frac{d^2\Phi}{d\phi^2} + m^2\Phi = 0 \tag{4-13a}$$

$$\frac{d^2 R}{\partial r^2} + \frac{2}{r}\frac{dR}{dr} - \frac{n(n+1)}{r^2}R = 0 \tag{4-13b}$$

$$\frac{d}{d\mu}\left[(1-\mu^2)\frac{dM}{d\mu}\right] + \left[n(n+1) - \frac{m^2}{1-\mu^2}\right]M = 0 \tag{4-13c}$$

and their elementary solutions are taken as

$$\Phi(\phi):\quad \sin m\phi \quad \text{and} \quad \cos m\phi \tag{4-13d}$$

$$R(r):\quad r^n \quad \text{and} \quad r^{-n-1} \tag{4-13e}$$

$$M(\mu):\quad P_n^m(\mu) \quad \text{and} \quad Q_n^m(\mu) \tag{4-13f}$$

For this special case the separated equation (4-13b) for the function $R(r)$ is an *Euler-Cauchy* type differential equation which has solutions r^n and r^{-n-1}.

4. *Temperature depends on* (r, μ). The governing equation is obtained from equation (4-3) by proper simplification to yield

$$\frac{\partial^2 T}{\partial r^2} + \frac{2}{r}\frac{\partial T}{\partial r} + \frac{1}{r^2}\frac{\partial}{\partial \mu}\left[(1-\mu^2)\frac{\partial T}{\partial \mu}\right] = 0 \tag{4-14}$$

where $T \equiv T(r, \mu)$.

The separation of this equation by setting $T(r,\mu) = R(r)M(\mu)$ leads to the following separated equations

$$\frac{d^2R}{dr^2} + \frac{2}{r}\frac{dR}{dr} - \frac{n(n+1)}{r^2}R = 0 \tag{4-15a}$$

$$\frac{d}{d\mu}\left[(1-\mu^2)\frac{dM}{d\mu}\right] + n(n+1)M = 0 \tag{4-15b}$$

and their elementary solutions are taken as

$$R(r)\colon r^n \qquad \text{and} \qquad r^{-n-1} \tag{4-15c}$$

$$M(\mu)\colon P_n(\mu) \qquad \text{and} \qquad Q_n(\mu) \tag{4-15d}$$

5. *Temperature depends on* (r, t). Equation (4-3) simplifies to

$$\frac{\partial^2 T}{\partial r^2} + \frac{2}{r}\frac{\partial T}{\partial r} = \frac{1}{\alpha}\frac{\partial T}{\partial t} \tag{4-16a}$$

which is written in the form

$$\frac{1}{r}\frac{\partial^2}{\partial r^2}(rT) = \frac{1}{\alpha}\frac{\partial T}{\partial t} \tag{4-16b}$$

A new dependent variable is defined as

$$U(r,t) = rT(r,t) \tag{4-16c}$$

Then equation (4-16b) is transformed into

$$\frac{\partial^2 U}{\partial r^2} = \frac{1}{\alpha}\frac{\partial U}{\partial t} \tag{4-17}$$

which is now the one-dimensional, time-dependent heat conduction equation in the rectangular coordinate system and the separation of which has already been considered in Chapter 2.

Once the elementary solutions of the heat-conduction equation are available, the general solution is constructed by the superposition of the elementary solutions.

4-2 LEGENDRE FUNCTIONS AND LEGENDRE'S ASSOCIATED FUNCTIONS

In this section we present a brief discussion of the properties of the Legendre functions and Legendre's associated functions. The reader should consult references 5-14 for detailed treatment of this subject.

Legendre Functions

It has been shown that the separation of the heat conduction equation for azimuthally symmetric temperature (i.e., temperature independent of ϕ) results in the Legendre's differential equation for $M(\mu)$ as

$$\frac{d}{d\mu}\left[(1-\mu^2)\frac{dM}{d\mu}\right]+n(n+1)M=0 \tag{4-18}$$

This differential equation is a special case of the Sturm–Liouville equation discussed in Chapter 2, with $p(\mu)=1-\mu^2, q(\mu)=0, w(\mu)=1$, and $\lambda=n(n+1)$. Clearly, the separation constants $n(n+1)$ are the eigenvalues, in which n, in general, is any number; depending on the nature of the problem, n can be a positive integer or fractional.

According to the theory of linear differential equations, equation (4-18) has two linearly independent solutions. These solutions, denoted by the symbols $P_n(\mu)$ and $Q_n(\mu)$, are called *Legendre functions of degree n, of the first and second kinds, respectively.*

For integer values of n the series defining the function $P_n(\mu)$ terminates at a finite number of terms, hence the Legendre function $P_n(\mu)$ becomes the *Legendre polynomial* $P_n(\mu)$, which is convergent in the interval $-1 \leqslant \mu \leqslant 1$. The first few of the Legendre polynomials are given as [5, p. 86; 7, p. 151]

$$\begin{aligned}
&P_0(\mu)=1 && P_1(\mu)=\mu\\
&P_2(\mu)=\tfrac{1}{2}(3\mu^2-1) && P_3(\mu)=\tfrac{1}{2}(5\mu^3-3\mu)\\
&P_4(\mu)=\tfrac{1}{8}(35\mu^4-30\mu^2+3) && P_5(\mu)=\tfrac{1}{8}(63\mu^5-70\mu^3+15\mu)\\
&P_6(\mu)=\tfrac{1}{16}(231\mu^6-315\mu^4+105\mu^2-5) && P_7(\mu)=\tfrac{1}{16}(429\mu^7-639\mu^5\\
&P_8(\mu)=\tfrac{1}{128}(6435\mu^8-12012\mu^6 && \qquad +315\mu^3-35\mu)\\
&\qquad +6930\mu^4-1260\mu^2+35)
\end{aligned} \tag{4-19}$$

Any other $P_n(\mu)$, when n is a positive integer, is obtainable from the following recurrence relation

$$(n+1)P_{n+1}(\mu)-(2n+1)\mu P_n(\mu)+nP_{n-1}(\mu)=0 \tag{4-20}$$

The Legendre polynomials $P_n(\mu)$ are also obtainable from the Rodrigues' formula [15]

$$P_n(\mu) = \frac{1}{2^n n!} \frac{d^n}{d\mu^n} (\mu^2 - 1)^n \tag{4-21}$$

This formula is useful to evaluate definite integrals involving Legendre polynomials. We present in Appendix V numerical values of the first seven of the Legendre polynomials $P_n(\mu)$.

The Legendre function $Q_n(\mu)$ being infinite at $\mu = \pm 1$ for all values of n, it is excluded from the solution on the physical grounds.

Figures 4-1 and 4-2 show a plot of the first four of $P_n(\mu)$ and $Q_n(\mu)$ functions. Clearly, $Q_n(\mu)$ functions become infinite at $\mu = \pm 1$.

Legendre's Associated Functions

We have seen that the separation of heat conduction equation for sphere resulted in a differential equation for the $M(\mu)$ variable in the form [see equation (4-7d)]

$$\frac{d}{d\mu}\left[(1-\mu^2)\frac{dM}{d\mu}\right] + \left[n(n+1) - \frac{m^2}{1-\mu^2}\right]M = 0 \tag{4-22}$$

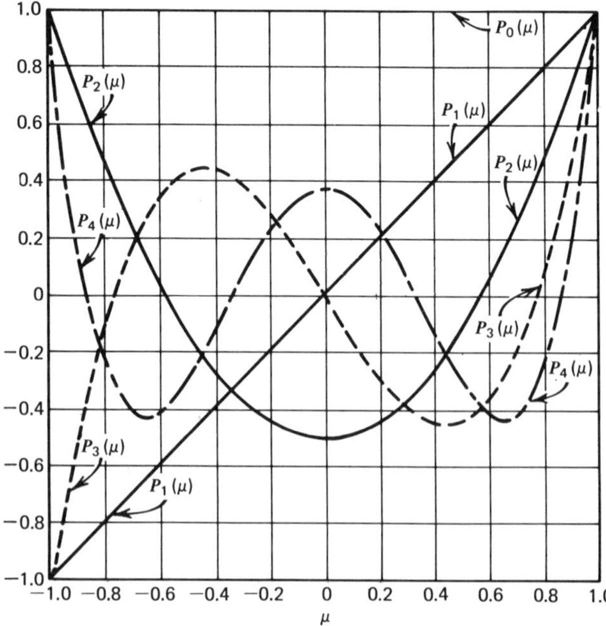

Fig. 4-1 Legendre polynomials $P_n(\mu)$ for $n = 0, 1, 2, 3, 4$.

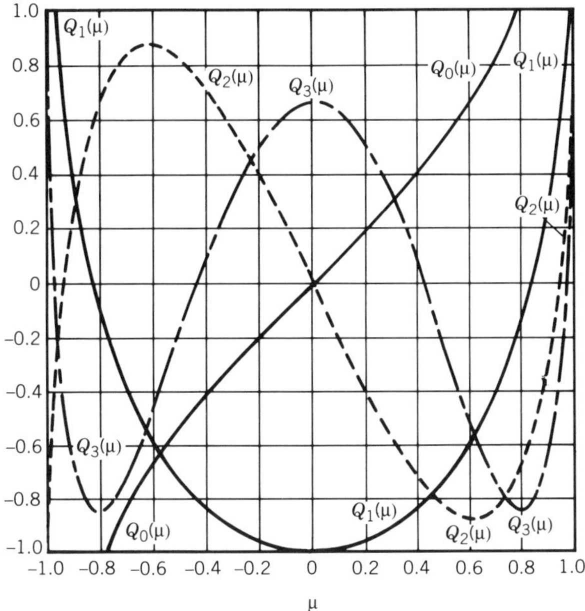

Fig. 4-2 Legendre functions of the second kind, $Q_n(\mu)$ for $n = 0, 1, 2, 3$.

which is called *Legendre's associated differential equation.* The two solutions $P_n^m(\mu)$ and $Q_n^m(\mu)$ of this differential equation are known as *the associated Legendre functions of degree n and order m,* of the first and second kinds, respectively. Here the order m of these functions has resulted from the separation constant associated with the separation of the ϕ variable; therefore its values depend on the range of the ϕ variable. When the range of ϕ is $0 \leqslant \phi \leqslant 2\pi$, the values of m are taken as positive integers ($m = 0, 1, 2, 3, \ldots$) to satisfy the physical requirement that the temperature remain periodic in ϕ with period 2π.

The first few of $P_n^m(\mu)$ functions for integer values of n and m over the range of $-1 \leqslant \mu \leqslant 1$ are given by

$$\begin{aligned} P_1^1(\mu) &= -(1-\mu^2)^{1/2} & P_2^1(\mu) &= -3(1-\mu^2)^{1/2}\mu \\ P_2^2(\mu) &= 3(1-\mu^2) & P_3^1(\mu) &= -\tfrac{3}{2}(1-\mu^2)^{1/2}(5\mu^2-1) \\ P_3^2(\mu) &= 15\mu(1-\mu^2) & P_3^3(\mu) &= -15(1-\mu^2)^{3/2} \end{aligned} \tag{4-23}$$

Here $P_n^0(\mu)$ are not included because they are the same as Legendre polynomials $P_n(\mu)$.

The recurrence formula among $P_n^m(\mu)$ functions for integer values of m and n is given by [5, p. 304; 11, p. 360; 12, p. 62]

$$(n-m+1)P_{n+1}^m(\mu) - (2n+1)P_n^m(\mu) + (n+m)P_{n-1}^m(\mu) = 0 \tag{4-24}$$

For $m=0$, this expression reduces to the recurrence relation (4-20) for the Legendre polynomials.

The $P_n^m(\mu)$ functions can also be expressed in terms of the $P_n(\mu)$ functions by the following differential relation [5, p. 116; 12, p. 53]:

$$P_n^m(\mu)=(-1)^m(1-\mu^2)^{m/2}\frac{d^m}{d\mu^m}P_n(\mu) \tag{4-25}$$

The $Q_n^m(\mu)$ functions become infinite at $\mu=\pm 1$ for all values of n, whether integer on not; hence they are inadmissible as solutions on the physical grounds when the region contains $\mu=\pm 1$.

4-3 ORTHOGONALITY OF LEGENDRE FUNCTIONS

In heat conduction problems on spheres involving the variation of temperature with μ or μ, ϕ variables, the Legendre functions will appear in the solution; hence the orthogonality property of the Legendre functions will be needed. We present below some of these orthogonality conditions for ready reference later in this chapter.

The Legendre polynomials $P_n(\mu)$ have the following orthogonality property over the range $-1 \leqslant \mu \leqslant 1$ [5, p. 88; 12, p. 51]

$$\int_{-1}^{1} P_n(\mu)P_{n'}(\mu)d\mu=\begin{cases}0 & \text{for } n \neq n' \\ N(n) & \text{for } n=n'\end{cases} \tag{4-26}$$

where

$$N(n)\equiv\int_{-1}^{1}[P_n(\mu)]^2 d\mu=\frac{2}{2n+1} \tag{4-27}$$

and n, n' are positive integers. This orthogonality condition is needed for the problem of full sphere (i.e., $-1 \leqslant \mu \leqslant 1$) when temperature varies with μ but not with the azimuth angle ϕ.

The orthogonality of the associated Legendre function $P_n^m(\mu)$ in the interval $-1 \leqslant \mu \leqslant 1$ is given by [5, p. 117; 6, p. 324; 12, p. 54; 14, p. 184]

$$\int_{-1}^{1} P_n^m(\mu)P_{n'}^m(\mu)d\mu=\begin{cases}0 & \text{for } n \neq n' \\ N(m,n) & \text{for } n=n'\end{cases} \tag{4-28}$$

where

$$N(m,n)=\int_{-1}^{1}[P_n^m(\mu)]^2 d\mu=\frac{2}{2n+1}\frac{(n+m)!}{(n-m)!} \tag{4-29}$$

and n, n', m are positive integers, zero being included, $m \leqslant n$. This orthogonality condition is needed for the problem of full sphere (i.e., $-1 \leqslant \mu \leqslant 1$) when temperature varies with both μ and ϕ. We note that for $m = 0$, equation (4-29) reduces to equation (4-27).

The orthogonality of Legendre polynomials $P_n(\mu)$ in the half range $0 \leqslant \mu \leqslant 1$ is more involved and given by [5, p. 109; 6, p. 306; 7, p. 172]

$$\int_0^1 P_n(\mu) P_{n'}(\mu) d\mu = \begin{cases} 0 & \text{if } n, n' \text{ both } even \text{ or both } odd, n \neq n' \quad \text{(4-30)} \\ \dfrac{1}{2n+1} & \text{if } n = n' \quad \text{(4-31)} \\ \dfrac{(-1)^{(n+n'+1)/2} n! n'!}{2^{n+n'-1}(n-n')(n+n'+1)\left(\dfrac{n}{2}!\right)^2 \left(\dfrac{n'-1}{2}!\right)^2} & \text{if } n \text{ } even, n' \text{ } odd \quad \text{(4-32)} \end{cases}$$

where n and n' are positive integers.

4-4 REPRESENTATION OF AN ARBITRARY FUNCTION IN TERMS OF LEGENDRE FUNCTIONS

In the solution of a heat conduction problem with temperature depending on the μ and/or ϕ variables, the representation of an arbitrary function $F(\mu)$ or $F(\mu, \phi)$ in terms of Legendre polynomials or the spherical harmonics is needed. Here we discuss such representations.

Representation in Region $-1 \leqslant \mu \leqslant 1$

This region is encountered in the problems of the full sphere; we consider the following two cases:

The Representation of $F(\mu)$. When temperature depends on μ but it is azimuthally symmetric, the representation of an arbitrary function $F(\mu)$ defined in the interval $-1 \leqslant \mu \leqslant 1$ is needed in terms of the Legendre polynomials $P_n(\mu), n = 0, 1, 2, \ldots,$ in the form

$$F(\mu) = \sum_{n=0}^{\infty} c_n P_n(\mu) \qquad \text{in} \qquad -1 \leqslant \mu \leqslant 1 \tag{4-33}$$

To determine the coefficients c_n we utilize the orthogonality of the Legendre polynomials given by equation (4-26). If it is assumed that the series on the right of equation (4-33) can be integrated term by term over the range $-1 \leqslant \mu \leqslant 1$, we operate on both sides of equation (4-33) by the operator $\int_{-1}^{1} P_{n'}(\mu) d\mu$ and utilize the above orthogonality relation to obtain

$$c_n = \frac{1}{N(n)} \int_{-1}^{1} P_n(\mu) F(\mu) d\mu \tag{4-34}$$

The substitution of equation (4-34) into (4-33) yields

$$F(\mu) = \sum_{n=0}^{\infty} \frac{1}{N(n)} P_n(\mu) \int_{-1}^{1} P_n(\mu') F(\mu') d\mu' \quad \text{in} \quad -1 \leqslant \mu \leqslant 1 \tag{4-35}$$

where

$$N(n) = \frac{2}{2n+1}, \qquad n = 0, 1, 2, 3 \ldots \tag{4-36}$$

The Representation of $F(\mu, \phi)$. The representation of this type is generally needed in the problems of full sphere when the temperature is a function of both μ and ϕ variables. Consider a function $F(\mu, \phi)$ defined in the interval $-1 \leqslant \mu \leqslant 1, 0 \leqslant \phi \leqslant 2\pi$ to be represented in terms of the elementary solutions

$$P_n^m(\mu) \qquad \text{and} \qquad (A \cos m\phi + B \sin m\phi)$$

where m, n are positive integers, zero being included, with $m \leqslant n$, in the form

$$F(\mu, \phi) = \sum_{n=0}^{\infty} \left[A_n P_n(\mu) + \sum_{m=1}^{n} (A_{nm} \cos m\phi + B_{nm} \sin m\phi) P_n^m(\mu) \right] \tag{4-37}$$

or

$$F(\mu, \phi) = \sum_{n=0}^{\infty} \sum_{m=0}^{n} (A_{nm} \cos m\phi + B_{nm} \sin m\phi) P_n^m(\mu)$$

$$\text{in} \qquad -1 \leqslant \mu \leqslant 1, \quad 0 \leqslant \phi \leqslant 2\pi \tag{4-38}$$

where

$$n, m = 0, 1, 2, \ldots \qquad \text{and} \qquad m \leqslant n$$

To determine the coefficients A_{nm} and B_{nm} we utilize the orthogonality of the associated Legendre functions $P_n^m(\mu)$ in the interval $-1 \leqslant \mu \leqslant 1$ given by equation (4-28).

To determine the coefficients B_{mm} we operate on both sides of equation (4-38) successively by the operators

$$\int_0^{2\pi} \sin m'\phi \, d\phi \qquad \text{and} \qquad \int_{-1}^{1} P_{n'}^m(\mu) d\mu$$

and utilize the orthogonality properties of trigonometric functions to obtain

$$B_{nm} = \frac{1}{\pi N(m,n)} \int_{\phi'=0}^{2\pi} \int_{\mu'=-1}^{1} \sin m\phi' P_n^m(\mu') F(\mu', \phi') d\mu' d\phi' \tag{4-39}$$

where

$$N(m,n) = \int_{-1}^{1} [P_n^m(\mu)]^2 d\mu = \frac{2}{2n+1} \frac{(n+m)!}{(n-m)!}, \quad m \leqslant n \tag{4-40}$$

To determine the coefficients A_{nm} we operate on both sides of equation (4-38) successively by the operators

$$\int_0^{2\pi} \cos m'\phi \, d\phi \qquad \text{and} \qquad \int_{-1}^{1} P_{n'}^m(\mu) d\mu$$

and utilize the orthogonality properties of trigonometric functions to obtain

$$A_{nm} = \frac{1}{\pi N(m,n)} \int_{\phi'=0}^{2\pi} \int_{\mu'=-1}^{1} \cos m\phi' P_n^m(\mu') F(\mu', \phi') d\mu' d\phi' \tag{4-41}$$

where π *should be replaced by* 2π *for* $m = 0$ and $N(m, n)$ is given by equation (4-40).

When the coefficients A_{nm} and B_{nm} as determined above are introduced into equation (4-38) and the trigonometric terms are combined as

$$\cos m\phi \cos m\phi' + \sin m\phi \sin m\phi' = \cos m(\phi - \phi') \tag{4-42}$$

the representation (4-38) becomes

$$F(\mu, \phi) = \frac{1}{\pi} \sum_{n=0}^{\infty} \sum_{m=0}^{n} \frac{P_n^m(\mu)}{N(m,n)} \int_{\phi'=0}^{2\pi} \int_{\mu'=-1}^{1} F(\mu', \phi') P_n^m(\mu') \cos m(\phi - \phi') d\mu' d\phi' \tag{4-43}$$

where π *should be replaced by* 2π *for* $m = 0$. By comparing equations (4-38) and (4-43) we write

$$[A_{nm} \cos m\phi + B_{nm} \sin m\phi]$$
$$\equiv \frac{1}{\pi N(m,n)} \int_{\phi'=0}^{2\pi} \int_{\mu'=-1}^{1} F(\mu', \phi') P_n^m(\mu') \cos m(\phi - \phi') d\mu' d\phi' \tag{4-44}$$

where π *should be replaced by* 2π *for* $m = 0$.

Equation (4-43) is now written more explicitly in the form

$$F(\mu, \phi) = \frac{1}{4\pi} \sum_{n=0}^{\infty} (2n+1) P_n(\mu) \int_{\phi'=0}^{2\pi} \int_{\mu'=-1}^{1} F(\mu', \phi') P_n(\mu') d\mu' \, d\phi'$$

$$+ \frac{1}{2\pi} \sum_{n=0}^{\infty} \sum_{m=1}^{n} (2n+1) \frac{(n-m)!}{(n+m)!} P_n^m(\mu)$$

$$\cdot \int_{\phi'=0}^{2\pi} \int_{\mu'=-1}^{1} F(\mu', \phi') P_n^m(\mu') \cos[m(\phi - \phi')] d\mu' \, d\phi' \tag{4-45}$$

in $-1 \leqslant \mu \leqslant 1, 0 \leqslant \phi \leqslant 2\pi$

This representation is valid for all values of μ and ϕ in the range $-1 \leqslant \mu \leqslant 1$, $0 \leqslant \phi \leqslant 2\pi$, provided that the function $F(\mu, \phi)$ satisfies the conditions that would have to be satisfied if it were to be developed into a Fourier's series.

Representation in Region $0 \leqslant \mu \leqslant 1$

This region is encountered in the problems of the *hemisphere* as illustrated in Fig. 4-3. When temperature is azimuthally symmetric but depends on the μ variable among other variables, it may be necessary to represent an arbitrary function $F(\mu)$ defined in the interval $0 \leqslant \mu \leqslant 1$ in terms of the $P_n(\mu)$ functions in the form

$$F(\mu) = \sum_n c_n P_n(\mu) \qquad \text{in} \qquad 0 < \mu \leqslant 1 \tag{4-46}$$

Here, the values of n should be so chosen that the boundary condition at $\mu = 0$ is satisfied. We consider this expansion for the following two different boundary conditions at $\mu = 0$, the base of the hemisphere.

1. For a boundary condition of the *first kind* at $\mu = 0$ we have

$$P_n(\mu) = 0 \qquad \text{at} \qquad \mu = 0 \tag{4-47}$$

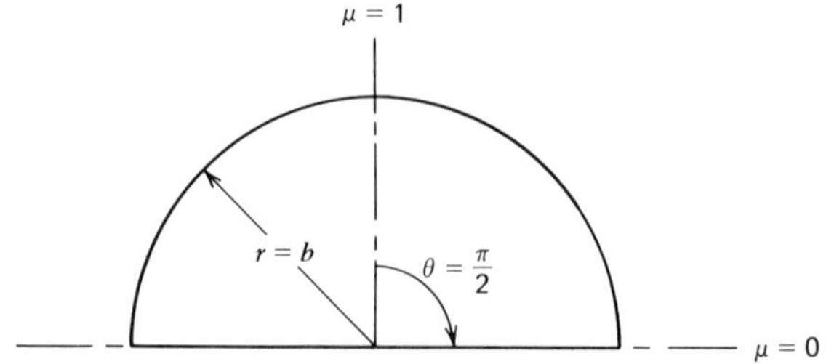

Fig. 4-3 Coordinates for a hemisphere.

This requirement is satisfied if $P_n(\mu)$ is chosen as the Legendre polynomials with n being *odd positive integer* (i.e., $n = 1, 3, 5 \ldots$). This is apparent from the definition of Legendre polynomials given by equation (4-19).

2. For a boundary condition of the *second kind* at $\mu = 0$ we have

$$\frac{dP_n(\mu)}{d\mu} = 0 \qquad \text{at} \qquad \mu = 0 \tag{4-48}$$

This requirement is satisfied if $P_n(\mu)$ is chosen as the Legendre polynomial with n being *even positive integers*, zero being included (i.e., $n = 0, 2, 4, 6 \ldots$). This is apparent from the definition of the Legendre polynomials given by equation (4-19).

To determine the coefficients c_n in equation (4-46), we utilize the orthogonality of Legendre polynomials in the interval $0 \leqslant \mu \leqslant 1$ given by equation (4-31).

The coefficients c_n are now determined by operating on both sides of equation (4-46) by the operator $\int_0^1 P_{n'}(\mu)d\mu$ and utilizing the orthogonality relations. We find

$$c_n = \frac{1}{N(n)} \int_0^1 F(\mu')P_n(\mu')d\mu' \tag{4-49}$$

where

$$N(n) \equiv \int_0^1 [P_n(\mu)]^2 d\mu = \frac{1}{2n+1} \tag{4-50}$$

and the values of n are chosen as

$n = 1, 3, 5 \ldots$ for boundary condition of the *first kind* at $\mu = 0$

$n = 0, 2, 4 \ldots$ for boundary condition of the *second kind* at $\mu = 0$

Introducting the coefficients c_n into equation (4-46), the representation of $F(\mu)$ becomes

$$F(\mu) = \sum_n (2n+1)P_n(\mu) \int_{\mu'=0}^1 F(\mu')P_n(\mu')d\mu' \qquad \text{in} \qquad 0 < \mu \leqslant 1 \tag{4-51}$$

where the values of n depend on the type of the boundary condition at the surface $\mu = 0$ as follows:

1. When the boundary condition at $\mu = 0$ is of the *first kind*, take $n = 1, 3, 5, 7, \ldots,$ that is, odd positive integers.
2. When the boundary condition at $\mu = 0$ is of the *second kind*, take $n = 0, 2, 4, 6, \ldots,$ that is, even positive integers, zero being included.

4-5 PROBLEMS IN (r, t) VARIABLES

The heat conduction problem for a sphere involving (r, t) variables can be transformed into a problem of a slab or a semiinfinite medium by the transformation of the dependent variable as in equation (4-16c). Then, the resulting problem can be solved readily by the techniques described in Chapter 2.

Solid Sphere $0 \leqslant r \leqslant b$

Consider the heat conduction problem in a solid sphere $0 \leqslant r \leqslant b$, with heat generation and subject to nonhomogeneous boundary condition of the third kind at the boundary surface $r = b$ as illustrated in Fig. 4-4. The mathematical formulation of the problem is given as

$$\frac{1}{r}\frac{\partial^2}{\partial r^2}(rT) + \frac{1}{k}g(r) = \frac{1}{\alpha}\frac{\partial T}{\partial t} \quad \text{in} \quad 0 \leqslant r < b, \quad t > 0 \tag{4-52}$$

$$\frac{\partial T}{\partial r} + HT = f \quad \text{at} \quad r = b, \quad t > 0 \tag{4-53}$$

$$T = F(r) \quad \text{for} \quad t = 0, \quad \text{in}\, 0 \leqslant r \leqslant b \tag{4-54}$$

A new dependent variable $U(r, t)$ is defined as

$$U(r, t) = rT(r, t) \tag{4-55}$$

Then the problem is transformed to

$$\frac{\partial^2 U}{\partial r^2} + \frac{rg(r)}{k} = \frac{1}{\alpha}\frac{\partial U}{\partial t} \quad \text{in} \quad 0 < r < b, \quad t > 0 \tag{4-56}$$

$$U = 0 \quad \text{at} \quad r = 0, \quad t > 0 \tag{4-57}$$

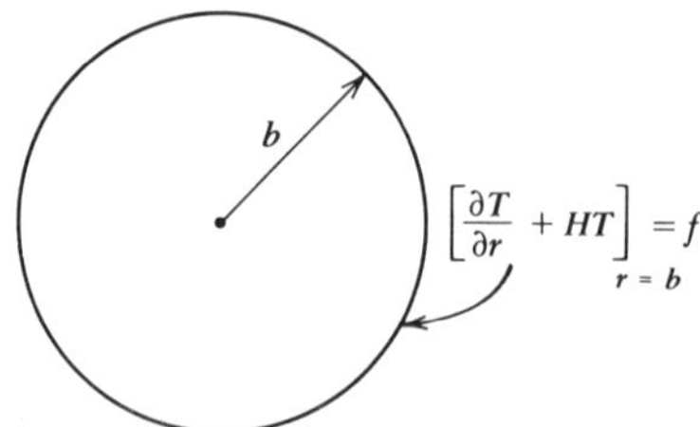

Fig. 4-4 Boundary condition for a solid sphere.

$$\frac{\partial U}{\partial r} + \left(H - \frac{1}{b}\right)U = bf \quad \text{at} \quad r = b, \quad t > 0 \tag{4-58}$$

$$U = rF(r) \quad \text{for} \quad t = 0, \quad \text{in} \quad 0 \leqslant r \leqslant b \tag{4-59}$$

This is a problem of heat conduction in a slab $0 \leqslant r \leqslant b$, which can readily be solved by the application of the techniques described in Chapter 2. We now illustrate the application with the examples given below.

Example 4-1

A solid sphere of radius $r = b$ is initially at temperature $F(r)$ and for times $t > 0$ the boundary surface at $r = b$ dissipates heat by convection into a medium at zero temperature. Obtain an expression for the temperature distribution $T(r, t)$ in the sphere for times $t > 0$.

Solution. The mathematical formulation of this problem is given as

$$\frac{1}{r}\frac{\partial^2}{\partial r^2}(rT) = \frac{1}{\alpha}\frac{\partial T}{\partial t} \quad \text{in} \quad 0 \leqslant r < b, \quad t > 0 \tag{4-60}$$

$$\frac{\partial T}{\partial r} + HT = 0 \quad \text{at} \quad r = b, \quad t > 0 \tag{4-61a}$$

$$T = F(r) \quad \text{for} \quad t = 0, \quad \text{in} \quad 0 \leqslant r \leqslant b \tag{4-61b}$$

When this problem is transformed by the transformation $U(r, t) = rT(r, t)$, the transformed system becomes

$$\frac{\partial^2 U}{\partial r^2} = \frac{1}{\alpha}\frac{\partial U}{\partial t} \quad \text{in} \quad 0 < r < b, \quad t > 0 \tag{4-62}$$

$$U = 0 \quad \text{at} \quad r = 0, \quad t > 0 \tag{4-63a}$$

$$\frac{\partial U}{\partial r} + \left(H - \frac{1}{b}\right)U = 0 \quad \text{at} \quad r = b, \quad t > 0 \tag{4-63b}$$

$$U = rF(r) \quad \text{for} \quad t = 0, \quad \text{in} \quad 0 \leqslant r \leqslant b \tag{4-63c}$$

This is a homogeneous heat conduction problem for a slab $0 \leqslant r \leqslant b$; its solution for $U(r, t)$ is readily obtainable by the approach described in Chapter 2. After the transformation of the solution for $U(r, t)$ to $T(r, t)$ we obtain

$$T(r, t) = \frac{2}{r}\sum_{m=1}^{\infty} e^{-\alpha\beta_m^2 t}\frac{\beta_m^2 + K^2}{b(\beta_m^2 + K^2) + K}\sin\beta_m r\int_{r'=0}^{b} r'F(r')\sin\beta_m r'\,dr' \tag{4-64}$$

where

$$K \equiv H - \frac{1}{b} \tag{4-65}$$

and the β_m values are the positive roots of

$$\beta_m b \cot \beta_m b + bK = 0 \tag{4-66}$$

The roots of this transcendental equation are real if $bK > -1$ (see Appendix II, the table for the roots of $\xi \cot \xi + c = 0$). When the value of K as defined above is introduced into this inequality we find $(bH - 1) > -1$, which implies that $H > 0$. This result is consistent with the requirement on the physical grounds that in the original sphere problem we should have $H > 0$. Therefore, in the pseudoproblem the coefficient $(Hb - 1)$ may be negative, but the quantity Hb is always positive.

Insulated Boundary

When the boundary at $r = b$ is insulated, we have $H = 0$. For this special case $\beta_0 = 0$ is also an eigenvalue. Then the term

$$\frac{3}{b^3} \int_0^b r^2 F(r) dr$$

resulting from the eigenvalue $\beta_0 = 0$ should be added on the right-hand side of equation (4-64). This term implies that, after the transients have passed, the steady-state temperature in the medium is the mean of the initial temperature distribution $F(r)$ over the volume of the insulated sphere (see note 1 at end of chapter for further discussion of this matter).

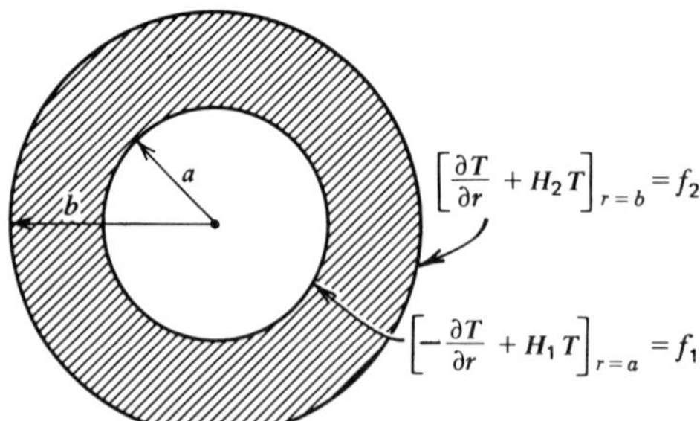

Fig. 4-5 Boundary conditions for a hollow sphere.

Hollow Sphere $a \leqslant r \leqslant b$

We now consider the problem of heat conduction in a hollow sphere $a \leqslant r \leqslant b$, with heat generation and subject to nonhomogeneous boundary conditions of the third kind at $r = a$ and $r = b$ as illustrated in Fig. 4-5. The mathematical formulation of the problem is given as

$$\frac{1}{r}\frac{\partial^2}{\partial r^2}(rT) + \frac{1}{k}g(r) = \frac{1}{\alpha}\frac{\partial T}{\partial t} \qquad \text{in} \quad a < r < b, \quad t > 0 \tag{4-67a}$$

$$-\frac{\partial T}{\partial r} + H_1 T = f_1 \qquad \text{at} \quad r = a, \quad t > 0 \tag{4-67b}$$

$$\frac{\partial T}{\partial r} + H_2 T = f_2 \qquad \text{at} \quad r = b, \quad t > 0 \tag{4-67c}$$

$$T = F(r) \qquad \text{for} \quad t = 0, \quad \text{in } a \leqslant r \leqslant b \tag{4-67d}$$

A new dependent variable is now defined as

$$U(r, t) = rT(r, t) \tag{4-68}$$

Then equations (4-67) are transformed into

$$\frac{\partial^2 U}{\partial r^2} + \frac{rg(r)}{k} = \frac{1}{\alpha}\frac{\partial U}{\partial t} \qquad \text{in} \quad a < r < b, \quad t > 0 \tag{4-69a}$$

$$-\frac{\partial U}{\partial r} + \left(H_1 + \frac{1}{a}\right)U = af_1 \qquad \text{at} \quad r = a, \quad t > 0 \tag{4-69b}$$

$$\frac{\partial U}{\partial r} + \left(H_2 - \frac{1}{b}\right)U = bf_2 \qquad \text{at} \quad r = b, \quad t > 0 \tag{4-69c}$$

$$U = rF(r) \qquad \text{for} \quad t = 0, \quad \text{in } a \leqslant r \leqslant b \tag{4-69d}$$

If a shift in the space coordinate is introduced as

$$x = r - a \tag{4-70}$$

the system (4-69) becomes

$$\frac{\partial^2 U}{\partial x^2} + \frac{(a + x)g(x + a)}{k} = \frac{1}{\alpha}\frac{\partial U}{\partial t} \qquad \text{in} \quad 0 < x < L, \quad t > 0 \tag{4-71a}$$

$$-\frac{\partial U}{\partial x} + K_1 U = af_1 \qquad \text{at} \quad x = 0, \quad t > 0 \tag{4-71b}$$

$$\frac{\partial U}{\partial x} + K_2 U = bf_2 \qquad \text{at} \qquad x = L, \qquad t > 0 \tag{4-71c}$$

$$U = (x + a)F(x + a) \qquad \text{for} \qquad t = 0, \qquad \text{in} \quad 0 \leqslant x \leqslant L \tag{4-71d}$$

where

$$K_1 = H_1 + \frac{1}{a}, \qquad K_2 = H_2 - \frac{1}{b}, \qquad \text{and} \qquad L = b - a \tag{4-72}$$

which is a problem of heat conduction for a slab, $0 \leqslant x \leqslant L$, and can be solved by the techniques described in Chapter 2. We illustrate the application with examples given below.

Example 4-2

A hollow sphere $a \leqslant r \leqslant b$ is initially at a temperature $F(r)$, for times $t > 0$ heat is dissipated by convection from the boundaries at $r = a$ and $r = b$ into an environment at zero temperature. Obtain an expression for the temperature distribution $T(r, t)$ in the sphere for times $t > 0$.

Solution. This problem is a special case of the problem (4-67) with $g(r) = 0$ and $f_1 = f_2 = 0$. Therefore the governing problem is obtained from equations (4-67) by setting

$$g(r) = 0, \qquad f_1 = 0, \qquad \text{and} \qquad f_2 = 0 \tag{4-73}$$

Then the transformation of the resulting problem by the transformations

$$U(r, t) = rT(r, t) \qquad \text{and} \qquad x = r - a \tag{4-74}$$

leads to a system similar to (4-71), with $g = 0$, $f_1 = 0$, and $f_2 = 0$, which is a homogeneous heat conduction problem for a slab and can readily be solved by the method described in Chapter 2. When the solution for $U(x, t)$ is obtained and transformed back to $T(r, t)$ by the transformations (4-70) and (4-68), the solution for the considered sphere problem becomes

$$T(r, t) = \frac{1}{r} \sum_{m=1}^{\infty} e^{-\alpha\beta_m^2 t} \frac{1}{N(\beta_m)} R(\beta_m, r) \int_{r'=a}^{b} r'F(r')R(\beta_m, r')dr' \tag{4-75}$$

where the $R(\beta_m, r)$, $N(\beta_m)$, and β_m values are obtained from Table 2-2, case 1, as

$$R(\beta_m, r) = \beta_m \cos \beta_m(r - a) + K_1 \sin \beta_m(r - a) \tag{4-76a}$$

$$\frac{1}{N(\beta_m)} = \frac{2}{(\beta_m^2 + K_1^2)[(b - a) + K_2/(\beta_m^2 + K_2^2)] + K_1} \tag{4-76b}$$

where

$$K_1 \equiv H_1 + \frac{1}{a}, \qquad K_2 = H_2 - \frac{1}{b} \tag{4-76c}$$

and the β_m values are the positive roots of

$$\tan \beta_m(b-a) = \frac{\beta_m(K_1 + K_2)}{\beta_m^2 - K_1 K_2} \tag{4-76d}$$

Insulated Boundaries

When both boundaries at $r = a$ and $r = b$ are insulated, we have $H_1 = H_2 = 0$. For this special case $\beta_0 = 0$ is also an eigenvalue. Then the term

$$\frac{3}{b^3 - a^3} \int_a^b r^2 F(r) dr$$

resulting from the zero eigenvalue should be added on the right-hand side of equation (4-75). This term implies that, after the transients have passed, the steady-state temperature in the medium is the mean of the initial temperature distribution $F(r)$ over the volume of the insulated sphere.

Example 4-3

A hollow sphere, $a \leqslant r \leqslant b$, is initially at temperature $F(r)$. For times $t > 0$, the boundaries at $r = a$ and $r = b$ are kept at zero temperature. Obtain an expression for the temperature distribution $T(r, t)$ in the sphere for times $t > 0$.

Solution. The mathematical formulation of the problem is given as

$$\frac{1}{r} \frac{\partial^2}{\partial r^2}(rT) = \frac{1}{\alpha} \frac{\partial T}{\partial t} \quad \text{in} \quad a < r < b, \quad t > 0 \tag{4-77a}$$

$$T = 0 \quad \text{at} \quad r = a \text{ and } r = b, \quad t > 0 \tag{4-77b}$$

$$T = F(r) \quad \text{for} \quad t = 0, \quad \text{in} \quad a \leqslant r \leqslant b \tag{4-77c}$$

This system is now transformed successively by the application of the transformations

$$U(r, t) = rT(r, t) \quad \text{and} \quad x = r - a \tag{4-78}$$

We obtain

$$\frac{\partial^2 U}{\partial x^2} = \frac{1}{\alpha} \frac{\partial U}{\partial t} \quad \text{in} \quad 0 < x < (b-a), \quad t > 0 \tag{4-79a}$$

$$U = 0 \qquad \text{at} \qquad x = 0 \text{ and } x = b - a, t > 0 \tag{4-79b}$$

$$U = (x + a)F(x + a) \qquad \text{for} \qquad t = 0, \qquad \text{in } 0 \leqslant x \leqslant b - a \tag{4-79c}$$

This equation for $U(x, t)$ is readily solved as described in Chapter 2 and after the transformation to $T(r, t)$ according to equation (4-78) we find

$$T(r, t) = \frac{2}{r(b-a)} \sum_{m=1}^{\infty} e^{-\alpha\beta_m^2 t} \sin \beta_m(r - a) \int_{r'=a}^{b} r'F(r') \sin \beta_m(r' - a)dr' \tag{4-80a}$$

where the β_m values are the roots of

$$\sin \beta_m(b - a) = 0 \tag{4-80b}$$

or

$$\beta_m = \frac{m\pi}{b - a}, \qquad m = 1, 2, 3 \ldots \tag{4-80c}$$

4-6 HOMOGENEOUS PROBLEMS IN (r, μ, t) VARIABLES

In this section we illustrate with examples the application of the method of the separation of variables to the solution of homogeneous heat conduction problems involving (r, μ, t) variables, for example, $T \equiv T(r, \mu, t)$.

Problem of a Full Sphere

Example 4-4

Obtain an expression for the temperature distribution $T(r, \mu, t)$ in a solid sphere, $-1 \leqslant \mu \leqslant 1, 0 \leqslant r \leqslant b$, which is initially at a temperature $F(r, \mu)$ and for times $t > 0$ boundary surface at $r = b$ is kept at zero temperature

Solution. The mathematical formulation of this problem is given as

$$\frac{\partial^2 T}{\partial r^2} + \frac{2}{r}\frac{\partial T}{\partial r} + \frac{1}{r^2}\frac{\partial}{\partial \mu}\left[(1 - \mu^2)\frac{\partial T}{\partial \mu}\right] = \frac{1}{\alpha}\frac{\partial T}{\partial t}$$

$$\text{in} \qquad 0 \leqslant r < b, \quad -1 \leqslant \mu \leqslant 1, \quad t > 0 \tag{4-81a}$$

$$T = 0 \qquad \text{at} \qquad r = b, \quad \text{for} \quad t > 0 \tag{4-81b}$$

$$T = F(r, \mu) \qquad \text{for} \qquad t = 0, \quad \text{in the sphere} \tag{4-81c}$$

Defining a new dependent variable $V(r, \mu, t)$ as

$$V = r^{1/2} T \tag{4-82}$$

the problem (4-81) is transformed into [see equation (4-9)]

$$\frac{\partial^2 V}{\partial r^2} + \frac{1}{r}\frac{\partial V}{\partial r} - \frac{1}{4}\frac{V}{r^2} + \frac{1}{r^2}\frac{\partial}{\partial \mu}\left[(1-\mu^2)\frac{\partial V}{\partial \mu}\right] = \frac{1}{\alpha}\frac{\partial V}{\partial t}$$

$$\text{in} \quad 0 \leqslant r < b, \quad -1 \leqslant \mu \leqslant 1, \quad t > 0 \tag{4-83a}$$

$$V = 0 \quad \text{at} \quad r = b, \quad t > 0 \tag{4-83b}$$

$$V = r^{1/2} F(r, \mu) \quad \text{for} \quad t = 0, \quad \text{in the sphere} \tag{4-83c}$$

The elementary solutions of equation (4-83a) are given by equations (4-11). The solutions $Q_n(\mu)$ become infinite at $\mu = \pm 1$ and $Y_{n+1/2}(\lambda r)$ [or $J_{-n-1/2}(\lambda r)$] become infinite at $r = 0$; therefore they are inadmissible as solutions on the physical grounds. Then, the elementary solutions that are admissible for this problem include

$$e^{-\alpha\lambda^2 t}, \qquad J_{n+1/2}(\lambda r), \qquad \text{and} \qquad P_n(\mu)$$

where $P_n(\mu)$ is the Legendre polynomial as defined by equation (4-19) with $n = 0, 1, 2, 3, \ldots$. The complete solution for $V(r, \mu, t)$ is constructed as

$$V(r, \mu, t) = \sum_{n=0}^{\infty} \sum_{p=1}^{\infty} c_{np} e^{-\alpha\lambda_{np}^2 t} J_{n+1/2}(\lambda_{np} r) P_n(\mu) \tag{4-84}$$

where the coefficients c_{np} and the eigenvalues λ_{np} are to be so determined that the boundary condition (4-83b) and the initial condition (4-83c) are satisfied. If the λ_{np} values are taken as the positive roots of

$$J_{n+1/2}(\lambda_{np} b) = 0 \tag{4-85}$$

the boundary condition at $r = b$ is satisfied. The application of the initial condition (4-83c) gives

$$r^{1/2} F(r, \mu) = \sum_{n=0}^{\infty} \sum_{p=1}^{\infty} c_{np} J_{n+1/2}(\lambda_{np} r) P_n(\mu) \quad \text{in} \quad 0 \leqslant r < b, \quad -1 \leqslant \mu \leqslant 1 \tag{4-86}$$

To determine the coefficients c_{np} we operate on both sides of equation (4-86) successively by the operators

$$\int_{-1}^{1} P_{n'}(\mu)d\mu \qquad \text{and} \qquad \int_{0}^{b} rJ_{n+1/2}(\lambda_{np'}r)dr$$

and utilize the orthogonality relations (4-26) for the Legendre polynomials and (3-19) for the Bessel functions. We obtain

$$c_{np} = \frac{1}{N(n)N(\lambda_{np})}\int_{r=0}^{b}\int_{\mu=-1}^{1} r^{3/2}J_{n+1/2}(\lambda_{np}r)P_n(\mu)F(r,\mu)d\mu dr \tag{4-87a}$$

where the norms are defined as

$$N(n) \equiv \int_{-1}^{1} [P_n(\mu)]^2 d\mu \qquad \text{and} \qquad N(\lambda_{np}) \equiv \int_{0}^{b} rJ_{n+1/2}^2(\lambda_{np}r)dr \tag{4-87b}$$

The coefficient c_{np} as given above is introduced into equation (4-84) and the resulting expression is transformed to $T(r, \mu, t)$ by the transformation (4-82). We find

$$T(r,\mu,t) = \sum_{n=0}^{\infty}\sum_{p=1}^{\infty} \frac{1}{N(n)N(\lambda_{np})} e^{-\alpha\lambda_{np}^2 t} r^{-1/2} J_{n+1/2}(\lambda_{np}r)P_n(\mu)$$
$$\cdot \int_{r'=0}^{b}\int_{\mu'=-1}^{1} r'^{3/2}J_{n+1/2}(\lambda_{np}r')P_n(\mu')F(r',\mu')d\mu' dr' \tag{4-88}$$

where the λ_{np} values are the positive roots of

$$J_{n+1/2}(\lambda_{np}b) = 0 \tag{4-89a}$$

the norm $N(n)$ is obtained from equation (4-27) as

$$N(n) = \frac{2}{2n+1} \tag{4-89b}$$

and the norm $N(\lambda_{np})$ is obtained from equations (25) of Appendix IV by utilizing the condition (4-89a) as

$$N(\lambda_{np}) = -\frac{b^2}{2}J_{n-1/2}(\lambda_{np}b)J_{n+3/2}(\lambda_{np}b) \qquad \text{if equation (25a) of Appendix IV is used} \tag{4-89c}$$

$$= \frac{b^2}{2}[J'_{n+1/2}(\lambda_{np}b)]^2 \qquad \text{if equation (25b) of Appendix IV is used} \tag{4-89d}$$

and the n values are positive integers, zero being included.

We note that the eigenfunctions $J_{n+1/2}(\lambda_{np}r)$, the norm $N(\lambda_{np})$ given by equation (4-89d), and the expression (4-89a) for the eigenvalues λ_{np} are the same as those obtainable from Table 3-1, case 3, with $v = n + \frac{1}{2}$. Therefore, Tables 3-1 and 3-3 are useful for the determination of the eigenvalues, the eigenfunctions, and the norms associated with the r variable in the solution of equation (4-83a) for a solid and hollow sphere, respectively.

Problem of a Hemisphere

Example 4-5

Obtain an expression for the temperature distribution $T(r, \mu, t)$ in a solid hemisphere, $0 \leqslant \mu \leqslant 1, 0 \leqslant r \leqslant b$, which is initially at a temperature $F(r, \mu)$ and for times $t > 0$ the boundary surfaces at $r = b$ and $\mu = 0$ are kept at zero temperature as illustrated in Fig. 4-6.

Solution. The mathematical formulation of this problem is given as

$$\frac{\partial^2 T}{\partial r^2} + \frac{2}{r}\frac{\partial T}{\partial r} + \frac{1}{r^2}\frac{\partial}{\partial \mu}\left[(1-\mu^2)\frac{\partial T}{\partial \mu}\right] = \frac{1}{\alpha}\frac{\partial T}{\partial t}$$

$$\text{in} \quad 0 \leqslant r < b, \quad 0 < \mu \leqslant 1, \quad \text{for} \quad t > 0 \tag{4-90a}$$

$$T = 0 \quad \text{at} \quad r = b \quad \text{and} \quad \mu = 0, \quad \text{for} \quad t > 0 \tag{4-90b}$$

$$T = F(r, \mu) \quad \text{for} \quad t = 0, \text{ in the hemisphere} \tag{4-90c}$$

A new variable $V(r, \mu, t)$ is defined as

$$V = r^{1/2} T \tag{4-91}$$

Then, the problem (4-90) is transformed into

$$\frac{\partial^2 V}{\partial r^2} + \frac{1}{r}\frac{\partial V}{\partial r} - \frac{1}{4}\frac{V}{r^2} + \frac{1}{r^2}\frac{\partial}{\partial \mu}\left[(1-\mu^2)\frac{\partial V}{\partial \mu}\right] = \frac{1}{\alpha}\frac{\partial V}{\partial t}$$

$$\text{in} \quad 0 \leqslant r < b, \quad 0 < \mu \leqslant 1, \quad \text{for} \quad t > 0 \tag{4-92a}$$

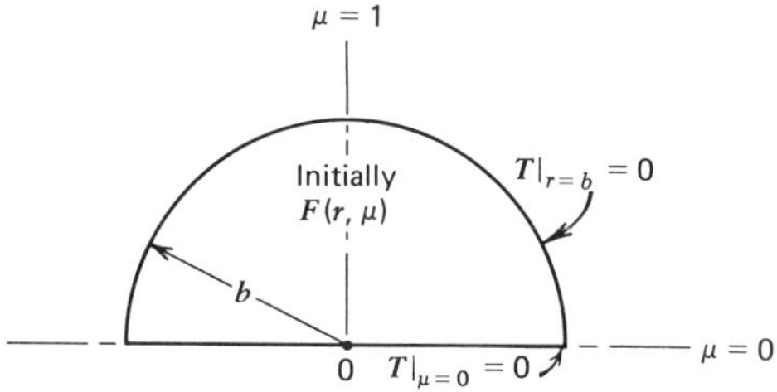

Fig. 4-6 Boundary and initial conditions for a hemisphere in Example 4-5.

$$V = 0 \qquad \text{at} \qquad r = b \quad \text{and} \quad \mu = 0, \quad \text{for} \quad t > 0 \tag{4-92b}$$

$$V = r^{1/2} F(r, \mu) \qquad \text{for} \qquad t = 0, \quad \text{in the hemisphere} \tag{4-92c}$$

The elementary solutions of equation (4-92a) that are admissible for this problem are

$$e^{-\alpha\lambda^2 t}, \qquad J_{n+1/2}(\lambda r), \qquad \text{and} \qquad P_n(\mu) \tag{4-93}$$

where $P_n(\mu)$ is the Legendre polynomial as defined by equation (4-19). To satisfy the boundary condition of the first kind at the surface $\mu = 0$ (i.e., the base of the hemisphere) the degree n should be taken as *odd positive integer* (i.e., $n = 1, 3, 5, \ldots$) for reasons discussed in Section 4-4. Then, the complete solution for $V(r, \mu, t)$ is constructed as

$$V(r, \mu, t) = \sum_{n=1,3,5,\ldots}^{\infty} \sum_{p=1}^{\infty} c_{np} e^{-\alpha\lambda_{np}^2 t} J_{n+1/2}(\lambda_{np} r) P_n(\mu) \tag{4-94}$$

which satisfies the boundary condition at $\mu = 0$. It will also satisfy the boundary condition at $r = b$ if the λ_{np} values are taken as the positive roots of

$$J_{n+1/2}(\lambda_{np} b) = 0 \tag{4-95}$$

The application of the initial condition (4-92c) to the solution (4-94) gives

$$r^{1/2} F(r, \mu) = \sum_{n=1,3,5,\ldots}^{\infty} \sum_{p=1}^{\infty} c_{np} J_{n+1/2}(\lambda_{np} r) P_n(\mu) \qquad \text{in} \qquad 0 \leqslant r < b, 0 < \mu \leqslant 1 \tag{4-96}$$

To determine the coefficients c_{np} we operate on both sides of equation (4-96) successively by the operators

$$\int_0^1 P_{n'}(\mu) d\mu \qquad \text{and} \qquad \int_0^b r J_{n+1/2}(\lambda_{np'} r) dr$$

where $n' = 1, 3, 5, \ldots$ and utilize the orthogonality relations (4-31) and (3-19). We obtain

$$c_{np} = \frac{1}{N(n) N(\lambda_{np})} \int_{r=0}^{b} \int_{\mu=0}^{1} r^{3/2} J_{n+1/2}(\lambda_{np} r) P_n(\mu) F(r, \mu) d\mu dr \tag{4-97a}$$

where the norms are defined as

$$N(n) \equiv \int_0^1 [P_n(\mu)]^2 d\mu \qquad \text{and} \qquad N(\lambda_{np}) \equiv \int_0^b r J_{n+1/2}^2(\lambda_{np} r) dr \tag{4-97b}$$

The coefficient c_{np} as given above is introduced into equation (4-94) and the resulting expression is transformed into $T(r, \mu, t)$ by the transformation (4-91). We obtain

$$T(r, \mu, t) = \sum_{n=1,3,5,\ldots}^{\infty} \sum_{p=1}^{\infty} \frac{1}{N(n)N(\lambda_{np})} e^{-\alpha\lambda_{np}^2 t} r^{-1/2} J_{n+1/2}(\lambda_{np} r) P_n(\mu) \cdot \int_{r'=0}^{b} \int_{\mu'=0}^{1} r'^{3/2} J_{n+1/2}(\lambda_{np} r') P_n(\mu') F(r', \mu') d\mu' dr' \quad \text{(4-98)}$$

where the eigenvalues λ_{np} are the positive roots of

$$J_{n+1/2}(\lambda_{np} b) = 0 \quad \text{(4-99a)}$$

$N(n)$ is determined from equation (4-31) as

$$N(n) = \frac{1}{2n+1} \quad \text{(4-99b)}$$

and $N(\lambda_{np})$ is obtained by using equation (25a) of Appendix IV and by utilizing the result (4-99a)

$$N(\lambda_{np}) = -\frac{b^2}{2} J_{n-1/2}(\lambda_{np} b) J_{n+3/2}(\lambda_{np} b) \quad \text{(4-99c)}$$

and the n values are odd positive integers, that is, $n = 1, 3, 5, \ldots$.

Example 4-6

Obtain an expression for the temperature distribution $T(r, \mu, t)$ in a hemisphere, $0 \leqslant \mu \leqslant 1, 0 \leqslant r \leqslant b$, which is initially at temperature $F(r, \mu)$ and for times $t > 0$ the boundary surface at $r = b$ is kept insulated and the boundary surface at $\mu = 0$ (i.e., the base of the sphere) is kept at zero temperature as illustrated in Fig. 4-7.

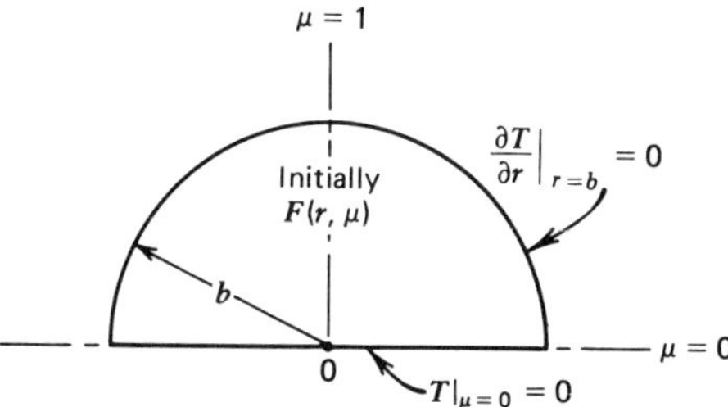

Fig. 4-7 Boundary and initial conditions for a hemisphere in Example 4-6.

Solution. This problem is similar to the one considered in Example 4-5 except the boundary surface at $r = b$ is now insulated. The differential equation is the same as that equation (4-90a); therefore we give only the boundary and initial conditions

$$T = 0 \quad \text{at} \quad \mu = 0 \,(\text{the base}), \quad t > 0 \tag{4-100a}$$

$$\frac{\partial T}{\partial r} = 0 \quad \text{at} \quad r = b \,(\text{spherical surface}), \quad t > 0 \tag{4-100b}$$

$$T = F(r, \mu) \quad \text{for} \quad t = 0, \quad \text{in the hemisphere} \tag{4-100c}$$

If this problem is transformed with the transformation (4-91), the differential equation is transformed to that given by equation (4-92a) and the boundary and initial conditions (4-99) are transformed into

$$V = 0 \quad \text{at} \quad \mu = 0 \,(\text{the base}), \quad t > 0 \tag{4-101a}$$

$$\frac{\partial V}{\partial r} - \frac{1}{2b} V = 0 \qquad r = b, \quad t > 0 \tag{4-101b}$$

$$V = r^{1/2} F(r, \mu) \quad \text{for} \quad t = 0, \quad \text{in the hemisphere} \tag{4-101c}$$

The elementary solutions of this problem are the same as those given by equation (4-93) and the complete solution for $V(r, \mu, t)$ is constructed as

$$V(r, \mu, t) = \sum_{n=1,3,5,\ldots}^{\infty} \sum_{p=1}^{\infty} c_{np} e^{-\alpha \lambda_{np}^2 t} J_{n+1/2}(\lambda_{np} r) P_n(\mu) \tag{4-102}$$

If the eigenvalues λ_{np} are taken as the positive roots of

$$\frac{d}{dr} J_{n+1/2}(\lambda_{np} b) - \frac{1}{2b} J_{n+1/2}(\lambda_{np} b) = 0 \tag{4-103a}$$

or

$$\lambda_{np} J'_{n+1/2}(\lambda_{np} b) - \frac{1}{2b} J_{n+1/2}(\lambda_{np} b) = 0 \tag{4-103b}$$

the boundary condition (4-101b) at $r = b$ is satisfied. The application of the initial condition (4-101c) to equation (4-102) gives

$$r^{1/2} F(r, \mu) = \sum_{n=1,3,5,\ldots}^{\infty} \sum_{p=1}^{\infty} c_{np} J_{n+1/2}(\lambda_{np} r) P_n(\mu) \tag{4-104}$$

To determine the coefficients c_{np} we operate on both sides of this equation successively by the operators

$$\int_0^1 P_{n'}(\mu)d\mu \qquad \text{and} \qquad \int_0^b rJ_{n+1/2}(\lambda_{np'}r)dr$$

where $n' = 1, 3, 5, \ldots$ and utilize the orthogonality relations (4-30), (4-31) and (3-19). We obtain

$$c_{np} = \frac{1}{N(n)N(\lambda_{np})} \int_{r=0}^b \int_{\mu=0}^1 r^{3/2} J_{n+1/2}(\lambda_{np}r)P_n(\mu)F(r,\mu)d\mu\, dr \qquad \text{(4-105a)}$$

where

$$N(n) \equiv \int_0^1 [P_n(\mu)]^2 d\mu \qquad \text{and} \qquad N(\lambda_{np}) \equiv \int_0^b rJ^2_{n+1/2}(\lambda_{np}r)dr \qquad \text{(4-105b)}$$

The coefficient c_{np} as given above is introduced into equation (4-102) and the resulting solution is transformed into $T(r, \mu, t)$ by the transformation (4-91). We obtain

$$T(r,\mu,t) = \sum_{n=1,3,5,\ldots}^{\infty} \sum_{p=1}^{\infty} \frac{1}{N(n)N(\lambda_{np})} e^{-\alpha\lambda_{np}^2 t} r^{-1/2} J_{n+1/2}(\lambda_{np}r)P_n(\mu)$$
$$\cdot \int_{r'=0}^b \int_{\mu'=0}^1 r'^{3/2} J_{n+1/2}(\lambda_{np}r')P_n(\mu')F(r',\mu')d\mu'\, dr' \qquad \text{(4-106)}$$

where the eigenvalues λ_{np} are the positive roots of

$$\lambda_{np}J'_{n+1/2}(\lambda_{np}b) - \frac{1}{2b} J_{n+1/2}(\lambda_{np}b) = 0 \qquad \text{(4-107a)}$$

the norm $N(n)$ is obtained from equation (4-31) as

$$N(n) = \frac{1}{2n+1} \qquad \text{(4-107b)}$$

The norm $N(\lambda_{np})$ is determined by using equation (25a) of Appendix IV as

$$N(\lambda_{np}) = \tfrac{1}{2}b^2[J^2_{n+1/2}(\lambda_{np}b) - J_{n-1/2}(\lambda_{np}b)J_{n+3/2}(\lambda_{np}b)] \qquad \text{(4-107c)}$$

and the n values are *odd positive integers* (i.e., $n = 1, 3, 5, \ldots$).

We note that the general form of the solution (4-106) is exactly the same as that of equation (4-98) of the previous example except the expressions defining the eigenvalues λ_{np} and the norm $N(\lambda_{np})$ are different. The eigenfunctions $J_{n+1/2}(\lambda_{np}r)$, the eigenvalues λ_{np}, and the norm $N(\lambda_{np})$ if determined by using equation (25b) of Appendix IV are the same as those obtainable from Table 3-1, case 1, by setting

$$\nu = n + \tfrac{1}{2}, H = -\frac{1}{2b} \qquad \text{and} \qquad \lambda_{np} = \beta_m.$$

All Boundaries Insulated

When all the boundary surfaces of the region are insulated, the analysis is performed in a manner similar to those illustrated above with other boundary conditions; but for this special case $\lambda_{0,0} = 0$ is also an eigenvalue. Then the term

$$\frac{\int_{\text{Region}} F(r,\mu) r^2 \, dr \, d\mu}{\int_{\text{Region}} r^2 \, dr \, d\mu}$$

that results from the zero eigenvalue should be added to the solution. This term implies that, after the transients have passed, the steady-state temperature in the medium is the mean of the initial distribution $F(r,\mu)$ over the volume of the insulated region.

4-7 HOMOGENEOUS PROBLEMS IN (r, μ, ϕ, t) VARIABLES

In this section we illustrate with examples the solution of the homogeneous problems of heat conduction involving (r, μ, ϕ, t) variables.

Example 4-7

A solid sphere of radius $r = b$ is initially at temperature $F(r, \mu, \phi)$. For times $t > 0$ the boundary surface at $r = b$ is kept at zero temperature. Obtain an expression for the temperature distribution $T(r, \mu, \phi, t)$ in the sphere for times $t > 0$.

Solution. The mathematical formulation of this problem is given as

$$\frac{\partial^2 T}{\partial r^2} + \frac{2}{r}\frac{\partial T}{\partial r} + \frac{1}{r^2}\frac{\partial}{\partial \mu}\left[(1-\mu^2)\frac{\partial T}{\partial \mu}\right] + \frac{1}{r^2(1-\mu^2)}\frac{\partial^2 T}{\partial \phi^2} = \frac{1}{\alpha}\frac{\partial T}{\partial t}$$

$$\text{in} \quad 0 \leqslant r < b, -1 \leqslant \mu \leqslant 1, 0 \leqslant \phi \leqslant 2\pi \text{ for } t > 0 \tag{4-108a}$$

$$T = 0 \qquad \text{at} \quad r = b, \text{ for } t > 0 \tag{4-108b}$$

$$T = F(r, \mu, \phi) \qquad \text{for} \quad t = 0, \text{ in the sphere} \tag{4-108c}$$

A new variable $V(r, \mu, \phi, t)$ is defined as

$$V = r^{1/2} T \tag{4-109}$$

The problem (4-108) is transformed into

$$\frac{\partial^2 V}{\partial r^2} + \frac{1}{r}\frac{\partial V}{\partial r} - \frac{1}{4}\frac{V}{r^2} + \frac{1}{r^2}\frac{\partial}{\partial \mu}\left[(1-\mu^2)\frac{\partial V}{\partial \mu}\right] + \frac{1}{r^2(1-\mu^2)}\frac{\partial^2 V}{\partial \phi^2} = \frac{1}{\alpha}\frac{\partial V}{\partial t}$$

$$\text{in} \quad 0 \leqslant r < b, \quad -1 \leqslant \mu \leqslant 1, \quad 0 \leqslant \phi \leqslant 2\pi, \quad \text{for} \quad t > 0 \tag{4-110a}$$

$$V = 0 \quad \text{at} \quad r = b, \text{ for } t > 0 \tag{4-110b}$$

$$V = r^{1/2} F(r, \mu, \phi) \quad \text{for} \quad t = 0, \text{ in the sphere} \tag{4-110c}$$

The elementary solution of equation (4-110a) that are admissible on the physical grounds are [see equations (4-8)]

$$e^{-\alpha\lambda^2 t}, \qquad J_{n+1/2}(\lambda r), \qquad P_n^m(\mu), \qquad (A\cos m\phi + B\sin m\phi) \tag{4-111}$$

where $P_n^m(\mu)$ is the associated Legendre function of the first kind defined by equation (4-25), with n and m being positive integers (i.e., $n, m = 0, 1, 2, 3, \ldots$) and $m \leqslant n$. The choice of m as positive integer satisfies the requirement that the temperature T (or V) is periodic with period 2π in the interval $0 \leqslant \phi \leqslant 2\pi$. Then the complete solution for $V(r, \mu, \phi, t)$ is constructed as

$$V(r, \mu, \phi, t) = \sum_{n=0}^{\infty} \sum_{p=1}^{\infty} \sum_{m=0}^{n} e^{-\alpha\lambda_{np}^2 t} J_{n+1/2}(\lambda_{np} r) P_n^m(\mu) \cdot (A_{mnp}\cos m\phi + B_{mnp}\sin m\phi) \tag{4-112}$$

This solution satisfies the differential equation (4-110a) and remains bounded in the region $-1 \leqslant \mu \leqslant 1, 0 \leqslant \phi \leqslant 2\pi$, and $0 \leqslant r \leqslant b$. If the eigenvalues λ_{np} are chosen as the positive roots of

$$J_{n+1/2}(\lambda_{np} b) = 0 \tag{4-113}$$

it also satisfies the boundary condition at $r = b$. The expansion coefficients A_{mnp} and B_{mnp} are to be determined so that the initial condition for the problem is satisfied. The application of the initial condition (4-110c) to equation (4-112) gives

$$r^{1/2} F(r, \mu, \phi) = \sum_{n=0}^{\infty} \sum_{p=1}^{\infty} \sum_{m=0}^{n} J_{n+1/2}(\lambda_{np} r) P_n^m(\mu)(A_{mnp}\cos m\phi + B_{mnp}\sin m\phi)$$

$$\text{in } -1 \leqslant \mu \leqslant 1, 0 \leqslant \phi \leqslant 2\pi, 0 \leqslant r \leqslant b \tag{4-114}$$

To determine the expansion coefficients we operate on both sides of equation (4-114) by the operator

$$\int_0^b rJ_{n+1/2}(\lambda_{np'}r)dr \tag{4-115}$$

and utilize the orthogonality relation (3-19). We obtain

$$f(\mu,\phi) = \sum_{n=0}^{\infty}\sum_{m=0}^{n} P_n^m(\mu)[A_{mnp}\cos m\phi + B_{mnp}\sin m\phi]N(\lambda_{np})$$

$$\text{in } -1 \leqslant \mu \leqslant 1, 0 \leqslant \phi \leqslant 2\pi \tag{4-116a}$$

where

$$f(\mu,\phi) \equiv \int_0^b r^{3/2}F(r,\mu,\phi)J_{n+1/2}(\lambda_{np}r)dr \tag{4-116b}$$

$$N(\lambda_{np}) \equiv \int_0^b rJ_{n+1/2}^2(\lambda_{np}r)dr = -\frac{b^2}{2}J_{n-1/2}(\lambda_{np}b)J_{n+3/2}(\lambda_{np}b) \tag{4-116c}$$

Here we utilized equation (25a) of Appendix IV together with equation (4-113) to evaluate the norm $N(\lambda_{np})$.

Equation (4-116a) is a representation of a function $f(\mu,\phi)$ defined in the interval $-1 \leqslant \mu \leqslant 1, 0 \leqslant \phi \leqslant 2\pi$ in terms of $P_n^m(\mu)$, $\sin m\phi$ and $\cos m\phi$. Such an expansion was considered previously in equation (4-38) and the expansion coefficients were given by equation (4-44). Therefore, the expansion coefficients in equation (4-116a) are obtained from equation (4-44) as

$$N(\lambda_{np})[A_{mnp}\cos m\phi + B_{mnp}\sin n\phi] \equiv \frac{1}{\pi N(m,n)}\int_{\phi'=0}^{2\pi}\int_{\mu'=-1}^{1} f(\mu',\phi')$$

$$\cdot P_n^m(\mu')\cos[m(\phi-\phi')]d\mu' d\phi' \tag{4-117a}$$

where π *should be replaced by* 2π *for* $m=0$, and the norm $N(m,n)$ is as given by equation (4-40):

$$N(m,n) \equiv \int_{-1}^{1} [P_n^m(\mu)]^2 d\mu = \left(\frac{2}{2n+1}\right)\frac{(n+m)!}{(n-m)!} \tag{4-117b}$$

The coefficients given by equations (4-117) are introduced into equation (4-112) and the resulting expression is transformed into $T(r,\mu,\phi,t)$ by the transformation

(4-109). We obtain

$$
\begin{aligned}
T(r,\mu,\phi,t) = {} & \frac{1}{\pi}\sum_{n=0}^{\infty}\sum_{p=1}^{\infty}\sum_{m=0}^{n}\frac{e^{-\alpha\lambda_{np}^2 t}}{N(m,n)N(\lambda_{np})} r^{-1/2}J_{n+1/2}(\lambda_{np}r)P_n^m(\mu) \\
& \cdot \int_{r'=0}^{b}\int_{\mu'=-1}^{1}\int_{\phi'=0}^{2\pi} r'^{3/2}J_{n+1/2}(\lambda_{np}r')P_n^m(\mu')\cos m(\phi-\phi') \\
& \cdot F(r',\mu',\phi')d\phi'\,d\mu'\,dr'
\end{aligned}
\tag{4-118}
$$

where π *should be replaced by* 2π *for* $m = 0$, and the eigenvalues λ_{np} are the positive roots of

$$
J_{n+1/2}(\lambda_{np}b) = 0 \tag{4-119}
$$

the norms $N(m,n)$ and $N(\lambda_{np})$ are given by

$$
N(m,n) = \left(\frac{2}{2n+1}\right)\frac{(n+m)!}{(n-m)!} \tag{4-120}
$$

$$
N(\lambda_{np}) = -\frac{b^2}{2}J_{n-1/2}(\lambda_{np}b)J_{n+3/2}(\lambda_{np}r) \tag{4-121}
$$

and n, m are positive integers, zero being included.

4-8 MULTIDIMENSIONAL STEADY-STATE PROBLEMS

In this section we illustrate with examples the solution of multidimensional, steady-state heat conduction equation with no heat generation subject to only one nonhomogeneous boundary condition by the method of separation of variables.

Example 4-8

Determine the steady-state temperature distribution $T(r,\mu,\phi)$ in a solid sphere of radius $r = b$ with its boundary surface at $r = b$ is kept at temperature $f(\mu,\phi)$.

Solution. The mathematical formulation of this problem is given as

$$
\frac{\partial^2 T}{\partial r^2} + \frac{2}{r}\frac{\partial T}{\partial r} + \frac{1}{r^2}\frac{\partial}{\partial \mu}\left[(1-\mu^2)\frac{\partial T}{\partial \mu}\right] + \frac{1}{r^2(1-\mu^2)}\frac{\partial^2 T}{\partial \phi^2} = 0
$$

$$
\text{in } 0 \leqslant r < b, \quad -1 \leqslant \mu \leqslant 1, \quad 0 \leqslant \phi \leqslant 2\pi \tag{4-122a}
$$

$$
T = f(\mu,\phi) \qquad \text{at} \qquad r = b \tag{4-122b}
$$

The elementary solutions of equation (4-122a) are given by equations (4-13d, e, f). The solutions r^{-n-1} and $Q_n^m(\mu)$ are to be excluded, because the former becomes infinite at $r = 0$ and the latter at $\mu = \pm 1$. Then the solutions that are admissible include

$$r^n, \quad P_n^m(\mu), \qquad \text{and} \qquad (A \cos m\phi + B \sin m\phi) \tag{4-123}$$

where $P_n^m(\mu)$ is the associated Legendre function of the first kind as defined by equation (4-25), with n and m being positive integers (i.e., $n, m = 0, 1, 2 \ldots$) and $m \leqslant n$. The choice of m as positive integer satisfies the requirement that the temperature T is periodic with period 2π in the region $0 \leqslant \phi \leqslant 2\pi$. The complete solution for $T(r, \mu, \phi)$ is constructed as

$$T(r, \mu, \phi) = \sum_{n=0}^{\infty} \sum_{m=0}^{n} r^n P_n^m(\mu)(A_{mn} \cos m\phi + B_{mn} \sin m\phi) \tag{4-124}$$

which satisfies the differential equation (4-122a) and remains finite in the region $-1 \leqslant \mu \leqslant 1$, $0 \leqslant \phi \leqslant 2\pi$, $0 \leqslant r < b$. The coefficients A_{mn} and B_{mn} are to be determined so that the boundary condition at $r = b$ is satisfied. The application of the boundary condition (4-122b) gives

$$f(\mu, \phi) = \sum_{n=0}^{\infty} \sum_{m=0}^{n} P_n^m(\mu)(A_{mn} \cos m\phi + B_{mn} \sin m\phi) b^n$$

$$\text{in} \quad -1 \leqslant \mu \leqslant 1, \; 0 \leqslant \phi \leqslant 2\pi \tag{4-125}$$

Equation (4-125) is a representation of a function $f(\mu, \phi)$ defined in the interval $-1 \leqslant \mu \leqslant 1, 0 \leqslant \phi \leqslant 2\pi$ in terms of the functions $P_n^m(\mu)$, $\sin m\phi$ and $\cos m\phi$. Such a representation was considered previously in equation (4-38), and the expansion coefficients were given by equation (4-44). Therefore, the coefficients in equation (4-125) are obtained from equation (4-44) as

$$[A_{mn} \cos m\phi + B_{mn} \sin m\phi] \equiv \frac{1}{\pi N(m, n)} \int_{\phi'=0}^{2\pi} \int_{\mu'=-1}^{1} \frac{f(\mu', \phi')}{b^n} P_n^m(\mu') \cdot \cos[m(\phi - \phi')] \, d\mu' \, d\phi' \tag{4-126a}$$

where π *should be replaced by* 2π *for* $m = 0$, and the norm $N(m, n)$ is as given by equation (4-40):

$$N(m, n) \equiv \int_{-1}^{1} [P_n^m(\mu)]^2 \, d\mu = \frac{2}{2n+1} \frac{(n+m)!}{(n-m)!} \tag{4-126b}$$

When the coefficients given by equation (4-126a) are introduced into equation

(4-124) the solution becomes

$$T(r, \mu, \phi) = \frac{1}{\pi} \sum_{n=0}^{\infty} \sum_{m=0}^{n} \frac{2n+1}{2} \frac{(n-m)!}{(n+m)!} \left(\frac{r}{b}\right)^n P_n^m(\mu)$$

$$\cdot \int_{\phi'=0}^{2\pi} \int_{\mu'=-1}^{1} P_n^m(\mu') \cos m(\phi - \phi') f(\mu', \phi') \, d\mu' \, d\phi' \quad (4\text{-}127)$$

where π *should be replaced by* 2π *for* $m = 0, m = 0, 1, 2, 3, \ldots, n = 0, 1, 2, 3, \ldots,$ $m \leqslant n$.

Example 4-9

Determine the steady-state temperature distribution $T(r, \mu)$ in a solid hemisphere of radius $r = b$, in the region $0 \leqslant r \leqslant b, 0 \leqslant \mu \leqslant 1$, with its spherical surface at $r = b$ kept at temperature $f(\mu)$ and its base at $\mu = 0$ is insulated as illustrated in Fig. 4-8.

Solution. The mathematical formulation of the problem is given as

$$\frac{\partial^2 T}{\partial r^2} + \frac{2}{r} \frac{\partial T}{\partial r} + \frac{1}{r^2} \frac{\partial}{\partial \mu}\left[(1 - \mu^2) \frac{\partial T}{\partial \mu}\right] = 0 \quad \text{in} \quad 0 \leqslant r < b, 0 < \mu \leqslant 1 \qquad (4\text{-}128\text{a})$$

$$\frac{\partial T}{\partial \mu} = 0 \qquad \text{at} \quad \mu = 0 \qquad (4\text{-}128\text{b})$$

$$T = f(\mu) \qquad \text{at} \quad r = b \qquad (4\text{-}128\text{c})$$

The elementary solutions of equation (4-128a) are given by equations (4-15c,d). The solutions r^{-n-1} and $Q_n(\mu)$ are inadmissible, because the former becomes infinite at $r = 0$ and the latter becomes infinite at $\mu = 1$. Then, the elementary solutions that are admissible are taken as

$$r^n \quad \text{and} \quad P_n(\mu) \qquad (4\text{-}129)$$

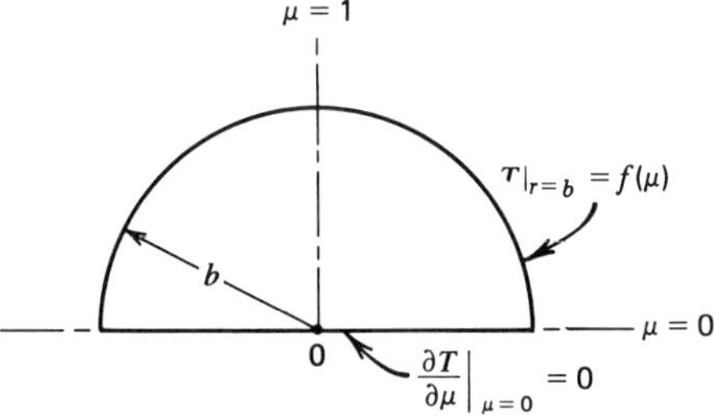

Fig. 4-8 Boundary conditions for a hemisphere in Example 4-9.

where $P_n(\mu)$ is the Legendre polynomial as defined by equations (4-19). The complete solution for $T(r, \mu)$ is constructed as

$$T(r, \mu) = \sum_n c_n r^n P_n(\mu) \tag{4-130}$$

The application of the boundary condition at $r = b$ gives

$$f(\mu) = \sum_n c_n b^n P_n(\mu) \qquad \text{in} \qquad 0 \leqslant \mu \leqslant 1 \tag{4-131}$$

This is a representation of a function $f(\mu)$ defined in the interval $0 \leqslant \mu \leqslant 1$ in terms of the Legendre polynomials. Such a representation was considered previously in equation (4-46) and the expansion coefficients were given by equation (4-49). Therefore, the coefficients c_n in equation (4-131) are readily obtained from equations (4-49) as

$$c_n = \frac{1}{N(n)} \int_0^1 \frac{f(\mu')}{b^n} P_n(\mu')\, d\mu' \tag{4-132a}$$

where

$$N(n) \equiv \int_0^1 [P_n(\mu)]^2\, d\mu = \frac{1}{2n+1} \tag{4-132b}$$

$$n = 0, 2, 4 \ldots \tag{4-132c}$$

for the boundary condition of the *second kind* at $\mu = 0$. Introducing c_n into equation (4-130) the solution becomes

$$T(r, \mu) = \sum_{n=0,2,4,\ldots}^{\infty} (2n+1)\left(\frac{r}{b}\right)^n P_n(\mu) \int_{\mu'=0}^1 f(\mu') P_n(\mu')\, d\mu' \tag{4-133}$$

or if n is replaced by $2n$, this result is written as

$$T(r, \mu) = \sum_{n=0}^{\infty} (4n+1)\left(\frac{r}{b}\right)^{2n} P_{2n}(\mu) \int_{\mu'=0}^1 f(\mu') P_{2n}(\mu')\, d\mu' \tag{4-134}$$

in equation (4-134) we have $n = 0, 1, 2, 3, \ldots$

4-9 TRANSIENT-TEMPERATURE CHARTS

Transient-temperature charts similar to those considered for a slab and a long solid cylinder can also be constructed for the case of a solid sphere by solving

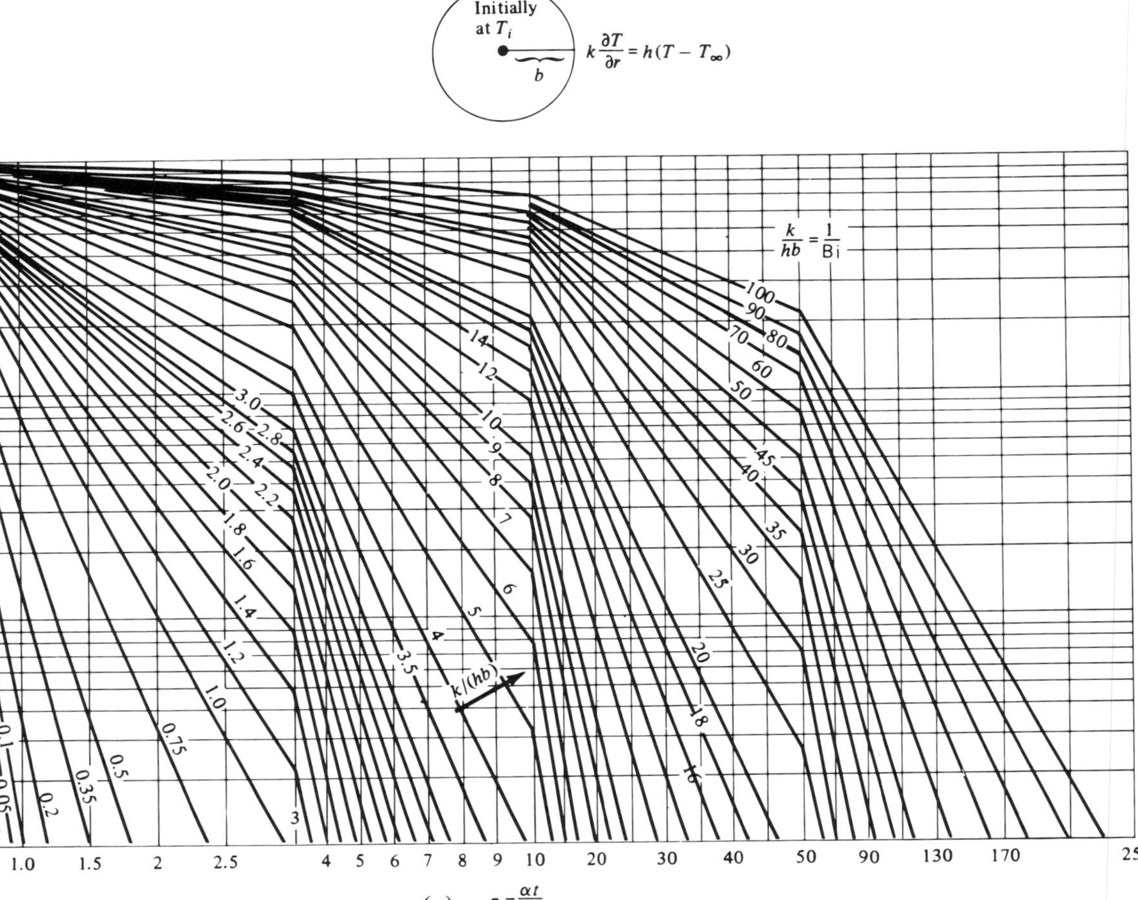

(*a*)

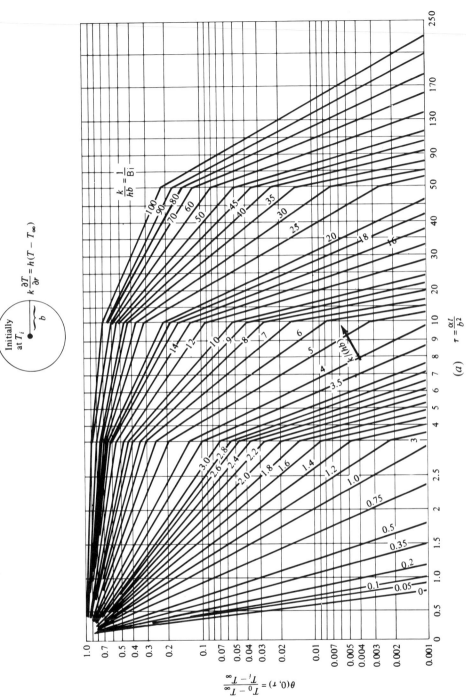

Fig. 4-9 Transient-temperature chart for a solid sphere subjected to convection. (From Heisler [15]). (*a*) Dimensionless temperature $\theta(0, \tau)$ at the center; (*b*) position correction chart for use with part *a*.

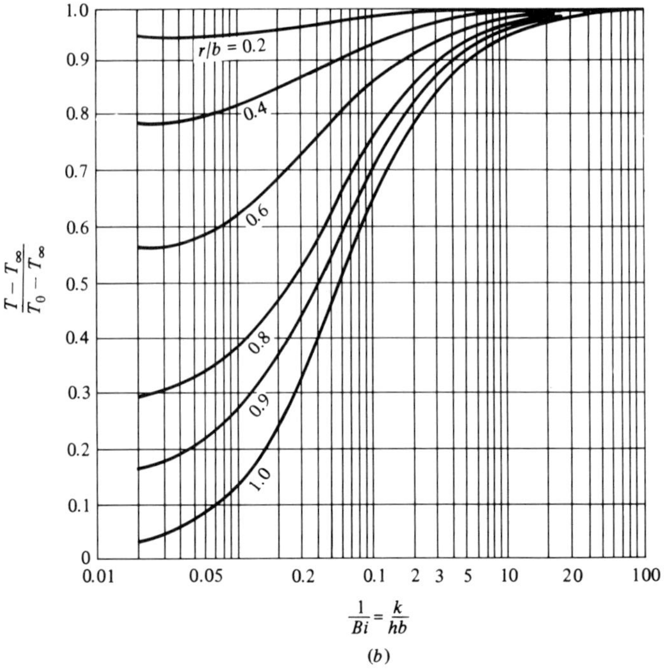

(*b*)

Fig. 4-9 (*Continued*)

the following transient heat conduction problem

$$\frac{1}{R^2}\frac{\partial}{\partial R}\left(R^2\frac{\partial \theta}{\partial R}\right) = \frac{\partial \theta}{\partial \tau} \quad \text{in} \quad 0 < R < 1, \quad \text{for} \quad \tau > 0 \tag{4-135a}$$

$$\frac{\partial \theta}{\partial R} = 0 \quad \text{at} \quad R = 0, \quad \text{for} \quad \tau > 0 \tag{4-135b}$$

$$\frac{\partial \theta}{\partial R} + \text{Bi}\,\theta = 0 \quad \text{at} \quad R = 1, \quad \text{for} \quad \tau = 0 \tag{4-135c}$$

$$\theta = 1 \quad \text{in} \quad 0 \leqslant R \leqslant 1, \quad \text{for} \quad \tau = 0 \tag{4-135d}$$

Here, the dimensionless parameters Bi, τ, θ, and R are as defined by equations 3-199a,b,c and d, respectively. In this case b is the sphere radius. The results are shown in Fig. 4-9*a*,*b*. Figure 4-9*a* shows the dimensionless center temperature $\theta(0, \tau)$ for the sphere as a function of dimensionless time τ for several different values of the parameter 1/Bi. Figure 4-9*b* relates the temperatures at different locations within the sphere to the center temperature $\theta(0, \tau)$. The use of these charts is similar to that described in Example 2-17 in Chapter 2.

REFERENCES

1. H. S. Carslaw and J. C. Jaeger, *Conduction of Heat in Solids*, Clarendon Press, London, 1959.
2. M. N. Özişik, *Boundary Value Problems of Heat Conduction*, International Textbook Co., Scranton, Pa., 1968.
3. A. V. Luikov, *Analytical Heat Diffusion Theory*, Academic Press, New York, 1968.
4. J. Crank, *The Mathematics of Diffusion*, Clarendon Press, London, 1957.
5. T. M. MacRobert, *Spherical Harmonics*, 3rd ed., Pergamon Press, New York, 1967.
6. E. T. Whitaker and G. N. Watson, *A Course of Modern Analysis*, Cambridge University Press London, 1965.
7. W. E. Byerly, *Fourier's Series and Spherical, Cylindrical and Ellipsoidal Harmonics*, Dover Publications, New York, 1959.
8. M. Abramowitz and I. A. Stegun, *Handbook of Mathematical Functions*, National Bureau of Standards, Applied Mathematic Series 55, U.S. Government Printing Office, Washington, D.C., 20402, 1964.
9. E. W. Hobson, *The Theory of Spherical and Ellipsoidal Harmonics*, Cambridge University Press, 1932.
10. G. Sansone, *Orthogonal Functions*, Interscience Publishers, New York, 1959.
11. H. Bateman, *Partial Differential Equations of Mathematical Physics*, Dover Publications, New York, 1944.
12. W. Magnus and F. Oberhettinger, *Formulas and Theorems for the Special Functions of Mathematical Physics*, Chelsea Publishing Co., New York, 1949.
13. I. N. Sneddon, *Special Functions of Mathematical Physics and Chemistry*, Oliver and Boyd, London, 1961.
14. E. W. Barnes, *Quart, J. Pure Apl. Math.* **39**, 97–204, 1908.
15. O. Rodrigues, *Corr. Ec. Roy. Polytech.* **III**, 1816.

PROBLEMS

4-1 By making use of the Rodrigues' formula given by equation (4-21) show that the integral

$$I \equiv \int_{-1}^{1} f(\mu) P_n(\mu)\, d\mu$$

when performed by repeated integrations by parts can be expressed in the form

$$I = (-1)^n \frac{1}{2^n n!} \int_{-1}^{1} (\mu^2 - 1)^n \frac{d^n f(\mu)}{d\mu^n}\, d\mu$$

4-2 Consider the heat conduction problem for a spherical cavity $a \leqslant r < \infty$

given in the form

$$\frac{1}{r}\frac{\partial^2}{\partial r^2}(rT)+\frac{g(r)}{k}=\frac{1}{a}\frac{\partial T}{\partial t} \quad \text{in} \quad a<r<\infty, \quad t>0$$

$$-\frac{\partial T}{\partial r}+HT=f_1 \quad \text{at} \quad r=a, \quad t>0$$

$$T=F(r) \quad \text{for} \quad t=0, \quad \text{in} \quad a\leqslant r<\infty$$

By utilizing the transformations $U=rT$ and $x=r-a$, transform the above problem to the problem of heat conduction in a semiinfinite medium in the rectangular coordinate system.

4-3 A hollow sphere $a\leqslant r\leqslant b$ is initially at temperature $F(r)$. For times $t>0$ the boundary surface at $r=a$ is kept insulated and the boundary at $r=b$ dissipates heat by convection into a medium at zero temperature. Obtain an expression for the temperature distribution $T(r,t)$ in the sphere.

4-4 A solid sphere of radius $r=b$ is initially at a temperature $F(r)$. For times $t>0$ the boundary surface at $r=b$ is kept at zero temperature. Obtain an expression for the temperature distribution $T(r,t)$ in the sphere for times $t>0$.

4-5 Obtain an expression for the temperature distribution $T(r,\mu,t)$ in a solid sphere of radius $r=b$ that is initially at temperature $F(r,\mu)$ and for times $t>0$ the boundary surface at $r=b$ is kept insulated.

4-6 Obtain an expression for the temperature distribution $T(r,\mu,t)$ in a solid hemisphere, $0\leqslant\mu\leqslant 1, 0\leqslant r\leqslant b$, which is initially at temperature $F(r,\mu)$ and for times $t>0$ the boundary at the spherical surface $r=b$ is kept at zero temperature and at the base $\mu=0$ is kept insulated.

4-7 A solid sphere of radius $r=b$ is initially at a temperature $F(r,\mu,\phi)$. For times $t>0$ the boundary surface $r=b$ dissipates heat by convection into a medium at zero temperature. Obtain an expression for the temperature distribution $T(r,\mu,\phi,t)$ in the sphere for times $t>0$.

4-8 Solve Problem 4-7 for the case when the boundary surface at $r=b$ is kept insulated.

4-9 A solid sphere of radius $r=b$ is initially at temperature $F(r)$. For times $t>0$ the boundary at $r=b$ is kept insulated. Develop an expression for the temperature distribution $T(r,t)$ in the sphere for times $t>0$.

4-10 By separating equation (4-6a), show that the resulting separated equations are as given by equations (4-7).

4-11 By separating equation (4-3), develop the resulting separated equations.

4-12 Consider a region $b \leqslant r < \infty$ (i.e., a region bounded internally by a sphere of radius $r = b$). Initially, the region is at a temperature $F(r)$. For times $t > 0$, the boundary surface at $r = b$ is kept at zero temperature. Develop an expression for the temperature distribution $T(r, t)$ in the medium for times $t > 0$.

4-13 Determine the steady-state temperature distribution $T(r, \mu)$ in a solid hemisphere of radius $r = b$, in the region $0 \leqslant r \leqslant b$, $0 \leqslant \mu \leqslant 1$, with its spherical surface at $r = b$ kept at temperature $f(\mu)$ and its base at $\mu = 0$ is kept at zero temperature.

4-14 Obtain an expression for the steady-state temperature $T(r, \mu)$ in a solid sphere of radius $r = b$ when the boundary surface at $r = b$ is kept at temperature $f(\mu)$.

4-15 A solid sphere of radius $r = b$ is initially at a uniform temperature $F(r)$. For times $t > 0$ the boundary surface at $r = b$ is kept at a constant temperature T_b. Obtain an expression for the temperature distribution $T(r, t)$ in the sphere.

4-16 A solid sphere of radius $r = b$ is initially at temperature $F(r)$. For times $t > 0$, the heat transfer at the boundary surface $r = b$ is given by $(\partial T/\partial r) + HT = f_b$, where f_b is constant. Obtain an expression for the temperature distribution $T(r, t)$ in the sphere.

4-17 A hollow sphere $a \leqslant r \leqslant b$ is initially at temperature $F(r)$. For times $t > 0$, heat is generated in the region at a constant rate of g_0 per unit volume and the boundary surfaces at $r = a$ and $r = b$ are kept at uniform temperatures T_a and T_b, respectively. Obtain an expression for the temperature distribution $T(r, t)$ in the sphere.

4-18 Repeat problem 4-12 for the case when the boundary surface at $r = b$ is kept insulated.

4-19 Consider a region $b < r < \infty$ (i.e., a region internally bounded by a sphere of radius $r = b$). Initially the region is at zero temperature. For times $t > 0$, the boundary surface at $r = b$ is kept at a constant temperature T_0. Develop an expression for the temperature distribution $T(r, t)$ in the region for times $t > 0$.

NOTE

1. Equations (4-32) for $H = 0$ become

$$\frac{\partial^2 U}{\partial r^2} = \frac{1}{\alpha}\frac{\partial U}{\partial t} \quad \text{in} \quad 0 < r < b, \quad t > 0 \tag{1a}$$

$$U = 0 \quad \text{at} \quad r = 0, \quad t > 0 \tag{1b}$$

$$\frac{\partial U}{\partial r}-\frac{1}{b}U=0 \quad \text{at} \quad r=b, \quad t>0 \tag{1c}$$

$$U=rF(r) \quad \text{for} \quad t=0, \quad \text{in} \quad 0\leqslant r\leqslant b \tag{1d}$$

Appropriate eigenvalue problem for the solution of this system is given as

$$\frac{d^2R_m}{dr^2}+\beta_m^2R_m=0 \quad \text{in} \quad 0<r<b \tag{2a}$$

$$R_m=0 \quad \text{at} \quad r=0 \tag{2b}$$

$$\frac{dR_m}{dr}-\frac{1}{b}R_m=0 \quad \text{at} \quad r=b \tag{2c}$$

The solution of system (1) is obtainable according to equation (2-13) of Chapter 2 as

$$U(r,t)=\sum_{m=0}^{\infty} e^{-\alpha\beta_m^2t}\frac{1}{N(\beta_m)}R(\beta_m,r)\int_0^b R(\beta_m,r')r'F(r')\,dr' \tag{3a}$$

where

$$N(\beta_m)=\int_0^b[R(\beta_m,r)]^2\,dr \tag{3b}$$

and $m=0,1,2,3,\ldots$. For $\beta_m\neq 0$, the eigenvalues β_m and the eigenfunctions $R(\beta_m,r)$ are obtainable from Table 2-2 of Chapter 2. However, for $\beta_m=0$, equations (2) have a solution $R(\beta_m,r)=r$. Then the solution (3) includes a term corresponding to the zero eigenvalue and takes the form

$$U(r,t)=\frac{r\int_0^b r'^2F(r')\,dr'}{\int_0^b r'^2\,dr'}+\sum_{m=1}^{\infty}\frac{e^{-\alpha\beta_m^2t}}{N(\beta_m)}R(\beta_m,r)\int_0^b R(\beta_m,r')r'F(r')\,dr' \tag{4}$$

where $U(r,t)$ is related to the temperature by $U(r,t)=rT(r,t)$; then equation (4) becomes

$$T(r,t)=\frac{3}{b^3}\int_0^b r^2F(r)\,dr+\frac{1}{r}\sum_{m=1}^{\infty}\frac{e^{-\alpha\beta_m^2t}}{N(\beta_m)}R(\beta_m,r)\int_0^b R(\beta_m,r')r'F(r')\,dr' \tag{5}$$

Thus, the first term on the right is the mean of the initial temperature distribution over the volume of the sphere.

5

THE USE OF DUHAMEL'S THEOREM

So far we considered the solution of heat conduction problems with time-independent boundary conditions and energy-generation term. However, there are many engineering problems such as heat transfer in internal-combustion engine walls and space reentry problems in which the boundary condition functions are *time-dependent*. In nuclear reactor fuel elements during power transients, the energy-generation rate in the fuel elements varies with time. Duhamel's theorem provides a convenient approach for developing solution to heat conduction problems with time-dependent boundary conditions and/or time-dependent energy generation by utilizing the solution to the same problem with time-independent boundary conditions and/or time-independent energy generation. The method is applicable to linear problems because it is based on the superposition principle. A proof of Duhamel's theorem can be found in several references [1; 2, p. 162; 3, p. 30]. The proof given in reference 1 considers a general convection-type boundary condition from which the cases of prescribed heat flux and prescribed temperature boundary conditions are obtainable as special cases. Here we present a statement of Duhamel's theorem and then illustrate its application in the solution of heat conduction problems with time-dependent boundary condition function and/or heat generation.

5-1 THE STATEMENT OF DUHAMEL'S THEOREM

Consider the three-dimensional, nonhomogeneous heat conduction problem in a region R with time-dependent boundary condition function and heat generation given in the form

$$\nabla^2 T(\mathbf{r},t) + \frac{1}{k} g(\mathbf{r},t) = \frac{1}{\alpha}\frac{\partial T(\mathbf{r},t)}{\partial t} \quad \text{in} \quad \text{region } R, \quad t>0 \tag{5-1a}$$

$$k_i \frac{\partial T}{\partial n_i} + h_i T = f_i(\mathbf{r},t) \quad \text{on} \quad \text{boundary } S_i, \quad t>0 \tag{5-1b}$$

$$T(\mathbf{r},t) = F(\mathbf{r}) \quad \text{for} \quad t=0, \quad \text{in region } R \tag{5-1c}$$

where $\partial/\partial n_i$ is the derivative along outward-drawn normal to the boundary surface $S_i, i = 1, 2, \ldots, N$ and N being the number of continuous boundary surfaces of the region R. Here k_i and h_i are coefficients that are assumed to be constant. By setting $k_i = 0$ we obtain boundary condition of the first kind, and by setting $h_i = 0$ we obtain boundary condition of the second kind.

The problem given by equations (5-1) cannot be solved by the techniques described in the previous chapters because the nonhomogeneous terms $g(\mathbf{r},t)$ and $f_i(\mathbf{r},t)$ depend on time. Therefore, instead of solving this problem directly, we express its solution in terms of the solution of the simpler auxiliary problem as now defined. Let $\Phi(\mathbf{r},t,\tau)$ be the solution of problem (5-1) on the assumption that the nonhomogeneous terms $g(\mathbf{r},\tau)$ and $f_i(\mathbf{r},\tau)$ do not depend on time; namely, the variable τ is merely a parameter but not a time variable. Then, $\Phi(\mathbf{r},t,\tau)$ is the solution of the following simpler auxiliary problem

$$\nabla^2 \Phi(\mathbf{r},t,\tau) + \frac{1}{k} g(\mathbf{r},\tau) = \frac{1}{\alpha}\frac{\partial \Phi(\mathbf{r},t,\tau)}{\partial t} \quad \text{in} \quad \text{region } R, \quad t>0 \tag{5-2a}$$

$$k_i \frac{\partial \Phi(\mathbf{r},t,\tau)}{\partial n_i} + h_i \Phi(\mathbf{r},t,\tau) = f_i(\mathbf{r},\tau) \quad \text{on} \quad \text{boundary } S_i, \quad t>0 \tag{5-2b}$$

$$\Phi(\mathbf{r},t,\tau) = F(\mathbf{r}) \quad \text{for} \quad t=0, \quad \text{in region } R \tag{5-2c}$$

where $\partial/\partial n_i$ and S_i as defined previously, and the function $\Phi(\mathbf{r},t,\tau)$ depends on τ because $g(\mathbf{r},\tau)$ and $f_i(\mathbf{r},\tau)$ depend on τ.

The problem (5-2) can be solved with the techniques described in the previous chapters because $g(\mathbf{r},\tau)$ and $f(\mathbf{r},\tau)$ do not depend on time. Suppose the solution $\Phi(\mathbf{r},t,\tau)$ of the auxiliary problem (5-2) is available. Then, Duhamel's theorem relates the solution $T(\mathbf{r},t)$ of the problem (5-1) to the solution $\Phi(\mathbf{r},t,\tau)$ of the auxiliary problem (5-2) by the following integral expression

$$T(\mathbf{r},t) = \frac{\partial}{\partial t}\int_{\tau=0}^{t} \Phi(\mathbf{r},t-\tau,\tau)\,d\tau \tag{5-3}$$

This result can be expressed in the alternative form by performing the differentiation under the integral sign; we obtain

$$T(\mathbf{r},t) = F(\mathbf{r}) + \int_{\tau=0}^{t} \frac{\partial}{\partial t}\Phi(\mathbf{r},t-\tau,\tau)d\tau \tag{5-4}$$

since

$$\Phi(\mathbf{r},t-\tau,\tau)|_{\tau=t} = \Phi(\mathbf{r},0,\tau) = F(\mathbf{r})$$

We now examine some special cases of Duhamel's theorem given by equation (5-4).

1. *Initial temperature zero.* For this special case we have $F(\mathbf{r}) = 0$ and equation (5-4) reduces to

$$T(\mathbf{r},t) = \int_{\tau=0}^{t} \frac{\partial}{\partial t}\Phi(\mathbf{r},t-\tau,\tau)d\tau \tag{5-5}$$

2. *Initial temperature zero, problem has only one nonhomogeneity.* The solid is initially at zero temperature and the problem involves only one nonhomogeneous term. Namely, if there is heat generation, all the boundary conditions for the problem are homogeneous; or, if there is no heat generation in the medium, only one of the boundary conditions is nonhomogeneous. For example, we consider a problem in which there is no heat generation, but one of the boundary conditions, say, the one at the boundary surface S_1 is nonhomogeneous.

$$\nabla^2 T(\mathbf{r},t) = \frac{1}{\alpha}\frac{\partial T(\mathbf{r},t)}{\partial t} \quad \text{in} \quad \text{region } R, \quad t>0 \tag{5-6a}$$

$$k_i\frac{\partial T}{\partial n_i} + h_i T = \delta_{1i} f_i(t) \quad \text{on} \quad \text{boundary } S_i, \quad t>0 \tag{5-6b}$$

$$T(\mathbf{r},t) = 0 \quad \text{for} \quad t=0, \quad \text{in region } R \tag{5-6c}$$

where $i = 1, 2, \ldots, N$ and δ_{1i} is the Kronecker delta defined as

$$\delta_{1i} = \begin{cases} 0 & i \neq 1 \\ 1 & i = 1 \end{cases}$$

The corresponding auxiliary problem is taken as

$$\nabla^2\Phi(\mathbf{r},t) = \frac{1}{\alpha}\frac{\partial\Phi(\mathbf{r},t)}{\partial t} \quad \text{in} \quad \text{region } R, \quad t>0 \tag{5-7a}$$

$$k_i \frac{\partial \Phi}{\partial n_i} + h_i \Phi = \delta_{1i} \qquad \text{on} \qquad \text{boundary } S_i, \quad t > 0 \tag{5-7b}$$

$$\Phi(\mathbf{r}, t) = 0 \qquad \text{for} \qquad t = 0, \qquad \text{in region } R \tag{5-7c}$$

Then, the solution $T(\mathbf{r}, t)$ of the problem (5-6) is related to the solution $\Phi(\mathbf{r}, t)$ of the problem (5-7) by

$$T(\mathbf{r}, t) = \int_{\tau=0}^{t} f(\tau) \frac{\partial \Phi(\mathbf{r}, t - \tau)}{\partial t} d\tau \tag{5-8}$$

The validity of this result is apparent from the fact that if $\Phi(\mathbf{r}, t, \tau)$ is the solution of the problem (5-7) for a boundary condition $\delta_{1i} f_i(\tau)$, then $\Phi(\mathbf{r}, t, \tau)$ is related to $\Phi(\mathbf{r}, t)$ by

$$\Phi(\mathbf{r}, t, \tau) = f(\tau) \Phi(\mathbf{r}, t) \tag{5-9}$$

When equation (5-9) is introduced into equation (5-5), the result (5-8) is obtained.

If the boundary condition function $f_i(t)$ has discontinuities, say, at times $t = \tau_j (j = 0, 1, 2, \ldots)$, then the time integral in equation (5-8) needs to be broken in parts at the points of discontinuities with proper cognizance of the effects of step changes in surface condition to the temperature in the medium. This matter will be discussed further in the next section.

The physical significance of the function $\Phi(\mathbf{r}, t)$ governed by the auxiliary problem (5-7) is dependent on the type of boundary condition considered for the physical problem (5-6). If the boundary condition is of the *first kind* [i.e., $T = \delta_{1i} f_i(t)$], then the boundary condition for the auxiliary problem is also of the first kind, [i.e., $\Phi = \delta_{1i}$]. Then, the function $\Phi(\mathbf{r}, t)$ represents the response function for a solid initially at zero temperature and for times $t > 0$, one of the boundary surfaces is subjected to a unit step change in the surface temperature. If the boundary condition for the physical problem is of the *second kind* [i.e., prescribed heat flux, $k_i(\partial T / \partial n_i) = \delta_{1i} f_i(t)$], then the boundary condition for the auxiliary problem is also of the second kind; hence $\Phi(\mathbf{r}, t)$ represents the response function for a unit step change in the applied surface heat flux.

5-2 TREATMENT OF DISCONTINUITIES

If the boundary condition function $f(t)$ has discontinuities resulting from step changes in the applied surface temperature, heat flux, or ambient temperature, then the integral appearing in Duhamel's theorem (5-8) needs to be broken into parts at the points of such discontinuities. Here we illustrate how to break the integral into parts at the points of discontinuities by integration by parts and come up with an alternative form of Duhamel's theorem.

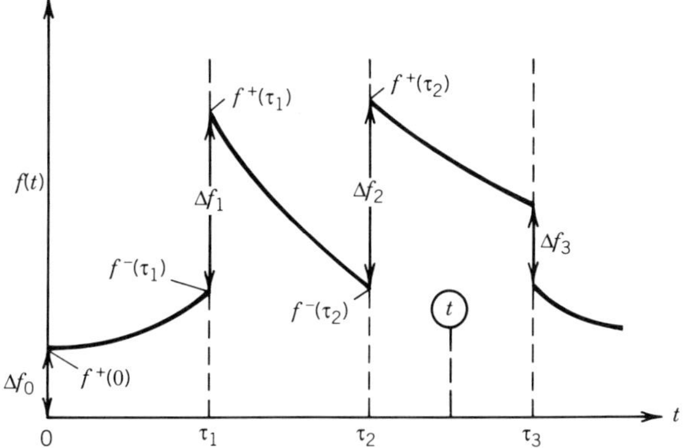

Fig. 5-1 Boundary-condition function $f(t)$ with discontinuities.

Consider, for example, the boundary condition function $f(t)$ that has three discontinuities at times $t = 0^+, \tau_1$, and τ_2 over the time domain $0^- < t < \tau_3$, where 0^+ denotes approaching the origin from the right and 0^- approaching from the left as illustrated in Fig. 5-1. In addition, f^- denotes the limiting value of f at the discontinuity as it is approached from the left and f^+ denotes the limiting value of f as it is approached from the right. It is assumed that the medium is initially at zero temperature.

We consider Duhamel's theorem given by equation (5-8) as

$$T(\mathbf{r}, t) = \int_{\tau=0}^{t} f(\tau) \frac{\partial \Phi(\mathbf{r}, t-\tau)}{\partial t} d\tau \tag{5-10}$$

By differentiating the function $\Phi(\mathbf{r}, t-\tau)$ with respect to t and τ and comparing the resulting expressions, we note that the following relationship holds

$$\frac{\partial \Phi(\mathbf{r}, t-\tau)}{\partial t} = -\frac{\partial \Phi(\mathbf{r}, t-\tau)}{\partial \tau} \tag{5-11}$$

Then equation (5-10) is written as

$$T(\mathbf{r}, t) = -\int_{\tau=0}^{t} f(\tau) \frac{\partial \Phi(\mathbf{r}, t-\tau)}{\partial \tau} d\tau \tag{5-12}$$

Suppose the function $f(t)$ has three discontinuities as illustrated in Fig. 5-1 and we wish to have the solution $T(\mathbf{r}, t)$ to be determined over the time interval $\tau_2 < t < \tau_3$. Then the integral is broken into parts at the discontinuities at $t = 0, \tau_1$, and τ_2 and equation (5-12) is written as

$$T(\mathbf{r}, t) = -\left\{ \int_{\tau=0}^{\tau_1} + \int_{\tau_1}^{\tau_2} + \int_{\tau_2}^{t} \right\} \left\{ f(\tau) \frac{\partial \Phi(\mathbf{r}, t-\tau)}{\partial \tau} d\tau \right\} \tag{5-13a}$$

where $\tau_2 < t < \tau_3$. Each of these integrals is performed by parts to yield

$$
\begin{aligned}
T(\mathbf{r}, t) &= f^{+}(0)\Phi(\mathbf{r}, t-0) - f^{-}(\tau_1)\Phi(\mathbf{r}, t-\tau_1) + \int_0^{\tau_1} \Phi(\mathbf{r}, t-\tau)\frac{df(\tau)}{d\tau}\,d\tau \\
&\quad + f^{+}(\tau_1)\Phi(\mathbf{r}, t-\tau_1) - f^{-}(\tau_2)\Phi(\mathbf{r}, t-\tau_2) + \int_{\tau_1}^{\tau_2} \Phi(\mathbf{r}, t-\tau)\frac{df(\tau)}{d\tau}\,d\tau \\
&\quad + f^{+}(\tau_2)\Phi(\mathbf{r}, t-\tau_2) - f^{-}(t)\Phi(\mathbf{r}, t-t) + \int_{\tau_2}^{t} \Phi(\mathbf{r}, t-\tau)\frac{df(\tau)}{d\tau}\,d\tau
\end{aligned}
\tag{5-13b}
$$

Collecting the terms and noting that $\Phi(\mathbf{r}, t-t) = \Phi(\mathbf{r}, 0) = 0$, equation (5-13b) takes the form

$$
\begin{aligned}
T(\mathbf{r}, t) &= \Phi(\mathbf{r}, t) f^{+}(0) + \Phi(\mathbf{r}, t-\tau_1)[f^{+}(\tau_1) - f^{-}(\tau_1)] \\
&\quad + \Phi(\mathbf{r}, t-\tau_2)[f^{+}(\tau_2) - f^{-}(\tau_2)] + \int_{\tau=0}^{t} \Phi(\mathbf{r}, t-\tau)\frac{df(\tau)}{d\tau}\,d\tau
\end{aligned}
\tag{5-14}
$$

which is written more compactly as

$$
T(\mathbf{r}, t) = \int_{\tau=0}^{t} \Phi(\mathbf{r}, t-\tau)\frac{df(\tau)}{d\tau}\,d\tau + \sum_{j=0}^{2} \Phi(\mathbf{r}, t-\tau_j)\Delta f_j \tag{5-15}
$$

where t lies in the interval $\tau_2 < t < \tau_3$ and the following definitions are used:

$$
\Delta f_j = f^{+}(\tau_j) - f^{-}(\tau_j) \qquad \text{with} \qquad f^{-}(0) = 0
$$

$\tau_j =$ the times at which a step change of magnitude Δf_j occurs in the surface condition

In equation (5-15), the integral term is for the contribution of the continuous portion of the boundary condition function $f(t)$ and the summation term is for the contribution of finite step changes Δf_j occurring in $f(t)$ at the discontinuities.

Generalization to *N* Discontinuities

In the preceding example, we considered only three discontinuities in the boundary condition function $f(t)$ over the time domain $0 < t < \tau_3$. Suppose the function $f(t)$ has N discontinuities over the time domain $0 < t < \tau_N$ and the temperature $T(\mathbf{r}, t)$ is required over the time interval $\tau_{N-1} < t < \tau_N$. The specific result given by equation (5-15) is generalized as

$$
T(\mathbf{r}, t) = \int_{\tau=0}^{t} \Phi(\mathbf{r}, t-\tau)\frac{df(\tau)}{d\tau}\,d\tau + \sum_{j=0}^{N-1} \Phi(\mathbf{r}, t-\tau_j)\Delta f_j \tag{5-16}
$$

where N is the number of discontinuities over the time domain $0 < t < \tau_N$ and the temperature $T(\mathbf{r}, t)$ is for times t in the interval

$$\tau_{N-1} < t < \tau_N$$

Equation (5-16) is the alternative form of Duhamel's theorem (5-8). In consists of the integral and summation terms and is called the *Stieltjes integral.*

All Step Changes

We now consider a situation in which the boundary condition function $f(t)$ consists of a series of step changes Δf_j occurring at times $\tau_j = j\Delta t$, but has no continuous parts as illustrated in Fig. 5-2. For this specific case the integral term drops out and the solution (5-16) reduces to

$$T(\mathbf{r}, t) = \sum_{j=0}^{N-1} \Phi(\mathbf{r}, t - j\Delta t)\Delta f_j \tag{5-17a}$$

where t is in the time interval $(N-1)\Delta t < t < N\Delta t$. This result can be written in a more general form as

$$T(\mathbf{r}, t) = \sum_{j=0}^{\infty} \Phi(\mathbf{r}, t - j\Delta t)\Delta f_j U(t - j\Delta t) \tag{5-17b}$$

where

$$U(t - j\Delta t) = \text{the unit step function}$$

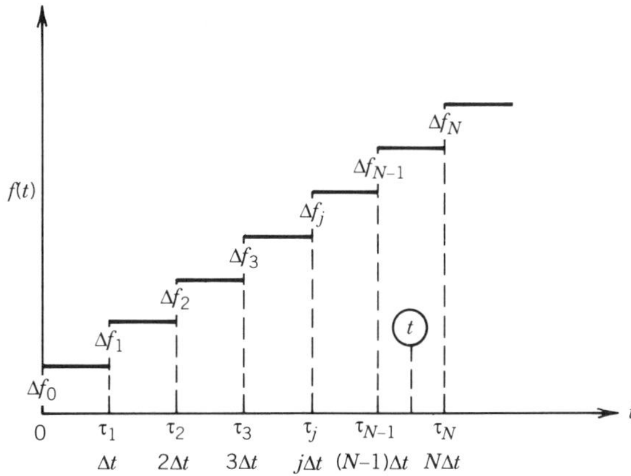

Fig. 5-2 Stepwise varying boundary condition function $f(t)$.

5-3 APPLICATIONS OF DUHAMEL'S THEOREM

We now illustrate with examples the application of Duhamel's theorem for the solution of heat-conduction problems with time-dependent boundary condition function and/or heat generation in terms of the solution of the same problem for time independent boundary condition function and/or heat generation.

Example 5-1

A slab of thickness L is initially at zero temperature. For times $t > 0$, the boundary surface at $x = 0$ is kept at zero temperature, while the surface at $x = L$ is subjected to a time varying temperature $f(t)$ defined by

$$f(t) = \begin{cases} bt & \text{for} \quad 0 < t < \tau_1 \qquad \text{(5-18a)} \\ 0 & \text{for} \quad t > \tau_1 \qquad \text{(5-18b)} \end{cases}$$

as illustrated in Fig. 5-3. Using Duhamel's theorem, develop an expression for the temperature distribution $T(x, t)$ in the slab for times (i) $t < \tau_1$ and (ii) $t > \tau_1$.

Solution. The mathematical formulation of this heat conduction problem is given by

$$\frac{\partial^2 T(x,t)}{\partial x^2} = \frac{1}{\alpha}\frac{\partial T(x,t)}{\partial t} \quad \text{in} \quad 0 < x < L, \quad t > 0 \tag{5-19a}$$

$$T(x,t) = 0 \quad \text{at} \quad x = 0, \quad t > 0 \tag{5-19b}$$

$$T(x,t) = f(t) \quad \text{at} \quad x = L, \quad t > 0 \tag{5-19c}$$

$$T(x,t) = 0 \quad \text{for} \quad t = 0 \tag{5-19d}$$

where $f(t)$ is defined by equation (5-18). The corresponding auxiliary problem

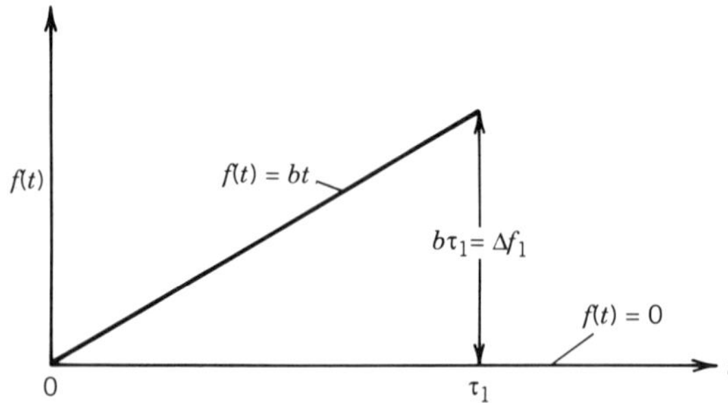

Fig. 5-3 Variation of surface temperature $f(t)$ with time for Example 5-1.

becomes

$$\frac{\partial^2 \Phi(x,t)}{\partial x^2} = \frac{1}{\alpha}\frac{\partial \Phi(x,t)}{\partial t} \qquad \text{in} \qquad 0 < x < L, \quad t > 0 \tag{5-20a}$$

$$\Phi(x,t) = 0 \qquad \text{at} \qquad x = 0, \qquad t > 0 \tag{5-20b}$$

$$\Phi(x,t) = 1 \qquad \text{at} \qquad x = L, \qquad t > 0 \tag{5-20c}$$

$$\Phi(x,t) = 0 \qquad \text{for} \qquad t = 0 \tag{5-20d}$$

The solution for the auxiliary problem (5-20) is determined as

$$\Phi(x,t) = \frac{x}{L} + \frac{2}{L}\sum_{m=1}^{\infty} e^{-\alpha\beta_m^2 t}(-1)^m \frac{1}{\beta_m}\sin\beta_m x \tag{5-21a}$$

where

$$\beta_m = \frac{m\pi}{L} \tag{5-21b}$$

The function $\Phi(x, t-\tau)$ is obtained by replacing t by $t-\tau$ in equation (5-21a).

$$\Phi(x,t-\tau) = \frac{x}{L} + \frac{2}{L}\sum_{m=1}^{\infty} e^{-\alpha\beta_m^2 (t-\tau)}\frac{(-1)^m}{\beta_m}\sin\beta_m x \tag{5-22}$$

Duhamel's theorem can now be applied either by using the form given by equation (5-10) or (5-16). Here, the latter is preferred since the contribution of discontinuity to the solution appears explicitly. The solutions for times $t < \tau_1$ and $t > \tau_1$ are considered below separately.

i. *Times* $t < \tau_1$. The boundary condition function $f(t)$ has no discontinuity; thus the summation term in equation (5-16) drops out to give

$$T(x,t) = \int_{\tau=0}^{t} \Phi(x,t-\tau)\frac{df(\tau)}{d\tau}\,d\tau \qquad \text{for} \qquad t < \tau_1 \tag{5-23}$$

where $[df(\tau)/d\tau] = b$ and $\Phi(x, t-\tau)$ is obtained from equation (5-22). Then equation (5-23) becomes

$$T(x,t) = \int_{\tau=0}^{t} \left\{ \frac{x}{L} + \frac{2}{L}\sum_{m=1}^{\infty} e^{-\alpha\beta_m^2 (t-\tau)}\frac{(-1)^m}{\beta_m}\sin\beta_m x \right\} b\,d\tau \tag{5-24}$$

The integration is performed to give

$$T(x,t) = b\frac{x}{L}t + b\frac{2}{L}\sum_{m=1}^{\infty}\frac{(-1)^m}{\alpha\beta_m^3}\left(1 - e^{-\alpha\beta_m^2 t}\right)\sin\beta_m x \tag{5-25}$$

for $t < \tau_1$.

ii. *Times* $t > \tau_1$. The boundary surface function $f(t)$ has only one discontinuity at time $t = \tau_1$ and the resulting step change in $f(t)$ is a decrease in temperature, that is, $\Delta f_1 = -b\tau_1$. Then equation (5-16) reduces to

$$T(x,t) = \int_{\tau=0}^{\tau_1} \Phi(x,t-\tau)\frac{df}{d\tau}d\tau + \int_{\tau_1}^{t} \Phi(x,t-\tau)\frac{df}{d\tau}d\tau + \Phi(x,t-\tau_1)\Delta f_1 \tag{5-26a}$$

where $(df/d\tau) = b$ for the first integral, $(df/d\tau) = 0$ for the second integral, and $\Delta f_1 = -b\tau_1$. Substituting these results into equation (5-26a), we find

$$T(x,t) = \int_{\tau=0}^{\tau_1} \Phi(x,t-\tau)(b)d\tau + 0 - \Phi(x,t-\tau_1)b\tau_1 \tag{5-26b}$$

where $\Phi(x,t-\tau)$ and $\Phi(x,t-\tau_1)$ are obtained from equation (5-22). Then equation (5-26b) takes the form

$$\begin{aligned} T(x,t) = &\int_{\tau=0}^{\tau_1} \left\{\frac{x}{L} + \frac{2}{L}\sum_{m=1}^{\infty} e^{-\alpha\beta_m^2(t-\tau)}\frac{(-1)^m}{\beta_m}\sin\beta_m x\right\} b\,d\tau \\ &- b\tau_1\left[\frac{x}{L} + \frac{2}{L}\sum_{m=1}^{\infty} e^{-\alpha\beta_m^2(t-\tau_1)}\frac{(-1)^m}{\beta_m}\sin\beta_m x\right] \end{aligned} \tag{5-27}$$

Clearly, the integral is similar to the one given by equation (5-24) except the upper limit is τ_1; hence it can be performed readily.

Example 5-2

A semiinfinite solid, $0 \leqslant x < \infty$, is initially at zero temperature. For times $t > 0$ the boundary surface at $x = 0$ is kept at temperature $f(t)$. Obtain an expression for the temperature distribution $T(x,t)$ in the solid for times $t > 0$ assuming that $f(t)$ has no discontinuities.

Solution. The mathematical formulation of this problem is given as

$$\frac{\partial^2 T(x,t)}{\partial x^2} = \frac{1}{\alpha}\frac{\partial T(x,t)}{\partial t} \quad \text{in} \quad 0 < x < \infty, \quad t > 0 \tag{5-28a}$$

$$T(x,t) = f(t) \quad \text{at} \quad x = 0, \quad t > 0 \tag{5-28b}$$

$$T(x,t) = 0 \quad \text{for} \quad t = 0, \quad \text{in } 0 \leqslant x < \infty \tag{5-28c}$$

The auxiliary problem is taken as

$$\frac{\partial^2 \Phi(x,t)}{\partial x^2} = \frac{1}{\alpha}\frac{\partial \Phi(x,t)}{\partial t} \quad \text{in} \quad 0 < x < \infty, \quad t > 0 \tag{5-29a}$$

$$\Phi(x,t) = 1 \qquad \text{at} \qquad x = 0, \qquad t > 0 \tag{5-29b}$$

$$\Phi(x,t) = 0 \qquad \text{for} \qquad t = 0, \qquad \text{in } 0 \leqslant x < \infty \tag{5-29c}$$

Then the solution of the problem (5-28) is given in terms of the solution of the auxiliary problem (5-29), by the Duhamel's theorem (5-8) as

$$T(x,t) = \int_{\tau=0}^{t} f(\tau) \frac{\partial \Phi(x, t-\tau)}{\partial t} d\tau \tag{5-30}$$

The solution $\Phi(x,t)$ of the auxiliary problem (5-29) is obtainable from the solution $T(x,t)$ given by equation (2-58e) by the relation $\Phi(x,t) = 1 - T(x,t)$, and setting in equation (2-58e) $T_0 = 1$. Thus we obtain

$$\Phi(x,t) = 1 - \operatorname{erf}\left(\frac{x}{\sqrt{4\alpha t}}\right) = \operatorname{erfc}\left(\frac{x}{\sqrt{4\alpha t}}\right) = \frac{2}{\sqrt{\pi}} \int_{x/\sqrt{4\alpha t}}^{\infty} e^{-\xi^2} d\xi \tag{5-31}$$

Then

$$\frac{\partial \Phi(x, t-\tau)}{\partial t} = \frac{x}{\sqrt{4\pi\alpha}(t-\tau)^{3/2}} \exp\left[-\frac{x^2}{4\alpha(t-\tau)}\right] \tag{5-32}$$

Introducing equation (5-32) into equation (5-30) the solution of the problem (5-28) becomes

$$T(x,t) = \frac{x}{\sqrt{4\pi\alpha}} \int_{\tau=0}^{t} \frac{f(\tau)}{(t-\tau)^{3/2}} \exp\left[-\frac{x^2}{4\alpha(t-\tau)}\right] d\tau \tag{5-33}$$

To express this result in an alternative form, a new independent variable η is defined as

$$\eta = \frac{x}{\sqrt{4\alpha(t-\tau)}} \tag{5-34}$$

Then

$$t - \tau = \frac{x^2}{4\alpha\eta^2} \qquad \text{and} \qquad d\tau = \frac{2}{\eta}(t-\tau) d\eta \tag{5-35a,b}$$

Introducing equations (5-34) and (5-35) into equation (5-33), we obtain

$$T(x,t) = \frac{2}{\sqrt{\pi}} \int_{x/\sqrt{4\alpha t}}^{\infty} e^{-\eta^2} f\left(t - \frac{x^2}{4\alpha\eta^2}\right) d\eta \tag{5-36}$$

Periodically Varying $f(t)$. We consider the surface temperature $f(t)$ be a periodic in time in the form

$$f(t) = T_0 \cos(\omega t - \beta) \tag{5-37}$$

The solution (5-36) becomes

$$\frac{T(x,t)}{T_0} = \frac{2}{\sqrt{\pi}} \int_{x/\sqrt{4\alpha t}}^{\infty} e^{-\eta^2} \cos\left[\omega\left(t - \frac{x^2}{4\alpha\eta^2}\right) - \beta\right] d\eta \tag{5-38}$$

or

$$\frac{T(x,t)}{T_0} = \frac{2}{\sqrt{\pi}} \int_0^{\infty} e^{-\eta^2} \cos\left[\omega\left(t - \frac{x^2}{4\alpha\eta^2}\right) - \beta\right] d\eta$$

$$- \frac{2}{\sqrt{\pi}} \int_0^{x/\sqrt{4\alpha t}} e^{-\eta^2} \cos\left[\omega\left(t - \frac{x^2}{4\alpha\eta^2}\right) - \beta\right] d\eta \tag{5-39}$$

The first definite integral can be evaluated [3, p. 65]; then

$$\frac{T(x,t)}{T_0} = \exp\left[-x\left(\frac{\omega}{2\alpha}\right)^{1/2}\right] \cos\left[\omega t - x\left(\frac{\omega}{2\alpha}\right)^{1/2} - \beta\right]$$

$$- \frac{2}{\sqrt{\pi}} \int_0^{x/\sqrt{4\alpha t}} e^{-\eta^2} \cos\left[\omega\left(t - \frac{x^2}{4\alpha\eta^2}\right) - \beta\right] d\eta \tag{5-40}$$

Here the second term on the right represents the transients that die away as $t \to \infty$, and the first term represents the steady oscillations of temperature in the medium after the transients have passed.

Example 5-3

A plate of thickness L is initially at zero temperature. For times $t > 0$, the boundary surface at $x = L$ is kept insulated while the surface at $x = 0$ is subjected to a heat flux $f(t)$ varying with time as

$$-k\frac{\partial T}{\partial x} = f(t) = \begin{cases} t & \text{for} \quad 0 < t < \tau_1 & \text{(5-41a)} \\ 0 & \text{for} \quad t > \tau_1 & \text{(5-41b)} \end{cases}$$

Using Duhamel's theorem, develop an expression for the temperature distribution $T(x, t)$ in the slab for times: (i) $t < \tau_1$ and (ii) $t > \tau_1$.

Solution. The mathematical formulation of this heat conduction problem is given by

$$\frac{\partial^2 T(x,t)}{\partial x^2} = \frac{1}{\alpha}\frac{\partial T}{\partial t} \quad \text{in} \quad 0 < x < L, \quad t > 0 \tag{5-42a}$$

$$-k\frac{\partial T}{\partial x}=f(t) \qquad \text{at} \qquad x=0, \qquad t>0 \tag{5-42b}$$

$$\frac{\partial T}{\partial x}=0 \qquad \text{at} \qquad x=L, \qquad t>0 \tag{5-42c}$$

$$T(x,t)=0 \qquad \text{for} \qquad t=0 \tag{5-42d}$$

where $f(t)$ is defined by equation (5-41). The corresponding auxiliary problem is given by

$$\frac{\partial^2\Phi(x,t)}{\partial x^2}=\frac{1}{\alpha}\frac{\partial\Phi(x,t)}{\partial t} \qquad \text{in} \qquad 0<x<L, \quad t>0 \tag{5-43a}$$

$$-k\frac{\partial\Phi}{\partial x}=1 \qquad \text{at} \qquad x=0 \qquad t>0 \tag{5-43b}$$

$$\frac{\partial\Phi}{\partial x}=0 \qquad \text{at} \qquad x=L, \qquad t>0 \tag{5-43c}$$

$$\Phi(x,t)=0 \qquad \text{for} \qquad t=0 \tag{5-43d}$$

The solution for the auxiliary problem is

$$\Phi(x,t)=\frac{\alpha}{Lk}t+\frac{2}{Lk}\sum_{m=1}^{\infty}\frac{\cos\beta_m x}{\beta_m^2}\left(1-e^{-\alpha\beta_m^2 t}\right) \tag{5-44}$$

where $\beta_m=(m\pi/L)$. Duhamel's theorem given by equation (5-16) is now applied.

i. *Times* $t<\tau_1$. The boundary-condition function $f(t)$ has no discontinuity; then the summation term in equation (5-16) drops out and we obtain

$$T(x,t)=\int_{\tau=0}^{t}\Phi(x,t-\tau)\frac{df(\tau)}{d\tau}d\tau \qquad \text{for} \qquad t<\tau_1 \tag{5-45}$$

where $[df(\tau)/d\tau]=1$ and $\Phi(x,t-\tau)$ is obtainable from equation (5-44) by replacing t by $(t-\tau)$. Then, equation (5-45) becomes

$$T(x,t)=\int_{\tau=0}^{t}\left[\frac{\alpha}{Lk}(t-\tau)+\frac{2}{Lk}\sum_{m=1}^{\infty}\frac{\cos\beta_m x}{\beta_m^2}-\frac{2}{Lk}\sum_{m=1}^{\infty}\frac{\cos\beta_m x}{\beta_m^2}e^{-\alpha\beta_m^2(t-\tau)}\right]d\tau \tag{5-46}$$

The integration is performed to give

$$T(x,t)=\frac{\alpha}{2Lk}t^2+\frac{2}{Lk}\sum_{m=1}^{\infty}\frac{\cos\beta_m x}{\beta_m^2}t-\frac{2}{\alpha Lk}\sum_{m=1}^{\infty}\frac{\cos\beta_m x}{\beta_m^4}(1-e^{-\alpha\beta_m^2 t})$$

$$\text{for}\qquad t<\tau_1 \tag{5-47}$$

ii. *Times* $t>\tau_1$. The boundary surface function $f(t)$ has only one discontinuity at time $t=\tau_1$ and the resulting change in $f(t)$ is a decrease in the amount $\Delta f_1=-\tau_1$. Then, Duhamel's theorem (5-16) reduces to

$$T(x,t)=\int_{\tau=0}^{\tau_1}\Phi(x,t-\tau)\frac{df}{d\tau}d\tau+\int_{\tau_1}^{t}\Phi(x,t-\tau)\frac{df}{d\tau}d\tau+\Phi(x,t-\tau_1)\Delta f_1 \tag{5-48}$$

where $(df/d\tau)=1$ for the first integral, $(df/d\tau)=0$ for the second integral, and $\Delta f_1=-\tau_1$.

Substituting these results into equation (5-48), we obtain

$$T(x,t)=\int_{\tau=0}^{\tau_1}\Phi(x,t-\tau)d\tau+0-\Phi(x,t-\tau_1)\tau_1 \tag{5-49}$$

where the functions $\Phi(x,t-\tau)$ and $\Phi(x,t-\tau_1)$ are obtainable from equation (5-44). Then equation (5-49) becomes

$$T(x,t)=\int_{\tau=0}^{\tau_1}\left[\frac{\alpha}{Lk}(t-\tau)+\frac{2}{Lk}\sum_{m=1}^{\infty}\frac{\cos\beta_m x}{\beta_m^2}-\frac{2}{Lk}\sum_{m=1}^{\infty}\frac{\cos\beta_m x}{\beta_m^2}e^{-\alpha\beta_m^2(t-\tau)}\right]d\tau$$
$$-\tau_1\left[\frac{\alpha}{Lk}(t-\tau_1)+\frac{2}{Lk}\sum_{m=1}^{\infty}\frac{\cos\beta_m x}{\beta_m^2}-\frac{2}{Lk}\sum_{m=1}^{\infty}\frac{\cos\beta_m x}{\beta_m^2}e^{-\alpha\beta_m^2(t-\tau_1)}\right] \tag{5-50}$$

The integral is similar to that in equation (5-46); hence can readily be performed.

Example 5-4

A solid cylinder, $0\leqslant r\leqslant b$, is initially at zero temperature. For times $t>0$ the boundary surface at $r=b$ is kept at temperature $T=f(t)$, which varies with time. Obtain an expression for the temperature distribution $T(r,t)$ in the cylinder for times $t>0$. Assume that $f(t)$ has no discontinuities.

Solution. The mathematical formulation of this problem is given by

$$\frac{\partial^2 T(r,t)}{\partial r^2}+\frac{1}{r}\frac{\partial T(r,t)}{\partial r}=\frac{1}{\alpha}\frac{\partial T(r,t)}{\partial t}\qquad\text{in}\qquad 0\leqslant r<b,\quad t>0 \tag{5-51a}$$

$$T = f(t) \qquad \text{at} \qquad r = b, \qquad t > 0 \tag{5-51b}$$

$$T = 0 \qquad \text{for} \qquad t = 0, \qquad \text{in } 0 \leqslant r \leqslant b \tag{5-51c}$$

and the auxiliary problem is taken as

$$\frac{\partial^2 \Phi(r,t)}{\partial r^2} + \frac{1}{r}\frac{\partial \Phi(r,t)}{\partial r} = \frac{1}{\alpha}\frac{\partial \Phi(r,t)}{\partial t} \qquad \text{in} \qquad 0 \leqslant r < b, \quad t > 0 \tag{5-52a}$$

$$\Phi = 1 \qquad \text{at} \qquad r = b, \qquad t > 0 \tag{5-52b}$$

$$\Phi = 0 \qquad \text{for} \qquad t = 0, \qquad \text{in } 0 \leqslant r \leqslant b \tag{5-52c}$$

Then, the solution of the problem (5-51) can be written in terms of the solution of the auxiliary problem (5-52) by Duhamel's theorem given by equation (5-8) as

$$T(r,t) = \int_{\tau=0}^{t} f(\tau) \frac{\partial \Phi(r, t-\tau)}{\partial t} d\tau \tag{5-53}$$

If $\psi(r,t)$ is the solution of the problem for a solid cylinder, $0 \leqslant r \leqslant b$, initially at temperature unity and for times $t > 0$, the boundary surface at $r = b$ is kept at zero temperature, then the solution for $\psi(r,t)$ is obtainable from the solution (3-68) by setting $T_0 = 1$ in that equation; we find

$$\psi(r,t) = \frac{2}{b}\sum_{m=1}^{\infty} e^{-\alpha\beta_m^2 t} \frac{J_0(\beta_m r)}{\beta_m J_1(\beta_m b)} \tag{5-54a}$$

where the β_m values are the positive roots of

$$J_0(\beta_m b) = 0 \tag{5-54b}$$

The solution $\Phi(r,t)$ of the auxiliary problem (5-52) is obtainable from the solution $\psi(r,t)$ given by equation (5-54) as

$$\Phi(r,t) = 1 - \psi(r,t) = 1 - \frac{2}{b}\sum_{m=1}^{\infty} e^{-\alpha\beta_m^2 t} \frac{J_0(\beta_m r)}{\beta_m J_1(\beta_m b)} \tag{5-55}$$

Introducing equation (5-55) into equation (5-53), the solution of the problem (5-51) becomes

$$T(r,t) = \frac{2\alpha}{b}\sum_{m=1}^{\infty} e^{-\alpha\beta_m^2 t} \beta_m \frac{J_0(\beta_m r)}{J_1(\beta_m b)} \int_0^t e^{\alpha\beta_m^2 \tau} f(\tau) d\tau \tag{5-56}$$

where the β_m values are the roots of $J_0(\beta_m b) = 0$. The solution (5-56) does not explicitly show that $T(r,t) \to f(t)$ for $r \to b$. To obtain alternative form of this solution, the time integration is performed by parts

$$T(r,t) = f(t) \cdot \frac{2}{b} \sum_{m=1}^{\infty} \frac{J_0(\beta_m r)}{\beta_m J_1(\beta_m b)} - \frac{2}{b} \sum_{m=1}^{\infty} \frac{J_0(\beta_m r)}{\beta_m J_1(\beta_m b)} \left[f(0) e^{-\alpha \beta_m^2 t} + \int_0^t e^{-\alpha \beta_m^2 (t-\tau)} df(\tau) \right] \tag{5-57}$$

We note that the solution (5-54a) for $t = 0$ should be equal to the initial temperature $\psi(r, 0) = 1$; thus we have

$$1 = \frac{2}{b} \sum_{m=1}^{\infty} \frac{J_0(\beta_m r)}{\beta_m J_1(\beta_m b)} \tag{5-58}$$

which gives the desired closed-form expression for first series on the right-hand side of equation (5-57). Then, the solution (5-57) is written as

$$T(r,t) = f(t) - \frac{2}{b} \sum_{m=1}^{\infty} \frac{J_0(\beta_m r)}{\beta_m J_1(\beta_m b)} \left[f(0) e^{-\alpha \beta_m^2 t} + \int_0^t e^{-\alpha \beta_m^2 (t-\tau)} df(\tau) \right] \tag{5-59}$$

The solution given in this form clearly shows that $T(r,t) = f(t)$ at $r = b$.

Example 5-5

A solid cylinder, $0 \leqslant r \leqslant b$, is initially at zero temperature. For times $t > 0$ heat is generated in the solid at a rate of $g(t)$ per unit volume and the boundary surface at $r = b$ is kept at zero temperature. Obtain an expression for the temperature distribution $T(r,t)$ in the cylinder for times $t > 0$. Assume that $g(t)$ has no discontinuities.

Solution. The mathematical formulation of this problem is given by

$$\frac{\partial^2 T(r,t)}{\partial r^2} + \frac{1}{r}\frac{\partial T(r,t)}{\partial r} + \frac{g(t)}{k} = \frac{1}{\alpha}\frac{\partial T(r,t)}{\partial t} \quad \text{in} \quad 0 \leqslant r < b, \quad t > 0 \tag{5-60a}$$

$$T = 0 \quad \text{at} \quad r = b, \quad t > 0 \tag{5-60b}$$

$$T = 0 \quad \text{for} \quad t = 0, \quad \text{in } 0 \leqslant r \leqslant b \tag{5-60c}$$

and the auxiliary problem is taken as

$$\frac{\partial^2 \Phi(r,t)}{\partial r^2} + \frac{1}{r}\frac{\partial \Phi(r,t)}{\partial r} + \frac{1}{k} = \frac{1}{\alpha}\frac{\partial \Phi(r,t)}{\partial t} \quad \text{in} \quad 0 \leqslant r < b, \quad t > 0 \tag{5-61a}$$

$$\Phi = 0 \quad \text{at} \quad r = b, \quad t > 0 \tag{5-61b}$$

$$\Phi = 0 \qquad \text{for} \quad t = 0, \qquad \text{in } 0 \leqslant r \leqslant b \tag{5-61c}$$

Then, the solution of the problem (5-60) is related to the solution of the auxiliary problem (5-61) by Duhamel's theorem as

$$T(r, t) = \int_{\tau=0}^{t} g(\tau) \frac{\partial \Phi(r, t-\tau)}{\partial t} d\tau \tag{5-62}$$

The solution of the auxiliary problem (5-61) is obtainable from equation (3-193) by setting $g_0 = 1$ and $F(r) = 0$; we find

$$\Phi(r, t) = \frac{b^2 - r^2}{4k} - \frac{2}{bk} \sum_{m=1}^{\infty} e^{-\alpha\beta_m^2 t} \frac{J_0(\beta_m r)}{\beta_m^3 J_1(\beta_m b)} \tag{5-63a}$$

where the β_m values are the positive roots of

$$J_0(\beta_m b) = 0 \tag{5-63b}$$

Introducing equation (5-63a) into (5-62) we obtain the solution as

$$T(r, t) = \frac{2\alpha}{bk} \sum_{m=1}^{\infty} e^{-\alpha\beta_m^2 t} \frac{J_0(\beta_m r)}{\beta_m J_1(\beta_m b)} \int_{\tau=0}^{t} g(\tau) e^{\alpha\beta_m^2 \tau} d\tau \tag{5-64}$$

REFERENCES

1. R. C. Bartels and R. V. Churchill, *Bull. Am. Math. Soc.* **48**, 276–282, 1942.
2. I. N. Sneddon, *Fourier Transforms*, McGraw-Hill, New York, 1951.
3. H. S. Carslaw and J. C. Jaeger, *Conduction of Heat in Solids*, Clarendon Press, London, 1959.

PROBLEMS

5-1 A slab, $0 \leqslant x \leqslant L$, is initially at zero temperature. For times $t > 0$, the boundary surface at $x = 0$ is subjected to a time-varying temperature $f(t) = b + ct$, while the boundary surface at $x = L$ is kept at zero temperature. Using Duhamel's theorem, develop an expression for the temperature distribution $T(x, t)$ in the slab for times $t > 0$.

5-2 A semiinfinite solid, $0 \leqslant x < \infty$, is initially at zero temperature. For times $t > 0$, the boundary surface at $x = 0$ is kept at temperature $T = T_0 t$, where T_0 is a constant. Using Duhamel's theorem obtain an expression for the temperature distribution $T(x, t)$ in the region for times $t > 0$.

5-3 A slab, $0 \leqslant x \leqslant L$, is initially at zero temperature. For times $t > 0$ the boundary at $x = 0$ is kept insulated and the convection boundary condition at $x = L$ is given as $(\partial T/\partial x) + HT = f(t)$, where $f(t)$ is a function of time. Obtain an expression for the temperature distribution $T(x, t)$ in the slab for times $t > 0$.

5-4 A solid cylinder, $0 \leqslant r \leqslant b$, is initially at zero temperature. For times $t > 0$ the boundary condition at $r = b$ is given as $\partial T/\partial r + HT = f(t)$, where $f(t)$ is a function of time. Obtain an expression for the temperature distribution $T(r, t)$ in the cylinder for times $t > 0$.

5-5 A solid sphere, $0 \leqslant r \leqslant b$, is initially at zero temperature, for times $t > 0$ the boundary surface $r = b$ is kept at temperature $f(t)$, which varies with time. Obtain an expression for the temperature distribution $T(r, t)$ in the sphere.

5-6 A solid cylinder, $0 \leqslant r \leqslant b$, is initially at zero temperature. For times $t > 0$, heat is generated in the solid at a rate of $g(t)$ per unit volume whereas the boundary surface at $r = b$ dissipates heat by convection into a medium at zero temperature. Obtain an expression for the temperature distribution $T(r, t)$ in the cylinder for times $t > 0$.

5-7 A rectangular region $0 \leqslant x \leqslant a, 0 \leqslant y \leqslant b$ is initially at zero temperature. For times $t > 0$ the boundaries at $x = 0$ and $y = 0$ are kept insulated, the boundaries at $x = a$ and $y = b$ are kept at zero temperature while heat is generated in the region at a rate of $g(t)$ per unit volume. Obtain an expression for the temperature distribution in the region using Duhamel's theorem.

5-8 A slab of thickness L is initially at zero temperature. For times $t > 0$, the boundary surface at $x = L$ is kept at zero temperature, while the boundary surface at $x = 0$ is subjected to a time varying temperature $f(t)$ defined by

$$f(t) = \begin{cases} ct & \text{for} \quad 0 < t < \tau_1 \\ 0 & \text{for} \quad t > \tau_1 \end{cases}$$

Using Duhamel's theorem, develop an expression for the temperature distribution $T(x, t)$ for times (i) $t < \tau_1$ and (ii) $t > \tau_1$.

5-9 A semiinfinite medium, $0 < x < \infty$, is initially at zero temperature. For times $t > 0$, the boundary surface at $x = 0$ is subjected to a time-varying temperature:

$$f(t) = \begin{cases} ct & \text{for} \quad 0 < t < \tau_1 \\ 0 & \text{for} \quad t > \tau_1 \end{cases}$$

Using Duhamel's theorem, develop an expression for the temperature distribution $T(x, t)$ for times (i) $t < \tau_1$ and (ii) $t > \tau_1$.

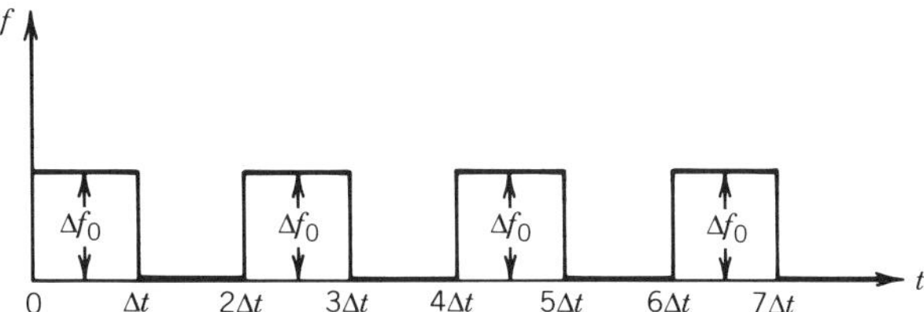

Fig. 5-4 Periodically varying surface temperature.

5-10 A slab of thickness L is initially at zero temperature. For times $t > 0$, the boundary surface at $x = 0$ is subjected to a time-varying temperature $f(t)$ defined by

$$f(t) = \begin{cases} a + bt & \text{for} \quad 0 < t < \tau_1 \\ 0 & \text{for} \quad t > \tau_1 \end{cases}$$

and the boundary at $x = L$ is kept insulated. Using Duhamel's theorem, develop an expression for the temperature distribution in the slab for times (i) $t < \tau_1$ and (ii) $t > \tau_1$.

5-11 A semiinfinite medium, $x > 0$, is initially at zero temperature. For times $t > 0$, the boundary surface at $x = 0$ is subjected to a periodically varying temperature as illustrated in Fig. 5-4. Develop an expression for the temperature distribution in the medium at times (i) $0 < t < \Delta t$, (ii) $\Delta t < t < 2\Delta t$, and (iii) $6\Delta t < t < 7\Delta t$.

5-12 Repeat Problem 5-5 for the case of surface temperature $f(t)$ varying with time as

$$f(t) = \begin{cases} bt & \text{for} \quad 0 < t < \tau_1 \\ 0 & \text{for} \quad t > \tau_1 \end{cases}$$

and determine the temperature distribution $T(r, t)$ in the sphere for times (i) $t < \tau_1$ and (ii) $t > \tau_1$.

6

THE USE OF GREEN'S FUNCTION

Green's function in the solution of partial differential equations of mathematical physics can be found in several references [1–11]. In this chapter we first discuss the physical significance of Green's function and then present sufficiently general expressions for the solution of inhomogeneous transient heat conduction problems with energy generation, inhomogeneous boundary conditions, and a given initial condition, in terms of Green's function. Application to one-, two-, and three-dimensional problems of finite, semiinfinite, and infinite regions is illustrated with representative examples in the rectangular, cylindrical, and spherical coordinate systems. Once Green's function is available for a given problem, the solution for the temperature distribution is determined immediately from the analytic expressions given in this chapter.

6-1 GREEN'S FUNCTION APPROACH FOR SOLVING NONHOMOGENEOUS TRANSIENT HEAT CONDUCTION

We consider the following three-dimensional nonhomogeneous boundary-value problem of heat conduction:

$$\nabla^2 T(\mathbf{r},t) + \frac{1}{k} g(\mathbf{r},t) = \frac{1}{\alpha}\frac{\partial T(\mathbf{r},t)}{\partial t} \qquad \text{in} \qquad \text{region } R, \quad t>0 \tag{6-1a}$$

$$k_i \frac{\partial T}{\partial n_i} + h_i T = h_i T_{\infty i} \equiv f_i(\mathbf{r},t) \qquad \text{on} \qquad S_i, \qquad t>0 \tag{6-1b}$$

$$T(\mathbf{r},t) = F(\mathbf{r}) \qquad \text{for} \qquad t=0, \qquad \text{in} \quad R \tag{6-1c}$$

where $\partial/\partial n_i$ denotes differentiation along the *outward-drawn normal* to the boundary surface $S_i, i = 1, 2, \ldots, N$, and N is the number of continuous boundary surfaces of the region. For generality it is assumed that the generation term $g(\mathbf{r}, t)$ and the boundary-condition function $f_i(\mathbf{r}, t)$ vary with both position and time. Here, k_i and h_i are to be treated as coefficients that are considered constants.

To solve the preceding heat conduction problem we consider the following auxiliary problem for the same region R:

$$\nabla^2 G(\mathbf{r}, t \,|\, \mathbf{r}', \tau) + \frac{1}{k}\delta(\mathbf{r} - \mathbf{r}')\delta(t - \tau) = \frac{1}{\alpha}\frac{\partial G}{\partial t} \qquad \text{in} \qquad \text{region } R, \quad t > \tau \qquad \text{(6-2a)}$$

$$k_i \frac{\partial G}{\partial \eta_i} + h_i G = 0 \qquad \text{on} \qquad S_i, \qquad t > \tau \qquad \text{(6-2b)}$$

obeying the causality requirement that Green's function G be zero for $t < \tau$ [2]. The source in equation (6-2a) is a unit impulsive source for the three-dimensional problem considered here, the delta function $\delta(\mathbf{r} - \mathbf{r}')$ represents a *point heat source* located at $\mathbf{r}'$, while the delta function $\delta(t - \tau)$ indicates that it is an instantaneous heat source releasing its energy spontaneously at time $t = \tau$.

In the case of two-dimensional problems, $\delta(\mathbf{r} - \mathbf{r}')$ is a two-dimensional delta function that characterizes a *line heat source* located at $\mathbf{r}'$, while for the one-dimensional problems $\delta(x - x')$ is a one-dimensional delta function which represents a *plain surface heat source* located at x'.

Three-Dimensional Problems

The physical significance of Green's function $G(\mathbf{r}, t \,|\, \mathbf{r}', \tau)$ for the three-dimensional problems is as follows: It *represents the temperature at the location* $\mathbf{r}$, *at time* t, *due to an instantaneous point source of unit strength, located at the point* $\mathbf{r}'$, *releasing its energy spontaneously at time* $t = \tau$. The auxiliary problem satisfied by Green's function is valid over the same region R as the original physical problem (6-1), but the boundary conditions (6-2b) is the homogeneous version of the boundary conditions (6-1b) and the initial condition is zero.

On the basis of this definition, the physical significance of Green's function may be interpreted as

$$G(\mathbf{r}, t \,|\, \mathbf{r}', \tau) \equiv G(\text{effect} \,|\, \text{impulse}) \qquad \text{(6-3)}$$

The first part of the argument, "$\mathbf{r}, t$," represents the "*effect*," that is, the temperature in the medium at the location $\mathbf{r}$ at time t, while the second part, "$\mathbf{r}', \tau$," represents the *impulse*, that is, the impulsive (instantaneous) point source located at $\mathbf{r}'$, releasing its heat spontaneously at time τ.

The usefulness of Green's function lies in the fact that the solution of the original problem (6-1) can be represented only in terms of Green's function.

Therefore, once the Green's function is known, the temperature distribution $T(\mathbf{r}, t)$ in the medium is readily computed. The mathematical proof for the developments of such expressions can be found in the texts [1, 2, 6]. Here we present only the resulting expressions, illustrate their use with representative examples and describe a very simple approach for the determination of Green's functions.

In the case of three-dimensional transient, nonhomogeneous heat conduction problem given by equation (6-1), the solution for $T(\mathbf{r}, t)$ is expressed in terms of the three-dimensional Green's function $G(\mathbf{r}, t|\mathbf{r}', \tau)$ as

$$T(\mathbf{r},t) = \int_R G(\mathbf{r},t|\mathbf{r}',\tau)|_{\tau=0} F(\mathbf{r}')dv' + \frac{\alpha}{k}\int_{\tau=0}^{t} d\tau \int_R G(\mathbf{r},t|\mathbf{r}',\tau) g(\mathbf{r}',\tau)dv' + \alpha \int_{\tau=0}^{t} d\tau \sum_{i=1}^{N} \int_{S_i} G(\mathbf{r},t|\mathbf{r}',\tau)|_{\mathbf{r}'=\mathbf{r}_i} \frac{1}{k_i} f_i(\mathbf{r}',\tau) ds_i' \tag{6-4}$$

where R refers to the entire volume of the region considered; S_i refers to the boundary surface S_i of the region $R, i = 1, 2, \ldots, N$ and N is the number of continuous boundary surfaces; and dv' and ds_i' refer to differential volume and surface elements, respectively, in the $\mathbf{r}'$ variable. The physical significance of various terms in the solution (6-4) is as follows:

The first term on the right-hand side of equation (6-4) is for the contribution of the initial condition function $F(\mathbf{r})$ on the temperature distribution; that is, Green's function evaluated for $\tau = 0$ is multiplied by $F(\mathbf{r})$ and integrated over the region R.

The second term is for the contribution of the energy generation $g(\mathbf{r}, t)$ on the temperature $T(\mathbf{r}, t)$; that is, Green's function $G(\mathbf{r}, t|\mathbf{r}', \tau)$ multiplied by the energy generation $g(\mathbf{r}, \tau)$, integrated over the region R and over the time from $\tau = 0$ to t.

The last term represents the contribution of the nonhomogeneous terms $f_i(\mathbf{r}', \tau)$ of the boundary conditions on the temperature. It consists of Green's function evaluated at the boundary, multiplied by $f_i(\mathbf{r}', \tau)$, integrated over the boundary surface and over the time from $\tau = 0$ to t.

For generality, the physical problem (6-1) is formulated by considering a boundary condition of the *third kind* (i.e., convection) for which $f_i(\mathbf{r}, \tau) \equiv h_i T_{\infty i}(\mathbf{r}, \tau)$, where $T_{\infty i}(\mathbf{r}, t)$ is the ambient temperature. The solution (6-4) is also applicable for the boundary condition of the *second kind* (i.e., prescribed heat flux if $f_i(\mathbf{r}, \tau)$ is interpreted as the prescribed boundary heat flux. For such a case, we first set $h_i T_{\infty i} \equiv f_i(\mathbf{r}, \tau)$ and then let $h_i = 0$ on the left-hand side. In the case of boundary condition of the *first kind*, some modification is needed in the third term on the right-hand side of the solution (6-4). The reason for this is that the boundary condition of the first kind is obtainable from equation (6-1b) by setting $k_i = 0$; then h_i cancels out and $f_i(\mathbf{r}, \tau) \equiv T_{\infty i}(\mathbf{r}, \tau)$ represents the ambient temperature. For such a case, difficulty arises in setting $k_i = 0$ in the solution (6-4), because k_i appears in the denominator. This difficulty can be alleviated by making the following change in the last term in the solution (6-4):

$$\text{Replace} \quad \left.\frac{1}{k_i} G\right|_{\mathbf{r}'=\mathbf{r}_i} \quad \text{by} \quad -\left.\frac{1}{h_i}\frac{\partial G}{\partial n_i}\right|_{\mathbf{r}'=\mathbf{r}_i} \tag{6-5}$$

The validity of this replacement is apparent if the boundary condition (6-2b) of the auxiliary problem is rearranged in the form

$$\frac{1}{k_i} G = -\frac{1}{h_i}\frac{\partial G}{\partial n_i} \quad \text{on surface} \quad S_i \tag{6-6}$$

We now examine the application of the general solution (6-4) for the cases of two- and one-dimensional problems.

Two-Dimensional Problems

The problems defined by equations (6-1) and (6-2) are also applicable for the two-dimensional case, if ∇^2 is treated as a two-dimensional Laplacian operator and $\delta(\mathbf{r} - \mathbf{r}')$ as a two-dimensional delta function, that is, $\delta(\mathbf{r} - \mathbf{r}') \equiv \delta(x - x')\delta(y - y')$ in the (x, y) coordinate system, and so forth.

For such a case, the physical significance of the two-dimensional Green's function is as follows: *It represents the temperature $T(\mathbf{r}, t)$ at the location $\mathbf{r}$, at time t, in the two-dimensional region R, due to an instantaneous line source of unit strength, located at $\mathbf{r}'$, releasing its energy spontaneously at time $t = \tau$.* This interpretation is similar to that for the three-dimensional problem considered previously, except the source is a *line heat source* of unit strength.

For the two-dimensional case, the solution (6-4) reduces to

$$\begin{aligned} T(\mathbf{r}, t) = & \int_A G(\mathbf{r}, t|\mathbf{r}', \tau)|_{\tau=0} F(\mathbf{r}') dA' \\ & + \frac{\alpha}{k}\int_{\tau=0}^{t} d\tau \int_A G(\mathbf{r}, t|\mathbf{r}', \tau) g(\mathbf{r}', \tau) dA' \\ & + \alpha \int_{\tau=0}^{t} d\tau \sum_{i=1}^{N} \int_{\substack{\text{Boundary} \\ \text{path } i}} G(\mathbf{r}, t|\mathbf{r}', \tau)|_{\mathbf{r}'=\mathbf{r}_i} \frac{1}{k_i} f_i dl_i \end{aligned} \tag{6-7}$$

where A is the area of the region under consideration, dl_i is the differential length along the boundary path of the boundary $i, i = 1, 2, \ldots, N$, and N is the number of continuous boundary paths of the region A. For a boundary condition of the *first kind* at the boundary, say, $i=j$, the term $(1/k_j)G|_{\mathbf{r}'=\mathbf{r}'_j}$ should be replaced by $-(1/h_j)(\partial G/\partial n_j)|_{\mathbf{r}'=\mathbf{r}'_j}$ for the boundary $i=j$ in accordance with equation (6-5).

We note that the space integrations over the initial condition function $F(\mathbf{r})$ and the energy-generation function $g(\mathbf{r},t)$ are surface integrals instead of the volume integrals, while the integration over the boundary-condition function f_i is a contour integral instead of a surface integral.

One-Dimensional Problems

For the one-dimensional temperature field, the problems defined by equations (6-1) and (6-2) are applicable if ∇^2 is considered as one-dimensional Laplacian operator and $\delta(\mathbf{r}-\mathbf{r}')$ as one-dimensional delta function, that is, $\delta(\mathbf{r}-\mathbf{r}')=\delta(x-x')$ for the (x) coordinate, and so on. Then, *Green's function* $G(x,t|x',\tau)$ *represents the temperature* $T(x,t)$, *at the location* x, *at time* t, *due to an instantaneous surface heat source of strength unity, located at* x', *releasing its energy spontaneously at time* $t=\tau$.

For the one-dimensional case, the solution (6-4) reduces to

$$
\begin{aligned}
T(x,t) = & \int_L x'^P G(x,t|x',\tau)|_{\tau=0} F(x')dx' \\
& + \frac{\alpha}{k}\int_{\tau=0}^{t} d\tau \int_L x'^P G(x,t|x',\tau)| g(x',\tau)dx' \\
& + \alpha \int_{\tau=0}^{t} d\tau \sum_{i=1}^{2} [x'^P G(x,t|x',\tau)]_{x'=x_i} \frac{1}{k_i} f_i
\end{aligned} \tag{6-8}
$$

where x'^P is the Sturm–Liouville weight function such that

$$
P = \begin{cases} 0 & \text{slab} \\ 1 & \text{cylinder} \\ 2 & \text{sphere} \end{cases}
$$

Here L refers to the thickness or radius of the one-dimensional region and $G(x,t|x',\tau)|_{x'=x_i}$ refers to the value of G evaluated at the boundary points $x'=x_i$. For a boundary condition of the *first kind* at the boundary, say, $i=j$, the term $(1/k_j)G|_{x'=x_j}$ should be replaced by $-(1/h_j)(\partial G/\partial n_j)|_{x'=x_j}$ for the boundary $i=j$ in accordance with equation (6-5).

We note that, in equation (6-8) the space integrations over the initial condition function $F(x)$ and the energy-generation function $g(x,t)$ are line integrations, while the boundary-condition functions f_i are evaluated at the two boundary points.

6-2 REPRESENTATION OF POINT, LINE, AND SURFACE HEAT SOURCES WITH DELTA FUNCTIONS

The energy source will be called an *instantaneous source* if it releases its energy spontaneously or a *continuous source* if it releases its energy continuously over time. In the definition of Green's function, we also refer to a *point* source, a *line source*, and a *surface source* of unit strength, in addition to the customarily used volume heat source that has the dimension W/m^3.

In order to identify such energy sources with a unified notation we introduce the symbol

$$g_B^A$$

where the superscript A refers to

$$\begin{aligned} A &\equiv \quad \text{i} \qquad\qquad \text{or} \qquad \text{c} \\ &\equiv (\text{instantaneous}) \quad \text{or} \quad (\text{continuous}) \end{aligned}$$

and the subscript B denotes

$$\begin{aligned} B &\equiv \quad \text{p}, \qquad \text{L}, \qquad \text{or} \qquad \text{s} \\ &\equiv (\text{point}), \quad (\text{line}), \quad \text{or} \quad (\text{surface}) \end{aligned}$$

and no subscript will be used for the volumetric source. Thus, based on the above notation, we write

$$\begin{aligned} g_\text{p}^\text{i} &= \text{instantaneous point source} \\ g_\text{p}^\text{c} &= \text{continuous point source} \\ g_\text{L}^\text{i} &= \text{instantaneous line source} \\ g_\text{s}^\text{c} &= \text{continuous surface source} \\ g^\text{i} &= \text{instantaneous volumetric source} \\ g &= \text{volumetric source} \end{aligned}$$

and so forth.

In the analytic solution of temperature $T(\mathbf{r}, t)$ in terms of Green's functions given by equations (6-4), (6-7) and (6-8), the energy-generation term $g(\mathbf{r}, t)$ appears under the integral sign. In order to perform the integration over a point source, surface source, instantaneous source, and so on, proper mathematical representations should be used to define such sources.

Here we describe a procedure for the identification of such sources with the delta-function notation and the determination of their dimensions.

Three-Dimensional Case

Rectangular Coordinates. Consider an instantaneous point heat source g_p^i located at the point (x', y', z') and releasing its entire energy spontaneously at time $t = \tau$. Such a source is related to the volumetric heat source $g(x, y, z, t)$ by

$$g_p^i \delta(x - x')\delta(y - y')\delta(z - z')\delta(t - \tau) \equiv g(x, y, z, t) \tag{6-9a}$$

where $\delta(.)$ is the Dirac delta function. A brief description of the properties of Dirac's delta function is given in Appendix VII.

When the dimensions are introduced into equation (6-9a), we obtain

$$\begin{array}{cccccc} g_p^i \delta(x - x') & \delta(y - y') & \delta(z - z') & \delta(t - \tau) & \equiv & g(x, y, z, t) \\ g_p^i \ \mathrm{m}^{-1} & \mathrm{m}^{-1} & \mathrm{m}^{-1} & \mathrm{s}^{-1} & & \mathrm{W\,m}^{-3} \end{array} \tag{6-9b}$$

Hence the dimension of an instantaneous point source g_p^i is Ws.

Cylindrical Coordinates. In the case of (r, ϕ, z) cylindrical coordinates, equation (6-9b) takes the form

$$\begin{array}{ccccc} g_p^i \dfrac{1}{r} & \delta(r - r')\delta(\phi - \phi') & \delta(z - z') & \delta(t - \tau) & \equiv g(r, \phi, z, t) \\ g_p^i \mathrm{m}^{-1}\mathrm{m}^{-1} & & \mathrm{m}^{-1} & \mathrm{s}^{-1} & \mathrm{W\,m}^{-3} \end{array} \tag{6-9c}$$

Hence g_p^i has the dimension Ws. The term r^{-1} appearing in equation (6-9c) is due to the scale factors associated with the transformation of the reciprocal of the volume element $(dV)^{-1}$ from the rectangular to a curvilinear coordinate system according to equation (1-25b): $(dV)^{-1} = (a_1 a_2 a_3\, du_1\, du_2\, du_3)^{-1}$. In the case of the cylindrical coordinate system, we have $(a_r a_\phi a_z)^{-1} = (1 \cdot r \cdot 1)^{-1} = r^{-1}$.

Spherical Coordinates. In the case of (r, ϕ, μ) spherical coordinate system, equation (6-9b) takes the form

$$\begin{array}{ccccc} g_p^i \dfrac{1}{r^2\sqrt{1-\mu^2}} & \delta(r - r') & \delta(\phi - \phi')\delta(\mu - \mu') & \delta(t - \tau) & \equiv g(r, \phi, \mu, t) \\ g_p^i \mathrm{m}^{-2} & \mathrm{m}^{-1} & & \mathrm{s}^{-1} & \mathrm{W\,m}^{-3} \end{array} \tag{6-9d}$$

Hence g_p^i has the dimension Ws. The term $(r^2\sqrt{1-\mu^2})^{-1}$ appearing in equation (6-9d), as stated above, is associated with the transformation of $(dV)^{-1}$ from the rectangular to the spherical coordinate system. That is, $(a_r a_\phi a_\theta)^{-1} = (1 \cdot r \sin\theta \cdot r)^{-1} = (r^2\sqrt{1-\mu^2})^{-1}$, where $\mu = \cos\theta$.

In the case of *continuous point source* g_p^c, the representation has no delta function with respect to time; hence the dimension of g_p^c is W.

One-Dimensional Case

We now examine the representation of an instantaneous energy source in the one-dimensional rectangular, cylindrical, and spherical coordinates.

Rectangular Coordinates. An instantaneous plane-surface heat source g_s^i is represented by

$$g_s^i \delta(x - x')\delta(t - \tau) \equiv g(x, t) \tag{6-10a}$$

$$g_s^i \mathrm{m}^{-1} \qquad \mathrm{s}^{-1} \qquad \mathrm{W\,m}^{-3}$$

Hence g_s^i has the dimension $(\mathrm{Ws})/\mathrm{m}^2$.

Cylindrical Coordinates. An instantaneous cylindrical-surface heat source g_s^i is represented by

$$g_s^i \frac{1}{2\pi r} \delta(r - r')\delta(t - \tau) \equiv g(r, t) \tag{6-10b}$$

$$g_s^i \mathrm{m}^{-1}\mathrm{m}^{-1} \qquad \mathrm{s}^{-1} \qquad \mathrm{Wm}^{-3}$$

Hence an instantaneous cylindrical-surface heat source has the dimension (Ws)/m. In equation (6-10b), the variable r appearing in the denominator is associated with the scale factor of the transformation. That is, g_s^i represents the strength of the cylindrical surface source per unit length and the quantity $g_s^i/2\pi r$ represents the source strength per unit area.

Spherical Coordinates. An instantaneous spherical-surface source g_s^i is represented by

$$g_s^i \frac{1}{4\pi r^2} \delta(r - r')\delta(t - \tau) \equiv g(r, t) \tag{6-10c}$$

$$g_s^i \mathrm{m}^{-2}\mathrm{m}^{-1} \qquad \mathrm{s}^{-1} \qquad \mathrm{Wm}^{-3}$$

Thus an instantaneous spherical-surface heat source has the dimension Ws. The variable r^2 appearing in the denominator is associated with the scale factor of the transformation. That is, $g_s^i/4\pi r^2$ represents the spherical surface source of strength per unit area.

6-3 DETERMINATION OF GREEN'S FUNCTIONS

Once Green's function is available, the temperature distribution $T(\mathbf{r}, t)$ in a medium is determined from the expressions given by equations (6-4), (6-7), and (6-8), respectively, for the three-, two-, and one-dimensional transient linear heat conduction problems. Therefore, the establishment of the proper Green's function for any given situation is an integral part of the solution methodology utilizing the Green's function approach. Reference 1 uses the Laplace transform technique, and reference 2 describes the method of images for the determination of Green's functions. Here we present a very simple, straightforward yet very general approach that utilizes the classical separation of variables technique for the determination of Green's functions.

We consider the following, three-dimensional, homogeneous transient heat conduction problem:

$$\nabla^2 T(\mathbf{r},t) = \frac{1}{\alpha}\frac{\partial T(\mathbf{r},t)}{\partial t} \qquad \text{in} \qquad \text{region } R, \quad t>0 \tag{6-11a}$$

$$\frac{\partial T}{\partial n_i} + H_i T = 0 \qquad \text{on} \qquad S_i, \qquad t>0 \tag{6-11b}$$

$$T(\mathbf{r},t) = F(\mathbf{r}) \qquad \text{for} \qquad t=0, \qquad \text{in region } R \tag{6-11c}$$

The solution of this problem has been extensively studied in the Chapters 2–4 by the method of separation of variables, and a large number of specific solutions has been already generated for a variety of situations. Suppose the solution of the homogeneous problem (6-11) is symbolically expressed in the form

$$T(\mathbf{r},t) = \int_R K(\mathbf{r},\mathbf{r}',t)\cdot F(\mathbf{r}')dv' \tag{6-12}$$

The physical significance of equation (6-12) implies that all the terms in the solution, except the initial condition function, are lumped into a single term $K(\mathbf{r},\mathbf{r}',t)$, that we shall call the kernel of the integration. The kernel $K(\mathbf{r},\mathbf{r}',t)$, multiplied by the initial condition function $F(\mathbf{r}')$ and integrated over the region R, gives the solution to Problem (6-11).

Now we consider the Green's function approach for the solution of the problem (6-11). It is obtained from the general solution (6-4) as

$$T(\mathbf{r},t) = \int_R G(\mathbf{r},t|\mathbf{r}',\tau)|_{\tau=0}\cdot F(\mathbf{r}')dv' \tag{6-13}$$

since the generation and the nonhomogeneous boundary condition functions are all zero.

A comparison of the solutions (6-12) and (6-13) implies that

$$G(\mathbf{r},t|\mathbf{r}',\tau)|_{\tau=0} = K(\mathbf{r},\mathbf{r}',t) \tag{6-14}$$

Then, we conclude that the kernel $K(\mathbf{r},\mathbf{r}',t)$, obtained by rearranging the homogeneous part of the transient heat conduction equation in the form given by equation (6-12), represents Green's function evaluated for $\tau = 0$: $G(\mathbf{r}, t|\mathbf{r}', 0)$.

Therefore, the solutions developed in Chapters 2–4 for the homogeneous transient heat conduction problems can readily be rearranged in the form given by equation (6-12) in order to obtain $G(\mathbf{r}, t|\mathbf{r}', 0)$. That is, to obtain $G(\mathbf{r}, t|\mathbf{r}', 0)$, the appropriate homogeneous problem is solved and rearranged in the form given by equation (6-12).

The general solution given by equation (6-4) requires that Green's function $G(\mathbf{r}, t|\mathbf{r}', \tau)$ should also be known in order to determine the contributions of the energy generation and nonhomogeneous boundary conditions on the solution.

It has been shown by Özişik [6] that *Green's function* $G(\mathbf{r}, t|\mathbf{r}', \tau)$ *for the transient heat conduction is obtainable from* $G(\mathbf{r}, t|\mathbf{r}', 0)$ *by replacing t by* $(t - \tau)$ *in the latter.*

The validity of this result will also be shown in Chapter 13.

We now illustrate the determination of Green's function from the solution of homogeneous problems with specific examples. In order to alleviate the details of the solution procedure, examples are chosen from those problems that have already been solved in the previous chapters.

Example 6-1

Determine the Green's function appropriate for the solution of the following nonhomogeneous heat conduction problem for a solid cylinder:

$$\frac{1}{r}\frac{\partial}{\partial r}\left(r\frac{\partial T}{\partial r}\right) + \frac{1}{k}g(r, t) = \frac{1}{\alpha}\frac{\partial T}{\partial t} \quad \text{in} \quad 0 \leqslant r < b, \quad t > 0 \tag{6-15a}$$

$$T = f(t) \quad \text{at} \quad r = b, \quad t > 0 \tag{6-15b}$$

$$T = F(r) \quad \text{for} \quad t = 0, \quad \text{in} \quad 0 \leqslant r \leqslant b \tag{6-15c}$$

Solution. To determine the desired Green's function we consider the homogeneous version of the problem defined by equations (6-15) for the same region given as

$$\frac{1}{r}\frac{\partial}{\partial r}\left(r\frac{\partial \psi}{\partial r}\right) = \frac{1}{\alpha}\frac{\partial \psi}{\partial t} \quad \text{in} \quad 0 \leqslant r < b, \quad t > 0 \tag{6-16a}$$

$$\psi = 0 \quad \text{at} \quad r = b, \quad t > 0 \tag{6-16b}$$

$$\psi = F(r) \quad \text{for} \quad t = 0, \quad \text{in} \quad 0 \leqslant r \leqslant b \tag{6-16c}$$

This homogeneous problem can readily be solved by the method of separation of variables; or its solution is immediately obtainable from equation (3-67) of Example 3-3. We write this solution in the same general form as given by equation (6-12), namely, as

$$\psi(\mathbf{r}, t) = \int_{r'=0}^{b} \left[\frac{2}{b^2} \sum_{m=1}^{\infty} e^{-\alpha\beta_m^2 t} \frac{1}{J_1^2(\beta_m b)} r' J_0(\beta_m r) J_0(\beta_m r') \right] F(r')dr' \tag{6-17}$$

where the β_m values are the roots of $J_0(\beta_m b) = 0$.

The solution of the homogeneous problem (6-16) in terms of Green's function is given, according to equation (6-13), as

$$\psi(r, t) = \int_{r'=0}^{b} r' G(r, t|r', \tau)|_{\tau=0} F(r')dr' \tag{6-18}$$

By comparing the two solutions (6-17) and (6-18) we find the Green's function for $\tau = 0$.

$$G(r, t|r', \tau)|_{\tau=0} = \frac{2}{b^2} \sum_{m=1}^{\infty} e^{-\alpha\beta_m^2 t} \frac{1}{J_1^2(\beta_m b)} J_0(\beta_m r) J_0(\beta_m r') \tag{6-19}$$

Then, replacing t by $(t - \tau)$ in equation (6-19) we obtain the desired Green's function as

$$G(r, t|r', \tau) = \frac{2}{b^2} \sum_{m=1}^{\infty} e^{-\alpha\beta_m^2(t-\tau)} \frac{1}{J_1^2(\beta_m b)} J_0(\beta_m r) J_0(\beta_m r') \tag{6-20}$$

Example 6-2

Determine the Green's function appropriate for the solution of the following nonhomogeneous transient heat conduction problem:

$$\frac{\partial^2 T}{\partial x^2} + \frac{1}{k} g(x, t) = \frac{1}{\alpha} \frac{\partial T}{\partial t} \quad \text{in} \quad 0 < x < \infty, \quad t > 0 \tag{6-21a}$$

$$T = f(t) \quad \text{at} \quad x = 0 \tag{6-21b}$$

$$T = F(x) \quad \text{for} \quad t = 0, \quad \text{in} \quad 0 < x < \infty \tag{6-21c}$$

Solution. We consider the homogeneous part of this problem given by

$$\frac{\partial^2 \psi}{\partial x^2} = \frac{1}{\alpha} \frac{\partial \psi}{\partial t} \quad \text{in} \quad 0 < x < \infty, \quad t > 0 \tag{6-22a}$$

$$\psi = 0 \quad \text{at} \quad x = 0, \quad t > 0 \tag{6-22b}$$

$$\psi = F(x) \quad \text{for} \quad t = 0, \quad \text{in} \quad 0 < x < \infty \tag{6-22c}$$

The solution of this problem is obtained from equation (2-58a) and rearranged in the form

$$\psi(x,t)=\int_{x'=0}^{\infty}\frac{1}{(4\pi\alpha t)^{1/2}}\left[\exp\left(-\frac{(x-x')^2}{4\alpha t}\right)-\exp\left(-\frac{(x+x')^2}{4\alpha t}\right)\right]F(x')dx' \tag{6-23}$$

By comparing this solution with equation (6-13) we conclude that $G(x,t|x',\tau)|_{\tau=0}$ is given by

$$G(x,t|x',0)=\frac{1}{(4\pi\alpha t)^{1/2}}\left[\exp\left(-\frac{(x-x')^2}{4\alpha t}\right)-\exp\left(-\frac{(x+x')^2}{4\alpha t}\right)\right] \tag{6-24}$$

Green's function $G(x,t|x',\tau)$ is determined by replacing t by $(t-\tau)$ in this equation:

$$G(x,t|x',\tau)=\frac{1}{[4\pi\alpha(t-\tau)]^{1/2}}\left[\exp\left(-\frac{(x-x')^2}{4\alpha(t-\tau)}\right)-\exp\left(-\frac{(x+x')^2}{4\alpha(t-\tau)}\right)\right] \tag{6-25}$$

Example 6-3

Determine the Green's function for the solution of the following nonhomogeneous transient heat conduction in a slab of thickness L.

$$\frac{\partial^2 T}{\partial x^2}+\frac{1}{k}g(x,t)=\frac{1}{\alpha}\frac{\partial T}{\partial t} \qquad \text{in} \qquad 0<x<L, \quad t>0 \tag{6-26a}$$

$$\frac{\partial T}{\partial x}=f_1(t) \qquad \text{at} \qquad x=0, \qquad t>0 \tag{6-26b}$$

$$\frac{\partial T}{\partial x}+HT=f_2(t) \qquad \text{at} \qquad x=L, \qquad t>0 \tag{6-26c}$$

$$T=F(x) \qquad \text{for} \qquad t=0, \qquad \text{in} \quad 0<x<L \tag{6-26d}$$

Solution. We consider only the homogeneous version of this problem given by

$$\frac{\partial^2 \psi}{\partial x^2}=\frac{1}{\alpha}\frac{\partial \psi}{\partial t} \qquad \text{in} \qquad 0<x<L, \quad t>0 \tag{6-27a}$$

$$\frac{\partial \psi}{\partial x}=0 \qquad \text{at} \qquad x=0, \qquad t>0 \tag{6-27b}$$

$$\frac{\partial \psi}{\partial x} + H\psi = 0 \qquad \text{at} \qquad x = L, \qquad t > 0 \tag{6-27c}$$

$$\psi = F(x) \qquad \text{for} \qquad t = 0, \qquad \text{in} \quad 0 < x < L \tag{6-27d}$$

The solution of this problem is obtained from equation (2-45a) and rearranged in the form

$$\psi(x,t) = 2\int_{x'=0}^{L}\left[\sum_{m=1}^{\infty} e^{\alpha\beta_m^2 t}\frac{\beta_m^2 + H^2}{L(\beta_m^2 + H^2) + H}\cos\beta_m x\cos\beta_m x'\right]F(x')dx' \tag{6-28}$$

By comparing this solution with equation (6-13) we conclude that $G(x,t|x',0)$ is given by

$$G(x,t|x',0) = 2\sum_{m=1}^{\infty} e^{-\alpha\beta_m^2 t}\frac{\beta_m^2 + H^2}{L(\beta_m^2 + H^2) + H}\cos\beta_m x\cos\beta_m x' \tag{6-29}$$

and Green's function $G(x,t|x',\tau)$ is obtained by replacing t by $(t-\tau)$ in the expression

$$G(x,t|x',\tau) = 2\sum_{m=1}^{\infty} e^{-\alpha\beta_m^2(t-\tau)}\frac{\beta_m^2 + H^2}{L(\beta_m^2 + H^2) + H}\cos\beta_m x\cos\beta_m x' \tag{6-30}$$

In the following sections we illustrate the application of Green's function technique for the solution of nonhomogeneous boundary-value problems of heat conduction in the rectangular, cylindrical, and spherical coordinate systems.

6-4 APPLICATIONS OF GREEN'S FUNCTION IN THE RECTANGULAR COORDINATE SYSTEM

In this section we illustrate with examples the application of the Green's function technique in the solution of nonhomogeneous boundary-value problems of heat conduction in the rectangular coordinate system. For convenience in the determination of Green's function we consider, as examples, those problems for which solutions are available in Chapter 2 for their homogeneous part.

Example 6-4

An infinite medium $-\infty < x < \infty$ is initially at temperature $F(x)$; for times $t > 0$ there is heat generation within the solid at a rate of $g(x,t)$ per unit time, per unit volume. Obtain an expression for the temperature distribution $T(x, t)$ for times $t > 0$ by the Green's function technique.

Solution. The mathematical formulation of this problem is given as

$$\frac{\partial^2 T(x,t)}{\partial x^2} + \frac{1}{k} g(x,t) = \frac{1}{\alpha} \frac{\partial T}{\partial t} \quad \text{in} \quad -\infty < x < \infty, \quad t > 0 \tag{6-31a}$$

$$T = F(x) \quad \text{for} \quad t = 0, \quad \text{in the region} \tag{6-31b}$$

To determine Green's function we consider the homogeneous version of this problem given as

$$\frac{\partial^2 \psi(x,t)}{\partial x^2} = \frac{1}{\alpha} \frac{\partial \psi(x,t)}{\partial t} \quad \text{in} \quad -\infty < x < \infty, \quad t > 0 \tag{6-32a}$$

$$\psi = F(x) \quad \text{for} \quad t = 0, \quad \text{in the region} \tag{6-32b}$$

The solution of this homogeneous problem is obtainable from equation (2-70) as

$$\psi(x,t) = \int_{x'=-\infty}^{\infty} \left[(4\pi\alpha t)^{-1/2} \exp\left(-\frac{(x-x')^2}{4\alpha t} \right) \right] F(x') dx'. \tag{6-33}$$

The solution of the problem (6-32) can be written in terms of Green's function, according to equation (6-13), as

$$\psi(x,t) = \int_{x'=-\infty}^{\infty} G(x,t|x',\tau)|_{\tau=0} F(x') dx' \tag{6-34}$$

A comparison of equations (6-33) and (6-34) yields

$$G(x,t|x',\tau)|_{\tau=0} = (4\pi\alpha t)^{-1/2} \exp\left(-\frac{(x-x')^2}{4\alpha t} \right) \tag{6-35}$$

The desired Green's function is obtained by replacing t by $(t-\tau)$ in equation (6-35); we find

$$G(x,t|x',\tau) = [4\pi\alpha(t-\tau)]^{-1/2} \exp\left(-\frac{(x-x')^2}{4\alpha(t-\tau)} \right) \tag{6-36}$$

Then the solution of the nonhomogeneous problem (6-31), according to equation (6-8), is given as

$$T(x,t) = (4\pi\alpha t)^{-1/2} \int_{x'=-\infty}^{\infty} \exp\left[-\frac{(x-x')^2}{4\alpha t}\right] F(x')dx'$$
$$+ \frac{\alpha}{k} \int_{\tau=0}^{t} d\tau \int_{x'=-\infty}^{\infty} [4\pi\alpha(t-\tau)]^{-1/2} \exp\left[-\frac{(x-x')^2}{4\alpha(t-\tau)}\right] g(x',\tau)dx'. \tag{6-37}$$

We now examine some special cases of the solution (6-37).

1. There is no heat generation. By setting $g(x',\tau) = 0$, equation (6-37) reduces to

$$T(x,t) = (4\pi\alpha t)^{-1/2} \int_{x'=-\infty}^{\infty} \exp\left[-\frac{(x-x')^2}{4\alpha t}\right] F(x')dx' \tag{6-38}$$

which is the same as that given by equation (2-70).

2. Medium is initially at zero temperature, an instantaneous distributed heat source of strength $g^i(x)$ Ws/m^3 releases its heat spontaneously at time $t = 0$. By setting

$$F(x) = 0, \qquad g(x,t) = g^i(x)\delta(t-0) \tag{6-39}$$

equation (6-37) reduces to

$$T(x,t) = (4\pi\alpha t)^{-1/2} \int_{x'=-\infty}^{\infty} \exp\left[-\frac{(x-x')^2}{4\alpha t}\right]\left[\frac{\alpha}{k} g^i(x')\right] dx' \tag{6-40}$$

A comparison of equations (6-38) and (6-40) reveals that

$$F(x') \equiv \frac{\alpha}{k} g^i(x') = \frac{1}{\rho c_p} g^i(x') \tag{6-41}$$

Equation (6-41) implies that the heat-conduction problem for an instantaneous distributed heat source $g^i(x)$ releasing its heat at time $t = 0$ is equivalent to an initial value problem with the initial temperature distribution as given by equation (6-41).

3. Medium is initially at zero temperature; for times $t > 0$ a plane surface heat source of strength $g_s^c(t)$ W/m^2 situated at $x = a$ releases its heat continuously. By setting

$$F(x) = 0, \qquad g(x,t) = g_s^c(t)\delta(x-a) \tag{6-42}$$

equation (6-37) reduces to

$$T(x,t)=\frac{\alpha}{k}\int_{\tau=0}^{t}[4\pi\alpha(t-\tau)]^{-1/2}\exp\left[-\frac{(x-a)^2}{4\alpha(t-\tau)}\right]g_s^c(\tau)d\tau \tag{6-43}$$

Example 6-5

A slab, $0\leqslant x\leqslant L$, is initially at temperature $F(x)$. For times $t>0$, the boundaries at $x=0$ and $x=L$ are maintained at temperatures $f_1(t)$ and $f_2(t)$ respectively, whereas heat is generated in the medium at a rate of $g(x,t)$ W/m^3. Obtain an expression for the temperature distribution $T(x,t)$ in the slab for times $t>0$.

Solution. The mathematical formulation of this problem is given as

$$\frac{\partial^2 T(x,t)}{\partial x^2}+\frac{1}{k}g(x,t)=\frac{1}{\alpha}\frac{\partial T(x,t)}{\partial t} \quad \text{in} \quad 0<x<L, \quad t>0 \tag{6-44a}$$

$$T=f_1(t) \quad \text{at} \quad x=0, \quad t>0 \tag{6-44b}$$

$$T=f_2(t) \quad \text{at} \quad x=L, \quad t>0 \tag{6-44c}$$

$$T=F(x) \quad \text{for} \quad t=0, \quad \text{in} \quad 0\leqslant x\leqslant L \tag{6-44d}$$

To determine the appropriate Green's function, we consider the homogeneous version of this problem as

$$\frac{\partial^2 \psi(x,t)}{\partial x^2}=\frac{1}{\alpha}\frac{\partial \psi(x,t)}{\partial t} \quad \text{in} \quad 0<x<L, \quad t>0 \tag{6-45a}$$

$$\psi=0 \quad \text{at} \quad x=0 \text{ and } x=L, \quad t>0 \tag{6-45b}$$

$$\psi=F(x) \quad \text{for} \quad t=0, \quad \text{in} \quad 0\leqslant x\leqslant L \tag{6-45c}$$

The problem (6-45) is exactly the same as that given by equations (2-151), and its solution is obtainable from equations (2-154) and (2-155a,b) as

$$\psi(x,t)=\int_{x'=0}^{L}\left[\frac{2}{L}\sum_{m=1}^{\infty}e^{-\alpha\beta_m^2 t}\sin\beta_m x\sin\beta_m x'\right]F(x')dx' \tag{6-46}$$

where

$$\beta_m=\frac{m\pi}{L}, \qquad m=1,2,3,\ldots$$

Also, the solution of problem (6-45) in terms of Green's function is given, according to equation (6-13), as

$$\psi(x,t)=\int_{x'=0}^{L} G(x,t|x',\tau)|_{\tau=0}F(x')dx' \tag{6-47}$$

A comparison of equations (6-46) and (6-47) gives

$$G(x,t|x',\tau)|_{\tau=0}=\frac{2}{L}\sum_{m=1}^{\infty} e^{-\alpha\beta_m^2 t}\sin\beta_m x\sin\beta_m x' \tag{6-48}$$

The desired Green's function is obtained by replacing t by $(t-\tau)$ in equation (6-48); we find

$$G(x,t|x',\tau)=\frac{2}{L}\sum_{m=1}^{\infty} e^{-\alpha\beta_m^2(t-\tau)}\sin\beta_m x\sin\beta_m x' \tag{6-49}$$

Then the solution of the nonhomogeneous problem (6-44) is given in terms of the Green's function, according to equation (6-8), as

$$\begin{aligned}T(x,t)=&\int_{x'=0}^{L} G(x,t|x',\tau)|_{\tau=0}F(x')dx'\\&+\frac{\alpha}{k}\int_{\tau=0}^{t}d\tau\int_{x'=0}^{L} G(x,t|x',\tau)g(x',\tau)dx'\\&+\alpha\int_{\tau=0}^{t}\left.\frac{\partial G(x,t|x',\tau)}{\partial x'}\right|_{x'=0} f_1(\tau)d\tau\\&-\alpha\int_{\tau=0}^{t}\left.\frac{\partial G(x,t|x',\tau)}{\partial x'}\right|_{x'=L} f_2(\tau)d\tau\end{aligned} \tag{6-50}$$

We note that in the problem (6-44) the boundary conditions are both of the first kind. Therefore, in the solution (6-50), we made replacements according to equation (6-5) in the terms involving the boundary-condition functions $f_1(\tau)$ and $f_2(\tau)$. Namely, we replaced $G|_{x'=0}$ by $+(\partial G/\partial x')|_{x'=0}$ for the terms involving $f_1(\tau)$ and $G|_{x'=L}$ by $-(\partial G/\partial x')|_{x'=L}$ for the term involving $f_2(\tau)$.

Introducing the above expression for Green's function into equation (6-50) we obtain the solution in the form

$$\begin{aligned}T(x,t)=&\frac{2}{L}\sum_{m=1}^{\infty} e^{-\alpha\beta_m^2 t}\sin\beta_m x\int_{x'=0}^{L}\sin\beta_m x' F(x')dx'\\&+\frac{\alpha}{k}\frac{2}{L}\sum_{m=1}^{\infty} e^{-\alpha\beta_m^2 t}\sin\beta_m x\int_{\tau=0}^{t} e^{\alpha\beta_m^2\tau}d\tau\int_{x'=0}^{L}\sin\beta_m x' g(x',\tau)dx'\end{aligned}$$

$$+\alpha\frac{2}{L}\sum_{m=1}^{\infty} e^{-\alpha\beta_m^2 t}\beta_m \sin\beta_m x\int_{\tau=0}^{t} e^{\alpha\beta_m^2\tau} f_1(\tau)d\tau$$

$$-\alpha\frac{2}{L}\sum_{m=1}^{\infty}(-1)^m e^{-\alpha\beta_m^2 t}\beta_m \sin\beta_m x\int_{\tau=0}^{t} e^{\alpha\beta_m^2\tau} f_2(\tau)d\tau \qquad (6\text{-}51)$$

where

$$\beta_m=\frac{m\pi}{L}, \qquad m=1,2,3,\ldots$$

The solution (6-51) appears to vanish at the two boundaries $x=0$ and $x=L$, instead of yielding the boundary conditions functions $f_1(t)$ and $f_2(t)$ at these locations. The reason for this is that these two terms involve series that are not uniformly convergent at the location $x=0$ and $x=L$. Therefore, the above solution is valid in the open interval $0<x<L$. Such phenomena occur when the solution derives its basis from the orthogonal expansion technique with the boundary condition being utilized to develop the eigencondition. Similar results are reported in pages 102 and 103 of reference 1. This difficulty can be alleviated by integrating by parts the last two integrals in equation (6-51), and replacing the resulting series expressions by their equivalent closed-form expressions. Another approach to avoid this difficulty is to remove the non-homogeneities from the boundary condition by a splitting-up procedure as described in Section 1–7 of Chapter 1. We now examine some special cases of solution (6-51):

1. The medium is initially at zero temperature. The boundaries at $x=0$ and $x=L$ are kept at zero temperature for times $t>0$, and a distributed heat source of strength $g^i(x)$ W/m^3 releases its heat spontaneously at time $t=0$. For this special case we set

$$F(x)=0, \qquad f_1(t)=0, \qquad f_2(t)=0, \qquad \text{and} \qquad g(x,t)=g^i(x)\delta(t-0) \qquad (6\text{-}52)$$

Then, the solution (6-51) reduces to

$$T(x,t)=\frac{2}{L}\sum_{m=1}^{\infty} e^{-\alpha\beta_m^2 t}\sin\beta_m x\int_{x'=0}^{L}\left[\frac{\alpha}{k}g^i(x')\right]\sin\beta_m x'\,dx'. \qquad (6\text{-}53)$$

A comparison of this solution with the first term in equation (6-51) reveals that the problem of heat conduction for an instantaneous distributed heat source $g^i(x)$ W/m^3 releasing its heat at time $t=0$ is equivalent to the problem in which the medium is initially at a temperature

$$F(x')\equiv\frac{\alpha}{k}g^i(x')=\frac{1}{\rho c_p}g^i(x') \qquad (6\text{-}54)$$

2. Medium is initially at zero temperature; for times $t > 0$ boundaries at $x = 0$ and $x = L$ are kept at zero temperature and a plane surface heat source of strength $g_s^c(t)\,\mathrm{W/m^2}$ situated at $x = a(< L)$ releases its heat continuously. For this case we set

$$F(x) = f_1(t) = f_2(t) = 0 \qquad \text{amd} \qquad g(x,t) = g_s^c(t)\delta(x - a) \tag{6-55}$$

Then equation (6-51) reduces to

$$T(x,t) = \frac{2\alpha}{kL}\sum_{m=1}^{\infty} e^{-\alpha\beta_m^2 t}\sin\beta_m x \sin\beta_m a \int_{\tau=0}^{t} e^{\alpha\beta_m^2\tau} g_s^c(\tau)\,d\tau \tag{6-56}$$

where

$$\beta_m = \frac{m\pi}{L}, \qquad m = 1, 2, 3, \ldots$$

Example 6-6

A rectangular parallelepiped, $0 \leqslant x \leqslant a, 0 \leqslant y \leqslant b, 0 \leqslant z \leqslant c$, is initially at temperature $F(x, y, z)$. For times $t > 0$ heat is generated in the medium at a rate of $g(x, y, z, t)\,\mathrm{W/m^3}$ while the boundary surfaces are kept at zero temperature. Obtain an expression for the temperature distribution in the solid for times $t > 0$.

Solution. The mathematical formulation of this problem is given as

$$\frac{\partial^2 T}{\partial x^2} + \frac{\partial^2 T}{\partial y^2} + \frac{\partial^2 T}{\partial z^2} + \frac{1}{k}g(x,y,z,t) = \frac{1}{\alpha}\frac{\partial T}{\partial t}$$

$$\text{in} \quad 0 < x < a, 0 < y < b, 0 < z < c, \quad \text{for } t > 0 \tag{6-57a}$$

$$T = 0 \quad \text{at} \quad \text{all boundaries,} \quad \text{for } t > 0 \tag{6-57b}$$

$$T = F(x,y,z) \quad \text{for} \quad t = 0, \text{ in the region} \tag{6-57c}$$

To determine the appropriate Green's function, we consider the homogeneous version of this problem as

$$\frac{\partial^2 \psi}{\partial x^2} + \frac{\partial^2 \psi}{\partial y^2} + \frac{\partial^2 \psi}{\partial z^2} = \frac{1}{\alpha}\frac{\partial \psi}{\partial t}$$

$$\text{in} \quad 0 < x < a, 0 < y < b, 0 < z < c, \quad \text{for } t > 0 \tag{6-58a}$$

$$\psi = 0 \quad \text{at} \quad \text{all boundaries,} \quad \text{for } t > 0 \tag{6-58b}$$

$$\psi = F(x,y,z) \quad \text{for} \quad t = 0, \text{ in the region.} \tag{6-58c}$$

The problem (6-58) is the same as that given by equations (2-87); its solution is obtainable from equation (2-92) as

$$\psi(x,y,z,t)=\int_{x'=0}^{a}\int_{y'=0}^{b}\int_{z'=0}^{c}\left[\frac{8}{abc}\sum_{m=1}^{\infty}\sum_{n=1}^{\infty}\sum_{p=1}^{\infty}e^{-\alpha(\beta_m^2+\gamma_n^2+\eta_p^2)t}\right.$$
$$\left.\cdot\sin\beta_m x\sin\gamma_n y\sin\eta_p z\sin\beta_m x'\sin\gamma_n y'\sin\eta_p z'\right]$$
$$\cdot F(x',y',z')dx'\,dy'\,dz' \tag{6-59}$$

where

$$\beta_m=\frac{m\pi}{a},\qquad \gamma_n=\frac{n\pi}{b},\qquad \eta_p=\frac{p\pi}{c},\qquad \text{with } (m,n,p)=1,2,3,4\ldots$$

Also the solution of the problem (6-58) in terms of Green's function is given, according to equation (6-13), as

$$\psi(x,y,z,t)=\int_{x'=0}^{a}\int_{y'=0}^{b}\int_{z'=0}^{c}G(x,y,z,t|x',y',z',\tau)|_{\tau=0}$$
$$\cdot F(x',y',z')dx'\,dy'\,dz' \tag{6-60}$$

A comparison of equations (6-59) and (6-60) gives

$$G(x,y,z,t|x',y',z',\tau)|_{\tau=0}=\frac{8}{abc}\sum_{m=1}^{\infty}\sum_{n=1}^{\infty}\sum_{p=1}^{\infty}e^{-\alpha(\beta_m^2+\gamma_n^2+\eta_p^2)t}$$
$$\cdot\sin\beta_m x\sin\gamma_n y\sin\eta_p z$$
$$\cdot\sin\beta_m x'\sin\gamma_n y'\sin\eta_p z' \tag{6-61}$$

The desired Green's function is obtained by replacing t by $(t-\tau)$ in equation (6-61); we find

$$G(x,y,z,t|x',y',z',\tau)=\frac{8}{abc}\sum_{m=1}^{\infty}\sum_{n=1}^{\infty}\sum_{p=1}^{\infty}e^{-\alpha(\beta_m^2+\gamma_n^2+\eta_p^2)(t-\tau)}$$
$$\cdot\sin\beta_m x\sin\gamma_n y\sin\eta_p z$$
$$\cdot\sin\beta_m x'\sin\gamma_n y'\sin\eta_p z'. \tag{6-62}$$

Then the solution of the nonhomogeneous problem (6-57) is given in terms of the above Green's function, according to equation (6-4), as

$$T(x,y,z,t)=\int_{x'=0}^{a}\int_{y'=0}^{b}\int_{z'=0}^{c}G(x,y,z,t|x',y',z',\tau)|_{\tau=0}$$
$$\cdot F(x',y',z')dx'\,dy'\,dz'$$

$$+\frac{\alpha}{k}\int_{\tau=0}^{t} d\tau \int_{x'=0}^{a}\int_{y'=0}^{b}\int_{z'=0}^{c}$$
$$\cdot G(x,y,z,t|x',y',z',\tau)g(x',y',z',\tau)dx'\,dy'\,dz' \tag{6-63}$$

where Green's function is defined above.

6-5 APPLICATIONS OF GREEN'S FUNCTION IN THE CYLINDRICAL COORDINATE SYSTEM

In this section we illustrate with examples the application of Green's function in the solution of nonhomogeneous boundary-value problems of heat conduction in the cylindrical coordinate system. For convenience in the determination of Green's function, we have chosen those problems for which solutions are available in Chapter 3 for their homogeneous part.

Example 6-7

A solid cylinder, $0 \leqslant r \leqslant b$, is initially at temperature $F(r)$. For times $t > 0$ there is heat generation in the medium at a rate of $g(r,t)$ W/m^3 while the boundary surface at $r = b$ is kept at temperature $f(t)$. Obtain an expression for the temperature distribution $T(r,t)$ in the cylinder for times $t > 0$.

Solution. The mathematical formulation of this problem is given as

$$\frac{\partial^2 T}{\partial r^2}+\frac{1}{r}\frac{\partial T}{\partial r}+\frac{1}{k}g(r,t)=\frac{1}{\alpha}\frac{\partial T}{\partial t} \quad \text{in} \quad 0\leqslant r<b, \quad t>0 \tag{6-64a}$$

$$T=f(t) \quad \text{at} \quad r=b, \quad \text{for} \quad t>0 \tag{6-64b}$$

$$T=F(r) \quad \text{for} \quad t=0, \quad \text{in} \quad 0\leqslant r\leqslant b \tag{6-64c}$$

To determine the appropriate Green's function, we consider the homogeneous version of this problem as

$$\frac{\partial^2 \psi}{\partial r^2}+\frac{1}{r}\frac{\partial \psi}{\partial r}=\frac{1}{\alpha}\frac{\partial \psi}{\partial t} \quad \text{in} \quad 0\leqslant r<b, \quad t>0 \tag{6-65a}$$

$$\psi=0 \quad \text{at} \quad r=b, \quad t>0 \tag{6-65b}$$

$$\psi=F(r) \quad \text{for} \quad t=0, \quad \text{in} \quad 0\leqslant r\leqslant b \tag{6-65c}$$

The problem (6-65) is the same as that considered in Example 3-3. Its solution is obtainable from equation (3-67a) as

$$\psi(r,t) = \int_{r'=0}^{b} r' \left[\frac{2}{b^2} \sum_{m=1}^{\infty} e^{-\alpha\beta_m^2 t} \frac{J_0(\beta_m r)}{J_1^2(\beta_m b)} J_0(\beta_m r') \right] F(r')\, dr' \tag{6-66}$$

where the β_m values are positive roots of $J_0(\beta_m b) = 0$. Also the solution of problem (6-65) in terms of Green's function is given, according to equation (6-8), as

$$\psi(r,t) = \int_{r'=0}^{b} r' G(r,t|r',\tau)|_{\tau=0} F(r')\, dr' \tag{6-67}$$

where r' is the Sturm–Liouville weight function. A comparison of equations (6-66) and (6-67) yields

$$G(r,t|r',\tau)|_{\tau=0} = \frac{2}{b^2} \sum_{m=1}^{\infty} e^{-\alpha\beta_m^2 t} \frac{J_0(\beta_m r)}{J_1^2(\beta_m b)} J_0(\beta_m r') \tag{6-68}$$

The desired Green's function is obtained by replacing t by $(t-\tau)$ in equation (6-68); we find

$$G(r,t|r',\tau) = \frac{2}{b^2} \sum_{m=1}^{\infty} e^{-\alpha\beta_m^2 (t-\tau)} \frac{J_0(\beta_m r)}{J_1^2(\beta_m b)} J_0(\beta_m r'). \tag{6-69}$$

Then the solution of the nonhomogeneous problem (6-64) in terms of the above Green's function is given, according to equation (6-8), as

$$T(r,t) = \int_{r'=0}^{b} r' G(r,t|r',\tau)|_{\tau=0} F(r')\, dr' + \frac{\alpha}{k} \int_{\tau=0}^{t} d\tau \int_{r'=0}^{b} r' G(r,t|r',\tau) g(r',\tau)\, dr' - \alpha \int_{\tau=0}^{t} \left[r' \frac{\partial G}{\partial r'} \right]_{r'=0} \cdot f(\tau)\, d\tau \tag{6-70}$$

Here, the boundary condition at $r = b$ being of the first kind, we replaced $[G]_{r'=b}$ by $-[\partial G/\partial r']_{r'=b}$ according to equation (6-5).

Introducing the above Green's function into equation (6-70) and noting that

$$\left[r' \frac{\partial G}{\partial r'} \right]_{r'=b} = -\frac{2}{b} \sum_{m=1}^{\infty} e^{-\alpha\beta_m^2 (t-\tau)} \beta_m \frac{J_0(\beta_m r)}{J_1(\beta_m b)} \tag{6-71}$$

we obtain

$$T(r,t) = \frac{2}{b^2} \sum_{m=1}^{\infty} e^{-\alpha\beta_m^2 t} \frac{J_0(\beta_m r)}{J_1^2(\beta_m b)} \int_{r'=0}^{b} r' J_0(\beta_m r') F(r')\, dr'$$

$$+\frac{2\alpha}{kb^2}\sum_{m=1}^{\infty} e^{-\alpha\beta_m^2 t}\frac{J_0(\beta_m r)}{J_1^2(\beta_m b)}\int_{\tau=0}^{t} e^{\alpha\beta_m^2\tau}\,d\tau\int_{r'=0}^{b} r' J_0(\beta_m r')g(r',\tau)\,dr'$$

$$+\frac{2\alpha}{b}\sum_{m=1}^{\infty} e^{-\alpha\beta_m^2 t}\beta_m\frac{J_0(\beta_m r)}{J_1(\beta_m b)}\int_{\tau=0}^{t} e^{\alpha\beta_m^2\tau}f(\tau)\,d\tau \tag{6-72}$$

where the β_m values are the positive roots of $J_0(\beta_m b)=0$. In this solution the first term on the right-hand side is for the effects of the initial condition function $F(r)$, and it is the same as that given by equation (3-67). The second term is for the effects of the heat generation function $g(r,t)$. The last term is for the effects of the boundary-condition function $f(t)$. This solution (6-72) appears to vanish at the boundary $r=b$ instead of yielding the boundary condition function $f(t)$. The reason for this is that the last term in equation (6-72) involves a series that is not uniformly convergent at $r=b$. This difficulty can be alleviated by integrating the last term by parts and replacing the resulting series by its closed-form expression. An alternative approach would be to split up the original problem as discussed in Section 1–8 of Chapter 1 in order to remove the nonhomogeneity from the boundary condition. We examine some special cases of the solution (6-72).

1. Cylinder has zero initial temperature, zero surface temperature, but heat is generated within the solid at a constant rate of g_0 W/m^3.

   By setting in equation (6-72), $F(r)=0, f(t)=0$, and $g(r,t)=g_0$, we obtain

$$T(r,t)=\frac{2g_0}{kb}\sum_{m=1}^{\infty}\frac{J_0(\beta_m r)}{\beta_m^3 J_1(\beta_m b)}-\frac{2g_0}{kb}\sum_{m=1}^{\infty}e^{-\alpha\beta_m^2 t}\frac{J_0(\beta_m r)}{\beta_m^3 J_1(\beta_m b)}. \tag{6-73}$$

   For $t\to\infty$, the second term on the right-hand side vanishes and the first term must be equal to the steady-state temperature distribution in the cylinder, namely

$$T(r,\infty)=\frac{2g_0}{kb}\sum_{m=1}^{\infty}\frac{J_0(\beta_m r)}{\beta_m^3 J_1(\beta_m b)}\equiv\frac{g_0(b^2-r^2)}{4k} \tag{6-74}$$

   Introducing (6-74) into (6-73), the solution becomes

$$T(r,t)=\frac{g_0(b^2-r^2)}{4k}-\frac{2g_0}{kb}\sum_{m=1}^{\infty}e^{-\alpha\beta_m^2 t}\frac{J_0(\beta_m r)}{\beta_m^3 J_1(\beta_m b)} \tag{6-75}$$

2. Cylinder has zero initial temperature, zero surface temperature, but there is a line heat source of strength $g_L^c(t)$ W/m situated along the centerline of the cylinder and releasing its heat continuously for times

$t > 0$. For this special case we set in equation (6-72)

$$F(r) = 0, \qquad f(t) = 0, \qquad \text{and} \qquad g(r', \tau) = g_L^c(\tau)\frac{1}{2\pi r'}\delta(r' - 0)$$

Then, equation (6-72) reduces to

$$T(r,t) = \frac{\alpha}{k\pi b^2}\sum_{m=1}^{\infty} e^{-\alpha\beta_m^2 t}\frac{J_0(\beta_m r)}{J_1^2(\beta_m b)}\int_{\tau=0}^{t} e^{-\alpha\beta_m^2\tau} g_L^c(\tau)\,d\tau \qquad (6\text{-}76)$$

3. Cylinder has zero initial temperature, zero surface temperature, but there is an instantaneous volume heat source of strength $g^i(r)$ Ws/m^3 which releases its heat spontaneously at time $t = 0$. For this case we set in equation (6-72)

$$F(r) = 0, \qquad f(t) = 0, \qquad \text{and} \qquad g(r', \tau) = g^i(r')\delta(\tau - 0)$$

Then equation (6-72) reduces to

$$T(r,t) = \frac{2}{b^2}\sum_{m=1}^{\infty} e^{-\alpha\beta_m^2 t}\frac{J_0(\beta_m r)}{J_1^2(\beta_m b)}\int_{r'=0}^{b} r' J_0(\beta_m r')\frac{\alpha g^i(r')}{k}\,dr' \qquad (6\text{-}77)$$

A comparison of this solution with the first term in equation (6-72) reveals that

$$\frac{\alpha g^i(r)}{k} \equiv F(r)$$

Namely, an instantaneous volume heat source of strength $g^i(r)$ Ws/m^3 releasing its heat spontaneously at time $t = 0$ is equivalent to an initial temperature distribution $\alpha g^i(r)/k$.

Example 6-8

A hollow cylinder, $a \leqslant r \leqslant b$, is initially at temperature $F(r)$. For times $t > 0$ there is heat generation in the medium at a rate of $g(r,t)$ W/m^3 while the boundary surfaces at $r = a$ and $r = b$ are kept at zero temperatures. Obtain an expression for the temperature distribution $T(r,t)$ in the cylinder for times $t > 0$.

Solution. The mathematical formulation of this problem is given as

$$\frac{\partial^2 T}{\partial r^2} + \frac{1}{r}\frac{\partial T}{\partial r} + \frac{1}{k}g(r,t) = \frac{1}{\alpha}\frac{\partial T}{\partial t} \quad \text{in} \quad a < r < b, \quad t > 0 \qquad (6\text{-}78a)$$

$$T = 0 \quad \text{at} \quad r = a, \quad r = b, \quad t > 0 \qquad (6\text{-}78b)$$

$$T = F(r) \qquad \text{for} \quad t=0, \qquad \text{in the region} \tag{6-78c}$$

To determine the appropriate Green's function, we consider the homogeneous version of this problem as

$$\frac{\partial^2 \psi}{\partial r^2} + \frac{1}{r}\frac{\partial \psi}{\partial r} = \frac{1}{\alpha}\frac{\partial \psi}{\partial t} \qquad \text{in} \quad a < r < b, \qquad t > 0 \tag{6-79a}$$

$$\psi = 0 \qquad \text{at} \quad r = a, \quad r = b, \quad t > 0 \tag{6-79b}$$

$$\psi = F(r) \qquad \text{for} \quad t = 0, \qquad \text{in the region} \tag{6-79c}$$

The problem (6-79) is the same as that considered in Example 3-5; the solution is obtainable from equation (3-78) as

$$\psi(r,t) = \int_{r'=a}^{b} r' \left[\frac{\pi^2}{2} \sum_{m=1}^{\infty} e^{-\alpha \beta_m^2 t} \frac{\beta_m^2 J_0^2(\beta_m a)}{J_0^2(\beta_m a) - J_0^2(\beta_m b)} R_0(\beta_m r) R_0(\beta_m r') \right] F(r')\, dr' \tag{6-80a}$$

where

$$R_0(\beta_m, r) = J_0(\beta_m r) Y_0(\beta_m b) - J_0(\beta_m b) Y_0(\beta_m r) \tag{6-80b}$$

and the β_m values are the positive roots of

$$J_0(\beta_m a) Y_0(\beta_m b) - J_0(\beta_m b) Y_0(\beta_m a) = 0 \tag{6-80c}$$

Also the solution of the problem (6-79) in terms of Green's function is given, according to equation (6-8), as

$$\psi(r,t) = \int_{r'=a}^{b} r' G(r,t|r',\tau)|_{\tau=0} F(r')\, dr' \tag{6-81}$$

A comparison of equations (6-80a) and (6-81) yields

$$G(r,t|r',\tau)|_{\tau=0} = \frac{\pi^2}{2} \sum_{m=1}^{\infty} e^{-\alpha \beta_m^2 t} \frac{\beta_m^2 J_0(\beta_m a)}{J_0^2(\beta_m a) - J_0^2(\beta_m b)} R_0(\beta_m r) R_0(\beta_m r'). \tag{6-82}$$

The desired Green's function is obtained by replacing t by $(t-\tau)$ in equation (6-82); we find

$$G(r,t|r',\tau) = \frac{\pi^2}{2} \sum_{m=1}^{\infty} e^{-\alpha \beta_m^2 (t-\tau)} \frac{\beta_m^2 J_0(\beta_m a)}{J_0^2(\beta_m a) - J_0^2(\beta_m b)} R_0(\beta_m r) R_0(\beta_m r'). \tag{6-83}$$

Then the solution of the above nonhomogeneous problem (6-78) in terms of this Green's function is given, according to equation (6-8), as

$$T(r,t)=\int_{r'=a}^{b} r'G(r,t|r',\tau)|_{\tau=0}F(r')\,dr' + \frac{\alpha}{k}\int_{\tau=0}^{t} d\tau \int_{r'=a}^{b} r'G(r,t|r',\tau)g(r',\tau)\,dr' \tag{6-84}$$

Introducing the foregoing Green's function into equation (6-84), the solution of the problem (6-78) becomes

$$\begin{aligned}T(r,t)=&\frac{\pi^2}{2}\sum_{m=1}^{\infty} e^{-\alpha\beta_m^2 t}\frac{\beta_m^2 J_0(\beta_m a)}{J_0^2(\beta_m a)-J_0^2(\beta_m b)}R_0(\beta_m r)\int_{r'=a}^{b} r'R_0(\beta_m r')F(r')\,dr'\\ &+\frac{\pi^2\alpha}{2k}\sum_{m=1}^{\infty} e^{-\alpha\beta_m^2 t}\frac{\beta_m^2 J_0(\beta_m a)}{J_0^2(\beta_m a)-J_0^2(\beta_m b)}R_0(\beta_m r)\\ &\cdot\int_{\tau=0}^{t} e^{\alpha\beta_m^2\tau}\,d\tau\int_{r'=a}^{b} r'R_0(\beta_m r')g(r',\tau)\,dr'\end{aligned} \tag{6-85}$$

where $R_0(\beta_m, r)$ as given by equation (6-80b) and the β_m values are the roots of the transcendental equation (6-80c). Clearly, several special cases are obtainable from the solution (6-85).

6-6 APPLICATIONS OF GREEN'S FUNCTION IN THE SPHERICAL COORDINATE SYSTEM

In this section we illustrate with examples the application of Green's function in the solution of nonhomogeneous boundary-value problems of heat conduction in the spherical coordinate system. For convenience in the determination of Green's function we have chosen those examples for which solutions are available in Chapter 4 for their homogeneous parts.

Example 6-9

A hollow sphere $a \leqslant r \leqslant b$, is initially at temperature $F(r)$. For time $t > 0$ heat is generated within the sphere at a rate of $g(r, t)$ W/m^3 while the boundaries at $r = a$ and $r = b$ are kept at zero temperature. Obtain an expression for the temperature distribution $T(r, t)$ in the sphere for times $t > 0$.

Solution. The mathematical formulation of this problem is given as

$$\frac{1}{r}\frac{\partial^2}{\partial r^2}(rT)+\frac{1}{k}g(r,t)=\frac{1}{\alpha}\frac{\partial T}{\partial t} \quad \text{in} \quad a<r<b, \quad t>0 \tag{6-86a}$$

$$T=0 \quad \text{at} \quad r=a \text{ and } r=b, \quad t>0 \tag{6-86b}$$

$$T=F(r) \quad \text{for} \quad t=0, \text{ in } \quad a\leqslant r\leqslant b \tag{6-86c}$$

To determine the Green's function we consider the homogeneous version of this problem as

$$\frac{1}{r}\frac{\partial^2}{\partial r^2}(r\psi) = \frac{1}{\alpha}\frac{\partial \psi}{\partial t} \qquad \text{in} \qquad a < r < b, \qquad t > 0 \tag{6-87a}$$

$$\psi = 0 \qquad \text{at} \qquad r = a \quad \text{and} \quad r = b, \quad t > 0 \tag{6-87b}$$

$$\psi = F(r) \qquad \text{for} \qquad t = 0, \quad \text{in} \qquad a \leqslant r \leqslant b \tag{6-87c}$$

This homogeneous problem is the same as that considered in Example 4-3; its solution is obtainable from equation (4-80) as

$$\psi(r,t) = \int_{r'=0}^{b} r'^2 \left[\frac{2}{r'r(b-a)} \sum_{m=1}^{\infty} e^{-\alpha\beta_m^2 t} \sin\beta_m(r'-a)\sin\beta_m(r-a) \right] F(r')\,dr' \tag{6-88a}$$

where the β_m values are the positive roots of

$$\sin\beta_m(b-a) = 0 \tag{6-88b}$$

or

$$\beta_m = \frac{m\pi}{b-a}, \qquad m = 1, 2, 3 \ldots \tag{6-88c}$$

The solution of the problem (6-87) in terms of Green's function is given, according to equation (6-13), as

$$\psi(r,t) = \int_{r'=a}^{b} r'^2 G(r,t\,|\,r',\tau)|_{\tau=0} F(r')\,dr' \tag{6-89}$$

where r'^2 is the Sturm–Liouville weight function. A comparison of equations (6-88a) and (6-89) gives

$$G(r,t\,|\,r',\tau)|_{\tau=0} = \frac{2}{r'r(b-a)} \sum_{m=1}^{\infty} e^{-\alpha\beta_m^2 t} \cdot \sin\beta_m(r'-a)\sin\beta_m(r-a). \tag{6-90}$$

The desired Green's function is obtained by replacing t by $(t-\tau)$ in equation (6-90); we find

$$G(r,t\,|\,r',\tau) = \frac{2}{r'r(b-a)} \sum_{m=1}^{\infty} e^{-\alpha\beta_m^2 (t-\tau)} \cdot \sin\beta_m(r'-a)\sin\beta_m(r-a). \tag{6-91}$$

Then the solution of the nonhomogeneous problem (6-86) in terms of Green's function is given, according to equation (6-8), as

$$T(r,t) = \int_{r'=a}^{b} r'^2 G(r,t|r',\tau)|_{\tau=0} F(r')\,dr' + \frac{\alpha}{k}\int_{\tau=0}^{t} d\tau \int_{r'=a}^{b} r'^2 G(r,t|r',\tau) g(r',\tau)\,dr' \tag{6-92}$$

Introducing the above Green's function into equation (6-92), the solution becomes

$$\begin{aligned} T(r,t) = {} & \frac{2}{r(b-a)} \sum_{m=1}^{\infty} e^{-\alpha\beta_m^2 t} \sin\beta_m(r-a) \int_{r'=a}^{b} r' \sin\beta_m(r'-a) F(r')\,dr' \\ & + \frac{\alpha}{k}\frac{2}{r(b-a)} \sum_{m=1}^{\infty} e^{-\alpha\beta_m^2 t} \sin\beta_m(r-a) \int_{\tau=0}^{t} e^{\alpha\beta_m^2 \tau}\,d\tau \\ & \cdot \int_{r'=a}^{b} r' \sin\beta_m(r'-a) g(r',\tau)\,dr' \end{aligned} \tag{6-93a}$$

where the β_m values are the positive roots of

$$\sin\beta_m(b-a) = 0 \tag{6-93b}$$

We now consider some special cases of the solution (6-93).

1. The medium is initially at zero temperature, the heat source is a spherical surface heat source of radius r_1 (i.e., $a < r_1 < b$) of total strength $g_s^c(t)$ W, which releases its heat continuously for times $t > 0$. In this case we set in equation (6-93a)

$$F(r') = 0, \qquad g(r',\tau) = g_s^c(\tau)\frac{1}{4\pi r'^2}\delta(r'-r_1) \tag{6-94}$$

and perform the integration with respect to the variable r'. We find

$$\begin{aligned} T(r,t) = {} & \frac{\alpha}{k}\frac{1}{2\pi r r_1(b-a)} \sum_{m=1}^{\infty} e^{-\alpha\beta_m^2 t} \sin\beta_m(r-a) \\ & \cdot \sin\beta_m(r_1-a) \int_{\tau=0}^{t} e^{\alpha\beta_m^2\tau} g_s^c(\tau)\,d\tau \end{aligned} \tag{6-95a}$$

where

$$\beta_m = \frac{m\pi}{b-a}, \qquad m = 1, 2, 3\ldots \tag{6-95b}$$

2. The medium is initially at zero temperature, the heat source is an

instantaneous spherical surface heat source of radius r_1 (i.e., $a < r_1 < b$) of total strength g_s^i Ws, which releases its heat spontaneously at time $t = 0$. In this case we set in equation (6-93a)

$$F(r') = 0, \qquad g(r', \tau) = g_s^i \frac{1}{4\pi r'^2} \delta(r' - r_1)\delta(\tau - 0) \tag{6-96}$$

and perform the integrations with respect to the variables r' and τ. We find

$$T(r, t) = \frac{\alpha}{k} \frac{1}{2\pi r r_1(b-a)} \sum_{m=1}^{\infty} e^{-\alpha\beta_m^2 t} \sin \beta_m(r-a) \sin \beta_m(r_1 - a) g_s^i \tag{6-97a}$$

where

$$\beta_m = \frac{m\pi}{b-a}, \qquad m = 1, 2, 3 \ldots \tag{6-97b}$$

Example 6-10

A solid sphere $0 \leqslant r \leqslant b$ is initially at zero temperature. For times $t > 0$ heat is generated within the sphere at a rate of $g(r, t)$ W/m^3 while the boundary surface at $r = b$ is kept at zero temperature. Obtain an expression for the temperature distribution in the sphere by the Green's function approach.

Solution. The mathematical formulation of this problem is given as

$$\frac{1}{r}\frac{\partial^2}{\partial r^2}(rT) + \frac{1}{k} g(r, t) = \frac{1}{\alpha}\frac{\partial T}{\partial t} \qquad \text{in} \quad 0 \leqslant r < b, \quad t > 0 \tag{6-98a}$$

$$T = 0 \qquad \text{at} \quad r = b, \quad t > 0 \tag{6-98b}$$

$$T = 0 \qquad \text{for} \quad t = 0, \quad \text{in} \quad 0 \leqslant r \leqslant b \tag{6-98c}$$

To determine the Green's function we consider the following homogeneous problem:

$$\frac{1}{r}\frac{\partial^2}{\partial r^2}(r\psi) = \frac{1}{\alpha}\frac{\partial \psi}{\partial t} \qquad \text{in} \quad 0 \leqslant r < b, \quad t > 0 \tag{6-99a}$$

$$\psi = 0 \qquad \text{at} \quad r = b, \quad t > 0 \tag{6-99b}$$

$$\psi = F(r) \qquad \text{for} \quad t = 0, \quad \text{in} \quad 0 \leqslant r \leqslant b \tag{6-99c}$$

The solution of the problem (6-99) is obtainable by converting it to a slab problem or directly from the solution (4-64) by setting $K \to \infty$. We find

$$\psi(r,t) = \int_{r'=0}^{b} r'^2 \left[\frac{2}{r'rb} \sum_{m=1}^{\infty} e^{-\alpha\beta_m^2 t} \sin\beta_m r' \sin\beta_m r \right] F(r')\,dr' \qquad \text{(6-100a)}$$

where the β_m values are the positive roots of

$$\sin\beta_m b = 0 \qquad \text{(6-100b)}$$

or

$$\beta_m = \frac{m\pi}{b}, \qquad m = 1, 2, 3 \ldots$$

Also the solution of the problem (6-99) in terms of Green's function is given, according to equation (6-13), as

$$\psi(r,t) = \int_{r'=0}^{b} r'^2 G(r,t|r',\tau)|_{\tau=0} F(r')\,dr' \qquad \text{(6-101)}$$

A comparison of equations (6-100a) and (6-101) gives

$$G(r,t|r',\tau)|_{\tau=0} = \frac{2}{r'rb} \sum_{m=1}^{\infty} e^{-\alpha\beta_m^2 t} \cdot \sin\beta_m r' \sin\beta_m r \qquad \text{(6-102)}$$

The desired Green's function is obtained by replacing t by $(t-\tau)$ in equation (6-102):

$$G(r,t|r',\tau) = \frac{2}{r'rb} \sum_{m=1}^{\infty} e^{-\alpha\beta_m^2 (t-\tau)} \sin\beta_m r' \sin\beta_m r \qquad \text{(6-103)}$$

Then the solution of the nonhomogeneous problem (6-98) in terms of Green's function is given, according to equation (6-8), as

$$T(r,t) = \frac{\alpha}{k} \int_{\tau=0}^{t} d\tau \int_{r'=0}^{b} r'^2 G(r,t|r',\tau) g(r',\tau)\,dr' \qquad \text{(6-104)}$$

Introducing the Green's function given by equation (6-103), into equation (6-104), the solution of the problem (6-98) becomes

$$T(r,t) = \frac{\alpha}{k}\frac{2}{br} \sum_{m=1}^{\infty} e^{-\alpha\beta_m^2 t} \sin\beta_m r \int_{\tau=0}^{t} e^{\alpha\beta_m^2 \tau}\,d\tau \int_{r'=0}^{b} r' \sin\beta_m r' g(r',\tau)\,dr' \qquad \text{(6-105)}$$

where

$$\beta_m = \frac{m\pi}{b}, \qquad m = 1, 2, 3 \ldots$$

We now consider some special cases of the solution (6-105).

1. The Heat source is an instantaneous volume heat source of strength $g^{\mathrm{i}}(r)\,(\mathrm{Ws})/\mathrm{m}^3$ that releases its heat spontaneously at time $t = 0$. By setting in equation (6-105)

$$G(r', \tau) = g^{\mathrm{i}}(r')\delta(\tau - 0) \tag{6-106}$$

and performing the integration with respect to the variable τ we obtain

$$T(r, t) = \frac{\alpha}{k}\frac{2}{br}\sum_{m=1}^{\infty} e^{-\alpha\beta_m^2 t}\sin\beta_m r \int_{r'=0}^{b} r' g^{\mathrm{i}}(r')\sin\beta_m r'\,dr' \tag{6-107}$$

2. The heat source is an instantaneous point heat source of strength $g_{\mathrm{p}}^{\mathrm{i}}$ Ws, which is situated at the center of the sphere and releases its heat spontaneously at time $t = 0$. By setting in equation (6-105).

$$g(r', \tau) = \frac{g_{\mathrm{p}}^{\mathrm{i}}}{4\pi r'^2}\delta(r' - 0)\delta(\tau - 0) \tag{6-108}$$

and performing the integrations with respect to the variables r' and τ, we obtain

$$T(r, t) = \frac{\alpha}{k}\frac{1}{2rb^2}\sum_{m=1}^{\infty} e^{-\alpha\beta_m^2 t} m \sin\beta_m r \cdot g_{\mathrm{p}}^{\mathrm{i}} \tag{6-109}$$

where $\beta_m = m\pi/b$.

Example 6-11

A solid sphere of radius $r = b$ is initially at temperature $F(r, \mu)$. For times $t > 0$ heat is generated in the sphere at a rate of $g(r, \mu, t)$ W/m^3, while the boundary surface at $r = b$ is kept at zero temperature. Obtain an expression for the temperature distribution $T(r, \mu, t)$ in the sphere for times $t > 0$.

Solution. The mathematical formulation of this problem is given as

$$\frac{\partial^2 T}{\partial r^2} + \frac{2}{r}\frac{\partial T}{\partial r} + \frac{1}{r^2}\frac{\partial}{\partial \mu}\left[(1 - \mu^2)\frac{\partial T}{\partial \mu}\right] + \frac{1}{k}g(r, \mu, t) = \frac{1}{\alpha}\frac{\partial T}{\partial t}$$

$$\text{in} \quad 0 \leqslant r < b, \quad -1 \leqslant \mu \leqslant 1, \quad t > 0 \tag{6-110a}$$

$$T = 0 \quad \text{at} \quad r = b, \quad t > 0 \tag{6-110b}$$

$$T = F(r, \mu) \quad \text{for} \quad t = 0, \quad \text{in the sphere} \tag{6-110c}$$

To determine the Green's function, we consider the homogeneous version of this problem as

$$\frac{\partial^2\psi}{\partial r^2}+\frac{2}{r}\frac{\partial\psi}{\partial r}+\frac{1}{r^2}\frac{\partial}{\partial\mu}\left[(1-\mu^2)\frac{\partial\psi}{\partial\mu}\right]=\frac{1}{\alpha}\frac{\partial\psi}{\partial t}$$

$$\text{in} \quad 0\leqslant r<b, \quad -1\leqslant\mu\leqslant 1, \quad \text{for} \quad t>0 \tag{6-111a}$$

$$\psi=0 \quad \text{at} \quad r=b, \quad t>0 \tag{6-111b}$$

$$\psi=F(r,\mu) \quad \text{for} \quad t=0 \quad \text{in the sphere} \tag{6-111c}$$

This homogeneous problem is the same as that considered in Example 4-4; the solution is obtainable from equation (4-88) as

$$\psi(r,\mu,t)=\int_{r'=0}^{b}\int_{\mu'=-1}^{1} r'^2\Bigg[\sum_{n=0}^{\infty}\sum_{p=1}^{\infty}\frac{1}{N(n)N(\lambda_{np})}e^{-\alpha\lambda_{np}^2 t} \cdot(r'r)^{-1/2}J_{n+1/2}(\lambda_{np}r)P_n(\mu)J_{n+1/2}(\lambda_{np}r')P_n(\mu')\Bigg] \cdot F(r',\mu')d\mu'\,dr' \tag{6-112a}$$

where the λ_{np} values are the positive roots of

$$J_{n+1/2}(\lambda_{np}b)=0 \tag{6-112b}$$

and the norms $N(n)$ and $N(\lambda_{np})$ are given as

$$N(n)=\frac{2}{2n+1} \tag{6-112c}$$

$$N(\lambda_{np})=-\frac{b^2}{2}J_{n-1/2}(\lambda_{np}b)J_{n+3/2}(\lambda_{np}b). \tag{6-112d}$$

The solution of the problem (6-111) in terms of Green's function is given, according to equation (6-13), as

$$\psi(r,\mu,t)=\int_{r'=0}^{b}\int_{\mu'=-1}^{1} r'^2 G(r,\mu,t|r',\mu',\tau)|_{\tau=0}F(r',\mu')d\mu'\,dr'. \tag{6-113}$$

A comparison of equations (6-112a) and (6-113) yields

$$G(r,\mu,t|r',\mu',\tau)|_{\tau=0}=\sum_{n=0}^{\infty}\sum_{p=1}^{\infty}\frac{1}{N(n)N(\lambda_{np})}e^{-\alpha\lambda_{np}^2 t} \cdot(r'r)^{-1/2}J_{n+1/2}(\lambda_{np}r)P_n(\mu)J_{n+1/2}(\lambda_{np}r')P_n(\mu') \tag{6-114}$$

The desired Green's function is obtained by replacing t by $(t-\tau)$ in equation (6-114); we find

$$G(r,\mu,t|r',\mu',\tau)=\sum_{n=0}^{\infty}\sum_{p=1}^{\infty}\frac{1}{N(n)N(\lambda_{np})}e^{-\alpha\lambda_{np}^2(t-\tau)}\cdot(r'r)^{-1/2}J_{n+1/2}(\lambda_{np}r)P_n(\mu)J_{n+1/2}(\lambda_{np}r')P_n(\mu') \quad (6\text{-}115)$$

Then the solution of the nonhomogeneous problem (6-110) in terms of this Green's function is given, according to equation (6-7), as

$$T(r,\mu,t)=\int_{r'=0}^{b}\int_{\mu'=-1}^{1}r'^2G(r,\mu,t|r',\mu',\tau)|_{\tau=0}F(r',\mu')d\mu'\,dr' + \frac{\alpha}{k}\int_{\tau=0}^{t}d\tau\int_{r'=0}^{b}\int_{\mu'=-1}^{1}r'^2G(r,\mu,t|r',\mu',\tau)\cdot g(r',\mu,\tau)d\mu'\,dr' \quad (6\text{-}116)$$

Introducing the above Green's function into (6-116), the solution becomes

$$\begin{aligned}T(r,\mu,t)={}&\sum_{n=0}^{\infty}\sum_{p=1}^{\infty}\frac{1}{N(n)N(\lambda_{np})}e^{-\alpha\lambda_{np}^2t}r^{-1/2}J_{n+1/2}(\lambda_{np}r)P_n(\mu)\\&\cdot\int_{r'=0}^{b}\int_{\mu'=-1}^{1}r'^{3/2}J_{n+1/2}(\lambda_{np}r')P_n(\mu')F(r',\mu')d\mu'\,dr'\\&+\frac{\alpha}{k}\sum_{n=0}^{\infty}\sum_{p=1}^{\infty}\frac{1}{N(n)N(\lambda_{np})}e^{-\alpha\lambda_{np}^2t}r^{-1/2}J_{n+1/2}(\lambda_{np}r)P_n(\mu)\\&\cdot\int_{\tau=0}^{t}e^{\alpha\lambda_{np}^2\tau}d\tau\int_{r'=0}^{b}\int_{\mu'=-1}^{1}r'^{3/2}J_{n+1/2}(\lambda_{np}r')P_n(\mu')\\&\cdot g(r',\mu',\tau)d\mu'\,dr'\end{aligned} \quad (6\text{-}117)$$

Clearly several special cases are obtainable from this solution.

6-7 PRODUCT OF GREEN'S FUNCTIONS

The multidimensional Green's functions can be obtained from the multiplication of one-dimensional Green's functions for all cases in the rectangular coordinate system and for some cases in the cylindrical coordinate system; but the multiplication procedure is not possible in the spherical coordinate system. We illustrate this matter with examples in the rectangular and cylindrical coordinates.

Rectangular Coordinates

The three-dimensional Green's function $G(x, y, z, t \mid x', y', z', \tau)$ can be obtained from the product of the three one-dimensional Green's functions as

$$G(x, y, z, t \mid x', y', z', \tau) = G_1(x, t \mid x', \tau) \cdot G_2(y, t \mid y', \tau) \cdot G_3(z, t \mid z', \tau) \qquad \text{(6-118)}$$

where each of the one-dimensional Green's functions G_1, G_2, and G_3 depends on the extent of the region (i.e., finite, semiinfinite, or infinite) and the boundary conditions associated with it (i.e., first, second, or third kind). We present below a tabulation of the three-dimensional Green's functions in the rectangular coordinates as the product of three one-dimensional Green's functions.

***Region:* $0 \leqslant x \leqslant a, 0 \leqslant y \leqslant b, 0 \leqslant z \leqslant c$**

$$\begin{aligned} G(x, y, z, t \mid x', y', z', \tau) = & \left[\sum_{m=1}^{\infty} e^{-\alpha \beta_m^2 (t-\tau)} \frac{1}{N(\beta_m)} X(\beta_m, x) X(\beta_m, x') \right] \\ & \cdot \left[\sum_{n=1}^{\infty} e^{-\alpha \gamma_n^2 (t-\tau)} \frac{1}{N(\gamma_n)} Y(\gamma_n, y) Y(\gamma_n, y') \right] \\ & \cdot \left[\sum_{p=1}^{\infty} e^{-\alpha \eta_p^2 (t-\tau)} \frac{1}{N(\eta_p)} Z(\eta_p, z) Z(\eta_p, z') \right] \end{aligned} \qquad \text{(6-119)}$$

where the eigenfunctions, eigenconditions, and normalization integrals are obtainable from Table 2-2. In each direction there are nine different combinations of boundary conditions; therefore the result given by equation (6-119) together with Table 2-2 represents $9 \times 9 \times 9 = 729$ different cases.

***Region:* $-\infty < x < \infty, 0 \leqslant y \leqslant b, 0 \leqslant z \leqslant c$**

$$\begin{aligned} G(x, y, z, t \mid x', y', z', \tau) = & \left[\{4\pi\alpha(t-\tau)\}^{-1/2} \cdot \exp\left\{ -\frac{(x-x')^2}{4\alpha(t-\tau)} \right\} \right] \\ & \cdot \left[\sum_{n=1}^{\infty} e^{-\alpha \gamma_n^2 (t-\tau)} \frac{1}{N(\gamma_n)} Y(\gamma_n, y) Y(\gamma_n, y') \right] \\ & \cdot \left[\sum_{p=1}^{\infty} e^{-\alpha \eta_p^2 (t-\tau)} \frac{1}{N(\eta_p)} Z(\eta_p, z) Z(\eta_p, z') \right] \end{aligned} \qquad \text{(6-120)}$$

where the infinite medium Green's function, shown inside the first bracket, is obtained from equation (6-36). The result given by equation (6-120) when used together with the Table 2-2 represents $9 \times 9 = 81$ different cases.

Region: $-\infty < x < \infty, -\infty < y < \infty, 0 \leqslant z \leqslant c$

$$G(x, y, z, t \mid x', y', z', \tau) = \left[\{4\pi\alpha(t-\tau)\}^{-1/2} \cdot \exp\left\{ -\frac{(x-x')^2}{4\alpha(t-\tau)} \right\} \right]$$
$$\cdot \left[\{4\pi\alpha(t-\tau)\}^{-1/2} \cdot \exp\left\{ -\frac{(y-y')^2}{4\alpha(t-\tau)} \right\} \right]$$
$$\cdot \left[\sum_{p=1}^{\infty} e^{-\alpha\eta_p^2(t-\tau)} \frac{1}{N(\eta_p)} Z(\eta_p, z) Z(\eta_p, z') \right] \tag{6-121}$$

This result, together with Table 2-2, represents nine different cases.

Region: $-\infty < x < \infty, -\infty < y < \infty, -\infty < z < \infty$

$$G(x, y, z, t \mid x', y', z', \tau) = \left[\{4\pi\alpha(t-\tau)\}^{-1/2} \cdot \exp\left\{ -\frac{(x-x')^2}{4\alpha(t-\tau)} \right\} \right]$$
$$\cdot \left[\{4\pi\alpha(t-\tau)\}^{-1/2} \cdot \exp\left\{ -\frac{(y-y')^2}{4\alpha(t-\tau)} \right\} \right]$$
$$\cdot \left[\{4\pi\alpha(t-\tau)\}^{-1/2} \cdot \exp\left\{ -\frac{(z-z')^2}{4\alpha(t-\tau)} \right\} \right] \tag{6-122}$$

Region: $0 \leqslant x < \infty$

In the foregoing expressions for the three-dimensional Green's functions, if any one of the region is semiinfinite, the Green's function for that region should be replaced by the semiinfinite medium Green's function given below.

1. Boundary condition at $x = 0$ is of the first kind [constructed from equation (2-58a)]:

$$G(x, t \mid x', \tau) = [4\pi\alpha(t-\tau)]^{-1/2} \left[\exp\left(-\frac{(x-x')^2}{4\alpha(t-\tau)} \right) - \exp\left(-\frac{(x+x')^2}{4\alpha(t-\tau)} \right) \right] \tag{6-123a}$$

2. Boundary condition at $x = 0$ is of the second kind [constructed from equation (2-63c)]:

$$G(x, t \mid x', \tau) = [4\pi\alpha(t-\tau)]^{-1/2} \left[\exp\left(-\frac{(x-x')^2}{4\alpha(t-\tau)} \right) + \exp\left(-\frac{(x+x')^2}{4\alpha(t-\tau)} \right) \right] \tag{6-123b}$$

3. Boundary condition at $x = 0$ is of the third kind [constructed from

equation (2-54)]:

$$G(x,t|x',\tau)=\int_{\beta=0}^{\infty}e^{-\alpha\beta^2(t-\tau)}\frac{1}{N(\beta)}X(\beta,x)X(\beta,x')d\beta \qquad \text{(6-123c)}$$

where

$$X(\beta,x)=\beta\cos\beta x+H_1\sin\beta x$$

$$\frac{1}{N(\beta)}=\frac{2}{\pi}\frac{1}{\beta^2+H_1^2}, \qquad H_1=\frac{h_1}{k}$$

Example 6-12

Consider a rectangular parallelepiped in the region $0\leqslant x\leqslant a, 0\leqslant y\leqslant b$, $0\leqslant z\leqslant c$ initially at zero temperature. For times $t>0$, all boundaries are maintained at zero temperature while energy is generated in the medium at a rate of $g(x,y,z,t)\,\text{W/m}^3$. Develop the three-dimensional Green's function needed for the solution of this heat conduction equation with the Green's function approach.

Solution. The Green's function for this problem is obtainable as a product of three one-dimensional finite-region Green's functions subjected to the boundary condition of the first kind. We use the formalism given by equation (6-119) together with case 9 of Table 2-2 to obtain

$$\begin{aligned}G(x,y,z,t|x',y',z',\tau)=&\left[\frac{2}{a}\sum_{m=1}^{\infty}e^{-\alpha\beta_m^2(t-\tau)}\sin\beta_m x\cdot\sin\beta_m x'\right]\\&\cdot\left[\frac{2}{b}\sum_{n=1}^{\infty}e^{-\alpha\gamma_n^2(t-\tau)}\sin\gamma_n y\cdot\sin\gamma_n y'\right]\\&\cdot\left[\frac{2}{c}\sum_{p=1}^{\infty}e^{-\alpha\eta_p^2(t-\tau)}\sin\eta_p z\cdot\sin\eta_p z'\right]\end{aligned} \qquad \text{(6-124a)}$$

where the eigenvalues β_m, γ_n, and η_p are the positive roots of the transcendental equations

$$\sin\beta_m a=0, \qquad \sin\gamma_n b=0, \qquad \sin\eta_p c=0 \qquad \text{(6-124b)}$$

Cylindrical Coordinates

The multiplication of one-dimensional Green's functions in order to get multi-dimensional Green's function is possible if the problem involves only the (r,z,t) variables, that is if the problem has azimuthal symmetry. When the problem involves (r, z, ϕ, t) variables, it is not possible to separate the Green's function associated with the r and ϕ variables.

We present below a tabulation of two-dimensional Green's functions in the (r, z, t) variables in the cylindrical coordinates developed by the multiplication of two one-dimensional Green's functions.

***Region:* $0 \leqslant r \leqslant b, 0 \leqslant z \leqslant c$**

$$G(r,z,t|r',z',\tau) = \left[\sum_{m=1}^{\infty} e^{-\alpha\beta_m^2(t-\tau)} \frac{1}{N(\beta_m)} r' R_0(\beta_m, r) R_0(\beta_m, r')\right] \cdot \left[\sum_{p=1}^{\infty} e^{-\alpha\eta_p^2(t-\tau)} \frac{1}{N(\eta_p)} Z(\eta_p, z) Z(\eta_p, z')\right] \tag{6-125}$$

where the eigenfunctions, eigenvalues, and normalization integrals for the r variable are obtained from Table 3-1 by setting $v = 0$ in the results given in this table and for the z-variable are obtained from Table 2-2. Table 3-1 involves three different cases and Table 2-2, nine different cases; hence the result given by equation (6-125) represents $3 \times 9 = 27$ different cases.

***Region:* $0 \leqslant r < \infty, 0 \leqslant z \leqslant c$**

$$G(r,z,t|r',z',\tau) = \left[(2\alpha(t-\tau))^{-1/2} r' \exp\left(-\frac{r^2 + r'^2}{4\alpha(t-\tau)}\right) I_0\left(\frac{rr'}{2\alpha(t-\tau)}\right)\right] \cdot \left[\sum_{p=1}^{\infty} e^{-\alpha\eta_p^2(t-\tau)} \frac{1}{N(\eta_p)} Z(\eta_p, z) Z(\eta_p, z')\right] \tag{6-126}$$

where the Green's function for the r variable is constructed from the solution given by equation (3-90) and the eigenfunctions, eigenvalues, and normalization integrals associated with the z variable are obtainable from Table 2-2. Therefore, the result given by equation (6-126) represents 9 different cases.

***Region:* $a \leqslant r < b, 0 \leqslant z \leqslant c$**

$$G(r,z,t|r',z',\tau) = \left[\sum_{m=1}^{\infty} e^{-\alpha\beta_m^2(t-\tau)} \frac{1}{N(\beta_m)} r' R_0(\beta_m, r) R_0(\beta_m, r')\right] \cdot \left[\sum_{p=1}^{\infty} e^{-\alpha\eta_p^2(t-\tau)} \frac{1}{N(\eta_p)} Z(\eta_p, z) Z(\eta_p, z')\right] \tag{6-127}$$

where the eigenfunctions, eigenvalues, and normalization integrals for the r variable are obtainable from Table 3-3 and for the z variable, from Table 2-2. In Table 3-3, only the boundary conditions of the first and second kind are considered, because the results for the boundary condition of the third kind are rather involved.

Region: $a \leqslant r < \infty, 0 \leqslant z \leqslant c$

$$G(r,z,t|r',z',\tau) = \left[\int_{\beta=0}^{\infty} e^{-\alpha\beta^2(t-\tau)} \frac{\beta}{N(\beta)} r' R_0(\beta,r) R_0(\beta,r')\right] \cdot \left[\sum_{p=1}^{\infty} e^{-\alpha\eta_p^2(t-\tau)} \frac{1}{N(\eta_p)} Z(\eta_p,z) Z(\eta_p,z')\right] \tag{6-128}$$

where the Green's function for the r variable given inside the first bracket is obtained from the rearrangement of equation (3-98). The eigenfunctions, eigenvalues, and the normalization integral associated with the r variable are given in Table 3-2 for three different boundary conditions at $r = a$, and those associated with the z variable are given in Table 2-2 for nine different combinations of boundary conditions. Therefore, the result given by equation (6-128) represents $3 \times 9 = 27$ different cases.

In the foregoing expression for the two dimensional Green's function in the cylindrical coordinates, we considered only a finite region $0 \leqslant z \leqslant c$ for the z variable. If it is semiinfinite or infinite, the corresponding Green's function is obtained from those discussed for the rectangular coordinate system.

REFERENCES

1. H. S. Carslaw and J. C. Jaeger, *Conduction of Heat in Solids*, Clarendon Press, London, 1959.
2. P. M. Morse and H. Feshback, *Methods of Theoretical Physics*, McGraw-Hill, New York, 1953.
3. I. N. Sneddon, *Partial Differential Equations*, McGraw-Hill, New York, 1957.
4. J. W. Dettman, *Mathematical Methods in Physics and Engineering*, McGraw-Hill, New York, 1962.
5. R. Courant and D. Hilbert, *Methods of Mathematical Physics*, Interscience Publishers, New York, 1953.
6. M. N. Özişik, *Boundary Value Problems of Heat Conduction*, International Textbook Company, Scranton, Pa., 1968.
7. M. D. Greenberg, *Applications of Green's Functions in Science and Engineering*, Prentice-Hall, Englewood Cliffs, N.J., 1971.
8. A. M. Aizen, I. S. Redchits, and I. M. Fedotkin, *J. Eng. Phys.* **26**, 453–458, 1974.
9. A. G. Butkovskiy, *Green's Functions and Transfer Functions Handbook*, Halstead Press: A Division of Wiley, New York, 1982.
10. I. Stakgold, *Green's Functions and Boundary Value Problems*, Wiley, New York, 1979.
11. J. V. Beck, K. D. Cole, A. Haji-Sheikh, and B. Litkouki, *Heat Conduction Using Green's Functions*, Hemisphere Publishing Co., Washington D.C., 1992.

PROBLEMS

6-1 A semiinfinite region $0 \leqslant x < \infty$ is initially at temperature $F(x)$. For times $t > 0$, boundary surface at $x = 0$ is kept at zero temperature and heat is generated within the solid at a rate of $g(x, t)$ W/m^3. Determine the Green's function for this problem, and using this Green's function obtain an expression for the temperature distribution $T(x, t)$ within the medium for times $t > 0$.

6-2 Repeat Problem 6-1 for the case when the boundary surface at $x = 0$ is kept insulated.

6-3 A slab, $0 \leqslant x \leqslant L$, is initially at temperature $F(x)$. For times $t > 0$, heat is generated within the slab at a rate of $g(x, t)$ W/m^3, boundary surface at $x = 0$ is kept insulated and the boundary surface at $x = L$ dissipates heat by convection into a medium at zero temperature. Using the Green's function approach, obtain an expression for the temperature distribution $T(x, t)$ in the slab for times $t > 0$.

6-4 Using the Green's function approach solve the following heat conduction problem for a rectangular region $0 \leqslant x \leqslant a, 0 \leqslant y \leqslant b$:

$$\frac{\partial^2 T}{\partial x^2} + \frac{\partial^2 T}{\partial y^2} + \frac{1}{k} g(x, y, t) = \frac{1}{\alpha} \frac{\partial T}{\partial t} \qquad \text{in} \qquad 0 < x < a, \quad 0 < y < b, \quad t > 0$$

$$\frac{\partial T}{\partial x} = 0 \qquad \text{at} \qquad x = 0, \qquad t > 0$$

$$\frac{\partial T}{\partial x} + H_2 T = 0 \qquad \text{at} \qquad x = a, \qquad t > 0$$

$$T = 0 \qquad \text{at} \qquad y = 0, \qquad t > 0$$

$$\frac{\partial T}{\partial y} + H_4 T = 0 \qquad \text{at} \qquad y = b, \qquad t > 0$$

$$T = F(x, y) \qquad \text{for} \qquad t = 0, \qquad \text{in the region}$$

6-5 Solve the following heat conduction problem by using Green's function approach:

$$\frac{\partial^2 T}{\partial x^2} + \frac{\partial^2 T}{\partial y^2} + \frac{1}{k} g(x, y, t) = \frac{1}{\alpha} \frac{\partial T}{\partial t} \qquad \text{in} \qquad 0 < x < \infty, \quad 0 < y < b, \quad t > 0$$

$$T = 0 \qquad \text{at} \qquad x = 0, \qquad t > 0$$

$$-\frac{\partial T}{\partial y}+H_1 T=0 \qquad \text{at} \qquad y=0, \qquad t>0$$

$$T=0 \qquad \text{at} \qquad y=b, \qquad t>0$$

$$T=F(x,y) \qquad \text{for} \qquad t=0, \qquad \text{in the region}$$

6-6 Solve the following heat conduction problem by using Green's function approach:

$$\frac{\partial^2 T}{\partial x^2}+\frac{1}{k}g(x,t)=\frac{1}{\alpha}\frac{\partial T}{\partial t} \qquad \text{in} \qquad 0<x<L, \quad t>0$$

$$T=0 \qquad \text{at} \qquad x=0, \qquad t>0$$

$$\frac{\partial T}{\partial x}+HT=0 \qquad \text{at} \qquad x=L, \qquad t>0$$

$$T=F(x) \qquad \text{for} \qquad t=0, \qquad \text{in} \quad 0\leqslant x\leqslant L$$

6-7 A rectangular region $0\leqslant x\leqslant a, 0\leqslant y\leqslant b$ is initially at temperature $F(x,y)$. For times $t>0$, heat is generated within the solid at a rate of $g(x,y,t)$ W/m^3, while all boundaries are kept at zero temperature. Obtain an expression for the temperature distribution $T(x,y,t)$ in the region for times $t>0$.

6-8 A three-dimensional infinite medium, $-\infty<x<\infty, -\infty<y<\infty, -\infty<z<\infty$, is initially at temperature $F(x,y,z)$. For times $t>0$ heat is generated in the medium at a rate of $g(x,y,z,t)$ W/m^3. Using Green's function approach obtain an expression for the temperature distribution $T(x,y,z,t)$ in the region for time $t>0$. Also consider the following special cases:

1. The heat source is a point heat source of strength $g_p^c(t)$ W situated at the location (x_1,y_1,z_1), that is, $g(x,y,z,t)=g_p^c(t)\delta(x-x_1)\delta(y-y_1)\delta(z-z_1)$ releases its heat for times $t>0$.
2. The heat source is an instantaneous point heat source of strength g_p^i Ws, which releases its heat spontaneously at time $t=0$, at the location (x_1,y_1,z_1), that is, $g(x,y,z,t)=g_p^i\delta(t-0)\delta(x-x_1)\delta(y-y_1)\,\delta(z-z_1)$.

6-9 Solve the following heat conduction problem for a solid cylinder $0\leqslant r\leqslant b$ by Green's function approach:

$$\frac{\partial^2 T}{\partial r^2}+\frac{1}{r}\frac{\partial T}{\partial r}+\frac{1}{k}g(r,t)=\frac{1}{\alpha}\frac{\partial T}{\partial t} \qquad \text{in} \qquad 0\leqslant r<b, \quad t>0$$

$$\frac{\partial T}{\partial r} + HT = 0 \qquad \text{at} \quad r = b, \qquad t > 0$$

$$T = F(r) \qquad \text{for} \quad t = 0, \qquad \text{in} \quad 0 \leqslant r \leqslant b$$

6-10 Solve the following heat conduction problem by Green's function approach:

$$\frac{\partial^2 T}{\partial r^2} + \frac{1}{r}\frac{\partial T}{\partial r} + \frac{1}{k} g(r, t) = \frac{1}{\alpha}\frac{\partial T}{\partial t} \qquad \text{in} \quad 0 \leqslant r < \infty, \quad t > 0$$

$$T = F(r) \qquad \text{for} \quad t = 0, \qquad \text{in the region}$$

Also consider the following special case: Medium initially at zero temperature, the heat source is an instantaneous line-heat source of strength g_L^i, Ws/m situated along the z axis and releases its heat spontaneously at time $t = 0$, that is,

$$g(r, t) = \frac{g_L^i}{2\pi r} \delta(r - 0)\delta(t - 0)$$

6-11 Solve the following heat conduction problem by Green's function approach:

$$\frac{\partial^2 T}{\partial r^2} + \frac{1}{r}\frac{\partial T}{\partial r} + \frac{1}{k} g(r, t) = \frac{1}{\alpha}\frac{\partial T}{\partial t} \qquad \text{in} \quad a < r < \infty, \quad t > 0$$

$$T = 0 \qquad \text{at} \quad r = a, \qquad t > 0$$

$$T = F(r) \qquad \text{for} \quad t = 0, \qquad \text{in the region}$$

6-12 Repeat problem (6-11) for the boundary condition:

$$-\frac{\partial T}{\partial r} + HT = 0 \qquad \text{at } r = a$$

6-13 Solve the following heat conduction problem for a hollow cylinder by Green's function approach:

$$\frac{\partial^2 T}{\partial r^2} + \frac{1}{r}\frac{\partial T}{\partial r} + \frac{\partial^2 T}{\partial z^2} + \frac{1}{k} g(r, z, t) = \frac{1}{\alpha}\frac{\partial T}{\partial t} \qquad \text{in} \quad a < r < b, \quad 0 < z < c, \quad t > 0$$

$$T = 0 \qquad \text{at} \quad r = a, \quad r = b, \quad t > 0$$

7

THE USE OF LAPLACE TRANSFORM

The method of Laplace transform has been widely used in the solution of time-dependent heat conduction problems, because the partial derivative with respect to the time variable can be removed from the differential equation of heat conduction by the Laplace transformation. Although the application of Laplace transform for the removal of the partial derivative is a relatively straightforward matter, the inversion of the transformed solution generally is rather involved unless the inversion is available in the standard Laplace transform tables.

In this chapter we present a brief description of the basic operational properties of the Laplace transformation and illustrate with numerous examples its application in the solution of one-dimensional transient heat conduction problems. The orthogonal expansion technique and the Green's function approach discussed previously provide a much easier and straightforward method for solving such problems, but the solutions converge very slowly for small times. The Laplace transformation has the advantage that, it allows for the making of small time approximation in order to obtain solutions that are strictly applicable for small times, but are very rapidly convergent. This aspect of Laplace transformation will be emphasized later in this chapter.

The reader should consult references 1–7 for a more detailed discussion of the Laplace transform theory and references 8–12 for further applications of the Laplace transformation in the solution of heat conduction problems.

7-1 DEFINITION OF LAPLACE TRANSFORMATION

The *Laplace transform* and the *inversion formula* of a function $F(t)$ is defined by

$$\text{Laplace transform:} \quad \mathscr{L}[F(t)] \equiv \bar{F}(s) = \int_{t'=0}^{\infty} e^{-st'} F(t')dt' \qquad \text{(7-1a)}$$

$$\text{Inversion formula:} \quad F(t) = \frac{1}{2\pi i} \int_{s=\gamma - i\infty}^{\gamma + i\infty} e^{st} \bar{F}(s) ds \tag{7-1b}$$

where s is the Laplace transform variable, $i = \sqrt{-1}$, γ is a positive number, and the bar denotes the transform.

Thus, the Laplace transform of a function $F(t)$ consists of multiplying the function $F(t)$ by e^{-st} and integrating it over t from 0 to ∞. The inversion formula consists of the complex integration as defined by equation (7-1b).

Some remarks on the existence of the Laplace transform of a function $F(t)$ as defined by equation (7-1a) might be in order to illustrate the significance of this matter. For example, the integral (7-1a) may not exist because, (1) $F(t)$ may have infinite discontinuities for some values of t, or (2) $F(t)$ may have singularity as $t \to 0$, or (3) $F(t)$ may diverge exponentially for large t. The conditions for the existence of the Laplace transform defined by equation (7-1a) may be summarized as follows:

1. Function $F(t)$ is continuous or piecewise continuous in any interval $t_1 \leqslant t \leqslant t_2$, for $t_1 > 0$.
2. $t^n |F(t)|$ is bounded as $t \to 0^+$ for some number n when $n < 1$.
3. Function F(t) is of exponential order, namely, $e^{-\gamma t}|F(t)|$ is bounded for some positive number γ as $t \to \infty$.

For example, $F(t) = e^{t^2}$ is *not* of exponential order, that is, $e^{-\gamma t} \cdot e^{t^2}$ is unbounded at $t \to \infty$ for all values of γ, hence its Laplace transform *does not* exist. The Laplace transform of a function $F(t) = t^n$, when $n \leqslant -1$, does not exist because of condition (2), that is $\int_0^\infty e^{-st} t^n dt$ for $n \leqslant -1$ diverges at the origin.

Example 7-1

Determine the Laplace transform of the following functions: $F(t) = 1, t, e^{\pm at}$ and t^n with $n > -1$ but not necessarily integer.

Solution. According to the definition of the Laplace transform given by equation (7-1a), the Laplace transforms of these functions are given as

$$F(t) = 1: \ \bar{F}(s) = \int_0^\infty 1 \cdot e^{-st} dt = -\frac{1}{s} e^{-st} \bigg|_0^\infty = \frac{1}{s} \tag{7-2}$$

$$F(t) = t: \ \bar{F}(s) = \int_0^\infty t e^{-st} dt = \frac{1}{s^2} \tag{7-3}$$

$$F(t) = e^{\pm at}: \ \bar{F}(s) = \int_0^\infty e^{\pm at} e^{-st} dt = \int_0^\infty e^{-(s \mp a)t} dt = \frac{1}{s \mp a} \tag{7-4}$$

$$F(t) = t^n, \quad n > -1: \ \bar{F}(s) = \int_0^\infty t^n e^{-st} dt \tag{7-5a}$$

Now let $\xi = st$ and $d\xi = sdt$; then

$$\bar{F}(s) = s^{-n-1}\int_0^\infty \xi^n e^{-\xi} d\xi = \frac{\Gamma(n+1)}{s^{n+1}} \tag{7-5b}$$

where the integral $\int_0^\infty \xi^n e^{-\xi} d\xi$ is the *gamma function*, $\Gamma(n+1)$. The gamma function has the property $\Gamma(n+1) = n\Gamma(n)$; if n is an integer, we have $\Gamma(n+1) = n!$.

7-2 PROPERTIES OF LAPLACE TRANSFORM

Here we present some of the properties of Laplace transform that are useful in the solution of heat-conduction problems with Laplace transformation.

Linear Property

If $\bar{F}(s)$ and $\bar{G}(s)$ are the Laplace transform of functions $F(t)$ and $G(t)$ with respect to the t variable respectively, we may write

$$\mathscr{L}[c_1 F(t) + c_2 G(t)] = c_1 \bar{F}(s) + c_2 \bar{G}(s) \tag{7-6}$$

where c_1 and c_2 are any constants.

Example 7-2

By utilizing the linear property of the Laplace transform and the Laplace transform of $e^{\pm at}$ given by equation (7-4), determine the Laplace transform of the functions $\cosh at$ and $\sinh at$.

Solution. For $\cosh at$ we write

$$F_1(t) \equiv \cosh at = \tfrac{1}{2}(e^{at} + e^{-at}) \tag{7-7}$$

$$\bar{F}_1(s) = \frac{1}{2}\left(\frac{1}{s-a} + \frac{1}{s+a}\right) = \frac{s}{s^2 - a^2} \tag{7-8}$$

Similarly

$$F_2(t) \equiv \sinh at = \tfrac{1}{2}(e^{at} - e^{-at}) \tag{7-9}$$

$$\bar{F}_2(s) = \frac{1}{2}\left(\frac{1}{s-a} - \frac{1}{s+a}\right) = \frac{s}{s^2 - a^2} \tag{7-10}$$

Laplace Transform of Derivatives

The Laplace transform of the first derivative $dF(t)/dt$ of a function $F(t)$ is readily obtained by utilizing the definition of the Laplace transform and integrating it by parts:

$$\mathscr{L}[F'(t)] = \int_0^\infty F'(t)e^{-st}\,dt = [F(t)e^{-st}]_0^\infty + s\int_0^\infty F(t)e^{-st}\,dt \tag{7-11a}$$

$$\mathscr{L}[F'(t)] = s\bar{F}(s) - F(0) \tag{7-11b}$$

where the prime denotes differentiation with respect to t and $F(0)$ indicates the value of $F(t)$ at $t = 0^+$, namely, as we approach zero from the positive side. Thus, the Laplace transform of the first derivative of a function is equal to multiplying the transform of the function by s and subtracting from it the value of this function at $t = 0^+$.

This result is now utilized to determine the Laplace transform of the second derivative of a function $F(t)$ as

$$\begin{aligned}\mathscr{L}[F''(t)] &= s\mathscr{L}[F'(t)] - F'(0) = s[s\bar{F}(s) - F(0)] - F'(0)\\ &= s^2\bar{F}(s) - sF(0) - F'(0)\end{aligned} \tag{7-12}$$

Similarly, the Laplace transform of the third derivative becomes

$$\mathscr{L}[F'''(t)] = s^3\bar{F}(s) - s^2F(0) - sF'(0) - F''(0) \tag{7-13}$$

In general, the Laplace transform of the nth derivative is given as

$$\mathscr{L}[F^{(n)}(t)] = s^n\bar{F}(s) - s^{n-1}F(0) - s^{n-2}F^{(1)}(0) - s^{n-3}F^{(2)}(0) - \cdots - F^{(n-1)}(0) \tag{7-14}$$

where

$$F^{(n)}(t) \equiv \frac{d^nF(t)}{dt^n}$$

Laplace Transform of Integrals

The Laplace transform of the integral $\int_0^t F(\tau)\,d\tau$ of a function $F(t)$ is determined as now described. Let

$$g(t) \equiv \int_0^t F(\tau)\,d\tau \tag{7-15a}$$

then

$$g'(t) = F(t) \tag{7-15b}$$

We take the Laplace transform of both sides of equation (7-15b) and utilize the result in equation (7-11b) to obtain

$$s\bar{g}(s) = \bar{F}(s) \tag{7-16}$$

since $g(0) = 0$. After rearranging, we find

$$\bar{g}(s) \equiv \mathscr{L}\left[\int_0^t F(\tau)\,d\tau\right] = \frac{1}{s}\bar{F}(s) \tag{7-17}$$

This procedure is repeated to obtain the Laplace transform of the double integration of a function $F(t)$

$$\mathscr{L}\left[\int_0^t \int_0^{\tau_2} F(\tau_1)\,d\tau_1\,d\tau_2\right] = \frac{1}{s^2}\bar{F}(s) \tag{7-18}$$

In general, the Laplace transform of the nth integral of a function $F(t)$ is given as

$$\mathscr{L}\left[\int_0^t \cdots \int_0^{\tau_n} F(\tau_1)\,d\tau_1 \cdots d\tau_n\right] = \frac{1}{s^n}\bar{F}(s) \tag{7-19}$$

Change of Scale

Let $\bar{F}(s)$ be the Laplace transform of a function $F(t)$. Then, the Laplace transforms of functions $F(at)$ and $F[(1/a)t]$, where a is a real, positive constant, are determined as

$$\mathscr{L}[F(at)] = \int_0^\infty F(at)e^{-st}\,dt = \frac{1}{a}\int_0^\infty F(u)e^{-(s/a)u}\,du = \frac{1}{a}\bar{F}\left(\frac{s}{a}\right) \tag{7-20a}$$

where we set $u = at$. Similarly.

$$\mathscr{L}\left[F\left(\frac{1}{a}t\right)\right] = \int_0^\infty F\left(\frac{t}{a}\right)e^{-st}\,dt = a\int_0^\infty F(u)e^{-asu}\,du = a\bar{F}(as) \tag{7-20b}$$

where we set $u = t/a$.

Example 7-3

The Laplace transform of cosh t is given as $\mathscr{L}[\cosh t] = s/(s^2 - 1)$. By utilizing the "change of scale" property, determine the Laplace transform of the functions cosh at and cosh (t/a).

Solution. By utilizing equation (7-20a) we obtain

$$\mathscr{L}[\cosh at] = \frac{1}{a}\frac{(s/a)}{(s/a)^2 - 1} = \frac{s}{s^2 - a^2} \tag{7-21a}$$

and by utilizing equation (7-20b) we find

$$\mathscr{L}\left[\cosh\frac{t}{a}\right] = a\frac{as}{(as)^2 - 1} = \frac{s^2}{s^2 - (1/a)^2} \tag{7-21b}$$

Shift Property

When the Laplace transform $\bar{F}(s)$ of a function $F(t)$ is known, the shift property enables us to write the Laplace transform of a function $e^{\pm at}F(t)$, where a is a constant; that is

$$\begin{aligned}\mathscr{L}[e^{\pm at}F(t)] &= \int_0^\infty e^{-st}e^{\pm at}F(t)\,dt = \int_0^\infty e^{-(s\mp a)t}F(t)\,dt \\ &= \bar{F}(s\mp a)\end{aligned} \tag{7-22}$$

Example 7-4

The Laplace transform of cos bt is given as $\mathscr{L}[\cos bt] = s/(s^2+b^2)$. By utilizing the "shift property" determine the Laplace transform of the function $e^{-at}\cdot\cos bt$.

Solution. By equation (7-22) we immediately write

$$\mathscr{L}[e^{-at}\cos bt] = \frac{s+a}{(s+a)^2+b^2} \tag{7-23}$$

Laplace Transform of Translated Function

The unit step function (or the Heaviside unit function) is useful in denoting the translation of a function. Figure 7-1 shows the physical significance of the unit step functions $U(t)$ and $U(t-a)$; namely

$$U(t) = \begin{cases} 1 & t>0 \\ 0 & t<0 \end{cases} \tag{7-24a}$$

$$U(t-a) = \begin{cases} 1 & t>a \\ 0 & t<a \end{cases} \tag{7-24b}$$

We now consider a function $F(t)$ defined for $t>0$ as illustrated in Fig. 7-2a and the translation of this function from $t=0$ to $t=a$ as illustrated in Fig. 7-2b. The translated function $U(t-a)\cdot F(t-a)$ represents the function $F(t)$ defined for $t>0$,

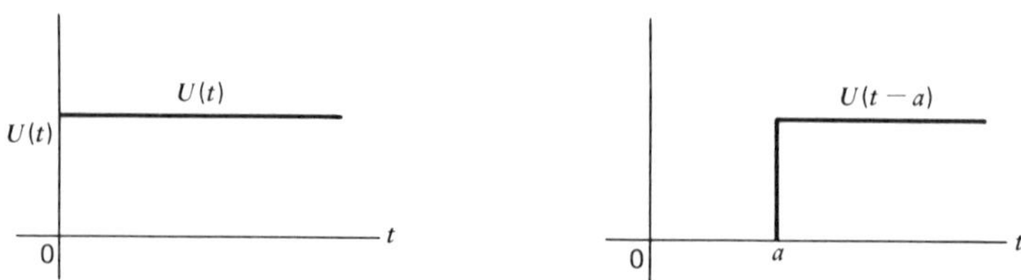

Fig. 7-1 Definition of the unit step functions $U(t)$ and $U(t-a)$.

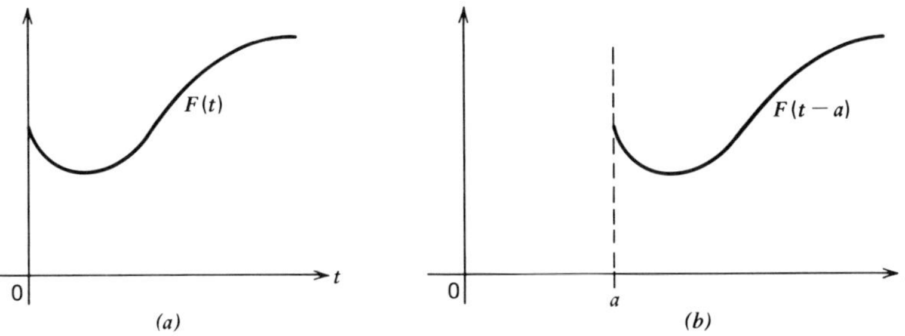

Fig. 7-2 The translation of a function $F(t)$: (a) the function $F(t)$ and (b) the translation of $F(t)$ from $t = 0$ to $t = a$.

translated by an amount $t = a$ in the positive t direction; namely

$$U(t-a)F(t-a) = \begin{cases} F(t-a) & \text{for} \quad t > a \\ 0 & \text{for} \quad t < a \end{cases} \tag{7-25}$$

The Laplace transform of this translated function is determined as

$$\begin{aligned} \mathscr{L}[U(t-a)F(t-a)] &= \int_0^\infty e^{-st} U(t-a)F(t-a)\, dt \\ &= \int_{t=a}^\infty e^{-st} F(t-a)\, dt \\ &= \int_{\eta=0}^\infty e^{-s(a+\eta)} F(\eta)\, d\eta = e^{-as} \int_{\eta=0}^\infty e^{-s\eta} F(\eta)\, d\eta \\ &= e^{-as} \bar{F}(s) \end{aligned} \tag{7-26}$$

where a new variable η is defined as $\eta = t - a$. This result shows that the Laplace

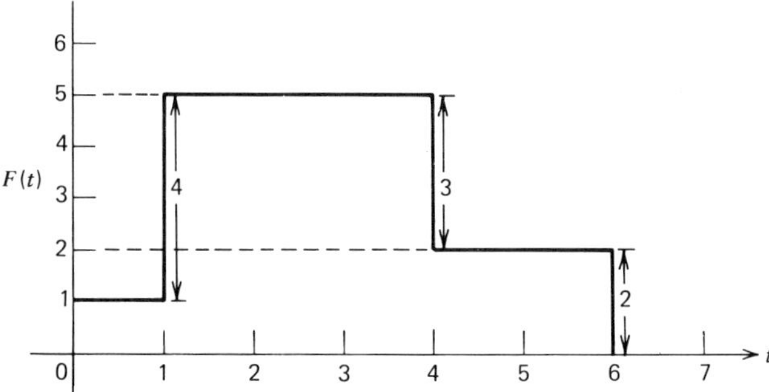

Fig. 7-3 The function defined by equation (7-28).

transform of a translated function $U(t-a)F(t-a)$ is equal to the Laplace transform $\bar{F}(s)$ of the function $F(t)$ multiplied by e^{-as}.

Similarly, the Laplace transform of a unit step function $U(t-a)$ is given by

$$\mathscr{L}[U(t-a)] = e^{-as}\frac{1}{s} \tag{7-27}$$

Example 7-5

Determine the Laplace transform of the following function

$$F(t) = \begin{cases} 0 & \text{for} \quad t<0 \\ 1 & \text{for} \quad 0<t<1 \\ 5 & \text{for} \quad 1<t<4 \\ 2 & \text{for} \quad 4<t<6 \\ 0 & \text{for} \quad t>6 \end{cases} \tag{7-28}$$

which is illustrated in Fig. 7-3.

Solution. The function given by equation (7-28) is represented in terms of the unit step functions as

$$F(t) = U(t-0) + 4U(t-1) - 3U(t-4) - 2U(t-6) \tag{7-29}$$

and the Laplace transform of this function becomes

$$\bar{F}(s) = \frac{1}{s} + 4e^{-s}\frac{1}{s} - 3e^{-4s}\frac{1}{s} - 2e^{-6s}\frac{1}{s} \tag{7-30}$$

Laplace Transform of Delta Function

The delta function $\delta(x)$ is defined to be zero everywhere except at $x = 0$ such that

$$\delta(x) = 0, \qquad x \neq 0 \tag{7-31}$$

and

$$\int_{-\infty}^{\infty} \delta(x)\,dx = 1 \tag{7-32}$$

The Laplace transform of the delta function $\delta(x)$ is given by

$$\mathscr{L}[\delta(x)] \equiv \bar{\delta}(s) = \int_{0}^{\infty} e^{-sx}\delta(x)\,dx = 1 \tag{7-33}$$

The properties of the delta function are given in Appendix VI.

Laplace Transform of Convolution

Let $f(t)$ and $g(t)$ be two functions of t defined for $t > 0$. The *convolution integral* or briefly the *convolution* of these two functions is denoted by the notation $f * g$ and defined by the equation

$$f * g = \int_0^t f(t-\tau) g(\tau)\, d\tau \tag{7-34}$$

$$= \int_0^t f(\tau) g(t-\tau)\, d\tau \tag{7-35}$$

Thus we have the relation $f * g = g * f$. The Laplace transform of the convolution $f * g$ is given by

$$\mathscr{L}[f * g] \equiv \overline{f * g} = \bar{f}(s)\bar{g}(s) \tag{7-36}$$

That is, the Laplace transform of the convolution is equal to the product of the Laplace transforms $\bar{f}(s)$ and $\bar{g}(s)$ of these two functions.

Derivatives of Laplace Transform

We now derive an expression for the derivative of the Laplace transform of a function. Consider the Laplace transform $\bar{F}(s)$ of a function $F(t)$ given by

$$\bar{F}(s) = \int_0^\infty e^{-st} F(t)\, dt \tag{7-37}$$

By differentiating both sides of equation (7-37) with respect to s we obtain

$$\frac{d\bar{F}(s)}{ds} \equiv \bar{F}'(s) = \int_0^\infty (-t) e^{-st} F(t)\, dt$$

$$\bar{F}'(s) = \mathscr{L}[(-t)F(t)] \tag{7-38a}$$

or by differentiating equation (7-37) n times we obtain

$$\frac{d^n \bar{F}(s)}{ds^n} \equiv \bar{F}^{(n)}(s) = \mathscr{L}[(-t)^n F(t)], \qquad n = 1, 2, 3 \ldots \tag{7-38b}$$

Thus, the nth differentiation of the Laplace transform $\bar{F}(s)$ is equal to the Laplace transform of $(-t)^n F(t)$. This relation is useful in finding the inverse transforms with the aid of partial fractions and in many other applications.

Example 7-6

The Laplace transform of $F(t) = \sin \beta t$ is given as $\bar{F}(s) = \beta/(s^2 + \beta^2)$. Determine the Laplace transform of the function "$t \sin \beta t$."

Solution. By applying the formula (7-38a) we write

$$\mathscr{L}[(-t)F(t)] = \frac{d}{ds}\bar{F}(s)$$

$$\mathscr{L}[-t \sin \beta t] = \frac{d}{ds}\left[\frac{\beta}{s^2 + \beta^2}\right] = -\frac{2s\beta}{(s^2 + \beta^2)^2} \tag{7-39}$$

or

$$\mathscr{L}[t \sin \beta t] = \frac{2\beta s}{(s^2 + \beta^2)^2} \qquad \text{for} \qquad s > 0 \tag{7-40}$$

The Integration of Laplace Transform

Consider the Laplace transform of a function $F(t)$ given by

$$\bar{F}(s) = \int_0^\infty e^{-st} F(t)\, dt \tag{7-41}$$

We integrate both sides of this equation with respect to s from s to b and obtain

$$\int_s^b \bar{F}(s')\, ds' = \int_s^b \int_0^\infty e^{-s't} F(t)\, dt\, ds'$$

$$= \int_0^\infty F(t)\left[\int_s^b e^{-s't}\, ds'\right] dt = \int_0^\infty \frac{F(t)}{t}(e^{-st} - e^{-bt})\, dt \tag{7-42}$$

If the function $F(t)$ is such that $F(t)/t$ exists at $t \to 0$, the integral uniformly converges. Then, letting $b \to \infty$, equation (7-42) becomes

$$\int_s^\infty \bar{F}(s')\, ds' = \int_0^\infty \left[\frac{F(t)}{t}\right] e^{-st}\, dt \equiv \mathscr{L}\left[\frac{F(t)}{t}\right] \tag{7-43}$$

Thus, the integration of the Laplace transform $\bar{F}(s)$ of a function $F(t)$ with respect to s from s to ∞, is equal to the Laplace transform of the function $F(t)/t$. This result is useful in the determination of the Laplace transform of the function $F(t)/t$ when the Laplace transform $\bar{F}(s)$ of the function $F(t)$ is known.

Example 7-7

The Laplace transform of $\sin \beta t$ is given as $\beta/(s^2 + \beta^2)$. Determine the Laplace transform of $(1/t) \sin \beta t$.

Solution. We utilize the formula (7-43)

$$\mathscr{L}\left[\frac{F(t)}{t}\right] = \int_s^\infty \bar{F}(s')\,ds'$$

Introducing the function as given above we obtain

$$\mathscr{L}\left[\frac{\sin \beta t}{t}\right] = \int_s^\infty \frac{\beta}{s'^2 + \beta^2}\,ds' = \left[\tan^{-1}\frac{s'}{\beta}\right]_s^\infty = \frac{\pi}{2} - \tan^{-1}\left(\frac{s}{\beta}\right) \tag{7-44}$$

7-3 THE INVERSION OF LAPLACE TRANSFORM USING THE INVERSION TABLES

In heat conduction problems, the Laplace transformation is generally applied to the time variable. Therefore, an important step in the final analysis is the inversion of the transformed function from the Laplace variable s domain to the actual time variable t domain. To facilitate such analysis comprehensive tables have been prepared for the inversion of the Laplace transform of a large class of functions [7]. We present in Table 7-1 the Laplace transform of various functions which are useful in the analysis of heat-conduction problems.

If the Laplace transform $\bar{F}(s)$ of a function $F(t)$ is expressible in the form

$$\bar{F}(s) = \frac{G(s)}{H(s)} \tag{7-45}$$

where $G(s)$ and $H(s)$ are polynomials with no common factor, with $G(s)$ being lower degree than $H(s)$, and the factors of $H(s)$ are all linear and distinct, then equation (7-45) can be expressed in the form

$$\bar{F}(s) = \frac{G(s)}{H(s)} = \frac{c_1}{s - a_1} + \frac{c_2}{s - a_2} + \cdots + \frac{c_n}{s - a_n} \tag{7-46}$$

Here the c_i values are independent of s. Then, by the theory of partial fractions c_i values are determined as

$$c_i = \lim_{s \to a_i} [(s - a_i)\bar{F}(s)] \tag{7-47}$$

Clearly, if a function $\bar{F}(s)$ is expressible in partial fractions as in equations (7-46), its inversion is readily obtained by the use of the Laplace transform table.

Also, there are many occasions that the transformed function $\bar{F}(s)$ will not appear in the standard transform tables. In such cases it will be necessary to use the inversion formula (7-1b) to determine the function. Such an inversion is

TABLE 7-1 A Table of Laplace Transform of Functions

No.	$\bar{F}(s)$	$F(t)$
1	$\frac{1}{s}$	1
2	$\frac{1}{s^2}$	t
3	$\frac{1}{s^n}(n = 1, 2, 3, \ldots)$	$\frac{t^{n-1}}{(n-1)!}$
4	$\frac{1}{\sqrt{s}}$	$\frac{1}{\sqrt{\pi t}}$
5	$s^{-3/2}$	$2\sqrt{t/\pi}$
6	$s^{-(n+1/2)}(n = 1, 2, 3, \ldots)$	$\frac{2^n}{[1 \cdot 3 \cdot 5 \cdot \ldots \cdot (2n-1)]\sqrt{\pi}} t^{n-1/2}$
7	$\frac{1}{s^n}(n > 0)$	$\frac{1}{\Gamma(n)} t^{n-1}$
8	$\frac{1}{s+a}$	e^{-at}
9	$\frac{1}{(s+a)^n}(n = 1, 2, 3, \ldots)$	$\frac{t^{n-1}e^{-at}}{(n-1)!}$
10	$\frac{\Gamma(k)}{(s+a)^k}(k > 0)$	$t^{k-1}e^{-at}$
11	$\frac{1}{(s+a)(s+b)}(a \neq b)$	$\frac{e^{-at} - e^{-bt}}{b-a}$
12	$\frac{s}{(s+a)(s+b)}(a \neq b)$	$\frac{ae^{-at} - be^{-bt}}{a-b}$
13	$\frac{1}{s^2+a^2}$	$\frac{1}{a}\sin at$
14	$\frac{s}{s^2+a^2}$	$\cos at$
15	$\frac{1}{s^2-a^2}$	$\frac{1}{a}\sinh at$
16	$\frac{s}{s^2-a^2}$	$\cosh at$

TABLE 7-1 (*Continued*)

No.	$\bar{F}(s)$	$F(t)$
17	$\dfrac{1}{s(s^2+a^2)}$	$\dfrac{1}{a^2}(1-\cos at)$
18	$\dfrac{1}{s^2(s^2+a^2)}$	$\dfrac{1}{a^3}(at-\sin at)$
19	$\dfrac{1}{(s^2+a^2)^2}$	$\dfrac{1}{2a^3}(\sin at - at\cos at)$
20	$\dfrac{s}{(s^2+a^2)^2}$	$\dfrac{t}{2a}\sin at$
21	$\dfrac{s^2}{(s^2+a^2)^2}$	$\dfrac{1}{2a}(\sin at + at\cos at)$
22	$\dfrac{s^2-a^2}{(s^2+a^2)^2}$	$t\cos at$
23	$\dfrac{1}{\sqrt{s}+a}$	$\dfrac{1}{\sqrt{\pi t}} - ae^{a^2t}\operatorname{erfc} a\sqrt{t}$
24	$\dfrac{\sqrt{s}}{s-a^2}$	$\dfrac{1}{\sqrt{\pi t}} + ae^{a^2t}\operatorname{erf} a\sqrt{t}$
25	$\dfrac{\sqrt{s}}{s+a^2}$	$\dfrac{1}{\sqrt{\pi t}} - \dfrac{2a}{\sqrt{\pi}}e^{-a^2t}\displaystyle\int_0^{a\sqrt{t}} e^{\lambda^2}\,d\lambda$
26	$\dfrac{1}{\sqrt{s}(s-a^2)}$	$\dfrac{1}{a}e^{a^2t}\operatorname{erf} a\sqrt{t}$
27	$\dfrac{1}{\sqrt{s}(s+a^2)}$	$\dfrac{2}{a\sqrt{\pi}}e^{-a^2t}\displaystyle\int_0^{a\sqrt{t}} e^{\lambda^2}\,d\lambda$
28	$\dfrac{b^2-a^2}{(s-a^2)(b+\sqrt{s})}$	$e^{a^2t}[b - a\operatorname{erf} a\sqrt{t}] - be^{b^2t}\operatorname{erfc} b\sqrt{t}$
29	$\dfrac{1}{\sqrt{s}(\sqrt{s}+a)}$	$e^{a^2t}\operatorname{erfc} a\sqrt{t}$
30	$\dfrac{1}{(s+a)\sqrt{s+b}}$	$\dfrac{1}{\sqrt{b-a}}e^{-at}\operatorname{erf}(\sqrt{b-a}\sqrt{t})$
31	$\dfrac{\sqrt{s+2a}}{\sqrt{s}} - 1$	$ae^{-at}[I_1(at) + I_0(at)]$

TABLE 7-1 (*Continued*)

No.	$\bar{F}(s)$	$F(t)$
32	$\dfrac{1}{\sqrt{s+a}\sqrt{s+b}}$	$e^{-(a+b)t/2}I_0\left(\dfrac{a-b}{2}t\right)$
33	$\dfrac{1}{\sqrt{s^2+a^2}}$	$J_0(at)$
34	$\dfrac{(\sqrt{s^2+a^2}-s)^\nu}{\sqrt{s^2+a^2}}(\nu > -1)$	$a^\nu J_\nu(at)$
35	$\dfrac{(s-\sqrt{s^2-a^2})^\nu}{\sqrt{s^2-a^2}}(\nu > -1)$	$a^\nu I_\nu(at)$
36	$\dfrac{1}{s}e^{-ks}$	$u(t-k)$
37	$\dfrac{1}{s^2}e^{-ks}$	$(t-k)u(t-k)$
38	$\dfrac{1}{s}e^{-k/s}$	$J_0(2\sqrt{kt})$
39	$\dfrac{1}{s^\mu}e^{-k/s}(\mu > 0)$	$\left(\dfrac{t}{k}\right)^{(\mu-1)/2}J_{\mu-1}(2\sqrt{kt})$
40	$\dfrac{1}{s^\mu}e^{-k/s}(\mu > 0)$	$\left(\dfrac{t}{k}\right)^{(\mu-1)/2}I_{\mu-1}(2\sqrt{kt})$
41	$e^{-k\sqrt{s}}(k > 0)$	$\dfrac{k}{2\sqrt{\pi t^3}}\exp\left(-\dfrac{k^2}{4t}\right)$
42	$\dfrac{1}{s}e^{-k\sqrt{s}}(k \geqslant 0)$	$\operatorname{erfc}\dfrac{k}{2\sqrt{t}}$
43	$\dfrac{1}{\sqrt{s}}e^{-k\sqrt{s}}(k \geqslant 0)$	$\dfrac{1}{\sqrt{\pi t}}\exp\left(-\dfrac{k^2}{4t}\right)$
44	$\dfrac{1}{s^{3/2}}e^{-k\sqrt{s}}(k \geqslant 0)$	$2\sqrt{\dfrac{t}{\pi}}\exp\left(-\dfrac{k^2}{4t}\right)-k\operatorname{erfc}\dfrac{k}{2\sqrt{t}}$ $=2\sqrt{t}\,i\operatorname{erfc}\dfrac{k}{2\sqrt{t}}$
45	$\dfrac{1}{s^{1+n/2}}e^{-k\sqrt{s}}(n=0,1,2,\ldots,k\geqslant 0)$	$(4t)^{n/2}i^n\operatorname{erfc}\dfrac{k}{2\sqrt{t}}$

TABLE 7-1 *(Continued)*

No.	$\bar{F}(s)$	$F(t)$
46	$\dfrac{e^{-k\sqrt{s}}}{a+\sqrt{s}}(k \geqslant 0)$	$\dfrac{1}{\sqrt{\pi t}}\exp\left(-\dfrac{k^2}{4t}\right) - ae^{ak}e^{a^2t}\operatorname{erfc}\left(a\sqrt{t}+\dfrac{k}{2\sqrt{t}}\right)$
47	$\dfrac{e^{-k\sqrt{s}}}{\sqrt{s}(a+\sqrt{s})}(k \geqslant 0)$	$e^{ak}a^{a^2t}\operatorname{erfc}\left(a\sqrt{t}+\dfrac{k}{2\sqrt{t}}\right)$
48	$\dfrac{e^{-k\sqrt{s(s+a)}}}{\sqrt{s(s+a)}}(k \geqslant 0)$	$e^{-at/2}I_0(\frac{1}{2}a\sqrt{t^2-k^2})u(t-k)$
49	$\dfrac{e^{-k\sqrt{s^2+a^2}}}{\sqrt{s^2+a^2}}(k \geqslant 0)$	$J_0(a\sqrt{t^2-k^2})u(t-k)$
50	$\dfrac{e^{-k\sqrt{s^2+a^2}}}{\sqrt{s^2-a^2}}(k \geqslant 0)$	$I_0(a\sqrt{t^2-k^2})u(t-k)$
51	$\dfrac{ae^{-k\sqrt{s}}}{s(a+\sqrt{s})}(k \geqslant 0)$	$-e^{ak}e^{a^2t}\operatorname{erfc}\left(a\sqrt{t}+\dfrac{k}{2\sqrt{t}}\right)+\operatorname{erfc}\dfrac{k}{2\sqrt{t}}$
52	$\dfrac{1}{s^2}e^{-k\sqrt{s}}$	$\left(t+\dfrac{k^2}{2}\right)\operatorname{erfc}\left(\dfrac{k}{2\sqrt{t}}\right)-k\left(\dfrac{t}{\pi}\right)^{1/2}\exp\left(-\dfrac{k^2}{4t}\right)$
53	$\dfrac{1}{s}\ln s$	$-\gamma-\ln t(\gamma = 0.57721\,56649\ldots$ Euler's constant)
54	$\ln\dfrac{s+a}{s+b}$	$\dfrac{1}{t}(e^{-bt}-e^{-at})$
55	$\ln\dfrac{s^2+a^2}{s^2}$	$\dfrac{2}{t}(1-\cos at)$
56	$\ln\dfrac{s^2-a^2}{s^2}$	$\dfrac{2}{t}(1-\cosh at)$
57	$K_0(ks)(k>0)$	$\dfrac{1}{\sqrt{t^2-k^2}}u(t-k)$
58	$K_0(k\sqrt{s})(k>0)$	$\dfrac{1}{2t}\exp\left(-\dfrac{k^2}{4t}\right)$
59	$\dfrac{1}{\sqrt{s}}K_1(k\sqrt{s})(k>0)$	$\dfrac{1}{k}\exp\left(-\dfrac{k^2}{4t}\right)$

generally performed by the method of contour integration and the calculus of residues that require rather elaborate analysis. Therefore, the use of inversion formula (7-1b) will not be considered here.

Example 7-8

Determine the function whose Laplace transform is

$$\bar{F}(s) = \frac{b^2}{s(s^2 + b^2)}. \tag{7-48}$$

Solution. This function is not available in the Laplace transform table in this form; but it can be expressible in partial fractions as

$$\bar{F}(s) = \frac{b^2}{s(s^2 + b^2)} = \frac{c_1}{s} + \frac{c_2 s + c_3}{s^2 + b^2} \tag{7-49}$$

Then

$$b^2 = c_1 b^2 + c_3 s + (c_1 + c_2)s^2 \tag{7-50}$$

Equating the coefficients of like powers of s, we obtain $c_1 = 1, c_2 = -1$, and $c_3 = 0$. Hence

$$\bar{F}(s) = \frac{1}{s} - \frac{s}{s^2 + b^2} \tag{7-51}$$

Each term on the right-hand side is readily inverted using Table 7-1, cases 1 and 14; we find

$$F(t) = 1 - \cos bt \tag{7-52}$$

7-4 APPLICATION OF LAPLACE TRANSFORM IN THE SOLUTION OF TIME-DEPENDENT HEAT CONDUCTION PROBLEMS

In this section we illustrate with representative examples the use of Laplace transform technique in the solution of time-dependent heat conduction problems. In this approach, the Laplace transform is applied to remove the partial derivative with respect to the time variable, the resulting equation is solved for the transform of temperature, and the transform is inverted to recover the solution for the temperature distribution. The approach is straightforward in principle, but generally the inversion is difficult unless the transform is available in the Laplace transform tables. In the following examples, typical heat-conduction problems are solved by using the Laplace transform table to invert the transform.

Example 7-9

A semiinfinite medium, $x \geqslant 0$, is initially at zero temperature. For times $t > 0$, the boundary surface at $x = 0$ is subjected to a temperature $T = f(t)$ that varies with time. Obtain an expression for the temperature distribution $T(x, t)$ in the medium for times $t > 0$.

Solution. The mathematical formulation of this problem is given as

$$\frac{\partial^2 T(x,t)}{\partial x^2} = \frac{1}{\alpha}\frac{\partial T(x,t)}{\partial t} \quad \text{in} \quad 0 < x < \infty, \quad t > 0 \tag{7-53a}$$

$$T(x,t) = f(t) \quad \text{at} \quad x = 0, \quad t > 0 \tag{7-53b}$$

$$T(x,t) = 0 \quad \text{as} \quad x \to \infty, \quad t > 0 \tag{7-53c}$$

$$T(x,t) = 0 \quad \text{for} \quad t = 0, \quad \text{in} \quad x \geqslant 0 \tag{7-53d}$$

We recall that this problem was solved in Example 5-2 by the application of Duhamel's method. Here the Laplace transform technique is used to solve the same problem, and the standard Laplace transform table is utilized to invert the resulting transform. Taking the Laplace transform of equations (7-53) we obtain

$$\frac{d^2 \bar{T}(x,s)}{dx^2} - \frac{s}{\alpha}\bar{T}(x,s) = 0 \quad \text{in} \quad 0 < x < \infty \tag{7-54a}$$

$$\bar{T}(x,s) = \bar{f}(s) \quad \text{at} \quad x = 0 \tag{7-54b}$$

$$\bar{T}(x,s) = 0 \quad \text{as} \quad x \to \infty \tag{7-54c}$$

The solution of equations (7-54) is given as

$$\bar{T}(x,s) = \bar{f}(s) \cdot \bar{g}(x,s) \tag{7-55a}$$

where

$$\bar{g}(x,s) \equiv e^{-x\sqrt{s/\alpha}} \tag{7-55b}$$

Since the functional form of $\bar{f}(s)$ is not explicitly specified, it is better to make use of the convolution property of the Laplace transform given by equation (7-36) to invert this transform. Namely, in view of equation (7-36), we write the result in equation (7-55a) as

$$\bar{T}(x,s) = \bar{f}(s) \cdot \bar{g}(x,s) = \mathscr{L}[f(t) * g(x,t)] \tag{7-56}$$

The inversion of this result gives

$$T(x,t) = f(t) * g(x,t) \tag{7-57a}$$

and utilizing the definition of the convolution $f*g$ given by equation (7-35) equation (7-57a) is written as

$$T(x,t) = \int_0^t f(\tau) g(x, t-\tau)\, d\tau \tag{7-57b}$$

To complete the solution of this problem we need to know the function $g(x,t)$. However, the Laplace transform $\bar{g}(x,s)$ of this function is given by equation (7-55b); then, the function $g(x,t)$ can be determined by the inversion of this transform. The transform $\bar{g}(x,s)$ is readily inverted by utilizing Table 7-1, case 41; we find

$$g(x,t) = \frac{x}{2\sqrt{\pi \alpha t^3}} e^{-x^2/4\alpha t} \tag{7-57c}$$

After replacing t by $(t-\tau)$ in this result, we introduce it into equation (7-57b) to obtain the desired solution as

$$T(x,t) = \frac{x}{\sqrt{4\pi\alpha}} \int_{\tau=0}^{t} \frac{f(\tau)}{(t-\tau)^{3/2}} \exp\left[\frac{-x^2}{4\alpha(t-\tau)}\right] d\tau \tag{7-58}$$

This result is the same as that given by equation (5-33) which was obtained by utilizing the Duhamel's theorem. The temperature $T(x,t)$ can be determined from equation (7-58) for any specified form of the function $f(t)$ by performing the integration. Sometimes it is easier to introduce the transform $\bar{f}(s)$ of the function $f(t)$ into equation (7-55a) and then invert the result rather than performing the integration in equation (7-58). This matter is now illustrated for some special cases of function $f(t)$.

1. $f(t) = T_0 =$ *constant*. Then, the transform of $f(t) = T_0$ is $\bar{f}(s) = T_0/s$. Introducing this result into equation (7-55) we obtain

$$\bar{T}(x,s) = \frac{T_0}{s} e^{-x\sqrt{s/\alpha}} \tag{7-59a}$$

The transform (7-59a) is readily inverted by utilizing Table 7-1, case 42. We obtain

$$T(x,t) = T_0 \operatorname{erfc}(x/\sqrt{4\alpha t}) \tag{7-59b}$$

2. $f(t) = T_0 t^{1/2}$. The transform of this function is obtained from Table 7-1, case 5 as $\bar{f}(s) = T_0(\sqrt{\pi}/2)s^{-3/2}$. Introducing this result into equation (7-55) we obtain

$$\bar{T}(x,s) = T_0 \frac{\sqrt{\pi}}{2} s^{-3/2} e^{-x\sqrt{s/\alpha}} \tag{7-60a}$$

This result is inverted by utilizing Table 7-1, case 44; we find

$$T(x,t) = T_0 \left[t^{1/2} e^{-x^2/4\alpha t} - \frac{x}{2}\sqrt{\frac{\pi}{\alpha}} \operatorname{erfc} \frac{x}{\sqrt{4\alpha t}} \right] \tag{7-60b}$$

Example 7-10

A semiinifinite medium, $0 \leqslant x < \infty$, is initially at zero temperature. For times $t > 0$, the boundary surface at $x = 0$ is subjected to convection with an environment at temperature T_∞. Obtain an expression for the temperature distribution $T(x,t)$ in the solid for times $t > 0$.

Solution. The mathematical formulation of the problem is given as

$$\frac{\partial^2 T(x,t)}{\partial x^2} = \frac{1}{\alpha} \frac{\partial T(x,t)}{\partial t} \quad \text{in} \quad 0 < x < \infty, \quad t > 0 \tag{7-61a}$$

$$-k\frac{\partial T}{\partial x} + hT = hT_\infty \quad \text{at} \quad x = 0, \quad t > 0 \tag{7-61b}$$

$$T = 0 \quad \text{as} \quad x \to \infty, \quad t > 0 \tag{7-61c}$$

$$T = 0 \quad \text{for} \quad t = 0, \quad \text{in} \quad 0 \leqslant x < 0 \tag{7-61d}$$

The Laplace transform of equations (7-61) becomes

$$\frac{d^2 \bar{T}(x,s)}{dx^2} - \frac{s}{\alpha} \bar{T}(x,s) = 0 \quad \text{in} \quad 0 < x < \infty \tag{7-62a}$$

$$-k\frac{d\bar{T}}{dx} + h\bar{T} = \frac{1}{s} hT_\infty \quad \text{at} \quad x = 0 \tag{7-62b}$$

$$\bar{T} = 0 \quad \text{as} \quad x \to \infty \tag{7-62c}$$

The solution of equations (7-62) is

$$\frac{\bar{T}(x,s)}{T_\infty} = H\sqrt{\alpha} \frac{e^{-(x/\sqrt{\alpha})\sqrt{s}}}{s(H\sqrt{\alpha} + \sqrt{s})} \tag{7-63a}$$

where $H \equiv h/k$. The inversion of this result is available in Table 7-1, case 51; then the solution becomes

$$\frac{T(x,t)}{T_\infty} = \operatorname{erfc}\left(\frac{x}{\sqrt{4\alpha t}}\right) - e^{Hx + H^2\alpha t} \operatorname{erfc}\left(H\sqrt{\alpha t} + \frac{x}{\sqrt{4\alpha t}}\right) \tag{7-63b}$$

7-5 APPROXIMATIONS FOR SMALL TIMES

The solutions of time-dependent heat conduction problems for finite regions, such as slabs or cylinders of finite radius, are in the form of series which converge rapidly for large values of t, but converge very slowly for the small values of t. Therefore, such solutions are not suitable for numerical computations for very small values of time. For example, the solution of the slab problem given by

$$\frac{T(x,t)}{T_1} = \left(1 - \frac{x}{L}\right) - \frac{2}{L}\sum_{n=1}^{\infty} e^{-\alpha\beta_n^2 t}\frac{\sin\beta_n x}{\beta_n} \qquad \text{where} \qquad \beta_n = \frac{n\pi}{L} \tag{7-64}$$

converges very slowly for the values of $\alpha t/L^2$ less than approximately 0.02. Therefore, for such cases, it is desirable to develop alternative forms of the solutions that will converge fast for small times.

When the Laplace transform is applied to the time variable, it transforms the equation in t into an equation in s. Therefore, it is instructive to examine the values of t in the time domain with the corresponding values of s in the Laplace transform domain. With this objective in mind we now examine the Laplace transform of some functions.

Consider a function $F(t)$ that is represented as a polynomial in t in the form

$$F(t) = \sum_{k=0}^{n} a_k \frac{t^k}{k!} = a_0 + a_1\frac{t}{1!} + a_2\frac{t^2}{2!} + \cdots + a_n\frac{t^n}{n!} \tag{7-65a}$$

Since the function has only a finite number of terms, we can take its Laplace transform term by term to obtain

$$\bar{F}(s) = \sum_{k=0}^{n} a_k \frac{1}{s^{k+1}} = a_0\frac{1}{s} + a_1\frac{1}{s^2} + \cdots + a_n\frac{1}{s^{n+1}} \tag{7-65b}$$

according to the transform Table 7-1, case 3.

The coefficients a_0 and a_n may be determined from equation (7-65a) and (7-65b) as

$$a_0 = \lim_{t\to 0} F(t) = \lim_{s\to\infty} s\bar{F}(s) \tag{7-66a}$$

$$a_n = n! \lim_{t \to \infty} \frac{F(t)}{t^n} = \lim_{s \to 0} s^{n+1} \bar{F}(s) \tag{7-66b}$$

The relations given by equations (7-66) indicate that the large values of s in the Laplace transform domain correspond to small values of t in the time domain. Although the results given above are derived for a function $F(t)$, which is a polynomial, they are also applicable for other types of functions. Consider, for example, the following function and its transform

$$F(t) = \cosh kt \qquad \text{and} \qquad \bar{F}(s) = \frac{s}{s^2 - k^2} \tag{7-67a}$$

which satisfies the relation

$$\lim_{t \to 0} \cosh kt = \lim_{s \to \infty} s \frac{s}{s^2 - k^2} \tag{7-67b}$$

and this result is similar to that given by equation (7-66a).

These facts can be utilized to obtain an approximate solution for the function $F(t)$ valid for small times from the knowledge of its transform evaluated for large values of s as illustrated in several references [16; 4, pp. 82–85; 6; 8]; that is the transform of the desired function can be expanded as an asymptotic series and then inverted term by term. For example, in the problems of slab of finite thickness, the transform of temperature $\bar{T}(x, s)$ contains hyperbolic functions of $\sqrt{s/\alpha}$. These hyperbolic functions may be expanded in a series of negative exponentials of $\sqrt{s/\alpha}$ and the resulting expression is then inverted term by term. The solution obtained in this manner will converge fast for small time. In the problems of a solid cylinder of finite radius, for example, the transform of temperature involves Bessel functions of $\sqrt{s/\alpha}$. Then, the procedure consists of using asymptotic expansion of Bessel functions in order to obtain a form involving negative exponentials of $\sqrt{s/\alpha}$ with coefficients that are series in $(1/\sqrt{s/\alpha})$. The resulting expression is then inverted term by term. The solutions obtained in this manner will converge fast. Many examples of this procedure is given in reference 16.

Example 7-11

A slab, $0 \leqslant x \leqslant L$, is initially of zero temperature. For times $t > 0$, the boundary at $x = 0$ is kept insulated and the boundary at $x = L$ is kept at constant temperature T_0. Obtain an expression for the temperature distribution $T(x, t)$ which is useful for small values of time.

$$\frac{\partial^2 T(x,t)}{\partial x^2} = \frac{1}{\alpha} \frac{\partial T(x,t)}{\partial t} \qquad \text{in} \qquad 0 < x < L, \quad t > 0 \tag{7-68a}$$

$$\frac{\partial T}{\partial x}=0 \qquad \text{at} \qquad x=0, \qquad t>0 \tag{7-68b}$$

$$T=T_0 \qquad \text{at} \qquad x=L, \qquad t>0 \tag{7-68c}$$

$$T=0 \qquad \text{for} \qquad t=0, \qquad \text{in} \quad 0\leqslant x\leqslant L \tag{7-68d}$$

The Laplace transform of the equations (7-68) is

$$\frac{d^2\bar{T}(x,s)}{dx^2}-\frac{s}{\alpha}\bar{T}(x,s)=0 \qquad \text{in} \qquad 0\leqslant x\leqslant L \tag{7-69a}$$

$$\frac{d\bar{T}}{dx}=0 \qquad \text{at} \qquad x=0 \tag{7-69b}$$

$$\bar{T}=\frac{T_0}{s} \qquad \text{at} \qquad x=L \tag{7-69c}$$

The solution of equations (7-69) is

$$\frac{\bar{T}(x,s)}{T_0}=\frac{\cosh(x\sqrt{s/\alpha})}{s\cosh(L\sqrt{s/\alpha})} \tag{7-70}$$

The inversion of this transform in this form yields a solution for $T(x,t)$ which is slowly convergent for small values of time. To obtain a solution applicable for very small times we expand this transform as an asymptotic series in negative exponentials of $\sqrt{s/\alpha}$ as given below.

$$\begin{aligned}\frac{\bar{T}(x,s)}{T_0}&=\frac{e^{x\sqrt{s/\alpha}}+e^{-x\sqrt{s/\alpha}}}{s[e^{L\sqrt{s/\alpha}}+e^{-L\sqrt{s/\alpha}}]}\\&=\frac{1}{s}[e^{-(L-x)\sqrt{s/\alpha}}+e^{-(L+x)\sqrt{s/\alpha}}][1+e^{-2L\sqrt{s/\alpha}}]^{-1}\end{aligned} \tag{7-71}$$

The last term is expanded in binomial series

$$\begin{aligned}\frac{\bar{T}(x,s)}{T_0}&=\frac{1}{s}[e^{-(L-x)\sqrt{s/\alpha}}+e^{-(L+x)\sqrt{s/\alpha}}]\left[\sum_{n=0}^{\infty}(-1)^n e^{-2Ln\sqrt{s/\alpha}}\right]\\&=\frac{1}{s}\sum_{n=0}^{\infty}(-1)^n e^{-[L(1+2n)-x]\sqrt{s/\alpha}}+\frac{1}{s}\sum_{n=0}^{\infty}(-1)^n e^{-[L(1+2n)+x]\sqrt{s/\alpha}}\end{aligned} \tag{7-72}$$

The inversion of this transform is available in Table 7-1, case 42. Inverting term by term we obtain

$$\frac{T(x,t)}{T_0}=\sum_{n=0}^{\infty}(-1)^n \operatorname{erfc}\left(\frac{L(1+2n)-x}{\sqrt{4\alpha t}}\right)+\sum_{n=0}^{\infty}(-1)^n \operatorname{erfc}\left(\frac{L(1+2n)+x}{\sqrt{4\alpha t}}\right) \tag{7-73}$$

which converges rapidly for small values of t.

Example 7-12

A slab, $0 \leqslant x \leqslant L$, is initially at uniform temperature T_0. For times $t>0$, the boundary surface at $x=0$ is kept insulated and the boundary at $x=L$ dissipates heat by convection into an environment of zero temperature. Obtain an expression for the temperature distribution $T(x,t)$ which is useful for small times.

Solution. The mathematical formulation of this problem is given as

$$\frac{\partial^2 T(x,t)}{\partial x^2}=\frac{1}{\alpha}\frac{\partial T(x,t)}{\partial t} \quad \text{in} \quad 0<x<L, \quad t>0 \tag{7-74a}$$

$$\frac{\partial T}{\partial x}=0 \quad \text{at} \quad x=0, \quad t>0 \tag{7-74b}$$

$$\frac{\partial T}{\partial x}+HT=0 \quad \text{at} \quad x=L, \quad t>0 \tag{7-74c}$$

$$T=T_0 \quad \text{for} \quad t=0, \quad \text{in} \quad 0\leqslant x\leqslant L \tag{7-74d}$$

The Laplace transform of these equations gives

$$\frac{d^2\bar{T}(x,s)}{dx^2}-\frac{s}{\alpha}\bar{T}(x,s)=-\frac{T_0}{\alpha} \quad \text{in} \quad 0<x<L \tag{7-75a}$$

$$\frac{d\bar{T}}{dx}=0 \quad \text{at} \quad x=0 \tag{7-75b}$$

$$\frac{d\bar{T}}{dx}+H\bar{T}=0 \quad \text{at} \quad x=L \tag{7-75c}$$

The solution of equations (7-75) is

$$\frac{\bar{T}(x,s)}{T_0}=\frac{1}{s}-H\frac{\cosh(x\sqrt{s/\alpha})}{s\left[\sqrt{\frac{s}{\alpha}}\sinh\left(L\sqrt{\frac{s}{\alpha}}\right)+H\cosh\left(L\sqrt{\frac{s}{\alpha}}\right)\right]} \tag{7-76}$$

Since the solution is required for small times, we need to expand this transform as an asymptotic series in negative exponentials and then invert it term by term. The procedure is as follows:

$$\frac{\bar{T}(x,s)}{T_0}=\frac{1}{s}-H\frac{e^{x\sqrt{s/\alpha}}+e^{-x\sqrt{s/\alpha}}}{s[\sqrt{s/\alpha}(e^{L\sqrt{s/\alpha}}-e^{-L\sqrt{s/\alpha}})+H(e^{L\sqrt{s/\alpha}}+e^{-L\sqrt{s/\alpha}})]}$$

$$=\frac{1}{s}-\frac{H}{s}\frac{e^{-(L-x)\sqrt{s/\alpha}}+e^{-(L+x)\sqrt{s/\alpha}}}{H+\sqrt{s/\alpha}}\left[1+\frac{H-\sqrt{s/\alpha}}{H+\sqrt{s/\alpha}}e^{-2L\sqrt{s/\alpha}}\right]^{-1} \tag{7-77}$$

Expanding the last term in the bracket in binomial series we obtain

$$\frac{\bar{T}(x,s)}{T_0}=\frac{1}{s}-\frac{H}{s}\frac{e^{-(L-x)\sqrt{s/\alpha}}+e^{-(L+x)\sqrt{s/\alpha}}}{H+\sqrt{s/\alpha}}\left[\sum_{n=0}^{\infty}(-1)^n\left(\frac{H-\sqrt{s/\alpha}}{H+\sqrt{s/\alpha}}\right)^n e^{-2Ln\sqrt{s/\alpha}}\right] \tag{7-78a}$$

or

$$\frac{\bar{T}(x,s)}{T_0}=\frac{1}{s}-\frac{H}{s}\frac{e^{-(L-x)\sqrt{s/\alpha}}+e^{-(L+x)\sqrt{s/\alpha}}}{H+\sqrt{s/\alpha}}$$

$$+\frac{H}{s}\frac{H+\sqrt{s/\alpha}}{(H+\sqrt{s/\alpha})^2}[e^{-(3L-x)\sqrt{s/\alpha}}+e^{-(3L+x)\sqrt{s/\alpha}}]-\cdots \tag{7-78b}$$

The first few terms can readily be inverted by the Laplace transform Table 7-1, case 1 and 51; we obtain

$$\frac{T(x,t)}{T_0}=1-\left[\operatorname{erfc}\frac{L-x}{\sqrt{4\alpha t}}-e^{H(L-x)+H^2\alpha t}\cdot\operatorname{erfc}\left(H\sqrt{\alpha t}+\frac{L-x}{\sqrt{4\alpha t}}\right)\right]$$

$$-\left[\operatorname{erfc}\frac{L+x}{\sqrt{4\alpha t}}-e^{H(L+x)+H^2\alpha t}\cdot\operatorname{erfc}\left(H\sqrt{\alpha t}+\frac{L+x}{\sqrt{4\alpha t}}\right)\right]+\cdots \tag{7-79}$$

This solution converges fast for small times.

Example 7-13

A solid sphere of radius $r=b$ is initially at a uniform temperature T_0. For times $t>0$, the boundary surface at $r=b$ is kept at zero temperature. Obtain an expression for the temperature distribution $T(r,t)$ which is useful for small times.

Solution. The mathematical formulation of this problem is given as

$$\frac{1}{r}\frac{\partial^2}{\partial r^2}(rT)=\frac{1}{\alpha}\frac{\partial T(r,t)}{\partial t} \qquad \text{in} \qquad 0\leqslant r<b,\quad t>0 \tag{7-80a}$$

$$T(r,t)=0 \qquad \text{at} \qquad r=b, \qquad t>0 \tag{7-80b}$$

$$T(r,t)=T_0 \qquad \text{for} \qquad t=0, \qquad \text{in} \quad 0\leqslant r\leqslant b \tag{7-80c}$$

The Laplace transform of equation (7-80) gives

$$\frac{1}{r}\frac{d^2}{dr^2}(r\bar{T})-\frac{s}{\alpha}\bar{T}(r,s)=-\frac{T_0}{\alpha} \qquad \text{in} \qquad 0\leqslant r\leqslant b \tag{7-81a}$$

$$\bar{T}(r,s)=0 \qquad \text{at} \qquad r=b \tag{7-81b}$$

The solution of equation (7-81) is

$$\frac{\bar{T}(r,s)}{T_0}=\frac{1}{s}-\frac{b}{sr}\frac{\sinh(r\sqrt{s/\alpha})}{\sinh(b\sqrt{s/\alpha})} \tag{7-82}$$

To obtain a solution that converges rapidly for small times, we expand this transform as an asymptotic series in negative exponentials, and then invert term by term. The procedure is as follows:

$$\begin{aligned}\frac{\bar{T}(r,s)}{T_0}&=\frac{1}{s}-\frac{b}{r}\frac{e^{r\sqrt{s/\alpha}}-e^{-r\sqrt{s/\alpha}}}{s[e^{b\sqrt{s/\alpha}}-e^{-b\sqrt{s/\alpha}}]}\\&=\frac{1}{s}-\frac{b}{r}\frac{1}{s}[e^{-(b-r)\sqrt{s/\alpha}}-e^{-(b+r)\sqrt{s/\alpha}}][1-e^{-2b\sqrt{s/\alpha}}]^{-1}\end{aligned} \tag{7-83}$$

The last term in the bracket is expanded in binomial series; we obtain

$$\begin{aligned}\frac{\bar{T}(r,s)}{T_0}&=\frac{1}{s}-\frac{b}{r}\frac{1}{s}[e^{-(b-r)\sqrt{s/\alpha}}-e^{-(b+r)\sqrt{s/\alpha}}]\left[\sum_{n=0}^{\infty}e^{-2bn\sqrt{s/\alpha}}\right]\\&=\frac{1}{s}-\frac{b}{r}\sum_{n=0}^{\infty}\left\{\frac{1}{s}e^{-[b(1+2n)-r]\sqrt{s/\alpha}}-\frac{1}{s}e^{-[b(1+2n)+r]\sqrt{s/\alpha}}\right\}\end{aligned} \tag{7-84}$$

This transform is readily inverted by utilizing the Laplace transform Table 7-1, case 42; we find

$$\frac{T(r,t)}{T_0}=1-\frac{b}{r}\sum_{n=0}^{\infty}\left\{\operatorname{erfc}\frac{b(1+2n)-r}{\sqrt{4\alpha t}}-\operatorname{erfc}\frac{b(1+2n)+r}{\sqrt{4\alpha t}}\right\} \tag{7-85}$$

This solution converges fast for small values of times.

REFERENCES

1. N. W. McLachlan, *Modern Operational Calculus*, Macmillan, New York, 1948.
2. R. V. Churchill, *Operational Mathematics*, McGraw-Hill, New York, 1958.
3. M. G. Smith, *Laplace Transform Theory*, Van Nostrand, New York, 1966.
4. N. W. McLachlan, *Complex Variable Theory and Transform Calculus*, Cambridge University Press, London, 1953.
5. W. Kaplan, *Operational Methods for Linear Systems*, Addison-Wesley, Reading, Mass., 1962.
6. H. S. Carslaw and J. C. Jaeger, *Operational Methods in Applied Mathematics*, Oxford University Press, London, 1948.
7. A. Erde'lyi, *Tables of Integral Transforms I*, McGraw-Hill, New York, 1954.
8. H. S. Carslaw and J. C. Jaeger, *Conduction of Heat in Solids*, 2 ed., Oxford University Press, London, 1959.
9. Ian N. Sneddon, *Use of Integral Transforms*, McGraw-Hill, New York, 1972.
10. J. Irving and N. Mullineaux, *Mathematics in Physics and Engineering*, Academic Press, New York, 1959.
11. V. S. Arpaci, *Conduction Heat Transfer*, Addison-Wesley, Reading, Mass., 1966.
12. A. V. Luikov, *Analytical Heat Diffusion Theory*, Academic Press, New York, 1968.
13. E. C. Titchmarsh, *Fourier Integrals*, 2 ed., Clarendon Press, London, 1962.
14. R. C. Bartels and R. V. Churchill, *Bull. Amer. Math. Soc.* **48**, 276–282, 1942.
15. H. B. Dwight, *Tables of Integrals and Other Mathematical Data*, 4th ed., MacMillan, New York, 1961.
16. S. Goldstein, *Proc. London Math. Soc.*, 2nd series, **34**, 51–88, 1932.

PROBLEMS

7-1 A semiinfinite medium, $0 \leqslant x < \infty$, is initially at uniform temperature T_0. For times $t > 0$ the boundary surface at $x = 0$ is maintained at zero temperature. Obtain an expression for the temperature distribution $T(x, t)$ in the medium for times $t > 0$ by solving this problem with the Laplace transformation.

7-2 A semiinfinite medium, $0 \leqslant x < \infty$, is initially at a uniform temperature T_0. For times $t > 0$ it is subjected to a prescribed heat flux at the boundary surface $x = 0$:

$$-k\frac{\partial T}{\partial x} = f_0 = \text{constant} \qquad \text{at } x = 0$$

Obtain an expression for the temperature distribution $T(x, t)$ in the medium for times $t > 0$ by using Laplace transformation.

7-3 A semiinfinite medium, $0 \leqslant x < \infty$, is initially at uniform temperature T_0.

For times $t > 0$, the boundary surface at $x = 0$ is kept at zero temperature while heat is generated in the medium at a constant rate of g_0 W/m^3. Obtain an expression for the temperature distribution $T(x, t)$ in the medium for times $t > 0$ by using Laplace transformation.

7-4 A slab, $0 \leqslant x \leqslant L$, is initially at uniform temperature T_0. For times $t > 0$, the boundary surface at $x = 0$ is kept insulated and the boundary surface at $x = L$ is kept at zero temperature. Obtain an expression for the temperature distribution $T(x, t)$ in the slab valid for very small times.

7-5 A slab, $0 \leqslant x \leqslant L$, is initially at zero temperature. For times $t > 0$, heat is generated in the slab at a constant rate of g_0 W/m^3 while the boundary surface at $x = 0$ is kept insulated and the boundary surface at $x = L$ is kept at zero temperature. Obtain an expression for the temperature distribution $T(x, t)$ in the slab for very small times.

7-6 A slab, $0 \leqslant x \leqslant L$, is initially at zero temperature. For times $t > 0$, the boundary surface at $x = 0$ is kept insulated while the boundary surface at $x = L$ is subjected to a heat flux:

$$k \frac{\partial T}{\partial x} = f_0 = \text{constant} \qquad \text{at } x = L$$

Obtain an expression for the temperature distribution $T(x, t)$ in the slab for very small times.

7-7 A solid cylinder, $0 \leqslant r \leqslant b$, is initially at a uniform temperature T_0. For times $t > 0$, the boundary surface at $r = b$ is kept at zero temperature. Obtain an expression for the temperature distribution $T(r, t)$ in the solid valid for very small times.

7-8 A solid cylinder, $0 \leqslant r \leqslant b$, is initially at a uniform temperature T_0. For times $t > 0$, the boundary surface at $r = b$ is subjected to convection boundary condition in the form

$$\frac{\partial T}{\partial r} + HT = 0 \qquad \text{at } r = b$$

Obtain an expression for the temperature distribution $T(r, t)$ in the solid valid for very small times.

7-9 A solid sphere, $0 \leqslant r \leqslant b$, is initially at a uniform temperature T_0. For times $t > 0$, the boundary surface at $r = b$ is kept at zero temperature. Obtain an expression for the temperature distribution $T(r, t)$ in the solid valid for very small times.

8

ONE-DIMENSIONAL COMPOSITE MEDIUM

The transient-temperature distribution in a composite medium consisting of several layers in contact has numerous applications in engineering. In this chapter, the mathematical formulation of one-dimensional transient heat conduction in a composite medium consisting of M parallel layers of slabs, cylinders, or spheres is presented. The transformation of the problem with nonhomogeneous boundary conditions into the one with homogeneous boundary conditions is described. The *orthogonal expansion* technique is used to solve the homogeneous problem of composite medium of finite thickness, while the *Laplace transformation* is used to solve the homogeneous problem of composite medium of infinite and semiinfinite thickness.

The Green's function approach is used for solving the nonhomogeneous problem with energy generation in the medium.

The reader should consult references 1–13 for the theory and the application of the generalized orthogonal expansion technique and the Green's function approach in the solution of heat conduction problems of composite media. The use of Laplace transform technique in the solution of composite media problems is given in references 14–17 and the application of the integral transform technique and various other approaches can be found in the references 18–38.

8-1 MATHEMATICAL FORMULATION OF ONE-DIMENSIONAL TRANSIENT HEAT CONDUCTION IN A COMPOSITE MEDIUM

We consider a composite medium consisting of M parallel layers of slabs, cylinders, or spheres as illustrated in Fig. 8-1. We assume the existence of contact conductance h_i at the interfaces $x = x_i, i = 2, 3, \ldots, M$. Initially each layer is at a

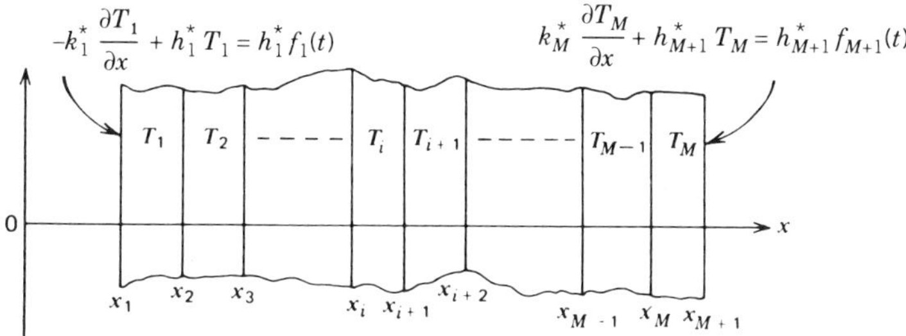

Fig. 8-1 M-layer composite region.

specified temperature $T_i(x,0) = F_i(x)$, in $x_i < x < x_{i+1}$, $i = 1, 2, 3, \ldots, M$, for $t = 0$. For times $t > 0$, energy is generated in each layer at a rate of $g_i(x,t)$, W/m^3, in $x_i < x < x_{i+1}, i = 1, 2, 3, \ldots, M$, while the energy is dissipated with convection from the two outer boundary surfaces $x = x_1$ and $x = x_{M+1}$, into ambients at temperatures $f_1(t)$ and $f_{M+1}(t)$, with heat transfer coefficients h_1^* and h_{M+1}^*, respectively.

The mathematical formulation of this heat conduction problem is given as follows.

The differential equations for each of the M layers are

$$\alpha_i \frac{1}{x^p} \frac{\partial}{\partial x}\left(x^p \frac{\partial T_i}{\partial x}\right) + \frac{\alpha_i}{k_i} g_i(x,t) = \frac{\partial T_i(x,t)}{\partial t} \qquad \text{in} \qquad x_i < x < x_{i+1}, \quad t > 0, \quad i = 1, 2, \ldots, M \tag{8-1}$$

where

$$p = \begin{cases} 0 & \text{slab} \\ 1 & \text{cylinder} \\ 2 & \text{sphere} \end{cases}$$

Subject to the boundary conditions

$$-k_1^* \frac{\partial T_1}{\partial x} + h_1^* T_1 = h_1^* f_1(t) \qquad \text{at the outer boundary } x = x_1, \quad t > 0 \tag{8-2a}$$

$$\left.\begin{aligned} -k_i \frac{\partial T_i}{\partial x} &= h_{i+1}(T_i - T_{i+1}) \qquad &(8\text{-}2\text{b}) \\ k_i \frac{\partial T_i}{\partial x} &= k_{i+1} \frac{\partial T_{i+1}}{\partial x} \qquad &(8\text{-}2\text{c}) \end{aligned}\right\} \quad \text{at the interfaces } x = x_{i+1}, \quad i = 1, 2, \ldots, M-1, \quad t > 0$$

$$k_M^* \frac{\partial T_M}{\partial x} + h_{M+1}^* T_M = h_{M+1}^* f_{M+1}(t) \qquad \text{at the outer boundary, } x = x_{M+1}, \quad t > 0 \tag{8-2d}$$

and the initial conditions:

$$T_i(x,t) = F_i(x) \qquad \text{for} \qquad t=0, \quad \text{in} \quad x_i < x < x_{i+1}, \quad i=1,2,\ldots,M \tag{8-3}$$

where $T_i(x,t)$ is the temperature of the layer i, $i = 1,2,\ldots,M$. The problem contains M partial differential equations, $2M$ boundary conditions and M initial conditions, hence it is mathematically well posed.

In order to distinguish the coefficients associated with the boundary conditions for the outer surfaces from those k and h for the medium and interfaces, an asterisk is used in the quantities h_1^*, h_{M+1}^*, k_1^*, and k_M^* appearing in the boundary conditions for the outer surfaces. The reason for this is that these quantities will be treated as coefficients, so that the boundary conditions of the first and second kind will be obtainable for the outer boundary surfaces by setting the values of these coefficients properly.

8-2 TRANSFORMATION OF NONHOMOGENEOUS BOUNDARY CONDITIONS INTO HOMOGENEOUS ONES

It is more convenient to solve the problems with homogeneous boundary conditions than with nonhomogeneous boundary conditions. The problem of time-dependent heat conduction for a M-layer composite medium with heat generation and nonhomogeneous outer boundary conditions can be transformed into a problem with heat generation but homogeneous boundary conditions by a procedure similar to that described in Chapter 1, Section 1-7 for the single-layer problem.

The problem defined by equations (8-1)–(8-3) has nonhomogeneous boundary conditions at the outer surfaces. In order to transform this time-dependent problem into a one with homogeneous boundary conditions, we consider $T_i(x,t)$ constructed by the superposition of three simpler problems in the form

$$T_i(x,t) = \theta_i(x,t) + \phi_i(x) f_1(t) + \psi_i(x) f_{M+1}(t)$$
$$\text{in} \qquad x_i < x < x_{i+1}, \quad i = 1,2,\ldots,M, \quad \text{for} \quad t>0 \tag{8-4}$$

Where the functions $\phi_i(x), \psi_i(x)$, and $\theta_i(x,t)$ are the solutions of the following subproblems:

1. The functions $\phi_i(x)$ are the solutions of the following steady-state problem for the same region, with no heat generation, but with one non-homogeneous boundary condition at $x = x_1$.

$$\frac{d}{dx}\left(x^p \frac{d\phi_i}{dx} \right) = 0 \qquad \text{in} \qquad x_i < x < x_{i+1}, \quad i = 1,2,\ldots,M \tag{8-5a}$$

subject to the boundary conditions

$$-k_1^* \frac{d\phi_1(x)}{dx} + h_1^* \phi_1(x) = h_1^* \qquad \text{at} \qquad x = x_1 \tag{8-5b}$$

$$-k_i \frac{d\phi_i}{dx} = h_{i+1}(\phi_i - \phi_{i+1}) \tag{8-5c}$$

$$k_i \frac{d\phi_i}{dx} = k_{i+1} \frac{d\phi_{i+1}}{dx} \tag{8-5d}$$

at the interfaces $x = x_{i+1}$, $i = 1, 2, \ldots, M-1$

$$k_M^* \frac{d\phi_M}{dx} + h_{M+1}^* \phi_M = 0 \qquad \text{at} \qquad x = x_{M+1} \tag{8-5e}$$

2. The functions $\psi_i(x)$ are the solutions of the following steady-state problem for the same region, with no heat generation, but with one non-homogeneous boundary condition at $x = x_{M+1}$

$$\frac{d}{dx}\left(x^p \frac{d\psi_i}{dx}\right) = 0 \qquad \text{in} \qquad x_i < x < x_{i+1}, \quad i = 1, 2, \ldots, M \tag{8-6a}$$

subject to the boundary conditions

$$-k_1^* \frac{d\psi_1}{dx} + h_1^* \psi_1 = 0 \qquad \text{at} \qquad x = x_1 \tag{8-6b}$$

$$-k_i \frac{d\psi_i}{dx} = h_{i+1}(\psi_i - \psi_{i+1}) \tag{8-6c}$$

$$k_i \frac{d\psi_i}{dx} = k_{i+1} \frac{d\psi_{i+1}}{dx} \tag{8-6d}$$

at the interfaces $x = x_{i+1}$, $i = 1, 2, \ldots, M-1$

$$k_M^* \frac{d\psi_M}{dx} + h_{M+1}^* \psi_M = h_{M+1}^* \qquad \text{at} \qquad x = x_{M+1} \tag{8-6e}$$

3. The functions $\theta_i(x, t)$ are the solutions of the following time-dependent heat conduction problem for the same region, with heat generation, but subject to homogeneous boundary conditions

$$\alpha_i \frac{1}{x^p} \frac{\partial}{\partial x}\left(x^p \frac{\partial \theta_i}{\partial x}\right) + g_i^*(x, t) = \frac{\partial \theta_i(x, t)}{\partial t},$$

$$\text{in} \qquad x_i < x < x_{i+1}, \quad i = 1, 2, \ldots, M, \quad \text{for} \quad t > 0 \tag{8-7a}$$

where

$$g_i^*(x,t) \equiv \frac{\alpha_i}{k_i} g_i(x,t) - \left[\phi_i(x) \frac{df_1(t)}{dt} + \psi_i(x) \frac{df_M(t)}{dt} \right]$$

Subject to the boundary conditions

$$-k_1^* \frac{d\theta_1}{\partial x} + h_1^* \theta_1 = 0 \qquad \text{at} \qquad x = x_1, \qquad t > 0 \qquad \text{(8-7b)}$$

$$-k_i \frac{\partial \theta_i}{\partial x} = h_{i+1}(\theta_i - \theta_{i+1}) \qquad \text{(8-7c)}$$

at the interfaces $x = x_{i+1}$, $i = 1, 2, \ldots, M-1$ for $t > 0$

$$k_i \frac{\partial \theta_i}{\partial x} = k_{i+1} \frac{\partial \theta_{i+1}}{\partial x} \qquad \text{(8-7d)}$$

$$k_M^* \frac{\partial \theta_M}{\partial x} + h_{M+1}^* \theta_M = 0 \qquad \text{at} \qquad x = x_{M+1}, \quad t > 0 \qquad \text{(8-7e)}$$

and the initial conditions

$$\theta_i(x,t) = F_i(x) - [\phi_i(x) f_1(0) + \psi_i(x) f_{M+1}(0)] \equiv F_i^*(x),$$

$$\text{for} \quad t = 0, \quad \text{in} \quad x_i < x < x_{i+1}, \quad i = 1, 2, \ldots, M \qquad \text{(8-7f)}$$

The validity of this superposition procedure can readily be verified by introducing the equation (8-4) into the original problem given by equations (8-1)–(8-3) and utilizing the above three subproblems defined by equations (8-5)–(8-7).

Example 8-1

A two-layer slab consists of the first layer in $0 \leqslant x \leqslant a$ and the second layer in $a \leqslant x \leqslant b$, which are in perfect thermal contact as illustrated in Fig. 8-2. Let k_1

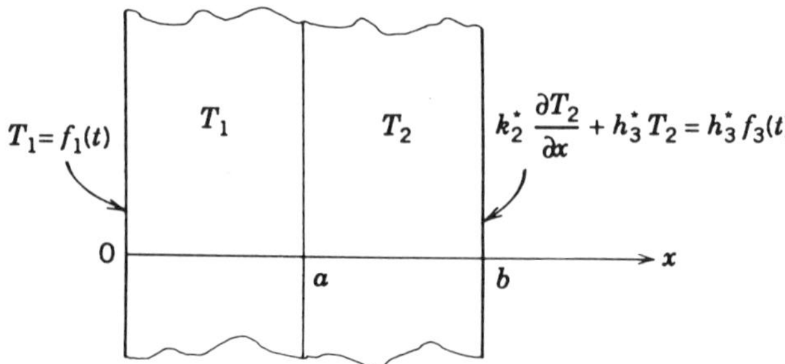

Fig. 8-2 Two-layer slab with perfect thermal contact at the interface.

and k_2 be the thermal conductivities, and α_1 and α_2 the thermal diffusivities for the first and second layer, respectively. Initially, the first region is at temperature $F_1(x)$ and the second region at $F_2(x)$. For times $t > 0$ the boundary surface at $x = 0$ is kept at temperature $f_1(t)$ and the boundary at $x = b$ dissipates heat by convection, with a heat transfer coefficient h_3^* into an ambient at temperature $f_3(t)$. By applying the splitting-up procedure described above, separate this problem into (i) two steady-state problems each with one nonhomogeneous boundary conditions, and (ii) one time-dependent problem with homogeneous boundary conditions and the initial condition. Figure 8-2 shows the geometry coordinates and the boundary conditions for the original problem.

Solution. The mathematical formulation of this problem is given as

$$\alpha_1 \frac{\partial^2 T_1}{\partial x^2} = \frac{\partial T_1(x,t)}{\partial t} \qquad \text{in} \qquad 0 < x < a, \quad t > 0 \tag{8-8a}$$

$$\alpha_2 \frac{\partial^2 T_2}{\partial x^2} = \frac{\partial T_2(x,t)}{\partial t} \qquad \text{in} \qquad a < x < b, \quad t > 0 \tag{8-8b}$$

subject to the boundary conditions

$$T_1(x,t) = f_1(t) \qquad \text{at} \qquad x = 0, \quad t > 0 \tag{8-8c}$$

$$T_1(x,t) = T_2(x,t) \qquad \text{at} \qquad x = a, \quad t > 0 \tag{8-8d}$$

$$k_1 \frac{\partial T_1}{\partial x} = k_2 \frac{\partial T_2}{\partial x} \qquad \text{at} \qquad x = a, \quad t > 0 \tag{8-8e}$$

$$k_2^* \frac{\partial T_2}{\partial x} + h_3^* T_2 = h_3^* f_3(t) \qquad \text{at} \qquad x = b, \quad t > 0 \tag{8-8f}$$

and the initial conditions

$$T_1(x,t) = F_1(x) \qquad \text{for} \qquad t = 0, \quad 0 < x < a \tag{8-8g}$$

$$T_2(x,t) = F_2(x) \qquad \text{for} \qquad t = 0, \quad a < x < b \tag{8-8h}$$

To transform this problem, we construct the solution of $T_i(x,t)$, $i = 1, 2$, by the superposition of the following three simpler problems in the form

$$T_i(x,t) = \theta_i(x,t) + \phi_i(x) f_1(t) + \psi_i(x) f_3(t) \qquad \text{in} \qquad x_i < x < x_{i+1}, \quad i = 1, 2 \tag{8-9}$$

where $x_1 = 0$, $x_2 = a$ and $x_3 = b$.

The functions $\phi_i(x), \psi_i(x)$, and $\theta_i(x,t)$, for $i = 1,2$ are the solutions of the following three simpler problems, respectively.

1. The functions $\phi_i(x), i = 1,2$ satisfy the steady-state heat conduction problem given as

$$\frac{d^2\phi_1(x)}{dx^2} = 0 \qquad \text{in} \qquad 0 < x < a \tag{8-10a}$$

$$\frac{d^2\phi_2(x)}{dx^2} = 0 \qquad \text{in} \qquad a < x < b \tag{8-10b}$$

subject to the boundary conditions

$$\phi_1(x) = 1 \qquad \text{at} \qquad x = 0 \tag{8-10c}$$

$$\left.\begin{aligned} &\phi_1(x) = \phi_2(x) && (8\text{-}10\text{d}) \\ &k_1 \frac{d\phi_1}{dx} = k_2 \frac{d\phi_2}{dx} && (8\text{-}10\text{e}) \end{aligned}\right\} \quad \text{at the interface } x = a$$

$$k_2^* \frac{d\phi_2}{dx} + h_3^* \phi_2 = 0 \qquad \text{at} \qquad x = b \tag{8-10f}$$

2. The functions $\psi_i(x), i = 1,2$ satisfy the steady-state heat conduction problem given as

$$\frac{d^2\psi_1(x)}{dx^2} = 0 \qquad \text{in} \qquad 0 < x < a \tag{8-11a}$$

$$\frac{d^2\psi_2(x)}{dx^2} = 0 \qquad \text{in} \qquad a < x < b \tag{8-11b}$$

subject to the boundary conditions

$$\psi_1(x) = 0 \qquad \text{at} \qquad x = 0 \tag{8-11c}$$

$$\left.\begin{aligned} &\psi_1(x) = \psi_2(x) && (8\text{-}11\text{d}) \\ &k_1 \frac{d\psi_1}{dx} = k_2 \frac{d\psi_2}{dx} && (8\text{-}11\text{e}) \end{aligned}\right\} \quad \text{at the interface } x = a$$

$$k_2^* \frac{d\psi_2}{dx} + h_3^* \psi_2 = h_3^* \qquad \text{at} \qquad x = b \tag{8-11f}$$

3. The functions $\theta_i(x,t), i=1,2$ are the solutions of the following homogeneous problem

$$\alpha_1 \frac{\partial^2 \theta_1}{\partial x^2} = \frac{\partial \theta_1(x,t)}{\partial t} \quad \text{in} \quad 0<x<a, \quad t>0 \tag{8-12a}$$

$$\alpha_2 \frac{\partial^2 \theta_2}{\partial x_2} = \frac{\partial \theta_2(x,t)}{\partial t} \quad \text{in} \quad a<x<b, \quad t>0 \tag{8-12b}$$

Subject to the boundary conditions

$$\theta_1(x,t)=0 \quad \text{at} \quad x=0, \quad t>0 \tag{8-12c}$$

$$\left.\begin{aligned} &\theta_1(x,t)=\theta_2(x,t) \qquad (8\text{-}12\text{d}) \\ &k_1\frac{\partial \theta_1}{\partial x} = k_2 \frac{\partial \theta_2}{\partial x} \qquad (8\text{-}12\text{e}) \end{aligned}\right\} \quad \text{at the interface } x=a, \quad t>0$$

$$k_2^* \frac{\partial \theta_2}{\partial x} + h_3^* \theta_2 = 0 \quad \text{at} \quad x=b \tag{8-12f}$$

and the initial conditions

$$\theta_1(x,t) = F_1(x) - f_1(0)\phi_1(x) - f_3(0)\psi_1(x) \equiv F_1^*(x)$$
$$\text{for} \quad t=0, \quad \text{in} \quad 0 \leqslant x < a \tag{8-12g}$$

$$\theta_2(x,t) = F_2(x) - f_1(0)\phi_2(x) - f_3(0)\psi_2(x) \equiv F_2^*(x)$$
$$\text{for} \quad t=0, \quad \text{in} \quad a \leqslant x < b \tag{8-12h}$$

The validity of this superposition procedure can be verified by introducing the transformation given by equation (8-9) into the original problem (8-8) and utilizing the definition of the subproblems given by equations (8-10)–(8-12).

Clearly, the time-dependent problem (8-12) has homogeneous boundary conditions.

Solving Steady-State Problem of *M*-Layer Slab, Cylinder, or Sphere

We consider a steady-state problem with no energy generation, one nonhomogeneous boundary condition of the type given by equation (8-5), but for a M-layer slab, cylinder, or sphere. The mathematical formulation is given by

$$\frac{d}{dx}\left(x^p \frac{d\phi_i}{dx}\right) = 0 \quad \text{in} \quad x_i < x < x_{i+1}, \quad i=1,2,\ldots,M \tag{8-13a}$$

subject to the boundary conditions

$$-k_1^*\frac{d\phi_1(x)}{dx}+h_1^*\phi_1(x)=h_1^* \qquad \text{at} \qquad x=x_1 \tag{8-13b}$$

$$\left.\begin{aligned} -k_i\frac{d\phi_i}{dx}&=h_{i+1}(\phi_i-\phi_{i+1}) \qquad (8\text{-}13\text{c})\\ k_i\frac{d\phi_i}{dx}&=k_{i+1}\frac{d\phi_{i+1}}{dx} \qquad (8\text{-}13\text{d})\end{aligned}\right\} \quad \begin{aligned}&\text{at the interfaces}\\ &x=x_{i+1},\\ &i=1,2,\ldots,M-1\end{aligned}$$

$$k_M^*\frac{d\phi_M}{dx}+h_{M+1}^*\phi_M=0 \qquad \text{at} \qquad x=x_{M+1} \tag{8-13e}$$

where

$$p=\begin{cases}0 & \text{slab}\\ 1 & \text{cylinder}\\ 2 & \text{sphere}\end{cases} \tag{8-13f}$$

The solution of the ordinary differential equation (8-13a), for any layer i is given in the form

$$\text{Slab:} \qquad \phi_i(x)=A_i+B_ix \tag{8-14a}$$

$$\text{Cylinder:} \qquad \phi_i(x)=A_i+B_i\ln x \tag{8-14b}$$

$$\text{Sphere:} \qquad \phi_i(x)=A_i+\frac{B_i}{x} \tag{8-14c}$$

The solution involves two unknown coefficients A_i and B_i for each layer i; then, for a M-layer problem, $2M$ unknown coefficients are to be determined. Substituting the solution given by equations (8-14) into the boundary conditions (8-13b,c,d,e), one obtains $2M$ equations for the determination of the $2M$ unknown coefficients A_i, B_i for $i=1,2,\ldots,M$.

The solution of the homogeneous transient heat conduction problems of the type given by equations (8-12), but for the M-layer medium, is described in the next section.

8-3 ORTHOGONAL EXPANSION TECHNIQUE FOR SOLVING *M*-LAYER HOMOGENEOUS PROBLEMS

We now consider the solution of the homogeneous problem of heat conduction in a composite medium consisting of M parallel layers of slabs, cylinder, or

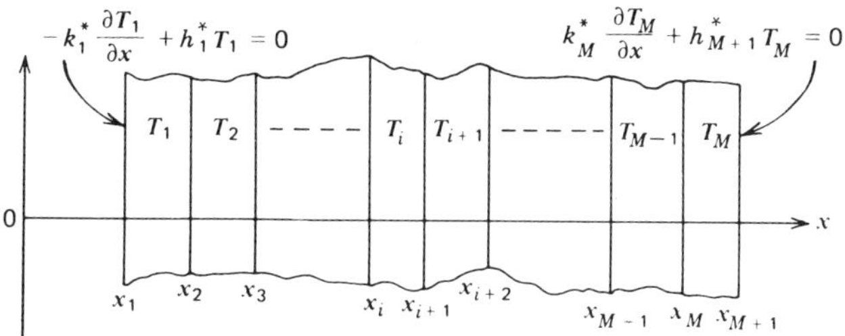

Fig. 8-3 M-layer composite region.

spheres in contact as illustrated in Fig. 8-3. For generality we assume contact resistance at the interfaces and convection from the outer boundaries. Let h_i be the arbitrary film coefficient at the interfaces $x = x_i, i = 2, 3, \ldots, M$, and h_1^* and h_{M+1}^* the heat transfer coefficients at the outer boundaries $x = x_1$ and $x = x_{M+1}$, respectively. Each layer is homogeneous and isotropic and has thermal properties (i.e., ρ, C_p, k) that are constant within the layer and different from those of the adjacent layers. Initially each layer is at a specified temperature $T_i(x, t) = F_i(x)$, in $x_i < x < x_{i+1}$, $i = 1, 2, \ldots, M$. For times $t > 0$, heat is dissipated by convection from the two outer boundaries into environments at zero temperature. There is no heat generation in the medium. We are interested in the determination of the temperature distribution $T_i(x, t)$, in the layers $i = 1, 2, \ldots, M$, for times $t > 0$. The mathematical formulation of this heat conduction problem is given by

$$\alpha_i \frac{1}{x^p} \frac{\partial}{\partial x}\left(x^p \frac{\partial T_i}{\partial x}\right) = \frac{\partial T_i(x,t)}{\partial t} \quad \text{in} \quad x_i < x < x_{i+1},$$

$$\text{for} \quad t > 0, \quad i = 1, 2, \ldots, M \tag{8-15}$$

where

$$p = \begin{cases} 0 & \text{slab} \\ 1 & \text{cylinder} \\ 2 & \text{sphere} \end{cases}$$

subject to the boundary conditions

$$-k_1^* \frac{\partial T_1}{\partial x} + h_1^* T_1 = 0 \quad \text{at the outer boundary } x = x_1, \quad \text{for} \quad t > 0 \tag{8-16a}$$

$$-k_i \frac{\partial T_i}{\partial x} = h_{i+1}(T_i - T_{i+1}) \quad \text{at the interfaces } x = x_{i+1}, \; i = 1, 2, \ldots, M-1 \tag{8-16b}$$

$$k_i \frac{\partial T_i}{\partial x} = k_{i+1} \frac{\partial T_{i+1}}{\partial x} \quad \text{for} \quad t > 0 \tag{8-16c}$$

$$k_M^* \frac{\partial T_M}{\partial x} + h_{M+1}^* T_M = 0 \qquad \text{at the outer boundary } x = x_{M+1}, \quad \text{for} \quad t > 0 \tag{8-16d}$$

and the initial conditions

$$T_i(x,t) = F_i(x) \qquad \text{for} \qquad t = 0, \qquad \text{in} \qquad x_i < x < x_{i+1}, \quad i = 1, 2, \ldots, M \tag{8-17}$$

The finite value of interface conductance h_{i+1}, in equations (8-16b) implies that the temperature is discontinuous at the interfaces. The boundary conditions (8-16c) represents the continuity of heat flux at the interfaces.

When the interface conductance $h_{i+1} \to \infty$, the boundary condition (8-16b) reduces to

$$T_i = T_{i+1} \qquad \text{at} \qquad x = x_{i+1}, \quad i = 1, 2, \ldots, M-1 \quad \text{for} \quad t > 0 \tag{8-18}$$

which implies the continuity of temperature or *perfect thermal contact* at the interfaces.

To solve the above heat conduction problem, the variables are separated in the form

$$T_i(x,t) = \psi_i(x)\Gamma(t) \tag{8-19a}$$

When equation (8-19a) is introduced into (8-15) we obtain

$$\alpha_i \frac{1}{x^p} \frac{1}{\psi_i(x)} \frac{d}{dx}\left(x^p \frac{d\psi_i}{dx}\right) = \frac{1}{\Gamma(t)} \frac{d\Gamma(t)}{dt} \equiv -\beta^2 \tag{8-19b}$$

where β is the separation constant. We recall that, in separating the variables for the case of a single-layer problem, the thermal diffusivity α was retained on the side of the equation where the time-dependent function $\Gamma(t)$ was collected. In the case of composite medium, α_i is retained on the left-hand side of equation (8-19b) where the space dependent function $\psi_i(x)$ are collected. The reason for this is to keep the solution of the time dependent function $\Gamma(t)$ independent of α_i since it is discontinuous at the interfaces.

The separation given by equations (8-19b) results in the following ordinary differential equations for the determination of the functions $\Gamma(t)$ and $\psi_i(\beta, x)$:

$$\frac{d\Gamma(t)}{dt} + \beta_n^2 \Gamma(t) = 0 \qquad \text{for} \qquad t > 0 \tag{8-20}$$

$$\frac{1}{x^p} \frac{d}{dx}\left(x^p \frac{d\psi_{in}}{dx}\right) + \frac{\beta_n^2}{\alpha_i} \psi_{in} = 0 \qquad \text{in} \qquad x_i < x < x_{i+1}, \quad i = 1, 2, \ldots, M \tag{8-21}$$

where $\psi_{in} \equiv \psi_i(\beta_n, x)$. The subscript n is included to imply that there are an infinite number of discrete values of the eigenvalues $\beta_1 < \beta_2 < \cdots < \beta_n < \cdots$ and the corresponding eigenfunctions ψ_{in}.

The boundary conditions for equations (8-21) are obtained by introducing equation (8-19a) into the boundary conditions (8-16); we find

$$-k_1^* \frac{d\psi_{1n}}{dx} + h_1^* \psi_{1n} = 0 \qquad \text{at the outer boundary } x = x_1 \tag{8-22a}$$

$$-k_i \frac{d\psi_{in}}{dx} = h_{i+1}(\psi_{in} - \psi_{i+1,n}) \tag{8-22b}$$

$$k_i \frac{d\psi_{in}}{dx} = k_{i+1} \frac{d\psi_{i+1,n}}{dx} \tag{8-22c}$$

at the interfaces $x = x_{i+1}$, where $i = 1, 2, \ldots, M-1$

$$k_M^* \frac{d\psi_{Mn}}{dx} + h_{M+1}^* \psi_{Mn} = 0 \qquad \text{at the outher boundary } x = x_{M+1} \tag{8-22d}$$

Equations (8-21) subject to the boundary conditions (8-22) constitute an eigenvalue problem for the determination of the eigenvalues β_n and the corresponding eigenfunction ψ_{in}.

The eigenfunctions ψ_{in} of the eigenvalue problem defined by equations (8-21) and (8-22) satisfy the following orthogonality relation [13]

$$\sum_{i=1}^{M} \frac{k_i}{\alpha_i} \int_{x=x_i}^{x_{i+1}} x^p \psi_{in}(x) \psi_{ir}(x)\, dx = \begin{cases} 0 & \text{for} \quad n \neq r \\ N_n & \text{for} \quad n = r \end{cases} \tag{8-23a}$$

where the norm N_n is defined as

$$N_n = \sum_{j=1}^{M} \frac{k_j}{\alpha_j} \int_{x_j}^{x_{j+1}} x^p \psi_{jn}^2(x)\, dx \tag{8-23b}$$

and ψ_{in}, ψ_{ir} are the two different eigenfunctions.

The solution for the time-variable function $\Gamma(t)$ is immediately obtained from equation (8-20) as

$$\Gamma(t) = e^{-\beta_n^2 t} \tag{8-24}$$

and the general solution for the temperature distribution $T_i(x, t)$, in any region i, is constructed as

$$T_i(x, t) = \sum_{n=1}^{\infty} c_n e^{-\beta_n^2 t} \psi_{in}(x), \qquad i = 1, 2, \ldots, M \tag{8-25}$$

where the summation is over all eigenvalues β_n. This solution satisfies the differential equations (8-15) and the boundary conditions (8-16). We now constrain this solution to satisfy the initial conditions (8-17), and obtain

$$F_i(x) = \sum_{n=1}^{\infty} c_n \psi_{in}(x) \quad \text{in} \quad x_i < x < x_{i+1}, \quad i = 1, 2, \ldots, M \tag{8-26}$$

The coefficients c_n can be determined by utilizing the above orthogonality relation as now described.

We operate on both sides of equation (8-26) by the operator

$$\frac{k_i}{\alpha_i} \int_{x_i}^{x_{i+1}} x^p \psi_{ir}(x)\, dx$$

and sum up the resulting expressions from $i = 1$ to M (i.e., over all regions) to obtain

$$\sum_{i=1}^{M} \frac{k_i}{\alpha_i} \int_{x_i}^{x_{i+1}} x^p \psi_{ir}(x) F_i(x)\, dx = \sum_{n=1}^{\infty} c_n \left[\sum_{i=1}^{M} \frac{k_i}{\alpha_i} \int_{x_i}^{x_{i+1}} x^p \psi_{ir}(x) \psi_{in}(x)\, dx \right] \tag{8-27}$$

In view of the orthogonality relation (8-23), the term inside the bracket on the right-hand side of equation (8-27) vanishes for $n \neq r$ and becomes equal to N_n for $n = r$. Then the coefficients c_n are determined as

$$c_n = \frac{1}{N_n} \sum_{i=1}^{M} \frac{k_i}{\alpha_i} \int_{x_i}^{x_{i+1}} x^p \psi_{in}(x) F_i(x)\, dx \tag{8-28}$$

Before introducing this result into equation (8-25), we change the summation indice from i to j, and the dummy integration variable from x to x' in equation (8-28) to avoid confusion with the index i and the space variable x in equation (8-25). Then, the solution for the temperature distribution $T_i(x, t)$ in any region i of the composite medium is determined as

$$T_i(x,t) = \sum_{n=1}^{\infty} e^{-\beta_n^2 t} \frac{1}{N_n} \psi_{in}(x) \sum_{j=1}^{M} \frac{k_j}{\alpha_j} \int_{x'=x_j}^{x_{j+1}} x'^p \psi_{jn}(x') F_j(x')\, dx'$$

$$\text{in} \quad x_i < x < x_{i+1}, \qquad i = 1, 2, \ldots, M \tag{8-29a}$$

where the norm N_n is defined as

$$N_n = \sum_{j=1}^{M} \frac{k_j}{\alpha_j} \int_{x_j}^{x_{j+1}} x'^p \psi_{jn}^2(x')\, dx' \tag{8-29b}$$

and

$$p = \begin{cases} 0 & \text{slab} \\ 1 & \text{cylinder} \\ 2 & \text{sphere} \end{cases} \tag{8-29c}$$

An examination of this solution reveals that, for $M = 1$, equation (8-29) reduce to the solution for the single-region problem considered in the previous chapters if we set $\beta_n^2 = \alpha\gamma_n^2$ where α is the thermal diffusivity.

Green's Function for Composite Medium

The solution given by equation (8-29) can be recast to define the composite medium Green's function. That is, the solution (8-29a) is rearranged as

$$T_i(x,t) = \sum_{j=1}^{M} \int_{x'=x_j}^{x_{j+1}} \frac{k_j}{\alpha_j}\left[\sum_{n=1}^{\infty} e^{-\beta_n^2 t} \frac{1}{N_n} \psi_{in}(x)\psi_{jn}(x')\right] x'^p F_j(x')\, dx'$$
$$\text{in} \quad x_i < x < x_{i+1}, \quad i = 1, 2, \ldots, M \tag{8-30}$$

This result is now written more compactly, by introducing the Green's function notation as

$$T_i(x,t) = \sum_{j=1}^{M} \int_{x'=x_j}^{x_{j+1}} x'^p G_{ij}(x,t|x',\tau)|_{\tau=0} F_j(x')\, dx' \tag{8-31a}$$

where x'^p is the Sturm–Liouville weight function and $G_{ij}(x,t|x',\tau)|_{\tau=0}$ is defined as

$$G_{ij}(x,t|x',\tau)|_{\tau=0} = \sum_{n=1}^{\infty} e^{-\beta_n^2 t} \frac{1}{N_n} \frac{k_j}{\alpha_j} \psi_{in}(x)\psi_{jn}(x') \tag{8-31b}$$

$$p = \begin{cases} 0 & \text{slab} \\ 1 & \text{cylinder} \\ 2 & \text{sphere} \end{cases} \tag{8-31c}$$

in the region $x_i < x < x_{i+1}, i = 1, 2, \ldots, M$. Thus $G_{ij}(x,t|x',\tau)|_{\tau=0}$ represents the Green's function evaluated for $\tau = 0$ associated with the solution of one-dimensional homogeneous composite medium problem defined by equations (8-15)–(8-17).

To solve the nonhomogeneous composite medium problem, such as the one with energy generation, the Green's function $G_{ij}(x,t|x',\tau)$ is needed. It is obtained from equation (8-31b) by replacing t by $(t - \tau)$. Thus the Green's function for the problem becomes

$$G_{ij}(x,t|x',\tau) = \sum_{n=1}^{\infty} e^{-\beta_n^2(t-\tau)} \frac{1}{N_n} \frac{k_j}{\alpha_j} \psi_{in}(x)\psi_{jn}(x') \tag{8-32}$$

in the region $x_i \leqslant x \leqslant x_{i+1}, i = 1, 2, \ldots, M$.

The use of Green's function in the solution of nonhomogeneous one-dimensional composite medium problems will be demonstrated later in this chapter.

8-4 DETERMINATION OF EIGENFUNCTIONS AND EIGENVALUES

The general solution $\psi_{in}(x)$ of the eigenvalue problem given by equations (8-21) and (8-22) can be written in the form

$$\psi_{in}(x) = A_{in}\phi_{in}(x) + B_{in}\theta_{in}(x) \qquad \text{in} \qquad x_i < x < x_{i+1}, \quad i = 1, 2, \dots, M \tag{8-33}$$

where $\phi_{in}(x)$ and $\theta_{in}(x)$ are the two linearly independent solutions of equations (8-21) and A_{in}, B_{in} are the coefficients. Table 8-1 lists the functions $\phi_{in}(x)$ and $\theta_{in}(x)$ for slabs, cylinders, and spheres. The heat conduction problem of an M-layer composite medium, in general, involves M solutions in the form given by equation (8-33); hence, there are $2M$ arbitrary coefficients, A_{in} and B_{in}, $i = 1, 2, \dots, M$ to be determined. The boundary conditions (8-22) provide a system of $2M$, linear, homogeneous equations for the determination of these $2M$ coefficients; but, because the resulting system of equations is homogeneous, the coefficients can be determined only in terms of any one of them (i.e., the nonvanishing one) or within a multiple of an arbitrary constant. This arbitrariness does not cause any difficulty, because the arbitrary constant will appear both in the numerator and denominator of equation (8-29) or equation (8-31); hence it will cancel out. Therefore, in the process of determining the coefficients A_{in} and B_{in} from the system of $2M$ homogeneous equations, any one of the nonvanishing coefficients, say, A_{in}, can be set equal to unity without loss of generality.

Finally, an additional relationship is needed for the determination of the eigenvalues β_n. This additional relationship is obtained from the requirement that the above system of $2M$ homogeneous equations has a nontrivial solution, that is, the determinant of the coefficients A_{in} and B_{in} vanishes. This condition leads to a *transcendental equation* for the determination of the eigenvalues

$$\beta_1 < \beta_2 < \beta_3 < \cdots < \beta_n < \cdots \tag{8-34}$$

TABLE 8-1 Linearly Independent Solutions $\phi_{in}(x)$ and $\theta_{in}(x)$ of Equation (8-21) for Slabs, Cylinders, and Spheres

Geometry	$\phi_{in}(x)$	$\theta_{in}(x)$
Slab	$\sin\left(\frac{\beta_n}{\sqrt{\alpha_i}}x\right)$	$\cos\left(\frac{\beta_n}{\sqrt{\alpha_i}}x\right)$
Cylinder	$J_0\left(\frac{\beta_n}{\sqrt{\alpha_i}}x\right)$	$Y_0\left(\frac{\beta_n}{\sqrt{\alpha_i}}x\right)$
Sphere	$\frac{1}{x}\sin\left(\frac{\beta_n}{\sqrt{\alpha_i}}x\right)$	$\frac{1}{x}\cos\left(\frac{\beta_n}{\sqrt{\alpha_i}}x\right)$

Clearly, for each of these eigenvalues there are the corresponding set of values of A_{in} and B_{in}, hence of the eigenfunctions $\psi_{in}(x)$. Once the eigenfunctions $\psi_{in}(x)$ and the eigenvalues β_n are determined by the procedure outlined above, the temperature distribution $T_i(x,t)$ in any region i of the composite medium is determined by equations (8-29).

Example 8-2

Consider transient heat conduction in a three-layer composite medium with perfect thermal contact at the interfaces and convection at the outer boundary surfaces. Give the eigenvalue problem and develop the equations for the determination of the coefficients A_{in}, B_{in} of equation (8-33) and the eigenvalues

$$\beta_1 < \beta_2 < \beta_3 < \cdots < \beta_n < \cdots$$

Solution. The eigenvalue problem for this transient heat conduction is similar to that given by equations (8-21) and (8-22), except $M = 3$, and the interface conductances are taken as infinite: $h_2 \to \infty, h_3 \to \infty$. Then, the eigenvalue problem becomes

$$\frac{1}{x^p}\frac{d}{dx}\left(x^p \frac{d\psi_{in}}{dx}\right) + \frac{\beta_n^2}{\alpha_i}\psi_{in} = 0 \quad \text{in} \quad x_i < x < x_{i+1}, \quad i = 1, 2, 3 \tag{8-35}$$

Subject to the boundary conditions,

$$-k_1^* \frac{d\psi_{1n}}{dx} + h_1^* \psi_{1n} = 0 \quad \text{at the outer boundary } x = x_1 \tag{8-36a}$$

$$\left.\begin{aligned} &\psi_{in} = \psi_{i+1,n} \\ &k_i \frac{d\psi_{in}}{dx} = k_{i+1}\frac{d\psi_{i+1,n}}{dx} \end{aligned}\right\} \quad \begin{aligned} &\text{at the interfaces } x = x_{i+1}, \quad (8\text{-}36\text{b}) \\ &i = 1, 2 \quad (8\text{-}36\text{c}) \end{aligned}$$

$$k_3^* \frac{d\psi_{3n}}{dx} + h_4^* \psi_{3n} = 0 \quad \text{at the outer boundary } x = x_4 \tag{8-36d}$$

where the eigenfunctions $\psi_{in}(x)$ are given by

$$\psi_{in}(x) = A_{in}\phi_{in}(x) + B_{in}\theta_{in}(x), \quad i = 1, 2, 3 \tag{8-37}$$

and $\phi_{in}(x)$ and $\theta_{in}(x)$ are as specified in Table 8-1.

The first step in the analysis is the determination of the six coefficients A_{in}, B_{in} with $i = 1, 2, 3$. Without loss of generality, we set of the nonvanishing coefficients, say, A_{1n} equal to unity:

$$A_{1n} = 1 \tag{8-38}$$

The eigenfunctions $\psi_{in}(x)$ given by equation (8-37) with $A_{1n} = 1$ are introduced into the boundary conditions (8-36). The resulting system of equations is expressed in the matrix form as

$$\begin{bmatrix} X_1 & Y_1 & 0 & 0 & 0 & 0 \\ \phi_{1n} & \theta_{1n} & -\phi_{2n} & -\theta_{2n} & 0 & 0 \\ k_1\phi'_{1n} & k_1\theta'_{1n} & -k_2\phi'_{2n} & -k_2\theta'_{2n} & 0 & 0 \\ 0 & 0 & \phi_{2n} & \theta_{2n} & -\phi_{3n} & -\theta_{3n} \\ 0 & 0 & k_2\phi'_{2n} & k_2\theta'_{2n} & -k_3\phi'_{3n} & -k_3\theta'_{3n} \\ 0 & 0 & 0 & 0 & X_3 & Y_3 \end{bmatrix} \begin{bmatrix} 1 \\ B_{1n} \\ A_{2n} \\ B_{2n} \\ A_{3n} \\ B_{3n} \end{bmatrix} = \begin{bmatrix} 0 \\ 0 \\ 0 \\ 0 \\ 0 \\ 0 \end{bmatrix} \tag{8-39}$$

where

$$X_1 = -k_1^*\phi'_{1n} + h_1^*\phi_{1n} \qquad Y_1 = -k_1^*\theta'_{1n} + h_1^*\theta_{1n} \tag{8-40}$$

$$X_3 = k_3^*\phi'_{3n} + h_4^*\phi_{3n} \qquad Y_3 = k_3^*\theta'_{3n} + h_4^*\theta_{3n} \tag{8-41}$$

and the primes denote differentiation with respect to x. Only five of these equations can be used to determine the coefficients. We choose the first five of them; the resulting system of equations for the determination of these five coefficients is given in the matrix form as

$$\begin{bmatrix} Y_1 & 0 & 0 & 0 & 0 \\ \theta_{1n} & -\phi_{2n} & -\theta_{2n} & 0 & 0 \\ k_1\theta'_{1n} & -k_2\phi'_{2n} & -k_2\theta'_{2n} & 0 & 0 \\ 0 & \phi_{2n} & \theta_{2n} & -\phi_{3n} & -\theta_{3n} \\ 0 & k_2\phi'_{2n} & k_2\theta'_{2n} & -k_3\phi'_{3n} & -k_3\theta'_{3n} \end{bmatrix} \begin{bmatrix} B_{1n} \\ A_{2n} \\ B_{2n} \\ A_{3n} \\ B_{3n} \end{bmatrix} = \begin{bmatrix} -X_1 \\ -\phi_{1n} \\ -k_1\phi'_{1n} \\ 0 \\ 0 \end{bmatrix} \tag{8-42}$$

Thus, the solution of equations (8-42) gives the five coefficients B_{1n}, A_{2n}, B_{2n}, A_{3n}, and B_{3n}. The transcendental equation for the determination of the eigenvalues $\beta_1 < \beta_2 < \cdots < \beta_n < \cdots$, is obtained from the requirement that the determinant of the coefficients in the system of equations (8-48) should vanish. This condition leads to the following transcendental equation for the determination of the eigenvalues, $\beta_1 < \beta_2 < \beta_3 < \cdots < \beta_n < \cdots$.

$$\begin{vmatrix} X_1 & Y_1 & 0 & 0 & 0 & 0 \\ \phi_{1n} & \theta_{1n} & -\phi_{2n} & -\theta_{2n} & 0 & 0 \\ k_1\phi'_{1n} & k_1\theta'_{1n} & -k_2\phi'_{2n} & -k_2\theta'_{2n} & 0 & 0 \\ 0 & 0 & \phi_{2n} & \theta_{2n} & -\phi_{3n} & -\theta_{3n} \\ 0 & 0 & k_2\phi'_{2n} & k_2\phi'_{2n} & -k_3\phi'_{3n} & -k_3\theta'_{3n} \\ 0 & 0 & 0 & 0 & X_3 & Y_3 \end{vmatrix} = 0 \tag{8-43}$$

8-5 APPLICATIONS OF ORTHOGONAL EXPANSION TECHNIQUE

In this section we illustrate the application of the orthogonal expansion technique described previously for the solution of transient homogeneous heat conduction problems of a two-layer cylinder and a two-layer slab.

Example 8-3

A two-layer solid cylinder as illustrated in Fig. 8-4 contains an inner region $0 \leqslant r \leqslant a$ and an outer region $a \leqslant r \leqslant b$ that are in perfect thermal contact; k_1 and k_2 are the thermal conductivities, and α_1 and α_2 are the thermal diffusivities of the inner and outer regions, respectively. Initially, the inner region is at temperature $\theta_1(r,t) = F_1(r)$ and the outer region at temperature $\theta_2(r,t) = F_2(r)$. For times $t > 0$, heat is dissipated by convection from the outer surface at $r = b$ into an environment at zero temperature. Develop an expression for the temperature distribution in the cylinders for times $t > 0$.

Solution. The mathematical formulation of the problem is given by

$$\frac{\alpha_1}{r}\frac{\partial}{\partial r}\left(r\frac{\partial \theta_1}{\partial r}\right) = \frac{\partial \theta_1(r,t)}{\partial t} \qquad \text{in} \qquad 0 \leqslant r < a, \quad t > 0 \tag{8-44a}$$

$$\frac{\alpha_2}{r}\frac{\partial}{\partial r}\left(r\frac{\partial \theta_2}{\partial r}\right) = \frac{\partial \theta_2(r,t)}{\partial t} \qquad \text{in} \qquad a < r < b, \quad t > 0 \tag{8-44b}$$

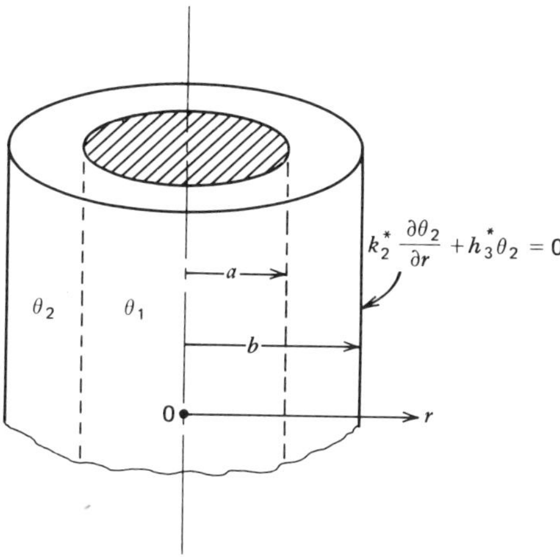

Fig. 8-4 Two-layer cylinder with perfect thermal contact at the interface.

Subject to the boundary conditions

$$\theta_1(r,t) = \text{finite} \qquad \text{at} \qquad r=0, \quad t>0 \tag{8-44c}$$

$$\theta_1(r,t) = \theta_2(r,t) \qquad \text{at} \qquad r=a, \quad t>0 \tag{8-44d}$$

$$k_1 \frac{\partial \theta_1}{\partial r} = k_2 \frac{\partial \theta_2}{\partial r} \qquad \text{at} \qquad r=a, \quad t>0 \tag{8-44e}$$

$$k_2^* \frac{\partial \theta_2}{\partial r} + h_3^* \theta_2 = 0 \qquad \text{at} \qquad r=b, \quad t>0 \tag{8-44f}$$

and to the initial conditions

$$\theta_1(r,t) = F_1(r) \qquad \text{for} \qquad t=0, \quad 0<r<a \tag{8-44g}$$

$$\theta_2(r,t) = F_2(r) \qquad \text{for} \qquad t=0, \quad a<r<b \tag{8-44h}$$

The corresponding eigenvalue problem is taken as

$$\frac{1}{r}\frac{d}{dr}\left(r\frac{d\psi_{1n}}{dr}\right) + \frac{\beta_n^2}{\alpha_1}\psi_{1n}(r) = 0 \qquad \text{in} \qquad 0 \leqslant r < a \tag{8-45a}$$

$$\frac{1}{r}\frac{d}{dr}\left(r\frac{d\psi_{2n}}{dr}\right) + \frac{\beta_n^2}{\alpha_2}\psi_{2n}(r) = 0 \qquad \text{in} \qquad a < r \leqslant b \tag{8-45b}$$

Subject to the boundary conditions

$$\psi_{1n}(r) = \text{finite} \qquad \text{at} \qquad r=0 \tag{8-45c}$$

$$\psi_{1n}(r) = \psi_{2n}(r) \qquad \text{at} \qquad r=a \tag{8-45d}$$

$$k_1 \frac{d\psi_{1n}}{dr} = k_2 \frac{d\psi_{2n}}{dr} \qquad \text{at} \qquad r=a \tag{8-45e}$$

$$k_2^* \frac{d\psi_{2n}}{dr} + h_3^* \psi_{2n} = 0 \qquad \text{at} \qquad r=b \tag{8-45f}$$

The general solution of the above eigenvalue problem 8-45, according to Table 8-1, is taken as

$$\psi_{in}(r) = A_{in} J_0\left(\frac{\beta_n}{\sqrt{\alpha_i}} r\right) + B_{in} Y_0\left(\frac{\beta_n}{\sqrt{\alpha_i}} r\right), \quad i=1,2 \tag{8-46}$$

The boundary condition (8-45c) requires that $B_{1n}=0$. Then the solutions $\psi_{in}(r)$ for the two regions become

$$\psi_{1n}(r)=J_0\left(\frac{\beta_n}{\sqrt{\alpha_1}}r\right) \qquad \text{in} \qquad 0\leqslant r\leqslant a \tag{8-47a}$$

$$\psi_{2n}(r)=A_{2n}J_0\left(\frac{\beta_n}{\sqrt{\alpha_2}}r\right)+B_{2n}Y_0\left(\frac{\beta_n}{\sqrt{\alpha_2}}r\right) \qquad \text{in} \qquad a<r<b \tag{8-47b}$$

where we have chosen $A_{1n}=1$ for the reason stated previously. The requirement that the solutions (8-47) should satisfy the remaining three boundary conditions (8-45d,e,f) leads, respectively, to the following equations for the determination of these coefficients

$$J_0\left(\frac{\beta_n a}{\sqrt{\alpha_1}}\right)=A_{2n}J_0\left(\frac{\beta_n a}{\sqrt{\alpha_2}}\right)+B_{2n}Y_0\left(\frac{\beta_n a}{\sqrt{\alpha_2}}\right) \tag{8-48a}$$

$$\frac{k_1}{k_2}\sqrt{\frac{\alpha_2}{\alpha_1}}J_1\left(\frac{\beta_n a}{\sqrt{\alpha_1}}\right)=A_{2n}J_1\left(\frac{\beta_n a}{\sqrt{\alpha_1}}\right)+B_{2n}Y_1\left(\frac{\beta_n a}{\sqrt{\alpha_2}}\right) \tag{8-48b}$$

$$-\left[A_{2n}J_1\left(\frac{\beta_n b}{\sqrt{\alpha_2}}\right)+B_{2n}Y_1\left(\frac{\beta_n b}{\sqrt{\alpha_2}}\right)\right]$$
$$+\frac{h_3^*\sqrt{\alpha_2}}{k_2^*\beta_n}\left[A_{2n}J_0\left(\frac{\beta_n b}{\sqrt{\alpha_2}}\right)+B_{2n}Y_0\left(\frac{\beta_n b}{\sqrt{\alpha_2}}\right)\right]=0 \tag{8-48c}$$

These equations are now written in the matrix form as

$$\begin{bmatrix} J_0(\gamma) & -J_0\left(\frac{a}{b}\eta\right) & -Y_0\left(\frac{a}{b}\eta\right) \\ KJ_1(\gamma) & -J_1\left(\frac{a}{b}\eta\right) & -Y_1\left(\frac{a}{b}\eta\right) \\ 0 & \frac{H}{\eta}J_0(\eta)-J_1(\eta) & \frac{H}{\eta}Y_0(\eta)-Y_1(\eta) \end{bmatrix}\begin{bmatrix} 1 \\ A_{2n} \\ B_{2n} \end{bmatrix}=\begin{bmatrix} 0 \\ 0 \\ 0 \end{bmatrix} \tag{8-49}$$

where we defined

$$\gamma\equiv\frac{a\beta_n}{\sqrt{\alpha_1}} \quad \eta\equiv\frac{b\beta_n}{\sqrt{\alpha_2}} \quad H\equiv\frac{bh_3^*}{k_2^*} \quad K\equiv\frac{k_1}{k_2}\sqrt{\frac{\alpha_2}{\alpha_1}} \tag{8-50}$$

Any two of these equations can be used to determine the coefficients A_{2n} and

B_{2n}. We choose the first two and write the resulting equations as

$$\begin{bmatrix} J_0\left(\frac{a}{b}\eta\right) & Y_0\left(\frac{a}{b}\eta\right) \\ J_1\left(\frac{a}{b}\eta\right) & Y_1\left(\frac{a}{b}\eta\right) \end{bmatrix} \begin{bmatrix} A_{2n} \\ B_{2n} \end{bmatrix} = \begin{bmatrix} J_0(\gamma) \\ KJ_1(\gamma) \end{bmatrix} \tag{8-51}$$

Then, A_{2n} and B_{2n} are obtained as

$$A_{2n} = \frac{1}{\Delta}\left[J_0(\gamma)Y_1\left(\frac{a}{b}\eta\right) - KJ_1(\gamma)Y_0\left(\frac{a}{b}\eta\right)\right] \tag{8-52a}$$

$$B_{2n} = \frac{1}{\Delta}\left[KJ_1(\gamma)J_0\left(\frac{a}{b}\eta\right) - J_0(\gamma)J_1\left(\frac{a}{b}\eta\right)\right] \tag{8-52b}$$

where

$$\Delta = J_0\left(\frac{a}{b}\eta\right)Y_1\left(\frac{a}{b}\eta\right) - J_1\left(\frac{a}{b}\eta\right)Y_0\left(\frac{a}{b}\eta\right) \tag{8-52c}$$

Finally, the equation for the determination of the eigenvalues is obtained from the requirement that in equation (8-49) the determinant of the coefficients should vanish. Then, the β_n values are the roots of the following transcendental equation

$$\begin{vmatrix} J_0(\gamma) & -J_0\left(\frac{a}{b}\eta\right) & -Y_0\left(\frac{a}{b}\eta\right) \\ KJ_1(\gamma) & -J_1\left(\frac{a}{b}\eta\right) & -Y_1\left(\frac{a}{b}\eta\right) \\ 0 & \frac{H}{\eta}J_0(\eta) - J_1(\eta) & \frac{H}{\eta}Y_0(\eta) - Y_1(\eta) \end{vmatrix} = 0 \tag{8-53}$$

Having established the relations for the determination of the coefficients A_{2n}, B_{2n} and the eigenvalues β_n, the eigenfunctions $\psi_{1n}(r)$ and $\psi_{2n}(r)$ are obtained according to equations (8-47). Then, the solution for the temperature $\theta_i(r, t)$, $i = 1, 2$ in any of the regions is given by equations (8-29) as

$$\theta_i(r, t) = \sum_{n=1}^{\infty} \frac{1}{N_n} e^{-\beta_n^2 t}\psi_{in}(r)\left[\frac{k_1}{\alpha_1}\int_0^a r'\psi_{1n}(r')F_1(r')\,dr' + \frac{k_2}{\alpha_2}\int_a^b r'\psi_{2n}(r')F_2(r')\,dr'\right], \quad i = 1, 2 \tag{8-54a}$$

where

$$N_n = \frac{k_1}{\alpha_1}\int_0^a r'\psi_{1n}^2(r')\,dr' + \frac{k_2}{\alpha_2}\int_a^b r'\psi_{2n}^2(r')\,dr' \tag{8-54b}$$

$$\psi_{1n}(r) = J_0\left(\frac{\beta_n}{\sqrt{\alpha_1}}r\right) \tag{8-54c}$$

$$\psi_{2n}(r) = A_{2n}J_0\left(\frac{\beta_n}{\sqrt{\alpha_2}}r\right) + B_{2n}Y_0\left(\frac{\beta_n}{\sqrt{\alpha_2}}r\right) \tag{8-54d}$$

This result is now written more compactly in terms of the Green's function as

$$\theta_i(r,t) = \int_0^a r'G_{i1}(r,t|r',\tau)|_{\tau=0}F_1(r')\,dr' + \int_a^b r'G_{i2}(r,t|r',\tau)|_{\tau=0}F_2(r')\,dr', \quad i=1,2 \tag{8-55a}$$

where $G_{ij}(r,t|r',\tau)|_{\tau=0}$ is defined as

$$G_{ij}(r,t|r',\tau)|_{\tau=0} = \sum_{n=1}^{\infty} e^{-\beta_n^2 t}\frac{1}{N_n}\frac{k_j}{\alpha_j}\psi_{in}(r)\psi_{jn}(r') \tag{8-55b}$$

Example 8-4

A two-layer slab consists of the first layer in $0 \leqslant x \leqslant a$ and the second layer in $a \leqslant x \leqslant b$, which are in perfect thermal contact as illustrated in Fig. 8-5. Let k_1 and k_2 be the thermal conductivities, and α_1 and α_2 the thermal diffusivities for the first and second layers, respectively. Initially, the first region is at temperature $F_1(x)$ and the second region at $F_2(x)$. For times $t>0$ the boundary surface at $x=0$ is kept at zero temperature and the boundary surface at $x=b$ dissipates heat by convection into a medium at zero temperature. Obtain an expression for the temperature distribution in the slab for times $t>0$.

Solution. The mathematical formulation of this problem is given as

$$\alpha_1\frac{\partial^2 T_1}{\partial x^2} = \frac{\partial T_1(x,t)}{\partial t} \quad \text{in} \quad 0<x<a, \quad t>0 \tag{8-56a}$$

$$\alpha_2\frac{\partial^2 T_2}{\partial x^2} = \frac{\partial T_2(x,t)}{\partial t} \quad \text{in} \quad a<x<b, \quad t>0 \tag{8-56b}$$

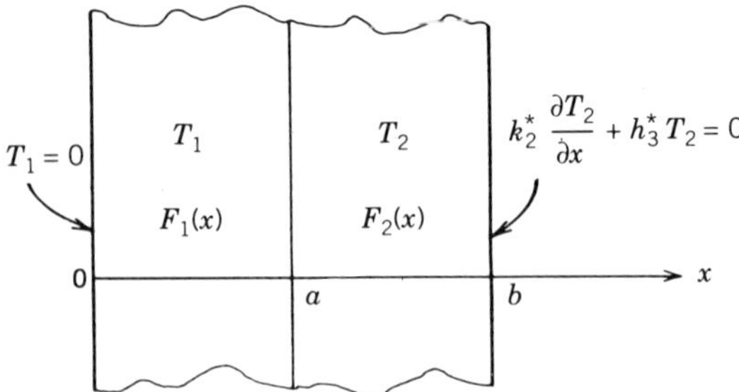

Fig. 8-5 Two-layer slab with perfect thermal contact at the interface.

subject to the boundary conditions

$$T_1(x,t) = 0 \qquad \text{at} \qquad x = 0, \quad t > 0 \tag{8-56c}$$

$$T_1(x,t) = T_2(x,t) \qquad \text{at} \qquad x = a, \quad t > 0 \tag{8-56d}$$

$$k_1 \frac{\partial T_1}{\partial x} = k_2 \frac{\partial T_2}{\partial x} \qquad \text{at} \qquad x = a, \quad t > 0 \tag{8-56e}$$

$$k_2^* \frac{\partial T_2}{\partial x} + h_3^* T_2 = 0 \qquad \text{at} \qquad x = b, \quad t > 0 \tag{8-56f}$$

and the initial conditions

$$T_1(x,t) = F_1(x) \qquad \text{for} \qquad t = 0, \quad 0 < x < a \tag{8-56g}$$

$$T_2(x,t) = F_2(x) \qquad \text{for} \qquad t = 0, \quad a < x < b \tag{8-56h}$$

The corresponding eigenvalue problem is taken as

$$\frac{d^2\psi_{1n}}{dx^2} + \frac{\beta_n^2}{\alpha_1}\psi_{1n}(x) = 0 \qquad \text{in} \qquad 0 < x < a \tag{8-57a}$$

$$\frac{d^2\psi_{2n}}{dx^2} + \frac{\beta_n^2}{\alpha_2}\psi_{2n}(x) = 0 \qquad \text{in} \qquad a < x < b \tag{8-57b}$$

subject to the boundary conditions

$$\psi_{1n}(x) = 0 \qquad \text{at} \qquad x = 0 \tag{8-57c}$$

$$\psi_{1n}(x) = \psi_{2n}(x) \qquad \text{at} \qquad x = a \tag{8-57d}$$

$$k_1 \frac{d\psi_{1n}}{dx} = k_2 \frac{d\psi_{2n}}{dx} \qquad \text{at} \qquad x = a \tag{8-57e}$$

$$k_2^* \frac{d\psi_{2n}}{dx} + h_3^* \psi_{2n} = 0 \qquad \text{at} \qquad x = b \tag{8-57f}$$

The general solution of the above eigenvalue problem, according to Table 8-1, is taken as

$$\psi_{in}(x) = A_{in} \sin\left(\frac{\beta_n}{\sqrt{\alpha_i}} x\right) + B_{in} \cos\left(\frac{\beta_n}{\sqrt{\alpha_i}} x\right), \qquad i = 1, 2 \tag{8-58}$$

The boundary condition (8-57c) requires that $B_{1n} = 0$. Then, the solutions ψ_{in} for the two regions are reduced to

$$\psi_{1n}(x) = \sin\left(\frac{\beta_n}{\sqrt{\alpha_1}} x\right) \qquad \text{in} \qquad 0 < x < a \tag{8-59a}$$

$$\psi_{2n}(x) = A_{2n} \sin\left(\frac{\beta_n}{\sqrt{\alpha_2}} x\right) + B_{2n} \cos\left(\frac{\beta_n}{\sqrt{\alpha_2}} x\right) \qquad \text{in} \qquad a < x < b \tag{8-59b}$$

where we have chosen $A_{1n} = 1$ for the reason stated previously. The requirement that the solutions (8-59) should satisfy the remaining boundary conditions (8-57d, e, f) yields the following equations for the determination of these coefficients.

$$\sin\left(\frac{\beta_n a}{\sqrt{\alpha_1}}\right) = A_{2n} \sin\left(\frac{\beta_n a}{\sqrt{\alpha_2}}\right) + B_{2n} \cos\left(\frac{\beta_n a}{\sqrt{\alpha_2}}\right) \tag{8-60a}$$

$$\frac{k_1}{k_2}\sqrt{\frac{\alpha_2}{\alpha_1}} \cos\left(\frac{\beta_n a}{\sqrt{\alpha_1}}\right) = A_{2n} \cos\left(\frac{\beta_n a}{\sqrt{\alpha_2}}\right) - B_{2n} \sin\left(\frac{\beta_n a}{\sqrt{\alpha_2}}\right) \tag{8-60b}$$

$$\left[A_{2n} \cos\left(\frac{\beta_n b}{\sqrt{\alpha_2}}\right) - B_{2n} \sin\left(\frac{\beta_n b}{\sqrt{\alpha_2}}\right)\right] + \frac{h_3^* \sqrt{\alpha_2}}{k_2^* \beta_n}\left[A_{2n} \sin\left(\frac{\beta_n b}{\sqrt{\alpha_2}}\right) + B_{2n} \cos\left(\frac{\beta_n b}{\sqrt{\alpha_2}}\right)\right] = 0 \tag{8-60c}$$

These equations are now written in the matrix from as

$$\begin{bmatrix} \sin\gamma & -\sin\left(\frac{a}{b}\eta\right) & -\cos\left(\frac{a}{b}\eta\right) \\ K\cos\gamma & -\cos\left(\frac{a}{b}\eta\right) & \sin\left(\frac{a}{b}\eta\right) \\ 0 & \frac{H}{\eta}\sin\eta + \cos\eta & \frac{H}{\eta}\cos\eta - \sin\eta \end{bmatrix} \begin{bmatrix} 1 \\ A_{2n} \\ B_{2n} \end{bmatrix} = \begin{bmatrix} 0 \\ 0 \\ 0 \end{bmatrix} \tag{8-61a}$$

where we defined

$$\gamma \equiv \frac{a\beta_n}{\sqrt{\alpha_1}} \qquad \eta \equiv \frac{b\beta_n}{\sqrt{\alpha_2}} \qquad H \equiv \frac{bh_3^*}{k_2^*} \qquad K \equiv \frac{k_1}{k_2}\sqrt{\frac{\alpha_2}{\alpha_1}} \tag{8-61b}$$

We choose the first two of these equations to determine the coefficients A_{2n} and B_{2n}; these two equations are written as

$$\begin{bmatrix} \sin\left(\frac{a}{b}\eta\right) & \cos\left(\frac{a}{b}\eta\right) \\ \cos\left(\frac{a}{b}\eta\right) & -\sin\left(\frac{a}{b}\eta\right) \end{bmatrix} \begin{bmatrix} A_{2n} \\ B_{2n} \end{bmatrix} = \begin{bmatrix} \sin\gamma \\ K\cos\gamma \end{bmatrix} \tag{8-62}$$

Then, A_{2n} and B_{2n} are determined as

$$A_{2n} = \frac{1}{\Delta}\left[-\sin\gamma\sin\left(\frac{a}{b}\eta\right) - K\cos\gamma\cos\left(\frac{a}{b}\eta\right)\right] \tag{8-63a}$$

$$B_{2n} = \frac{1}{\Delta}\left[K\cos\gamma\sin\left(\frac{a}{b}\eta\right) - \sin\gamma\cos\left(\frac{a}{b}\eta\right)\right] \tag{8-63b}$$

where

$$\Delta = -\sin^2\left(\frac{a}{b}\eta\right) - \cos^2\left(\frac{a}{b}\eta\right) = -1 \tag{8-63c}$$

Finally, the equation for the determination of the eigenvalues β_n is obtained from the requirement that in equation (8-61a) the determinant of the coefficients should vanish. This condition yields the following transcendental equation for the determination of the eigenvalues β_n

$$\begin{bmatrix} \sin\gamma & -\sin\left(\frac{a}{b}\eta\right) & -\cos\left(\frac{a}{b}\eta\right) \\ K\cos\gamma & -\cos\left(\frac{a}{b}\eta\right) & \sin\left(\frac{a}{b}\eta\right) \\ 0 & \frac{H}{\eta}\sin\eta + \cos\eta & \frac{H}{\eta}\cos\eta - \sin\eta \end{bmatrix} = 0 \tag{8-64}$$

The formal solution of this problem is now complete. That is, the coefficients A_{2n} and B_{2n} are given by equations (8-63), the eigenvalues β_n by equation (8-64). Then, the eigenfunctions $\psi_{1n}(x)$ and $\psi_{2n}(x)$ defined by equations (8-59) are known and the temperature distribution $T_i(x,t), i=1,2$ in any one of the two regions is determined according to equation (8-29) as

$$T_i(x,t)=\sum_{n=1}^{\infty}\frac{1}{N_n}e^{-\beta_n^2 t}\psi_{in}(x)\left[\frac{k_1}{\alpha_1}\int_{x'=0}^{a}\psi_{1n}(x')F_1(x')dx' + \frac{k_2}{\alpha_2}\int_{a}^{b}\psi_{2n}(x')F_2(x')dx'\right], \qquad i=1,2 \tag{8-65a}$$

where

$$N_n=\frac{k_1}{\alpha_1}\int_0^a \psi_{1n}^2(x')dx'+\frac{k_2}{\alpha_2}\int_a^b \psi_{2n}^2(x')dx \tag{8-65b}$$

$$\psi_{1n}(x)=\sin\left(\frac{\beta_n}{\sqrt{\alpha_1}}x\right) \tag{8-65c}$$

$$\psi_{2n}(x)=A_{2n}\sin\left(\frac{\beta_n}{\sqrt{\alpha_2}}x\right)+B_{2n}\cos\left(\frac{\beta_n}{\sqrt{\alpha_2}}x\right) \tag{8-65d}$$

This result can be written more compactly in terms of the Green's function as

$$T_i(x,t)=\int_0^a G_{i1}(x,t|x',\tau)|_{\tau=0}F_1(x')dx' + \int_a^b G_{i2}(x,t|x',\tau)|_{\tau=0}F_2(x')dx', \qquad i=1,2 \tag{8-66a}$$

where $G_{ij}(x,t|x',\tau)|_{\tau=0}$, the Green's function evaluated at $\tau=0$, is given by

$$G_{ij}(x,t|x',\tau)|_{\tau=0}=\sum_{n=1}^{\infty}e^{-\beta_n^2 t}\frac{1}{N_n}\frac{k_j}{\alpha_j}\psi_{in}(x)\psi_{jn}(x') \tag{8-66b}$$

where N_n, $\psi_{in}(x)$, $i=1,2$ are defined by equations (8-65b, c, d).

8-6 GREEN'S FUNCTION APPROACH FOR SOLVING NONHOMOGENEOUS PROBLEMS

The use of Green's function is a very convenient approach for solving nonhomogeneous problems of transient heat conduction in a composite medium, if the general expression relating the solution for the temperature $T_i(x,t)$ to the Green's function is known and the appropriate Green's function is available. The general procedure is similar to that described for the case of a single-region medium, except the functional forms of the general solution and the Green's function are different.

In the following analysis, we assume that the nonhomogeneity associated with the boundary conditions is removed by a splitting up procedure described previously; hence the energy generation term is the only nonhomogeneity in the problem.

We consider the following transient heat conduction problem for a M-layer composite medium with energy generation, homogeneous boundary conditions at the outer surfaces and contact conductance at the interfaces

$$\alpha_i \frac{1}{x^p}\frac{\partial}{\partial x}\left(x^p \frac{\partial T_i}{\partial x}\right) + \frac{\alpha_i}{k_i} g_i(x,t) = \frac{\partial T_i(x,t)}{\partial t} \quad \text{in} \quad x_i < x < x_{i+1}, \quad t > 0, \quad i = 1,2,\ldots,M \tag{8-67a}$$

where

$$p = \begin{cases} 0 & \text{slab} \\ 1 & \text{cylinder} \\ 2 & \text{sphere} \end{cases} \tag{8-67b}$$

Subject to the boundary conditions

$$-k_1^* \frac{\partial T_1}{\partial x} + h_1^* T_1 = 0 \qquad \text{at the outer boundary } x = x_1, \quad t > 0 \tag{8-67c}$$

$$-k_i \frac{\partial T_i}{\partial x} = h_{i+1}(T_i - T_{i+1}) \tag{8-67d}$$

$$k_i \frac{\partial T_i}{\partial x} = k_{i+1} \frac{\partial T_{i+1}}{\partial x} \tag{8-67e}$$

at the interfaces $x = x_{i+1}$, $i = 1,2,\ldots,M-1$ for $t > 0$

$$k_M^* \frac{\partial T_M}{\partial x} + h_{M+1}^* T_M = 0 \qquad \text{at the outer boundary } x = x_{M+1}, \quad t > 0 \tag{8-67f}$$

and the initial conditions

$$T_i(x,t) = F_i(x) \quad \text{for} \quad t = 0, \quad \text{in} \quad x_i < x < x_{i+1}, \quad i = 1,2,\ldots,M \tag{8-67g}$$

Appropriate eigenvalue problem for the solution of the above heat conduction problem is taken as

$$\frac{1}{x^p}\frac{d}{dx}\left(x^p \frac{d\psi_{in}}{dx}\right) + \frac{\beta_n^2}{\alpha_i}\psi_{in}(x) = 0 \quad \text{in} \quad x_i < x < x_{i+1}, \quad i = 1,2,\ldots,M \tag{8-68a}$$

Subject to the boundary conditions

$$-k_1^* \frac{d\psi_{1n}}{dx} + h_1^* \psi_{1n} = 0 \qquad \text{at the outer boundary,} \quad x = x_1 \tag{8-68b}$$

$$-k_i \frac{d\psi_{in}}{dx} = h_{i+1}(\psi_{in} - \psi_{i+1,n}) \tag{8-68c}$$

$$k_i \frac{d\psi_{in}}{dx} = k_{i+1} \frac{d\psi_{i+1,n}}{dx} \tag{8-68d}$$

at the interfaces $x = x_{i+1}$, $i = 1, 2, \ldots, M-1$

$$k_M^* \frac{d\psi_{Mn}}{dx} + h_{M+1}^* \psi_{Mn} = 0 \qquad \text{at the outer boundary,} \quad x = x_{M+1} \tag{8-68e}$$

The solution of this multilayer transient heat conduction problem in terms of the composite medium Green's function $G_{ij}(x, t|x', \tau)$ can be obtained by proper rearrangement of the general solution given by Yener and Özişik [25] and Özişik [8]. We write the resulting expression in the form

$$\begin{aligned} T_i(x,t) = \sum_{j=1}^{M} \Bigg\{ & \int_{x_j}^{x_{j+1}} x'^p |G_{ij}(x,t|x',\tau)|_{\tau=0} F_j(x') dx' \\ & + \int_{\tau=0}^{t} d\tau \int_{x_j}^{x_{j+1}} x'^p G_{ij}(x,t|x',\tau) \left[\frac{\alpha_j}{k_j} g_j(x',\tau) \right] dx' \Bigg\} \\ & \text{in} \quad x_i < x < x_{i+1}, \quad i = 1, 2, \ldots, M \end{aligned} \tag{8-69a}$$

where the composite medium Green's function $G_{ij}(x, t|x', \tau)$ is defined as

$$G_{ij}(x,t|x',\tau) = \sum_{n=1}^{\infty} e^{-\beta_n^2 (t-\tau)} \frac{1}{N_n} \frac{k_j}{\alpha_j} \psi_{in}(x) \psi_{jn}(x') \tag{8-69b}$$

and

$$p = \begin{cases} 0 & \text{slab} \\ 1 & \text{cylinder} \\ 2 & \text{sphere} \end{cases} \tag{8-69c}$$

The norm N_n is given by

$$N_n = \sum_{j=1}^{M} \frac{k_j}{\alpha_j} \int_{x_j}^{x_{j+1}} x'^p \psi_{jn}^2(x') dx' \tag{8-69d}$$

where $\psi_{in}(x)$ and $\psi_{jn}(x)$ are the eigenfunctions, the β_n values are the eigenvalues of the eigenvalue problem (8-68), $G_{ij}(x, t|x', \tau)|_{\tau=0}$ is the composite medium Green's function evaluated at $\tau = 0$, and $G_{ij}(x, t|x', \tau)$ is the composite medium Green's function.

The function $G_{ij}(x, t|x', \tau)|_{\tau=0}$ is obtainable by rearranging the solution given by equations (8-29), of the homogeneous problem defined by equations (8-15)–(8-17), in the form

$$T_i(x, t) = \sum_{j=1}^{M} \int_{x_j}^{x_{j+1}} x'^p \left[\sum_{n=1}^{\infty} e^{-\beta_n^2 t} \frac{1}{N_n} \frac{k_j}{\alpha_j} \psi_{in}(x) \psi_{jn}(x') \right] F_j(x') dx' \qquad \text{(8-70a)}$$

where x'^p is the Sturm–Liouville weight function with $p = 0, 1$ and 2 for slab, cylinder, and sphere, respectively. Then, the function inside the bracket in equation (8-70a) is $G_{ij}(x, t|x', \tau)|_{\tau=0}$, that is

$$G_{ij}(x, t|x', \tau)|_{\tau=0} = \sum_{n=1}^{\infty} e^{-\beta_n^2 t} \frac{1}{N_n} \frac{k_j}{\alpha_j} \psi_{in}(x) \psi_{jn}(x') \qquad \text{(8-70b)}$$

and the Green's function is obtained by replacing t by $(t - \tau)$ in this expression:

$$G_{ij}(x, t|x', \tau) = \sum_{n=1}^{\infty} e^{-\beta_n^2 (t-\tau)} \frac{1}{N_n} \frac{k_j}{\alpha_j} \psi_{in}(x) \psi_{jn}(x') \qquad \text{(8-70c)}$$

We now illustrate the use of Green's function approach for developing solutions for the nonhomogeneous transient heat conduction problems of composite medium with specific examples. In order to alleviate the details of developing solutions for the corresponding homogeneous problems, we have chosen the examples from those considered in the previous sections for which solutions are already available for the homogeneous part.

Example 8-5

A two-layer solid cylinder contains an inner region $0 \leqslant r \leqslant a$ and an outer region $a \leqslant r \leqslant b$ that are in perfect thermal contact. Initially, the inner and outer regions are at temperatures $F_1(r)$ and $F_2(r)$, respectively. For times $t > 0$, heat is generated in the inner and outer regions at rates $g_1(r, t)$ and $g_2(r, t)$ W/m^3, respectively, while the heat is dissipated by convection from the outer boundary surface at $r = b$ into a medium at zero temperature. Obtain an expression for the temperature distribution in the cylinder for times $t > 0$.

Solution. The mathematical formulation of this problem is given as

$$\alpha_1 \frac{1}{r} \frac{\partial}{\partial r} \left(r \frac{\partial T_1}{\partial r} \right) + \frac{\alpha_1}{k_1} g_1(r, t) = \frac{\partial T_1(r, t)}{\partial t} \quad \text{in} \quad 0 < r < a, \quad t > 0 \qquad \text{(8-71a)}$$

$$\alpha_2 \frac{1}{r} \frac{\partial}{\partial r} \left(r \frac{\partial T_2}{\partial r} \right) + \frac{\alpha_2}{k_2} g_2(r, t) = \frac{\partial T_2(r, t)}{\partial t} \quad \text{in} \quad a < r < b, \quad t > 0 \qquad \text{(8-71b)}$$

subject to the boundary conditions

$$T_1(r,t) = T_2(r,t) \quad \text{at} \quad r=a, \quad t>0 \tag{8-71c}$$

$$k_1 \frac{\partial T_1}{\partial r} = k_2 \frac{\partial T_2}{\partial r} \quad \text{at} \quad r=a, \quad t>0 \tag{8-71d}$$

$$k_2^* \frac{\partial T_2}{\partial r} + h_3^* T_2 = 0 \quad \text{at} \quad r=b, \quad t>0 \tag{8-71e}$$

and the initial conditions

$$T_1(r,t) = F_1(r) \quad \text{for} \quad t=0 \ \text{ in } \ 0 \leqslant r < a \tag{8-71f}$$

$$T_2(r,t) = F_2(r) \quad \text{for} \quad t=0 \ \text{ in } \ a < r < b \tag{8-71g}$$

The solution of this problem is written in terms of the Green's function, according to the general solution given by equations (8-69), in the form

$$\begin{aligned} T_i(r,t) = &\int_{r'=0}^{a} r'[G_{i1}(r,t|r',\tau)]_{\tau=0} \cdot F_1(r')dr' + \int_{r'=a}^{b} r'[G_{i2}(r,t|r',\tau)]_{\tau=0} F_2(r')dr' \\ &+ \int_{\tau=0}^{t} d\tau \left[\int_{r'=0}^{a} r' G_{i1}(r,t|r',\tau) \cdot g_1(r',\tau)dr' \right. \\ &+ \left. \int_{r'=a}^{b} r' G_{i2}(r,t|r',\tau) g_2(r',\tau)dr' \right], \qquad i=1,2 \end{aligned} \tag{8-72a}$$

where the Green's function $G_{ij}(r,t|r',\tau)$ is obtainable from the solution of the homogeneous version of the problem. The homogeneous version of this problem is already considered in Example 8-3; hence the $G(r,t|r',\tau)|_{\tau=0}$ is obtainable from equation (8-55b) and $G(r,t|r',\tau)$ is obtained by replacing in the expression, $G(r,t|r',\tau)|_{\tau=0}$, t by $(t-\tau)$. Thus the Green's functions become

$$G_{ij}(r,t|r',\tau) = \sum_{n=1}^{\infty} e^{-\beta_n^2(t-\tau)} \frac{1}{N_n} \frac{k_j}{\alpha_j} \psi_{in}(r) \psi_{jn}(r') \tag{8-72b}$$

$$G_{ij}(r,t|r',\tau)|_{\tau=0} = \sum_{n=1}^{\infty} e^{-\beta_n^2 t} \frac{1}{N_n} \frac{k_j}{\alpha_j} \psi_{in}(r) \psi_{jn}(r') \tag{8-72c}$$

where the norm N_n is obtained from equation (8-54b) as

$$N_n = \frac{k_1}{\alpha_1} \int_0^a r' \psi_{1n}^2(r')dr' + \frac{k_2}{\alpha_2} \int_a^b r' \psi_{2n}^2(r')dr' \tag{8-72d}$$

The eigenfunctions $\psi_{1n}(r)$ and $\psi_{2n}(r)$ are obtained from equations (8-54c) and (8-54d), respectively:

$$\psi_{1n}(r) = J_0\left(\frac{\beta_n}{\sqrt{\alpha_1}} r\right) \tag{8-72e}$$

$$\psi_{2n}(r) = A_{2n} J_0\left(\frac{\beta_n}{\sqrt{\alpha_2}} r\right) + B_{2n} Y_0\left(\frac{\beta_n}{\sqrt{\alpha_2}} r\right) \tag{8-72f}$$

The coefficients A_{2n} and B_{2n} are given by equations (8-52a) and (8-52b), respectively. The eigenvalues β_n are the roots of the transcendental equation (8-53).

Example 8-6

In a two-layer slab, the first $(0 < x < a)$ and the second $(a < x < b)$ layers are in perfect thermal contact. Initially, the first layer is at temperature $F_1(x)$ and the second layer is at temperature $F_2(x)$. For times $t > 0$, the boundary at $x = 0$ is kept at zero temperature, the boundary at $x = b$ dissipates heat by convection into a medium at zero temperature, while heat is generated in the first layer at a rate of $g_1(x, t)$ W/m^3. Obtain an expression for the temperature distribution in the medium for times $t > 0$.

Solution. The mathematical formulation of this problem is given as

$$\alpha_1 \frac{\partial^2 T_1}{\partial x^2} + \frac{\alpha_1}{k_1} g_1(x, t) = \frac{\partial T_1(x, t)}{\partial t} \quad \text{in} \quad 0 < x < a, \quad t > 0 \tag{8-73a}$$

$$\alpha_2 \frac{\partial^2 T_2}{\partial x^2} = \frac{\partial T_2(x, t)}{\partial t} \quad \text{in} \quad a < x < b, \quad t > 0 \tag{8-73b}$$

Subject to the boundary conditions

$$T_1(x, t) = 0 \quad \text{at} \quad x = 0, \quad t > 0 \tag{8-73c}$$

$$T_1(x, t) = T_2(x, t) \quad \text{at} \quad x = a, \quad t > 0 \tag{8-73d}$$

$$k_1 \frac{\partial T_1}{\partial x} = k_2 \frac{\partial T_2}{\partial x} \quad \text{at} \quad x = a, \quad t > 0 \tag{8-73e}$$

$$k_2^* \frac{\partial T_2}{\partial x} + h_3^* T_2 = 0 \quad \text{at} \quad x = b, \quad t > 0 \tag{8-73f}$$

and the initial conditions

$$T_1(x,t) = F_1(x) \qquad \text{for} \qquad t=0, \quad \text{in} \qquad 0<x<a \tag{8-73g}$$

$$T_2(x,t) = F_2(x) \qquad \text{for} \qquad t=0, \quad \text{in} \qquad a<x<b \tag{8-73h}$$

This heat conduction problem is a special case of the general problem given by equations (8-67). Therefore, its solution is immediately obtainable in terms of the Green's functions from the general solution (8-69) as

$$\begin{aligned} T_i(x,t) = {} & \int_{x'=0}^{a} [G_{i1}(x,t|x',\tau)]_{\tau=0} F_1(x')dx' + \int_{x'=a}^{b} [G_{i2}(x,t|x',\tau)]_{\tau=0} F_2(x')dx' \\ & + \int_{\tau=0}^{t} d\tau \int_{x'=0}^{a} G_{i1}(x,t|x',\tau) \frac{\alpha_1}{k_1} g_1(x',\tau)dx', \qquad i=1,2 \end{aligned} \tag{8-74}$$

where the Green's function is obtainable from the solution of the homogeneous version of the heat conduction problem given by equations (8-73). Actually, the homogeneous version of this problem is exactly the same as that considered in Example 8-1 given by equations (8-8). Therefore, the desired Green's function is obtainable from the result given by equation (8-66b) by replacing t by $(t-\tau)$ in this expression. We find

$$G_{ij}(x,t|x',\tau) = \sum_{n=1}^{\infty} e^{-\beta_n^2(t-\tau)} \frac{1}{N_n} \frac{k_j}{\alpha_j} \psi_{in}(x)\psi_{jn}(x') \tag{8-75a}$$

where the norm N_n is obtained from equation (8-65b) as

$$N_n = \frac{k_1}{\alpha_1} \int_0^a \psi_{1n}^2(x')dx' + \frac{k_2}{\alpha_2} \int_a^b \psi_{2n}^2(x')dx' \tag{8-75b}$$

The eigenfunctions $\psi_{1n}(x)$ and $\psi_{2n}(x)$ are obtained from equations (8-65c) and (8-65d), respectively, as

$$\psi_{1n}(x) = \sin\left(\frac{\beta_n}{\sqrt{\alpha_1}} x\right) \tag{8-75c}$$

$$\psi_{2n}(x) = A_{2n} \sin\left(\frac{\beta_n}{\sqrt{\alpha_2}} x\right) + B_{2n} \cos\left(\frac{\beta_n}{\sqrt{\alpha_2}} x\right) \tag{8-75d}$$

The coefficients A_{2n} and B_{2n} are given by equations (8-63) and the eigenvalues β_n are the positive roots of the transcendental equation (8-64).

8-7 USE OF LAPLACE TRANSFORM FOR SOLVING SEMIINFINITE AND INFINITE MEDIUM PROBLEMS

The Laplace transform technique is convenient for the solution of composite medium problems involving regions of semiinfinite or infinite in extent. In this approach, the partial derivatives with respect to time is removed by the application of the Laplace transform, the resulting system of ordinary differential equations is solved and the transforms of temperatures are inverted; but the principal difficulty lies in the inversion of the resulting transform. In this section, we examine the solution of two-layer composite medium problems of semiinfinite and infinite extend by the Laplace transform technique and consider only those problems for which the inversion of the transforms can be performed by using the standard Laplace transform inversion tables.

Example 8-7

Two semiinfinite regions, $x > 0$ and $x < 0$, illustrated in Fig. 8-6 are in perfect thermal contact. Initially, the region 1 (i.e., $x > 0$) is at a uniform temperature T_0, and the region 2 (i.e., $x < 0$) is at zero temperature. Obtain an expression for the temperature distribution in the medium for times $t > 0$.

Solution. For convenience in the analysis, we define a dimensionless temperature $\theta_i(x, t)$ as

$$\theta_i(x,t) = \frac{T_i(x,t)}{T_0} \qquad i = 1, 2 \tag{8-76}$$

Then, the mathematical formulation of the problem, in terms of $\theta_i(x, t)$, is given as

$$\frac{\partial^2 \theta_1}{\partial x^2} = \frac{1}{\alpha_1}\frac{\partial \theta_1(x,t)}{\partial t} \qquad \text{in} \qquad x > 0, \quad t > 0 \tag{8-77a}$$

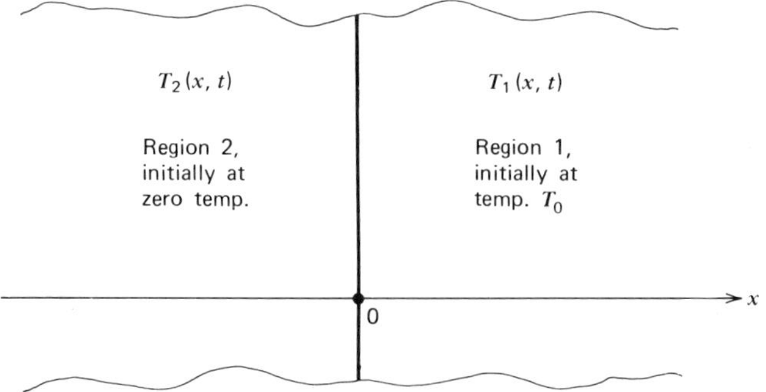

Fig. 8-6 Two semiinfinite regions in perfect thermal contact.

$$\frac{\partial^2 \theta_2}{\partial x^2} = \frac{1}{\alpha_2} \frac{\partial \theta_2(x,t)}{\partial t} \qquad \text{in} \qquad x < 0, \quad t > 0 \tag{8-77b}$$

Subject to the boundary conditions

$$\theta_1(x,t)|_{x=0^+} = \theta_2(x,t)|_{x=0^-} \qquad t > 0 \tag{8-78a}$$

$$-k_1 \left.\frac{\partial \theta_1}{\partial x}\right|_{x=0^+} = k_2 \left.\frac{\partial \theta_2}{\partial x}\right|_{x=0^-} \qquad t > 0 \tag{8-78b}$$

$$\left.\frac{\partial \theta_1}{\partial x}\right|_{x\to\infty} = \left.\frac{\partial \theta_2}{\partial x}\right|_{x\to-\infty} = 0 \qquad t > 0 \tag{8-78c}$$

and the initial conditions

$$\theta_1(x,t) = 1 \qquad \text{for} \qquad t = 0, \quad x > 0 \tag{8-78d}$$

$$\theta_2(x,t) = 0 \qquad \text{for} \qquad t = 0, \quad x < 0 \tag{8-78e}$$

The Laplace transform of equations (8-77) is

$$\frac{d^2 \bar{\theta}_1(x,s)}{dx^2} = \frac{1}{\alpha_1}[s\bar{\theta}_1(x,s) - 1] \qquad \text{in} \qquad x > 0 \tag{8-79a}$$

$$\frac{d^2 \bar{\theta}_2(x,s)}{dx^2} = \frac{1}{\alpha_2} s\bar{\theta}_2(x,s) \qquad \text{in} \qquad x < 0 \tag{8-79b}$$

and the Laplace transform of the boundary conditions gives

$$\bar{\theta}_1(0^+,s) = \bar{\theta}_2(0^-,s) \tag{8-80a}$$

$$-k_1 \left.\frac{d\bar{\theta}_1}{dx}\right|_{x=0^+} = k_2 \left.\frac{d\bar{\theta}_2}{dx}\right|_{x=0^-} \tag{8-80b}$$

$$\left.\frac{d\bar{\theta}_1}{dx}\right|_{x\to\infty} = \left.\frac{d\bar{\theta}_2}{dx}\right|_{x\to-\infty} = 0 \tag{8-80c}$$

The solutions of equations (8-79) subject to the boundary conditions (8-80) are

$$\bar{\theta}_1(x,s) = \frac{1}{s} - \frac{1}{1+\beta}\frac{1}{s} e^{-(s/\alpha_1)^{1/2} x} \qquad \text{for} \qquad x > 0 \tag{8-81a}$$

$$\bar{\theta}_2(x,s) = \frac{\beta}{1+\beta}\frac{1}{s} e^{-(s/\alpha_2)^{1/2}|x|} \qquad \text{for} \qquad x < 0 \tag{8-81b}$$

where

$$\beta \equiv \frac{k_1}{k_2}\left(\frac{\alpha_2}{\alpha_1}\right)^{1/2} \tag{8-81c}$$

These transforms can be inverted using the Laplace transform Table 7-1, cases 1 and 42. The resulting expressions for the temperature distribution in the medium become

$$\theta_1(x,t) \equiv \frac{T_1(x,t)}{T_0} = 1 - \frac{1}{1+\beta}\operatorname{erfc}\left(\frac{x}{2\sqrt{\alpha_1 t}}\right), \qquad \text{for} \qquad x > 0 \tag{8-82a}$$

$$\theta_2(x,t) \equiv \frac{T_2(x,t)}{T_0} = \frac{\beta}{1+\beta}\operatorname{erfc}\left(\frac{|x|}{2\sqrt{\alpha_2 t}}\right), \qquad \text{for} \qquad x < 0 \tag{8-82b}$$

Example 8-8

A two-layer medium illustrated in Fig. 8-7 is composed of region $1, 0 < x < L$, and region $2, x > L$, which are in perfect thermal contact. Initially, region 1 is at a uniform temperature T_0 and region 2 is at zero temperature. For times $t > 0$, the boundary surface at $x = 0$ is kept insulated. Obtain an expression for the temperature distribution in the medium for times $t > 0$

Solution. We define a dimensionless temperature $\theta_i(x,t)$ as

$$\theta_i(x,t) = \frac{T_i(x,t)}{T_0} \qquad i = 1, 2 \tag{8-83}$$

Then, the mathematical formulation of the problem in terms of $\theta_i(x,t)$ is

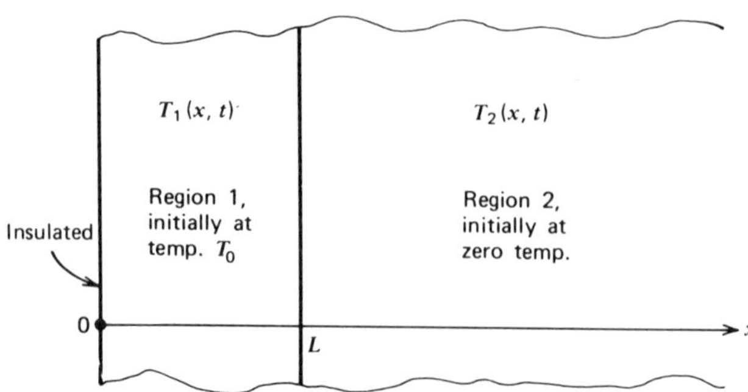

Fig. 8-7 A finite region and a semiinfinite region in perfect thermal contact.

given as

$$\frac{\partial^2 \theta_1}{\partial x^2} = \frac{1}{\alpha_1}\frac{\partial \theta_1(x,t)}{\partial t} \quad \text{in} \quad 0 < x < L, \quad t > 0 \tag{8-84a}$$

$$\frac{\partial^2 \theta_2}{\partial x^2} = \frac{1}{\alpha_2}\frac{\partial \theta_2(x,t)}{\partial t} \quad \text{in} \quad x > L, \quad t > 0 \tag{8-84b}$$

subject to the boundary conditions

$$\frac{\partial \theta_1}{\partial x} = 0 \quad \text{at} \quad x = 0, \quad t > 0 \tag{8-85a}$$

$$\theta_1(x,t) = \theta_2(x,t) \quad \text{at} \quad x = L, \quad t > 0 \tag{8-85b}$$

$$k_1 \frac{\partial \theta_1}{\partial x} = k_2 \frac{\partial \theta_2}{\partial x} \quad \text{at} \quad x = L, \quad t > 0 \tag{8-85c}$$

$$\theta_2(x,t) \to 0 \quad \text{as} \quad x \to \infty, \quad t > 0 \tag{8-85d}$$

and the initial conditions

$$\theta_1(x,t) = 1 \quad \text{for} \quad t = 0, \quad \text{in} \quad 0 < x < L \tag{8-85e}$$

$$\theta_2(x,t) = 0 \quad \text{for} \quad t = 0, \quad \text{in} \quad x > L \tag{8-85f}$$

The Laplace transform of equations (8-84) is

$$\frac{d^2 \bar{\theta}_1(x,s)}{dx^2} = \frac{1}{\alpha_1}[s\bar{\theta}_1(x,s) - 1] \quad \text{in} \quad 0 < x < L \tag{8-86a}$$

$$\frac{d^2 \bar{\theta}_2(x,s)}{dx^2} = \frac{1}{\alpha_2} s\bar{\theta}_2(x,s) \quad \text{in} \quad x > L \tag{8-86b}$$

The Laplace transform of the boundary conditions gives

$$\frac{d\bar{\theta}_1}{dx} = 0 \quad \text{at} \quad x = 0 \tag{8-87a}$$

$$\bar{\theta}_1 = \bar{\theta}_2 \quad \text{at} \quad x = L \tag{8-87b}$$

$$k_1 \frac{d\bar{\theta}_1}{dx} = k_2 \frac{d\bar{\theta}_2}{dx} \quad \text{at} \quad x = L \tag{8-87c}$$

$$\bar{\theta}_2 \to 0 \quad \text{as} \quad x \to \infty \tag{8-87d}$$

The solution of equation (8-86a) that satisfies the boundary condition (8-87a) is taken in the form

$$\bar{\theta}_1(x,s) = \frac{1}{s} + A\cosh\left(x\sqrt{\frac{s}{\alpha_1}}\right) \quad \text{in} \quad 0 \leqslant x < L \tag{8-88a}$$

and the solution of equation (8-86b) satisfying the boundary condition (8-87d) as

$$\bar{\theta}_2(x,s) = Be^{-x(\sqrt{s/\alpha_2})} \quad \text{in} \quad x > L \tag{8-88b}$$

The constants A and B are determined by the application of the remaining boundary conditions (8-87b,c); we find

$$A = -\frac{1-\gamma}{s}\frac{e^{-\sigma L}}{1-\gamma e^{-2\sigma L}} \tag{8-89a}$$

$$B = \frac{1+\gamma}{2s}e^{\sigma\mu L}\frac{1-e^{-2\sigma L}}{1-\gamma e^{-2\sigma L}} \tag{8-89b}$$

where

$$\sigma \equiv \sqrt{\frac{s}{\alpha_1}}, \qquad \mu \equiv \sqrt{\frac{\alpha_1}{\alpha_2}} \tag{8-89c}$$

$$\gamma \equiv \frac{\beta-1}{\beta+1}, \qquad \beta \equiv \frac{k_1}{k_2}\frac{1}{\mu} \tag{8-89d}$$

Introducing equations (8-89) into (8-88) we obtain

$$\bar{\theta}_1(x,s) = \frac{1}{s} - \frac{1-\gamma}{2s}\frac{e^{-\sigma(L-x)} + e^{-\sigma(L+x)}}{1-\gamma e^{-2\sigma L}}, \quad \text{in} \quad 0 < x < L \tag{8-90a}$$

$$\bar{\theta}_2(x,s) = \frac{1+\gamma}{2s}\frac{e^{-\sigma\mu(x-L)} - e^{-\sigma(2L+\mu x-\mu L)}}{1-\gamma e^{-2\sigma L}}, \quad x > L \tag{8-90b}$$

Here we note that $|\gamma| < 1$. Therefore, the term $[1-\gamma\exp(-2\sigma L)]^{-1}$ can be expanded as a binomial series, and equations (8-90) become

$$\bar{\theta}_1(x,s) = \frac{1}{s} - \frac{1-\gamma}{2}\sum_{n=0}^{\infty}\gamma^n\left[\frac{e^{-\sigma[(2n+1)L-x]}}{s} + \frac{e^{-\sigma[(2n+1)L+x]}}{s}\right],$$

$$\text{in} \quad 0 < x < L \tag{8-91a}$$

$$\bar{\theta}_2(x,s) = \frac{1+\gamma}{2}\sum_{n=0}^{\infty}\gamma^n\left[\frac{e^{-\sigma[2nL+\mu(x-L)]}}{s} - \frac{e^{-\sigma[(2n+2)L+\mu(x-L)]}}{s}\right], \quad \text{in} \quad x > L \tag{8-91b}$$

The inversion of these transforms are available in the Laplace transform Table 7-1 as cases 1 and 42. After the inversion, the temperature distribution in the medium becomes

$$\theta_1(x,t) \equiv \frac{T_1(x,t)}{T_0} = 1 - \frac{1-\gamma}{2}\sum_{n=0}^{\infty}\gamma^n\left\{\operatorname{erfc}\left[\frac{(n+1)L-x}{2\sqrt{\alpha_1 t}}\right] + \operatorname{erfc}\left[\frac{(2n+1)L+x}{2\sqrt{\alpha_1 t}}\right]\right. \quad \text{in} \quad 0 < x < L \tag{8-92a}$$

$$\theta_2(x,t) \equiv \frac{T_2(x,t)}{T_0} = \frac{1+\gamma}{2}\sum_{n=0}^{\infty}\gamma^n\left\{\operatorname{erfc}\left[\frac{2nL+\mu(x-L)}{2\sqrt{\alpha_1 t}}\right] - \operatorname{erfc}\left[\frac{(2n+2)L+\mu(x-L)}{2\sqrt{\alpha_1 t}}\right]\right. \quad \text{in} \quad x > L \tag{8-92b}$$

where γ and μ are defined by equations (8-89).

REFERENCES

1. V. Vodicka, *Schweizer Arch.* **10**, 297–304, 1950.
2. V. Vodicka, *Math. Nach.* **14**, 47–55, 1955.
3. P. E. Bulavin and V. M. Kascheev, *Int. Chem. Eng.* **1**, 112–115, 1965.
4. C. W. Tittle, *J. Appl. Phys.* **36**, 1486–1488, 1965.
5. C. W. Tittle and V. L. Robinson, Analytical Solution of Conduction Problems in Composite Media, ASME Paper 65-WA-HT-52, 1965.
6. H. L. Beach, The Application of the Orthogonal Expansion Technique to Conduction Heat Transfer Problems in Multilayer Cylinders, M. S. Thesis, Mech. and Aerospace Eng. Dept., N.C. State University, Releigh, N.C. 27607, 1967.
7. C. H. Moore, Heat Transfer Across Surfaces in Contact: Studies of Transients in One-Dimensional Composite Systems, Ph.D. Dissertation, Mechanical Eng. Dept., Southern Methodist University, Dallas, Texas, 1967.
8. M. N. Özişik, *Boundary Value Problems of Heat Conduction*, International Textbook Co., Scranton, Pa., 1968; Dover Publications, New York, 1989.
9. M. H. Cobble, *J. Franklin Inst.* **290** (5), 453–465, 1970.
10. G. P. Mulholland and M. N. Cobble, *Int. J. Heat Mass Transfer* **15**, 147–160, 1972.
11. C. A. Chase, D. Gidaspow, and R. E. Peck, Diffusion of Heat and Mass in Porous Medium with Bulk Flow: Part I. Solution by Green's Functions, Chem. Eng. Progr. Symposium Ser. No. 92, Vol. 65, 91–109, 1969.

12. B. S. Baker, D. Gidaspow, and D. Wasan, in *Advances in Electrochemistry and Electrochemical Engineering*, Wiley-Interscience, New York, 1971, pp. 63–156.
13. M. D. Mikhailov and M. N. Özişik, *Unified Analysis and Solutions of Heat and Mass Diffusion*, Wiley, New York, 1984.
14. J. Crank, *The Mathematics of Diffusion*, 2nd ed., Clarendon Press, London, 1975.
15. H. S. Carslaw and J. C. Jaeger, *Conduction of Heat in Solids*, Clarendon Press, London, 1959.
16. V. S. Arpaci, *Conduction Heat Transfer*, Addison-Wesley, Reading, Mass., 1966.
17. A. V. Luikov, *Analytical Heat Diffusion Theory*, Academic Press, New York, 1968.
18. N. Y. Ölçer, *Ingenieur-Arch.* **36**, 285–293, 1968.
19. N. Y. Ölçer, *Quart. Appl. Math.* **26**, 355–371, 1968.
20. N. Y. Ölçer, *Nucl. Eng. Design* **7**, 92–112, 1968.
21. Kanae Senda, A Family of Integral Transforms and Some Applications to Physical Problems, Technology Reports of the Osaka University, Osaka, Japan, No. 823, Vol. 18, 261–286, 1968.
22. J. D. Lockwood and G. P. Mulholland, *J. Heat Transfer* **95c**, 487–491, 1973.
23. M. D. Mikhailov, *Int. J. Eng. Sci.* **11**, 235–241, 1973.
24. M. D. Mikhailov, *Int. J. Heat Mass Transfer* **16**, 2155–2164, 1973.
25. Y. Yener and M. N. Özişik, *Proceedings of the 5th International Heat Transfer Conference*, Tokyo, Sept. 1974.
26. J. Padovan, *AIAA J.* **12**, 1158–1160, 1974.
27. M. Ben-Amoz, *Int. J. Eng. Sci.*, **12**, 633, 1974.
28. G. Horvay, R. Mani, M. A. Veluswami, and G. E. Zinsmeister, *J. Heat Transfer* **95**, 309, 1973.
29. M. Ben-Amoz, *ZAMP*, **27**, 335–345, 1976.
30. S. C. Huang and Y. P. Chang, *J. Heat Transfer* **102**, 742–748, 1980.
31. M. D. Mikhailov and M. N. Özişik, *Int. J. Heat Mass Transfer* **28**, 1039–1045, 1985.
32. J. Baker-Jarvis and R. Inguva, *J. Heat Transfer* **107**, 39–43, 1985.
33. S. C. Huang and J. P. Chang, *J. Heat Transfer* **102**, 742–748, 1980.
36. A. Haji-Sheikh and M. Mashena, *J. Heat Transfer*, **109**, 551–556, 1987.
37. M. D. Mikhailov and M. N. Özişik, *Int. J. Heat Mass Transfer* **29**, 340–342, 1988.
38. A. Haji-Sheikh and J. V. Beck, *J. Heat Transfer* **112**, 28–34, 1990.

PROBLEMS

8-1 A two-layer solid cylinder contains the inner region, $0 \leqslant r \leqslant a$, and the outer region, $a \leqslant r \leqslant b$, which are in perfect thermal contact. Initially, the inner region is at temperature $F_1(r)$ and the outer region at temperature $F_2(r)$. For times $t > 0$, the boundary surface at $r = b$ is kept at zero temperature. Obtain an expression for the temperature distribution in the medium. Also, express the solution in terms of Green's function and determine the Green's function for this problem.

8-2 A two-layer slab consists of the first layer $0 \leqslant x \leqslant a$ and the second layer $a \leqslant x \leqslant b$, which are in perfect thermal contact. Initially, the first region is at temperature $F_1(x)$ and the second region is at temperature $F_2(x)$. For times $t > 0$, the outer boundaries at $x = 0$ and $x = b$ are kept at zero temperatures. Obtain an expression for the temperature distribution in the medium. Also determine the Green's function for this problem.

8-3 A two-layer hollow cylinder consists of the first layer $a \leqslant r \leqslant b$ and the second layer $b \leqslant r \leqslant c$, which are in perfect thermal contact. Initially the first region is at temperature $F_1(r)$ and the second region at temperature $F_2(r)$. For times $t > 0$, the outer boundaries at $r = a$ and $r = c$ are kept at zero temperature. Obtain an expression for the temperature distribution in the medium. Also, determine the Green's function for this problem.

8-4 Repeat Problem 8-2 for the case when boundary surface at $x = 0$ is kept insulated and the boundary surface at $x = b$ dissipates heat by convection into an environment at zero temperature. Also determine the Green's function for this problem.

8-5 Repeat Problem 8-3 for the case when the boundary surface at $r = a$ is kept insulated and the boundary surface at $r = c$ dissipates heat by convection into an environment at zero temperature. Also determine the Green's function for this problem.

8-6 A two-layer solid cylinder contains the inner region, $0 \leqslant r \leqslant a$, and the outer region, $a \leqslant r \leqslant b$, which are in perfect thermal contact. Initially the inner region is at temperature $F_1(r)$, the outer region at temperature $F_2(r)$. For times $t > 0$, heat is generated in the inner region at a rate of $g_1(r, t)$ W/m^3 while the boundary surface at $r = b$ is kept at temperature $f(t)$. By following an approach discussed in Example 8-5 transform this problem into a one with homogeneous boundary condition at $r = b$.

8-7 A two-layer slab consists of the first layer $0 \leqslant x \leqslant a$ and the second layer $a \leqslant x \leqslant b$, which are in perfect thermal contact. Initially the first region is at temperature $F_1(x)$ and the second region at temperature $F_2(x)$. For times $t > 0$, heat is generated in the first region at a rate of $g_1(x, t)$, W/m^3, and in the second region at a rate of $g_2(x, t)$, W/m^3, while the outer boundary surfaces at $x = 0$ and $x = b$ are kept at temperatures $f_1(t)$ and $f_2(t)$ respectively. Split up this problem into a steady-state problem and a time-dependent problem with heat generation, subject to homogeneous boundary conditions by following the procedure discussed in Section 8-2.

8-8 Solve Problem 8-1 with the additional condition that heat is generated in the inner region, $0 \leqslant r \leqslant a$, at a rate of $g_1(r, t)$, W/m^3. Utilize the Green's function constructed in Problem 8-1 to solve this nonhomogeneous problem.

8-9 Solve Problem 8-2 with the additional condition that heat is generated in the first and second layers at a rate of $g_1(x, t)$ and $g_2(x, t)$, W/m^3 respectively. Utilize the Green's function constructed in Problem 8-2 to solve this problem.

8-10 Solve Problem 8-3 with the additional condition that heat is generated in the first and second regions at a rate of $g_1(r, t)$ and $g_2(r, t)$, W/m^3, respectively.

9

APPROXIMATE ANALYTIC METHODS

Analytic solutions, whether exact or approximate, are always useful in engineering analysis, because they provide a better insight into the physical significance of various parameters affecting the problem. When exact analytic solutions are impossible or too difficult to obtain or the resulting analytic solutions are too complicated for computational purposes, approximate analytic solutions provide a powerful alternative approach to handle such problems.

There are numerous approximate analytic methods for solving the partial differential equations governing the engineering problems. In this chapter we present the *integral method*, the *Galerkin method*, and the method of *partial integration*, and illustrate their applications with representative examples. The accuracy of an approximate solution cannot be assessed unless the results are compared with the exact solution. Therefore, in order to give some idea of the accuracy of the approximate analysis, simple problems for which exact solutions are available are first solved with the approximate methods and the results are compared with the exact solutions. The applications to the solution of more complicated, nonlinear problems are then considered.

9-1 THE INTEGRAL METHOD—BASIC CONCEPTS

The use of integral method for the solution of partial differential equations dates back to von Kármán and Pohlhausen, who applied the method for the approximate analysis of boundary-layer momentum and energy equations of fluid mechanics [1]. Landahl [2] used it in the field of biophysics to solve the diffusion equation in connection with the spread of a concentrate. Merk [3] applied this approach to solve a two-dimensional steady-state melting problem, and Goodman

[4, 5] used it for the solution of a one-dimensional transient melting problem. Since then, this method has been applied in the solution of various types of one-dimensional transient heat conduction problems [6–16], melting and solidification problems [16–25], and heat and momentum transfer problems involving melting of ice in seawater, melting and extrusion of polymers [26–29].

The method is simple, straightforward, and easily applicable to both linear and nonlinear one-dimensional transient boundary value problems of heat conduction for certain boundary conditions. The results are approximate, but several solutions obtained with this method when compared with the exact solutions have confirmed that the accuracy is generally acceptable for many engineering applications. In this section we first present the basic concepts involved in the application of this method by solving a simple transient heat conduction problem for a semiinfinite medium. The method is then applied to the solution of various one-dimensional, time-dependent heat conduction problems. The application to the solution of melting, solidification, and ablation problems is considered in a later chapter on moving-boundary problems (Chapter 11).

When the differential equation of heat conduction is solved exactly in a given region subject to specified boundary and initial conditions, the resulting solution is satisfied at each point over the considered region; but with the integral method the solution is satisfied only on the average over the region. We now summarize the basic steps in the analysis with the integral method when it is applied to the solution of one-dimensional, transient heat-conduction problem in a semiinfinite medium subject to some prescribed boundary and uniform initial conditions but no heat generation.

1. The differential equation of heat conduction is integrated over a phenomenologic distance $\delta(t)$, called the *thermal layer* in order to remove from the differential equation the derivative with respect to the space variable. The thermal layer is defined as the distance beyond which, for practical purposes, there is no heat flow; hence the initial temperature distribution remains unaffected beyond $\delta(t)$. The resulting equation is called the *energy integral equation* (i.e., it is also called the *heat-balance integral*).

2. A suitable profile is chosen for the temperature distribution over the thermal layer. A polynomial profile is generally preferred for this purpose; experience has shown that there is no significant improvement in the accuracy of the solution to choose a polynomial greater than the fourth degree. The coefficients in the polynomial are determined in terms of the thermal layer thickness $\delta(t)$ by utilizing the actual (if necessary derived) boundary conditions.

3. When the temperature profile thus constructed is introduced into the energy integral equation and the indicated operations are performed, an ordinary differential equation is obtained for the thermal-layer thickness $\delta(t)$ with time as the independent variable. The solution of this differential equation subject to the appropriate initial condition [i.e., in this case $\delta(t) = 0$ for $t = 0$] gives $\delta(t)$ as a function of time.

4. Once $\delta(t)$ is available from step 3, the temperature distribution $T(x, t)$ is known as a function of time and position in the medium, and the heat flux at the surface is determined. Experience has shown that the method is more accurate for the determination of heat flux than the temperature profile.

9-2 INTEGRAL METHOD—APPLICATION TO LINEAR TRANSIENT HEAT CONDUCTION IN A SEMIINFINITE MEDIUM

To illustrate the mathematical details of the basic steps discussed above for the application of the integral method, in the following example we consider a problem of transient heat conduction in a semiinfinite medium with no energy generation.

Example 9-1

A semiinfinite medium $x \geqslant 0$ is initially at a uniform temperature T_i. For times $t > 0$ the boundary surface is kept at constant temperature T_0 as illustrated in Fig. 9-1. Develop expressions for the temperature distribution and the surface heat flux with the integral method by using a cubic polynomial approximation for the temperature profile.

Solution. The mathematical formulation of this problem is given as

$$\frac{\partial^2 T(x,t)}{\partial x^2} = \frac{1}{\alpha}\frac{\partial T(x,t)}{\partial t} \quad \text{in} \quad x > 0, \quad t > 0 \tag{9-1a}$$

$$T(x,t) = T_0 \quad \text{at} \quad x = 0, \quad t > 0 \tag{9-1b}$$

$$T(x,t) = T_i \quad \text{for} \quad t = 0, \quad \text{in} \quad x \geqslant 0 \tag{9-1c}$$

We solve this problem with the integral method by following the basic steps discussed above.

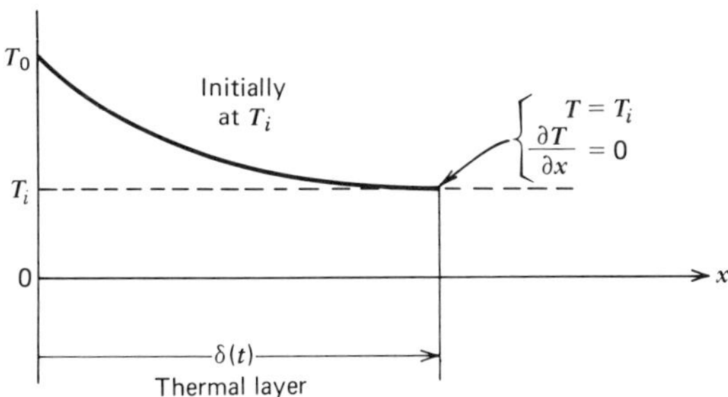

Fig. 9-1 Definition of thermal layer for heat conduction in a semiinfinite region.

1. We integrate equation (9-1a) with respect to the space variable from $x = 0$ to $x = \delta(t)$

$$\left.\frac{\partial T}{\partial x}\right|_{x=\delta(t)} - \left.\frac{\partial T}{\partial x}\right|_{x=0} = \frac{1}{\alpha}\int_{x=0}^{\delta(t)} \frac{\partial T}{\partial t}\,dx \tag{9-2a}$$

when the integral on the right-hand side is performed by the rule of differentiation under the integral sign we obtain

$$\left.\frac{\partial T}{\partial x}\right|_{x=\delta} - \left.\frac{\partial T}{\partial x}\right|_{x=0} = \frac{1}{\alpha}\left[\frac{d}{dt}\left(\int_{x=0}^{\delta} T\,dx\right) - \left.T\right|_{x=\delta}\frac{d\delta}{dt}\right] \tag{9-2b}$$

By the definition of thermal layer as illustrated in Fig. 9-1 we have

$$\left.\frac{\partial T}{\partial x}\right|_{x=\delta} = 0 \qquad \text{and} \qquad \left.T\right|_{x=\delta} = T_i \tag{9-3a}$$

and for convenience in the analysis we define

$$\theta \equiv \int_{x=0}^{\delta(t)} T(x,t)\,dx \tag{9-3b}$$

Introducing equations (9-3) into (9-2b) we obtain

$$-\alpha\left.\frac{\partial T}{\partial x}\right|_{x=0} = \frac{d}{dt}(\theta - T_i\delta) \tag{9-4}$$

which is called the *energy integral equation* for the problem considered here.

2. We choose a cubic polynomial representation for $T(x, t)$ in the form

$$T(x,t) = a + bx + cx^2 + dx^3 \qquad \text{in} \qquad 0 \leqslant x \leqslant \delta(t) \tag{9-5}$$

where the coefficients are in general functions of time. Four conditions are needed to determine these four coefficients in terms of $\delta(t)$. Three of these conditions are obtained from the boundary conditions at $x = 0$ and at the edge of the thermal layer $x = \delta(t)$, as

$$\left.T\right|_{x=0} = T_0, \quad \left.T\right|_{x=\delta} = T_i, \quad \left.\frac{\partial T}{\partial x}\right|_{x=\delta} = 0 \tag{9-6a}$$

The fourth condition may be derived by evaluating the differential equation (9-1a) at $x = 0$ and by making use of the fact that $T = T_0 =$ constant at $x = 0$; then the derivative of temperature with respect to the time vanishes at $x = 0$ and we obtain

$$\left.\frac{\partial^2 T}{\partial x^2}\right|_{x=0} = 0 \tag{9-6b}$$

Clearly, the fourth condition could also be derived by evaluating the differential equation (9-1a) at $x = \delta(t)$ and utilizing the fact that $T = T_i =$ constant, by definition, at $x = \delta$. This matter will be discussed later in the analysis. The application of the four conditions (9-6) to equation (9-5) yields the temperature profile in the form

$$\frac{T(x,t) - T_i}{T_0 - T_i} = 1 - \frac{3}{2}\frac{x}{\delta} + \frac{1}{2}\left(\frac{x}{\delta}\right)^3 \tag{9-7}$$

3. When the temperature profile (9-7) is introduced into the energy integral equation (9-4) and the indicated operations are performed, we obtain the following ordinary differential equation for $\delta(t)$

$$4\alpha = \delta\frac{d\delta}{dt} \qquad \text{for} \qquad t > 0 \tag{9-8a}$$

subject to

$$\delta = 0 \qquad \text{for} \qquad t = 0 \tag{9-8b}$$

The solution of equations (9-8) gives

$$\delta = \sqrt{8\alpha t} \tag{9-9}$$

4. Knowing $\delta(t)$, we determine the temperature distribution $T(x,t)$ according to equation (9-7) and the heat flux $q(0,t)$ at the surface $x = 0$ from its definition

$$q(0,t) = -k\left.\frac{\partial T}{\partial x}\right|_{x=0} = \frac{3k}{2\delta}(T_0 - T_i) \tag{9-10a}$$

where

$$\delta = \sqrt{8\alpha t} \tag{9-10b}$$

Other Profiles

In the foregoing example we considered a cubic polynomial representation for $T(x,t)$ that involved four unknown coefficients and required four conditions for their determination. Three of these conditions given by equation (9-6a) are the *natural conditions* for the problem, and the fourth condition equation (9-6b) is a *derived condition* obtained by evaluating the differential equation at $x = 0$. It is also possible to derive an alternative fourth condition by evaluating the differential equation at $x = \delta$ yielding

$$\left.\frac{\partial^2 T}{\partial x^2}\right|_{x=\delta} = 0 \tag{9-6b'}$$

Therefore, it is also possible to use the above three natural conditions (9-6a) together with the alternative derived condition (9-6b′) to obtain an *alternative cubic temperature profile* in the form

$$\frac{T(x,t) - T_i}{T_0 - T_i} = \left(1 - \frac{x}{\delta}\right)^3 \tag{9-11a}$$

where

$$\delta = \sqrt{24\alpha t} \tag{9-11b}$$

If a fourth-degree polynomial representation is used for $T(x, t)$, the resulting five coefficients are determined by the application of the five conditions given by equations (9-6a), (9-6b), and (9-6b′), and the following temperature profile is obtained

$$\frac{T(x,t) - T_i}{T_0 - T_i} = 1 - 2\left(\frac{x}{\delta}\right) + 2\left(\frac{x}{\delta}\right)^3 - \left(\frac{x}{\delta}\right)^4 \tag{9-12a}$$

where

$$\delta = \sqrt{\frac{40}{3}\alpha t} \tag{9-12b}$$

Comparison with Exact Solution. In the foregoing analysis we developed two different cubic temperature profiles given by equations (9-7) and (9-11) and a fourth-degree profile given by equation (9-12). One can also develop another approximate solution by utilizing a second-degree polynomial representation. The question regarding which one of these approximate solutions is more accurate cannot be answered until each of these solutions are compared with the exact solution of the problem given by

$$\frac{T(x,t) - T_i}{T_0 - T_i} = 1 - \operatorname{erf}\frac{x}{\sqrt{4\alpha t}} \tag{9-13a}$$

Figure 9-2 shows a comparison of these approximate temperature distributions with the exact solution. The agreement is better for small values of the parameter

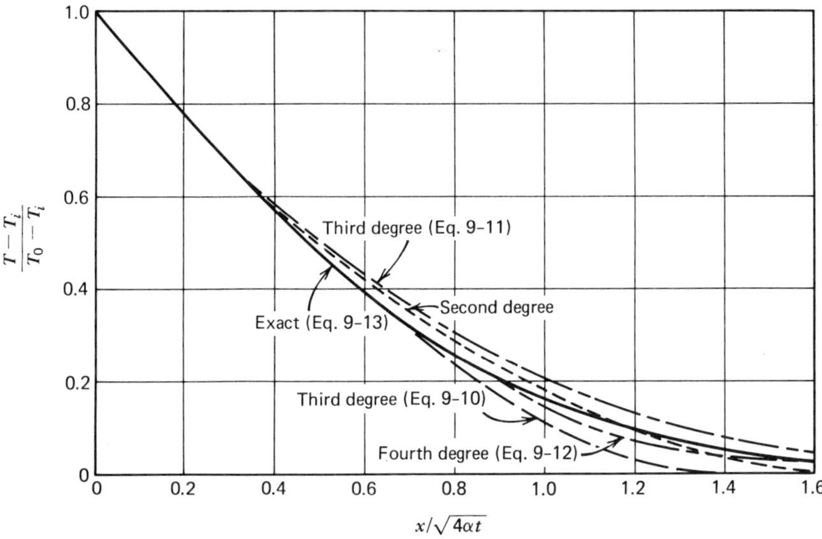

Fig. 9-2 Comparison of exact and approximate solutions for a semiinfinite region.

$x/\sqrt{4\alpha t}$. The fourth-degree polynomial approximation agrees better with the exact solution. The cubic polynomial representation utilizing the condition at $x = 0$ seems to agree with the exact solution better than the one utilizing the condition at $x = \delta$.

The heat flux at the boundary surface $x = 0$ is a quantity of practical interest, and for the various temperature profiles considered above it may be expressed in the form

$$q(t) = -k \left. \frac{\partial T}{\partial x} \right|_{x=0} = C \frac{k(T_0 - T_i)}{\sqrt{\alpha t}} \tag{9-13b}$$

Table 9-1 gives the values of the constant C as calculated from the above exact and approximate solutions. The fourth-degree polynomial approximation represents the heat flux with an error of approximately 3%, which is acceptable for most engineering applications.

Cylindrical and Spherical Symmetry. The use of polynomial representation for temperature, although giving reasonably good results in the rectangular coordinate system, will yield significant error in the problems of cylindrical and spherical symmetry [11]. This is to be expected since the volume into which the heat diffuses does not remain the same for equal increments of r in the cylindrical and spherical coordinate systems. This situation may be remedied by modifying the

TABLE 9-1 Error Involved in the Surface Heat Flux

Temperature Profile	C as Defined by Equation (9-13b)	Percent Error Involved
Exact (equation 9-13)	$\dfrac{1}{\sqrt{\pi}} = 0.565$	0
Cubic approximation (equation 9-7)	$\dfrac{3}{2\sqrt{8}} = 0.530$	6
Cubic approximation (equation 9-11)	$\dfrac{3}{\sqrt{24}} = 0.612$	8
Fourth-degree approximation (equation 9-12)	$\dfrac{2}{\sqrt{\dfrac{40}{3}}} = 0.548$	3

temperature profiles as

$$\text{Cylindrical symmetry:} \qquad T(r,t) = (\text{polynomial in } r)(\ln r) \tag{9-14a}$$

$$\text{Spherical symmetry:} \qquad T(r,t) = \frac{\text{polynomial in } r}{r} \tag{9-14b}$$

Since the problems with spherical symmetry can be transformed into a problem in the rectangular coordinate system as discussed in Chapter 4, one needs to be concerned with such a modification only for the cylindrical symmetry.

Problems with Energy Generation

The integral method is also applicable for the solution of one-dimensional transient heat conduction problems with a uniform energy generation that may be constant or time-dependent over the region. The following example illustrates the application to a problem with energy generation in the medium.

Example 9-2

A semiinfinite region, $x > 0$, is initially at a constant temperature T_i. For times $t > 0$ heat is generated within the solid at a rate of $g(t)$ W/m^3 while the boundary at $x = 0$ is kept at a constant temperature T_0. Obtain an expression for the temperature distribution $T(x,t)$ in the medium using the integral method.

Solution. The mathematical formulation of this problem is given as

$$\alpha \frac{\partial^2 T}{\partial x^2} + \frac{\alpha}{k} g(t) = \frac{\partial T}{\partial t} \qquad \text{in} \qquad x > 0, \quad t > 0 \tag{9-15a}$$

$$T = T_0 \qquad \text{at} \qquad x = 0, \quad t > 0 \tag{9-15b}$$

$$T = T_i \qquad \text{for} \qquad t = 0, \quad x \geqslant 0 \tag{9-15c}$$

The integration of equation (9-15a) from $x = 0$ to $x = \delta(t)$ gives

$$-\alpha \frac{\partial T}{\partial x}\bigg|_{x=0} + \frac{\alpha}{k} g(t)\delta(t) = \frac{d\theta}{dt} - T\bigg|_{x=\delta} \frac{d\delta}{dt} \tag{9-16}$$

where we utilized the condition $dT/dx = 0$ at $x = \delta$, and defined

$$\theta \equiv \int_{x=0}^{\delta(t)} T(x,t)\, dx \tag{9-17}$$

We note that equation (9-16) is similar to equation (9-2b) except for the generation term. The term $T|_{x=\delta}$ is now determined by evaluating the differential equation (9-15a) at $x = \delta(t)$ where $\partial T/\partial x = 0$, and then integrating the resulting ordinary differential equation from $t = 0$ to t subject to the condition $T = T_i$ for $t = 0$. We find

$$T|_{x=\delta} = T_i + \frac{\alpha}{k} G(t) \tag{9-18a}$$

where we defined

$$G(t) \equiv \int_0^t g(t')\, dt' \tag{9-18b}$$

We also note that the term $g(t)\delta(t)$ on the left-hand side equation (9-16) can be written as

$$g(t)\delta(t) = \delta(t)\frac{dG(t)}{dt} \tag{9-19}$$

Equations (9-18) and (9-19) are introduced into equation (9-16)

$$-\alpha \frac{\partial T}{\partial x}\bigg|_{x=0} = \frac{d\theta}{dt} - \frac{\alpha}{k}\left(G\frac{d\delta}{dt} + \delta\frac{dG}{dt}\right) - T_i\frac{d\delta}{dt}$$

or

$$-\alpha \frac{\partial T}{\partial x}\bigg|_{x=0} = \frac{d}{dt}\left[\theta - \frac{\alpha}{k}G\delta - T_i\delta\right] \tag{9-20}$$

which is the energy integral equation for the considered problem. Now, we assume a cubic polynomial representation for $T(x,t)$ in the form

$$T(x,t) = a_1 + a_2 x + a_3 x^2 + a_4 x^3 \tag{9-21}$$

and choose the four conditions needed to evaluate these four coefficients as

$$T\Big|_{x=0} = T_0, \qquad \frac{\partial T}{\partial x}\Big|_{x=\delta} = 0, \qquad \frac{\partial^2 T}{\partial x^2}\Big|_{x=\delta} = 0, \qquad T\Big|_{x=\delta} = T_i + \frac{\alpha}{k}G(t) \tag{9-22}$$

Then, the temperature profile is determined as

$$T(x,t) = T_0 + \left[1 - \left(1 - \frac{x}{\delta}\right)^3\right]F \quad \text{in} \quad 0 \leqslant x \leqslant \delta \tag{9-23a}$$

where

$$F \equiv (T_i - T_0) + \frac{\alpha}{k}G \tag{9-23b}$$

Introducing equation (9-23) into equation (9-20) and performing the indicated operations, we obtain the following differential equation for δ

$$12\alpha F^2 = (F\delta)\frac{d(F\delta)}{dt} \quad \text{for} \quad t > 0 \tag{9-24}$$

The solution of equation (9-24) subject to $\delta = 0$ for $t = 0$ gives

$$\delta^2 = 24\alpha \frac{\int_0^t F^2\,dt}{F^2} \tag{9-25}$$

Equation (9-23) together with equation (9-25) gives the temperature distribution in the medium as a function of time and position. For the special case of no heat generation, equations (9-23) and (9-25), respectively, reduce to

$$\frac{T(x,t) - T_i}{T_0 - T_i} = \left(1 - \frac{x}{\delta}\right)^3 \tag{9-26a}$$

$$\delta = \sqrt{24\alpha t} \tag{9-26b}$$

which are exactly the same as equations (9-11a,b).

9-3 INTEGRAL METHOD—APPLICATION TO NONLINEAR TRANSIENT HEAT CONDUCTION

Another advantage of the integral method is that it can handle the nonlinear problems quite readily. In the following two examples we illustrate the application of the integral method to the solution of nonlinear heat conduction problems. In the first example the nonlinearity is due to the boundary condition, in the second due to the differential equation.

Example 9-3

A semiinfinite medium is initially at uniform temperature T_i. For times $t > 0$, the boundary surface at $x = 0$ is subjected to a heat flux that is a prescribed function of time and surface temperature. Obtain an expression for the surface temperature $T_s(t)$ for times $t > 0$.

Solution. The mathematical formulation of this problem is given as

$$\frac{\partial^2 T}{\partial x^2} = \frac{1}{\alpha}\frac{\partial T}{\partial t} \quad \text{in} \quad x > 0, \quad \text{for} \quad t > 0 \tag{9-27a}$$

$$-\frac{\partial T}{\partial x} = f(T_s, t) \quad \text{at} \quad x = 0, \quad \text{for} \quad t > 0 \tag{9-27b}$$

$$T = T_i \quad \text{for} \quad t = 0, \quad \text{in} \quad x \geqslant 0 \tag{9-27c}$$

Here the boundary-condition function $f(T_s, t)$ is a function of time t and the boundary-surface temperature $T_s(t) \equiv T_s$ at $x = 0$.

The integration of the differential equation (9-27a) over the thermal layer $\delta(t)$ gives

$$\left.\frac{\partial T}{\partial x}\right|_{x=\delta} - \left.\frac{\partial T}{\partial x}\right|_{x=0} = \frac{1}{\alpha}\left[\frac{d}{dt}\left(\int_0^\delta T\,dx\right) - \left.T\right|_{x=\delta} \cdot \frac{d\delta}{dt}\right] \tag{9-28}$$

In view of the conditions

$$\left.\frac{\partial T}{\partial x}\right|_{x=\delta} = 0, \qquad \left.T\right|_{x=\delta} = T_i, \qquad -\left.\frac{\partial T}{\partial x}\right|_{x=0} = f(T_s, t) \tag{9-29}$$

Equation (9-28) becomes

$$\alpha f(T_s, t) = \frac{d}{dt}(\theta - T_i\delta) \tag{9-30a}$$

where

$$\theta \equiv \int_{x=0}^{\delta} T\,dx \tag{9-30b}$$

which is the energy integral equation for the considered problem. To solve this equation we choose a cubic polynomial representation for $T(x, t)$ as

$$T(x, t) = a_1 + a_2 x + a_3 x^2 + a_4 x^3 \tag{9-31}$$

These four coefficients are determined by utilizing the three conditions (9-29) together with the derived condition

$$\left.\frac{\partial^2 T}{\partial x^2}\right|_{x=\delta} = 0 \tag{9-32}$$

The resulting temperature profile becomes

$$T(x, t) - T_i = \frac{\delta f(T_s, t)}{3}\left(1 - \frac{x}{b}\right)^3 \quad \text{in} \quad 0 \leqslant x \leqslant \delta \tag{9-33}$$

and for $x = 0$, this relation gives

$$T_s(t) - T_i = \frac{\delta f(T_s, t)}{3} \tag{9-34}$$

From equations (9-33) and (9-34) we write

$$\frac{T(x, t) - T_i}{T_s(t) - T_i} = \left(1 - \frac{x}{\delta}\right)^3 \tag{9-35}$$

Introducing equation (9-35) into equation (9-30), performing the indicated operations, and eliminating δ from the resulting expression by means of equation (9-34) we obtain the following first-order ordinary differential equation for the determination of the surface temperature T_s.

$$\frac{4}{3}\alpha f(T_s, t) = \frac{d}{dt}\left[\frac{(T_s - T_i)^2}{f(T_s, t)}\right] \quad \text{for} \quad t > 0 \tag{9-36}$$

with

$$T_s = T_i \quad \text{for} \quad t = 0 \tag{9-37}$$

Equation (9-36) can be integrated numerically if the boundary condition function $f(T_s, t)$ depends on both the surface temperature and the time. For the special case of $f(T_s, t)$ being a function of surface temperature only, namely

$$f(T_s, t) = f(T_s) \tag{9-38}$$

equation (9-36) is written as

$$\frac{4}{3}\alpha f(T_s) = \frac{d}{dT_s}\left[\frac{(T_s - T_i)^2}{f(T_s)}\right]\cdot\frac{dT_s}{dt}$$

or

$$\frac{4}{3}\alpha = \frac{2(T_s - T_i)f(T_s) - f'(T_s)\cdot(T_s - T_i)^2}{f^3(T_s)}\frac{dT_s}{dt} \qquad \text{for} \qquad t > 0 \quad \text{(9-39a)}$$

with

$$T_s = T_i \qquad \text{for} \qquad t = 0 \quad \text{(9-39b)}$$

The integration of equation (9-39) establishes the relation between the surface temperature $T_s(t)$ and the time t as

$$\frac{4}{3}\alpha t = \int_{T_i}^{T_s} \frac{2(T_s - T_i)f(T_s) - f'(T_s)\cdot(T_s - T_i)^2}{f^3(T_s)}\, dT_s \quad \text{(9-40)}$$

where f' denotes differentiation with respect to T_s.

Example 9-4

A semiinfinite medium, $x > 0$, is initially at zero temperature. For times $t > 0$, the boundary surface is subjected to a prescribed heat flux that varies with time. The thermal properties $k(T)$, $C_p(T)$, and $\rho(T)$ are all assumed depend on temperature. Obtain an expression for the temperature distribution in the medium.

Solution. The mathematical formulation of this problem is given as

$$\frac{\partial}{\partial x}\left(k\frac{\partial T}{\partial x}\right) = \rho C_p \frac{\partial T}{\partial t} \qquad \text{in} \qquad x > 0, \quad t > 0 \quad \text{(9-41a)}$$

$$-k\frac{\partial T}{\partial x} = f(t) \qquad \text{at} \qquad x = 0, \quad t > 0 \quad \text{(9-41b)}$$

$$T = 0 \qquad \text{for} \qquad t = 0, \quad x \geqslant 0 \quad \text{(9-41c)}$$

where $k \equiv k(T)$, $C_p \equiv C_p(T)$, and $\rho \equiv \rho(T)$. By applying the transformation

$$U = \int_0^T \rho C_p\, dT \quad \text{(9-42)}$$

The system (9-41) is transformed into

$$\frac{\partial}{\partial x}\left(\alpha\frac{\partial U}{\partial x}\right) = \frac{\partial U}{\partial t} \qquad \text{in} \quad x > 0, \quad t > 0 \quad \text{(9-43a)}$$

$$-\frac{\partial U}{\partial x} = \frac{1}{\alpha_s} f(t) \qquad \text{at} \quad x = 0, \quad t > 0 \quad \text{(9-43b)}$$

$$U = 0 \qquad \text{for} \quad t = 0, \quad x \geqslant 0 \tag{9-43c}$$

where $\alpha \equiv \alpha(U)$ and α_s refers to the value of α at the boundary surface $x = 0$.

Equation (9-43a) is integrated over the thermal layer $\delta(t)$

$$\left[\alpha \frac{\partial U}{\partial x}\right]_{x=0}^{\delta} = \frac{d}{dt}\left[\int_0^{\delta} U\,dx - U\Big|_{\delta} \cdot \delta\right] \tag{9-44}$$

In view of the conditions

$$\frac{\partial U}{\partial x}\bigg|_{x=\delta} = 0, \quad U\bigg|_{x=\delta} = 0, \quad \left[\alpha \frac{\partial U}{\partial x}\right]_{x=0} = -f(t) \tag{9-45}$$

Equation (9-44) becomes

$$\frac{d\theta}{dt} = f(t) \tag{9-46a}$$

where

$$\theta \equiv \int_0^{\delta} U\,dx \tag{9-46b}$$

which is the energy integral equation for the considered problem. To solve this equation we choose a cubic polynomial representation for $U(x, t)$ as

$$U(x, t) = a_1 + a_2 x + a_3 x^2 + a_4 x^3 \tag{9-47}$$

The four coefficients are determined by utilizing the following four conditions

$$U\bigg|_{x=\delta} = 0, \quad \frac{\partial U}{\partial x}\bigg|_{x=\delta} = 0, \quad \frac{\partial U}{\partial x}\bigg|_{x=0} = -\frac{f(t)}{\alpha_s}, \quad \frac{\partial^2 U}{\partial x^2}\bigg|_{x=\delta} = 0 \tag{9-48}$$

Then, the corresponding profile becomes

$$U(x, t) = \frac{\delta f(t)}{3\alpha_s}\left(1 - \frac{x}{\delta}\right)^3 \qquad \text{in} \qquad 0 \leqslant x \leqslant \delta \tag{9-49}$$

By substituting of equation (9-49) into equations (9-46) and performing the indicated operations we obtain the following differential equation for the determination of the thermal-layer thickness $\delta(t)$:

$$\frac{d}{dt}\left[\frac{\delta^2 f(t)}{12\alpha_s}\right] = f(t) \qquad \text{for} \qquad t > 0 \tag{9-50a}$$

with

$$\delta = 0 \qquad \text{for} \qquad t = 0 \tag{9-50b}$$

The solution of equation (9-50) is

$$\delta = \left[\frac{12\alpha_s}{f(t)} \int_0^t f(t')\, dt' \right]^{1/2} \tag{9-51}$$

This equation cannot yet be used to calculate the thermal layer thickness δ directly, because it involves α_s, the thermal diffusivity evaluated at the surface temperature, U_s, which is still unknown. To circumvent this difficulty an additional relationship is needed between α_s and U_s; such a relationship is obtained as now described.

For $x = 0$ equation (9-49) gives

$$U_s = \frac{\delta f(t)}{3\alpha_s} \tag{9-52}$$

Eliminating δ between equations (9-51) and (9-52), we obtain

$$U_s \sqrt{\alpha_s} = \left[\frac{4}{3} f(t) \int_0^t f(t')\, dt' \right]^{1/2} \tag{9-53}$$

The computational procedure is as follows:

1. α_s is known as a function of T_s and hence of U_s. Then the left-hand side of equation (9-53) can be regarded to depend on α_s only.
2. Then use Eq. (9-53) to compute α_s as a function of time.
3. Knowing α_s at each time, use equation (9-51) to calculate δ.
4. Knowing δ, calculate U from (9-49).
5. Knowing U, determine the actual temperature $T(x, t)$ from equation (9-42).

9-4 INTEGRAL METHOD—APPLICATION TO A FINITE REGION

In the previous examples we considered the application of the integral method for the solution of transient heat conduction in a semiinfinite medium in which the thickness of the thermal layer $\delta(t)$ would increase indefinitely. However, for the problem of a slab, in $0 \leqslant x \leqslant L$, with the surface $x = L$ insulated, the analysis is exactly the same as that described for a semiinfinite region so long as the thickness of the thermal layer remains less than the thickness of the plate; but, as soon as the thickness of the thermal layer becomes equal to that of the slab, that is, $\delta(t) = L$, the thermal layer has no physical significance. A different analysis is needed for $\delta(t) > L$. This matter is illustrated with the following example.

Example 9-5

A slab, $0 \leqslant x \leqslant L$, is initially at a uniform temperature T_i. For times $t > 0$, the boundary surface at $x = 0$ is kept at a constant temperature T_0 and the boundary at $x = L$ is kept insulated. Obtain an expression for the temperature distribution in the slab for times $t > 0$ by using the integral method.

Solution. The mathematical formulation of this problem is given as

$$\frac{\partial^2 T}{\partial x^2} = \frac{1}{\alpha}\frac{\partial T}{\partial t} \quad \text{in} \quad 0 < x < L, \quad t > 0 \tag{9-54a}$$

$$T(x,t) = T_0 \quad \text{at} \quad x = 0, \quad t > 0 \tag{9-54b}$$

$$\frac{\partial T}{\partial x} = 0 \quad \text{at} \quad x = L, \quad t > 0 \tag{9-54c}$$

$$T(x,t) = T_i \quad \text{for} \quad t = 0, \quad \text{in} \quad 0 \leqslant x \leqslant L \tag{9-54d}$$

For the reasons discussed above the analysis is now performed in *two stages*:

1. The *first stage*, during which the thermal-layer thickness is less than the slab thickness (i.e., $\delta \leqslant L$).
2. The *second stage*, during which δ exceeds the slab thickness L.

The First Stage. For the case $\delta(t) < L$, we integrate equation (9-54a) over the thermal layer thickness and obtain

$$-\alpha \left.\frac{\partial T}{\partial x}\right|_{x=0} = \frac{d}{dt}(\theta - T_i \delta) \tag{9-55a}$$

where

$$\theta \equiv \int_{x=0}^{\delta} T(x,t)\,dx \tag{9-55b}$$

The energy integral equation thus obtained is exactly the same as that given by equation (9-4) for the semiinfinite region. We choose a cubic profile for the temperature as given by equation (9-5), apply the conditions given by equations (9-6) to determine the coefficients, and utilize equation (9-55) to obtain the thermal-layer thickness as discussed for the semiinfinite region. The resulting temperature profile becomes

$$\frac{T(x,t) - T_i}{T_0 - T_i} = 1 - \frac{3}{2}\left(\frac{x}{\delta}\right) + \frac{1}{2}\left(\frac{x}{\delta}\right)^3 \tag{9-56a}$$

where

$$\delta = \sqrt{8\alpha t} \tag{9-56b}$$

This solution is valid for $0 \leqslant x \leqslant \delta$, so long as $\delta \leqslant L$. The time t_L when $\delta = L$ is obtained from equation (9-56b) by setting $\delta = L$, that is

$$t_L = \frac{L^2}{8\alpha} \tag{9-57}$$

Clearly, the solution (9-56) is not applicable for times $t > t_L$.

The Second Stage. For times $t > t_L$, the concept of thermal layer has no physical significance. The analysis for the second stage may be performed in the following manner. We integrate the differential equation (9-54a) from $x = 0$ to $x = L$; we find

$$-\alpha \left.\frac{\partial T}{\partial x}\right|_{x=0} = \frac{d}{dt}(\theta - T_i L) \tag{9-58a}$$

where

$$\theta \equiv \int_0^L T(x,t)\,dx \tag{9-58b}$$

A comparison of equation (9-58) with equation (9-55a) reveals that in the latter, the plate thickness L has replaced $\delta(t)$, hence there is no thermal layer. The temperature $T(x,t)$ is again expressed by a polynomial. Suppose we choose a cubic polynomial representation in the form

$$T(x,t) = a + bx + cx^2 + dx^3 \qquad \text{in} \qquad 0 < x < L, \quad t > t_L \tag{9-59}$$

where the coefficients are generally function of time. In this case we have no thermal layer to be determined from the solution of the differential equation (9-58). Therefore, we choose only three conditions

$$\left.T\right|_{x=0} = T_0, \quad \left.\frac{\partial T}{\partial x}\right|_{x=L} = 0 \quad \text{and} \quad \left.\frac{\partial^2 T}{\partial x^2}\right|_{x=0} = 0 \tag{9-60a,b,c}$$

These conditions are applied to the cubic profile given by equation (9-59) and all the coefficients are expressed in terms of one of them, say $b \equiv b(t)$. We find the following profile

$$T(x,t) = T_0 + bL\left[\frac{x}{L} - \frac{1}{3}\left(\frac{x}{L}\right)^3\right] \quad \text{in} \quad 0 \leqslant x \leqslant L, \quad \text{for} \quad t \geqslant t_L \tag{9-61}$$

which is expressed in the dimensionless form as

$$\frac{T(x,t) - T_i}{T_0 - T_i} = 1 + \eta(t)\left[\frac{x}{L} - \frac{1}{3}\left(\frac{x}{L}\right)^3\right] \quad \text{in} \quad 0 \leqslant x \leqslant L \tag{9-62a}$$

where

$$\eta(t) \equiv \frac{Lb(t)}{T_0 - T_i} \tag{9-62b}$$

When the profile (9-62a) is introduced into equation (9-58a) and the indicated operations are performed, the following ordinary differential equation is obtained for the determination of $\eta(t)$:

$$\frac{d\eta(t)}{dt} + \frac{12\alpha}{5L^2}\eta(t) = 0 \quad \text{for} \quad t \geqslant t_L \tag{9-63}$$

The initial condition needed to solve this differential equation is determined from the requirement that the temperature defined by equation (9-62a) at

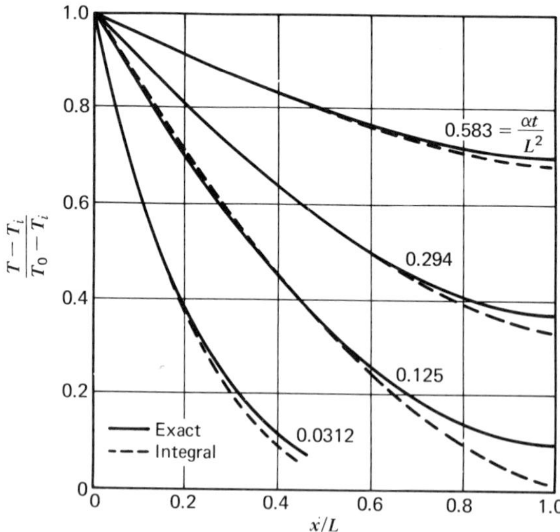

Fig. 9-3 Comparison of exact and approximate temperature profiles for a slab of thickness L. (From Reynolds and Dolton [7].)

$x = L$ at the time $t = t_L = L^2/8\alpha$ should be equal to the initial condition $T = T_i$; we find

$$\eta(t) = -\frac{3}{2} \qquad \text{for} \qquad t = t_L = \frac{L^2}{8\alpha} \tag{9-64}$$

The solution of the differential equation (9-63) subject to the initial condition (9-64) gives

$$\eta(t) = -\frac{3}{2}\exp\left[-\left(\frac{12\alpha}{5L^2}t - \frac{3}{10}\right)\right] \tag{9-65}$$

Thus equation (9-62a) with $\eta(t)$ as given by equation (9-65) represents the temperature distribution in the slab for times $t > t_L \equiv L^2/8\alpha$. Figure 9-3 shows a comparison of the exact and approximate solutions.

9-5 APPROXIMATE ANALYTIC METHODS OF RESIDUALS

When the exact solution T cannot be obtained for an ordinary or partial differential equation, a *trial family* of approximate solution $\tilde{T}$ containing a finite number of undetermined coefficients $c_1, c_2, \ldots, c_n$ can be constructed by the superposition of some *basis functions* such as polynomials, trigonometric functions, and similar. The *trial solution* is so selected that it satisfies the essential boundary conditions for the problem; but, when it is introduced into the differential equation it does not satisfy it and leads to a *residual* R, because it is not the exact solution. For the true solution the residuals vanish identically. Therefore, the problem of constructing an approximate solution becomes one of determining the unknown coefficients $c_1, c_2, \ldots, c_n$ so that the residual stays "*close*" to zero throughout the domain of the solution. Depending on the number of terms taken for the trial solution, the type of base functions used and the way the unknown coefficients are determined, several different approximate solutions are possible for a given problem.

Different schemes have been proposed for the determination of the unknown coefficients $c_1, c_2, \ldots, c_n$ associated with the construction of the trial family of approximate solution $\tilde{T}$. To illustrate the basic approaches followed in various approximate methods of solution, we examine the following simple problem considered in reference 55:

$$\frac{dT(t)}{dt} + T(t) = 0 \qquad \text{for} \qquad t > 0 \tag{9-66a}$$

$$T(t) = 1 \qquad \text{at} \qquad t = 0 \tag{9-66b}$$

The exact solution of this problem is given by

$$T(t) = e^{-t} \tag{9-67}$$

We wish to obtain an approximate solution for this problem in the interval $0 < t < 1$. To construct a trial solution $\tilde{T}(t)$, we choose the *basis functions* to be polynomial in t (i.e., $t, t^2, \ldots$). The *trial solution* that contains only two undetermined coefficients c_1 and c_2, and satisfies the condition (9-66b) for all values of c_1 and c_2, is taken as

$$\tilde{T}(t) = 1 + c_1 t + c_2 t^2 \tag{9-68}$$

Here, the first term on the right-hand side is included in order to satisfy the nonhomogeneous part of the boundary condition (9-66b).

The trial solution (9-68) satisfies the boundary condition (9-66b) for all values of c_1 and c_2; but, when it is introduced into the differential equation (9-66a) it yields a *residual* $R(c_1, c_2, t)$ as

$$R(c_1, c_2 t) = 1 + (1 + t)c_1 + (2t + t^2)c_2 \tag{9-69}$$

This residual vanishes only with the exact solution for the problem. Now, the problem of finding an approximate solution for the problem (9-66) in the interval $0 < t < 1$ becomes one of adjusting the values of c_1 and c_2 so that the residual $R(c_1, c_2, t)$ stays "close" to zero throughout the interval $0 < t < 1$.

Various schemes have been proposed for the determination of these unknown coefficients; when c_1 and c_2 are known, the trial solution $\tilde{T}(t)$ given by equation (9-68) becomes the approximate solution for the problem.

We briefly describe below some of the popular schemes for the determination of the unknown coefficients.

1. *Collocation Method.* If the trial solution contains n undetermined coefficients, n different locations are selected where the residual $R(t)$ is forced to vanish, thus providing n simultaneous algebraic equations for the determination of the coefficients $c_1, c_2, \ldots, c_n$. The basic assumption is that the residual does not deviate much from zero between the collocation locations. For the specific example considered previously, suppose we select the collocation locations $\frac{1}{3}$ and $\frac{2}{3}$. Introducing these values into the residual equation (9-69) we obtain

$$R(c_1, c_2, \tfrac{1}{3}) = 1 + \tfrac{4}{3}c_1 + \tfrac{7}{9}c_2 = 0 \tag{9-70a}$$

$$R(c_1, c_2, \tfrac{2}{3}) = 1 + \tfrac{5}{3}c_1 + \tfrac{16}{9}c_2 = 0 \tag{9-70b}$$

Thus we have two algebraic equations for the determination of the two unknown coefficients c_1 and c_2; or simultaneous solution gives

$$c_1 = -0.9310, \qquad c_2 = 0.3103$$

Introducing these coefficients into equation (9-68), we obtain the approximate solution for the problem (9-66) in the interval $0 < t < 1$, based on the collocation method as

$$\tilde{T}(t) = 1 - 0.9310t + 0.3103t^2 \tag{9-70c}$$

2. *Least-Squares Method.* Referring to the simple example considered above, the coefficients c_1 and c_2 are determined from the requirement that the integral of the square of the residual $R(c_1, c_2, t)$ given by equation (9-69) is minimized over the interval $0 < t < 1$. That is, we set

$$\frac{1}{2}\frac{\partial}{\partial c_1}\int_0^1 R^2(c_1, c_2, t)\,dt = \int_0^1 R\frac{\partial R}{\partial c_1}\,dt = \frac{3}{2} + \frac{7}{3}c_1 + \frac{9}{4}c_2 = 0 \tag{9-71a}$$

$$\frac{1}{2}\frac{\partial}{\partial c_2}\int_0^1 R^2(c_1, c_2, t)\,dt = \int_0^1 R\frac{\partial R}{\partial c_2}\,dt = \frac{4}{3} + \frac{9}{4}c_1 + \frac{38}{15}c_2 = 0 \tag{9-71b}$$

and again we have two algebraic equations for the determination of the two unknown constants c_1 and c_2. A simultaneous solution gives

$$c_1 = -0.9427, \qquad c_2 = 0.3110$$

Introducing these values into equation (9-68), we obtain the approximate solution for the problem (9-66) in the interval $0 < t < 1$ as

$$\tilde{T}(t) = 1 - 0.9427t + 0.3110t^2 \tag{9-71c}$$

3. *Rayleigh-Ritz Method.* This method requires the variational formulation of the differential equation so that the boundary conditions for the problem are incorporated into the variational form. Once the variational form is available, the trial solution given by equation (9-68) is introduced into the variational expression $J(c_1, c_2)$ and this result is minimized as

$$\frac{\partial J(c_1, c_2)}{\partial c_1} = 0 \tag{9-72a}$$

$$\frac{\partial J(c_1, c_2)}{\partial c_2} = 0 \tag{9-72b}$$

Thus, we have two algebraic equations for the determination of the two unknown coefficients c_1 and c_2. Generally, the difficulty with this approach is the determination of the corresponding variational expression. If the variational form cannot be found, the scheme is not applicable. The principles of variational calculus are discussed in several texts [32–36], and the application to the solution of heat conduction problems can be found in several other references [37–56]. Next, we consider the Galerkin method, which leads to the same approximate solution as the Rayleigh–Ritz method without requiring the variational form of the problem.

4. *Galerkin Method.* The method requires that the weighted averages of the residual $R(c_1, c_2, t)$ should vanish over the interval considered. The weight functions $w_1(t)$ and $w_2(t)$ are taken as the basis functions used to construct the trial solution $\tilde{T}(t)$. For the specific problem considered here, t and t^2 are the weight functions to be used for the integration of the residual $R(c_1, c_2, t)$ over the interval $0 < t < 1$. Thus, the Galerkin method becomes

$$\int_0^1 tR(c_1, c_2, t)\,dt = \int_0^1 t[1 + (1 + t)c_1 + (2t + t^2)c_2]\,dt$$

$$= \tfrac{1}{2} + \tfrac{5}{6}c_1 + \tfrac{11}{12}c_2 = 0 \tag{9-73a}$$

$$\int_0^1 t^2 R(c_1, c_2, t)\,dt = \int_0^1 t^2[1 + (1 + t)c_1 + (2t + t^2)c_2]\,dt$$

$$= \tfrac{1}{3} + \tfrac{7}{12}c_1 + \tfrac{9}{20}c_2 = 0 \tag{9-73b}$$

Equations (9-73) provide two algebraic equations for the determination of the unknown coefficients c_1 and c_2. A simultaneous solution gives

$$c_1 = -0.9143, \qquad c_2 = 0.2857$$

Introducing these coefficients into equation (9-68), the approximate solution becomes

$$\tilde{T}(t) = 1 - 0.9143t + 0.2857t^2 \tag{9-73c}$$

We also note that, the weight functions $w_1(t)$ and $w_2(t)$ can be interpreted as

$$w_i(t) = \frac{\partial \tilde{T}(t)}{\partial c_i}, \qquad i = 1, 2 \tag{9-74}$$

The Galerkin method does not require the variational form of the problem, and yields the same result as the Rayleigh–Ritz method, therefore the problem setup is easier and more direct. We now present further applications of the Galerkin method and discuss the construction of the basis functions.

9-6 THE GALERKIN METHOD

In the previous section, we illustrated with a simple example the basic concepts in the application of some of the popular approximate analytic methods for the solution of differential equations. Here we focus attention to the Galerkin method, especially to its application to more general problems and the methods of determining the trial functions.

The method is perfectly universal; it can be applied to elliptic, hyperbolic and parabolic equations, nonlinear problems as well as complicated boundary conditions. The reader should consult to references [16, 37, 53–55] for a discussion of the theory and application of the Galerkin method and to references 58–69 for its application in the solution of various types of boundary-value problems.

Application to Steady-State Heat Conduction

We consider a steady-state heat conduction problem given in the form

$$\nabla^2 T(\mathbf{r}) + AT(\mathbf{r}) + \frac{1}{k} g(\mathbf{r}) = 0 \qquad \text{in} \qquad R \tag{9-75a}$$

$$k\frac{\partial T}{\partial n} + hT = f(\mathbf{r}_s) \qquad \text{on boundary } S \tag{9-75b}$$

where $\partial/\partial n$ denotes derivative along the outward drawn normal to the boundary surface S.

Clearly, the problem defined by equations (9-75) covers a wide range of steady-state heat conduction problems as special cases.

Let $\phi_j(\mathbf{r}), j = 1, 2, 3, \ldots$, be a set of *basis functions*. We construct the n-term trial solution $\tilde{T}_n(\mathbf{r})$ in the form

$$\tilde{T}_n(\mathbf{r}) = \psi_0(\mathbf{r}) + \sum_{j=1}^{n} c_j \phi_j(\mathbf{r}) \tag{9-76}$$

where the function $\psi_0(\mathbf{r})$ is included to satisfy the nonhomogeneous part of the boundary condition (9-75b) and the basis functions $\phi_j(\mathbf{r})$ satisfy the homogeneous part. When all the boundary conditions are homogeneous, the function $\psi_0(\mathbf{r})$ is not neded. The subscript n in the trial solution $\tilde{T}_n(\mathbf{r})$ denotes that it is n-term trial solution.

When the trial solution (9-76) is substituted into the differential equation (9-75a), a residual $R(c_1, c_2, \ldots, c_n; \mathbf{r})$ is left, because $\tilde{T}_n(\mathbf{r})$ it is not an exact solution. We obtain

$$\nabla^2 \tilde{T}_n(\mathbf{r}) + A\tilde{T}_n(\mathbf{r}) + \frac{1}{k} g(\mathbf{r}) \equiv R(c_1, c_2, \ldots, c_n; \mathbf{r}) \neq 0 \tag{9-77}$$

Then the Galerkin method for the determination of the n unknown coefficients $c_1, c_2, \ldots, c_n$ is given by

$$\int_R \phi_j(\mathbf{r}) \left[\nabla^2 \tilde{T}_n(\mathbf{r}) + A\tilde{T}_n(\mathbf{r}) + \frac{1}{k} g(\mathbf{r}) \right] dv = 0; \qquad j = 1, 2, \ldots, n \tag{9-78a}$$

which is written more compactly in the form

$$\int_R \phi_j(\mathbf{r}) R(c_1, c_2, \ldots, c_n; \mathbf{r})\, dv = 0; \qquad j = 1, 2, \ldots, n \tag{9-78b}$$

Equations (9-78) provide n algebraic equations for the determination of n unknown coefficients $c_1, c_2, \ldots, c_n$.

If the problem can be solved by the separation of variables and the basis functions $\phi_j(\mathbf{r})$ are taken to be the eigenfunctions for the problem, then the solution generated by the Galerkin method becomes the exact solution for the problem as the number of terms n approaches infinity. However, in general, the eigenfunctions for the problem are not available, hence the question arises regarding what kind of functions should be chosen as the basis functions to construct the trial solution.

The functions $\phi_j(\mathbf{r})$, $(j = 1, 2, \ldots, n)$ should satisfy the homogeneous part of the boundary conditions and should be linearly independent over the given region R. The functions $\phi_j(\mathbf{r})$, $j = 1, 2, \ldots, n$, if possible, should belong to a class of functions that are complete in the considered region. They should be continuous in the region and should have continuous first and second derivatives. They may be polynomials, trigonometric, circular, or spherical functions, but they should satisfy the homogeneous part of the boundary conditions for the problem.

Construction of Function $\phi_j(\mathbf{r})$ When Boundary Conditions Are All of the First Kind

In regions having simple geometries, such as a slab, cylinder, sphere, or rectangle, the functions $\phi_j(\mathbf{r})$ can be taken as the eigenfunctions obtained by the separation of variables that are available in tabulated form in Chapters 2–4. Thus, the functions $\phi_j(\mathbf{r})$ can be used as the basis functions to construct the trial solution $\tilde{T}(\mathbf{r})$ for the problem. However, there are many situations in which the boundaries of the region are irregular; as a result, it becomes very difficult to find basis functions $\phi_j(\mathbf{r})$ that will satisfy the homogeneous boundary conditions. Here we present a methodology to construct the basis functions $\phi_j(\mathbf{r})$ for such situations in two-dimensional problems.

Let a function $\omega(x, y)$ be a continuous function and have continuous derivatives with respect to x and y within the region, and in addition satisfy the homogeneous boundary condition of the first kind at the boundaries of the region:

$$\omega(x, y) > 0 \qquad \text{in} \qquad R \tag{9-79a}$$

$$\omega(x, y) = 0 \qquad \text{on boundary } S \tag{9-79b}$$

Once the functions $\omega(x, y)$ are available, the basis functions $\phi_j(x, y)$ can be constructed by the products of $\omega(x, y)$ with various powers of x and y in the form

$$\phi_1 = \omega, \qquad \phi_2 = \omega x, \qquad \phi_3 = \omega y, \qquad \phi_4 = \omega x^2, \qquad \phi_5 = \omega x y, \ldots \tag{9-80}$$

The functions $\phi_j(x, y), j = 1, 2, \ldots, n$ constructed in this manner satisfy the homogeneous part of the boundary conditions for the problem, have continuous derivatives in x and y, and it is proved in reference 53, [p. 276] that they constitute a *complete* system of functions. Then, the problem becomes one of determining the auxiliary functions $\omega(x, y)$. These functions can be determined by utilizing the equations for the contour of the boundary as now described.

1. *Region Having a Single Continuous Contour.* If the region has a single continuous contour such as a circle, the equation of the boundary can be expressed in the form

$$F(x, y) = 0 \qquad \text{on boundary } S \tag{9-81}$$

Clearly, the function $F(x, y)$ is continuous, has partial derivatives with respect to x and y, and vanishes at the boundary of the region R. Then, the function $\omega(x, y)$ can be chosen as

$$\omega(x, y) = \pm F(x, y) \tag{9-82}$$

For example, for a circular region of radius R with center at the origin, the equation for the contour satisfies the equation

$$F(x, y) = R^2 - x^2 - y^2 = 0 \tag{9-83a}$$

and the function $\omega(x, y)$ is taken as

$$\omega(x, y) = R^2 - x^2 - y^2 \tag{9-83b}$$

2. *Region Having a Contour as Convex Polynomial.* Consider a region in the form of a convex polynomial and let the equations for the sides be given in the form

$$F_1 \equiv a_1 x + b_1 y + d_1 = 0 \tag{9-84a}$$

$$F_2 \equiv a_2 x + b_2 y + d_2 = 0 \tag{9-84b}$$

$$\vdots$$

$$F_n \equiv a_n x + b_n y + d_n = 0 \tag{9-84c}$$

Then, the function $\omega(x, y)$ chosen in the form

$$\omega(x, y) = \pm F_1(x, y) F_2(x, y) \ldots F_n(x, y) \tag{9-85}$$

vanishes at every point on the boundary and satisfies the homogeneous part of the boundary conditions of the first kind for the region.

3. *Region Having a Contour as Nonconvex Polynomial.* The construction of the function $\omega(x, y)$ for this case is more involved, because the function

$\omega(x, y)$ has to be assigned piecewise in different parts of the region. Further discussion of this matter can be found in reference 53 [p. 278].

Example 9-6

Construct the functions $\omega(x, y)$ as discussed above for the four different geometries shown in Fig. 9-4.

Solution. The equations of the contours for each of the four geometries shown in Fig. 9-4*a*,*b*,*c*,*d* are given, respectively, as

$$a - x = 0, \quad a + x = 0, \quad b - y = 0, \quad b + y = 0 \tag{9-86a}$$

$$y - \alpha x = 0, \quad y + \beta x = 0, \quad L - x = 0 \tag{9-86b}$$

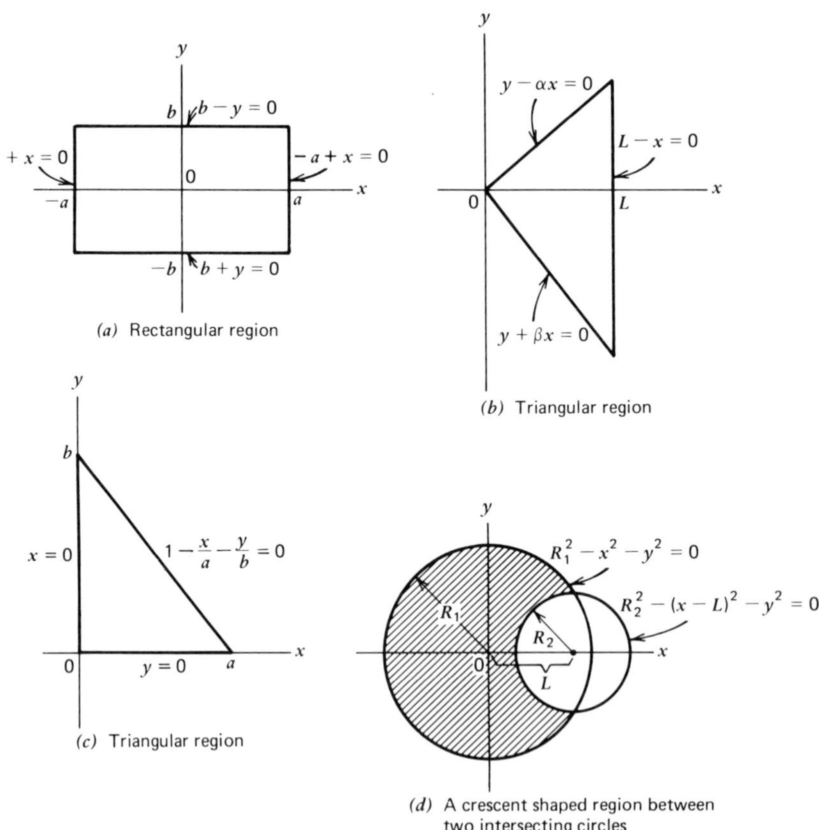

Fig. 9-4 Regions having boundary contour in the form of a convex polygon and a region bounded by two circles: (*a*) rectangular region; (*b*) triangular region; (*c*) triangular region; (*d*) a crescent-shaped region between two intersecting circles.

$$x = 0, \quad y = 0, \quad 1 - \frac{x}{a} - \frac{y}{b} = 0 \tag{9-86c}$$

$$R_1^2 - x^2 - y^2 = 0, \quad R_2^2 - (x - L)^2 - y^2 = 0 \tag{9-86d}$$

Then the corresponding functions $\omega(x, y)$ for each of these geometries shown in Fig. 9-4*a*,*b*,*c*,*d* are given respectively as

$$\omega(x, y) = (a^2 - x^2)(b^2 - y^2) \tag{9-87a}$$

$$\omega(x, y) = (y - \alpha x)(y + \beta x)(L - x) \tag{9-87b}$$

$$\omega(x, y) = xy\left(1 - \frac{x}{a} - \frac{y}{b}\right) \tag{9-87c}$$

$$\omega(x, y) = (R_1^2 - x^2 - y^2)[R_2^2 - (x - L)^2 - y^2] \tag{9-87d}$$

Construction of Functions $\phi_j(r)$ for Boundary Conditions of the Third Kind

We consider one-dimensional steady-state heat conduction in a slab of thickness L subjected to convection into a medium at zero temperature. The boundary conditions at $x = 0$ and $x = L$ are

$$\left[-\frac{dT}{dx} + h_1 T\right]_{x=0} = 0 \tag{9-88a}$$

$$\left[\frac{dT}{dx} + h_2 T\right]_{x=L} = 0 \tag{9-88b}$$

If such a heat conduction problem is to be solved by the Galerkin method, the first two trial functions $\phi_1(x)$ and $\phi_2(x)$ must be chosen as

$$\phi_1 = x^2\left(x - L - \frac{L}{2 + h_2 L}\right) \tag{9-89a}$$

$$\phi_2 = (L - x)^2\left(x + \frac{L}{2 + h_1 L}\right) \tag{9-89b}$$

and the remaining $\phi_3, \phi_4, \ldots$ can be taken as

$$\phi_j = x^j (L - x)^2, \qquad j = 3, 4, 5 \ldots \tag{9-89c}$$

Then, the trial solution constructed as

$$\tilde{T}_n(x) = \phi_1 + \phi_2 + \sum_{j=3}^{n} c_j\phi_j \qquad \text{in} \qquad 0 \leqslant x \leqslant L \tag{9-89d}$$

satisfies the boundary conditions (9-88a,b) for all values of $c_j, j \geqslant 3$.

For other combinations of the boundary conditions of the first, second, and third kinds, functions ϕ_j are constructed with similar considerations.

Integration Formula

In performing computations associated with the application of the Galerkin method, the following integration formula are useful [53, p. 269]

$$\int_0^L x^k(L-x)^m dx = \frac{k!\,m!}{(k+m+1)!} L^{k+m+1} \tag{9-90}$$

Example 9-7

Consider the following one-dimensional steady-state heat conduction problem

$$\frac{d^2T}{dx^2} + AT + Bx = 0 \qquad \text{in} \qquad 0 < x < 1 \tag{9-91a}$$

$$T = 0 \qquad \text{at} \qquad x = 0 \quad \text{and} \quad x = 1 \tag{9-91b}$$

where A and B are constants. Solve this problem by the Galerkin method using one and two term trial solutions and compare the approximate results with the exact solution of the problem for the case $A = B = 1$.

Solution. The application of the Galerkin method gives

$$\int_{x=0}^{1} \left[\frac{d^2\tilde{T}}{dx^2} + A\tilde{T} + Bx \right] \phi_i(x)dx = 0, \qquad i = 1, 2, \ldots. \tag{9-92}$$

where $\phi_i(x)$ are a set of basis functions and $\tilde{T}(x)$ is the trial solution. The basis functions are chosen as

$$\phi_1 = x(1-x), \qquad \phi_2 = x^2(1-x), \ldots \tag{9-93}$$

which satisfy the homogeneous boundary conditions (9-91b) for the problem.

1. *One-Term Trial Solution.* We choose the trial solution as

$$\tilde{T}_1(x) = c_1 \phi_1(x) \tag{9-94a}$$

where the basis function $\phi_1(x)$, satisfying the homogeneous boundary conditions (9-91b), is taken as

$$\phi_1(x) = x(1 - x) = x - x^2 \tag{9-94b}$$

Then we have

$$\tilde{T}_1(x) = c_1 x - c_1 x^2 \tag{9-94c}$$

$$\frac{d^2 \tilde{T}_1}{dx^2} = -2c_1 \tag{9-94d}$$

Introducing equations (9-94c,d) into equation (9-92), the one-term Galerkin method of solution gives

$$\int_{x=0}^{1} [-2c_1 + A(c_1 x - c_1 x^2) + Bx](x - x^2)dx = 0 \tag{9-95a}$$

Performing this integration and solving for c_1, we obtain

$$c_1 = \frac{B}{4[1 - (A/10)]} \tag{9-95b}$$

Then the one-term trial solution becomes

$$\tilde{T}_1(x) = \frac{B}{4[1 - (A/10)]} x(1 - x) \tag{9-96a}$$

For the case $A = B = 1$, this result reduces to

$$\tilde{T}_1(x) = \frac{1}{3.6} x(1 - x) \tag{9-96b}$$

2. *Two-Term Trial Solution.* The trial solution is taken as

$$\tilde{T}_2(x) = c_1 \phi_1(x) + c_2 \phi_2(x) \tag{9-97a}$$

where the basis functions $\phi_1(x)$ and $\phi_2(x)$ are chosen as

$$\phi_1(x) = x(1 - x), \qquad \phi_2(x) = x^2(1 - x) \tag{9-97b}$$

Then we have

$$\tilde{T}_2(x) = c_1(x - x^2) + c_2(x^2 - x^3) \tag{9-98a}$$

$$\frac{d^2 T_2(x)}{dx^2} = -2c_1 + 2c_2 - 6c_2 x \tag{9-98b}$$

Introducing equations (9-98a,b) into equations (9-92a) we obtain

$$\int_0^1 \left[\frac{d^2 \tilde{T}_2}{dx^2} + A\tilde{T}_2(x) + Bx \right] (x - x^2) dx = 0 \qquad \text{for} \quad i = 1 \tag{9-99a}$$

$$\int_0^1 \left[\frac{d^2 \tilde{T}_2}{dx^2} + A\tilde{T}_2(x) + Bx \right] (x^2 - x^3) dx = 0 \qquad \text{for} \quad i = 2 \tag{9-99b}$$

where

$$\frac{d^2 \tilde{T}_2}{dx^2} + A\tilde{T}_2(x) + Bx = (-2c_1 + 2c_2 - 6c_2 x) + A(c_1 x - c_1 x^2 + c_2 x^2 - c_2 x^3) + Bx \tag{9-99c}$$

When the integrations are performed, equations (9-99a,b) provides two algebraic equations for the determination of the two unknown coefficients c_1 and c_2:

$$\left(1 - \frac{A}{10}\right) c_1 + \frac{1}{2}\left(1 - \frac{A}{10}\right) c_2 = \frac{B}{4} \tag{9-99d}$$

$$\left(1 - \frac{A}{10}\right) c_1 + \frac{6}{15}\left(2 - \frac{A}{7}\right) c_2 = \frac{3}{10} B \tag{9-99e}$$

For the case of $A = B = 1$, these coefficients are

$$c_1 = \frac{71}{369}, \qquad c_2 = \frac{7}{41}$$

then, the two-term trial solution becomes

$$\tilde{T}_2(x) = x(1 - x)\left(\frac{71}{369} + \frac{7}{41} x\right) \tag{9-100}$$

TABLE 9-2 A Comparison of Approximate and Exact Solutions of Example 9-7 for $A = B = 1$

x	T Exact	T Approx.	% Error	T_2 Approx.	% Error
0.25	0.04400	0.0521	+18.4	0.04408	+0.18
0.50	0.06974	0.0694	− 0.48	0.06944	−0.43
0.75	0.06005	0.0521	−13.2	0.06009	+0.06
0.85	0.04282	0.0354	−17.3	0.04302	+0.46

3. *The Exact Solution.* The exact solution of the problem is given by

$$T(x) = \frac{B}{A}\left[\frac{\sin A^{1/2}x}{\sin A^{1/2}} - x\right] \tag{9-101a}$$

For $A = B = 1$, the solution becomes

$$T(x) = 1.1884 \sin x - x \tag{9-101b}$$

We present in Table 9-2 a comparison of the one- and two-term approximate solutions with the exact result. Clearly, the accuracy is significantly improved using a two-term solution.

Example 9-8

Consider the following steady-state heat conduction problem for a solid cylinder:

$$\frac{1}{r}\frac{d}{dr}\left(r\frac{dT}{dr}\right) + \left(1 - \frac{1}{r^2}\right)T = 0 \qquad \text{in} \qquad 1 < r < 2 \tag{9-102a}$$

$$T = 4 \qquad \text{at} \qquad r = 1 \tag{9-102b}$$

$$T = 8 \qquad \text{at} \qquad r = 2 \tag{9-102c}$$

Solve this problem by the Galerkin method using one-term trial solution and compare this approximate result with the exact solution of the problem.

Solution. The application of the Galerkin method gives

$$\int_{r=1}^{2}\left[\frac{1}{r}\frac{d}{dr}\left(r\frac{d\tilde{T}}{dr}\right) + \left(1 - \frac{1}{r^2}\right)\tilde{T}\right]\phi_i(r)\,dr = 0 \tag{9-103}$$

The one-term trial solution is taken in the form

$$\tilde{T}_1(r) = \psi_0(r) + c_1\phi_1(r) \tag{9-104a}$$

where the function $\psi_0(r)$ that satisfies the nonhomogeneous part of the boundary conditions (9-102b, c) is taken as

$$\psi_0(r) = 4r \tag{9-104b}$$

and the first basis function $\phi_1(r)$ that satisfies the homogeneous parts of the boundary conditions is taken as

$$\phi_1(r) = (r-1)(r-2) \tag{9-104c}$$

Substituting the trial solution (9-104) into equation (9-103) for $i = 1$, performing the integration and solving the result for c_1 we obtain $c_1 = -3.245$. Then, the one-term approximate solution becomes

$$\tilde{T}_1(r) = 3.245(r-1)(2-r) + 4r \tag{9-105}$$

The exact solution of this problem is

$$T(r) = 14.43 J_1(r) + 3.008\, Y_1(r) \tag{9-106}$$

where $J_1(r)$ and $Y_1(r)$ are the Bessel functions. A comparison of the approximate and exact solutions at the locations $r = 1.2$, 1.5, and 1.8 shows that the agreement is within 0.03%. Therefore, in this example even the one-term approximation gives very good result.

Example 9-9

Solve the steady-state heat conduction problem in a rectangular region $(-a, a; -b, b)$ with heat generation at a constant rate of g W/m^3 and the boundaries kept at zero temperature using the Galerkin method and compare the result with the exact solution.

Solution. The mathematical formulation of the problem is

$$\frac{\partial^2 T}{\partial x^2} + \frac{\partial^2 T}{\partial y^2} + \frac{1}{k} g = 0 \qquad \text{in} \qquad -a < x < a, \quad -b < y < b \tag{9-107a}$$

$$T = 0 \qquad \text{at} \qquad x = \pm a \qquad \text{and} \quad y = \pm b \tag{9-107b}$$

The solution of this problem by the Galerkin method is written as

$$\int_{x=-a}^{a} \int_{y=-b}^{b} \left[\frac{\partial^2 \tilde{T}}{\partial x^2} + \frac{\partial^2 \tilde{T}}{\partial y^2} + \frac{1}{k} g \right] \phi_i(x, y)\, dx\, dy = 0 \tag{9-108a}$$

We consider one-term trial solution taken as

$$\tilde{T}_1(x, y) = c_1 \phi_1(x, y) \tag{9-108b}$$

where the function ϕ_1 is obtained from equation (9-87a) as

$$\phi_1(x, y) = (a^2 - x^2)(b^2 - y^2) \tag{9-108c}$$

Introducing this trial solution into equation (9-108a) and performing the integrations we obtain

$$c_1 = \frac{5}{8}\frac{g/k}{a^2 + b^2}$$

Hence, the one-term approximate solution becomes

$$T_1(x, y) = \frac{5}{8}\frac{g/k}{a^2 + b^2}(a^2 - x^2)(b^2 - y^2) \tag{9-109}$$

The exact solution of this problem is

$$T(x, y) = \frac{g}{k}\left[\frac{a^2 - x^2}{2}\right] - 2a^2 \sum_{n=0}^{\infty} \frac{(-1)^n}{\beta_n^3} \frac{\cosh\left(\beta_n \frac{y}{b}\right) \cdot \cos\left(\beta_n \frac{x}{a}\right)}{\cosh\left(\beta_n \frac{b}{a}\right)} \tag{9-110}$$

where

$$\beta_n = \frac{(2n + 1)\pi}{2}$$

To compare these two results we consider the center temperature (i.e., $x = 0$, $y = 0$) for the case $a = b$, and obtain

$$\text{Approximate:} \quad T_1(0, 0) = \frac{5}{16}\frac{ga^2}{k} = 0.3125\frac{ga^2}{k} \tag{9-111a}$$

$$\text{Exact:} \quad T(0, 0) = \frac{ga^2}{k}\left[\frac{1}{2} - 2\sum_{n=0}^{\infty} \frac{(-1)^n}{\beta_n^3 \cdot \cosh \beta_n}\right] = 0.293\frac{ga^2}{k} \tag{9-111b}$$

The error involved with one-term solution is about 6.7%. For a two-term trial solution, the temperature distribution may be taken in the form

$$T_2(x, y) = (c_1 + c_2 x^2)(a^2 - x^2)(b^2 - y^2) \tag{9-112}$$

and the calculations are performed in a similar manner to determine the coefficients c_1 and c_2.

9-7 PARTIAL INTEGRATION

In the previous section, the Galerkin method has been applied to the solution of two-dimensional steady-state heat conduction problems by using a trial solution $\tilde{T}(x, y)$ in the x and y variables; as a result, the problem has been reduced to the solution of a set of algebraic equations for the determination of the unknown coefficients $c_1, c_2, \ldots, c_n$. A more accurate approximation is obtainable if a one-dimensional trial function is used either in the x variable $\tilde{T}(x)$ or the y variable $\tilde{T}(y)$, and the problem is reduced to the solution of an ordinary differential equation for the determination of a function $Y(y)$ or $X(x)$. One advantage of such an approach is that, in situations when the functional form of the temperature profile cannot be chosen a priori in one direction, it is left to be determined according to the character of the problem for the solution of the resulting ordinary differential equation.

The partial integration approach is also applicable for the approximate solution of transient heat conduction problems.

We illustrate the application of the partial integration technique with some representative examples.

Example 9-10

Solve the steady-state heat conduction problem considered in Example 9-9 with the Galerkin method using partial integration with respect to the y variable and solving the resulting ordinary differential equation in the x variable.

Solution. The Galerkin method when applied to the differential equation (9-107a) by partial integration with respect to the y variable, gives

$$\int_{y=-b}^{b} \left[\frac{\partial^2 \tilde{T}}{\partial x^2} + \frac{\partial^2 \tilde{T}}{\partial y^2} + \frac{g}{k} \right] \phi_i(y)\, dy = 0 \qquad \text{in} \qquad -a < x < a \tag{9-113}$$

We consider only a one-term trial solution $\tilde{T}_1(x, y)$ choosen as

$$\tilde{T}_1(x, y) = \phi_1(y) X(x) \tag{9-114a}$$

where

$$\phi_1(y) = b^2 - y^2 \tag{9-114b}$$

This trial solution satisfies the boundary conditions at $y = \pm b$; the function $X(x)$ is yet to be determined. Introducing the trial solution (9-114) into equation (9-113) and performing the indicated operations we obtain

$$X''(x) - \frac{5}{2b^2} X(x) = -\frac{5g}{4b^2 k} \qquad \text{in} \qquad -a < x < a \tag{9-115a}$$

Subject to the boundary conditions

$$X(x) = 0 \qquad \text{at} \qquad x = \pm a \tag{9-115b}$$

where the prime shows differentiation with respect to x. The solution of the problem given by equations (9-115) is

$$X(x) = \frac{g}{2k}\left[1 - \frac{\cosh\left(\sqrt{2.5}\,\frac{x}{b}\right)}{\cosh\left(\sqrt{2.5}\,\frac{a}{b}\right)}\right] \tag{9-116}$$

Then the one-term trial solution becomes

$$\tilde{T}_1(x, y) = \frac{g}{2k}(b^2 - y^2)\left[1 - \frac{\cosh\left(\sqrt{2.5}\,\frac{x}{b}\right)}{\cosh\left(\sqrt{2.5}\,\frac{a}{b}\right)}\right] \tag{9-117}$$

and the temperature at the center (i.e., $x = y = 0$) for $a = b$ becomes

$$\tilde{T}_1(x, y) = 0.3026\frac{ga^2}{k} \tag{9-118}$$

This result involves an error of only approximately 3.6%, whereas the one-term approximation obtained in the previous example by the application of the Galerkin method for both x and y variables involves an error of approximately 6.7%. Thus the solution by partial integration improves the accuracy.

Example 9-11

Consider the following steady-state heat conduction problem for a segment of a cylinder, $0 \leqslant r \leqslant 1, 0 \leqslant \theta \leqslant \theta_0$, in which heat is generated at a constant rate of g W/m^3 and all the boundary surfaces are kept at zero temperature.

$$\frac{1}{r}\frac{\partial}{\partial r}\left(r\frac{\partial T}{\partial r}\right) + \frac{1}{r^2}\frac{\partial^2 T}{\partial \theta^2} + \frac{g}{k} = 0 \qquad \text{in} \qquad 0 \leqslant r < 1, \quad 0 < \theta < \theta_0 \tag{9-119a}$$

$$T = 0 \qquad \text{at} \qquad r = 1 \qquad \theta = 0, \quad \theta = \theta_0 \tag{9-119b}$$

Solve this problem using the Galerkin method by partial integration with respect to the θ variable. Compare the approximate result with the exact solution.

Solution. The Galerkin method is now applied to the differential equation (9-119a) by partial integration with respect to the variable θ.

$$\int_{\theta=0}^{\theta_0}\left[\frac{1}{r}\frac{\partial}{\partial r}\left(r\frac{\partial \tilde{T}}{\partial r}\right)+\frac{1}{r^2}\frac{\partial^2 \tilde{T}}{\partial \theta^2}+\frac{g}{k}\right]\phi_i(\theta)d\theta=0 \qquad \text{in} \qquad 0\leqslant r<1 \tag{9-120}$$

we consider a one-term trial solution taken as

$$\tilde{T}(r,\theta)=F(r)\phi_1(\theta) \tag{9-121a}$$

where

$$\phi_1(\theta)=\sin\left(\frac{\pi\theta}{\theta_0}\right) \tag{9-121b}$$

The trial solution thus chosen satisfies the boundary conditions at $\theta=0$ and $\theta=\theta_0$; but the function $F(r)$ is yet to be determined. Introducing the trial solution (9-121) into (9-120) and performing the integration we obtain

$$\frac{1}{r}\frac{d}{dr}\left(r\frac{dF}{dr}\right)-\frac{\beta^2}{r^2}F(r)=-\frac{4g}{k\pi} \qquad \text{in} \qquad 0\leqslant r<1 \tag{9-122a}$$

$$F(r)=0 \qquad \text{at} \qquad r=1 \tag{9-122b}$$

where

$$\beta\equiv\frac{\pi}{\theta_0} \tag{9-122c}$$

A particular solution of equation (9-122a) is

$$F_p=\frac{4g}{\pi k}\frac{r^2}{\beta^2-4}$$

and the complete solution for $F(r)$ is constructed as

$$F(r)=c_1r^{\beta}+c_2r^{-\beta}+\frac{4g}{\pi k}\frac{r^2}{\beta^2-4} \tag{9-123}$$

Here, $c_2=0$ from the requirement that the solution should remain finite at $r=0$; c_1 is determined by the application of the boundary condition at $r=1$ to give

$$c_1=-\frac{4g}{\pi k}\frac{1}{\beta^2-4}$$

Then the solution for $F(r)$ is obtained as

$$F(r) = \frac{4g}{\pi k}\frac{r^2 - r^\beta}{\beta^2 - 4} \tag{9-124}$$

and the one-term trial solution $\tilde{T}(r, \theta)$ becomes

$$\tilde{T}(r, \theta) = \frac{4g}{\pi k}\frac{r^2 - r^{(\pi/\theta_0)}}{(\pi/\theta_0)^2 - 4}\sin\left(\frac{\pi\theta}{\theta_0}\right) \tag{9-125}$$

The exact solution of the problem (9-119) is

$$T(r, 0) = \frac{4g}{\pi k}\sum_{n=1,3,5\ldots}\frac{1}{n}\frac{r^2 - r^{(n\pi/\theta_0)}}{(n\pi/\theta_0)^2 - 4}\sin\left(\frac{n\pi\theta}{\theta_0}\right) \tag{9-126}$$

The one-term approximate solution obtained above represents the first term in the series of the exact solution.

Example 9-12

Solve the steady-state heat conduction problem with constant rate of heat generation for a region bounded by $x = 0, x = a, y = 0$, and $y = f(x)$ for the boundary conditions as shown in Fig. 9-5, using the Galerkin method by partial integration with respect to the y variable.

Solution. The mathematical formulation of this problem is given as

$$\frac{\partial^2 T}{\partial x^2} + \frac{\partial^2 T}{\partial y^2} + \frac{g}{k} = 0 \quad \text{in} \quad 0 < x < a, \quad 0 < y < f(x) \tag{9-127a}$$

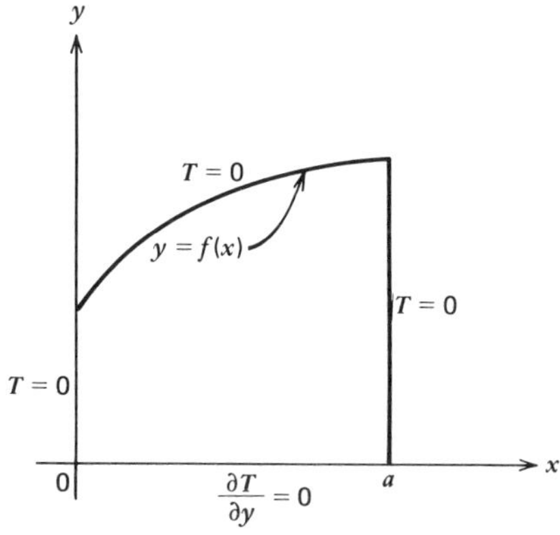

Fig. 9-5 Region considered in Example 9-12.

$$T = 0 \quad \text{at} \quad x = 0, \quad x = a, \quad \text{and} \quad y = f(x) \tag{9-127b}$$

$$\frac{\partial T}{\partial y} = 0 \quad \text{at} \quad y = 0 \tag{9-127c}$$

The Galerkin method is now applied to the differential equation (9-127a) by partial integration with respect to the y variable. We obtain

$$\int_{y=0}^{f(x)} \left[\frac{\partial^2 \tilde{T}}{\partial x^2} + \frac{\partial^2 \tilde{T}}{\partial y^2} + \frac{g}{k} \right] \phi_i(y) dy = 0 \tag{9-128}$$

We consider one-term trial solution taken as

$$\tilde{T}_1(x, y) = X(x) \cdot \phi_1(y) \tag{9-129a}$$

where

$$\phi_1(y) = [y^2 - f^2(x)] \tag{9-129b}$$

Clearly, this trial solution satisfies the boundary conditions at $y = 0$ and $y = f(x)$; but the function $X(x)$ is yet to be determined. Introducing the trial solution (9-129) into equation (9-128) and performing the indicated operations we obtain the following ordinary differential equation for the determination of the function $X(x)$.

$$\tfrac{2}{5} f^2 X'' + 2ff'X' + (ff'' + f'^2 - 1)X = -\frac{g}{2k} \quad \text{in} \quad 0 < x < a \tag{9-130}$$

subject to

$$X = 0 \quad \text{at} \quad x = 0 \quad \text{and} \quad x = a \tag{9-131}$$

Once the function $f(x)$ defining the form of the boundary arc is specified, this equation can be solved and the function $X(x)$ can be determined. For example, the case $y = f(x) = b$, corresponds to a rectangular region and the equation (9.130) reduces to

$$X'' - \frac{5}{2b^2} X = -\frac{5g}{4b^2 k} \quad \text{in} \quad 0 < x < a \tag{9-132}$$

$$X = 0 \quad \text{at} \quad x = 0 \quad \text{and} \quad x = a \tag{9-133}$$

which is the same as that given by equations (9-115); and the one term approximate solution becomes

$$\tilde{T}_1(x, y) = (y^2 - b^2) X(x) \tag{9-134}$$

where $X(x)$ is as given by equation (9-116).

9-8 APPLICATION TO TRANSIENT PROBLEMS

We now illustrate the application of the Galerkin method to the solution of time dependent problems with the following two examples.

Example 9-13

A slab in $0 \leqslant x \leqslant 1$ is initially at a temperature $T(x,t) = T_0(1 - x^2)$. For times $t > 0$, the boundary at $x = 0$ is kept insulated and the boundary at $x = 1$ is kept at zero temperature. Using the Galerkin method combined with partial integration, obtain an approximate solution for the temperature distribution $\tilde{T}(x,t)$ in the slab and compare it with the exact solution $T(x,t)$.

Solution. The mathematical formulation of this problem is given as

$$\frac{\partial^2 T}{\partial x^2} = \frac{1}{\alpha}\frac{\partial T(x,t)}{\partial t} \quad \text{in} \quad 0 < x < 1, \quad t > 0 \tag{9-135a}$$

$$\frac{\partial T}{\partial x} = 0 \quad \text{at} \quad x = 0, \quad t > 0 \tag{9-135b}$$

$$T = 0 \quad \text{at} \quad x = 1, \quad t > 0 \tag{9-135c}$$

$$T = T_0(1 - x^2) \quad \text{for} \quad t = 0, \quad \text{in} \quad 0 \leqslant x \leqslant 1 \tag{9-135d}$$

We apply the Galerkin method to equation (9-135a) with partial integration with respect to x and obtain

$$\int_{x=0}^{1} \left[\frac{\partial^2 \tilde{T}}{\partial x^2} - \frac{1}{\alpha}\frac{\partial \tilde{T}}{\partial t} \right] \phi_i(x)\,dx = 0 \tag{9-136}$$

and choose a one-term trial solution $\tilde{T}_1(x,t)$ as

$$\tilde{T}_1(x,t) = T_0 f(t)\phi_1(x) \tag{9-137a}$$

where

$$\phi_1(x) = 1 - x^2 \tag{9-137b}$$

$$f(t) = 1 \quad \text{for} \quad t = 0 \tag{9-137c}$$

and the function $f(t)$ is yet to be determined. Clearly, the trial solution chosen as above satisfies the initial condition and the two boundary conditions for the problem. Substituting the trial solution (9-137) into equation (9-136) and performing the indicated operations we obtain the differential equation for $f(t)$ as

$$\frac{\partial f(t)}{dt} + \frac{5\alpha}{2} f(t) = 0 \qquad \text{for} \qquad t > 0 \tag{9-138a}$$

$$f(t) = 1 \qquad \text{for} \qquad t = 0 \tag{9-138b}$$

The solution for $f(t)$ is

$$f(t) = e^{-(5\alpha/2)t} \tag{9-139}$$

and the one-term approximate solution $\tilde{T}_1(x, t)$ becomes

$$\frac{\tilde{T}(x, t)}{T_0} = (1 - x^2) e^{-(5\alpha/2)t} \tag{9-140}$$

The exact solution of the problem (9-135) is obtained as

$$\frac{T(x, t)}{T_0} = 4 \sum_{n=0}^{\infty} (-1)^n \frac{1}{\beta_n^3} e^{-\alpha\beta_n^2 t} \cos \beta_n x \tag{9-141a}$$

where

$$\beta_n = \frac{(2n+1)\pi}{2} \tag{9-141b}$$

We list in Table 9-3 a comparison of this approximate solution with the exact solution. Even the one-term approximate solution is in reasonably good agreement with the exact solution. Improved approximations can be obtained by choosing a higher-order trial solution in the form

$$\tilde{T}_n(x, t) = T_0 \sum_{i=1}^{n} f_i(t) \phi_i(x) \tag{9-142}$$

where the functions $\phi_i(x)$ satisfy the boundary conditions for the problem and

TABLE 9-3 A Comparison of Approximate and Exact Solutions of Example 9-13

	$[(\tilde{T}_1 - T)/T] \times 100$		
x	$\alpha t = 0.01$	$\alpha t = 0.1$	$\alpha t = 1$
0.2	+1	−1	+4.4
0.6	+2	+5.5	+3.1

the function $f_i(t)$ with $f_i(0)=1$ is determined from the resulting ordinary differential equations obtained after the application of the Galerkin method with partial integration with respect to the x variable.

Example 9-14

The transient heat conduction problem for a solid cylinder, $0 \leqslant r \leqslant 1$, with heat generation within the medium is given in the dimensionless form as

$$\frac{1}{r}\frac{\partial}{\partial r}\left(r\frac{\partial T}{\partial r}\right)+G(r)=\frac{\partial T(r,t)}{\partial t} \quad \text{in} \quad 0<r<1, \quad t>0 \tag{9-143a}$$

$$T=\text{finite} \quad \text{at} \quad r=0, \quad t>0 \tag{9-143b}$$

$$T=0 \quad \text{at} \quad r=1, \quad t>0 \tag{9-143c}$$

$$T=0 \quad \text{for} \quad t=0, \quad 0\leqslant r\leqslant 1 \tag{9-143d}$$

Solve this problem by the combined application of the Laplace transform and the Galerkin method.

Solution. The Laplace transform of this problem with respect to the time variable is

$$L(\bar{T})\equiv\frac{d}{dr}\left(r\frac{d\bar{T}}{dr}\right)-sr\bar{T}+\frac{1}{s}rG(r)=0 \quad \text{in} \quad 0<r<1 \tag{9-144a}$$

$$\bar{T}(r,s)=\text{finite} \quad \text{at} \quad r=0 \tag{9-144b}$$

$$\bar{T}(r,s)=0 \quad \text{at} \quad r=1 \tag{9-144c}$$

where $\bar{T}(r,s)$ is the Laplace transform of $T(r,t)$ and s is the Laplace transform variable.

The application of the Galerkin method to equation (9-144a) is written as

$$\int_{r=0}^{1} L[\bar{\bar{T}}(r,s)]\phi_i(r)dr=0 \tag{9-145}$$

where $\bar{\bar{T}}(r,s)$ is the trial solution for $\bar{T}(r,s)$ and $\phi_i(r)$ are the functions that satisfy the boundary conditions for the problem and from which the trial solution is constructed. In this example we show that if the proper function is chosen for $\phi_i(r)$ and sufficient number of $\phi_i(r)$ are included to construct the trial solution, it is possible to obtain the exact solution for the problem.

We choose $\phi_i(r)$ as

$$\phi_i(r)=J_0(\beta_i r) \tag{9-146a}$$

and the β_i values are the roots of

$$J_0(\beta_i) = 0 \tag{9-146b}$$

Then, each of the functions $\phi_i(r)$ satisfies the boundary conditions (9-144b, c) for the problem. We construct the trial solution $\bar{T}(r, s)$ in terms of the $\phi_i(r)$ functions as

$$\tilde{\bar{T}} = \sum_j c_j \phi_j(r) = \sum_j c_j J_0(\beta_j r) \tag{9-147}$$

where the summation is taken over the permissible values of β_j as defined by equation (9-146b). Introducing equations (9-147) and (9-146a) into equation (9-145) we obtain

$$\int_{r=0}^{1} L\left[\sum_j c_j J_0(\beta_j r)\right] \cdot J_0(\beta_i r)\, dr = 0, \qquad i = 1, 2, \ldots \tag{9-148a}$$

or

$$\sum_j c_j \int_{r=0}^{1} \left[\frac{d}{dr}\left(r\frac{dJ_0(\beta_j r)}{dr}\right) - srJ_0(\beta_j r)\right] J_0(\beta_i r)\, dr + \frac{1}{s}\int_{r=0}^{1} rG(r) J_0(\beta_i r)\, dr = 0 \tag{9-148b}$$

or

$$-\sum_j c_j(\beta_j^2 + s) \int_{r=0}^{1} rJ_0(\beta_j r) J_0(\beta_i r)\, dr + \frac{1}{s}\int_{r=0}^{1} rG(r) J_0(\beta_i r)\, dr = 0 \tag{9-148c}$$

The first integral is evaluated as

$$\int_{r=0}^{1} rJ_0(\beta_j r) J_0(\beta_i r)\, dr = \begin{cases} 0 & i \neq j \\ \frac{1}{2} J_1^2(\beta_i) & i = j \end{cases} \tag{9-149}$$

Introducing (9-149) into (9-148c), the summation drops out and we obtain

$$c_i = \frac{2}{s(s + \beta_i^2)} \frac{1}{J_1^2(\beta_i)} \int_{r=0}^{1} rG(r) J_0(\beta_i r)\, dr \tag{9-150}$$

We introduce equation (9-150) into (9-147) after changing i to j and r to r' to obtain

$$\tilde{\bar{T}}(r, s) = 2\sum_j \frac{1}{s(s + \beta_j^2)} \frac{J_0(\beta_j r)}{J_1^2(\beta_j)} \int_{r'=0}^{1} r' G(r') J_0(\beta_j r')\, dr' \tag{9-151a}$$

or

$$\tilde{\bar{T}}(r,s) = 2\sum_j \frac{1}{\beta_j^2}\left(\frac{1}{s} - \frac{1}{s+\beta_j^2}\right)\frac{J_0(\beta_j r)}{J_1^2(\beta_j)}\int_{r'=0}^{1} r'G(r')J_0(\beta_j r')dr' \quad (9\text{-}151b)$$

The Laplace transform can be inverted by means of the Laplace transform Table 7-1, cases 1 and 8; we obtain

$$\tilde{T}(r,t) = 2\sum_j \frac{1}{\beta_j^2}(1 - e^{-\beta_j^2 t})\frac{J_0(\beta_j r)}{J_1^2(\beta_j)}\int_{r'=0}^{1} r'G(r')J_0(\beta_j r')dr' \quad (9\text{-}152a)$$

where the summation is over all eigenvalues β_j's which are the positive roots of

$$J_0(\beta_j) = 0 \quad (9\text{-}152b)$$

We note that, the solution obtained in this manner is in fact the exact solution of this problem.

REFERENCES

1. H. Schlichting, *Boundary Layer, Theory*, 6th ed., McGraw-Hill, New York, 1968, Chapter 13.
2. H. D. Landahl, *Bull. Math. Biophys.* **15**, 49–61, 1953.
3. H. J. Merk, *Appl. Sci. Res.* **4**, Section A, 435–452, 1954.
4. T. R. Goodman, *Trans. ASME* **80**, 335–342, 1958.
5. T. R. Goodman, *J. Heat Transfer* **83c**, 83–86, 1961.
6. T. R. Goodman, in T. F. Irvine and J. P. Hartnett, eds., *Advances in Heat Transfer*, Vol. 1, Academic Press, New York, 1964, pp. 52–120.
7. W. C. Reynolds and T. A. Dolton, The Use of Integral Methods in Transient Heat Transfer Analysis, Department of Mechanical Engineering, Report No. 36, Stanford University, Stanford, Calif., Sept. 1, 1958.
8. K. T. Yang, *Trans. ASME* **80**, 146–147, 1958.
9. K. T. Yang, *International Developments in Heat Transfer*, Vol. 1, ASME, New York, 1963, pp. 18–27.
10. P. J. Schneider, *J. Aerospace Sci.* **27**, 546–549, 1960.
11. T. J. Lardner and F. B. Pohle, *J. Appl. Mech.* 310–312, June 1961.
12. R. Thorsen and F. Landis, *Int. J. Heat Mass Transfer* **8**, 189–192, 1965.
13. B. Persson and L. Persson, Calculation of the Transient Temperature Distribution in Convectively Heated Bodies With Integral Method, ASME Paper No. 64-HT-19, 1964.
14. F. A. Castello, An Evaluation of Several Methods of Approximating Solutions to the Heat Conduction Equation, ASME Paper No. 63-HT-44, 1963.
15. H. H. Bengston and F. Kreith, *J. Heat Transfer* **92c**, 182–184, 1970.

16. M. N. Özişik, *Boundary Value Problems of Heat Conduction*, International Textbook Co., Scranton, P., 1968.
17. M. Altman, *Chem. Eng. Prog. Symp. Series*, **57**, 16–23, Buffalo, 1961.
18. G. Poots, *Int. J. Heat Mass Transfer* **5**, 339–348, 1962.
19. G. Poots, *Int. J. Heat Mass Transfer* **5**, 525, 1962.
20. R. H. Tien and G. E. Geiger, *J. Heat Transfer* **89c**, 230–234, 1967.
21. R. H. Tien and G. E. Geiger, *J. Heat Transfer* **90c**, 27–31, 1968.
22. J. C. Muehlbauer, J. D. Hatcher, D. W. Lyons, and J. E. Sunderland, *J. Heat Transfer* **95c**, 324–331, 1973.
23. S. Cho and J. E. Sunderland, Melting or Freezing of Finite Slabs, ASME Paper 68-WA/HT-37, Dec. 1968.
24. S. H. Cho and J. E. Sunderland, *J. Heat Transfer* **91c**, 421–426, 1969.
25. K. Mody and M. H. Özişik, *Lett. Heat Mass Transfer* **2**, 487–493, 1975.
26. O. M. Griffin, *J. Heat Transfer* **95c**, 317–322, 1973.
27. O. M. Griffin, *Poly. Eng. Sci.* **12**, 140–149, 1972.
28. O. M. Griffin, *Proceedings of the 5th International Heat Transfer Conference*, Vol. 1, Tokyo, 1974, pp. 211–215.
29. O. M. Griffin, *Int. J. Heat Mass Transfer* **20**, 675–683, 1977.
30. W. F. Ames, *Nonlinear Partial Differential Equations in Engineering*, Academic Press, New York, 1965.
31. W. F. Ames, *Nonlinear Ordinary Differential Equations in Transport Process*, Academic Press, New York, 1968.
32. R. Weinstock, *Calculus of Variations*, McGraw-Hill, New York, 1952.
33. I. M. Gelfand and S. V. Fomin, *Calculus of Variations*, Prentice-Hall, Englewood Cliffs, N.J., 1963.
34. P. M. Morse and H. Feshbach, *Methods of Theoretical Physics*, McGraw-Hill, New York, 1953.
35. A. R. Forsythe, *Calculus of Variations*, Dover Publications, New York, 1960.
36. S. G. Mikhlin, *Variational Methods in Mathematical Physics*, Macmillan, New York, 1964.
37. R. S. Schechter, *The Variational Method in Engineering*, McGraw-Hill, New York, 1967.
38. M. A. Biot, *J. Aeronaut. Sci.* **24**, 857–873, 1957.
39. M. A. Biot, *J. Aeronaut. Sci.* **26**, 367–381, 1959.
40. P. D. Richardson, *J. Heat Transfer* **86c**, 298–299, 1964.
41. V. S. Arpaci and C. M. Vest, Variational Formulation of Transformed Diffusion Problems, ASME Paper No. 67-HT-77.
42. T. J. Lardner, *AIAA J.* **1**, 196–206, 1963.
43. M. E. Gurtin, *Quart. Appl. Math.* **22**, 252–256, 1964.
44. B. A. Finlayson and L. E. Scriven, *Int. J. Heat Mass Transfer* **10**, 799–821, 1967.
45. D. F. Hays and H. N. Curd, *Int. J. Heat Mass Transfer* **11**, 285–295, 1968.
46. P. Rafalski and W. Zyszkowski, *AIAA J.* **6**, 1606, 1968.
47. W. Zyszkowski, *J. Heat Transfer* **91c**, 77–82, 1969.

48. D. Djukic and B. Vujanovic, *Z. Angew. Math. Mech.* **51**, 611–616, 1971.
49. B. Vujanovic and A. M. Straus, *AIAA J.* **9**, 327–330, 1971.
50. B. Krajewski, *Int. J. Heat Mass Transfer* **18**, 495–502, 1975.
51. W. Ritz, *J. Reine Angewandte Math.* **135**, 1–61, 1908.
52. N. Kryloff, *Memorial Sci. Math. Paris* **49**, 1931.
53. L. V. Kantorovich and V. I. Krylov, *Approximate Methods of Higher Analysis*, Wiley, New York, 1964.
54. L. Collatz, *The Numerical Treatment of Differential Equations* (English trans.), Springer-Verlag, Heidelberg, 1960.
55. S. H. Crandall, *Engineering Analysis*, McGraw-Hill, New York, 1956.
56. V. S. Arpaci, *Conduction Heat Transfer*, Addison-Wesley, Reading, Mass., 1966.
57. G. B. Galerkin, *Vestnik Inzhenerov Teckhnikov*, p. 879, 1915.
58. N. W. McLachlan, *Phil. Mag.* **36**, 600, 1945.
59. D. Dicker and M. B. Friedman, *J. Appl. Mech.* **30**, 493–499, 1963.
60. F. Erdogan, *J. Heat Transfer* **85c**, 203–208, 1963.
61. L. J. Snyder, T. W. Spiggs, and W. E. Stewart, *AICHE J.* **10**, 535–540, 1964.
62. B. A. Finlayson and L. E. Scriven, *Chem. Eng. Sci.* **20**, 395–404, 1965.
63. P. A. Laura and A. J. Faulstich, *Int. J. Heat Mass Transfer* **11**, 297–303, 1968.
64. A. H. Eraslan, *J. Heat Transfer* **91c**, 212–220, 1969.
65. A. H. Eraslan, *AIAA J.* **10**, 1759–1766, 1966.
66. M. Mashena and A. Haji-Sheikh, *Int. J. Heat Mass Transfer* **29**, 317–329, 1986.
67. A. Haji-Sheikh and M. Mashena, *J. Heat Transfer* **109**, 551–556, 1987.
68. A. Haji-Sheikh and R. Lakshminarayanan, *J. Heat Transfer* **109**, 557–562, 1987.
69. S. Nomura and A. Haji-Sheikh, *J. Heat Transfer* **110**, 110–112, 1988.

PROBLEMS

9-1 A semiinfinite region $x > 0$ is initially at zero temperature. For times $t > 0$, the convection boundary condition at the surface $x = 0$ is given as $-k(\partial T/\partial x) + hT = f_1$, where $f_1 =$ constant. Obtain an expression for the temperature distribution $T(x, t)$ in the medium using the integral method with a cubic polynomial representation for temperature.

9-2 A semiinfinite medium $x > 0$ is initially at a uniform temperature T_i. For times $t > 0$, the boundary surface at $x = 0$ is subjected to a prescribed heat flux, that is, $-k(\partial T \partial x) = f(t)$ at $x = 0$, where $f(t)$ varies with time. Obtain an expression for the temperature distribution $T(x, t)$ in the medium using the integral method and a cubic polynomial representation for $T(x, t)$.

9-3 A region exterior to a cylindrical hole of radius $r = b$ (i.e., $r > b$) is initially at zero temperature. For times $t > 0$ the boundary surface at $r = b$ is kept at a constant temperature T_0. Obtain an expression for the temperature distribution in the medium using the integral method with a second-degree polynomial representation modified by $\ln r$ for $T(x, t)$.

9-4 A simiinfinite medium $x > 0$ is initially at zero temperature. For times $t > 0$ heat is generated in the medium at a constant rate of g W/m^3, while heat is removed from the boundary surface at $x = 0$ as $k(\partial T/\partial x) = f =$ constant. Obtain an expression for the temperature distribution $T(x, t)$ in the medium for times $t > 0$, using the integral method and a cubic polynomial representation for $T(x, t)$.

9-5 Consider a heat conduction problem for a semiinfinite medium $x > 0$ with the fourth-power radiative heat transfer at the boundary surface $x = 0$ defined as

$$\frac{\partial^2 T}{\partial x^2} = \frac{1}{\alpha}\frac{\partial T}{\partial t} \qquad \text{in} \qquad x > 0, \quad t > 0$$

$$k\frac{\partial T}{\partial x} = \sigma\varepsilon(T_s^4 - T_\infty^4) \qquad \text{at} \qquad x = 0, \quad t > 0$$

$$T = T_i \qquad \text{for} \qquad t = 0, \quad x \geqslant 0$$

where T_s is the surface temperature. Apply the formal solution given by equation (9-40) for the solution of this problem. For the case of $T_\infty = 0$, by performing the resulting integration analytically obtain an expression for the surface temperature T_s as a function of time.

9-6 Consider the following steady-state heat conduction problem for a rectangular region $0 < x < a, 0 < y < b$:

$$\frac{\partial^2 T}{\partial x^2} + \frac{\partial^2 T}{\partial y^2} = 0 \qquad \text{in} \qquad 0 < x < a, \quad 0 < y < b$$

$$T = 0 \qquad \text{at} \qquad y = 0, \qquad y = b$$

$$\frac{\partial T}{\partial x} = 0 \qquad \text{at} \qquad x = 0$$

$$T = T_0 \sin\left(\frac{3\pi y}{b}\right) \qquad \text{at} \qquad x = a$$

Solve this problem by the Galerkin method using partial integration with respect to the y variable for a trial function chosen in the form $\tilde{T}_1(x, y) = f(x)\sin(3\pi y/b)$ and compare this result with the exact solution.

9-7 Solve the following steady-state heat conduction problem:

$$\frac{\partial^2 T}{\partial x^2} + \frac{\partial^2 T}{\partial y^2} + \frac{1}{k}g = 0 \qquad \text{in the region shown in Fig. 9-4}b$$

$$T = 0 \qquad \text{on the boundaries}$$

using the Galerkin method and a one-term trial solution chosen in the form

$$\tilde{T}_1(x, y) = c_1(y - \alpha x)(y + \beta x)(L - x)$$

9-8 Solve the following steady-state heat conduction problem

$$\frac{\partial^2 T}{\partial x^2} + \frac{\partial^2 T}{\partial y^2} + \frac{1}{k} g = 0 \qquad \text{in the region shown in Fig. 9-4}c \quad \text{for} \quad a = b = 1$$

$$T = 0 \qquad \text{on the boundaries}$$

using the Galerkin method and a one-term trial solution.

10

MOVING HEAT SOURCE PROBLEMS

There are numerous engineering applications, such as welding, grinding, metal cutting, firing a bullet in a gun barrel, flame or laser hardening of metals, and many others in which the calculation of temperature field in the solid is modeled as a problem of heat conduction involving a moving heat source. Following the pioneering works of Rosenthal [1–4] on the determination of temperature distribution in a solid resulting from arc welding, numerous papers appeared on the subject of heat transfer in solids with moving heat sources [5–29].

In machining, grinding, cutting, and sliding of surfaces, the energy generated as a result of friction heating can be modeled as a moving heat source. The determination of temperature field around such heat sources has been studied by several investigators [14–23].

More recently lasers—because of their ability to produce high-power beams—have found applications in welding, drilling, cutting, machining of brittle materials, and surface hardening of metallic alloys. For example, in surface hardening, a high-power laser beam scans over the surface and unique metallurgical structures may be produced by rapid cooling that occurs subsequent to the laser heating. The determination of temperature field around a moving laser beam has been studied in several references [24–32].

The objective of this chapter is to introduce the mathematical formulation and the method of solution of heat conduction problems involving a moving heat source by considering simple, representative examples for which analytic solutions are obtainable by the method of separation of variables under quasi-stationary conditions.

10-1 MATHEMATICAL MODELING OF MOVING HEAT SOURCE PROBLEMS

A moving heat source, depending on the physical nature of the problem, can be modeled as a *point*, *line*, *surface*, or *ring* heat source that may release its energy either continuously over the time or spontaneously at specified times. As discussed in Chapter 6, we use the following notation to identify various types of *continuous* heat sources:

$$g_{\mathrm{p}}^{\mathrm{c}} = \text{point source, W}$$
$$g_{\mathrm{L}}^{\mathrm{c}} = \text{line source, W/m}$$
$$g_{\mathrm{s}}^{\mathrm{c}} = \text{surface source, W/m}^2$$

where the superscript c refers to a continuous source. For an instantaneous source we change the superscript c to i and alter the units of the source accordingly as discussed in Chapter 6.

The spatial distribution of the strength of the heat source depends on the physical nature of the source. For example, the energy distribution in a laser beam generally is not uniform spatially. It may have a Gaussian distribution (i.e., intensity decreasing exponentially from the center of the beam with the square of the radial distance) or a doughnut shape or a combination of these two shapes. Also, it may be a continuous source over the time or activated as pulses for short periods of time.

In this section we present the mathematical modeling of the determination of temperature fields in solids resulting from a moving *point*, *line*, and *surface* heat sources under the *quasi-stationary state* conditions.

A Moving Point Heat Source

We consider a point heat source of constant strength $g_{\mathrm{p}}^{\mathrm{c}}$ watts, releasing its energy continuously over the time while moving along the x axis in the positive x direction with a constant velocity u, in a stationary medium that is initially at zero temperature. Figure 10-1*a* illustrates the geometry and the coordinates.

The three-dimensional heat conduction equation in the fixed x, y, z coordinate system, assuming constant properties, is taken as

$$\frac{\partial^2 T}{\partial x^2} + \frac{\partial^2 T}{\partial y^2} + \frac{\partial^2 T}{\partial z^2} + \frac{1}{k} g(x, y, z, t) = \frac{1}{\alpha} \frac{\partial T}{\partial t} \tag{10-1a}$$

where $T \equiv T(x, y, z, t)$.

Let the heat source be a point heat source of constant strength $g_{\mathrm{p}}^{\mathrm{c}}$ watts, located at $y = 0, z = 0$ and releasing its energy continuously as it moves along the x axis in the positive x direction with a constant velocity u. Such a point heat source is related to the volumetric source $g(x, y, z, t)$ by the delta function

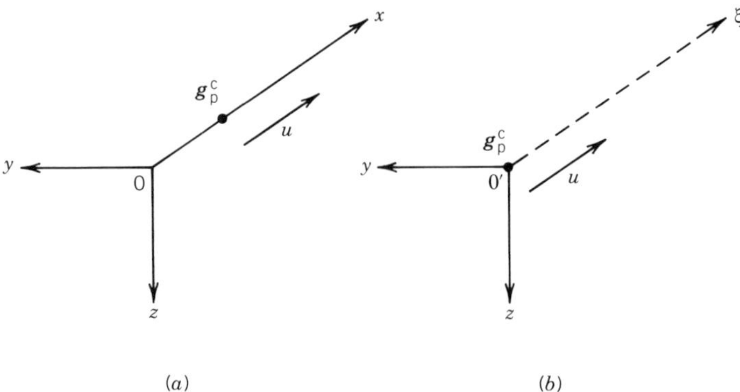

Fig. 10-1 A moving point heat source: (*a*) fixed coordinates x, y, z; (*b*) moving coordinates ξ, y, z.

notation as

$$g(x, y, z, t) \equiv g_p^c \delta(y-0)\delta(z-0)\delta(x-ut) \tag{10-1b}$$

$$\frac{\text{W}}{\text{m}^3} \qquad \text{W} \qquad \frac{1}{\text{m}} \qquad \frac{1}{\text{m}} \qquad \frac{1}{\text{m}}$$

where $\delta(*)$ denotes the Dirac delta function.

Transformation of the Origin. In the solution of moving heat source problems, it is convenient to let the coordinate system move with the source. This is achieved by introducing a new coordinate ξ defined by

$$\xi = x - ut \tag{10-2}$$

Figure 10-1*b* illustrates the new coordinate system ξ, y, z that moves with the source. The heat conduction equation (10-1a) is transformed from the fixed x, y, z coordinate system with fixed origin 0 to the moving coordinate system ξ, y, z with moving origin 0′ by the application of the chain rule of differentiation given by

$$\frac{\partial T(\xi, y, z, t)}{\partial t} = \frac{\partial T}{\partial \xi}\frac{\partial \xi}{\partial t} + \frac{\partial T}{\partial y}\frac{dy}{dt} + \frac{\partial T}{\partial z}\frac{dz}{dt} + \frac{\partial T}{\partial t}\frac{dt}{dt} = -u\frac{\partial T}{\partial \xi} + \frac{\partial T}{\partial t} \tag{10-3a}$$

since $(\partial \xi/\partial t) = u$. The derivatives with respect to x becomes

$$\frac{\partial T}{\partial x} = \frac{\partial T}{\partial \xi}\frac{\partial \xi}{\partial x} + \frac{\partial T}{\partial t}\frac{dt}{dx} + \cdots = \frac{\partial T}{\partial \xi} \tag{10-3b}$$

$$\frac{\partial^2 T}{\partial x^2} = \frac{\partial^2 T}{\partial \xi^2} \tag{10-3c}$$

but the partial derivatives with respect to y and z remain unaltered. Then, the heat conduction equation (10-1a) in the coordinate system ξ, y, z moving with the source is given by

$$\frac{\partial^2 T}{\partial \xi^2} + \frac{\partial^2 T}{\partial y^2} + \frac{\partial^2 T}{\partial z^2} + \frac{1}{k} g_{\mathrm{p}}^{\mathrm{c}} \delta(\xi - 0)\delta(y - 0)\delta(z - 0) = \frac{1}{\alpha}\left(\frac{\partial T}{\partial t} - u\frac{\partial T}{\partial \xi}\right) \tag{10-4}$$

We note that this equation is a special case of the heat conduction equation for a moving solid given by equation (1-56) in Chapter 1. In equation (10-4), the solid is moving with a velocity u in the negative ξ direction with respect to an observer located at the source; this is the reason for the negative sign in front of the velocity in equation (10-4).

Quasi-Stationary Condition. Experiments have shown that, if the solid is long enough compared to the penetration depth to heat transfer field, the temperature distribution around the heat source soon becomes independent of time. That is, an observer stationed at the moving origin 0′ of the ξ, y, z coordinate system fails to notice any change in the temperature distribution around him/her as the source moves on. This is identified as the quasi-stationary condition [3] and mathematically defined by setting $\partial T/\partial t = 0$. Therefore, the quasi-stationary form of equation (10-4) is obtained by setting $\partial T/\partial t = 0$ as

$$\frac{\partial^2 T}{\partial \xi^2} + \frac{\partial^2 T}{\partial y^2} + \frac{\partial^2 T}{\partial z^2} + \frac{1}{k} g_{\mathrm{p}}^{\mathrm{c}} \delta(\xi)\delta(y)\delta(z) = -\frac{u}{\alpha}\frac{\partial T}{\partial \xi} \tag{10-5}$$

where $\delta(y) \equiv \delta(y - 0)$, and so forth.

Equation (10-5) can be transformed into a more convenient form by introducing a new dependent variable $\theta(\xi, y, z)$ defined as

$$T(\xi, y, z) = \theta(\xi, y, z) e^{-(u/2\alpha)\xi} \tag{10-6}$$

Then equation (10-5) takes the form

$$\frac{\partial^2 \theta}{\partial \xi^2} + \frac{\partial^2 \theta}{\partial y^2} + \frac{\partial^2 \theta}{\partial z^2} - \left(\frac{u}{2\alpha}\right)^2 \theta + \frac{1}{k} g_{\mathrm{p}}^{\mathrm{c}} \delta(\xi)\delta(y)\delta(z) e^{(u/2\alpha)\xi} = 0 \tag{10-7}$$

Here, the exponential $e^{(u/2\alpha)\xi}$ appearing in the source term can be omitted, since the term vanishes for $\xi \neq 0$ because of the delta function $\delta(\xi)$ and the exponential becomes unity for $\xi = 0$.

A Moving Line Heat Source

We now consider a line heat source of constant strength $g_{\mathrm{L}}^{\mathrm{c}}$ W/m, located at the x axis and oriented parallel to the z axis as illustrated in Fig. 10-2a. The source releases its energy continuously over the time as it moves with a constant velocity u in the positive x direction. The medium is initially at zero temperature. We assume $(\partial T/\partial z) = 0$ everywhere in the medium. Then the two-dimensional heat conduction equation in the x, y coordinates is taken as

$$\frac{\partial^2 T}{\partial x^2} + \frac{\partial^2 T}{\partial y^2} + \frac{1}{k} g(x, y, t) = \frac{1}{\alpha}\frac{\partial T}{\partial t} \tag{10-8a}$$

where $T \equiv T(x, y, t)$. The line heat source g_L^c W/m is related to the volumetric source $g(x, y, t)$ by the delta function notation as

$$\underset{\frac{\text{W}}{\text{m}^3}}{g(x, y, t)} = \underset{\frac{\text{W}}{\text{m}}}{g_L^c}\,\underset{\frac{1}{\text{m}}}{\delta(y-0)}\,\underset{\frac{1}{\text{m}}}{\delta(x-ut)} \tag{10-8b}$$

Transformation of the Origin. This heat conduction problem is now transformed from the x, y fixed coordinates to new ξ, y coordinates moving with the line-heat source by the transformation

$$\xi = x - ut \tag{10-9}$$

as illustrated in Fig. 10-2.

By following the procedure described previously, we transform the heat conduction equation (10-8) to the moving coordinate system (ξ, y) as

$$\frac{\partial^2 T}{\partial \xi^2} + \frac{\partial^2 T}{\partial y^2} + \frac{1}{k} g_L^c \delta(\xi)\delta(y) = \frac{1}{\alpha}\left(\frac{\partial T}{\partial t} - u\frac{\partial T}{\partial \xi}\right) \tag{10-10}$$

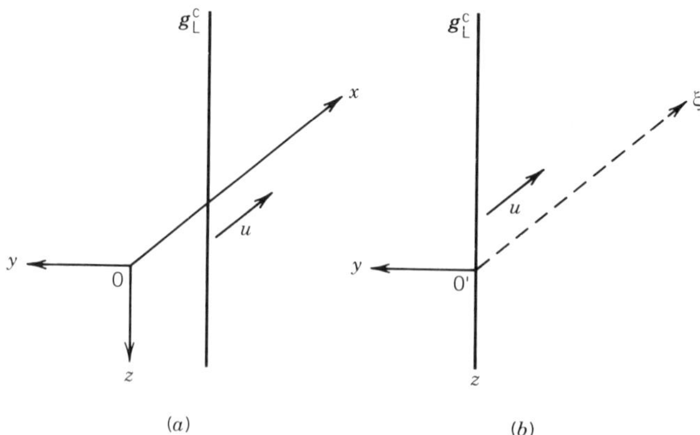

Fig. 10-2 A moving line heat source: (a) fixed coordinates x, y; (b) moving coordinates ξ, y.

Quasi-Stationary Condition. As discussed previously, the quasi-stationary form of equation (10-10) is obtained by setting $(\partial T/\partial t) = 0$. We find

$$\frac{\partial^2 T}{\partial \xi^2} + \frac{\partial^2 T}{\partial y^2} + \frac{1}{k} g_L^c \delta(\xi)\delta(y) = -\frac{u}{\alpha}\frac{\partial T}{\partial \xi} \tag{10-11}$$

This equation is transformed into a more convenient form by introducing a new dependent variable $\theta(\xi, y)$ defined as

$$T(\xi, y) = \theta(\xi, y)e^{-(u/2\alpha)\xi} \tag{10-12}$$

Then equation (10-11) takes the form

$$\frac{\partial^2 \theta}{\partial \xi^2} + \frac{\partial^2 \theta}{\partial y^2} - \left(\frac{u}{2\alpha}\right)^2 \theta + \frac{1}{k} G_L = 0 \tag{10-13a}$$

where

$$G_L \equiv g_L^c \delta(\xi)\delta(y) \tag{10-13b}$$

We note that the term $e^{(u/2\alpha)\xi}$ that would have appeared on the right-hand side of equation (10-13b) is omitted for the reason stated previously.

A Moving Plane Surface Heat Source

We now consider a plane surface heat source of constant strength g_s^c W/m^2, oriented perpendicular to the x axis, as illustrated in Fig. 10-3. The source releases its energy continuously over the time as it moves with a constant velocity u in the positive x direction. For the one-dimensional case considered here we assume $(\partial T/\partial y) = (\partial T/\partial z) = 0$ everywhere, hence the differential equation of heat

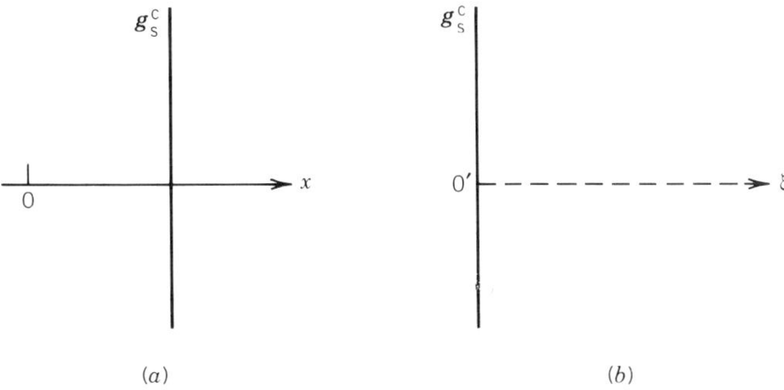

Fig. 10-3 A moving plane surface heat source: (*a*) fixed coordinate x; and (*b*) moving coordinate ξ.

conduction reduces to

$$\frac{\partial^2 T}{\partial x^2} + \frac{1}{k} g(x, t) = \frac{1}{\alpha} \frac{\partial T}{\partial t} \tag{10-14a}$$

where $T \equiv T(x, t)$. The moving continuous surface heat source g_s^c is related to the volumetric source $g(x, t)$ by

$$\underset{\frac{\mathrm{W}}{\mathrm{m}^3}}{g(x,t)} = \underset{\frac{\mathrm{W}}{\mathrm{m}^2}}{g_s^c}\,\underset{\frac{1}{\mathrm{m}}}{\delta(x - ut)} \tag{10-14b}$$

Transformation of the Origin. The heat conduction equation (10-14) is transformed from the fixed x coordinate to the moving ξ coordinate by the transformation

$$\xi = x - ut \tag{10-15}$$

By following the procedure described previously, the heat conduction equation (10-14) is transformed to the moving ξ coordinate as

$$\frac{\partial^2 T}{\partial \xi^2} + \frac{1}{k} g_s^c \delta(\xi) = \frac{1}{\alpha}\left(\frac{\partial T}{\partial t} - u \frac{\partial T}{\partial \xi}\right) \tag{10-16}$$

Quasi-Stationary Condition. Assuming quasi-stationary condition, equation (10-16) reduces to

$$\frac{d^2 T}{d\xi^2} + \frac{1}{k} g_s^c \delta(\xi) = -\frac{u}{\alpha} \frac{dT}{d\xi} \tag{10-17}$$

and with the application of the transformation

$$T(\xi) = \theta(\xi) e^{-(u/2\alpha)\xi} \tag{10-18}$$

equation (10-17), takes the form

$$\frac{d^2\theta}{d\xi^2} - \left(\frac{u}{2\alpha}\right)^2 \theta + \frac{1}{k} g_s^c \delta(\xi) = 0 \tag{10-19}$$

where the exponential $\exp(u\xi/2\alpha)$, which would appear as a multiplier to the source, is omitted for the reason stated previously.

10-2 ONE-DIMENSIONAL QUASI-STATIONARY PLANE HEAT SOURCE PROBLEM

In the problem of arc welding, the energy generated by the arc causes the electrode to melt, hence the problem of temperature distribution around the arc can be modeled as a problem of moving heat source. If the electrode is long enough with respect to its diameter, the heat transfer in the first few inches of the electrode can be envisioned as being of a quasi-stationary nature. If we assume there are no surface losses from the electrode (i.e., electrode is partially insulated), the corresponding heat transfer problem can be modeled as a one-dimensional moving heat source problem governed by the heat conduction equation (10-14). If we further assume that the quasi-stationary condition exists, the governing differential equation for this problem is taken as

$$\frac{d^2T}{d\xi^2}+\frac{1}{k}g_s^c\delta(\xi)=-\frac{u}{\alpha}\frac{dT}{d\xi}, \qquad \text{in} \qquad -\infty<\xi<\infty \tag{10-20}$$

subject to the boundary condition

$$\frac{dT}{d\xi}\to 0 \qquad \text{as} \qquad \xi\to\pm\infty \tag{10-21}$$

Applying the transformation (10-18), equation (10-20) takes the form

$$\frac{d^2\theta}{d\xi^2}-\left(\frac{u}{2\alpha}\right)^2\theta+\frac{1}{k}g_s^c\delta(\xi)=0 \qquad \text{in} \qquad -\infty<\xi<\infty \tag{10-22}$$

which is the same as that given by equation (10-19). The solution of this equation for $\xi\neq 0$, where the source term drops out, is taken as

$$\theta(\xi)=C_1e^{-(u/2\alpha)\xi}+C_2e^{(u/2\alpha)\xi} \qquad \text{for} \qquad -\infty<\xi<\infty \tag{10-23}$$

Introducing this result into equation (10-18), we obtain

$$T(\xi)=C_1e^{-(u/\alpha)\xi}+C_2 \qquad \text{for} \qquad -\infty<\xi<\infty \tag{10-24}$$

This solution is now considered separately for the regions $\xi<0$ and $\xi>0$ in the form

$$T^-(\xi)=C_1^-e^{-(u/\alpha)\xi}+C_2^- \qquad \text{for} \qquad \xi<0 \tag{10-25a}$$

$$T^+(\xi)=C_1^+e^{-(u/\alpha)\xi}+C_2^+ \qquad \text{for} \qquad \xi>0 \tag{10-25b}$$

and the unknown coefficients are determined by the application of the following boundary conditions:

$$\frac{dT^{\pm}}{d\xi} \to 0 \qquad \text{as} \qquad \xi \to \pm\infty \tag{10-26a,b}$$

$$T^{-} = T^{+} \qquad \text{at} \qquad \xi = 0 \quad \text{(continuity of temperature)} \tag{10-26c}$$

$$k\frac{dT^{-}}{d\xi} - k\frac{dT^{+}}{d\xi} = g_s^c \qquad \text{at} \qquad \xi = 0 \quad \text{(jump condition)} \tag{10-26d}$$

The last condition is obtained by integrating equation (10-20) with respect to ξ from $\xi = -\varepsilon$ to $\xi = +\varepsilon$ and then letting $\varepsilon \to 0$.

The application of the boundary conditions (10-26a, b) and the fact that $T^{+} \to 0$ as $\xi \to +\infty$ gives $C_1^{-} = 0, C_2^{+} = 0$; then

$$T^{-}(\xi) = C_2^{-} \qquad \text{for} \qquad \xi < 0 \tag{10-27a}$$

$$T^{+}(\xi) = C_1^{+} e^{-(u/\alpha)\xi} \qquad \text{for} \qquad \xi > 0 \tag{10-27b}$$

The requirement of continuity of temperature (10-26c) and the fact that $T^{+} \to 0$ as $\xi \to +\infty$, gives $C_2^{-} = C_1^{+} \equiv C$. The unknown constant C is determined by the application of the boundary condition (10-26d) to give

$$C_2^{-} = C_1^{+} \equiv C = \frac{\alpha}{uk} g_s^c \tag{10-28}$$

Then the solution for the problem becomes

$$T^{-}(\xi) = \frac{\alpha}{uk} g_s^c \qquad \text{for} \qquad \xi < 0 \tag{10-29}$$

$$T^{+}(\xi) = \frac{\alpha}{uk} g_s^c e^{-(u/\alpha)\xi} \qquad \text{for} \qquad \xi > 0 \tag{10-30}$$

Figure 10-4 shows a plot of the temperature profiles given by equations (10-29) and (10-30). Here, the rate of melting of the electrode is equivalent to the speed at which the arc moves along the electrode. The term $T^{+}(\xi)$ represents the temperature of a point at a distance ξ from the arc. The maximum value of temperature occurs at the moving source: $\xi = 0$. The medium remains at this maximum temperature after the source has moved further, because no surface losses have been allowed in the problem.

Effects of Surface Heat Losses. To illustrate the modeling of this problem for the case allowing for heat losses from the surfaces, we consider the same problem

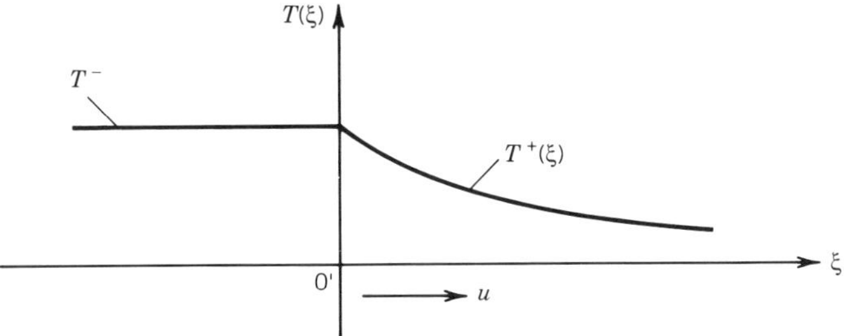

Fig. 10-4 Quasi-stationary temperature distribution around a moving surface heat source.

except assume that heat is lost by convection from the lateral surfaces of the rod into an ambient at zero temperature with a heat transfer coefficient h. If the rod has a uniform cross section A and perimeter P, then the governing one-dimensional heat conduction equation allowing for convection losses from the lateral surfaces is given by

$$\frac{\partial^2 T}{\partial x^2}+\frac{1}{k}g(x,t)=\frac{1}{\alpha}\frac{\partial T}{\partial t}+\frac{Ph}{Ak}T \tag{10-31a}$$

where the last term on the right-hand side represents convection heat losses from the lateral surfaces of the rod. The surface heat source g_s^c is related to the volumetric source $g(x,t)$ by

$$g(x,t)=g_s^c\delta(x-ut) \tag{10-31b}$$

The equation is transformed from the fixed coordinate x to the moving coordinate ξ by the transformation

$$\xi=x-ut \tag{10-32}$$

Then the heat conduction (10-31) takes the form

$$\frac{\partial^2 T}{\partial \xi^2}+\frac{1}{k}g_s^c\delta(\xi)=\frac{1}{\alpha}\left(\frac{\partial T}{\partial t}-u\frac{\partial T}{\partial \xi}\right)+\frac{Ph}{Ak}T \tag{10-33}$$

and for the quasi-stationary condition we have

$$\frac{d^2 T}{d\xi^2}+\frac{1}{k}g_s^c\delta(\xi)=-\frac{u}{\alpha}\frac{dT}{d\xi}+\frac{Ph}{Ak}T \tag{10-34a}$$

$$\frac{dT}{d\xi}\rightarrow 0 \qquad \text{for} \qquad \xi\rightarrow\pm\infty \tag{10-34b}$$

Applying the transformation

$$T(\xi) = \phi(\xi) e^{-(u/2\alpha)\xi} \tag{10-35}$$

equation (10-34) takes the form

$$\frac{d^2\phi(\xi)}{d\xi^2} - m^2\phi(\xi) + \frac{1}{k} g_s^c \delta(\xi) = 0 \qquad \text{in} \qquad -\infty < \xi < \infty \tag{10-36a}$$

where

$$m = \left[\left(\frac{u}{2\alpha} \right)^2 + \frac{Ph}{Ak} \right]^{1/2} \tag{10-36b}$$

The solution of equation (10-36a) for $\xi \neq 0$ where the source term drops out is given by

$$\theta(\xi) = C_1 e^{-m\xi} + C_2 e^{+m\xi} \qquad \text{for} \qquad -\infty < \xi < \infty \quad \text{and} \quad \xi \neq 0 \tag{10-37}$$

Introducing this result into equation (10-35), we obtain

$$T(\xi) = C_1 e^{-[m+(u/2\alpha)]\xi} + C_2 e^{[m-(u/2\alpha)]\xi} \qquad \text{for} \qquad -\infty < \xi < \infty \tag{10-38}$$

It is convenient to consider this solution for the regions $\xi < 0$ and $\xi > 0$, separately, as

$$T^-(\xi) = C_1^- e^{-[m+(u/2\alpha)]\xi} + C_2^- e^{[m-(u/2\alpha)]\xi} \qquad \text{for} \qquad \xi < 0 \tag{10-39a}$$

$$T^+(\xi) = C_1^+ e^{-[m+(u/2\alpha)]\xi} + C_2^+ e^{[m-(u/2\alpha)]\xi} \qquad \text{for} \qquad \xi > 0 \tag{10-39b}$$

and the boundary conditions for the determination of these four unknown coefficients are taken as

$$\frac{dT^\pm}{d\xi} = 0 \qquad \text{as} \qquad \xi \to \pm\infty \tag{10-40a,b}$$

$$T^- = T^+ \qquad \text{at} \qquad \xi = 0 \quad \text{(continuity of temperature)} \tag{10-40c}$$

$$k\frac{dT^-}{d\xi} - k\frac{dT^+}{d\xi} = g_s^c \qquad \text{at} \qquad \xi = 0 \quad \text{(jump condition)} \tag{10-40d}$$

Here, the last condition is obtained by integrating equation (10-34a) from $\xi = -\epsilon$ to $\xi = +\epsilon$ and then letting $\epsilon \to 0$. The application of the boundary conditions

(10-40) to equations (10-39) gives the solution as

$$T^{-}(\xi) = \frac{g_s^c}{2km} e^{[m-(u/2\alpha)]\xi} \qquad \text{for} \qquad \xi < 0 \tag{10-41a}$$

$$T^{+}(\xi) = \frac{g_s^c}{2km} e^{-[m+(u/2\alpha)]\xi} \qquad \text{for} \qquad \xi > 0 \tag{10-41b}$$

where

$$m = \sqrt{\left(\frac{u}{2\alpha}\right)^2 + \frac{Ph}{kA}} \tag{10-41c}$$

Clearly, equations (10-41) reduce to equation (10-30) for $h = 0$.

10-3 TWO-DIMENSIONAL QUASI-STATIONARY LINE HEAT SOURCE PROBLEM

We now examine a two-dimensional situation in which heat flows in the x and y directions while a line heat source of constant strength g_L^c W/m oriented parallel to the z axis moves along the x axis in the positive x-direction with a constant velocity u. We assume $(\partial T/\partial z) = 0$ everywhere.

Assuming quasi-stationary conditions, the transformed energy equation is the two-dimensional version of equation (10-19); that is

$$\frac{\partial^2\theta}{\partial\xi^2} + \frac{\partial^2\theta}{\partial y^2} - \left(\frac{u}{2\alpha}\right)^2 \theta + \frac{1}{k} g_L^c \delta(\xi)\delta(y) = 0 \tag{10-42a}$$

where

$$\xi = x - ut, \qquad \theta \equiv \theta(\xi, y) \tag{10-42b}$$

and $\theta(\xi, y)$ is related to the temperature $T(\xi, y)$ by

$$T(\xi, y) = \theta(\xi, y) e^{-(u/2\alpha)\xi} \tag{10-42c}$$

Since the boundary conditions for $T(\xi, y)$ at infinity are given by

$$\frac{\partial T}{\partial \xi} \to 0 \qquad \text{for} \qquad \xi \to \pm\infty \tag{10-43a}$$

$$\frac{\partial T}{\partial y} \to 0 \qquad \text{for} \qquad y \to \pm\infty \tag{10-43b}$$

and equation (10-42a) is symmetric with respect to the ξ and y variables; then, the function $\theta(\xi, y)$ depends only on the distance r from the heat source. To solve this problem, we write the homogeneous portion of the differential equation (10-42a) in the polar coordinates in the r variable as

$$\frac{1}{r}\frac{d}{dr}\left(r\frac{d\theta}{dr}\right)-\left(\frac{u}{2\alpha}\right)^2\theta=0 \quad \text{in} \quad 0<r<\infty \tag{10-44a}$$

and treat the source term as a boundary effect at $r=0$. To obtain a boundary condition at the origin, a circle of radius r is drawn around the line heat source, the heat released by the source is equated to the heat conducted away, and then r is allowed to go to zero. We find

$$\lim_{r\to 0}\left(-2\pi rk\frac{d\theta}{dr}\right)=g_{\mathrm{L}}^{\mathrm{c}} \quad \text{as} \quad r\to 0 \tag{10-44b}$$

$$\frac{d\theta}{dr}\to 0 \quad \text{as} \quad r\to\infty \tag{10-44c}$$

Equation (10-44a) is a modified Bessel equation of order zero and its solution satisfying the boundary condition (10-44c) is taken as

$$\theta(r)=CK_0\left(\frac{u}{2\alpha}r\right) \tag{10-45}$$

where K_0 is the modified Bessel function of order zero.

Introducing the solution (10-45) into the boundary condition (10-44b) we find

$$-C\,2\pi k\lim_{r\to 0}\left[r\frac{d}{dr}K_0\left(\frac{u}{2\alpha}r\right)\right]=g_{\mathrm{L}}^{\mathrm{c}} \tag{10-46a}$$

$$(C\;2\pi k)(1)=g_{\mathrm{L}}^{\mathrm{c}} \quad \text{or} \quad C=\frac{1}{2\pi k}g_{\mathrm{L}}^{\mathrm{c}} \tag{10-46b,c}$$

since

$$K_0\left(\frac{u}{2\alpha}r\right)\to -\ln\left(\frac{u}{2\alpha}r\right)$$

for small arguments and

$$r\frac{d}{dr}\left[\ln\left(\frac{u}{2\alpha}r\right)\right]=(r)\left[\frac{2\alpha}{ur}\frac{u}{2\alpha}\right]=1$$

Hence the solution for $\theta(r)$ becomes

$$\theta(r) = \frac{1}{2\pi k} g_{\mathrm{L}}^{\mathrm{c}} K_0\left(\frac{u}{2\alpha} r\right) \tag{10-47}$$

and $T(r, \xi)$ is determined according to equation (10-42c) as

$$T(r, \xi) = \frac{1}{2\pi k} g_{\mathrm{L}}^{\mathrm{c}} K_0\left(\frac{u}{2\alpha} r\right) e^{-(u/2\alpha)\xi} \tag{10-48}$$

The two-dimensional temperature field given by equation (10-48) can have application in the arc welding of thin plates along the edges.

For large values of r, equation (10-48) can be simplified by using the asymptotic value of $K_0(z)$ for large arguments:

$$K_0(z) \simeq \sqrt{\frac{\pi}{2z}}\, e^{-z} \qquad \text{for large } z \tag{10-49}$$

10-4 TWO-DIMENSIONAL QUASI-STATIONARY RING HEAT SOURCE PROBLEM

There are many engineering applications in which the moving heat source can be modeled as a *moving ring heat source*. Consider, for example, the turning operation for a cylindrical workpiece on a lathe in order to reduce its diameter. The thermal energy released from the cutting process will cause the heating of both the tool and the workpiece. In such turning operations the relative velocity of the tool with respect to the workpiece is large in the circumferential direction. Therefore, the heat generated during the turning operation can be regarded as a *ring heat source* moving along the outer boundary in the negative z direction as illustrated in Fig. 10-5. We assume azimuthal symmetry and a ring heat source of constant strength Q_0 watts, releasing its energy continuously as it moves with of constant velocity u along the outer surface of the cylinder. We allow for convection from the outer surface of the cylinder into an ambient at zero temperature and choose the initial temperature of the solid as zero.

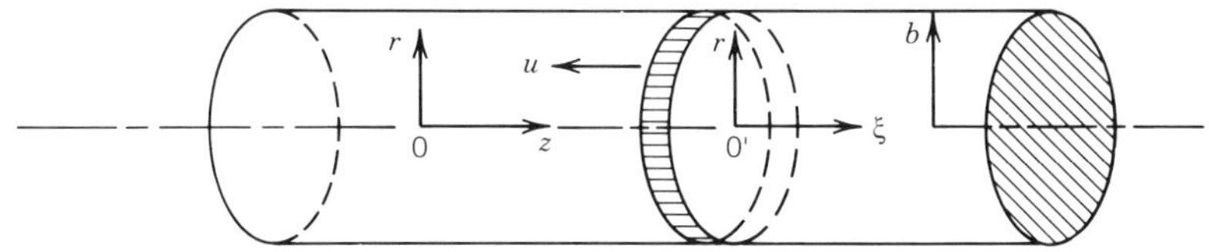

Fig. 10-5 Moving ring heat source.

The mathematical formulation of this problem is given as

$$\frac{1}{r}\frac{\partial}{\partial r}\left(r\frac{\partial T}{\partial r}\right)+\frac{\partial^2 T}{\partial z^2}+\frac{1}{k}\frac{Q_0}{2\pi b}\delta(r-b)\delta(z+ut)=\frac{1}{\alpha}\frac{\partial T}{\partial t}$$

$$\text{in} \quad 0<r<b, \quad -\infty<z<\infty \tag{10-50}$$

subject to the boundary conditions

$$\frac{\partial T}{\partial r}=0 \qquad \text{at} \qquad r=0 \tag{10-51a}$$

$$k\frac{\partial T}{\partial r}+hT=0 \qquad \text{at} \qquad r=b \tag{10-51b}$$

$$\frac{\partial T}{\partial r}\rightarrow 0 \qquad \text{as} \qquad z\rightarrow\pm\infty \tag{10-51c}$$

$$T=0 \qquad \text{for} \qquad t=0 \tag{10-51d}$$

This problem has been solved in reference 23 by using the integral transform technique. Here we describe its solution by the classical separation of variables technique. In equation (10-50), the delta function $\delta(r-b)$ denotes that the source is located at the outer surface of the cylinder and $\delta(z+ut)$ shows its position at time t along the z axis.

The fixed coordinate system r, z is now allowed to move with the source by introducing the transformation

$$\xi = z + ut \tag{10-52}$$

In the moving ξ coordinate system assuming *quasi-stationary condition*, equation (10-50) reduces to

$$\frac{1}{r}\frac{\partial}{\partial r}\left(r\frac{\partial T}{\partial r}\right)+\frac{\partial^2 T}{\partial \xi^2}+\frac{1}{k}\frac{Q_0}{2\pi b}\delta(r-b)\delta(\xi)=\frac{u}{\alpha}\frac{\partial T}{\partial \xi} \qquad \text{in} \qquad 0<r<b, \quad -\infty<\xi<\infty \tag{10-53a}$$

subject to the boundary conditions

$$\frac{\partial T}{\partial r}=0 \qquad \text{at} \qquad r=0 \tag{10-53b}$$

$$k\frac{\partial T}{\partial r}+hT=0 \qquad \text{at} \qquad r=b \tag{10-53c}$$

$$\frac{\partial T}{\partial \xi} \to 0 \qquad \text{as} \qquad \xi \to \pm\infty \tag{10-53d}$$

This problem is now expressed in the dimensionless form as

$$\frac{1}{R}\frac{\partial}{\partial R}\left(R\frac{\partial \psi}{\partial R}\right) + \frac{\partial^2 \psi}{\partial \eta^2} + \delta(R-1)\delta(\eta) = \frac{\text{Pe}}{2}\frac{\partial \psi}{\partial \eta} \qquad \text{in} \qquad 0 < R < 1, \quad -\infty < \eta < \infty \tag{10-54a}$$

subject to the boundary conditions

$$\frac{\partial \psi}{\partial R} = 0 \qquad \text{at} \qquad R = 0 \tag{10-54b}$$

$$\frac{\partial \psi}{\partial R} + \frac{\text{Bi}}{2}\psi = 0 \qquad \text{at} \qquad R = 1 \tag{10-54c}$$

$$\frac{\partial \psi}{\partial \eta} \to 0 \qquad \text{as} \qquad \eta \to \pm\infty \tag{10-54d}$$

where various dimensionless quantities are defined as

$$\left.\begin{aligned} &\eta = \frac{\xi}{b}, \qquad R = \frac{r}{b}, \qquad \psi = \frac{T}{A}, \qquad A = \frac{Q_0}{k2\pi b} \\ &\text{Pe} = \frac{u \cdot 2b}{\alpha} = \text{Peclét number}, \qquad \text{Bi} = \frac{h \cdot 2b}{k} = \text{Biot number} \end{aligned}\right\} \tag{10-55}$$

With the application of the transformation

$$\psi(R, \eta) = \theta(R, \eta) e^{-(\text{Pe}/4)\eta} \tag{10-56}$$

the differential equation (10-54a) is transformed to

$$\frac{1}{R}\frac{\partial}{\partial R}\left(R\frac{\partial \theta}{\partial R}\right) + \frac{\partial^2 \theta}{\partial \eta^2} - \left(\frac{\text{Pe}}{4}\right)^2 \theta + \delta(R-1)\delta(\eta) e^{-(\text{Pe}/4)\eta} = 0 \tag{10-57}$$

where $e^{-(\text{Pe}/4)\eta}$ appearing in the source term can be omitted because the source term vanishes for $\eta \neq 0$ and $\exp[-(\text{Pe}/4)\eta]$ becomes unity for $\eta = 0$. Therefore we need to consider the solution of the homogeneous equation

$$\frac{1}{R}\frac{\partial}{\partial R}\left(R\frac{\partial \theta}{\partial R}\right) + \frac{\partial^2 \theta}{\partial \eta^2} - \left(\frac{\text{Pe}}{4}\right)^2 \theta = 0 \qquad \text{in} \qquad 0 < R < 1, \quad -\infty < \eta < \infty \tag{10-58a}$$

subject to the boundary conditions

$$\frac{\partial \theta}{\partial R} = 0 \qquad \text{at} \qquad R = 0 \tag{10-58b}$$

$$\frac{\partial \theta}{\partial R} + \frac{\text{Bi}}{2}\theta = 0 \qquad \text{at} \qquad R = 1 \tag{10-58c}$$

$$\frac{\partial \theta}{\partial \eta} \rightarrow 0 \qquad \text{as} \qquad \eta \rightarrow \pm\infty \tag{10-58d}$$

in the regions $\eta < 0$ and $\eta > 0$. Let $\theta \equiv \theta^-$ be the solution for the region $\eta < 0$ and $\theta \equiv \theta^+$ be the solution for the region $\eta > 0$. The unknown coefficients associated with these solutions are determined from the requirement of continuity of temperature

$$\theta^- = \theta^+ \qquad \text{at} \qquad \eta = 0 \tag{10-59a}$$

and the jump condition

$$\frac{\partial \theta^-}{\partial \eta} - \frac{\partial \theta^+}{\partial \eta} = \delta(R - 1) \qquad \text{at} \qquad \eta = 0 \tag{10-59b}$$

This jump condition is obtained by integrating equation (10-57) from $\eta = -\epsilon$ to $\eta = +\epsilon$ and then letting $\epsilon \rightarrow 0$.

Once $\theta^\pm$ are determined, the dimensionless temperature $\psi^\pm$ is determined according to the transformation given by equation (10-56).

Finally, the solution for the dimensionless temperatures $\psi^\pm(R, \eta)$ are determined as

$$\psi^\pm(R, \eta) = \sum_{n=1}^{\infty} \frac{\beta_n^2}{\left(\frac{\text{Bi}}{2}\right)^2 + \beta_n^2} \frac{J_0(\beta_n R)}{J_0(\beta_n)} \frac{\exp\left(\frac{\text{Pe}}{4} \pm F\right)\eta}{F} \tag{10-60a}$$

where the β_n values are the roots of

$$-\beta_n J_1(\beta_n) + \frac{\text{Bi}}{2} J_0(\beta_n) = 0 \tag{10-60b}$$

and F is defined by

$$F \equiv \sqrt{\left(\frac{\text{Pe}}{4}\right)^2 + \beta_n^2} \tag{10-60c}$$

with plus and minus signs denoting the regions $\eta > 0$ and $\eta < 0$, respectively.

Here the Peclét number is a measure of the ratio of convective diffusion (i.e., due to the velocity of the moving source) to the conduction diffusion. Therefore, for the smaller Peclét number, the temperature field penetrates considerably farther "upstream" from the source than with the larger Peclét numbers.

REFERENCES

1. D. Rosenthal, *2-ene Congres National des Sciences*, Brussels, pp. 1277–1292, 1935 (in French).
2. D. Rosenthal and R. Schmerber, *Welding J.* **17**, 2–8, 1938.
3. D. Rosenthal, *ASME Trans.* **68**, 849–866, 1946.
4. D. Rosenthal and R. H. Cameron, *Trans. ASME* **69**, 961–968, 1947.
5. E. M. Mahla, M. C. Rawland, C. A. Shook, and G. E. Doan, *Welding J.* **20**, (10) (Res. Suppl.), 459, 1941.
6. W. A. Bruce, *J. Appl. Phys.* **10**, 578–585, 1939.
7. H. S. Carslaw and J. C. Jaeger, *Conduction of Heat in Solids*, Oxford at the Clarendon Press, London, 1959.
8. J. C. Jaeger, *J. Proc. Roy. Soc. N. S. W.* **76**, 203–224, 1942.
9. R. J. Grosh, E. A. Trabant, and G. A. Hawkins, *Q. Appl. Math.* **13**, 161–167, 1955.
10. R. Weichert and K. Schönert, *Q. J. Mech. Appl. Math.* **31** (3), 363–379, 1978.
11. Y. Terauchi, H. Nadano, and M. Kohno, *Bull. JSME* **28** (245), 2789–2795, 1985.
12. N. R. DeRuisseaux and R. D. Zerkle, *Trans. ASME, J. Heat Transfer* **92**, 456–464, 1970.
13. W. Y. D. Yuen, *Math. Engg. Ind.* **1**, 1–19, 1987.
14. J. O. Outwater and M. C. Shaw, *Trans. ASME* **74**, 73, 1952.
15. E. G. Loewen and M. C. Shaw, *Trans. ASME* **76**, 217, 1954.
16. B. T. Chao and K. J. Trigger, *Trans. ASME* **80**, 311, 1958.
17. A. Cameron, A. N. Gordon, and G. T. Symm, *Proc. Roy. Soc.* **A286**, 45–61, 1965.
18. G. T. Symm, *Q. J. Mech. Appl. Math.* **20**, 381–391, 1967.
19. J. R. Barber, *Int. J. Heat Mass Transfer* **13**, 857–869, 1970.
20. I. L. Ryhming, *Acta Mechanica* **32**, 261–274, 1979.
21. O. A. Tay, M. G. Stevenson, and G. de Vahl Davis, *Proc. Inst. Mech. Eng.* **188**, 627–638, 1974.
22. W. M. Mansour, M. O. N. Osman, T. S. Sansar, and A. Mazzawi, *Int. J. Prod. Rev.* **11**, 59–68, 1973.
23. R. G. Watts, *ASME Paper* No. 68-WA/HT-11, 1968.
24. M. F. Modest and H. Abakians, *J. Heat Transfer* **108**, 597–601, 1986.
25. J. F. Ready, *Effects of High Power Laser Radiation*, Academic Press, New York, 1971.
26. H. E. Cline and T. R. Anthony, *J. Appl. Phys.* **48**, 3895–3900, 1977.
27. Y. I. Nissim, A. Lietoila, R. B. Gold, and J. F. Gibbons, *J. Appl. Phys.* **51**, 274–279, 1980.
28. K. Brugger, *J. Appl. Physics* **43**, 577–583, 1972.

29. M. F. Modest and H. Abakians, *J. Heat Transfer* **108**, 602–607, 1986.
30. M. F. Modest and H. Abakians, *J. Heat Transfer* **108**, 597–601, 1986.
31. S. Biyikli and M. F. Modest, *J. Heat Transfer* **110**, 529–532, 1988.
32. S. Roy and M. F. Modest, *J. Thermophys.* **4**, 199–203, 1990.

PROBLEMS

10-1 Consider the three-dimensional quasi-stationary temperature field $T(\xi, y, z)$ governed by the differential equation (10-5) with the boundary conditions at infinity taken as

$$\frac{\partial T}{\partial \xi} \to 0 \quad \text{for} \quad \xi \to \pm\infty, \qquad \frac{\partial T}{\partial y} \to 0 \quad \text{for} \quad y \to \pm\infty,$$

$$\frac{\partial T}{\partial z} \to 0 \quad \text{for} \quad z \to \pm\infty$$

and the transformed equation (10-7) for the temperature field $\theta(\xi, y, z)$. In equation (10-7), which is symmetric with respect to the variables ξ, y, and z, the function $\theta(\xi, y, z)$ depends only on the distance r from the point heat source. Then

1. Equation (10-7) can be written in the polar coordinates with respect to the r variable only; write this equation without the source term.
2. Develop the boundary condition at $r = 0$ for this equation by drawing a sphere of radius r around the point heat source, then equating the heat released by the source to the heat conducted away and letting $r \to 0$.
3. By solving this equation in the polar coordinates, develop an expression for the quasi-stationary temperature field $T(r, \xi)$ around the moving point heat source.

10-2 Develop equation (10-31a) by writing an energy balance for a bar of uniform cross section with energy generation in the solid and heat dissipation from the lateral surfaces by convection with a heat transfer coefficient h into an ambient at zero temperature.

10-3 The temperature distribution in the gun barrel resulting from the firing of a bullet can be regarded as a problem of a point heat source moving with a constant velocity u along the axis of a solid cylinder of radius b if the base of the barrel is small enough compared to the outside radius of the barrel.

Assuming (1) constant speed and the rate of heat release by the point source, (2) no heat losses from the outer surface of the cylinder, and (3) cylinder long enough with respect to the diameter so that quasi-stationary state is established, develop the governing differential equations and the boundary conditions needed for the solution of the quasi-stationary temperature distribution in the cylinder.

10-4 Consider a boring process in order to increase the inside diameter of a hollow cylindrical workpiece. Such a problem can be modeled as a *moving ring heat source* advancing axially along the interior surface of a hollow cylinder. Assume a source of constant strength Q_0 watts, releasing its energy continuously as it moves with a constant speed u along the inner surface of the cylinder and heat loss by convection from the outer surface of the cylinder with a heat transfer coefficient h into an ambient at zero temperature. Initially the solid is also at zero temperature. Give the governing differential equations and the boundary conditions for the determination of the quasi-stationary temperature field in the cylinder. Note that this problem is analogous to that considered in Section 10-4, except the ring heat source is moving along the inside surface of the cylinder. Assume negligible heat loss from the inner surface of the hollow cylinder.

11

PHASE-CHANGE PROBLEMS

Transient heat transfer problems involving melting or solidification generally referred to as "phase-change" or "moving-boundary" problems are important in many engineering applications such as in the making of ice, the freezing of food, the solidification of metals in casting, the cooling of large masses of igneous rock, thermal energy storage, processing of chemicals and plastics, crystal growth, aerodynamic ablation, casting and welding of metals and alloys, and numerous others. The solution of such problems is inherently difficult because the interface between the solid and liquid phases is moving as the latent heat is absorbed or released at the interface; as a result, the location of the solid–liquid interface is not known *a priori* and must follow as a part of the solution. In the solidification of pure substances, like water, the solidification takes place at a discrete temperature, and the solid and liquid phases are separated by a *sharp moving interface*. On the other hand, in the solidification of mixtures, alloys, and impure materials the solidification takes place over an extended temperature range, and as a result the solid and liquid phases are separated by a *two-phase moving* region.

Early analytic work on the solution of phase-change problems include those by Lamé and Clapeyron [1] in 1831 and by Stefan [2] in 1891 in relation to the ice formation. The fundamental feature of this type of problem is that the location of the boundary is both unknown and moving, and that the parabolic heat conduction equation is to be solved in a region whose boundary is also to be determined. Although references 1, 2 are the early published works on this subject, the exact solution of a more general phase-change problem was discussed by F. Neumann in his lectures in the 1860s, but his lecture notes containing these solutions were not published until 1912. Since then, many phase-change problems have appeared in the literature, but the exact solutions are limited to a number of idealized situations involving semiinfinite or infinite regions and subject to simple boundary and initial conditions [3]. Because of the nonlinearity of such problems, the superposition principle is not applicable and each case must be treated separately. When exact solutions are not available, approximate, semi-analytic, and numerical methods can be used to solve the phase-change problems. We now present a brief discussion of various methods of solution of phase-change problems.

The *integral method*, which dates back to von Kármán and Pohlhausen, who used it for the approximate analysis of boundary-layer equations, was applied by Goodman [5, 6] to solve a one-dimensional transient melting problem, and subsequently by many other investigators [7–15] to solve various types of one-dimensional transient phase-change problems. This method provides a relatively straightforward and simple approach for approximate analysis of one-dimensional transient phase-change problems. The *variational formulation* derived by Biot [16] on the basis of an irreversible thermodynamic argument, was used in the solution of one-dimensional, transient phase-change problems [17–21]. The *moving heat source* (or the *integral equation*) *method*, originally applied by Lightfoot [22] to solve Neuman's problem, is based on the concept of representing the liberation (or absorption) of latent heat by a moving plane heat source (or sink) located at the solid–liquid interface. A general formulation of moving heat source approach is given in reference 23, and various application can be found in references 24–28. The *perturbation method* has been used by several investigators [29–34]; however, the analysis becomes very complicated if higher-order solutions are to be determined; also it is difficult to use this method for problems involving more than one dimension. The *embedding technique*, first introduced by Boley [35] to solve the problem of melting of a slab, has been applied to solve various phase-change problems [36–41]. The method appears to be versatile to obtain solutions for one, two, or three dimensions and to develop general starting solutions. A *variable eigenvalue approach* developed in connection with the solution of heat conduction problems involving time-dependent boundary condition parameters [42, 43] has been applied to solve one-dimensional transient phase-change problems [44]. The method is applicable to solve similar problems in the cylindrical or spherical symmetry. The *electrical network analog method* often used in early applications [45–49] has now been replaced by purely numerical methods of solution because of the availability of high-speed digital computers. A large number of purely numerical solutions of phase-change problems has been reported [50–81].

Reviews of phase-change problems up to 1965 can be found in references 82–84. Extensive list of references and treatments of the fundamentals of solidification can be found in standard texts [85–89].

Experimental investigation of phase-change problems is important in order to check the validity of various analytic models, but only a limited number of experimental studies are available in the literature [90–94].

11-1 MATHEMATICAL FORMULATION OF PHASE-CHANGE PROBLEMS

To illustrate the mathematical formulation of phase-change problems, we consider first a one-dimensional solidification problem and then a melting problem.

Interface Condition for One- and Multidimensional Phase Change Problems

Solidification Problem. A liquid having a single phase-change temperature T_m is confined to a semiinfinite region $0 < x < \infty$. Initially, the liquid is at a uniform temperature T_i which is higher than the phase-change temperature T_m. At time $t = 0$, the temperature of the boundary surface $x = 0$ is suddenly lowered to a temperature T_0, which is less than the melt temperature T_m and is maintained at that temperature for times $t > 0$. The solidification starts at the boundary surface $x = 0$ and the solid–liquid interface $x = s(t)$ moves in the *positive x direction.*

Figure 11-1*a* shows the geometry and coordinates for such a one-dimensional solidification problem. The temperatures $T_s(x, t)$ and $T_l(x, t)$ for the solid and liquid phases, respectively, are governed by the standard diffusion equations given by

$$\frac{\partial^2 T_s(x,t)}{\partial x^2} = \frac{1}{\alpha_s}\frac{\partial T_s(x,t)}{\partial t} \quad \text{in} \quad 0 < x < s(t), \quad t > 0 \qquad \text{(11-1a)}$$

$$\frac{\partial^2 T_l(x,t)}{\partial x^2} = \frac{1}{\alpha_l}\frac{\partial T_l(x,t)}{\partial t} \quad \text{in} \quad s(t) < x < \infty, \quad t > 0 \qquad \text{(11-1b)}$$

where we assumed constant properties for the solid and liquid phases. Here, $s(t)$

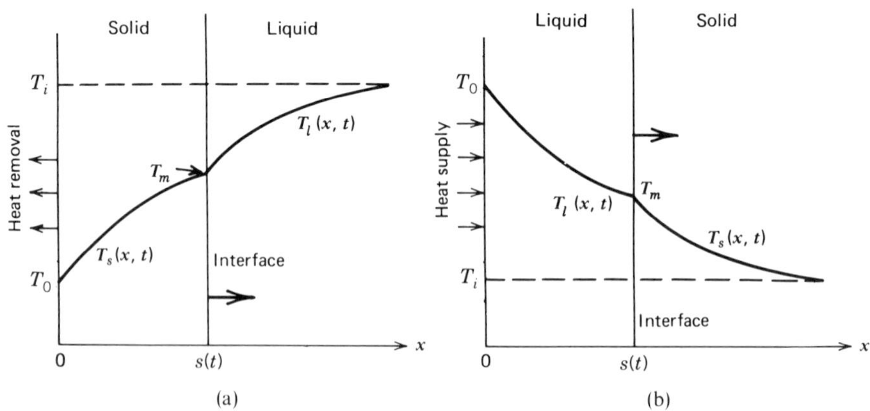

Fig. 11-1 Geometry and coordinates for one-dimensional (*a*) solidification and (*b*) melting problems.

is the location of the solid–liquid interface which is not known *a priori*, hence must be determined as a part of the solution. The subscripts s and l refer, respectively, to the solid and liquid phases. Therefore, the problem involves three unknowns, namely, $T_s(x,t)$, $T_l(x,t)$, and $s(t)$. An additional equation governing $s(t)$ is determined by considering an interface energy balance at $x = s(t)$, stated as

$$\begin{bmatrix}\text{Conduction}\\ \text{heat flux in}\\ \text{the negative } x\\ \text{direction through}\\ \text{the solid phase}\end{bmatrix} - \begin{bmatrix}\text{conduction}\\ \text{heat flux}\\ \text{in the negative } x\\ \text{direction through}\\ \text{the liquid phase}\end{bmatrix} = \begin{bmatrix}\text{rate of heat}\\ \text{liberated during}\\ \text{solidification}\\ \text{per unit area of}\\ \text{interface}\end{bmatrix} \tag{11-2a}$$

or

$$k_s \frac{\partial T_s}{\partial x} - k_l \frac{\partial T_l}{\partial x} = \rho L \frac{ds(t)}{dt} \qquad \text{at} \qquad x = s(t), \quad t > 0 \tag{11-2b}$$

where L is the latent heat per unit mass [i.e., (Ws)/kg] associated with the phase change. For the time being we neglected the density difference for the solid and liquid phases and assumed $\rho_l = \rho_s = \rho$ at the solid–liquid interface.

The continuity of temperature at the solid liquid interface is given by

$$T_s(x,t) = T_l(x,t) = T_m \qquad \text{at} \qquad x = s(t) \tag{11-3}$$

where $T_s(x,t)$ and $T_l(x,t)$ are the solid and liquid phase temperature at $x = s(t)$, respectively, and T_m is the phase-change temperature.

Summarizing, equations (11-1a), (11-1b), and (11-2b) provide three differential equations that govern the temperature distributions in the solid and liquid phases and the position $s(t)$ of the solid–liquid interface.

Equation (11-3) provides two boundary conditions. Other boundary conditions and the initial conditions are specified depending on the nature of the physical conditions at the boundary surfaces. This matter will be illustrated later in this chapter with specific examples.

Melting Problem. We now consider a solid having a single phase-change temperature T_m confined to a semiinfinite region $0 < x < \infty$. Initially, the solid is at a uniform temperature T_i that is lower than the phase-change temperature T_m. At time $t = 0$, the temperature of the boundary surface $x = 0$ is suddenly raised to a temperature T_0, which is higher than the melting temperature T_m and maintained at that temperature for times $t > 0$. We assume that the coordinate system for this melting problem is arranged as illustrated in Fig. 11-1*b*, so that the solid–liquid interface moves in the *positive x direction* as in the case of the solidification problem. The governing differential equations for this problem, assuming constant properties for each phase, are given by

$$\frac{\partial^2 T_l(x,t)}{\partial x^2} = \frac{1}{\alpha_l}\frac{\partial T_l(x,t)}{\partial t} \quad \text{in} \quad 0 < x < s(t), \quad t > 0 \tag{11-4a}$$

$$\frac{\partial^2 T_s(x,t)}{\partial x^2} = \frac{1}{\alpha_s}\frac{\partial T_s(x,t)}{\partial t} \quad \text{in} \quad s(t) < x < \infty, \quad t > 0 \tag{11-4b}$$

and an energy balance at the solid–liquid interface $x = s(t)$ shows that the resulting interface energy balance equation is exactly the same as given by equation (11-2b); hence we have

$$k_s\frac{\partial T_s}{\partial x} - k_l\frac{\partial T_l}{\partial x} = \rho L\frac{ds(t)}{dt} \quad \text{at} \quad x = s(t), \quad t > 0 \tag{11-4c}$$

Thus, equations (11-4a,b,c) provide three differential equations for the determination of the three unknowns $T_s(x,t)$, $T_l(x,t)$, and $s(t)$ for the melting problem considered here. Appropriate boundary and initial conditions need to be specified for their solution.

We note that, in the interface energy-balance equation (11-2b) or (11-4c), the term $ds(t)/dt$ represents the velocity of the interface in the positive x direction, hence we write

$$\frac{ds(t)}{dt} \equiv v_x(t) \tag{11-5a}$$

Then the interface energy-balance equation can be written as

$$k_s\frac{\partial T_s}{\partial x} - k_l\frac{\partial T_l}{\partial x} = \rho L v_x(t) \quad \text{at} \quad x = s(t) \tag{11-5b}$$

Effects of Density Change. The difference in the density of phases at the interface during phase change gives rise to liquid motion across the interface. Usually $\rho_s > \rho_l$, except for water, bismuth, and antimony, for which $\rho_s < \rho_l$.

To illustrate the effects of density change, we consider the one-dimensional solidification problem illustrated in Fig. 11-1*a*. Let $\rho_s > \rho_l$ and

v_x = velocity of the interface

v_l = velocity of the liquid at the interface

H_s, H_l = enthalpies per unit mass of the material for the solid and liquid phases at the interface

In the physical situation considered in Fig. 11-1*a*, the interface velocity v_x is in the positive x direction, and for $\rho_s > \rho_l$ the motion of the liquid is in the opposite direction. Then, the energy balance at the interface allowing for the contributions of various diffusive and convective energy transfer becomes

$$k_s \frac{\partial T_s}{\partial x} - k_l \frac{\partial T_l}{\partial x} = (\rho_l H_l - \rho_s H_s) v_x - \rho_l H_l v_l \qquad \text{at} \qquad x = s(t) \qquad \text{(11-6)}$$

The mass-conservation equation at the interface may be written as

$$(\rho_l - \rho_s) v_x = \rho_l v_l \qquad \text{(11-7a)}$$

$$v_l = -\frac{\rho_s - \rho_l}{\rho_l} v_x \qquad \text{(11-7b)}$$

Eliminating v_l from equation (11-6) by means of equation (11-7b) we obtain

$$k_s \frac{\partial T_s}{\partial x} - k_l \frac{\partial T_l}{\partial x} = \rho_s L v_x \qquad \text{(11-8a)}$$

since

$$H_l - H_s = L = \text{the latent heat} \qquad \text{(11-8b)}$$

which is similar to equation (11-4c) except ρ is now replaced by ρ_s.

Effects of Convection. Consider the solidification problem illustrated in Fig. 11-1*a*. If the heat transfer from the liquid phase to the solid–liquid interface is controlled by convection, and hence diffusion in the liquid phase is neglected, the interface energy-balance equation (11-2b) takes the form

$$k_s \frac{\partial T_s}{\partial x} - h(T_\infty - T_m) = \rho L \frac{ds(t)}{dt} \qquad \text{at} \qquad x = s(t) \qquad \text{(11-9)}$$

where h is the heat transfer coefficient for the liquid side, T_∞ is the bulk temperature of the liquid phase, and T_m is the melting-point temperature at the interface.

In the case of the melting problem illustrated in Fig. 11-1*b*, if convection is dominant in the liquid phase, equation (11-9) is applicable if the minus sign before h is changed to the plus sign.

Nonlinearity of Interface Condition. The interface boundary conditions given by equations (11-2b) and (11-9) are nonlinear. To show the nonlinearity of these equations, we need to relate $ds(t)/dt$ to the derivative of temperatures. This is done by taking the total derivative of the interface equation (11-3)

$$\left[\frac{\partial T_s}{\partial x} dx + \frac{\partial T_s}{\partial t} dt\right]_{x=s(t)} = \left[\frac{\partial T_l}{\partial x} dx + \frac{\partial T_l}{\partial t} dt\right]_{x=s(t)} = 0 \qquad \text{(11-10a)}$$

or

$$\frac{\partial T_s}{\partial x}\frac{ds(t)}{dt}+\frac{\partial T_s}{\partial t}=\frac{\partial T_l}{\partial x}\frac{ds(t)}{dt}+\frac{\partial T_l}{\partial t}=0 \qquad \text{at} \qquad x=s(t) \tag{11-10b}$$

which can be rearranged as

$$\frac{ds(t)}{dt}=-\frac{\partial T_s/\partial t}{\partial T_s/\partial x} \qquad \text{and} \qquad \frac{ds(t)}{dt}=-\frac{\partial T_l/\partial t}{\partial T_l/\partial x} \tag{11-10c}$$

Introducing these results, for example, into equation (11-2b) we obtain

$$k_s\frac{\partial T_s}{\partial x}-k_l\frac{\partial T_l}{\partial x}=-\rho L\frac{\partial T_s/\partial t}{\partial T_s/\partial x}=-\rho L\frac{\partial T_l/\partial t}{\partial T_s/\partial x} \tag{11-10d}$$

The nonlinearity of this equation is now apparent.

Generalization to Multidimension. The interface energy-balance equation developed above for the one-dimensional case is now generalized for the multidimensional situations. Figure 11-2 illustrates a solidification in a three-dimensional region. The solid and liquid phases are separated by a sharp interface defined by the equation

$$F(x,y,z,t)=0 \tag{11-11}$$

The requirement of the continuity of temperatures at the interface becomes

$$T_s(x,y,z,t)=T_l(x,y,z,t)=T_m \qquad \text{at} \qquad F(x,y,z,t)=0 \tag{11-12}$$

The interface energy-balance equation is written as

$$k_s\frac{\partial T_s}{\partial n}-k_l\frac{\partial T_l}{\partial n}=\rho L v_n \qquad \text{at} \qquad F(x,y,z,t)=0 \tag{11-13}$$

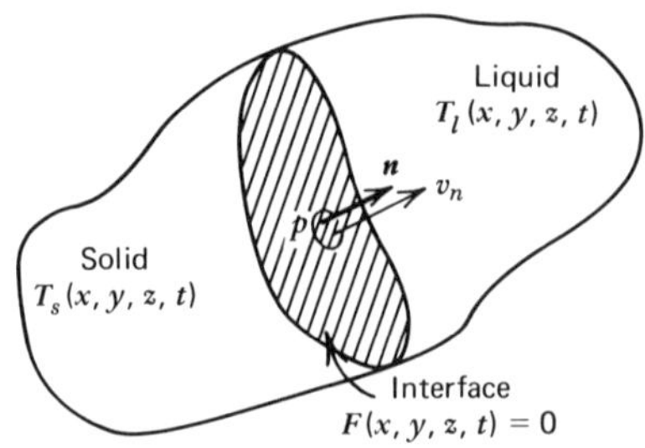

Fig. 11-2 Solidification in three dimensions. Interface is moving in the direction **n**.

where $\partial/\partial n$ denotes the derivative at the interface along the normal direction vector $\mathbf{n}$ at any location P on the interface and pointing toward the liquid region and v_n is the velocity of this interface at the location P in the direction $\mathbf{n}$. Here we assumed that the densities of the solid and liquid phases are the same.

The interface energy-balance equation (11-13) is not in a form suitable for developments of analytic or numerical solutions of the phase-change problems. An alternative form of this equation is given by [95]

$$\left[1+\left(\frac{\partial s}{\partial x}\right)^2+\left(\frac{\partial s}{\partial y}\right)^2\right]\left[k_s\frac{\partial T_s}{\partial z}-k_l\frac{\partial T_l}{\partial z}\right]=\rho L\frac{\partial s}{\partial t} \quad \text{at} \quad z=s(x,y,t) \tag{11-14}$$

This form of the interface energy balance equation is analogous to the form given by equation (11-2b) for the one-dimensional case; therefore, it is more suitable for numerical or analytic purposes. We now examine some special cases of equation (11-14).

For the two-dimensional problem involving (x, z, t) variables, if the location of the solid–liquid interface is specified by the relation $F(x,z,t)=z-s(x,t)=0$, then equation (11-14) reduces to

$$\left[1+\left(\frac{\partial s}{\partial x}\right)^2\right]\left[k_s\frac{\partial T_s}{\partial z}-k_l\frac{\partial T_l}{\partial z}\right]=\rho L\frac{\partial s}{\partial t} \quad \text{at} \quad z=s(x,t) \tag{11-15}$$

This equation is the same as that used in references 25, 38, and 39 for interface boundary condition in the analysis of two-dimensional phase-change problems.

For the one-dimensional problem involving (z, t) variables, if the location of the solid–liquid interface is given by $F(z,t)=z-s(t)=0$, equation (11-14) reduces to

$$k_s\frac{\partial T_s}{\partial z}-k_l\frac{\partial T_l}{\partial z}=\rho L\frac{ds}{dt} \quad \text{at} \quad z=s(t) \tag{11-16}$$

which is identical to equation (11-2b) if z is replaced by x.

In the cylindrical coordinate system involving (r, ϕ, t) variables, if the location of the solid–liquid interface is given by $F(r,\phi,t)=r-s(\phi,t)=0$, then the corresponding form of equation (11-14) becomes

$$\left[1+\frac{1}{s^2}\left(\frac{\partial s}{\partial \phi}\right)^2\right]\left[k_s\frac{\partial T_s}{\partial r}-k_l\frac{\partial T_l}{\partial r}\right]=\rho L\frac{\partial s}{\partial t} \quad \text{at} \quad r=s(\phi,t) \tag{11-17}$$

In the cylindrical coordinate system involving (r, z, t) variables, if the location of the solid–liquid interface is given as $F(r,z,t)=z-s(r,t)=0$, the interface equation takes the form

$$\left[1+\left(\frac{\partial s}{\partial r}\right)^2\right]\left[k_s\frac{\partial T_s}{\partial z}-k_l\frac{\partial T_l}{\partial z}\right]=\rho L\frac{\partial s}{\partial t} \quad \text{at} \quad z=s(r,t) \tag{11-18}$$

Dimensionless Variables of Phase-Change Problem

The role of dimensionless variables in phase-change problems is envisioned better if the interface energy-balance equation (11-2b) is expressed in the dimensionless form as

$$\frac{\partial \theta_s}{\partial \eta}-\frac{k_l}{k_s}\frac{\partial \theta_l}{\partial \eta}=\frac{1}{Ste}\frac{d\delta(\tau)}{d\tau} \tag{11-19}$$

$$\theta_i(\tau,\eta)=\frac{T_i(x,t)-T_m}{T_m-T_0}, \qquad i=\text{s or l}; \qquad \eta=x/b;$$

$$\delta(\tau)=\frac{s(t)}{b}; \qquad \tau=\frac{\alpha_s t}{b^2}; \qquad \text{Ste}=\frac{C_{ps}(T_m-T_0)}{L} \tag{11-20}$$

Here, b is a reference length, L is the latent heat, C_{ps} is the specific heat, T_m is the melting temperature, T_0 is a reference temperature, $s(t)$ is the location of the solid–liquid interface, and Ste is the *Stefan number*, named after J. Stefan. The above dimensionless variables, other than the Stefan number, are similar to those frequently used in the standard heat conduction problems; the Stefan number is associated with the phase-change process.

The Stefan number signifies the importance of sensible heat relative to the latent heat. If the Stefan number is small, say, less than approximately 0.1, the heat released or absorbed by the interface during phase change is affected very little as a result of the variation of the sensible heat content of the material during the propagation of heat through the medium. For materials such as aluminum, copper, iron, lead, nickel, and tin, the Stefan number based on a temperature difference between the melting temperature and the room temperature varies from 1 to 3. For melting or solidification processes taking place with much smaller temperature differences, the Stefan number is much smaller. For example, in phase-change problems associated with thermal energy storage, the temperature differences are small; as a result the Stefan number is generally smaller than 0.1.

11-2 EXACT SOLUTION OF PHASE-CHANGE PROBLEMS

The exact solution of phase-change problems is limited to few idealized situations for the reasons stated previously. They are mainly for the cases of one-dimensional infinite or semiinfinite regions and simple boundary conditions, such as the prescribed temperature at the boundary surface. Exact solutions are obtainable if the problem admits a similarity solution allowing the two independent variables x and t merge into a single similarity variable $x/t^{1/2}$. Some exact solutions can be found in references 3 and 4. We present below some of the exact solutions of phase-change problems.

Example 11.1

Solidification of a Supercooled Liquid in a Half-Space (*One-Phase Problem*). A supercooled liquid at a uniform temperature T_i which is lower than the solidification (or melting) temperature T_m of the solid phase is confined to a half-space $x > 0$. It is assumed that the solidification starts at the surface $x = 0$ at time $t = 0$ and the solid–liquid interface moves in the positive x direction. Figure 11-3 illustrates the geometry, the coordinates, and the temperature profiles. The solid phase being at the uniform temperature T_m throughout, there is no heat transfer through it; the heat released during the solidification process is transferred into the super-cooled liquid and raises its temperature. The temperature distribution is unknown only in the liquid phase; hence the problem is a one-phase problem. In the following analysis we determine the temperature distribution in the liquid phase and the location of the solid–liquid interface as a function of time.

Solution. Before presenting the analysis for the solution of this problem we discuss the implications of the supercooling of a liquid. If a liquid is cooled very slowly, the bulk temperature may be lowered below the solidification temperature and the liquid in such a state is called a *supercooled liquid.* After supercooling reaches some critical temperature, the solidification starts, and heat released during freezing raises the temperature of the supercooled liquid. Little is known about the actual condition of the solid–liquid interface during the solidification of a supercooled liquid. During the solidification of super-cooled water the interface may grow as a dentritic surface consisting of thin, plate-like crystals of ice interspersed in water rather than moving as a sharp interface [96]. As a result, it is a very complicated matter to include in the analysis the effects of irregular surface conditions. Therefore, in the following solution only an idealized situation is considered. Namely, it is assumed that

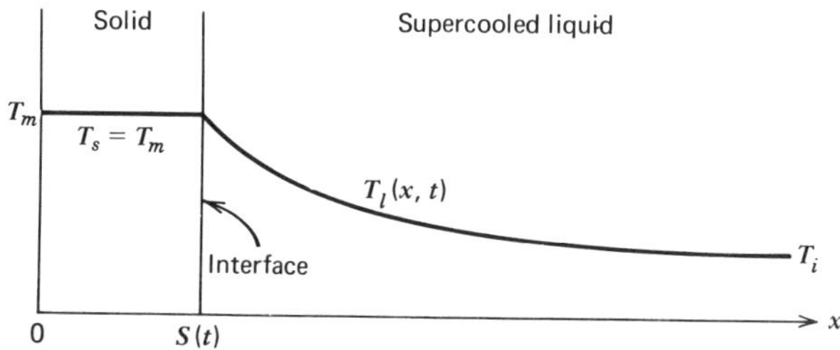

Fig. 11-3 Solidification of supercooled liquid in a half-space. One-phase problem.

the solid–liquid interface is a sharp surface whose motion is similar to that encountered in the normal solidification process. The mathematical formulation for the liquid phase is given as

$$\frac{\partial^2 T_l}{\partial x^2} = \frac{1}{\alpha_l}\frac{\partial T_l(x,t)}{\partial t} \quad \text{in} \quad s(t) < x < \infty, \quad t > 0 \tag{11-21}$$

$$T_l(x,t) \to T_i \quad \text{as} \quad x \to \infty, \quad t > 0 \tag{11-22a}$$

$$T_l(x,t) = T_i \quad \text{for} \quad t = 0, \quad \text{in} \quad x > 0 \tag{11-22b}$$

and for the interface as

$$T_l(x,t) = T_m \quad \text{at} \quad x = s(t), \quad t > 0 \tag{11-23a}$$

$$-k_l \frac{\partial T_l(x,t)}{\partial x} = \rho L \frac{ds(t)}{dt} \quad \text{at} \quad x = s(t), \quad t > 0 \tag{11-23b}$$

The interface equation (11-23b) states that the heat liberated at the interface as a result of solidification is equal to the heat conducted into the supercooled liquid. No equations are needed for the solid phase because it is at uniform temperature T_m. Recalling that $\operatorname{erfc}[x/2(\alpha_l t)^{1/2}]$ is a solution of the heat-conduction equation (11-21), we choose a solution for $T_l(x,t)$ in the form

$$T_l(x,t) = T_i + B \operatorname{erfc}[x/2(\alpha_l t)^{1/2}] \tag{11-24}$$

where B is an arbitrary constant. This solution satisfies the differential equation (11-21), the boundary condition (11-22a), and the initial condition (11-22b) since $\operatorname{erfc}(\infty) = 0$. If we require that the solution (11-24) should also satisfy the interface condition (11-23a), we find

$$T_m = T_i + B \operatorname{erfc}(\lambda) \tag{11-25a}$$

where

$$\lambda = \frac{s(t)}{2(\alpha_l t)^{1/2}} \tag{11-25b}$$

Since equation (11-25a) should be satisfied for all times, the *parameter* λ must be a constant. Equation (11-25a) is solved for the coefficient B

$$B = \frac{T_m - T_i}{\operatorname{erfc}(\lambda)} \tag{11-26}$$

and this result is int [illegible] obtain

$$\frac{[illegible]}{T_m - T_i} = \frac{[illegible]}{\operatorname{erfc}(\lambda)} \tag{11-27}$$

Finally, the interface energy-balance equation (11-23b) provides the additional relationship for the determination of the parameter λ. Namely, substituting $s(t)$ and $T_l(x, t)$ from equations (11-25b) and (11-27), respectively, into equation (11-23b) and after performing the indicated operations, we obtain the following transcendental equation for the determination of λ

$$\lambda e^{\lambda^2} \operatorname{erfc}(\lambda) = \frac{C_p(T_m - T_i)}{L\sqrt{\pi}} \tag{11-28}$$

TABLE 11-1 Tabulation of Equation (11-28)

λ	$\lambda e^{\lambda^2} \operatorname{erfc}(\lambda) = \frac{C_p(T_m - T_i)}{L\sqrt{\pi}}$
0.00	0.00000E + 00
0.10	8.96457E − 02
0.20	1.61804E − 01
0.30	2.20380E − 01
0.40	2.68315E − 01
0.50	3.07845E − 01
0.60	3.40683E − 01
0.70	3.68151E − 01
0.80	3.91280E − 01
0.90	4.10878E − 01
1.00	4.27584E − 01
1.10	4.41904E − 01
1.20	4.54245E − 01
1.30	4.64935E − 01
1.40	4.74241E − 01
1.50	4.82378E − 01
1.60	4.89525E − 01
1.70	4.95828E − 01
1.80	5.01408E − 01
1.90	5.06368E − 01
2.00	5.10791E − 01
2.50	5.27016E − 01
3.00	5.37003E − 01
3.50	5.43528E − 01
4.00	5.47998E − 01

and λ is the root of this equation. Knowing λ, we can determine the location of the solid–liquid interface $s(t)$ from equation (11-25b) and the temperature distribution $T_l(x, t)$ in the liquid phase from equation (11-27).

In Table 11-1 we present the values of $\lambda e^{\lambda^2}\,\mathrm{erfc}(\lambda)$ against λ. Thus, knowing the Stefan number, λ is determined from this table.

Example 11-2

Melting in a Half-Space (One-Phase Problem). A solid at the solidification (or melting) temperature T_m is confined to a half-space $x > 0$. At time $t = 0$, the temperature of the boundary surface at $x = 0$ is raised to T_0, which is higher than T_m and maintained at that temperature for times $t > 0$. As a result melting starts at the surface $x = 0$ and the solid–liquid interface moves in the positive x direction. Figure 11-4 shows the coordinates and the temperature profiles. The solid phase being at a constant temperature T_m throughout, the temperature is unknown only in the liquid phase, hence the problem is a one-phase problem. In the following analysis the temperature distribution in the liquid phase and the location of the solid–liquid interface are determined, as a function of time.

Solution. The mathematical formulation for the liquid-phase is given as

$$\frac{\partial^2 T_l(x,t)}{\partial x^2} = \frac{1}{\alpha_l}\frac{\partial T_l(x,t)}{\partial t} \quad \text{in} \quad 0 < x < s(t), \quad t > 0 \tag{11-29a}$$

$$T_l(x,t) = T_0 \quad \text{at} \quad x = 0, \quad t > 0 \tag{11-29b}$$

and for the interface as

$$T_l(x,t) = T_m \quad \text{at} \quad x = s(t), \quad t > 0 \tag{11-30a}$$

$$-k_l\frac{\partial T_l}{\partial x} = \rho L\frac{ds(t)}{dt} \quad \text{at} \quad x = s(t), \quad t > 0 \tag{11-30b}$$

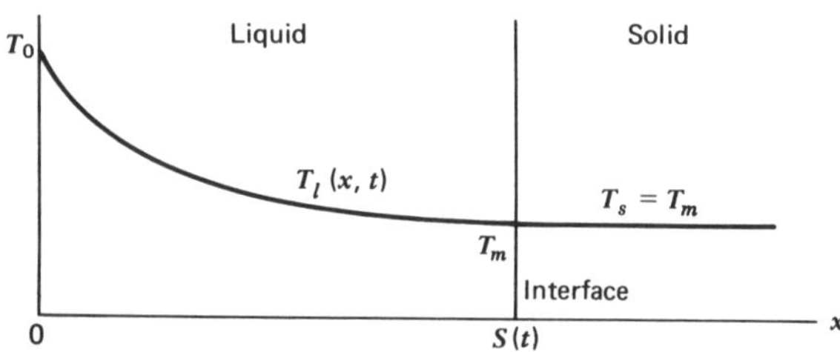

Fig. 11-4 Melting in a half-space. One-phase problem.

No equations are needed for the solid phase because it is at the melting temperature T_m throughout. If we assume a solution in the form

$$T_l(x,t) = T_0 + B\,\text{erf}\,[x/2(\alpha_l t)^{1/2}] \tag{11-31}$$

where B is an arbitrary constant, the differential equation (11-29a) and the boundary condition (11-29b) are satisfied since $\text{erf}(0) = 0$. If we impose the condition that this solution should also satisfy the boundary condition (11-30a) at $x = s(t)$, we obtain

$$T_m = T_0 + B\,\text{erf}(\lambda) \tag{11-32a}$$

where

$$\lambda = \frac{s(t)}{2(\alpha_l t)^{1/2}} \qquad \text{or} \qquad s(t) = 2\lambda(\alpha_l t)^{1/2} \tag{11-32b}$$

Equation (11-32a) implies that λ should be a constant. Then the coefficient B is determined from equation (11-32a) as

$$B = \frac{T_m - T_0}{\text{erf}(\lambda)} \tag{11-33}$$

Introducing equation (11-33) into (11-31) we obtain

$$\frac{T_l(x,t) - T_0}{T_m - T_0} = \frac{\text{erf}\,[x/2(\alpha_l t)^{1/2}]}{\text{erf}(\lambda)} \tag{11-34}$$

Finally, we utilize the interface condition (11-30b) to obtain an additional relationship for the determination of the parameter λ. When $s(t)$ and $T_l(x,t)$ from equations (11-32b) and (11-34), respectively, are introduced into equation (11-30b), the following transcendental equation, similar to equation (11-28), is obtained for the determination of λ

$$\lambda e^{\lambda^2}\,\text{erf}(\lambda) = \frac{C_p(T_0 - T_m)}{L\sqrt{\pi}} \tag{11-35}$$

and λ is the root of this equation. Knowing λ, $s(t)$ is determined from equation (11-32b) and $T_l(x,t)$ from equation (11-34).

In Table 11-2 we present the values of $\lambda\, e^{\lambda^2}\,\text{erf}(\lambda)$ against λ. Thus, knowing the Stefan number, λ is determined from this table.

Example 11-3

Solidification in a Half-Space (*Two-phase Problem*). A liquid at a uniform temperature T_i that is higher than the melting temperature T_m of the solid

TABLE 11-2 Tabulation of Equation (11-35)

λ	$\lambda e^{\lambda^2} \operatorname{erf}(\lambda) = \dfrac{C_p(T_0 - T_m)}{L\sqrt{\pi}}$
0.00	0.00000E + 00
0.10	1.13593E − 02
0.20	4.63583E − 02
0.30	1.07872E − 01
0.40	2.01089E − 01
0.50	3.34168E − 01
0.60	5.19315E − 01
0.70	7.74470E − 01
0.80	1.12590E + 00
0.90	1.61224E + 00
1.00	2.29070E + 00
1.10	3.24693E + 00
1.20	4.61059E + 00
1.30	6.58039E + 00
1.40	9.46482E + 00
1.50	1.37492E + 01
1.60	2.02078E + 01
1.70	3.00928E + 01
1.80	4.54593E + 01
1.90	6.97291E + 01
2.00	1.08686E + 02
2.50	1.29451E + 03
3.00	2.43087E + 04
3.50	7.31434E + 05
4.00	3.55444E + 07

phase is confined to a half-space $x > 0$. At time $t = 0$ the boundary surface at $x = 0$ is lowered to a temperature T_0 below T_m and maintained at that temperature for times $t > 0$. As a result, the solidification starts at the surface $x = 0$ and the solid–liquid interface moves in the positive x direction. Figure 11-5 illustrates the coordinates and the temperatures. This problem is a two-phase problem because the temperatures are unknown in both the solid and liquid phases. In the following analysis we determine the temperature distributions in both phases and the location of the solid–liquid interface. This problem is more general than the ones considered in the previous examples; its solution is known as *Neumann's solution.*

Solution. The mathematical formulation of this problem for the solid phase is given as

$$\frac{\partial^2 T_s}{\partial x^2} = \frac{1}{\alpha_s} \frac{\partial T_s(x,t)}{\partial t} \quad \text{in} \quad 0 < x < s(t), \quad t > 0 \tag{11-36a}$$

$$T_s(x,t) = T_0 \quad \text{at} \quad x = 0, \quad t > 0 \tag{11-36b}$$

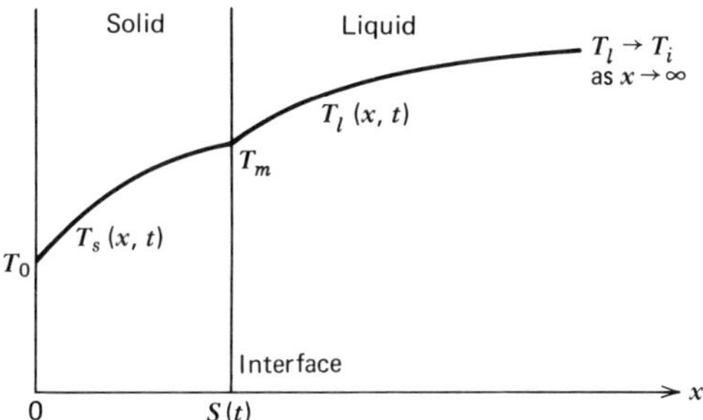

Fig. 11-5 Solidification in a half-space. Two-phase problem.

for the liquid phase as

$$\frac{\partial^2 T_1}{\partial x^2} = \frac{1}{\alpha_1} \frac{\partial T_1(x,t)}{\partial t} \qquad \text{in} \qquad s(t) < x < \infty, \quad t > 0 \tag{11-37a}$$

$$T_1(x,t) \rightarrow T_i \qquad \text{as} \qquad x \rightarrow \infty, \qquad t > 0 \tag{11-37b}$$

$$T_1(x,t) = T_i \qquad \text{for} \qquad t = 0, \qquad \text{in} \quad x > 0 \tag{11-37c}$$

and the coupling conditions at the interface $x = s(t)$ as

$$T_s(x,t) = T_1(x,t) = T_m \qquad \text{at} \quad x = s(t), \quad t > 0 \tag{11-38a}$$

$$k_s \frac{\partial T_s}{\partial x} - k_1 \frac{\partial T_1}{\partial x} = \rho L \frac{ds(t)}{dt} \qquad \text{at} \quad x = s(t), \quad t > 0 \tag{11-38b}$$

If we choose a solution for $T_s(x,t)$ in the form

$$T_s(x,t) = T_0 + A \operatorname{erf}[x/2(\alpha_s t)^{1/2}] \tag{11-39}$$

the differential equation (11-36a) and the boundary condition (11-36b) are satisfied.

If we choose a solution for $T_i(x,t)$ in the form

$$T_1(x,t) = T_i + B \operatorname{erfc}[x/2(\alpha_1 t)^{1/2}] \tag{11-40}$$

the differential equation (11-37a), the boundary condition (11-37b), and the initial condition (11-37c) are satisfied. The constants A and B are yet to be determined.

Equations (11-39) and (11-40) are introduced into the interface condition (11-38a); we find

$$T_0 + A\,\text{erf}(\lambda) = T_i + B\,\text{erfc}\left[\lambda\left(\frac{\alpha_s}{\alpha_l}\right)^{1/2}\right] = T_m \tag{11-41a}$$

where

$$\lambda = \frac{s(t)}{2(\alpha_s t)^{1/2}} \qquad \text{or} \qquad s(t) = 2\lambda(\alpha_s t)^{1/2} \tag{11-41b}$$

Equation (11-41a) implies that λ should be a constant. The coefficients A and B are determined from equations (11-41) as

$$A = \frac{T_m - T_0}{\text{erf}(\lambda)}, \qquad B = \frac{T_m - T_i}{\text{erfc}\,[\lambda(\alpha_s/\alpha_l)^{1/2}]} \tag{11-42}$$

Introducing the coefficients A and B into equations (11-39) and (11-40), we obtain the temperatures for the solid and liquid phases as

$$\frac{T_s(x,t) - T_0}{T_m - T_0} = \frac{\text{erf}\,[x/2(\alpha_s t)^{1/2}]}{\text{erf}(\lambda)} \tag{11-43a}$$

$$\frac{T_l(x,t) - T_i}{T_m - T_i} = \frac{\text{erfc}\,[x/2(\alpha_l t)^{1/2}]}{\text{erfc}\,[\lambda(\alpha_s/\alpha_l)^{1/2}]} \tag{11-43b}$$

The interface energy-balance equation (11-38b) is now used to determine the relation for the evaluation of the parameter λ. That is, when $s(t)$, $T_s(x,t)$ and $T_l(x,t)$ from equations (11-41b), (11-43a), and (11-43b), respectively, are substituted into equation (11-38b), we obtain the following transcendental equation for the determination of λ:

$$\frac{e^{-\lambda^2}}{\text{erf}(\lambda)} + \frac{k_l}{k_s}\left(\frac{\alpha_s}{\alpha_l}\right)^{1/2} \frac{T_m - T_i}{T_m - T_0} \frac{e^{-\lambda^2(\alpha_s/\alpha_l)}}{\text{erfc}\,[\lambda(\alpha_s/\alpha_l)^{1/2}]} = \frac{\lambda L\sqrt{\pi}}{C_{ps}(T_m - T_0)} \tag{11-44}$$

Once λ is known from the solution of this equation, $s(t)$ is determined from equation (11-41b), $T_s(x,t)$ from equation (11-43a) and $T_l(x,t)$ from equation (11-43b).

Example 11-4

Solidification by a Line Heat Sink in an Infinite Medium with Cyclindrical Symmetry (Two-Phase Problem). A line heat sink of strength Q, W/m is located at $r = 0$ in a large body of liquid at a uniform temperature T_i higher than the melting (or solidification) temperature T_m of the medium. The heat sink is

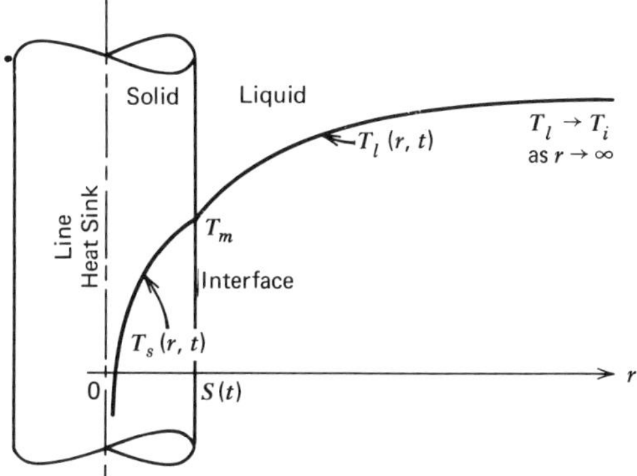

Fig. 11-6 Solidification by a line heat sink in an infinite medium with cylindrical symmetry. Two-phase problem.

activated at time $t = 0$ to absorb heat continuously for times $t > 0$. As a result, the solidification starts at the origin $r \to 0$ and the solid–liquid interface moves in the positive r direction. Figure 11-6 shows the coordinates and the temperature profiles. The problem has cylindrical symmetry, and the temperatures being unknown in both regions, it is a two-phase problem. In this example, the temperature distributions in the solid and liquid phases, and the location of the solid–liquid interface as a function of time will be determined.

Solution. Paterson [97] has shown that the exact solution to the above problem is obtainable if the solution of the heat conduction equation is chosen as an *exponential integral function* in the form $Ei(-r^2/4\alpha t)$. The function $-Ei(-x)$ is also denoted by $E_1(x)$ [99]. A tabulation of $E_1(x)$ function is given in Table 11-3 and a brief discussion of its properties is given in the note at the end of this chapter.

The mathematical formulation of this problem is given for the solid phase as

$$\frac{1}{r}\frac{\partial}{\partial r}\left(r\frac{\partial T_s}{\partial r}\right) = \frac{1}{\alpha_s}\frac{\partial T_s(r,t)}{\partial t} \qquad \text{in} \qquad 0 < r < s(t), \quad t > 0 \tag{11-45}$$

for the liquid phase as

$$\frac{1}{r}\frac{\partial}{\partial r}\left(r\frac{\partial T_l}{\partial r}\right) = \frac{1}{\alpha_l}\frac{\partial T_l(r,t)}{\partial t} \qquad \text{in} \qquad s(t) < r < \infty, \quad t > 0 \tag{11-46}$$

$$T_l(r,t) \to T_i \qquad \text{as} \qquad r \to \infty, \qquad t > 0 \tag{11-47a}$$

$$T_l(r,t) = T_i \qquad \text{for} \qquad t = 0, \qquad \text{in} \quad r > 0 \tag{11-47b}$$

TABLE 11-3 $E_1(x)$ or $-Ei(-x)$ Function[a]

x	$E_1(x)$	x	$E_1(x)$	x	$E_1(x)$	x	$E_1(x)$
0.00	∞	0.25	1.0442826	0.50	0.5597736	1.60	0.0863083
0.01	4.0379296	0.26	0.0138887	0.55	0.5033641	1.65	0.0802476
0.02	3.3547078	0.27	0.9849331				
0.03	2.9591187	0.28	0.9573083	0.60	0.4543795	1.70	0.0746546
0.04	2.6812637	0.29	0.9309182	0.65	0.4115170	1.75	0.0694887
0.05	2.4678985	0.30	0.9056767	0.70	0.3737688	1.80	0.0647131
0.06	2.2953069	0.31	0.8815057	0.75	0.3403408	1.85	0.0602950
0.07	2.1508382	0.32	0.8583352				
0.08	2.0269410	0.33	0.8361012	0.80	0.3105966	1.90	0.0562044
0.09	1.9187448	0.34	0.8147456	0.85	0.2840193	1.95	0.0524144
0.10	1.8229240	0.35	0.7942154	0.90	0.2601839	2.0	4.89005(−2)
0.11	1.7371067	0.36	0.7744622	0.95	0.2387375	2.1	4.26143
0.12	1.6595418	0.37	0.7554414				
0.13	1.5888993	0.38	0.7371121	1.00	0.2193839	2.2	3.71911
0.14	1.4241457	0.39	0.7194367	1.05	0.2018728	2.3	3.25023
0.15	1.4644617	0.40	0.7023801	1.10	0.1859909	2.4	2.84403
0.16	1.4091867	0.41	0.6859103	1.15	0.1715554	2.6	2.18502
0.17	1.3577806	0.42	0.6699973				
0.18	1.3097961	0.43	0.6546134	1.20	0.1584084	2.8	1.68553
0.19	1.2648584	0.44	0.6397328	1.25	0.1464134	3.0	1.30484
0.20	1.2226505	0.45	0.6253313	1.30	0.1354510	3.5	6.97014(−3)
0.21	1.1829020	0.46	0.6113865	1.35	0.1254168	4.0	3.77935
0.22	1.1453801	0.47	0.5978774	1.40	0.1162193	4.5	2.07340
0.23	1.1098831	0.48	0.5847843	1.45	0.1077774	5.0	1.14830
0.24	1.0762354	0.49	0.5720888	1.50	0.1000196	∞	0

[a]The figures in parentheses indicate the power of 10 by which the numbers to the left, and those below in the same column, are to be multiplied.

and for the solid–liquid interface as

$$T_s(r, t) = T_l(r, t) = T_m \qquad \text{at} \qquad r = s(t), \quad t > 0 \tag{11-48a}$$

$$k_s \frac{\partial T_s}{\partial r} - k_l \frac{\partial T_l}{\partial r} = \rho L \frac{ds(t)}{dt} \qquad \text{at} \qquad r = s(t), \quad t > 0 \tag{11-48b}$$

We now choose the solutions for the solid and liquid phases in the forms

$$T_s(r, t) = A - B Ei\left(\frac{-r^2}{4\alpha_s t}\right) \qquad \text{in} \qquad 0 < r < s(t) \tag{11-49a}$$

$$T_l(r, t) = T_i - C Ei\left(\frac{-r^2}{4\alpha_l t}\right) \qquad \text{in} \qquad s(t) < r < \infty \tag{11-49b}$$

and the derivatives of these solutions with respect to r are given as

$$\frac{\partial T_s(r,t)}{\partial r} = -\frac{2B}{r} e^{-r^2/4\alpha_s t} \tag{11-49c}$$

$$\frac{\partial T_l(r,t)}{\partial r} = -\frac{2C}{r} e^{-r^2/4\alpha_l t} \tag{11-49d}$$

The solution (11-49a) for $T_s(r,t)$ satisfies the differential equation (11-45), while the solution (11-49b) for $T_l(r,t)$ satisfies the differential equation (11-46), the boundary condition (11-47a), and the initial condition (11-47b) since $Ei(-\infty) = 0$. The remaining conditions are used to determine the coefficients A, B, and C as now described. The energy balance around the line-heat sink is written as

$$\lim_{r \to 0} \left[2\pi r k_s \frac{\partial T_s}{\partial r} \right] = Q \tag{11-50a}$$

Introducing equation (11-49c) into (11-50a) we find

$$B = -Q/4\pi k_s \tag{11-50b}$$

Equations (11-49a), (11-49b), and (11-50b) are introduced into the interface condition (11-48a)

$$A + \frac{Q}{4\pi k_s} Ei(-\lambda^2) = T_i - C Ei\left(\frac{-\lambda^2 \alpha_s}{\alpha_l} \right) = T_m \tag{11-51a}$$

where

$$\lambda = \frac{s(t)}{2(\alpha_s t)^{1/2}} \tag{11-51b}$$

Since equation (11-51a) should be valid for all values of time, we conclude that *λ must be a constant*. The coefficients A and C are solved from equations (11-51a); we find

$$A = T_m - \frac{Q}{4\pi k_s} Ei(-\lambda^2) \tag{11-52a}$$

$$C = \frac{T_i - T_m}{Ei(-\lambda^2 \alpha_s/\alpha_l)} \tag{11-52b}$$

The derivative of $s(t)$ is obtained from equation (11-51b) as

$$\frac{ds(t)}{dt} = \frac{2\alpha_s \lambda^2}{s} \tag{11-52c}$$

Introducing equations (11-52a) and (11-52b) into equations (11-49a,b) the solutions for the temperatures in the solid and liquid phases become

$$T_s(r,t) = T_m + \frac{Q}{4\pi k_s}\left[Ei\left(-\frac{r^2}{4\alpha_s t}\right) - Ei(-\lambda^2)\right] \quad \text{in} \quad 0 < r < s(t) \tag{11-53a}$$

$$T_l(r,t) = T_i - \frac{T_i - T_m}{Ei(-\lambda^2 \alpha_s/\alpha_l)} Ei\left(-\frac{r^2}{4\alpha_l t}\right), \quad \text{in} \quad s(t) < r < \infty \tag{11-53b}$$

Finally, when equations (11-52c) and (11-53) are introduced into the interface energy-balance equation (11-48b) the following transcendental equation is obtained for the determination of λ

$$\frac{Q}{4\pi} e^{-\lambda^2} + \frac{k_l(T_i - T_m)}{Ei(-\lambda^2 \alpha_s/\alpha_l)} e^{-\lambda^2 \alpha_s/\alpha_l} = \lambda^2 \alpha_s \rho L \tag{11-54}$$

and λ is the root of this equation. Once λ is known, the location of the solid–liquid interface is determined from equation (11-51b); and the temperatures in the solid and liquid phase, from equations (11-53a) and (11-53b), respectively.

A scrutiny of the foregoing exact analyses reveals that in the rectangular coordinate system exact solutions are obtained for some half-space problems when the solution of the heat conduction equation is chosen as a function of $xt^{-1/2}$, namely, as $\operatorname{erf}[x/2(\alpha t)^{1/2}]$ or $\operatorname{erfc}[x/2(\alpha t)^{1/2}]$. In the cylindrical symmetry the corresponding solutions are in the form

$$-Ei\left(-\frac{r^2}{4\alpha t}\right)$$

which is again a function of $rt^{-1/2}$. Paterson [97] has shown that the corresponding solution of the heat conduction equation in spherical symmetry is given in the form

$$\frac{(\alpha t)^{1/2}}{r} e^{-r^2/4\alpha t} - \tfrac{1}{2}\pi^{1/2} \operatorname{erfc}\left(\frac{r}{2(\alpha t)^{1/2}}\right)$$

11-3 INTEGRAL METHOD OF SOLUTION OF PHASE-CHANGE PROBLEMS

The integral method provides a relatively simple and straightforward approach for the solution of one-dimensional transient phase-change problems and has been used for this purpose by several investigators [5–15]. The basic theory of this method has already been described in the chapter on approximate solution of heat conduction problems. When it is applied to the solution of phase-change problems, the fundamental steps in the analysis remain essentially the same, except some modifications are needed in the construction of the temperature profile. In this section we illustrate the use of the integral method in the solution of phase-change problems with simple examples.

Example 11-5

Melting in a Half-Space (One-phase Problem). To give some idea on the accuracy of the integral method of solution of one-dimensional, time-dependent phase-change problems, we consider the one-phase melting problem for which exact solution is available in Example 11-2. The problem considered is the melting of a solid confined to a half-space $x > 0$, initially at the melting temperature T_m. For times $t > 0$ the boundary surface at $x = 0$ is kept at a constant temperature T_0, which is higher than the melting temperature T_m of the solid. The melting starts at the surface $x = 0$ and the solid–liquid interface moves in the positive x direction as illustrated in Fig. 11-4. In the following analysis we determine the location of the solid–liquid interface as a function of time.

Solution. The mathematical formulation of this problem is exactly the same as those given by equations (11-29) and (11-30). Namely, for the liquid phase the equations are given as

$$\frac{\partial^2 T_l}{\partial x^2} = \frac{1}{\alpha_l}\frac{\partial T_l(x,t)}{\partial t} \quad \text{in} \quad 0 < x < s(t), \quad t > 0 \tag{11-55a}$$

$$T_l(x,t) = T_0 \quad \text{at} \quad x = 0, \quad t > 0 \tag{11-55b}$$

and for the interface as

$$T_l(x,t) = T_m \quad \text{at} \quad x = s(t), \quad t > 0 \tag{11-56a}$$

$$-k_l\frac{\partial T_l}{\partial x} = \rho L\frac{ds(t)}{dt} \quad \text{at} \quad x = s(t), \quad t > 0 \tag{11-56b}$$

We recall that the first step in the analysis with the integral method is to define a thermal layer thickness beyond which the temperature gradient is considered zero for practical purposes. Referring to Fig. 11-4, we note that the location of the solid–liquid interface $x = s(t)$ is identical to the definition of the thermal layer, since the temperature gradient in the solid phase is zero for $x > s(t)$. Hence, we choose the region $0 \leqslant x \leqslant s(t)$ as the thermal layer appropriate for this problem and integrate the heat conduction equation from $x = 0$ to $x = s(t)$ to obtain

$$\left.\frac{\partial T}{\partial x}\right|_{x=s(t)} - \left.\frac{\partial T}{\partial x}\right|_{x=0} = \frac{1}{\alpha}\frac{d}{dt}\left[\left(\int_0^{s(t)} T\,dx\right) - \left.T\right|_{x=s(t)} s(t)\right] \tag{11-57}$$

For simplicity we omitted the subscript "l" and it will be done so in the following analysis. We note that equation (11-57) is similar to equation (9-2b) considered in Chapter 9. In view of the boundary conditions (11-56a) and (11-56b) the equation (11-57) reduces to

$$-\frac{\rho L}{k}\frac{ds(t)}{dt}-\left.\frac{\partial T}{\partial x}\right|_{x=0}=\frac{1}{\alpha}\frac{d}{dt}[\theta-T_m s(t)] \tag{11-58a}$$

where

$$\theta \equiv \int_0^{s(t)} T(x,t)dx \tag{11-58b}$$

Equation (11-58) is the *energy-integral equation* for this problem. To solve this equation we choose a second-degree polynomial approximation for the temperature in the form

$$T(x,t)=a+b(x-s)+c(x-s)^2 \tag{11-59}$$

where $s \equiv s(t)$. Three conditions are needed to determine these three coefficients. Equations (11-55b) and (11-56a) provide two conditions; but, the relation given by equation (11-56b) is not suitable for this purpose, because if it is used, the resulting temperature profile will involve the $ds(t)/dt$ term. When such a profile is substituted into the energy integral equation, a second-order ordinary differential equation will result for $s(t)$ instead of the usual first-order equation. To alleviate this difficulty an alternative relation is now developed [5]. The boundary condition (11-56a) is differentiated

$$dT \equiv \left[\frac{\partial T}{\partial x}dx+\frac{\partial T}{\partial t}dt\right]_{x=s(t)}=0 \tag{11-60a}$$

or

$$\frac{\partial T}{\partial x}\frac{ds(t)}{dt}+\frac{\partial T}{\partial t}=0 \tag{11-60b}$$

where we omitted the subscript l for simplicity. The term $ds(t)/dt$ is eliminated between equations (11-56b) and (11-60b)

$$\left(\frac{\partial T}{\partial x}\right)^2=\frac{\rho L}{k}\frac{\partial T}{\partial t} \quad \text{at} \quad x=s(t) \tag{11-61}$$

and eliminating $\partial T/\partial t$ between equations (11-55a) and (11-61) we obtain

$$\left(\frac{\partial T}{\partial x}\right)^2=\frac{\alpha\rho L}{k}\frac{\partial^2 T}{\partial x^2} \quad \text{at} \quad x=s(t) \tag{11-62}$$

This relation, together with the boundary conditions at $x = 0$ and $x = s(t)$

$$T = T_0 \qquad \text{at} \qquad x = 0 \tag{11-63a}$$

$$T = T_m \qquad \text{at} \qquad x = s(t) \tag{11-63b}$$

provide three independent relations for the determination of three unknown coefficients in equation (11-59); the resulting temperature profile becomes

$$T(x, t) = T_m + b(x - s) + c(x - s)^2 \tag{11-64a}$$

where

$$b = \frac{\alpha \rho L}{ks}[1 - (1 + \mu)^{1/2}] \tag{11-64b}$$

$$c = \frac{bs + (T_0 - T_m)}{s^2} \tag{11-64c}$$

$$\mu = \frac{2k}{\alpha \rho L}(T_0 - T_m) = \frac{2C_p(T_0 - T_m)}{L} \tag{11-64d}$$

Substituting the temperature profile (11-64) into the energy-integral equation (11-58) and performing the indicated operations we obtain the following ordinary differential equation for the determination of the location of the solid–liquid interface $s(t)$

$$s\frac{ds}{dt} = 6\alpha \frac{1 - (1 + \mu)^{1/2} + \mu}{5 + (1 + \mu)^{1/2} + \mu} \tag{11-65a}$$

with

$$s = 0 \qquad \text{for} \qquad t = 0 \tag{11-65b}$$

The solution of equation (11-65) is

$$s(t) = 2\lambda\sqrt{\alpha t} \tag{11-66a}$$

where

$$\lambda \equiv \left[3 \frac{1 - (1 + \mu)^{1/2} + \mu}{5 + (1 + \mu)^{1/2} + \mu} \right]^{1/2} \tag{11-66b}$$

We note that the approximate solution (11-66a) for $s(t)$ is of the same form as

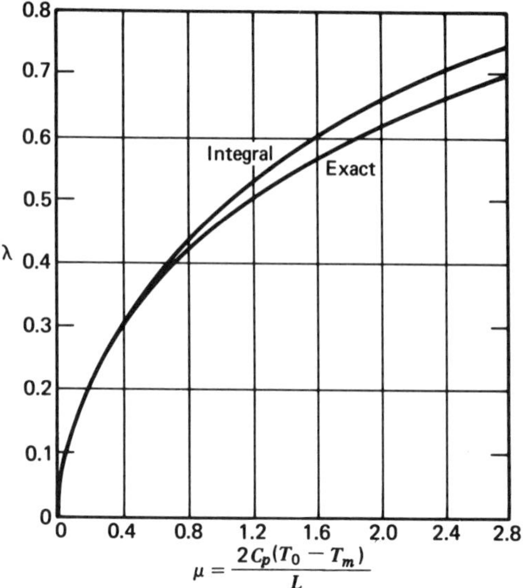

Fig. 11-7 A comparison of exact and approximate solutions of the melting problem in half-space. (From Goodman [5].)

the exact solution of the same problem given previously by equation (11-32b); but the parameter λ is given by equation (11-66b) for the approximate solution, whereas it is the root of the transcendental equation (11-35), that is,

$$\lambda e^{\lambda^2} \operatorname{erf}(\lambda) = \frac{C_p(T_0 - T_m)}{L\sqrt{\pi}} = \frac{\mu}{2\sqrt{\pi}} \tag{11-67}$$

for the exact solution. Therefore, the accuracy of the approximate analysis can be determined by comparing the exact and approximate values of λ as a function of the quantity μ. Now, recalling the definition of the Stefan number given by equation (11-20) we note that the parameter μ is actually twice the Stefan number. Figure 11-7 shows a comparison of the exact and approximate values of λ as a function of the parameter μ. The agreement between the exact and approximate analysis is reasonably good for the second-degree profile used here. If a cubic polynomial approximation were used, the agreement would be much closer [5].

11-4 VARIABLE-TIME-STEP METHOD FOR SOLVING PHASE-CHANGE PROBLEMS—A NUMERICAL SOLUTION

When analytic methods of solution are not possible or impractical, numerical techniques, such as finite differences or finite element is used for solving phase-change problems. The numerical methods of solving phase-change problems can be categorized as follows:

Fixed-grid methods, in which the space–time domain is subdivided into a finite number of equal grids $\Delta x, \Delta t$ for all times. Then the moving solid–liquid interface will in general lie somewhere between two grid points at any given time. The methods of Crank [101] and Ehrlich [102] are the examples for estimating the location of the interface by a suitable interpolation formula as a part of the solution.

Variable-grid methods, in which the space–time domain is subdivided into equal intervals in one direction only and the corresponding grid side in the other direction is determined so that the moving boundary always remains at a grid point. For example, Murray and Landis [52] chose equal steps Δt in the time domain and kept the number of space intervals fixed which in turn allowed the size of the space interval Δx changed (decreased or increased) as the interface moved. In an alternative approach, the space domain is subdivided into fixed equal intervals Δx, but time step is varied such that the interface moves a distance Δx during the time interval Δt, hence always remains at a grid point at the end of each time interval Δt. Several variations of such a variable time step approach have been reported by [52, 66, 79, 80].

Front-fixing method, used in one-dimensional problems. This is essentially a coordinate transformation scheme which immobilizes the moving front hence alleviates the need for tracking the moving front at the expense of solving a more complicated problem by the numerical scheme [77, 101].

Enthalpy method, which has been used by several investigators to solve phase-change problems in situations in which the material does not have a distinct solid–liquid interface. Instead, the melting or solidification takes place over an extended range of temperatures. The solid and liquid phases are separated by a two-phase moving region. In this approach, an enthalpy function, $H(T)$, which is the total heat content of the substance, is used as a dependent variable along with the temperature. The method is also applicable for phase-change problems involving a single phase-change temperature [73–76].

In this section we present the *modified variable-time-step* (MVTS) method described by Gupta and Kumar [79].

We consider the solidification of a liquid initially at the melting temperature T_m^*, confined to the region $0 \leqslant x \leqslant B$. For times $t > 0$, the boundary surface at $x = 0$ is subjected to convective cooling into an ambient at a constant temperature T_∞ with a heat transfer coefficient h, while the boundary surface at $x = B$ is kept insulated or satisfies the symmetry condition. The solidification starts at the boundary surface $x = 0$, and the solid–liquid interface moves in the x direction as illustrated in Fig. 11-8.

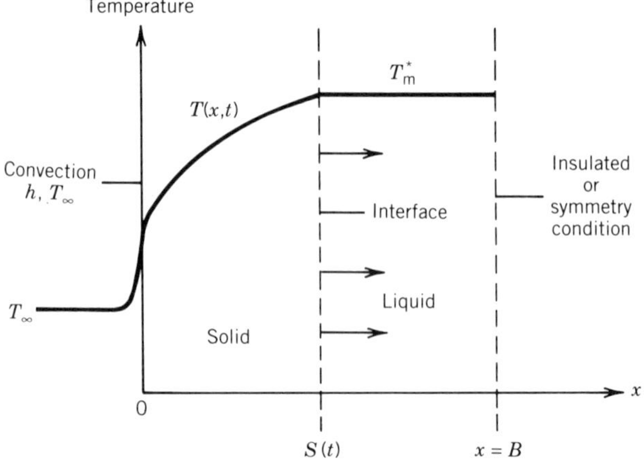

Fig. 11-8 Geometry and coordinate for single-phase solidification.

Temperature $T(x, t)$ varies only in the solid phase, since the liquid region is at the melting temperature T_m^*. We are concerned with the determination of the temperature distribution $T(x, t)$ in the solid phase and location of the interface as a function of time. The mathematical formulation of this solidification problem is given as follows:

Solid region: $$\frac{\partial^2 T}{\partial x^2} = \frac{1}{\alpha}\frac{\partial T}{\partial t} \quad \text{in} \quad 0 < x < s(t), \quad t > 0 \tag{11-68a}$$

$$-k\frac{\partial T}{\partial x} + hT = hT_\infty \quad \text{at} \quad x = 0, \quad t > 0 \tag{11-68b}$$

Interface: $$T(x, t) = T_m^* \quad \text{at} \quad x = s(t), \quad t > 0 \tag{11-68c}$$

$$k\frac{\partial T}{\partial x} = \rho L\frac{ds(t)}{dt} \quad \text{at} \quad x = s(t), \quad t > 0 \tag{11-68d}$$

where h is the heat transfer coefficient, $s(t)$ is the location of solid–liquid interface, ρ is the density, L is the latent heat of solidification (or melting), k is the thermal conductivity, and α is the thermal diffusivity.

To solve the above problem with finite differences, the "$x - t$" domain is subdivided into small intervals of constant Δx in space and variable Δt in time as illustrated in Fig. 11-9. The variable time step approach requires that at each time level t_n the time step Δt_n is so chosen that the interface moves exactly a distance Δx during the time interval Δt_n, hence always stays on the node. Therefore, we are concerned with the determination of the time step $\Delta t_n = t_{n+1} - t_n$ such that, in the time interval from t_n to t_{n+1}, the interface moves from the position

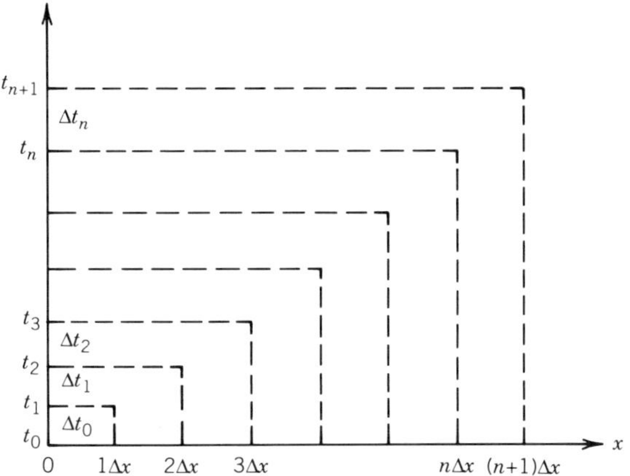

Fig. 11-9 Subdivision of $x - t$ domain using constant Δx, variable Δt.

$n\Delta x$ to the next position $(n+1)\Delta x$. We describe below first the finite-difference approximation of this solidification problem, and then the determination of the time step Δt_n.

The finite-difference approximation of equations (11-68) is described below:

Differential Equation (11-68a). This differential equation can be approximated with finite differences by using either the implicit scheme or the combined method. For simplicity we prefer the implicit method and write equation (11-68a) in finite-difference form as

$$\frac{T_{i-1}^{n+1} - 2T_i^{n+1} + T_{i+1}^{n+1}}{(\Delta x)^2} = \frac{1}{\alpha} \frac{T_i^{n+1} - T_i^n}{\Delta t_n} \tag{11-69a}$$

where the following notation is adopted

$$T(x, t_n) = T(i\Delta x, t_n) \equiv T_i^n \tag{11-69b}$$

Equation (11-69a) is rearranged as

$$[-r_n T_{i-1}^{n+1} + (1 + 2r_n)T_i^{n+1} - r_n T_{i+1}^{n+1}]^{(p)} = T_i^n \tag{11-70a}$$

where the superscript p over the bracket refers to the pth iteration, and the parameter r_n is defined as

$$r_n = \frac{\alpha \Delta t_n}{(\Delta x)^2}, \qquad i = 1, 2, 3, \ldots, \qquad \Delta t_n = t_{n+1} - t_n \tag{11-70b,c}$$

Boundary Condition at $x = 0$. The convection boundary condition (11-68b) is rearranged

$$\frac{\partial T}{\partial x} = HT - HT_\infty \qquad \text{where} \qquad H = \frac{h}{k} \tag{11-71a}$$

and then discretized as

$$\frac{T_1^{n+1} - T_0^{n+1}}{\Delta x} = HT_0^{n+1} - HT_\infty \tag{11-71b}$$

This result is rearranged in the form

$$[T_1^{n+1} - (1 + H\,\Delta x)T_0^{n+1}]^{(p)} = -H\,\Delta x T_\infty \tag{11-71c}$$

where superscript p over the bracket denotes the pth iteration. The finite-difference equation (11-71c) is first-order accurate.

Interface Conditions. The condition of continuity of temperature at the interface, equation (11-68c), is written as

$$T_{n+1}^{n+1} = T_{\text{m}}^* = \text{melting temperature} \tag{11-72a}$$

which is valid for all times. The interface energy balance equation (11-68d) is discretized as

$$\frac{T_{n+1}^{n+1} - T_n^{n+1}}{\Delta x} = \frac{\rho L}{k}\frac{\Delta x}{\Delta t_n} \tag{11-72b}$$

which is rearranged in the form

$$[\Delta t_n]^{(p+1)} = \frac{\rho L}{k}\left[\frac{(\Delta x)^2}{T_{\text{m}}^* - T_n^{n+1}}\right]^{(p)} \tag{11-72c}$$

since $T_{n+1}^{n+1} = T_{\text{m}}^* =$ melting temperature.

Determination of Time Steps

We now describe the algorithms for the determination of time step Δt_n such that during this time step, the interface moves exactly a distance Δx.

Starting Time Step Δt_0. An explicit expression can be developed for the calculation of the first step Δt_0 as follows. Set $n = 0$ in equations (11-71c) and (11-72c), and eliminate T_0^1 between the resulting two equations and note that $T_1^1 \equiv T_{\text{m}}^*$.

The following explicit expression is obtained for Δt_0.

$$\Delta t_0 = \frac{\rho L}{k} \frac{\Delta x(1 + H\,\Delta x)}{H(T_m^* - T_\infty)} \tag{11-73}$$

where $\Delta t_0 \equiv t_1 - t_0$.

Time Step Δt_1. We set $i = 1, n = 1$ in equation (11-70a) and note that $T_1^1 = T_2^2 = T_m^*$. Then equation (11-70a) becomes

$$[-r_1 T_0^2 + (1 + 2r_1)T_1^2]^{(p)} = (1 + r_1^{(p)})T_m^* \tag{11-74a}$$

and from the boundary condition (11-71c) for $n = 1$, we obtain

$$[-(1 + H\,\Delta x)T_0^2 + T_1^2]^{(p)} = -H\,\Delta x T_\infty \tag{11-74b}$$

To solve equations (11-74a) and (11-74b) for T_0^2 and T_1^2, the value of $r_1^{(p)}$ is needed; but $r_1^{(p)}$ defined by equation (11-70b) depends on $\Delta t_1^{(p)}$. Therefore, iteration is needed for their solution. To start iterations, we set

$$\Delta t_1^{(0)} = \Delta t_0$$

Then, $r_1^{(0)}$ is determined from equation (11-70b); using this value of $r_1^{(0)}$, equations (11-74a,b) are solved for T_0^2 and T_1^2. Knowing T_1^2, we can compute $\Delta t_1^{(1)}$ from equation (11-72c). Iterations are continued until the difference between two consecutive time steps

$$|\Delta t_1^{(p+1)} - \Delta t_1^{(p)}|$$

satisfies a specified convergence criteria.

Time Step Δt_n. The above results are now used in the following algorithm to calculate the time steps Δt_n at each time level $t_n, n = 2, 3, \ldots$.

1. The starting time step Δt_0 at the time level t_0 is calculated directly from the explicit expression (11-73) since all the quantities on the right-hand side of this equation are known.
2. The time steps Δt_n at the time levels $t_n, n = 2, 3, \ldots$ are calculated by iteration. A guess value Δt_n^0 is chosen as

$$\Delta t_n^{(0)} = \Delta t_{n-1}, \qquad n = 2, 3, \ldots \tag{11-75a}$$

The system of finite-difference equations (11-70), (11-71c), together with the condition (11-72a) are solved for $i = 1, 2, 3, \ldots, n$ by setting $p = 0$ and a first estimate is obtained for the nodal temperatures

$$[T_i^{n+1}]^{(0)}, \qquad \text{for} \qquad i = 1, 2, \ldots, n \tag{11-75b}$$

We note that the system of equations is tridiagonal, and hence readily solved.

3. The values of $[T_i^{n+1}]^{(0)}$ obtained from equation (11-75b) are introduced into equation (11-72c) for $p = 0$ and a first estimate for the time step $\Delta t_n^{(1)}$ is determined.
4. $\Delta t_n^{(1)}$ is used as a guess value and steps 2 and 3 are repeated to calculate a second estimate for the time step $\Delta t_n^{(2)}$.
5. The steps 2, 3, and 4 are repeated until the difference between two consecutive time steps

$$|\Delta t_n^{(p+1)} - \Delta t_n^{(p)}|$$

satisfies a specified convergence criteria.

Example 11-6

Consider a single-phase solidification problem for a liquid initially at the melting temperature T_m^*, confined to the region $0 \leqslant x \leqslant B$. Solidification takes place as a result of convective cooling at the boundary surface $x = 0$, while the boundary surface at $x = B$ is kept insulated. The mathematical formulation of this problem is given in the dimensionless form as follows:

$$\text{Solid region:} \quad \frac{\partial^2 T}{\partial x^2} = \frac{\partial T}{\partial t} \qquad \text{in} \qquad 0 < x < s(t), \quad t > 0$$

$$-\frac{\partial T}{\partial x} + 10T = 0 \qquad \text{at} \qquad x = 0, \qquad t > 0$$

$$\text{Interface:} \quad T(x, t) = 1 \qquad \text{at} \qquad x = s(t), \qquad t > 0$$

$$\frac{\partial T}{\partial x} = \frac{ds}{dt} \qquad \text{at} \qquad x = s(t), \qquad t > 0$$

Calculate the time step Δt required for the solid–liquid interface $s(t)$ to move one space interval $\Delta x = 0.1$ and the temperature of the boundary surface at $x = 0$ for the interface positions $s(t) = 0.1, 0.2, 0.3, \ldots, 1.0$.

Solution. This problem has been solved [79] by using the variable time step approach described above, and their results are listed in Table 11-4. For example, the first time step Δt_0, needed for the interface to move from $s(t) = 0$ to $s(t) = 0.1$, is determined directly from equation (11-73). The numerical

TABLE 11-4 Time Step Δt Required for the Interface Position to Move by One Space Interval Δx and Temperature of the Boundary Surface at $x = 0$

Interface Position $s(t)$	Time Step Δt	$T(0, t)$	Number of Iterations
0.1	0.0200	0.5000	0
0.2	0.0356	0.3596	4
0.3	0.0494	0.2770	4
0.4	0.0627	0.2242	4
0.5	0.0759	0.1879	4
0.6	0.0890	0.1616	4
0.7	0.1021	0.1416	4
0.8	0.1152	0.1260	4
0.9	0.1282	0.1135	4
1.0	0.1413	0.1032	4

values of various parameters appearing in this equation are determined by comparing the mathematical formulation of this example with that given by equations (11-68). We find

$$T_m^* = 1, \quad H = \frac{h}{k} = 10, \quad T_\infty = 0, \quad \frac{\rho L}{k} = 1$$

and the space step is chosen as $\Delta x = 0.1$. Introducing these numerical values into equation (11-73), the starting time step Δt_0 is determined as

$$\Delta t_0 = \frac{\rho L}{k} \frac{\Delta x(1 + H\,\Delta x)}{H(T_m^* - T_\infty)} = \frac{0.1(1 + 10 \times 0.1)}{10 \times (1 - 0)} = 0.020$$

The next time step Δt_1 needed for the interface to move from the position $s(t) = 0.1$ to the position $s(t) = 0.2$ is determined by an iterative procedure described previously. According to Table 11-4, a value of $\Delta t_1 = 0.0356$ is obtained with a maximum error of 0.05%. The remaining time steps are determined iteratively and listed in Table 11-4. Also included in this Table is $T(0, t)$, the temperature of the boundary surface at $x = 0$.

11-5 ENTHALPY METHOD FOR SOLUTION OF PHASE-CHANGE PROBLEMS—A NUMERICAL SOLUTION

In the solution of phase-change problems considered previously, the temperature has been the sole dependent variable. That is, the energy equation has been written separately for the solid and liquid phases and the temperatures have been coupled through the interface energy balance condition. Such a formulation gives rise to the tracking of the moving interface, and it is a difficult matter if the problem is to be solved with finite differences.

An alternative approach is the use of the enthalpy form of the energy equation along with the temperature. The advantage of the enthalpy method is that a single energy equation becomes applicable in both phases; hence there is no need to consider liquid and solid phases separately. Therefore, any numerical scheme such as the finite-difference or finite-element method can readily be adopted for the solution. In addition, the enthalpy method is capable of handling phase change problems in which the phase change occurs over an extended temperature range rather than at a single phase-change temperature.

Figure 11-10 shows enthalpy–temperature relations for (*a*) pure crystalline substances and eutectics and (*b*) glassy substances and alloys. For pure substances the phase change takes place at a discrete temperature, and hence is associated with the latent heat L. Therefore, in Fig. (11-10*a*) a jump discontinuity occurs at the melting temperature T_m^*; hence $\partial H/\partial T$ becomes infinite and the energy equation apparently is not meaningful at this point. However, it has been shown that [73] the enthalpy form of the energy equation given by

$$\nabla \cdot (k \nabla T) = \rho \frac{\partial H(T)}{\partial t} \tag{11-76}$$

is equivalent to the usual temperature form in which the heat conduction equation is written separately for the liquid and solid regions and coupled with the energy balance equation at the solid–liquid interface. Therefore, the enthalpy method is applicable for the solution of phase-change problems involving both a distinct phase change at a discrete temperature as well as phase change taking place over an extended range of temperatures.

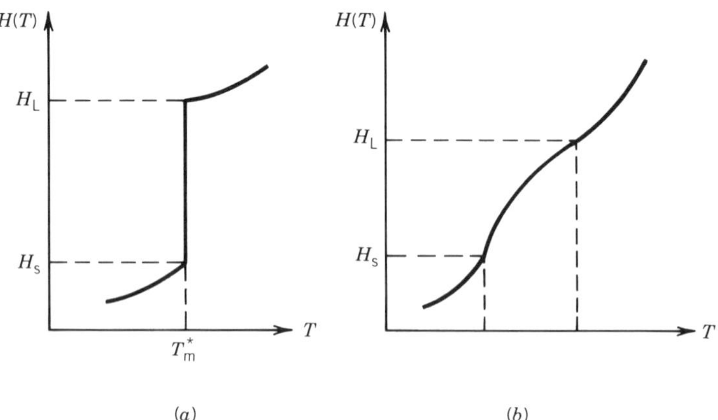

Fig. 11-10 Enthalpy–temperature relationship for (*a*) pure crystalline substances and eutectics and (*b*) glassy substances and alloys.

Figure 11-10*b* shows that for alloys and glassy substances there is no single melting-point temperature T_m^* because the phase change takes place over an extended temperature range from T_s to T_l, and a mushy zone exists between the all solid and all liquid regions.

To illustrate the physical significance of the enthalpy function $H(T)$, J/kg (joules per kilogram), in relation to the case of pure substances having a single melting-point temperature T_m^*, we refer to the plot of $H(T)$ as a function of temperature as illustrated in Fig. 11-10*a*. When the substance is in solid form at temperature T, the substance contains a sensible heat per unit mass $C_p(T - T_m^*)$, where the melting-point temperature T_m^* is taken as the reference temperature. In the liquid form, it contains latent heat L per unit mass in addition to the sensible heat, that is, $C_p(T - T_m^*) + L$. For the specific case considered here, the enthalpy is related to temperature by

$$H = \begin{cases} C_p(T - T_m^*) & \text{for} \quad T < T_m^* \qquad (11\text{-}77a) \\ C_p(T - T_m^*) + L & \text{for} \quad T > T_m^* \qquad (11\text{-}77b) \end{cases}$$

Conversely, given the enthalpy of the substance, the corresponding temperature is determined from

$$T = \begin{cases} T_m^* + \dfrac{H}{C_p} & \text{for} \quad H < 0 \qquad (11\text{-}78a) \\ T_m^* & \text{for} \quad 0 \leqslant H \leqslant L \qquad (11\text{-}78b) \\ T_m^* + \dfrac{H - L}{C_p} & \text{for} \quad H > L \qquad (11\text{-}78c) \end{cases}$$

In the case of glassy substances and alloys, there is no discrete melting-point temperature, because the phase change takes place over an extended range of temperatures as illustrated in Fig. 11-10*b*. Such relationship between $H(T)$ and T is obtained from either experimental data or standard physical tables. In general, enthalpy is a nonlinear function of temperature. Therefore an enthalpy versus temperature variation need to be available. Assuming linear release of latent heat over the mushy region, the variation of $H(T)$ with temperature can be taken as

$$H = \begin{cases} C_p T & \text{for} \quad T < T_s \quad \text{solid region} \qquad (11\text{-}79a) \\ C_p T + \dfrac{T - T_s}{T_l - T_s} L & \text{for} \quad T_s \leqslant T \leqslant T_l \quad \text{mushy region} \qquad (11\text{-}79b) \\ C_p T + L & \text{for} \quad T > T_l \quad \text{liquid region} \qquad (11\text{-}79c) \end{cases}$$

where L is the latent heat, and T_s and T_l are the solid- and liquid-phase temperatures, respectively.

To solve the phase-change problem with the enthalpy method, an explicit or an implicit finite-difference scheme can be used. The implicit scheme is generally preferred because of its ability to accommodate a wide range of time steps without the restriction of the stability criteria. We present below, the *implicit enthalpy* method for solving one-dimensional, two-phase solidification problem for a substance having a single phase-change temperature T_m^*.

Implicit Enthalpy Method for Solidification at a Single Phase-Change Temperature

We consider one-dimensional solidification of a liquid having a single melting-point temperature T_m^* and confined to the region $0 \leqslant x \leqslant B$. Initially, the liquid is at a uniform temperature T_0 that is higher than the melting temperature T_m^* of the liquid. For times $t > 0$, the boundary surface at $x = 0$ is kept at a temperature f that is lower than the melting temperature T_m^* of the substance. The boundary condition at $x = B$ satisfies the symmetry requirement. For simplicity, the properties are assumed to be constant.

The enthalpy formulation of this phase-change problem is given by

$$\rho \frac{\partial H}{\partial t} = k \frac{\partial^2 T}{\partial x^2}, \qquad \text{in} \quad 0 < x < B, \quad t > 0 \tag{11-80a}$$

$$T = f \qquad \text{at} \quad x = 0, \quad t > 0 \tag{11-80b}$$

$$\frac{\partial T}{\partial x} = 0 \qquad \text{at} \quad x = B, \quad t > 0 \tag{11-80c}$$

$$T = T_0 (\text{or } H = H_0) \qquad \text{for} \quad t = 0, \quad 0 \leqslant x \leqslant B \tag{11-80d}$$

To approximate this problem with finite differences, the region $0 \leqslant x \leqslant B$ is subdivided into M equal parts each of width $\Delta x = B/M$.

The finite-difference approximation of the differential equation (11-80a) using the implicit scheme is given by

$$\rho \frac{H_i^{n+1} - H_i^n}{\Delta t} = k \frac{T_{i-1}^{n+1} - 2T_i^{n+1} + T_{i+1}^{n+1}}{(\Delta x)^2} \tag{11-81}$$

where the subscript $i = 1, 2, \ldots, M - 1$ denotes the spatial discretization and the superscript $n = 1, 2, \ldots$ denotes the time discretization. The solution of equation (11-81) for the enthalpy H_i^{n+1} gives

$$H_i^{n+1} = H_i^n + \frac{k\Delta t}{\rho(\Delta x)^2}[F^*(H_{i-1}^{n+1}) - 2F^*(H_i^{n+1}) + F^*(H_{i+1}^{n+1})], \qquad i = 1, 2, 3, \ldots, M - 1 \tag{11-82}$$

where the notation

$$T = F^*(H) \tag{11-83}$$

denotes that the temperature T is related to the enthalpy H. The system of equations (11-82) can be written more compactly in the vector form as

$$\mathbf{H}^{n+1} = \mathbf{H}^n + \Delta t \mathbf{F}(\mathbf{H}^{n+1}) \tag{11-84a}$$

where $\mathbf{H}$ is a vector whose components are the nodal enthalpies H_i and $\mathbf{F}$ is a function with ith component given by

$$F_i(\mathbf{H}) = \frac{k}{\rho(\Delta x)^2}[F^*(H_{i-1}) - 2F^*(H_i) + F^*(H_{i+1})] \tag{11-84b}$$

For a substance having a single phase-change temperature T_m^*, the temperature is related to the enthalpy by

$$T = \begin{cases} \dfrac{H}{C_p} & H < C_p T_m^* \quad (11\text{-}85a) \\ T_m^* & C_p T_m^* \leqslant H \leqslant (C_p T_m^* + L) \quad (11\text{-}85b) \\ \dfrac{H - L}{C_p} & H > (C_p T_m^* + L) \quad (11\text{-}85c) \end{cases}$$

Equivalently, equation (11-85) can be written as

$$H(T) = \begin{cases} C_p T & T < T_m^* \\ C_p T + L & T > T_m^* \end{cases} \tag{11-86}$$

The difference between these equations and that given by equations (11-78) is that in the latter temperature T_m^* is used as the reference temperature. The finite-difference equations (11-84), together with the appropriate boundary and initial conditions for the problem and the "temperature–enthalpy" relations given by equations (11-85), constitute a set of equations for the determination of nodal enthalpies H_i^{n+1} at the time level $n + 1$, from the knowledge of the enthalpies H_i^n in the previous time level. These equations being nonlinear, an iterative scheme is needed to solve for $\mathbf{H}^{n+1}$. Furthermore, if it is required that the solid–liquid interface move one and only one spatial step Δx during each consecutive time step Δt, iteration becomes necessary to establish the magnitude of each time step accordingly. Voller and Cross [74] used enthalpy formulation for a one-dimensional solidification problem with a single phase-change temperature which led to very accurate solutions. We present below, the equations needed to perform such iterations.

Equation (11-84) is written in the form

$$\mathbf{G}(\mathbf{H}^{n+1}) \equiv \mathbf{H}^{n} + \Delta t \mathbf{F}(\mathbf{H}^{n+1}) - \mathbf{H}^{n+1} = \mathbf{0} \tag{11-87}$$

To calculate $\mathbf{H}^{n+1}$, the modification of Newton's method is applied

$$\mathbf{H}^{n+1,k+1} = \mathbf{H}^{n+1,k} - \omega \frac{\mathbf{G}(\mathbf{H}^{n+1,k})}{\mathbf{G}'(\mathbf{H}^{n+1,k})} \tag{11-88a}$$

where ω is the relaxation parameter, the superscript k denotes the number of iterations, and n is the number of discretization steps on time. The derivative $\mathbf{G}'$ with respect to $\mathbf{H}^{n+1,k}$ is determined as

$$\mathbf{G}'(\mathbf{H}^{n+1,k}) = \Delta t \frac{\partial \mathbf{F}(\mathbf{H}^{n+1,k})}{\partial \mathbf{H}^{n+1,k}} - \mathbf{I} \tag{11-88b}$$

$$\equiv \mathbf{J} - \mathbf{I} \tag{11-88c}$$

where **I** is the identity matrix and **J** is the Jacobian matrix whose components are given by

$$\mathbf{J}_{i,l} = \Delta t \left. \frac{\partial F_i}{\partial H_l} \right|_{\mathbf{H} = \mathbf{H}^{n+1,k}} \tag{11-89}$$

where $F_i(\mathbf{H})$ is as defined by equation (11-84b).

Then the equation for the determination of the ith component of enthalpy $H_i^{n+1,k+1}$ becomes

$$H_i^{n+1,k+1} = H_i^{n+1,k} + \omega \frac{H_i^{n,k} - H_i^{n+1,k} + \Delta t F_i(\mathbf{H}^{n+1,k})}{1 - J_{ii}} \tag{11-90a}$$

where

$$J_{ii} = \Delta t \left. \frac{\partial F_i}{\partial H_i} \right|_{\mathbf{H} = \mathbf{H}^{n+1,k}} \tag{11-90b}$$

The Algorithm

To start the iterations on **H**, an initial estimate on the components of enthalpy is chosen as

$$H_i^{n+1,0} = H_i^{n} + \Delta t F_i(\mathbf{H}^{n}) \tag{11-91}$$

where $F_i(\mathbf{H}^n)$ is as defined by equation (11-84b). Iterations are carried out by using equations (11-90) until a specified convergence criterion is achieved.

Also we need to perform iterations on the size of the time step Δt_k such that the interface will move one and only one grid point over the duration of this time step. This requirement can be satisfied by noting that at each time step one and only one nodal enthalpy takes the value $(CT_{\mathrm{m}}^* + \frac{1}{2}L)$.

Suppose the calculations are carried out up to the time level n and that the nodal enthalpies H_i^n are determined for all nodal points i at time t, namely, the time level n. Let, Δt_i denote the time step during which the interface moves by one spatial step Δx, from the node i to the node $i+1$. Then

$$\Delta t_i = t_{i+1} - t_i$$

Then the iterations on the size of the time step Δt is performed in the following manner:

1. The initial guess for the size of the time step Δt_k^0 is taken as

$$\Delta t_i^0 = \Delta t_{i-1}$$

2. The enthalpy distribution $\mathbf{H}^{t+\Delta t_i^m}$, where the superscript m on Δt refers to the mth iteration on the time step, is determined from the solution of equations (11-90) and (11-91). Here, the mth time step Δt_i^m is computed using an iterative scheme given by

$$\Delta t_i^{m+1} = \Delta t_i^m + \omega^* \Delta t_i^m \left(\frac{H_{i+1}^{t+\Delta t_i^m}}{CT_{\mathrm{m}}^* + (L/2)} - 1 \right) \tag{11-92}$$

where ω^* is the relaxation parameter associated with the time step iterations.

3. When the value of $H_{i+1}^{t+\Delta t_i^m}$ converges to $[CT_{\mathrm{m}}^* + (L/2)]$, the corresponding enthalpy values at all nodes are considered to be the solution for the time $t + \Delta t_i^m$.
4. Once the enthalpy values are available at the nodes, the corresponding values of node temperatures T_i are determined from the temperature–enthalpy relation given by equations (11-85).

Implicit Enthalpy Method for Solidification Over an Extended Temperature Range

If the phase change takes place over an extended temperature range, there is a mushy zone between the solidus and liquidus regions. In such a case, the enthalpy $H(T)$ is a smooth continuous function or piecewise continuous function. Assuming a linear variation of latent heat over the mushy region, the variation of $H(T)$ with temperature can be taken as that given by equations (11-79):

$$H = \begin{cases} C_p T & \text{for} \quad T < T_s \quad \text{solid region} & \text{(11-93a)} \\ C_p T + \dfrac{T - T_s}{T_l - T_s} L & \text{for} \quad T_s \leqslant T \leqslant T_l \quad \text{mushy region} & \text{(11-93b)} \\ C_p T + L & \text{for} \quad T > T_l \quad \text{liquid region} & \text{(11-93c)} \end{cases}$$

and the corresponding relations for temperature as a function of enthalpy becomes

$$T = \begin{cases} \dfrac{H}{C_p} & \text{for} \quad H < C_p T_s & \text{(11-94a)} \\ \dfrac{H(T_l - T_s) + L T_s}{C_p(T_l - T_s) + L} & \text{for} \quad C_p T_s \leqslant H \leqslant (C_p T_l + L) & \text{(11-94b)} \\ \dfrac{H - L}{C_p} & \text{for} \quad H > (C_p T_l + L) & \text{(11-94c)} \end{cases}$$

Then the algorithm described previously is applicable if equation (11-84b) is used together with equations (11-94).

Readers should consult reference 74 for a comparison of *explicit enthalpy* and *implicit enthalpy* methods of solution for phase change at a single temperature and constant properties with the *exact analytic solution* of a one-dimensional solidification problem.

REFERENCES

1. G. Lamé and B. P. Clapeyron, *Ann, Chem. Phys.* **47**, 250–256, 1831.
2. J. Stefan, *Ann. Phys. Chemie (Wiedemannsche Annalen)* **42**, 269–286, 1891.
3. H. S. Carslaw and J. C. Jaeger, *Conduction of Heat in Solids*, 2nd ed., Clarendon Press, London, 1959.
4. J. Crank, *Free and Moving Boundary Problems*, Oxford University Press, New York, 1984.
5. T. R. Goodman, *Trans. Am. Soc. Mech. Eng.* **80**, 335–342, 1958.
6. T. R. Goodman, *J. Heat Transfer* **83c**, 83–86, 1961.
7. T. R. Goodman and J. Shea, *J. Appl. Mech.* **32**, 16–24, 1960.
8. G. Poots, *Int. J. Heat Mass Transfer* **5**, 339–348, 1962.
9. G. Poots, *Int. J. Heat Mass Transfer* **5**, 525, 1962.
10. R. H. Tien, *Trans. Metall. Soc. AIME* **233**, 1887–1891, 1965.

11. R. H. Tien and G. E. Geiger, *J. Heat Transfer* **89c**, 230–234, 1967.
12. R. H. Tien and G. E. Geiger, *J. Heat Transfer* **90c**, 27–31, 1968.
13. S. H. Cho and J. E. Sunderland, *J. Heat Transfer* **91c**, 421–426, 1969.
14. J. C. Muehlbauer, J. D. Hatcher, D. W. Lyons, and J. E. Sunderland, *J. Heat Transfer* **95c**, 324–331, 1973.
15. K. Mody and M. N. Özışık, *Lett. Heat Mass Transfer* **2**, 487–493, 1975.
16. M. A. Biot, *J. Aeronaut. Sci.* **24**, 857–873, 1957.
17. T. J. Lardner, *AIAA J.* **1**, 196–206, 1963.
18. W. Zyskowski, The Transient Temperature Distribution in One-Dimensional Heat-Conduction Problems with Nonlinear Boundary Conditions, ASME Paper No. 68-HT-6, 1968.
19. M. A. Biot and H. Daughaday, *J. Aerospace Sci.* **29**, 227–229, 1962.
20. M. A. Biot, *Variational Principles in Heat Transfer*, Oxford University Press, London, 1970.
21. A. Prasad and H. C. Agrawal, *AIAA J.* **12**, 250–252, 1974.
22. N. M. H. Lightfoot, *Proc. London Math. Soc.* **31**, 97–116, 1929.
23. M. N. Özışık, *J. Heat Transfer*, **100C**, 370–371, 1978.
24. Y. K. Chuang and J. Szekely, *Int. J. Heat Mass Transfer* **14**, 1285–1295, 1971.
25. K. A. Rathjen and L. M. Jiji, *J. Heat Transfer* **93c**, 101–109, 1971.
26. Y. K. Chuang and J. Szekely, *Int. J. Heat Mass Transfer* **15**, 1171–1175, 1972.
27. H. Budhia and F. Kreith, *Int. J. Heat Mass Transfer* **16**, 195–211, 1973.
28. L. T. Rubenstein, *The Stefan Problem*, Trans. Math. Monographs Vol. 27, Am. Math. Soc., Providence, R. I., 1971, pp. 94–181.
29. K. A. Rathjen and L. M. Jiji, Transient Heat Transfer in Fins Undergoing Phase Transformation, *Fourth International Heat Transfer Conference, Paris-Versailles*, **2**, 1970.
30. R. I. Pedroso and G. A. Domoto, *J. Heat Transfer* **95**, 42, 1973.
31. R. I. Pedroso and G. A. Domoto, *Int. J. Heat Mass Transfer* **16**, 1037, 1973.
32. L. M. Jiji and S. Weimbaum, A Nonlinear Singular Perturbation Theory for Non-Similar Melting or Freezing Problems, Conduction Cu-3, 5th International Heat Transfer Conference, Tokyo, 1974.
33. D. S. Riley, F. T. Smith and G. Poots, *Int. J. Heat Mass Transfer* **17**, 1507, 1974.
34. C. L. Hwang and Y. P. Shih, *Int. J. Heat Mass Transfer* **18**, 689–695, 1975.
35. B. A. Boley, *J. Math. Phys.* **40**, 300–313, 1961.
36. B. A. Boley, *Int. J. Eng. Sci.* **6**, 89–111, 1968.
37. J. M. Lederman and B. A. Boley, *Int. J. Heat Mass Transfer* **13**, 413–427, 1970.
38. D. L. Sikarshie and B. A. Boley, *Int. J. Solids and Structures* **1**, 207–234, 1965.
39. B. A. Boley and H. P. Yagoda, *Q. Appl. Math.* **27**, 223–246, 1969.
40. Y. F. Lee and B. A. Boley, *Int. J. Eng. Sci.* **11**, 1277, 1973.
41. A. N. Güzelsu and A. S. Çakmak, *Int. J. Solids and Structures* **6**, 1087, 1970.
42. M. N. Özışık and R. L. Murray, *J. Heat Transfer* **96c**, 48–51, 1974.
43. Y. Yener and M. N. Özışık, On the Solution of Unsteady Heat Conduction in Multi-Region Finite Media with Time Dependent Heat Transfer Coefficient, Proceeding of 5th International Heat Transfer Conference, Tokyo, September 3–7, 1974.

44. M. N. Özışık and S. Güçeri, *Ca. J. Chem. Eng.* **55**, 145–148, 1977.
45. F. Kreith and F. E. Romie, *Proc. Phys. Soc. Lond.* **68(B)**, 283, 1955.
46. D. R. Otis, Solving the Melting Problem Using the Electric Analog to Heat Conduction, *Heat Transfer and Fluid Mechanics Institute*, Stanford Univ., 1956.
47. G. Liebmann, *ASME Trans.* **78**, 1267, 1956.
48. D. C. Baxter, The Fusion Times of Slabs and Cylinders, ASME Paper No. 61-WA-179, 1961.
49. C. F. Bonilla and A. L. Strupczewski, *Nucl. Struc. Eng.* **2**, 40–47, 1965.
50. J. Douglas and T. M. Gallie, *Duke Math. J.* **22**, 557, 1955.
51. J. Crank, *Quart. J. Mech. Appl. Math.* **10**, 220, 1957.
52. W. D. Murray and F. Landis, *Trans. Am. Soc. Mech. Eng.* **81**, 106, 1959.
53. G. S. Springer and D. R. Olson, Method of Solution of Axisymmetric Solidification and Melting Problems, ASME Paper No. 63-WA-246, 1962.
54. G. S. Springer and D. R. Olson, Axisymmetric Solidification and Melting of Materials, ASME Paper No. 63-WA-185, 1963.
55. D. N. de G. Allen and R. T. Severn, *Q. J. Mech. Appl. Math.* **15**, 53, 1962.
56. W. D. Seider and S. W. Churchill, The Effect of Insulation on Freezing Motion, 7th National Heat Transfer Conference, Cleveland, 1964.
57. C. Bonacina, G. Comini, A. Fasano, and M. Primicerio, *Int. J. Heat Mass Transfer* **16**, 1825–1832, 1973.
58. G. M. Dusinberre, A Note on Latent Heat in Digital Computer Calculations, ASME Paper No. 58-HT-7.
59. N. R. Eyres, D. R. Hartree, J. Ingham, R. Jackson, R. J. Jarjant, and J. B. Wagstaff, *Phil. Trans. Roy. Soc.* **A240**, 1–57, 1948.
60. D. C. Baxter, *J. Heat Transfer* **84c**, 317–326, 1962.
61. L. C. Tao, *AIChE J.* **13**, 165, 1967.
62. L. C. Tao, *AIChE J.* **14**, 720, 1968.
63. G. S. Springer, *Int. J. Heat Mass Transfer* **12**, 521, 1969.
64. S. H. Cho and J. E. Sunderland, *Int. J. Heat Mass Transfer* **13**, 123, 1970.
65. A. Lazaridis, *Int. J. Heat Mass Transfer* **13**, 1459–1477, 1970.
66. J. Crank and R. S. Gupta, *J. Int. Maths. Appl.* **10**, 296–304, 1972.
67. G. G. Sackett, *SIAM J. Num. Anal.* **8**, 80–96, 1971.
68. G. H. Meyer, *Num. Math.* **16**, 248–267, 1970.
69. J. Crank and R. D. Phahle, *Bull. Inst. Maths. Appl.* **9**, 12–14, 1973.
70. J. Szekely and M. J. Themelis, *Rate Phenomena in Process Metallurgy*, Wiley-Interscience, New York, 1971, Chapter 10.
71. R. D. Atthey, *J. Inst. Math. Appl.* **13**, 353–366, 1974.
72. G. H. Meyer, *SIAM J. Num. Anal.* **10**, 522–538, 1973.
73. N. Shamsunder and E. M. Sparrow, *J. Heat Transfer* **97c**, 333–340, 1975.
74. V. Voller and M. Cross, *Int. J. Heat Mass Transfer* **24**, 545–556, 1981.
75. V. Voller and M. Cross, *Int. J. Heat Mass Transfer* **26**, 147–150, 1983.
76. K. H. Tacke, *Int. J. Num. Meth. Eng.* **21**, 543–554, 1985.
77. R. M. Furzeland, *J. Inst. Math. Appl.* **26**, 411–429, 1980.
78. R. S. Gupta and D. Kumar, *Comp. Mech. Appl. Mech. Eng.* **23**, 101–109, 1980.

79. R. S. Gupta and D. Kumar, *Int. J. Heat Mass Transfer* **24**, 251–259, 1981.
80. Q. Pham, *Int. J. Heat Mass Transfer* **28**, 2079–2084, 1985.
81. D. Poirier and M. Salcudean, *J. Heat Transfer* **110**, 562–570, 1988.
82. B. A. Boley, in *Proceedings of the 3rd Symposium on Naval Structural Mechanics*, Pergamon Press, New York, 1963.
83. S. B. Bankoff, in *Advances in Chemical Engineering*, Vol. 5, Academic Press, New York, 1964.
84. J. C. Muehlbauer and J. E. Sunderland, *Appl. Mech. Rev.* **18**, 951–959, 1965.
85. W. Kurz and D. J. Fisher, *Fundamentals of Solidification*, Trans. Tech. Publications, Aerdermannsdorf, Switzerland, 1989.
86. M. C. Flemings, *Solidification Processing*, McGraw-Hill, New York, 1974.
87. A. Ohno, *Solidification*, Springer-Verlag, Berlin, 1988.
88. I. Minkoff, *Solidification and Cast Structure*, Wiley, New York, 1986.
89. M. Rappaz, Modelling of Microstructure Formation of Solidification Processes, *Int. Math. Rev.* **34**, 93–123, 1989.
90. L. T. Thomas and J. W. Westwater, *Chem. Eng. Prog. Symp.* **59**m 155–164, 1963.
91. D. V. Boger and J. E. Westwater, *J. Heat Transfer* **89c**, 81–89, 1967.
92. L. M. Jiji, K. A. Rathjen, and T. Drezewiecki, *Int. J. Heat Mass Transfer* **13**, 215–218, 1970.
93. J. A. Bailey and J. R. Davila, *Appl. Sci. Res.* **25**, 245–261, 1971.
94. J. Szekely and A. S. Jassal, *Met. Trans. B* **9B**, 389–398, 1978.
95. P. D. Patel, *AIAA J.* **6**, 2454, 1968.
96. R. R. Gilpin, *Int. J. Heat Mass Transfer* **20**, 693–699, 1977.
97. S. Paterson, *Proc. Glasgow Math. Assoc* **1**, 42–47, 1952–53.
98. N. N. Levedev, *Special Functions and Their Applications*, Prentice-Hall, Englewood Cliffs, N.J., 1965.
99. M. Abramowitz and I. A. Stegun, *Handbook of Mathematical Functions*, NBS Applied Mathematics Series 55, U.S. Government Printing Office, Washington, D.C., 1964, p. 239.
100. S. M. Selby, *Standard Mathematical Tables*, The Chemical Rubber Company, Ohio, 1971, p. 515.
101. J. Crank, *J. Mech. Appl. Math.* **10**, 220–231, 1957.
102. L. W. Ehrlick, *J. Assn. Comp. Math.* **5**, 161–176, 1958.
103. N. Shamsundar and E. Rooz, in *Handbook of Numerical Heat Transfer*, W. J. Minkowycz, E. M. Sparrow, G. E. Schneider, and R. H. Pletcher, eds., pp. 747–786, Wiley, New York, 1988.

PROBLEMS

11-1 Verify that the interface energy-balance equation (11-2c) is also applicable for the melting problem illustrated in Fig. 11-1b.

11-2 In the melting problem illustrated in Fig. 11-1b, if the heat transfer on the liquid side is by convection and on the solid phase is by pure conduction, derive the interface energy-balance equation. Take the bulk temperature of the liquid side as T_∞ and the heat transfer coefficient as h.

11-3 Solve exactly the phase-change problem considered in Example 11-2 for the case of solidification in a half-space $x > 0$. That is, a liquid at the melting temperature T_m^* is confined to a half-space $x > 0$. At time $t = 0$ the boundary at $x = 0$ is lowered to a temperature T_0 below T_m^* and maintained at that temperature for times $t > 0$. Determine the temperature distribution in the solid phase and the location of the solid–liquid interface as a function of time.

11-4 Solve exactly the problem considered in Example 11-3 for the case of melting. That is, a solid in $x > 0$ is initially at a uniform temperature T_i lower than the melting temperature T_m^*. For times $t > 0$ the boundary surface at $x = 0$ is kept at a constant temperature T_0, which is higher than the melting temperature T_m^*. Determine the temperature distribution in the liquid and solid phases, and the location of the solid–liquid interface as a function of time.

11-5 Solve exactly the problem considered in Example 11-4 for the case of melting. That is, a line heat source of strength Q, W/m is situated at $r = 0$ in an infinite medium that is at a uniform temperature T_i lower than the melting temperature T_m. The melting will start at $r \to 0$, and the solid–liquid interface will move in the positive r direction. Determine the temperature distribution in the solid and liquid phases, and the location of the solid-liquid interface as a function of time.

11-6 Using the integral method of solution, solve the solidification Problem 11-3 and obtain an expression for the location of the solid–liquid interface. Compare this result with that obtained in Example 11-5 for the case of melting.

11-7 A solid confined in a half-space $x > 0$ is initially at the melting temperature T_m^*. For times $t > 0$ the boundary surface at $x = 0$ is subjected to a heat flux in the form

$$-k\left.\frac{\partial T}{\partial x}\right|_{x=0} = H \equiv \text{constant}$$

Using the integral method of solution and a second-degree polynomial approximation for the temperature, obtain an expression for the location of the solid–liquid interface as a function of time.

11-8 Consider one-dimensional solidification of a liquid having a single melting-point temperature T_m^*, confined to the region $0 \leqslant x \leqslant B$. Initially the liquid is at a uniform temperature T_0 that is higher than the melting temperature T_m^* of the liquid. For times $t > 0$, the boundary surface at $x = 0$ is kept at a temperature $T = f$ that is lower than the melting temperature T_m^* of the substance. The boundary at $x = B$ satisfies the symmetry requirement. The properties are assumed to be constant. Develop the finite-difference formulation of this problem by using the *explicit enthalpy method*; that is, use the explicit finite difference scheme to discretize the differential equations.

11-9 Repeat Problem 11-8 for the case of a material having phase change over an extended temperature range.

11-10 A liquid having a single phase-change temperature T_m^* is confined to a semiinfinite region $0 < x < \infty$. Initially, the liquid is at a temperature $T_i(> T_m^*)$. At time $t = 0$, the temperature of the boundary surface at $x = 0$ is suddenly lowered to a temperature $T_0(< T_m^*)$ and maintained at that temperature for times $t > 0$. Determine the location of the solid–liquid interface and the temperature at a position $x = x_0$ as a function of time by using (a) exact analytic solution, (b) explicit enthalpy method, and (c) implicit enthalpy method. Numerical values for various quantities are given as

$$T_m = 0°\text{C}, \qquad T_i = 2°\text{C}, \qquad T_0 = -10°\text{C}$$

$$L = 100\ \text{MJ/kg}, \qquad (\rho C_p)_l = 2.5\ \text{MJ/(kg·°C)}, \qquad (\rho C_p)_s = 1.5\ \text{MJ/(kg·°C)}$$

$$K_l = 1.75\ \text{W/(m·°C)}, \qquad k_s = 2.25\ \text{W/(m·°C)}, \qquad X_0 = 50\ \text{cm}$$

(See Figures 3 and 7 of reference 74 to compare your results.)

NOTE

The exponential-integral function $-Ei(-x)$ or $E_1(x)$ is defined as

$$-Ei(-x) \equiv E_1(x) = \int_x^\infty \frac{e^{-u}}{u}\,du = \int_1^\infty \frac{e^{-xt}}{t}\,dt \qquad \text{for} \qquad x > 0 \tag{1}$$

The function $-Ei(-x)$, which is also denoted by $E_1(x)$, decreases monotonically from the value $E_1(0) = \infty$ to $E_1(\infty) = 0$ as x is varied from $x = 0$ to $x \to \infty$ as shown in Table 11-1. The derivative of $-Ei(-x)$ with respect to x is given as

$$\frac{d}{dx}\left[-Ei(-x)\right] = \frac{d}{dx}\left[\int_x^\infty \frac{e^{-u}}{u}\,du\right] = -\frac{e^{-x}}{x} \tag{2}$$

The notation $E_1(x)$ has been used for $-Ei(-x)$ function in reference 99 [p. 228], and its polynomial approximations are given for $0 \leqslant x \leqslant 1$ and $1 \leqslant x < \infty$ [99, p. 231]. A tabulation of $E_1(x)$ function is given in references 99 [p. 239] and 100 [p. 515].

12

FINITE-DIFFERENCE METHODS

Numerical methods are useful for solving fluid dynamics, heat and mass transfer problems, and other partial-differential equations of mathematical physics when such problems cannot be handled by the exact analysis because of nonlinearities, complex geometries, and complicated boundary conditions. The development of the high-speed digital computers significantly enhanced the use of numerical methods in various branches of science and engineering. Many complicated problems can now be solved at a very little cost and in a very short time with the available computing power.

Presently, two major approaches used in the numerical solution of partial-differential equations of heat, mass, and momentum transport include the *finite-difference method* (FDM) and the *finite-element method* (FEM). Extensive amount of literature exist on the application of FDMs [1–27] and FEMs [28–35]. Each method has its advantages depending on the nature of the physical problem to be solved. Finite-difference methods are simple to formulate, can be readily extended to two or three-dimensional problems, and are very easy to learn and apply to the solution of partial-differential equations encountered in the modeling of engineering problems. More recently, with the advent of *numerical grid generation* [36–41] approach, the FDM have become comparable to FEM in dealing with irregular geometries, while still maintaining the simplicity of the standard FDMs. Here we consider the application of FDMs to the solution of heat conduction problems. Despite the simplicity of the finite-difference representation of governing partial-differential equations, it requires considerable experience and knowledge to select appropriate finite-differencing scheme for a specific problem in hand. The type of partial differential equations, the number of physical dimensions, the type of coordinate system involved, whether the governing equations and boundary conditions are linear or nonlinear, and whether the pı are among the factors that affect the type of th from a large number of available methods. Tl hod for a specific problem in hand is an important first step in numerical solution with a finite-difference method. Therefore, we also present a classification of partial-differential equations encountered in the mathematical formulation of heat, mass, and momentum transfer problems and discuss the physical significance of such a classification in relation to the numerical solution of the problem.

12-1 CLASSIFICATION OF SECOND-ORDER PARTIAL-DIFFERENTIAL EQUATIONS

In the solution of partial-differential equations with finite differences, the choice of a particular finite-differencing scheme also depends on the type of the partial-differential equation considered. Generally, the partial-differential equations are classified into three categories, called *elliptic*, *parabolic*, and *hyperbolic*. To illustrate this matter we consider the following most general second-order partial-differential equation in two independent variables x, y given by Forsythe and Wasow [6]:

$$A\frac{\partial^2\phi}{\partial x^2}+B\frac{\partial^2\phi}{\partial x\,\partial y}+C\frac{\partial^2\phi}{\partial y^2}+D\frac{\partial\phi}{\partial x}+E\frac{\partial\phi}{\partial y}+F\phi+G(x,y)=0 \tag{12-1}$$

Here we assume a linear equation (this restriction is not essential), that is, the coefficients A, B, C, D, E, and F are functions of the two independent variables x, y, but not of the dependent variable ϕ.

The classification is made on the basis of the coefficients A, B, and C of the highest derivatives in equation (12-1), according to whether the determinant

$$-\begin{vmatrix} A & B \\ B & C \end{vmatrix}$$

is negative, zero, or positive. The differential equation is called

$$\text{Elliptic} \quad \text{if} \quad B^2-4AC<0 \tag{12-2a}$$

$$\text{Parabolic} \quad \text{if} \quad B^2-4AC=0 \tag{12-2b}$$

$$\text{Hyperbolic} \quad \text{if} \quad B^2-4AC>0 \tag{12-2c}$$

For example, the steady-state heat conduction equation with no energy generation

$$\frac{\partial^2 T}{\partial x^2}+\frac{\partial^2 T}{\partial y^2}=0 \tag{12-3a}$$

is *elliptic*. The steady-state heat conduction equation with energy generation

$$\frac{\partial^2 T}{\partial x^2}+\frac{\partial^2 T}{\partial y^2}+\frac{1}{k}g(x, y)=0 \tag{12-3b}$$

is also *elliptic*. The characteristic of the elliptic equation is that it requires the specification of appropriate boundary conditions at all boundaries.

The one-dimensional time-dependent heat conduction equation

$$\frac{\partial^2 T}{\partial x^2}=\frac{1}{\alpha}\frac{\partial T}{\partial t} \tag{12-3c}$$

is *parabolic*.

The second-order wave equation

$$\frac{\partial^2 \phi}{\partial x^2}=\frac{1}{c^2}\frac{\partial^2 \phi}{\partial t^2} \tag{12-3d}$$

where t is the time, x is the space variable, and c is the wave propagation speed, is *hyperbolic*.

The non-Fourier heat conduction equation

$$\frac{\partial^2 T}{\partial x^2}=\frac{1}{c^2}\frac{\partial^2 T}{\partial t^2}+\frac{1}{\alpha}\frac{\partial T}{\partial t} \tag{12-3e}$$

which is a second-order damped wave equation, is *hyperbolic*.

Physical Significance of Parabolic, Elliptic, and Hyperbolic Systems

In the foregoing discussion we considered a purely mathematical criterion given by equations (12-2) to classify the second-order partial-differential equation (12-1) into categories called *parabolic, elliptic, and hyperbolic*. We now discuss the physical significance of such a classification in computational domain.

Consider, for example, the steady-state heat conduction equation (12-3a) or (12-3b), which has second-degree partial derivatives in both x and y variables. The conditions at any given location are influenced by changes in conditions at both sides of that location, whether the changes are in the x variable or the y variable. Thus, the steady-state heat conduction equation is elliptic in both x and y space coordinates, and simply called *elliptic*.

Now let us consider the one-dimensional time-dependent heat conduction equation (12-4), which has a second-degree partial derivative in the x variable and a first-degree partial derivative in the time variable. The conditions at any given location x are influenced by changes in conditions at both sides of that location; hence the equation is regarded *elliptic* in the x variable. However, in the time variable t, the conditions at any instant are influenced only by changes taking place in conditions at times *earlier* than that time; hence the equation is parabolic in time and called *parabolic*.

The computational advantage of the *parabolic* system lies in the fact that significant reduction in computer time and computer storage can be realized. For example, in finite-difference solution of two-dimensional time-dependent heat conduction equation, which is parabolic in time, one needs to consider only a two-dimensional network for the temperature field. As the temperature field at any time is not affected by temperature field at *later* times, one starts with a given *initial* temperature field and *marches* forward to compute the temperature fields at successive time steps.

In the case of *hyperbolic* equation, however, it does not seem to be possible to relate it to be some distinct computational advantages in finite-difference solutions as in the case of parabolic systems. As it will be apparent later, the solution of hyperbolic heat conduction equation exhibits a wave–like propagation of the temperature field with a finite speed in contrast to the infinite speed of propagation associated with the parabolic heat conduction equation. Therefore, the solution of hyperbolic equations with finite differences requires special considerations and special schemes.

12-2 FINITE-DIFFERENCE APPROXIMATION OF DERIVATIVES THROUGH TAYLOR'S SERIES

The idea of finite-difference representation of a derivative can be introduced by recalling the definition of the derivative of the function $F(x, y)$ at $x = x_0, y = y_0$ with respect to x:

$$\frac{\partial F}{\partial x} = \lim_{\Delta x \to 0} \frac{F(x_0 + \Delta x, y_0) - F(x_0, y_0)}{\Delta x} \tag{12-4}$$

Clearly, if the function $F(x, y)$ is continuous, the right-hand side of equation (12-4) can be a reasonable approximation to $\partial F/\partial x$ for a *sufficiently small* but finite Δx.

We consider the Taylor series expansion of the functions $f(x+h)$ and $f(x-h)$ about the point x, as illustrated in Fig. 12-1, given by

$$f(x+h) = f(x) + hf'(x) + \frac{h^2}{2!}f''(x) + \frac{h^3}{3!}f'''(x) \tag{12-5a}$$

$$f(x-h) = f(x) - hf'(x) + \frac{h^2}{2!}f''(x) - \frac{h^3}{3!}f'''(x) \tag{12-5b}$$

where primes denote derivatives with respect to x. The first- and second-order derivatives $f'(x)$ and $f''(x)$ can be represented in the finite difference form in many

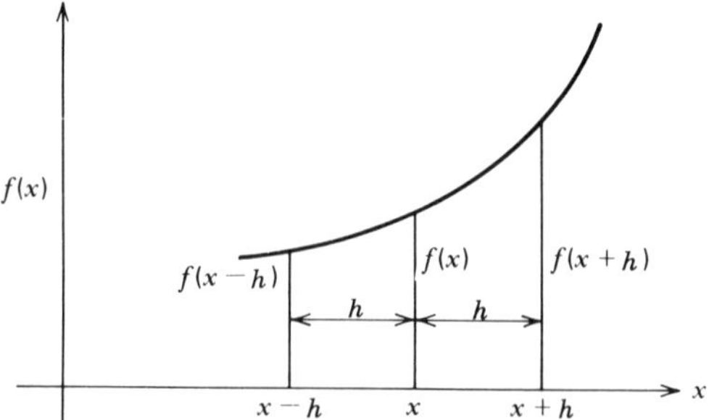

Fig. 12-1 Nomenclature for a Taylor series representation.

different ways by utilizing Taylor series expansions given by equations (12-5) as now described.

First Derivatives

The forward and backward first-order derivatives of $f(x)$ are obtained by solving equations (12-5a) and (12-5b) for $f'(x)$:

$$f'(x) = \frac{f(x+h) - f(x)}{h} - \frac{h}{2} f''(x) - \frac{h^2}{6} f'''(x) \cdots \qquad \text{(forward)} \qquad (12\text{-}6a)$$

$$f'(x) = \frac{f(x) - f(x-h)}{2} + \frac{h}{2} f''(x) - \frac{h^2}{6} f'''(x) + \cdots \qquad \text{(backward)} \qquad (12\text{-}6b)$$

Subtracting equations (12-5a) and (12-5b), the first-order central-difference approximation is determined as

$$f'(x) = \frac{f(x+h) - f(x-h)}{2h} - \frac{h^2}{6} f'''(x) - \cdots \qquad \text{(central)} \qquad (12\text{-}6c)$$

These three results are written more compactly as

$$f'(x) = \frac{f(x+h) - f(x)}{h} + O(h) \qquad \text{(forward)} \qquad (12\text{-}7a)$$

$$f'(x) = \frac{f(x) - f(x-h)}{h} + O(h) \qquad \text{(backward)} \qquad (12\text{-}7b)$$

$$f'(x) = \frac{f(x+h) - f(x-h)}{2h} + O(h^2) \qquad \text{(central)} \qquad (12\text{-}7c)$$

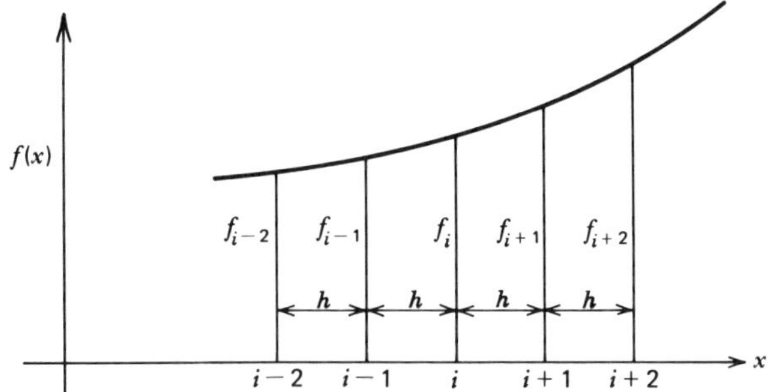

Fig. 12-2 Nomenclature for finite-difference representation of $f(x)$.

Here the notation $O(h)$ is used to show that the error involved is of the order of h; similarly, $O(h^2)$ is for the error of the order of h^2.

If we now introduce the notation

$$x = ih, \qquad x + h = (i+1)h, \qquad x - h = (i-1)h, \qquad \text{etc.} \tag{12-8a}$$

$$f(x) = f_i, \qquad f(x+h) = f_{i+1}, \qquad f(x-h) = f_{i-1}, \qquad \text{etc.} \tag{12-8b}$$

as illustrated in Fig. 12-2, then the finite-difference representation of the first derivative of function $f(x)$ about x, given by equations (12-7a,b,c), are written, respectively, as

$$f'_i = \frac{f_{i+1} - f_i}{h} + O(h) \qquad \text{(forward)} \tag{12-9}$$

$$f'_i = \frac{f_i - f_{i-1}}{h} + O(h) \qquad \text{(backward)} \tag{12-10}$$

$$f'_i = \frac{f_{i+1} - f_{i-1}}{2h} + O(h^2) \qquad \text{(central)} \tag{12-11}$$

Second Derivatives

We now proceed to the finite-difference representation of the second derivative $f''(x)$ of a function $f(x)$ about the point x. To obtain such results we consider a Taylor series expansion of functions $f(x+2h)$ and $f(x-2h)$ about x as

$$f(x+2h) = f(x) + 2hf'(x) + 2h^2 f''(x) + \tfrac{4}{3}h^3 f'''(x) + \cdots \tag{12-12a}$$

$$f(x-2h) = f(x) - 2hf'(x) + 2h^2 f''(x) - \tfrac{4}{3}h^3 f'''(x) + \cdots \tag{12-12b}$$

Eliminating $f'(x)$ between equations (12-5a) and (12-12a) we obtain

$$f''(x) = \frac{f(x) + f(x+2h) - 2f(x+h)}{h^2} - hf'''(x) \tag{12-13}$$

similarly, eliminating $f'(x)$ between equations (12-5b) and (12-12b) we find

$$f''(x) = \frac{f(x-2h) + f(x) - 2f(x-h)}{h^2} + hf'''(x) \tag{12-14}$$

Eliminating $f'(x)$ between equations (12-5a) and (12-5b) we obtain

$$f''(x) = \frac{f(x-h) + f(x+h) - 2f(x)}{h^2} - \frac{1}{12}h^2 f''''(x) \tag{12-15}$$

The results given by equations (12-13)–(12-15) are written more compactly as

$$f_i'' = \frac{f_i - 2f_{i+1} + f_{i+2}}{h^2} + O(h) \qquad \text{forward difference} \tag{12-16}$$

$$f_i'' = \frac{f_{i-2} - 2f_{i-1} + f_i}{h^2} + O(h) \qquad \text{backward difference} \tag{12-17}$$

$$f_i'' = \frac{f_{i-1} - 2f_i + f_{i+1}}{h^2} + O(h^2) \qquad \text{central difference} \tag{12-18}$$

where

$$f_i'' \equiv \left.\frac{d^2 f(x)}{dx^2}\right|_i$$

We note that the central-difference representation is accurate to $O(h^2)$ whereas the forward and backward differences to $O(h)$.

In the foregoing finite-difference expressions, two-point formulas are used for the first derivatives and three-point formulas for the second derivatives. It is possible to use more points in order to obtain more accurate finite-difference expressions.

Summary of First Derivatives

In the following formulas, the symbols B, C, and F denote backward, central, and forward, respectively.

where the subscripts i and j denote the grid points associated with the discretization in the x and y variables, respectively. Applying the central difference formula once more to discretize the partial derivatives with respect to the y variable on the right-hand side of equation (12-23) we obtain

$$\frac{\partial}{\partial x}\left(\frac{\partial f}{\partial y}\right) = \frac{1}{2\Delta x}\left(\frac{f_{i+1,j+1} - f_{i+1,j-1}}{2\Delta y} - \frac{f_{i-1,j+1} - f_{i-1,j-1}}{2\Delta y}\right) + O[(\Delta x)^2, (\Delta y)^2] \tag{12-24}$$

which is the finite-difference approximation of the mixed partial derivative $\partial^2 f/\partial x\,\partial y$ using central differences for both x and y variables. The order of differentiation is immaterial if the derivatives are continuous; that is $\partial^2 f/\partial x\,\partial y$ and $\partial^2 f/\partial y\,\partial x$ are equal.

In the above illustration we applied central differences for both derivatives in x and y. If all possible combinations of forward, backward, and central differences are considered, nine different cases arise for finite difference approximation of $\partial^2 f/\partial x\,\partial y$. Table 12-1 lists the finite-difference approximations for each of these nine different cases.

12-3 ERRORS INVOLVED IN NUMERICAL SOLUTIONS

In the numerical solution of partial-differential equations with finite differences, errors are involved in the discretization process as well as during the solution of the resulting algebraic equations with a computer. These errors can be classified as the *truncation*, *discretization*, and *round-off* errors.

The round-off error, as the name implies, is caused by rounding off of the numbers by the computer during the solution process. The discretization error is caused by replacing the continuous problem satisfied by the PDE by a discrete problem satisfied by the finite-difference approximation, including the contributions of the differential equation and boundary conditions, but without the round-off errors.

Consider, for example, the steady-state heat conduction equation

$$L(T) \equiv \frac{\partial^2 T}{\partial x^2} + \frac{\partial^2 T}{\partial y^2} = 0 \tag{12-25a}$$

and its finite-difference approximation given by

$$L_{\text{FD}}(T) \equiv \frac{T_{i-1,j} - 2T_{i,j} + T_{i+1,j}}{(\Delta x)^2} + \frac{T_{i,j-1} - 2T_{i,j} + T_{i,j+1}}{(\Delta y)^2} \tag{12-25b}$$

to be solved over a domain subject to some specified boundary conditions.

If the heat conduction problem is solved over the domain exactly by using the PDE and by using the finite-difference equations without any round-off error, the results will not be equal. The difference is called the *discretization* error caused by the truncation errors associated with finite-difference approximation of the *differential equation* and the *boundary conditions* for the problem.

The terminology, *truncation error* is used to identify the error resulting from the discretization of the PDE only:

$$\underbrace{\begin{pmatrix}\text{Exact solution}\\ \text{of PDE}\end{pmatrix}}_{L(T)} - \underbrace{\begin{pmatrix}\text{solution of finite-}\\ \text{difference equation}\\ \text{without}\\ \text{round-off error}\end{pmatrix}}_{L_{\mathrm{FD}}(T)} = \underbrace{\begin{pmatrix}\text{truncation}\\ \text{error}\end{pmatrix}}_{TE} \qquad (12\text{-}26)$$

Clearly, the truncation error is the difference between the exact solution of PDE and its finite-difference solution without the round-off error, and hence is a measure of the accuracy of representing the partial differential equation in the finite-difference form.

Table 12-2 lists *truncated leading error terms* in the finite differencing of the first and second derivatives using forward, backward, and central differencing schemes. Clearly, the leading error is of the order $O(\Delta x)$ for the forward and backward differences, while of the order $O[(\Delta x)^2]$ for central difference.

The *total error* involved in finite-difference calculations consists of the *discretization error* plus the *round-off error*. The discretization error increases with increasing mesh size while the round-off error decreases with increasing mesh

TABLE 12-2 Various Differencing Schemes and the Corresponding Truncated Leading Error Terms

Derivative	Finite-Difference Form	Truncated Leading Error Terms[a]
$\frac{df(x)}{dx}$	$\frac{f(x+\Delta x)-f(x)}{\Delta x}$ (forward)	$-\frac{\Delta x}{2}f'' - \frac{(\Delta x)^2}{6}f'''$
$\frac{df(x)}{dx}$	$\frac{f(x)-f(x-\Delta x)}{\Delta x}$ (backward)	$+\frac{\Delta x}{2}f'' - \frac{(\Delta x)^2}{6}f'''$
$\frac{df(x)}{dx}$	$\frac{f(x+\Delta x)-f(x-\Delta x)}{2\Delta x}$ (central)	$-\frac{(\Delta x)^2}{6}f'''$
$\frac{d^2f(x)}{dx^2}$	$\frac{f(x-\Delta x)-2f(x)+f(x+\Delta x)}{(\Delta x)^2}$ (central)	$-\frac{(\Delta x)^2}{12}f''''$

[a] Primes denote differentiation with respect to x.

size. Therefore, the total error is expected to exhibit a minimum as the mesh size is decreased.

Example 12-1

The following numerical representation of $f(x)$ is given at equally spaced intervals $\Delta x \equiv h = 1$.

x	0	1	2	3	4	5
$f(x)$	15	18	12	10	1	-6

Using finite differences, determine $f'_0 \equiv f'(x=0), f'_5 \equiv f'(x=5)$ and $f''_0 \equiv f''(x=0)$ accurate to the order h^2.

Solution. The first-order derivatives accurate to $O(h)^2$ are given by equations (12-20a) and (12-20b). The forward difference formula (12-20a) is used to determine f'_0 because no points are available in the backward direction and the backward-difference formula (12-20b) is used to determine f'_5 because no points are available in the forward direction. Hence we have

$$f'_0 = \tfrac{1}{2}(-3f_0 + 4f_1 - f_2) = \tfrac{1}{2}(-45 + 72 - 12) = 7.5 \quad 0(1)^2 \tag{12-27a}$$

$$f'_5 = \tfrac{1}{2}(f_3 - 4f_4 + 3f_5) = \tfrac{1}{2}(10 - 4 - 18) = -6 \quad 0(1)^2 \tag{12-27b}$$

Formula (12-23a) is used to determine f''_0; we find

$$f''_0 = \tfrac{1}{2}(2f_0 - 5f_1 + 4f_2 - f_3) = \tfrac{1}{2}(30 - 90 + 48 - 10) = -11 \quad 0(1)^2 \tag{12-27c}$$

12-4 CHANGING THE MESH SIZE

In most engineering applications, one will often have some idea of the general shape of the solution, especially of the locations where the profile will exhibit a sudden change in the first derivative. Therefore, to obtain higher resolution in the region where the gradients are expected to vary rapidly, it is desirable to use a finer mesh over that particular region rather than refining the mesh over the entire domain. To illustrate this matter we consider the simplest situation involving a change in mesh spacing only in one direction at some point in the region.

Figure 12-3 shows a change of the mesh size from Δx_1 to Δx_2 at some node i. A Taylor series expansion about the node i can be used to develop finite

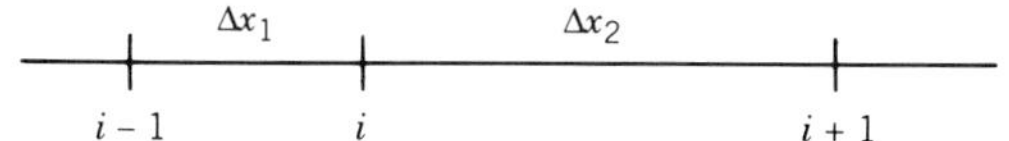

Fig. 12-3 Change of mesh size from Δx_1 to Δx_2 at node i.

difference approximation. That is, the function $f(x)$ is expanded about the node i in *forward* and *backward* Taylor series, respectively, as

$$f_{i+1} = f_i + \Delta x_2 \left.\frac{df}{dx}\right|_i + \frac{(\Delta x_2)^2}{2!}\left.\frac{d^2 f}{dx^2}\right|_i + \frac{(\Delta x_2)^3}{3!}\left.\frac{d^3 f}{dx^3}\right|_i + O[(\Delta x_2)^2] \tag{12-28a}$$

$$f_{i-1} = f_i - \Delta x_1 \left.\frac{df}{dx}\right|_i + \frac{(\Delta x_1)^2}{2!}\left.\frac{d^2 f}{dx^2}\right|_i - \frac{(\Delta x_1)^3}{3!}\left.\frac{d^3 f}{dx^3}\right|_i + O[(\Delta x_1)^2] \tag{12-28b}$$

To obtain a difference approximation for the second derivative at the node i, equation (12-28b) is multiplied by $(\Delta x_2/\Delta x_1)^2$ and the resulting expression is added to equation (12-28a) to give

$$f_{i+1} + \varepsilon^2 f_{i-1} = (1+\varepsilon^2) f_i + (1-\varepsilon)\Delta x_2 \left.\frac{df}{dx}\right|_i + (\Delta x_2)^2 \left.\frac{d^2 f}{dx^2}\right|_i + \frac{1}{6}(\Delta x_2 - \Delta x_1)(\Delta x_2)^2 \left.\frac{d^3 f}{dx^3}\right|_i + O[(\Delta x)^4] \tag{12-29a}$$

where

$$\varepsilon = \frac{\Delta x_2}{\Delta x_1} \tag{12-29b}$$

and $O[(\Delta x)^4]$ means the largest of $O[(\Delta x_1)^4]$ or $O[(\Delta x_2)^4]$. The finite-difference approximation for the second derivative is obtained by solving equation (12-29a) for $(d^2 f/dx^2)|_i$ to yield

$$\left.\frac{d^2 f}{dx^2}\right|_i = \frac{f_{i+1} - (1-\varepsilon^2) f_i + \varepsilon^2 f_{i-1}}{(\Delta x_2)^2} - \frac{1-\varepsilon}{\Delta x_2}\left.\frac{df}{dx}\right|_i + O[(\Delta x_2 - \Delta x_1)] \tag{12-30}$$

This expression is accurate to first order at i if $1 - \varepsilon = O[(\Delta x_1)^2]$. The above finite-difference approximations imply that, unless the mesh spacing is changed slowly, the truncation error deteriorates rather than improves.

12-5 CONTROL-VOLUME APPROACH

In the previous section it was assumed that the given partial-differential equation was the correct and appropriate form of the physical conservation law governing the problem and Taylor series approach was used as a purely mathematical procedure to develop the finite difference approximation to the derivatives.

In the alternative *control-volume* approach, the finite-difference equations are developed by constraining the partial-differential equation to a finite control-volume and conserving the specific physical quantity such as mass, momentum, or energy over the control volume. The basic concept is analogous to heat balance over a small volume surrounding a grid point commonly used by elementary textbooks on heat transfer to derive the finite-difference form of the heat conduction equation.

To develop the control-volume statement for a small finite region, it is instructive to work backward from the partial-differential equation governing the specific physical quantity. For illustration purposes we consider the transient heat conduction equation with energy generation given in the form

$$\rho C_p \frac{\partial T}{\partial t} = -\nabla \cdot \mathbf{q} + g \tag{12-31a}$$

where the heat flux vector $\mathbf{q}$ is related to the temperature $T(r, t)$ by the Fourier law

$$\mathbf{q} = -k\nabla T \tag{12-31b}$$

and g is the volumetric energy generation rate.

We integrate equation (12-31a) over a small fixed volume V

$$\int_V \rho C_p \frac{\partial T}{\partial t} dV = -\int_V \nabla \cdot \mathbf{q}\, dV + \int_V g\, dV \tag{12-31c}$$

The integral on the left-hand side can be removed by means of the mean value theorem for integrals. Similarly, the integral over g is also removed. The volume integral over the divergence of the heat flux vector is transformed to a surface integral by means of the divergence theorem. Then equation (12-31c) becomes

$$V\rho C_p \frac{\partial \bar{T}}{\partial t} = -\int_S \mathbf{q} \cdot \mathbf{n}\, dS + V\bar{g} \tag{12-32}$$

where S is the surface area of the control volume. Introducing the heat flux vector $\mathbf{q}$ from equation (12-31b) into equation (12-32) we find

$$V\rho C_p \frac{\partial \bar{T}}{\partial t} = \int_S k \frac{\partial T}{\partial n} dS + V\bar{g} \tag{12-33a}$$

since

$$\nabla T \cdot \mathbf{n} = \frac{\partial T}{\partial n} \tag{12-33b}$$

Here V is a small control volume; $\mathbf{n}$ and $(\partial/\partial n)$ are, respectively, the *outward-drawn* normal unit vector and derivative along the outward-drawn normal to the surface of the control volume; $\bar{T}$ and $\bar{g}$ are suitable averages over the control volume of temperature and the energy generation rate, respectively.

We have developed above a control-volume energy conservation equation for physical phenomena involving transient heat conduction. Similar conservation expressions can be developed for the conservation of mass or momentum and include situations involving convective transport.

Once the control-volume conservation equation is available, the corresponding finite-difference equation over the control volume is readily obtained by discretizing the derivative terms in this conservation equation.

The control-volume approach for the development of finite-difference equations has distinct advantages for being readily applicable to multidimensional problems, complicated boundary conditions and to situations involving variable mesh and physical properties. On the other hand, the accuracy estimates with the control-volume approach are difficult compared to that with the Taylor series expansion method which provides information on the order of the truncation error involved.

When applying the control-volume approach to develop the finite-difference equations, the finite difference nodes must be established first and then the control volumes must be identified.

Example 12-2

Consider the following one-dimensional steady-state heat conduction problem with variable thermal conductivity:

$$\frac{d}{dx}\left(k\frac{dT}{dx}\right)+g(x)=0 \qquad \text{in} \qquad 0<x<L \tag{12-34a}$$

$$-k\frac{dT}{dx}+h_0 T=h_0 T_{\infty 0} \qquad \text{at} \qquad x=0 \tag{12-34b}$$

$$k\frac{dT}{dx}+h_L T=h_L T_{\infty L} \qquad \text{at} \qquad x=L \tag{12-34c}$$

Using the control volume approach write the finite-difference form of this problem.

Solution. Figure 12-4 shows the finite-difference nodes constructed over the region. The integration of equation (12-34a) over the control volume about

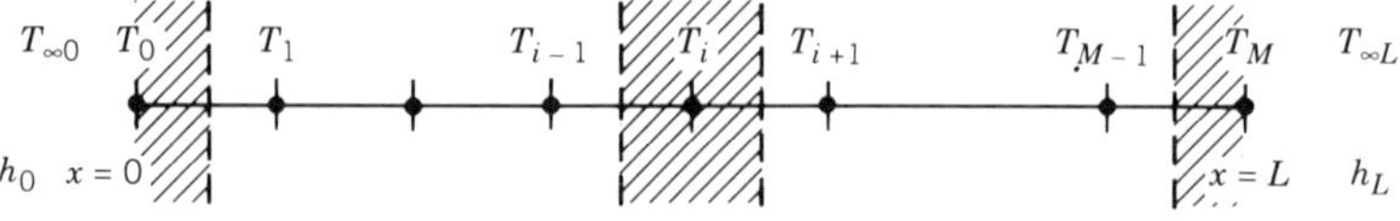

Fig. 12-4 Control volumes for one-dimensional case.

an internal node i gives

$$\left(k\frac{dT}{dx}\right)_{i+1/2}-\left(k\frac{dT}{dx}\right)_{i-1/2}+\Delta x\,\bar{g}_i=0 \tag{12-35}$$

and expressing the derivatives in the discrete form we obtain

$$k_{i+1/2}\frac{T_{i+1}-T_i}{\Delta x}-k_{i-1/2}\frac{T_i-T_{i-1}}{\Delta x}+\Delta x\,\bar{g}_i=0, \qquad i=1,2,\ldots,M-1 \tag{12-36}$$

where $\bar{g}_i$ is a suitable average value of g over the control volume about i. The integration of equation (12-34a) over the control volumes associated with the boundary nodes at $x=0$ and $x=L$, respectively, gives

$$\left(k\frac{dT}{dx}\right)_{1/2}-\left(k\frac{dT}{dx}\right)_0+\frac{1}{2}\Delta x\,\bar{g}_0=0 \tag{12-37a}$$

$$\left(k\frac{dT}{dx}\right)_M-k\left(\frac{dT}{dx}\right)_{M-1/2}+\frac{1}{2}\Delta x\,\bar{g}_M=0 \tag{12-37b}$$

Utilizing the boundary conditions (12-34b,c), the finite-difference representation of the boundary conditions at $x=0$ and $x=L$ is obtained from equations (12-37a) and (12-37b), respectively, as

$$k_{1/2}\frac{T_1-T_0}{\Delta x}+h_0(T_{\infty 0}-T_0)+\frac{1}{2}\Delta x\,\bar{g}_0=0 \tag{12-38a}$$

$$h_L(T_{\infty L}-T_M)-k_{M-1/2}\frac{T_M-T_{M-1}}{\Delta x}+\frac{1}{2}\Delta x\,\bar{g}_m=0 \tag{12-38b}$$

Thus equations (12-36) together with the boundary-condition equations (12-38a,b) provide $M+1$ algebraic equations for the determination of $M+1$ unknown node temperatures, $T_i, (i=0,1,\ldots,M)$. For constant thermal conductivity, this system of equations reduces to

$$2T_1-2\beta_0T_0=-(2\gamma_0+G_0), \qquad i=0 \tag{12-39a}$$

$$T_{i-1}-2T_i+T_{i+1}=-G_i, \qquad i=1,2,\ldots,M-1 \tag{12-39b}$$

$$2T_{M-1}-2\beta_LT_M=-(2\gamma_L+G_M), \qquad i=M \tag{12-39c}$$

where

$$\left.\begin{aligned}
\beta_0 &\equiv 1 + \frac{\Delta x\, h_0}{k}, \quad \gamma_0 \equiv \frac{\Delta x h_0 T_{\infty,0}}{k} \\
\beta_L &\equiv 1 + \frac{\Delta x\, h_L}{k}, \quad \gamma_L \equiv \frac{\Delta x h_L T_{\infty,L}}{k} \\
G_i &= \frac{(\Delta x)^2 \bar{g}_i}{k}
\end{aligned}\right\} \tag{12-40}$$

This system of equations can be expressed in the matrix form as

$$[\mathbf{A}]\{\mathbf{T}\} = \{\mathbf{B}\} \tag{12-41}$$

where

$$[\mathbf{A}] = \begin{bmatrix} -2\beta_0 & 2 & 0 & \cdots & 0 & 0 & 0 \\ 1 & -2 & 1 & & 0 & 0 & 0 \\ 0 & 1 & -2 & 1 & 0 & 0 & 0 \\ \cdot & & & & & & \\ 0 & & & & 1 & -2 & 1 \\ 0 & 0 & \cdots & \cdots & 0 & 2 & -2\beta_L \end{bmatrix} = \begin{matrix}\text{known} \\ \text{coefficient} \\ \text{matrix}\end{matrix} \tag{12-42a}$$

$$\{\mathbf{T}\} = \begin{Bmatrix} T_0 \\ T_1 \\ \cdot \\ \cdot \\ T_M \end{Bmatrix} = \begin{matrix}\text{unknown} \\ \text{vector,}\end{matrix} \qquad \{\mathbf{B}\} = \begin{Bmatrix} -(G_0 + 2\gamma_0) \\ -G_1 \\ \cdot \\ \cdot \\ -G_{M-1} \\ -(G_M + 2\gamma_L) \end{Bmatrix} = \begin{matrix}\text{known} \\ \text{vector}\end{matrix} \tag{12-42b,c}$$

It is to be noted that the system of algebraic equations (12-39) or (12-42) are second-order accurate, namely $0[(\Delta x)^2]$. If the boundary conditions (12-34b,c) were discretized by using first-order accurate, two-point forward and backward differencing, the resulting equations would be first-order accurate, specifically, $0[\Delta x]$.

12-6 FICTITIOUS NODE CONCEPT FOR DISCRETIZING BOUNDARY CONDITIONS

An alternative approach for developing second-order accurate finite-difference form of the convection boundary conditions (12-34b,c) is through the use of fictitious node concept.

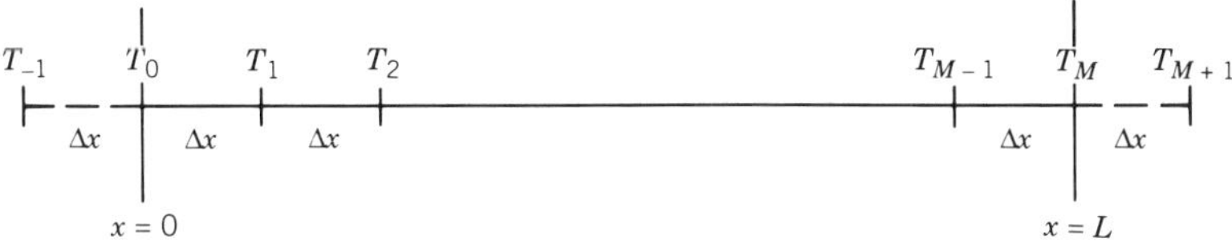

Fig. 12-5 Fictitious nodes -1 and $M+1$ at fictitious temperatures T_{-1} and T_{M+1}.

Figure 12-5 shows that the region $0 \leqslant x \leqslant L$ is extended outward by a distance Δx to the left and to the right giving rise to two fictitious nodes -1 and $M+1$ at fictitious temperatures T_{-1} and T_{M+1}, respectively. Then the application of the second-order accurate central-difference formula (12-19c) to discretize the first derivatives in the boundary conditions (12-34b,c), respectively, yields

$$-k\frac{T_1 - T_{-1}}{2\Delta x} + h_0 T_0 = h_0 T_{\infty,0} \tag{12-43a}$$

$$k\frac{T_{M+1} - T_{M-1}}{2\Delta x} + h_L T_M = h_L T_{\infty,L} \tag{12-43b}$$

The finite-difference equation (12-39b) is evaluated for $i = 0$ and $i = M$ to give, respectively

$$T_{-1} - 2T_0 + T_1 + \frac{(\Delta x)^2 g_0}{k} = 0 \tag{12-44a}$$

$$T_{M-1} - 2T_M + T_{M+1} + \frac{(\Delta x)^2 g_0}{k} = 0 \tag{12-44b}$$

The elimination of T_{-1} and T_{M+1} between equations (12-43) and (12-44) results in the following two finite-difference equations

$$2T_1 - 2\beta_0 T_0 + (2\gamma_0 + G_0) = 0 \qquad \text{at} \qquad x = 0 \qquad (i = 0) \tag{12-45a}$$

$$2T_{M-1} - 2\beta_L T_M + (2\gamma_L + G_M) = 0 \qquad \text{at} \qquad x = L \qquad (i = M) \tag{12-45b}$$

where the coefficients $\beta_0, \beta_{0L}, \gamma_0, \gamma_L$ are as defined by equations (12-40). We note that equations (12-45a,b) are the same as equations (12-39a,c).

12-7 METHODS OF SOLVING SIMULTANEOUS ALGEBRAIC EQUATIONS

So far we illustrated the basic steps in the transformation of a partial-differential equation and its boundary conditions into a system of algebraic equations. The methods of solving such a system of algebraic equations can be put into one of the two categories: (1) the *direct* methods in which a finite number of operations is involved in the solution and (2) the *iterative* techniques in which answers become progressively more accurate as the number of iterations is increased provided that the convergence criteria related to the *diagonal dominance* of the coefficient matrix is satisfied.

The reader should consult reference 42 for detailed description of various methods for solving systems of algebraic equations and the FORTRAN programs and the subroutines associated with them.

We present here briefly some of the *direct* and *iterative* methods of solving systems of algebraic equations.

Direct Methods

Generally, the direct methods are preferred for systems having banded matrix coefficients and for problems involving relatively simple geometries and boundary conditions. They are very efficient, but require large computer storage and give rise to the accumulation of round-off error if the number of equations is large. There is a wealth of literature on the subject of solving systems of simultaneous algebraic equations because of the importance of this subject in scientific computing. Here we present a brief discussion of some of these direct methods.

Cramer's Rule. One of the most elementary methods of solving a set of algebraic equations is by employing *Cramer's rule.* The method is not practical to use for large number of equations because it involves large number of operations. To solve a set of N equations, the number of basic operations needed is to the order of $O(N^4)$. It implies that doubling the number of equations to be solved would increase the computer time on the order of 2^4, or 16 times. Even if the computer time were available, the accuracy would be destroyed by round-off errors. Therefore, in comparison to other methods discussed below, this method is completely impractical and should not be used in the solution of finite difference equations; it is mentioned here in order to bring into attention its such a shortcoming.

Gauss Elimination Method. This is a commonly used direct method for solving simultaneous algebraic equations. In this method, the coefficient matrix is transformed into an upper triangular matrix by systematic application of some algebraic operations under which the solution to the system of equations remains invariant. Two principal operations applied include (1) multiplication or division of any equation by a constant and (2) replacement of any equation by the sum (or difference) of that equation with any other equation. Once the system is transformed into upper diagonal form, the solution starts from the last equation and proceeds upward by back substitutions.

To illustrate the procedure we consider the following simple example involving three unknowns, T_1, T_2, and T_3:

$$a_{11}T_1 + a_{12}T_2 + a_{13}T_3 = d_1 \tag{12-46a}$$

$$a_{21}T_1 + a_{22}T_2 + a_{23}T_3 = d_2 \tag{12-50b}$$

$$a_{31}T_1 + a_{32}T_2 + a_{33}T_3 = d_3 \tag{12-50c}$$

where $a_{ii} \neq 0$ for $i = 1$–3.

Equations are successively solved for the main diagonal unknowns:

$$T_1 = \frac{1}{a_{11}}(d_1 - a_{12}T_2 - a_{13}T_3) \tag{12-51a}$$

$$T_2 = \frac{1}{a_{22}}(d_2 - a_{21}T_1 - a_{23}T_3) \tag{12-51b}$$

$$T_3 = \frac{1}{a_{33}}(d_3 - a_{31}T_1 - a_{32}T_2) \tag{12-51c}$$

Initial guess values are chosen as

$$T_1^{(0)}, \quad T_2^{(0)}, \quad T_3^{(0)} \tag{12-52}$$

These guess values are used together with the most recently computed values to complete the *first round* of iterations as

$$T_1^{(1)} = \frac{1}{a_{11}}(d_1 - a_{12}T_2^{(0)} - a_{13}T_3^{(0)}) \tag{12-53a}$$

$$T_2^{(1)} = \frac{1}{a_{22}}(d_2 - a_{21}T_1^{(1)} - a_{13}T_3^{(0)}) \tag{12-53b}$$

$$T_3^{(1)} = \frac{1}{a_{33}}(d_3 - a_{31}T_1^{(1)} - a_{32}T_2^{(1)}) \tag{12-53c}$$

These first approximations are used together with the most recently computed values to complete the second round of iterations as

$$T_1^{(2)} = \frac{1}{a_{11}}(d_1 - a_{12}T_2^{(1)} - a_{13}T_3^{(1)}) \tag{12-54a}$$

$$T_2^{(2)} = \frac{1}{a_{22}}(d_2 - a_{21}T_1^{(2)} - a_{23}T_3^{(1)}) \tag{12-54b}$$

$$T_3^{(2)} = \frac{1}{a_{33}}(d_3 - a_{21}T_1^{(2)} - a_{32}T_2^{(2)}) \tag{12-54c}$$

The iteration procedure is continued in a similar manner.

A general expression for the $(n+1)th$ *round* of iterations of the above system is written as

$$T_1^{(n+1)} = \frac{1}{a_{11}}[d_1 - a_{12}T_2^{(n)} - a_{13}T_3^{(n)}] \tag{12-55a}$$

$$T_2^{(n+1)} = \frac{1}{a_{22}}[d_2 - a_{21}T_1^{(n+1)} - a_{23}T_3^{(n)}] \tag{12-55b}$$

$$T_3^{(n+1)} = \frac{1}{a_{33}}[d_3 - a_{31}T_1^{(n+1)} - a_{32}T_2^{(n+1)}] \tag{12-55c}$$

In the general case of M equations, the $(n+1)th$ *round* of iterations can be written as

$$T_i^{(n+1)} = \frac{1}{a_{ii}}\left\{d_i - \sum_{j=1}^{i-1} a_{ij}T_j^{(n+1)} - \sum_{j=i+1}^{M} a_{ij}T_j^{(n)}\right\} \qquad \text{for} \qquad i = 1 \text{ to } M \tag{12-56}$$

The criterion for convergence can be specified either as the *absolute convergence crierion* in the form

$$|T_i^{(n+1)} - T_i^{(n)}| \leqslant \epsilon \tag{12-57}$$

or as *relative convergence criterion* in the form

$$\left|\frac{T_i^{(n+1)} - T_i^{(n)}}{T_i^{(n+1)}}\right| \leqslant \epsilon \tag{12-58}$$

which should be satisfied for all T_i.

Successive Overrelaxation. The Gauss–Seidel method described previously, generally does not converge sufficiently fast. The successive overrelaxation is a method that can accelerate the convergence.

In this method, the iteration procedure is written as

$$T_i^{(n+1)} = \omega\left\{\begin{matrix}\text{RHS (right-hand side) of Gauss–Seidel} \\ \text{iteration given by equation (12-56)}\end{matrix}\right\} + (1-\omega)T_i^{(n)} \qquad \text{for} \qquad i = 1 \text{ to } M \tag{12-59}$$

Here ω is the *relaxation parameter.*

Clearly, the case $\omega = 1$ corresponds to Gauss–Seidel iteration. The choice of the relaxation parameter effects the speed of convergence, but the determination of the optimal value of ω is a difficult matter. Some numerical experimentation is necessary for selecting proper value of ω for a given problem. With the proper choice of ω, it may be possible to reduce the computation time by an order of magnitude; therefore, when the number of equations is large and reduction of the computation time is important, some experimentation with different values of ω is worthwhile.

The physical significance of the relaxation parameter ω is as follows. For $\omega = 1$, the Gauss–Seidel computed value of the unknown is stored as the current value. For underrelaxation, $0 < \omega < 1$, a weighted average of the Gauss–Seidel value and the value from the previous iteration are stored as the current value. For overrelaxation, $1 < \omega < 2$, the current stored value is essentially extrapolated beyond the Gauss– Seidel value. For $\omega > 2$, the calculations diverge.

Iterative techniques are used to solve very large systems of equations, because round-off error is much smaller with iterative techniques than with direct solution techniques. With direct techniques, round-off errors occur with each mathematical operation and accumulate until final answers are obtained. With iterative techniques, for all practical purposes, the round-off error in the final converged solution is due to that accumulated in the final iteration.

12-8 ONE-DIMENSIONAL, STEADY-STATE HEAT CONDUCTION IN CYLINDRICAL AND SPHERICAL SYMMETRY

We now examine the finite-difference representation of one-dimensional, constant-property, steady-state heat conduction equation with cylindrical and spherical symmetry. The governing heat conduction equation is given by

$$\frac{1}{r^p}\frac{d}{dr}\left(r^p\frac{dT}{dr}\right)+\frac{1}{k}g(r)=0 \qquad r\neq 0 \tag{12-60a}$$

or

$$\frac{d^2T}{dr^2}+\frac{p}{r}\frac{dT}{dr}+\frac{1}{k}g(r)=0 \qquad r\neq 0 \tag{12-60b}$$

where

$$p=\begin{cases}0 & \text{rectangular}\\ 1 & \text{cylindrical}\\ 1 & \text{spherical}\end{cases}$$

and the term $g(r)$ represents the volumetric energy generation rate (i.e., W/m^3).

Solid Cylinder and Sphere. For a solid cylinder and sphere, equation (12-60b) has an apparent singularity at the origin $r = 0$. However, an examination of equation (12-60b) reveals that both r and dT/dr becomes zero for $r = 0$; hence we have $\frac{0}{0}$ ratio at the origin. By the application of L'Hospital's rule it can be shown that this ratio has the following determinate form:

$$\left(\frac{1}{r}\frac{dT}{dr}\right)_{r=0} = \frac{(d/dr)\left(\dfrac{dT}{dr}\right)}{(d/dr)(r)}_{r=0} = \frac{d^2T}{dr^2}\bigg|_{r=0} \tag{12-61}$$

Then, equation (12-60b), at $r = 0$, becomes

$$(1+p)\frac{d^2T(r)}{dr^2} + \frac{1}{k}g(r) = 0, \qquad r = 0 \tag{12-62}$$

To approximate this equation in finite differences, a network of mesh size δ, as illustrated in Fig. 12-6, is constructed over the region. Then by using the second-order accurate finite-difference formula, the first and the second derivatives are directly discretized. The resulting finite-difference approximation to equation (12-60b) becomes

$$\frac{T_{i-1} - 2T_i + T_{i+1}}{\delta^2} + \frac{p}{i\delta}\frac{T_{i+1} - T_{i-1}}{2\delta} + \frac{1}{k}g_i = 0$$

$$\text{for} \qquad i = 1, 2, \ldots, M-1, \quad O(\delta^2) \tag{12-63}$$

This system provides $M - 1$ algebraic equations for the $M + 1$ unknown node temperatures $T_0, T_1, \ldots, T_{M-1}, T_M$. Two more relations are needed.

An additional relationship is obtained by discretizing equation (12-62) at $r = 0$. In order to use a second-order accurate central-difference formula at $r = 0$, a node is needed to the left of the origin $r = 0$. This is achieved by considering a fictitious node "-1" at a fictitious temperature T_{-1} located at a distance δ to the left of the r axis. The resulting finite-difference approximation of equation

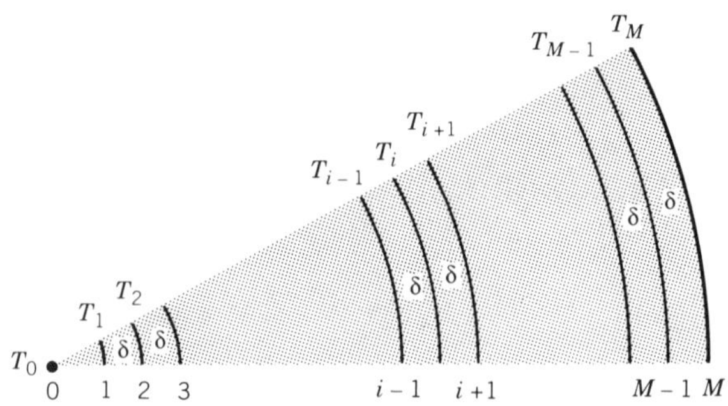

Fig. 12-6 Nomenclature for finite-difference representation for cylindrical and spherical symmetry.

(12-62) at $r = 0$ becomes

$$(1+p)\frac{T_{-1} - 2T_0 + T_1}{\delta^2} + \frac{1}{k} g_0 = 0, \qquad i = 0 \tag{12-64a}$$

where the fictitious temperature T_{-1} is determined by utilizing the symmetry condition at the node $i = 0$;

$$\left.\frac{dT}{dr}\right|_{r=0} = \frac{T_{-1} - T_1}{2\delta} = 0, \qquad \text{giving} \qquad T_{-1} = T_1 \tag{12-64b}$$

Introducing equation (12-64b) into (12-64a), the additional finite difference equation is determined as

$$2(1+p)\frac{T_1 - T_0}{\delta^2} + \frac{1}{k} g_0 = 0 \qquad \text{for} \qquad i = 0 \tag{12-65}$$

Equations (12-65) and (12-63) are now rearranged, respectively, as

$$2(1+p)(T_1 - T_0) + \frac{\delta^2 g_0}{k} = 0, \qquad \text{for} \qquad i = 0 \tag{12-66}$$

$$\left(1 - \frac{p}{2i}\right) T_{i-1} - 2T_i + \left(1 + \frac{p}{2i}\right) T_{i+1} + \frac{\delta^2 g_i}{k} = 0, \qquad \text{for} \qquad i = 1, 2, \ldots, M-1 \tag{12-67}$$

where

$$p = \begin{cases} 1 & \text{cylinder} \\ 2 & \text{sphere} \end{cases}$$

One more equation is needed to make the number of equations equal to the number of unknowns. It is obtained by considering the boundary condition at the node $i = M$ (i.e., $r = b$). The following possibilities can be considered at the node M:

1. The temperature T_b is specified at the boundary $r = b$. Then we have

$$T_M = T_b = \text{known} \tag{12-68}$$

and the system of equations (12-66), (12-67) and (12-68) provide $M + 1$ relations for the determination of $(M + 1)$ unknown node temperatures.

2. The boundary condition at $r = b$ is convection into an ambient at a constant temperature $T_{\infty,b}$ with a heat transfer coefficient h_b. The boundary condition is

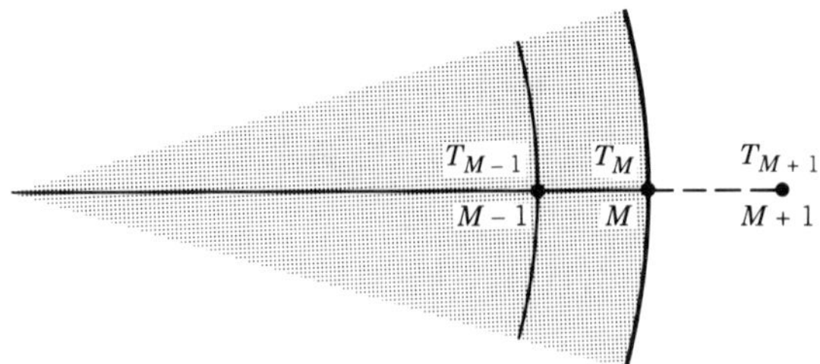

Fig. 12-7 Fictitious node $M+1$ at fictitious temperature T_{M+1}.

given by

$$k\frac{dT(r)}{dr}+h_bT(r)=h_bT_{\infty,b}=\text{known} \qquad \text{at} \qquad r=b \tag{12-69}$$

To discretize this equation about the boundary node M with a second-order central-difference formula, an additional node is needed to the right of the node M. This is obtained by considering an extension of the region by a distance δ to the right of the node M, giving rise to a fictitious node $M+1$ at a fictitious temperature T_{M+1} as illustrated in Fig. 12-7. Then the discretization of equation (12-69) about the node M with the central difference formula gives

$$k\frac{T_{M+1}-T_{M-1}}{2\delta}+h_bT_M=h_bT_{\infty,b} \tag{12-70}$$

An additional relationship needed to eliminate T_{M+1} is determined by evaluating equation (12-67) for $i=M$. We obtain

$$\left(1-\frac{p}{2M}\right)T_{M-1}-2T_M+\left(1+\frac{p}{2M}\right)T_{M+1}+\frac{\delta^2 g_M}{k}=0 \tag{12-71}$$

The elimination of T_{M+1} between equations (12-70) and (12-71) gives

$$2T_{M-1}-2\beta_M T_M+2\gamma_M+G_M=0 \qquad \text{for} \qquad i=M \tag{12-72a}$$

where

$$\beta_M=1+\left(1+\frac{p}{2M}\right)\frac{\delta h_b}{k} \tag{12-72b}$$

$$\gamma_M=\left(1+\frac{p}{2M}\right)\frac{\delta}{k}h_bT_{\infty,b} \tag{12-72c}$$

$$G_M=\frac{\delta^2 g_M}{k} \tag{12-72d}$$

which is accurate $0(\delta^2)$. Equations (12-66), (12-67), and (12-72) provide $M+1$ relations for the determination of $M+1$ unknown node temperatures for convection boundary conditions at $r=b$.

3. Boundary condition at $r=b$ is a prescribed heat flux boundary condition. For this case, the steady-state solution does not exist unless the energy generated in the medium equals to the total heat removal rate from the boundaries. Even for such a case, the steady-state solution for a solid cylinder or sphere is not unique; such a situation will not be considered.

Example 12-3

A 10-cm-diameter solid steel bar of thermal conductivity $k = 40$ W/(m·°C) is heated electrically by the passage of electric current which generates energy within the rod at a rate of $g = 4 \times 10^6$ W/m^3. Heat is dissipated from the surface of the rod by convection with a heat transfer coefficient $h = 400$ W/(m^2·°C) into an ambient at temperature $T_\infty = 20$°C. By dividing the radius into Five equal parts, develop the finite-difference equations for this heat conduction problem. Compare the finite-difference solution with the exact analytic solution for the cases when the *first-order* and the *second-order* accurate differencing are used for the convection boundary condition.

Solution. The problem involves six unknown node temperatures, T_i, $i = 0, 1, \ldots, 5$, since the region $0 \leqslant r \leqslant b$ is divided into five equal parts. The six finite-difference equations needed for their determination are obtained as

$$4(T_1 - T_0) + 10 = 0, \qquad i = 0 \tag{12-73}$$

$$\left(1 - \frac{1}{2i}\right)T_{i-1} - 2T_i + \left(1 + \frac{1}{2i}\right)T_{i+1} + 10 = 0, \qquad i = 1, 2, 3, 4 \tag{12-74}$$

For the boundary condition at $i = M = 5$, one can use either the first-order accurate formula

$$T_5 = \frac{1}{1.1}(T_4 + 2), \qquad i = 5 \tag{12-75a}$$

or the second-order accurate formula (c)

$$T_4 - 1.11T_5 + 7.2 = 0, \qquad i = 5 \tag{12-75b}$$

The exact solution of this problem is given by

$$T(r) = T_\infty + \frac{gb}{2h} + \frac{gb^2}{4k}\left[1 - \left(\frac{r}{b}\right)^2\right] \tag{12-76a}$$

TABLE 12-3 Comparison of Results with Exact Solution for Example 12-3

		$M = 5$		$M = 10$
$\frac{r}{b}$	Exact	1st-Order Accurate	2nd-Order Accurate	1st-Order Accurate
0.0	332.50	307.50	332.50	320.00
0.2	330.00	305.00	330.00	317.50
0.4	322.50	297.50	322.50	310.00
0.6	310.00	285.00	310.00	297.00
0.8	292.50	267.50	292.50	280.00
1.0	270.00	245.00	270.00	257.50

or

$$T(r) = 20 + 250 + 62.5\left[1 - \left(\frac{r}{b}\right)^2\right] \tag{12-76b}$$

Table 12-3 shows a comparison of finite-difference solutions with the exact results for the cases when the *first-order* and *second-order* accurate formulas are used for the convection boundary condition. Gauss elimination method is used to solve the resulting algebraic equations. The numerical results obtained with the second-order accurate formula are in excellent agreement with the exact solution; but the solution with the first-order formula is not so good; it underpredicts temperature from about 7 to 9%. Increasing the number of subdivisions from $M = 5$ to $M = 10$ improves the accuracy of the results with the first-order formula to about 4%.

Hollow Cylinder and Sphere. We now consider heat conduction in a hollow cylinder and sphere of inner radius $r = a$, outer radius $r = b$. To solve this problem with finite differences, a finite-difference network is constructed over the region

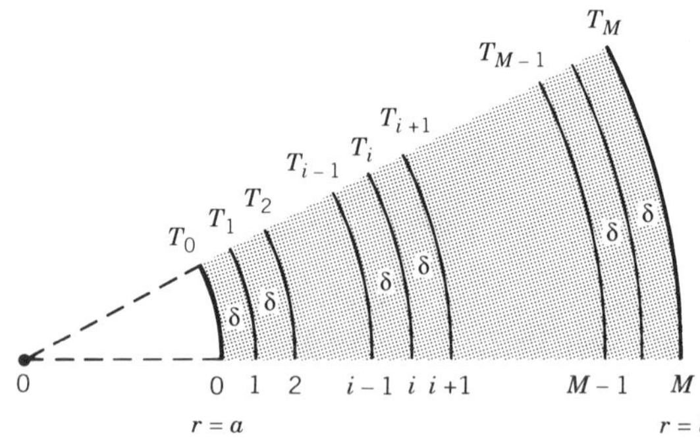

Fig. 12-8 Nomenclature for finite-difference representation for hollow sphere or cylinder.

as illustrated in Fig. 12-8. The governing heat conduction equation is given by

$$\frac{d^2T}{dr^2} + \frac{p}{r}\frac{dT}{dr} + \frac{1}{k}g(r) = 0 \qquad \text{in} \qquad a < r < b \tag{12-77}$$

For finite-difference representation of this equation, the region $a \leqslant r \leqslant b$ is divided into M subregions each of thickness δ given by

$$\delta = \frac{b-a}{M} \tag{12-78}$$

The differential equation is discretized by using the second-order accurate central-difference formula for both the second and the first derivatives. We obtain

$$\frac{T_{i-1} - 2T_i + T_{i+1}}{\delta^2} + \frac{p}{a + i\delta}\frac{T_{i+1} - T_{i-1}}{2\delta} + \frac{1}{k}g_i = 0 \tag{12-79}$$

which is rearranged in the form

$$\left[1 - \frac{p}{2[(a/\delta) + i]}\right]T_{i-1} - 2T_i + \left[1 + \frac{p}{2[(a/\delta) + i]}\right]T_{i+1} + \frac{\delta^2 g_i}{k} = 0 \tag{12-80}$$

$$\text{for} \qquad i = 1, 2, \ldots, M-1$$

where

$$p = \begin{cases} 1 & \text{cylinder} \\ 2 & \text{sphere} \end{cases}$$

Equations (12-80) provide $M - 1$ algebraic equations, but involve $(M + 1)$ unknown node temperatures T_i, $i = 0, 1, 2, \ldots, M$. The additional two relationships are obtained from the boundary conditions at $r = a$ and $r = b$. The following possibilities are considered for the boundary conditions:

1. Temperatures T_a and T_b are prescribed at the boundaries $r = a$ and $r = b$. Then the system of equations (12-80) provide $M - 1$ relations for the determination of $M - 1$ internal node temperatures, since $T_0 = T_a$ and $T_M = T_b$ are known.
2. The boundary conditions at $r = a$ and $r = b$ are convection into ambients at temperatures $T_{\infty,a}$ and $T_{\infty,b}$ with heat transfer coefficients h_a and h_b, respectively:

$$-k\frac{dT}{dr} + h_a T = h_a T_{\infty,a} = \text{known}, \qquad r = a \tag{12-81a}$$

$$k\frac{dT}{dr} + h_b T = h_b T_{\infty,b} = \text{known}, \qquad r = b \tag{12-81b}$$

These boundary conditions are discretized with the second-order accurate central-difference formula by using the fictitious node concept. The resulting finite-difference form of equations (12-81) becomes

$$2T_1 - 2\beta_0 T_0 + 2\gamma_0 + G_0 = 0, \qquad \text{for} \qquad i = 0 \tag{12-82a}$$

$$2T_{M-1} - 2\beta_M T_M + 2\gamma_M + G_M = 0, \qquad \text{for} \qquad i = M \tag{12-82b}$$

where

$$\beta_0 = 1 + \left[1 - \frac{p}{2(a/\delta)}\right]\frac{\delta h_a}{k}, \qquad \beta_M = 1 + \left[1 + \frac{p}{2[(a/\delta) + M]}\right]\frac{\delta h_b}{k} \tag{12-83a}$$

$$\gamma_0 = \left[1 - \frac{p}{2(a/\delta)}\right]\frac{\delta}{k}(h_a T_{\infty,a}), \qquad \gamma_M = \left[1 + \frac{p}{2[(a/\delta) + M]}\right]\frac{\delta}{k}(h_b T_{\infty,b}) \tag{12-83b}$$

$$G_0 = \frac{\delta^2 g_0}{k}, \qquad G_M = \frac{\delta^2 g_M}{k} \tag{12-84}$$

and $a > \delta$.

Summarizing, equations (12-80) and (12-82) provide $M + 1$ algebraic equations for the determination of $M + 1$ unknown node temperatures $T_i, i = 0, 1, 2, \ldots, M$.

12-9 MULTIDIMENSIONAL STEADY-STATE HEAT CONDUCTION

The extension of one-dimensional finite-differencing scheme for the discretization of multidimensional steady-state heat conduction equation is a straightforward matter which is now illustrated with examples.

Example 12-4

Consider the following steady-state heat conduction problem.

$$\frac{\partial^2 T}{\partial x^2} + \frac{\partial^2 T}{\partial y^2} + \frac{1}{k} g(x, y) = 0 \qquad \text{in} \qquad 0 \leqslant x \leqslant a, \qquad 0 \leqslant y \leqslant b \tag{12-85a}$$

$$T = f(y) \qquad \text{at} \qquad x = 0 \tag{12-85b}$$

$$k\frac{\partial T}{\partial x} + h_1 T = h_1 T_{\infty 1} \qquad \text{at} \qquad x = a \tag{12-85c}$$

$$-k\frac{\partial T}{\partial y} + h_2 T = h_2 T_{\infty 2} \qquad \text{at} \qquad y = 0 \tag{12-85d}$$

$$T = 0 \qquad \text{at} \qquad y = b \tag{12-85e}$$

Write the finite-difference form of this problem using second-order accurate differencing scheme.

Solution. Figure 12-9 shows the finite-difference net drawn over the region. We let

$$T(x, y) = T(i\Delta x, j\Delta y) \equiv T_{i,j} \tag{12-86}$$

where $i = 0, 1, 2, \ldots, M$ and $j = 0, 1, 2, \ldots, N$. We assume $\Delta x = \Delta y \equiv l$.

The finite-difference equations for various grid points are determined as follows:

1. *Internal Nodes,* $1 \leqslant i \leqslant M - 1$; $1 \leqslant j \leqslant N - 1$. The differential equation is discretized to yield the following finite-difference equations:

$$\frac{T_{i-1,j} - 2T_{i,j} + T_{i+1,j}}{(\Delta x)^2} + \frac{T_{i,j-1} - 2T_{i,j} + T_{i,j+1}}{(\Delta y)^2} + \frac{1}{k} g_{i,j} = 0 \tag{12-87a}$$

For the case $\Delta x = \Delta y = l$, these equations reduce to

$$(T_{i-1,j} + T_{i+1,j} + T_{i,j-1} + T_{i,j+1} - 4T_{i,j}) + \frac{l^2}{k} g_{i,j} = 0 \tag{12-87b}$$

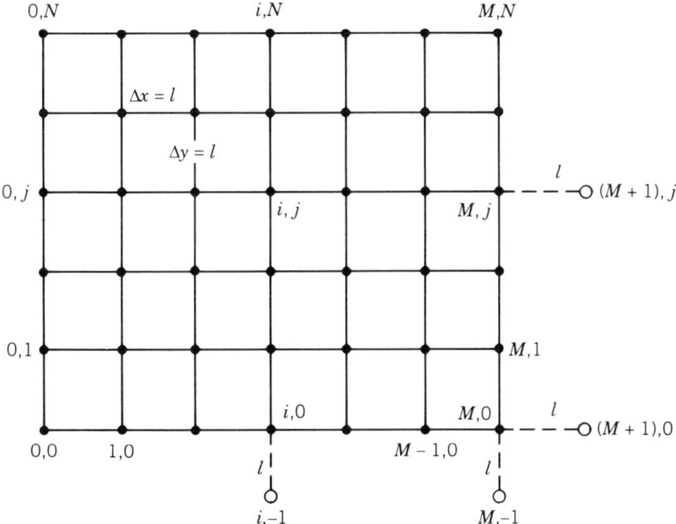

Fig. 12-9 Finite-difference network for Example 12-4.

2. *Boundary Nodes* $0,j$ *for* $j=0$ *to* $N-1$ *and* i,N *for* $i=0$ *to* M. Because temperatures are prescribed at these boundaries, no equations are needed.

3. *Boundary Nodes* M,j *for* $j=1$ *to* $N-1$. This boundary is subjected to convection. The finite-difference equations are developed either by writing an energy-balance equation for a control volume about the node M,j or by considering a fictitious node $M+1,j$ at a distance l outside the region at a fictitious temperature $T_{M+1,j}$ as illustrated in Fig. 2-9. We prefer the latter, and discretize the boundary condition (12-85c) using central differences as

$$k\frac{T_{M+1,j}-T_{M-1,j}}{2l}+h_1 T_{M,j}=h_1 T_{\infty 1} \tag{12-88a}$$

To eliminate $T_{M+1,j}$, an additional relationship is obtained by evaluating equation (12-87b) for $i=M$:

$$T_{M-1,j}+T_{M+1,j}+T_{M,j-1}+T_{M,j+1}-4T_{M,j}+\frac{l^2}{k}g_{M,j}=0 \tag{12-88b}$$

Eliminating $T_{M+1,j}$ between equations (12-88a) and (12-88b), the finite-difference equations for the nodes on this boundary becomes

$$2T_{M-1,j}+T_{M,j-1}+T_{M,j+1}-\left(4+\frac{2h_1 l}{k}\right)T_{M,j}=-\frac{2h_1 l}{k}T_{\infty,1}-\frac{l^2}{k}g_{M,j}$$

$$\text{for}\quad j=1 \text{ to } N-1 \tag{12-88c}$$

4. *Boundary Nodes* $i,0$ *for* $i=1,2,\ldots,M-1$. By following a procedure similar to that in case 3, the finite-difference equations for nodes on this boundary are determined as

$$2T_{i,1}+T_{i-1,0}+T_{i+1,0}-\left(4+\frac{2h_2 l}{k}\right)T_{i,0}=-\frac{2h_2 l}{k}-\frac{l^2}{k}g_{i,0} \tag{12-89}$$

5. *Node* $M,0$ *at Intersection of Two Convection Boundaries.* The finite-difference equation (12-87b) is evaluated for $i=M, j=0$. The resulting equation contains fictitious temperatures $T_{M,-1}$ and $T_{(M+1),0}$ at the fictitious nodes $M,-1$ and $(M+1),0$. Two additional relation needed to eliminate these fictitious temperatures are obtained by discretizing the boundary conditions (12-85c) and (12-85d) at the node $M,0$ by central differences, using these fictitious nodes. After the fictitious temperatures are eliminated, the resulting finite-difference equation for the node $M,0$

becomes

$$2T_{M-1,0} + 2T_{M,1} - \left(4 + \frac{2lh_1}{k} + \frac{2lh_2}{k}\right)T_{M,0}$$

$$= -\left(\frac{2lh_1}{k}T_{\infty 1} + \frac{2lh_2}{k}T_{\infty 2} + \frac{l^2}{k}g_{M,0}\right) \tag{12-90}$$

Example 12-5

Consider the following two-dimensional steady-state heat conduction for a solid cylinder of radius b in the r, ϕ variables

$$\frac{\partial^2 T}{\partial r^2} + \frac{1}{r}\frac{\partial T}{\partial r} + \frac{1}{r^2}\frac{\partial^2 T}{\partial \phi^2} + \frac{1}{k}g(r,\phi) = 0 \quad \text{in} \quad 0 \leqslant r < b, \quad 0 \leqslant \phi \leqslant 2\pi \tag{12-91a}$$

$$k\frac{\partial T}{\partial r} + hT = hT_\infty \quad \text{at} \quad r = b \tag{12-91b}$$

and T is periodic in ϕ with a period 2π (12-91c)

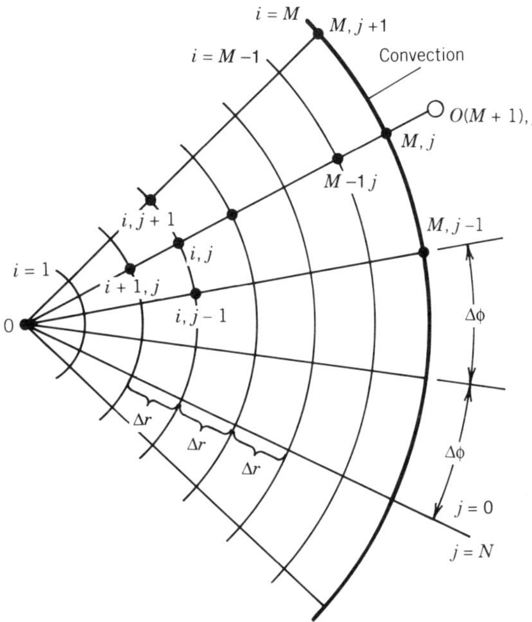

Fig. 12-10. An (r, ϕ) network in cylindrical coordinate system and the fictitious node "$M + 1, j$"

Figure 12-10 shows the finite-difference network in $\Delta r, \Delta\phi$. Write the finite-difference form of this problem using second-order accurate differencing scheme.

Solution. Temperature $T(r,\phi)$ at a grid point (i,j) is denoted by

$$T(r,\phi) = T(i\Delta r, j\Delta\phi) \equiv T_{ij} \tag{12-92a}$$

$$\text{for} \qquad i = 0, 1, 2, \ldots, M; \quad j = 0, 1, 2, \ldots, N \tag{12-92b}$$

Various derivatives in this problem are discretized by using second-order accurate central difference formula. The resulting finite-difference equations for various grid points are determined as follows:

1. *Internal Nodes,* $1 \leqslant i \leqslant M-1$; $0 \leqslant j \leqslant N$. Discretization of equation (12-91a) gives

$$\frac{T_{i-1,j} - 2T_{i,j} + T_{i,j+1}}{(\Delta r)^2} + \frac{1}{i\Delta r} \cdot \frac{T_{i+1,j} - T_{i-1,j}}{2\Delta r} + \frac{1}{i^2(\Delta r)^2} \frac{(T_{i,j-1} - 2T_{i,j} + T_{i,j+1})}{(\Delta\phi)^2} + \frac{1}{k} g_{i,j} = 0 \tag{12-93a}$$

After rearranging we find

$$\frac{1}{(\Delta r)^2}\left[\left(1 - \frac{1}{2i}\right)T_{i-1,j} - 2T_{i,j} + \left(1 + \frac{1}{2i}\right)T_{i+1,j}\right] + \frac{1}{i^2(\Delta r\,\Delta\phi)^2}(T_{i,j-1} - 2T_{i,j} + T_{i,j+1}) + \frac{1}{k} g_{i,j} = 0 \tag{12-93b}$$

and the condition that temperature is periodic in ϕ with a period 2π requires

$$T_{i,0} = T_{i,N} \tag{12-93c}$$

2. *The Center Node,* $T_j^0 \equiv T_0$. Equation (12-91a) appears to have singularity at the origin $r = 0$. To deal with this situation, equation (12-91a) is replaced by its Cartesian equivalent:

$$\frac{\partial^2 T}{\partial x^2} + \frac{\partial^2 T}{\partial y^2} + \frac{1}{k} g = 0 \qquad \text{as} \qquad r \to 0 \tag{12-94a}$$

We construct a circle of radius, Δr, center at $r = 0$. Let T_0 be the temperature at $r = 0$ and T_1, T_2, T_3, T_4 be the temperatures at the four nodes this circle intersects the x and y axes. Then the finite difference form of this equation about $r=0$ becomes

$$\frac{T_1+T_2+T_3+T_4-4T_0}{(\Delta r)^2}+\frac{1}{k}g_0=0 \tag{12-94b}$$

with a truncation error of the order of $(\Delta r)^2$. The rotation of the $0x$ and $0y$ axes about $r=0$ also leads to a similar difference equation. If we now denote $\tilde{T}_1$ as the arithmetic mean of the temperatures around the circle of radius Δr_1, then equation (12-94b) becomes

$$4\frac{\tilde{T}_1-T_0}{(\Delta r)^2}+\frac{1}{k}g_0=0 \qquad \text{at} \qquad r=0 \tag{12-94c}$$

where $\tilde{T}_1$ is the arithmetic mean of the values of $T_{1,j}$ around the circle of radius Δr with center at $r=0$, T_0 is the value of temperature at $r=0$. Thus, equation (12-94c) is the finite-difference form of equation (12-91a) for the central node at $r=0$.

3. *Boundary Nodes* (M,j), *for* $j=0$ *to* N. The finite-difference equation (12-93b) is evaluated for $i=M$:

$$\left(1-\frac{1}{2M}\right)T_{M-1,j}+\left(1+\frac{1}{2M}\right)T_{(M+1),j}-2\left(1+\frac{1}{(M\Delta\phi)^2}\right)T_{M,j}$$
$$+\frac{1}{(M\Delta\phi)^2}T_{M,j-1}+\frac{1}{M(\Delta\phi)^2}T_{M,j+1}+\frac{(\Delta r)^2 g_{M,j}}{k}=0 \tag{12-95a}$$

An additional relation needed to eliminate the *fictitious temperature* $T_{(m+1),j}$ is obtained by discretizing the boundary condition (12-91b) about the node (M,j) with central differences using the fictitious node $(M+1),j$. We obtain

$$k\frac{T_{(M+1),j}-T_{M-1,j}}{2\Delta r}+hT_{M,j}=hT_\infty$$

or solving for the fictitious temperature $T_{M+1,j}$:

$$T_{(M+1),j}=T_{M-1,j}-\frac{2h\,\Delta r}{k}T_{M,j}+\frac{2h\,\Delta r}{k}T_\infty \tag{12-95b}$$

Equation (12-95a), together with equation (12-95b), provides a second-order accurate finite-difference equations for the nodes M,j for $j=0,1,\ldots,N$ on the convection boundary.

12-10 ONE-DIMENSIONAL TIME-DEPENDENT HEAT CONDUCTION

In this section we present the finite-difference representation of one-dimensional transient heat conduction equation. There are several schemes available to express the time-dependent heat conduction equation in finite-difference form; for example, 13 different schemes are listed in reference 5. Each of these schemes has its advantages and limitations. We shall discuss some of these schemes with particular emphasis to the finite-difference approximation in the rectangular coordinates. Applications for the cylindrical and spherical symmetry will be presented afterward.

Explicit Method

We consider the one-dimensional, time-dependent heat conduction problem for a finite region $0 \leqslant x \leqslant L$ given as

$$\frac{\partial T}{\partial t} = \alpha \frac{\partial^2 T}{\partial x^2} \quad \text{in} \quad 0 < x < L, \quad t > 0 \tag{12-96}$$

Subject to the boundary and initial conditions

$$T(x,t) = T_a = \text{known} \quad \text{at} \quad x = 0, \quad t > 0 \tag{12-97a}$$

$$T(x,t) = T_L = \text{known} \quad \text{at} \quad x = L, \quad t > 0 \tag{12-97b}$$

$$T(x,t) = F(x) = \text{known} \quad \text{for} \quad t = 0 \tag{12-97c}$$

The differential equation (12-96) is represented in finite-difference form by using central differences to discretize $\partial^2 T/\partial x^2$ and forward differences to discretize $\partial T/\partial t$. We obtain

$$\frac{T_i^{n+1} - T_i^n}{\Delta t} = \alpha \frac{T_{i-1}^n - 2T_i^n + T_{i+1}^n}{(\Delta x)^2} + 0[\Delta t, (\Delta x)^2] \tag{12-98a}$$

where

$$T(x,t) = T(i\Delta x, n\Delta t) \equiv T_i^n \tag{12-98b}$$

Equation (12-98a) is rearranged as

$$T_i^{n+1} = rT_{i-1}^n + (1 - 2r)T_i^n + rT_{i+1}^n \tag{12-99a}$$

where

$$r = \frac{\alpha \Delta t}{(\Delta x)^2} \tag{12-99b}$$

$$n = 0, 1, 2, \ldots \qquad \text{and} \qquad i = 1, 2, \ldots, M - 1$$

with a truncation error of order $O[\Delta t, (\Delta x)^2]$.

The finite-difference representation given by equations (12-99) is called the *explicit* form because the unknown temperature T_i^{n+1} at time step $(n + 1)$ can be explicitly determined from the knowledge of the temperatures T_{i-1}^n, T_i^n, and T_{i+1}^n at the previous time step n according to equation (12-99a). The only disadvantage of this method is that, once α and Δx are fixed, there is a maximum permissible time-step size Δt which should not exceed the value imposed on by the following stability criterion:

$$0 < r \equiv \frac{\alpha \Delta t}{(\Delta x)^2} \leqslant \frac{1}{2} \tag{12-100}$$

That is, for given values of α and Δx, if the time step Δt exceeds the limit imposed on by the above criteria, the numerical calculations become unstable resulting from the amplification of errors. Figure 12-11 illustrates what happens to the numerical calculations when the above stability criterion is violated. In this figure, the numerical calculations performed with a time step satisfying the condition $r = \frac{5}{11} < \frac{1}{2}$ is in good agreement with the exact solution; whereas the numerical solution of the same problem with slightly larger time step which violates the above stability criterion (i.e., $r = \frac{5}{9} > \frac{1}{2}$), results in an unstable solution.

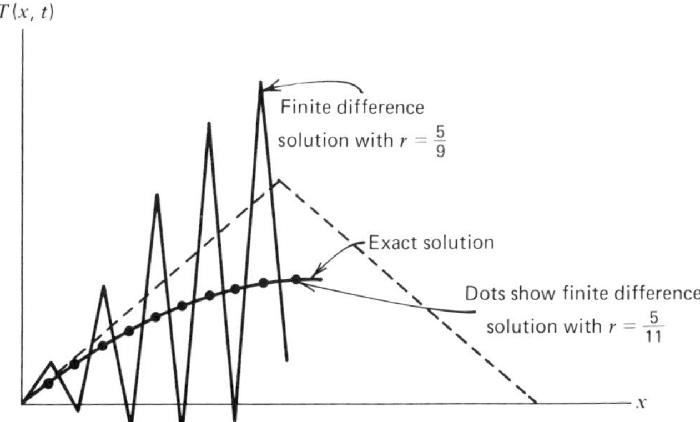

Fig. 12-11 Effects of parameter $r = \alpha\Delta t/(\Delta x)^2$ on the stability of finite-difference solution of the one-dimensional time-dependent heat conduction equation.

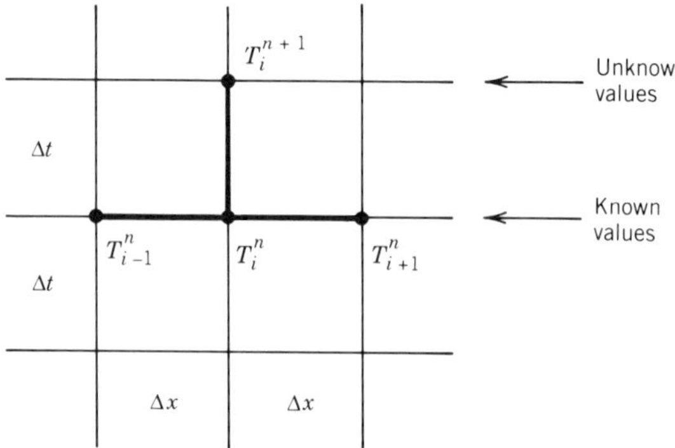

Fig. 12-12 The finite-difference molecules for the simple explicit scheme.

When the boundary conditions are prescribed temperatures at both boundaries, as is the case of the problem defined by equations (12-96) an (12-97), then the number of finite-difference equations (12-98) become equal to the number of unknown node temperatures. Figure 12-12 schematically illustrates the finite-difference molecules associated with the explicit scheme.

The computational procedure is as follows:

1. Start the calculations with $n = 0$. Compute the T_i^1, $i = 1, 2, \ldots, M - 1$, at the end of the first time step from equation (12-99a), since the right-hand side of this equation is known from the initial condition.
2. Set $n = 1$ and calculate T_i^2, $i = 1, 2, \ldots, M - 1$, at the end of the second time step from equation (12-99a), because the right-hand side of this equation is known from the previous time step.
3. Repeat the procedure for each subsequent time step and continue calculations until a specified time or some specified value of the temperature is reached.

Convection Boundary Conditions. Consider the boundary surfaces at $x = 0$ and $x = L$ are subjected to convection with heat transfer coefficients h_0 and h_L into ambients at temperatures $T_{\infty,0}$ and $T_{\infty,L}$, respectively. We have

$$-k\frac{\partial T}{\partial x} + h_0 T = h_0 T_{\infty,0} = \text{known}, \qquad \text{at} \qquad x = 0 \tag{12-101a}$$

$$k\frac{\partial T}{\partial x} + h_L T = h_L T_{\infty,L} = \text{known}, \qquad \text{at} \qquad x = L \tag{12-101b}$$

where the temperatures at the boundary nodes $i = 0$ and $i = M$ are unknown. Two additional relations are obtained by discretizing these two boundary conditions.

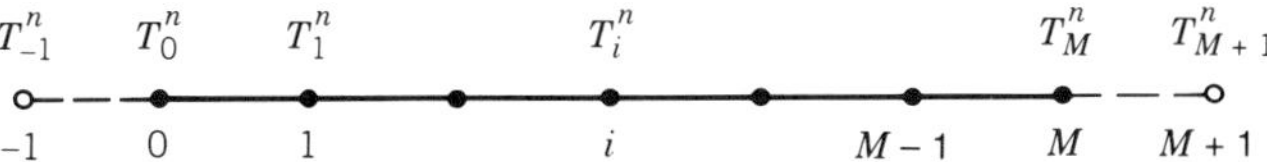

Fig. 12-13 Fictitious nodes -1 and $M+1$.

A very simple approach to discretize these boundary conditions is to use forward differencing for equation (12-101a) and backward differencing for equation (12-101b); but the results are only first-order accurate, $O(\Delta x)$.

A second-order accurate, that is, $O[(\Delta x)^2]$, differencing of these boundary conditions is possible if central differencing is used to discretize the first derivatives in these boundary conditions. To apply the central differencing, we consider a fictitious node "$-$ 1" at a fictitious temperature T_1^n and a fictitious node $M+1$ at a fictitious temperature T_{M+1}^n obtained by extending the region by Δx to the left and right, respectively, as illustrated in Fig. 12-13.

Equation (12-99a) is evaluated for $i=0$ and $i=M$, the resulting fictitious temperatures T_1^n and T_{M+1}^n are eliminated by utilizing the equations obtained by discretizing the boundary conditions (12-101a) and (12-101b) with central differences about the nodes 0 and $M+1$, respectively. Then the following, second-order finite-difference equations are obtained for the convection boundary conditions

$$T_0^{n+1} = (1-2r\beta_0)T_0^n + 2rT_1^n + 2r\gamma_0, \qquad \text{for} \quad i=0 \tag{12-102a}$$

$$T_M^{n+1} = 2rT_{M-1}^n + (1-2r\beta_L)T_M^n + 2r\gamma_L, \qquad \text{for} \quad i=M \tag{12-102b}$$

where

$$\beta_0 = 1 + \frac{\Delta x\, h_0}{k}, \quad \gamma_0 = \frac{\Delta x\, h_0}{k} T_{\infty,0} \tag{12-103a}$$

$$\beta_L = 1 + \frac{\Delta x\, h_L}{k}, \quad \gamma_L = \frac{\Delta x\, h_L}{k} T_{\infty,L} \tag{12-103b}$$

$$r = \frac{\alpha \Delta t}{(\Delta x)^2} \tag{12-103c}$$

Thus, the finite-difference equations (12-99) together with equations (12-102) provide $M+1$ expressions for the determination of $M+1$ unknown node temperatures at each time step.

For the second-order accurate finite-differencing of the convection boundary conditions considered here the stability criteria (12-100) should be modified as follows

$$1 - 2r\beta_0 \geqslant 0 \quad \text{or} \quad 0 < r \leqslant \frac{1}{2\beta_0} = \frac{1}{2 + 2[(\Delta x\, h_0)/k]} \tag{12-104a}$$

for the boundary condition at $x = 0$, and

$$1 - 2r\beta_L \geqslant 0 \quad \text{or} \quad 0 < r \leqslant \frac{1}{2\beta_L} = \frac{1}{2 + 2[(\Delta x\, h_L)/k]} \tag{12-104b}$$

for the boundary condition at $x = L$.

Clearly, the stability criterion imposed on by equations (12-104a,b) is more restrictive than that based on $r \leqslant \frac{1}{2}$. The smaller value of r obtained from equations (12-104a,b) should be used as the stability criteria for the solution of the problem.

Heuristic Argument of Stability. Computers cannot perform calculations to infinite accuracy. Therefore, in the numerical solution of finite-difference equations with a digital computer, round-off errors are introduced during calculations. The mathematical analysis of stability is concerned with the examination of the growth of errors while the computations are being performed. For an unstable system the error grows larger without bound, but for a stable system it should not grow without a bound.

Before presenting a rigorous analysis of the stability of the solution of finite-difference equations, it is instructive to give a heuristic discussion of the stability requirements.

We consider the explicit finite-difference equations (12-99a)

$$T_i^{n+1} = rT_{i-1}^n + (1 - 2r)T_i^n + rT_{i+1}^n \tag{12-105a}$$

Suppose at any time level n, the temperature T_{i-1}^n and T_{i+1}^n at the nodes $i - 1$ and $i + 1$ are equal. Equation (12-105a) is arranged as

$$T_i^{n+1} = T_{i-1}^n + (1 - 2r)(T_i^n - T_{i-1}^n) \tag{12-105b}$$

For illustration purposes, let $T_{i-1}^n = T_{i+1}^n = 0°\text{C}$ and $T_i^n = 100°\text{C}$. Equation (12-105b) is now used to calculate the temperature T_i^{n+1} of the node i at the next time level $n + 1$ as

$$T_i^{n+1} = 0 + (1 - 2r)(100 - 0) = (1 - 2r)100 \tag{12-106}$$

The physical situation requires that the temperature T_i^{n+1} cannot go below the temperature of the two neighboring nodes, 0°C. An examination of equation (12-106) revels that a negative value of $(1 - 2r)$ violates such a requirement. Therefore, to obtain meaningful results from the solution of the finite-difference equation (12-99a) the coefficient $(1 - 2r)$ should satisfy the following criterion

$$1 - 2r \geqslant 0 \quad \text{or} \quad r = \frac{\alpha \Delta t}{(\Delta x)^2} \leqslant \frac{1}{2} \tag{12-107a}$$

which is the same as given previously by equations (12-100). Such a restriction on the maximum value of r imposes the following limitation to the maximum size of the time step

$$\Delta t \leqslant \frac{(\Delta x)^2}{2\alpha} \tag{12-107b}$$

A similar physical argument can be applied to examine the stability conditions for the solution of finite-difference equations (12-102a). Consider equation (12-102a) for $\gamma_0 = 0$, which corresponds to convection into an ambient at zero temperature. We obtain

$$T_0^{n+1} = 2rT_1^n + (1 - 2r\beta_0)T_0^n \tag{12-108a}$$

Suppose at any time level, the temperatures of the nodes 0 and 1 be, respectively, $T_0^n = 100°\text{C}$ and $T_1^n = 0°\text{C}$. Equation (5-108a), to be used for predicting the temperature T_0^{n+1} of the node 0 at the next time level $n + 1$, becomes

$$T_0^{n+1} = 2r \times 0 + (1 - 2r\beta_0)100 = (1 - 2r\beta_0)100 \tag{12-108b}$$

The physically meaningful situation for the problem requires that the temperature T_0^{n+1} can assume values between 0°C and 100°C, but cannot go below the 0°C temperature of the neighboring node and of the ambient. An examination of equation (12-108b) reveals that a negative value of the parameter $(1 - 2r\beta_0)$ violates this requirement. Therefore, to obtain physically meaningful results from the solution of the finite-difference equation (12-108) the following criteria should be satisfied

$$1 - 2r\beta_0 \geqslant 0 \tag{12-109}$$

which is the same as that given by equation (12-104a).

Fourier Method of Stability Analysis. We now present rather straightforward but more rigorous analysis of the stability of finite-difference equations by using the Fourier (or Neumann) method of stability analysis.

In the Fourier method, the errors are expressed in a finite Fourier series, and then the propagation of growth of errors with time are examined. The method does not accommodate the effects of boundary conditions; but it is simple, straightforward and can readily be extended to multidimensional problems.

Consider the one-dimensional transient heat conduction equation (12-96) expressed in finite-difference form by using the explicit method

$$\frac{T_j^{n+1} - T_j^n}{\Delta t} = \alpha \frac{T_{j-1}^n - 2T_j^n + T_{j+1}^n}{(\Delta x)^2} \tag{12-110}$$

where the subscript j is the discretization index for the space variable (i.e., $x = j\Delta x$) and the superscript n for the time variable (i.e., $t = n\Delta t$). The numerical solution, T_N, of the problem can be written as the sum of the *exact solution*, T_E, of the problem and an error term ε in the form

$$T_N = T_E + \epsilon \tag{12-111}$$

where the numerical solution must satisfy the difference equation (12-110). Substitution of equation (12-111) into the difference equation (12-110) and noting that T_E should also satisfy the difference equation, we obtain

$$\frac{\epsilon_j^{n+1} - \epsilon_j^n}{\Delta t} = \alpha \frac{\epsilon_{j-1}^n - 2\epsilon_j^n + \epsilon_{j+1}^n}{(\Delta x)^2} \tag{12-112}$$

Machine computations introduce error almost at every stage of the calculations. Assume that the errors introduced at pivotal points along the initial (i.e., $t = 0$) line could be expressed in a finite Fourier series in terms of sine–cosine or complex exponentials. Here we prefer to use the latter. To examine the propagation of errors as time increases, one needs to consider only a single term in the series, because the finite-difference equations are linear. With these considerations one needs to examine the propagation of error due to a single term expressed in the form

$$\epsilon(j\Delta x, n\Delta t) \equiv \epsilon_j^n = e^{\gamma n\Delta t} \cdot e^{i\beta_m j\Delta x} \tag{12-113}$$

where $i = \sqrt{-1}$, β_m are the Fourier modes, γ is in general a complex quantity, $n\Delta t = t$, and $j\Delta x = x$. This equation is expressed in the form

$$\epsilon_j^n = \xi^n e^{i\beta_m j\Delta x} \tag{12-114a}$$

Similarly, we write

$$\epsilon_{j\pm 1}^n = \xi^n e^{i\beta_m (j\pm 1)\Delta x} \tag{12-114b}$$

$$\epsilon_j^{n+1} = \xi^{n+1} e^{i\beta_m j\Delta x} \tag{12-114c}$$

where

$$\xi = e^{\gamma\Delta t} \tag{12-115}$$

For this definition of ξ, the error term ε_j^n will not increase without a bound as t increases if

$$|\xi| \leqslant 1 \tag{12-116}$$

We substitute the error terms given by equations (12-114) into equation (12-112)

$$e^{i\beta_m j\Delta x}(\xi^{n+1} - \xi^n) = \frac{\alpha\Delta t}{(\Delta x)^2} e^{i\beta_m j\Delta x}\xi^n(e^{i\beta_m \Delta x} - 2 + e^{-i\beta_m\Delta x}) \tag{12-117a}$$

and after cancellation and some rearrangement we obtain

$$\xi - 1 = 2\frac{\alpha\Delta t}{(\Delta x)^2}\left(\frac{e^{i\beta_m\Delta x} + e^{-i\beta_m\Delta x}}{2} - 1\right) \tag{12-117b}$$

Noting that

$$\cos(\beta_m\Delta x) = \frac{e^{i\beta_m\Delta x} + e^{-i\beta_m\Delta x}}{2} \tag{12-118}$$

Equation (12-117) is written as

$$\xi = 1 - 2r(1 - \cos\beta_m\Delta x) \tag{12-119a}$$

where

$$r = \frac{\alpha\Delta t}{(\Delta x)^2} \tag{12-119b}$$

Here the parameter ξ is called the *amplification factor*. Recalling its definition by equation (12-115), the initial errors will not be amplified and the finite-difference calculations remain stable if the condition $|\xi| \leqslant 1$ is satisfied for all values of β_m. Applying this restriction to equation (12-119a) we obtain

$$|1 - 2r(1 - \cos\beta_m\Delta x)| \leqslant 1 \tag{12-120a}$$

or

$$-1 \leqslant \{1 - 2r(1 - \cos\beta_m\Delta x)\} \leqslant 1 \tag{12-120b}$$

which must be satisfied for all possible Fourier modes β_m. The right-hand side of this inequality is satisfied for all possible values of β_m. To satisfy the left-hand side under most strict conditions we must have $1 - \cos \beta_m \Delta x = 2$. Then we have

$$-1 \leqslant (1 - 4r) \quad \text{or} \quad r = \frac{\alpha \Delta t}{(\Delta x)^2} \leqslant \frac{1}{2} \tag{12-121a,b}$$

which is the criteria for stable solution of the explicit finite-difference equation (12-99a)

Implicit Method

The explicit method discussed previously is simple computationally, but the maximum size of the time step is restricted by stability considerations. If calculations are to be performed over a large period of time, the number of steps, hence the number of calculations needed may become prohibitively large. To alleviate this difficulty, finite-difference schemes that are not restrictive to the size of the time step Δt have been developed. One such method is the implicit method. We consider the one-dimensional diffusion equation

$$\frac{\partial T}{\partial t} = \alpha \frac{\partial^2 T}{\partial x^2} \tag{12-122}$$

The finite-difference representation of this equation with the implicit scheme is given by

$$\frac{T_i^{n+1} - T_i^n}{\Delta t} = \alpha \frac{T_{i-1}^{n+1} - 2T_i^{n+1} + T_{i+1}^{n+1}}{(\Delta x)^2} \tag{12-123}$$

which is accurate to $O[(\Delta x)^2, \Delta t]$ and unconditionally stable. This is an implicit scheme, because at each time level algebraic equations are to be solved simultaneously in order to determine the nodal temperatures at the next time level.

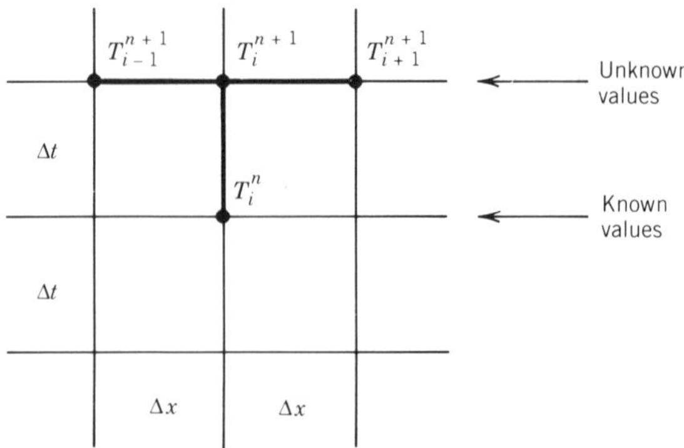

Fig. 12-14 The finite-difference molecules for the simple implicit scheme.

Figure 12-14 illustrates the expansion point $(i, n+1)$ and the surrounding finite-difference molecules. If the problem involves M unknown node temperatures, a simultaneous solution of M equations is required at each time step. Such a solution procedure is more involved computationally than that of the explicit scheme; but the method is advantageous in that there is no restriction on the size of the time step Δt by the stability considerations.

Stability Analysis. We apply the Fourier method of stability analysis to demonstrate that the simple implicit scheme is unconditionally stable.

As discussed previously, the numerical solution T_{N} is the sum of the exact solution, T_{E}, of the problem, plus an error term ϵ, given in the form

$$T_{\mathrm{N}} = T_{\mathrm{E}} + \epsilon \tag{12-124}$$

We introduce equation (12-124) into (12-123) and note that T_{E} should also satisfy the difference equation. We obtain

$$\frac{\epsilon_j^{n+1} - \epsilon_j^n}{\Delta t} = \alpha \frac{\epsilon_{j-1}^{n+1} - 2\epsilon_j^{n+1} + \epsilon_{j+1}^{n+1}}{(\Delta x)^2} \tag{12-125}$$

where we replaced the space variable index i by j. The error terms ϵ_j^n are represented as given by equations (12-114). Introducing the values ϵ_j from equation (12-114) into (12-125) and after cancellations and some rearrangement, we obtain

$$\xi - 1 = \frac{2\alpha\,\Delta t}{(\Delta x)^2}\,\xi\left(\frac{e^{i\beta_m\Delta x} + e^{-i\beta_m\Delta x}}{2} - 1\right) \tag{12-126}$$

where $i = \sqrt{-1}$. Noting that

$$\cos(\beta_m \Delta x) = \frac{e^{i\beta_m\Delta x} + e^{-i\beta_m\Delta x}}{2} \tag{12-127}$$

equation (12-126) is written as

$$\xi - 1 = 2r\xi(\cos\beta_m\Delta x - 1) \tag{12-128}$$

or

$$\xi - 1 = -4r\xi\sin^2\left(\frac{\beta_m\Delta x}{2}\right) \tag{12-129a}$$

where

$$r = \frac{\alpha\Delta t}{(\Delta x)^2} \tag{12-129b}$$

Equation (12-129a) is solved for ξ:

$$\xi = \left[1 + 4r \sin^2\left(\frac{\beta_m \Delta x}{2} \right) \right]^{-1} \tag{12-130}$$

For stability we need $|\xi| \leqslant 1$ and this condition is satisfied for all positive values of r. Therefore, the simple implicit finite-difference approximation is stable for all values of the time step Δt. However, the time step Δt must be kept reasonably small to obtain results sufficiently close to the exact solution of the partial-differential equation.

Combined Method

A combination of the explicit method given by equation (12-110) and the implicit method given by equation (12-123) is written as

$$\frac{T_i^{n+1} - T_i^n}{\Delta t} = \alpha \left[\theta \frac{T_{i-1}^{n+1} - 2T_i^{n+1} + T_{i+1}^{n+1}}{(\Delta x)^2} + (1-\theta) \frac{T_{i-1}^n - 2T_i^n + T_{i+1}^n}{(\Delta x)^2} \right] \tag{12-131}$$

where the constant $\theta (0 \leqslant \theta \leqslant 1)$ is the weight factor which represents the degree of implicitness. That is, equation (12-131) reduces to the simple explicit form for $\theta = 0$, to the Crank–Nicolson method for $\theta = \frac{1}{2}$ and to the simple implicit form for $\theta = 1$.

The order of accuracy of various difference schemes corresponding to specific values of θ are given by:

1. $\theta = 0$, the explicit method: $O[\Delta t, (\Delta x)^2]$
2. $\theta = 1$, the implicit method: $O[\Delta t, (\Delta x)^2]$
3. $\theta = \frac{1}{2}$, the Crank–Nicolson method: $O[(\Delta t)^2, (\Delta x)^2]$
4. $\theta = \dfrac{1}{2} - \dfrac{(\Delta x)^2}{12\alpha\,\Delta t}$: $O[(\Delta t)^2, (\Delta x)^4]$

Clearly, finite-difference schemes of various degree of accuracy are obtainable from the combined method by proper choice of the value of the weight factor θ.

The stability criterion for the combined method depends on the value of the weight factor θ as given below:

$$\tfrac{1}{2} \leqslant \theta \leqslant 1: \text{ unconditionally stable for all values of } r \tag{12-132a}$$

$$0 \leqslant \theta < \tfrac{1}{2}: \text{ stable only if } 0 \leqslant r \leqslant \frac{1}{2 - 4\theta} \tag{12-132b}$$

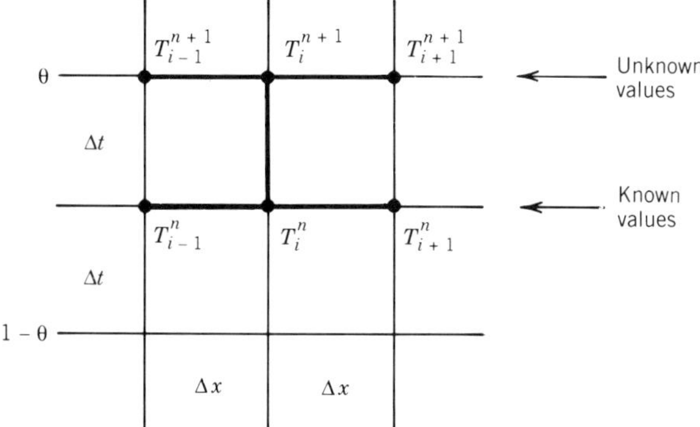

Fig. 12-15 The finite-difference molecules for the combined scheme.

where $r = (\alpha \Delta t)/(\Delta x)^2$. Figure 12-15 shows the finite-difference molecules for the combined method.

To solve equation (12-131), all the unknown temperatures T^{n+1} are moved on one side and all the known temperatures T^n on the other side. We obtain

$$\begin{aligned} -r\theta T_{i-1}^{n+1} &+ (1+2r\theta)T_i^{n+1} - r\theta T_{i+1}^{n+1} \\ &= r(1-\theta)T_{i-1}^{n} + [1-2r(1-\theta)]T_i^n + r(1-\theta)T_{i+1}^n \end{aligned} \tag{12-133}$$

where $r = (\alpha \Delta t)/(\Delta x)^2$. The resulting system of equations (12-133) have a tridiagonal linear coefficient matrix, hence can be solved with any one of the algorithms discussed previously.

When temperatures are prescribed at all boundaries, the system (12-133) provides complete set of algebraic equations for the determination of all the unknown internal node temperatures. With convection or prescribed heat flux boundary conditions, the temperatures at the boundary nodes are not known. Additional equations are obtained by either discretizing the boundary condition directly about the boundary node or the application of conservation principle for a control volume about the boundary node.

12-11 MULTIDIMENSIONAL TIME-DEPENDENT HEAT CONDUCTION

The finite-difference schemes such as the *explicit*, *implicit*, *Crank–Nicolson*, and *combined* methods presented previously with applications for the solution of one-dimensional transient heat conduction problems can readily be generalized for the solution of multidimensional transient heat conduction problems. Consider, for example, that a three-dimensional transient heat conduction problem is to be solved with an implicit method in order to alleviate the restriction imposed on the size of the permissible time step and suppose that there are N

interior nodes in each direction. Then the $N^3 \times N^3$ matrix must be solved for each time level and the procedure becomes impractical for large N. To alleviate such difficulties various alternative approaches have been proposed for the solution of multidimensional transient heat conduction problems. They include, among others, *alternating-direction-implicit* (ADI) methods advanced by Peaceman and Rachford [4], Douglas and Gun [43], Douglas [44], and a closely related method described by Yanenko [45].

Alternating-direction-explicit (ADE) methods have been proposed by Saul'yev [47], Barakat and Clark [46], Larkin [12], and Allada and Quon [13]. Several alternative schemes have also been proposed [48–50].

In this section we first illustrate the generalization of the *explicit* and *combined* methods for the solution of multidimensional transient heat conduction, and then present the ADI method applied for the solution of two-dimensional transient heat conduction.

Explicit Method Applied to Two-Dimensional Heat Conduction

Consider two-dimensional transient heat conduction equation with energy generation in the rectangular coordinate system taken as

$$\frac{\partial T}{\partial t} = \alpha\left(\frac{\partial^2 T}{\partial x^2} + \frac{\partial^2 T}{\partial y^2} + \frac{1}{k}g\right) \tag{12-134}$$

where $T \equiv T(x, y, t)$ and $g \equiv (x, y, t)$, subject to some specified boundary and initial conditions. To discretize this differential equation we introduce the notation

$$T(x, y, t) = T(i\Delta x, j\Delta y, n\Delta t) \equiv T_{i,j}^{n} \tag{12-135}$$

Then, the finite-difference approximation of the differential equation (12-134) at a grid point (x, y) by the simple explicit method using forward-time–central-space (FTCS) discretization gives

$$\frac{T_{i,j}^{n+1} - T_{i,j}^{n}}{\Delta t} = \alpha\left[\frac{T_{i-1,j}^{n} - 2T_{i,j}^{n} + T_{i+1,j}^{n}}{(\Delta x)^2} + \frac{T_{i,j-1}^{n} - 2T_{i,j}^{n} + T_{i,j+1}^{n}}{(\Delta y)^2} + \frac{1}{k}g_{i,j}^{n}\right] \tag{12-136}$$

This expression is rearranged in the form

$$T_{i,j}^{n+1} = T_{i,j}^{n} + r_x(T_{i-1,j}^{n} - 2T_{i,j}^{n} + T_{i+1,j}^{n}) + r_y(T_{i,j-1}^{n} - 2T_{i,j}^{n} + T_{i,j+1}^{n}) + \frac{\alpha\Delta t}{k}g_{i,j}^{n} \tag{12-137a}$$

where

$$r_x \equiv \frac{\alpha\Delta t}{(\Delta x)^2}, \quad r_y \equiv \frac{\alpha\Delta t}{(\Delta y)^2} \tag{12-137b}$$

For a square mesh $\Delta x = \Delta y = \delta$, equation (12-137a) reduces to

$$T_{i,j}^{n+1} = r(T_{i-1,j}^n + T_{i+1,j}^n + T_{i,j-1}^n + T_{i,j+1}^n) + (1-4r)T_{i,j}^n + rG_{ij}^n \qquad \text{(12-138a)}$$

where

$$r \equiv \frac{\alpha \Delta t}{\delta^2}, \quad G_{ij}^n \equiv \frac{\delta^2 g_{i,j}^n}{k} \qquad \text{(12-138b)}$$

Equation (12-137) or (12-138) provides explicit expression for the determination of $T_{i,j}^{n+1}$ at the time level $n+1$ from the knowledge of grid-point temperatures at the previous time level n. If temperature is prescribed at all boundaries, the number of equations are equal to the number of unknown grid temperatures; hence the problem is soluble.

For derivative boundary conditions, such as convection or prescribed heat flux, the temperatures at the boundary nodes are not known. For such cases, additional relations are developed by discretizing the boundary conditions. The discretization of the derivative term in the boundary condition can be made either by one-sided differences by using a backward or forward formula that is only first-order accurate. A second-order accurate discretization of the boundary condition is possible by introducing a fictitious node and using a central-difference formula. Alternatively, the control volume approach and conservation principle can be used to develop finite-difference approximation for the boundary conditions.

Stability. To obtain meaningful results from the solution of the difference equations (12-137a), the stability criterion associated with them should be established. We rewrite equation (12-137a) in the form

$$T_{j,k}^{n+1} = T_{j,k}^n + r_x(T_{j-1,k}^n - 2T_{j,k}^n + T_{j+1,k}^n) + r_y(T_{j,k-1}^n - 2T_{j,k}^n + T_{j,k+1}^n) \qquad \text{(12-139a)}$$

where

$$r_x = \frac{\alpha \Delta t}{(\Delta x)^2}, \quad r_y = \frac{\alpha \Delta t}{(\Delta y)^2} \qquad \text{(12-139b)}$$

Here the generation term is neglected because it does not influence the growth and propagation of errors, and the subscript i is replaced by j in order to distinguish the subscript from $i = \sqrt{-1}$, which will appear in the analysis.

The Fourier stability analysis described previously is now generalized for the two-dimensional case considered here by choosing the error term in the form

$$\varepsilon_{j,k}^n = \xi^n e^{i\beta_m j \Delta x} e^{i\eta_n k \Delta y} \quad \text{where} \quad \xi \equiv e^{\gamma \Delta t} \qquad \text{(12-140)}$$

and $i=\sqrt{-1}$, β_m and η_n are the Fourier modes. In view of the definition of ξ, the error term $\varepsilon_{j,k}^n$ will not increase without bound as t increases, provided that

$$|\xi| \leqslant 1 \tag{12-141}$$

The error term should also satisfy the finite defference equation (12-139a). Therefore, equation (12-140) is substituted into equation (12-139a) and after cancellations we obtain

$$\xi = 1 + r_x(e^{-i\beta_m\Delta x} + e^{i\beta_m\Delta x} - 2) + r_y(e^{-i\eta_n\Delta y} + e^{i\eta_n\Delta y} - 2) \tag{12-142}$$

which can be written as

$$\xi = 1 - 2r_x(1 - \cos\beta_m\Delta x) - 2r_y(1 - \cos\eta_n\Delta y) \tag{12-143}$$

since $\cos z = \frac{1}{2}(e^{-iz} + e^{iz})$.

The application of the stability criterion equation (12-141) to equation (12-143) yields

$$-1 \leqslant [1 - 2r_x(1 - \cos\beta_m\Delta x) - 2r_y(1 - \cos\eta_n\Delta y)] \leqslant 1$$

which must be satisfied for all values of β_m and η_n. The right-hand side is satisfied always. To satisfy the left-hand side under most strict conditions we must have $1 - \cos\beta_m\Delta x = 2$ and $1 - \cos\eta_n\Delta y = 2$, yielding

$$-1 \leqslant [1 - 4r_x - 4r_y] \tag{12-144a}$$

or

$$(r_x + r_y) \leqslant \tfrac{1}{2} \tag{12-144b}$$

or

$$\left[\frac{\alpha\Delta t}{(\Delta x)^2} + \frac{\alpha\Delta t}{(\Delta y)^2}\right] \leqslant \frac{1}{2} \tag{12-144c}$$

For the case $\Delta x = \Delta y = \delta$, the stability criterion becomes

$$r = \frac{\alpha\Delta t}{\delta^2} \leqslant \frac{1}{4} \tag{12-145}$$

which is as twice restrictive as the one-dimensional constraint $r \leqslant \frac{1}{2}$.

Example 12-6

Develop the stability criterion for the finite-difference approximation by the simple explicit method of the three-dimensional linear diffusion equation in the x, y, z rectangular coordinates.

Solution. The finite-difference equation (12-139) and the corresponding error term equation (12-140) are readily generalized to the three-dimensional case. The error term is substituted into the finite-difference equation and a procedure similar to that described previously is applied. The stability criterion

$$(r_x + r_y + r_z) \leqslant \tfrac{1}{2} \tag{12-146a}$$

or

$$\left[\frac{\alpha \Delta t}{(\Delta x)^2} + \frac{\alpha \Delta t}{(\Delta y)^2} + \frac{\alpha \Delta t}{(\Delta z)^2}\right] \leqslant \frac{1}{2} \tag{12-146b}$$

results. For the case $\Delta x = \Delta y = \Delta z = \delta$, the stability criterion becomes

$$r = \frac{\alpha \Delta t}{\delta^2} \leqslant \frac{1}{6} \tag{12-146c}$$

which is thrice as restrictive as the one-dimensional constraint $r \leqslant \frac{1}{2}$.

Combined Method Applied to Three-Dimensional Diffusion

We consider a three-dimensional linear diffusion problem in an isotropic solid governed by the partial-differential equation

$$\frac{\partial T}{\partial t} = \alpha\left(\frac{\partial^2 T}{\partial x^2} + \frac{\partial^2 T}{\partial y^2} + \frac{\partial^2 T}{\partial z^2}\right) \tag{12-147}$$

with appropriate boundary and initial conditions. To discretize this equation we introduce the notation

$$T(x, y, z, t) = T(i\Delta x, j\Delta y, k\Delta z, n\Delta t) \equiv T^n_{i,j,k} \tag{12-148}$$

Then, the finite-difference approximation of the differential equation (12-147) with the combined method, by using FTCS, becomes

$$\begin{aligned}\frac{T^{n+1}_{i,j,k} - T^n_{i,j,k}}{\alpha \Delta t} &= \theta[\Delta_{xx} T^{n+1}_{i,j,k} + \Delta_{yy} T^{n+1}_{i,j,k} + \Delta_{zz} T^{n+1}_{i,j,k}] \\ &\quad + (1-\theta)[\Delta_{xx} T^n_{i,j,k} + \Delta_{yy} T^n_{i,j,k} + \Delta_{zz} T^n_{i,j,k}]\end{aligned} \tag{12-149}$$

where the weight factor θ assumes values $0 \leqslant \theta \leqslant 1$, and the finite-difference operators Δ_{xx}, Δ_{yy}, and Δ_{zz} are defined as

$$\Delta_{xx} T^n_{i,j,k} \equiv \frac{1}{(\Delta x)^2}[T^n_{i+1,j,k} - 2T^n_{i,j,k} + T^n_{i-1,j,k}] \tag{12-150a}$$

$$\Delta_{yy} T^n_{i,j,k} \equiv \frac{1}{(\Delta y)^2}[T^n_{i,j+1,k} - 2T^n_{i,j,k} + T^n_{i,j-1,k}] \tag{12-150b}$$

$$\Delta_{zz} T^n_{i,j,k} \equiv \frac{1}{(\Delta z)^2}[T^n_{i,j,k+1} - 2T^n_{i,j,k} + T^n_{i,j,k-1}] \tag{12-150c}$$

Clearly, depending on the value chosen for the weight factor θ, the simple explicit, the simple implicit, and the Crank–Nicolson methods are readily obtained as special cases:

$\theta = 0$: The simple explicit scheme. The truncation error is $0[\Delta t, (\Delta x)^2, (\Delta y)^2, (\Delta z)^2]$ and the stability constraint on the time step Δt is

$$\left[\frac{\alpha \Delta t}{(\Delta x)^2} + \frac{\alpha \Delta t}{(\Delta y)^2} + \frac{\alpha \Delta t}{(\Delta z)^2}\right] \leqslant \frac{1}{2} \tag{12-151}$$

$\theta = \frac{1}{2}$: The Crank–Nicolson scheme. The truncation error is $O[(\Delta t)^2, (\Delta x)^2, (\Delta y)^2, (\Delta z)^2]$ and the finite-difference equations are unconditionally stable.

$\theta = 1$: The simple implicit scheme. The truncation error is $O[\Delta t, (\Delta x)^2, (\Delta y)^2, (\Delta z)^2]$ and the finite-difference equations are unconditionally stable.

For values of $0.5 \leqslant \theta \leqslant 1$, the scheme is unconditionally stable.

ADI Method Applied to Two-Dimensional Heat Conduction

We now present the alternating-direction-implicit (ADI) method for the solution of two-dimensional transient heat conduction in the rectangular coordinates. The principal advantage of the method lies in the fact that, the size of the matrix to be solved in each time level is reduced at the expense of solving a reduced matrix many times.

We consider the following transient heat conduction equation

$$\frac{1}{\alpha}\frac{\partial T}{\partial t} = \frac{\partial^2 T}{\partial x^2} + \frac{\partial^2 T}{\partial y^2} + \frac{1}{k} g(x, y, t) \tag{12-152}$$

subject to appropriate boundary and initial conditions, and introduce the notation

$$T(x, y, t) = T(i\Delta x, j\Delta y, n\Delta t) \equiv T^n_{i,j} \tag{12-153}$$

The finite-difference approximation of the differential equation (12-152) with the ADI method is based on the following concepts.

Suppose the computations are to be advanced from the (n)th time level to the $(n+1)$th time level. The simple *implicit method* is used for one of the directions, say, x, and the simple *explicit method* is used for the other direction, y. Then, the advancement from the $(n+1)$th level to the $(n+2)$th level is done by reversing the directions of the implicit and explicit methods. The computational procedure is continued by alternatively changing the directions of the explicit and implicit methods.

We now illustrate the application of the ADI method for the discretization of equation (12-152). Suppose the implicit scheme is used in the x direction and the explicit scheme in the y direction to advance from the nth to the $(n+1)$th time level. The finite-difference approximation of equation (12-152) is given by

$$\frac{T_{i,j}^{n+1}-T_{i,j}^{n}}{\alpha\Delta t}=\frac{T_{i-1,j}^{n+1}-2T_{i,j}^{n+1}+T_{i+1,j}^{n+1}}{(\Delta x)^2}+\frac{T_{i,j-1}^{n}-2T_{i,j}^{n}+T_{i,j+1}^{n}}{(\Delta y)^2}+\frac{1}{k}\bar{g}_{i,j} \tag{12-154a}$$

where $\bar{g}_{i,j}$ is the average of $g_{i,j}$ for the current and next time steps.

For the next time level, an *explicit* formulation is used for the x direction and an *implicit* formulation for the y direction. Then, the finite-difference approximation for equation (12-152) from the $(n+1)$th to the $(n+2)$nd time step becomes

$$\frac{T_{i,j}^{n+2}-T_{i,j}^{n+1}}{\alpha\Delta t}=\frac{T_{i-1,j}^{n+1}-2T_{i,j}^{n+1}+T_{i+1,j}^{n+1}}{(\Delta x)^2}+\frac{T_{i,j-1}^{n+2}-2T_{i,j}^{n+2}+T_{i,j+1}^{n+2}}{(\Delta y)^2}+\frac{1}{k}\bar{g}_{i,j} \tag{12-154b}$$

This equation utilizes the results from the previous time level $n+1$ to calculate the temperatures at the time level $n+2$.

For computational purposes, it is convenient to rearrange equations (12-154a) and (12-154b) such that at each time level, the unknown quantities appear on one side of the equality, say, on the left and the known quantities on the other side, on the right. Equations (12-154a) and (12-154b), respectively, become

$$\begin{aligned}-r_x T_{i-1,j}^{n+1}+(1+2r_x)T_{i,j}^{n+1}-r_x T_{i+1,j}^{n+1}&=r_y T_{i-1,j}^{n}+(1-2r_y)T_{i,j}^{n}\\&\quad+r_y T_{i+1,j}^{n}+\frac{\alpha\Delta t}{k}\bar{g}_{i,j}\end{aligned} \tag{12-155a}$$

for the time level $n+1$ and

$$\begin{aligned}-r_y T_{i,j-1}^{n+2}+(1+2r_y)T_{i,j}^{n+2}-r_y T_{i,j+1}^{n+2}&=r_x T_{i-1,j}^{n+1}+(1-2r_x)T_{i,j}^{n+1}\\&\quad+r_x T_{i+1,j}^{n+1}+\frac{\alpha\Delta t}{k}\bar{g}_{i,j}\end{aligned} \tag{12-155b}$$

for the time level $n + 2$, where

$$r_x = \frac{\alpha \Delta t}{(\Delta x)^2} \quad \text{and} \quad r_y = \frac{\alpha \Delta t}{(\Delta y)^2} \tag{12-156}$$

When solving the problem, equations (12-155a) and (12-155b) are repeated alternatively.

The advantage of this approach over the fully implicit or Crank–Nicolson methods is that, each equation, although implicit, is only tridiagonal. That is, equation (12-155a) contains implicit unknowns $T_{i,j}^{n+1}$, $T_{i-1,j}^{n+1}$, and $T_{i+1,j}^{n+1}$, while equation (12-155b) contains implicit unknowns $T_{i,j}^{n+2}$, $T_{i,j-1}^{n+2}$, and $T_{i,j+1}^{n+2}$. Therefore, the coefficient matrix is tridiagonal for each equation; hence the computation scheme is more efficient than those that are not tridiagonal.

If temperatures are prescribed at all boundaries, equations (12-155a,b) are sufficient to determine the unknown internal node temperatures.

For convection, prescribed heat flux boundary conditions, temperatures at the boundary nodes are unknown. For such cases, additional relations are obtained by applying the conservation principle to a control volume about each boundary node at which the node temperature is not known.

12-12 NONLINEAR HEAT CONDUCTION

In principle, there is no difficulty in applying finite-difference methods to nonlinear parabolic systems; but the resulting finite-difference equations become nonlinear and difficult to solve. The diffusion-type problems become nonlinear due to the nonlinearity of the governing differential equation, or the boundary condition or both. Consider, for example, the heat conduction equation

$$\nabla \cdot [k(T) \nabla T] + g(T) = \rho C_p(T) \frac{\partial T}{\partial t} \quad \text{in region } R, \quad t > 0 \tag{12-157}$$

which is nonlinear because the thermal properties and the energy-generation term depend on temperature. Consider the boundary condition given in the form

$$\frac{\partial T}{\partial n} = f(T_s) \quad \text{on boundary} \tag{12-158}$$

where $\partial/\partial n$ is the derivative along the outward-drawn normal to the boundary surface and T_s is the boundary surface temperature. This boundary condition becomes nonlinear if the function $f(T_s)$ involves a power of T_s, as in the case of radiation boundary condition

$$k \frac{\partial T}{\partial n} = \epsilon \sigma (T^4 - T_\infty^4) \tag{12-159}$$

or the natural-convection boundary condition

$$k\frac{\partial T}{\partial n} = c(|T - T_{\infty}|)^{1/4}(T - T_{\infty}) \tag{12-160}$$

where ϵ is the emissivity, σ is the Stefan–Boltzmann constant, and T_{∞} is the ambient temperature with which radiation or free convection takes place.

Various schemes are available for finite-difference approximation of nonlinear diffusion problems as a system of linear algebraic equations. They include, among others, the lagging of temperature-dependent properties by one time step, the use of three-time-level finite-differencing, and the linearization procedures. Here we consider the simplest procedure the lagging of temperature dependent properties by one time step.

Lagging of Properties by One Time Step and Extrapolation Schemes

We consider the nonlinear diffusion equation given in the form

$$\rho C_p(T)\frac{\partial T}{\partial t} = \frac{\partial}{\partial x}\left[k(T)\frac{\partial T}{\partial x}\right] \tag{12-161}$$

where the specific heat $C_p(T)$ and the thermal conductivity $k(T)$ vary with temperature. This equation can be discretized by using any one of the finite-difference schemes described previously. Here we prefer to use the *combined method* because of its versatility to yield the simple explicit, simple implicit, Crank–Nicolson, and other methods merely by the adjustment of a coefficient. The finite-difference representation of equation (12-161) with the combined method is given by

$$\begin{aligned}(\rho C_p)_i\frac{T_i^{n+1} - T_i^n}{\Delta t} &= \theta\left[k_{i-1/2}\frac{T_{i-1}^{n+1} - T_i^{n+1}}{(\Delta x)^2} + k_{i+1/2}\frac{T_{i+1}^{n+1} - T_i^{n+1}}{(\Delta x)^2}\right] \\ &\quad + (1-\theta)\left[k_{i-1/2}\frac{T_{i-1}^{n} - T_i^{n}}{(\Delta x)^2} + k_{i+1/2}\frac{T_{i+1}^{n} - T_i^{n}}{(\Delta x)^2}\right]\end{aligned} \tag{12-162}$$

where the constant $\theta (0 \leqslant \theta \leqslant 1)$ is the weight factor that represents the degree of implicitness. The values of $\theta = 0, \frac{1}{2}$, and 1, respectively, correspond to the explicit, Crank–Nicolson, and implicit schemes.

We note that the thermal properties $(\rho C_p)_i$ and $k_{i \pm 1/2}$ depend on temperature; but at this stage of the analysis it is not yet specified how they will be computed. This matter will be discussed later on.

Equation (12-162) can be written more compactly in the form

$$\begin{aligned}T_i^{n+1} - T_i^n &= \theta[A_i T_{i-1}^{n+1} - 2B_i T_i^{n+1} + D_i T_{i+1}^{n+1}] \\ &\quad + (1-\theta)[A_i T_{i-1}^{n} - 2B_i T_i^{n} + D_i T_{i+1}^{n}]\end{aligned} \tag{12-163}$$

where

$$A_i = \frac{k_{i-1/2}}{(\rho C_p)_i} \frac{\Delta t}{(\Delta x)^2} \tag{12-164a}$$

$$D_i = \frac{k_{i+1/2}}{(\rho C_p)_i} \frac{\Delta t}{(\Delta x)^2} \tag{12-164b}$$

$$B_i = \frac{1}{2}(A_i + D_i) = \frac{1}{2} \frac{k_{i-1/2} + k_{i+1/2}}{(\rho C_p)_i} \frac{\Delta t}{(\Delta x)^2} \tag{12-164c}$$

Equation (12-163) is now rearranged so that all unknown temperatures (i.e., those at the time level $n+1$) appear on one side and all the known temperatures (i.e., those at the time level n) on the other side.

$$\begin{aligned} &-\theta A_i T_{i-1}^{n+1} + (1 + 2\theta B_i) T_i^{n+1} - \theta D_i T_{i+1}^{n+1} \\ &\quad = (1-\theta) A_i T_{i-1}^n + [1 - 2(1-\theta)B_i] T_i^n + (1-\theta) D_i T_{i+1}^n \end{aligned} \tag{12-165}$$

We note that for the case of constant thermal properties we have

$$A_i = B_i = D_i = \frac{k}{\rho C_p} \frac{\Delta t}{(\Delta x)^2} = \frac{\alpha \Delta t}{(\Delta x)^2} \equiv r \tag{12-166}$$

and equation (12-165) reduces to the linear case given by equation (12-133). Assuming that the coefficients A_i, D_i, and B_i are available, the system (12-165) provides a complete set of equations for the determination of the unknown internal node temperatures when the temperatures at the boundary surfaces are prescribed. For the case of prescribed heat flux or convection boundary conditions, the temperatures at the boundaries are unknown; additional relations are developed by discretizing such boundary conditions. Since equations (12-165) have a tridiagonal coefficient matrix, any one of the algorithms discussed in Chapter 3 can be used for their solution provided that the coefficients A_i, B_i, and D_i are known at the time level $n+1$. The following approaches can be used to compute the properties.

The simplest, but less accurate, method for computing these coefficients is to lag the evaluation of the temperature-dependent properties by one time step. That is, to perform the computations at the time level $n+1$, the coefficients are evaluated at the previous time level n:

$$A_i \equiv A_i^n, \quad B_i \equiv B_i^n, \quad D_i \equiv D_i^n \tag{12-167a}$$

The thermal conductivity $k_{i \pm 1/2}$ can be evaluated at the time level n, either at the average temperature $(T_{i \pm 1}^n + T_i^n)/2$, or as the average of the thermal conductivities at the nodes i and $i \pm 1$:

$$k_{i \pm 1/2} = \frac{k_i^n + k_{i \pm 1}^n}{2} \tag{12-167b}$$

A more accurate approach for calculating the temperature dependent properties is the use of an extrapolation scheme as described below.

Consider, for example, that the thermal conductivity k^{n+1} at the time level $n+1$ is expanded in term of that at the time level n in the form

$$\begin{aligned} k^{n+1} &\cong k^n + \left(\frac{\partial k}{\partial t}\right)^n \Delta t \\ &\cong k^n + \left(\frac{\partial k}{\partial T}\right)^n \left(\frac{\partial T}{\partial t}\right)^n \Delta t \end{aligned} \tag{12-168a}$$

The time derivative of temperature is approximated by

$$\left(\frac{\partial T}{\partial t}\right)^n \cong \frac{T^n - T^{n-1}}{\Delta t} \tag{12-168b}$$

Introducing equation (12-168b) into (12-168a), the following expression is obtained for the determination of thermal conductivity at the time level $n+1$ from the knowledge of k^n:

$$k^{n+1} = k^n + \left(\frac{\partial k}{\partial T}\right)^n (T^n - T^{n-1}) \tag{12-169}$$

A similar expression can be written for the specific heat:

$$C_p^{n+1} = C_p^n + \left(\frac{\partial C_p}{\partial T}\right)^n (T^n - T^{n-1}) \tag{12-170}$$

Clearly, if the second terms on the right-hand side of equation (12-169) and (12-170) are neglected, the result is equivalent to the lagging of the coefficients.

The computation time for solving the system of equations (12-165) resulting from the nonlinear differential equation is longer than that resulting from the linear system, because the temperature-dependent thermal properties $k(T)$ and $C_p(T)$ need to be evaluated at each time step $n+1$ from their known values at the time level n according to equations (12-169) and (12-170).

REFERENCES

1. R. V. Southwell, *Relaxation Methods in Engineering Science*, Oxford University Press, London, 1940.

2. J. Crank and P. Nicolson, *Proc. Camb. Phil. Soc.* **43**, 50–67, 1947.
3. G. G. O'Brien, M. A. Hyman, and Kaplan, *J. Math. Phys.* **29**, 223–251, 1951.
4. D. W. Peaceman and H. H. Rachford, *J. Soc. Indust. Appl. Math.* **3**, 28–41, 1955.
5. R. D. Richtmeyer and K. W. Morton, *Difference Methods for Initial Value Problems*, 2nd ed., Interscience Publishers, New York, 1965.
6. G. E. Forsythe and W. R. Wasov, *Finite Difference Methods for Partial Differential Equations*, Wiley, New York, 1960.
7. G. M. Dusinberre, *Heat Transfer Calculations by Finite Differences*, International Textbook Company, Scranton, Pa., 1961.
8. L. Fox, *Numerical Solution of Ordinary and Partial Differential Equations*, Addison-Wesley, Reading, Mass., 1962.
9. G. D. Smith, *Numerical Solution of Partial Differential Equations with Exercises and Worked Solutions*, Oxford University Press, London, 1965.
10. J. K. Reid, ed., *Large Sparse Sets of Linear Equations*, Academic Press, New York, 1971.
11. D. J. Rose and R. A. Willoughby, eds., *Sparce Matrices and Their Applications*, Plenum Press, New York, 1972.
12. B. K. Larkin, *Chem. Eng. Prog. Sym. Ser.* **61**, 59, 1965.
13. S. R. Allada and D. Quon, *Heat Transfer Los Angeles, Chem. Eng. Prog. Sym. Ser.* **62**, (64), 151–156, 1966.
14. J. H. Wilkinson and C. Reinsch, *Handbook for Automatic Computation*, Vol. 2, *Linear Algebra*, Springer-Verlag, New York, 1971.
15. G. J. Trezek and J. G. Witwer, *J. Heat Transfer* **94c**, 321–323, 1972.
16. Y. Jaluria and K. E. Torrance, *Computational Heat Transfer*, Hemisphere Publishing, New York, 1986.
17. R. W. Hornbeck, *Numerical Methods*, Quantum Publishers, New York, 1975.
18. A. K. Aziz, ed., *Numerical Solutions of Boundary Value Problems for Ordinary Differential Equations*, Academic Press, New York, 1975.
19. G. E. Schneider, A. B. Strong, and M. M. Yovanovich, A. Physical Approach to the Finite Difference Solution of the Conduction Equation in Orthogonal Curvilinear Coordinates, ASME Paper No. 75-WA/HT-94, 1975.
20. N. D'Souza, Numerical Solution of One-Dimensional Inverse Transient Heat Conduction by Finite-Difference Method, ASME Paper No. 75-WA/HT-81, 1975.
21. M. L. James, G. M. Smith, and J. C. Wolford, *Applied Numerical Methods for Digital Computation*, IEP-Dun-Donnelley Publishers, New York, 1977.
22. N. D. Anderson, J. C. Tannehill, and R. H. Pletcher, *Computational Fluid Mechanics and Heat Transfer*, Hemisphere Publishing, New York, 1984.
23. W. J. Minkowycz, E. M. Sparrow, G. E. Schneider, and R. H. Pletcher, *Handbook of Numerical Heat Transfer*, Wiley-Interscience, New York, 1988.
24. T. M. Shih, *Numerical Heat Transfer*, Hemisphere Publishing, New York, 1984.
25. P. A. Roache, *Computational Fluid Dynamics*, Hermosa Publishers, Albuquerque, N.M., 1976.
26. R. Peyret and T. D. Taylor, *Computational Methods for Fluid Dynamics*, Springer-Verlag, New York, 1983.
27. S. V. Patankar, *Numerical Heat Transfer in Fluid Flow*, Hemisphere Publishing, New York, 1980.

28. A. J. Baker, *Finite Element Computational Fluid Mechanics*, Hemisphere Publishing, New York, 1983.
29. L. J. Segerlind, *Applied Finite Element Analysis*, Wiley, New York, 1976.
30. G. A. Ramirez and J. T. Oden, Finite-Element Technique Applied to Heat Conduction in Solids with Temperature Dependent Thermal Conductivity, ASME Paper No. 69-WA/HT-34, 1969.
31. E. F. Rybicki and A. T. Hopper, Higher Order Finite Element Method for Transient Temperature Analysis of Inhomogeneous Problems, ASME Paper No. 69-WA/HT-33, 1969.
32. O. C. Zienkiewicz and I. K. Cheung, *The Finite Element Method in Engineering Science*, McGraw-Hill, New York, 1971.
33. O. Ural, *Finite Element Method: Basic Concepts and Applications*, International Textbook Company, Scranton, Pa., 1973.
34. H. C. Martin and G. F. Carey, *Introduction of Finite-Element Analysis*, McGraw-Hill, New York, 1973.
35. K. H. Huebner, *Finite Element Method for Engineers*, Wiley, New York, 1975.
36. J. F. Thompson, *Elliptic Grid Generation*, Elsevier, New York, pp. 79–105, 1982.
37. J. F. Thompson, AIAA Paper No. 83-0447, AIAA 21st Aerospace Sciences Meeting, 1983.
38. J. F. Thompson, *AIAA J.* **22**, 1505–1523, 1984.
39. J. F. Thompson, *Appl. Num. Math.* **1**, 3, 1985.
40. J. F., Thompson, Z. U. A. Varsi, and C. W. Mastin, Boundary-Fitted Coordinate Systems for Numerical Solution of Partial Differential Equations—A Review, *J. Comp. Phys.* **48**, 1–108, 1982.
41. J. E. Thompson, Z. U. A. Varsi, and C. Mastin, *Numerical Grid Generation Foundations and Applications*. North-Holland, Elsevier Science Publishers, Amsterdam, Netherlands, 1985.
42. D. V. Griffiths and I. M. Smith, *Numerical Methods for Engineers*, CRC Press, Boston, 1991.
43. J. Douglas and J. E. Gun, *Numerische Mathematic* **6**, 428–453, 1964.
44. J. Douglas, *Numerische Mathematic* **4**, 41–63, 1962.
45. N. N. Yanenko, in *The Method of Fractional Steps: The Solution of Problem of Mathematical Physics in Several Variables*, M. Hold (Ed.), Springer-Verlag, New York, 1971.
46. H. Z. Barakat and J. A. Clark, *J. Heat Transfer*, **88**, 421–427, 1966.
47. V. K. Saul'yev, *Dokl. Akad. Nauk*, SSSR, **115**, 1077–1079, 1957 (Russian).
48. W. W. Yuen and L. W. Wong, *Num. Heat Transfer* **3**, 373–380, 1980.
49. B. P. P. J. Sommeljer, van der Houwen and J. G. Verwer, *Int. J. Num. Methods Eng.* **17**, 335–346, 1981.
50. D. J. Evans and G. Avdelas, *Itern. J. Computer Math, Section B* **6**, 335–358, 1978.

PROBLEMS

12-1 Using a Taylor's series expansion, show that a forward-difference representation of df/dx, which is accurate to the order of $O(h^3)$ is given in subscript notation as

$$f'_i = \frac{2f_{i+3} - 9f_{i+2} + 18f_{i+1} - 11f_i}{6h} + O(h^3)$$

12-2 Consider the function $f(x) = 2e^x$. Using a mesh size $\Delta x \equiv h = 0.1$, determine $f'(x)$ at $x = 2$ with the *forward* formulas (12-19a) accurate to $O(h)$ and the *central*-difference formula accurate to $O(h)^2$ and compare the results with the exact value.

12-3 Write the finite-difference form of the heat conduction equation

$$\frac{\partial^2 T}{\partial x^2} + \frac{\partial^2 T}{\partial y^2} + \frac{1}{k} g(x, y) = 0$$

for the 12 nodes, $i = 1, 2, \ldots, 12$, subject to the boundary conditions shown in the accompanying figure. Here, the temperatures f_1, f_2, and f_3 are prescribed.

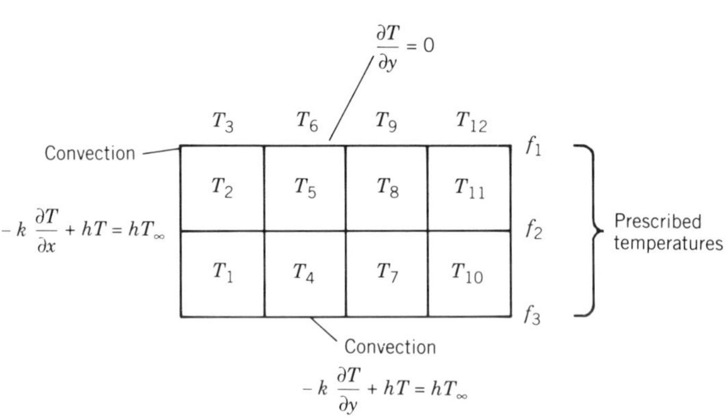

12-4 Solve the following set of algebraic equations by the Gauss elimination method:

$$\begin{aligned} T_1 + 3T_2 + T_3 &= 10 \\ 2T_1 - 2T_2 + 3T_3 + T_4 &= 11 \\ +4T_2 - 2T_3 - 2T_4 &= -2 \\ +4T_3 + 2T_4 &= 20 \end{aligned}$$

12-5 Solve the following steady-state heat conduction equation by finite differences using mesh sizes $\Delta x = \Delta y = 0.25$ and 0.1; compare the center temperature with the exact solution:

$$\frac{\partial^2 T}{\partial x^2} + \frac{\partial^2 T}{\partial y^2} = 0 \quad \text{in} \quad 0 < x < 1, \quad 0 < y < 1$$

$$T = 0 \quad \text{at} \quad x = 0, \quad x = 1$$

$$\frac{\partial T}{\partial y} = 0 \quad \text{at} \quad y = 0$$

$$T = \sin \pi x \quad \text{at} \quad y = 1$$

12-6 Consider the two-dimensional, time-dependent heat conduction equation given in the form

$$\frac{\partial^2 T}{\partial x^2} + \frac{\partial^2 T}{\partial y^2} = \frac{1}{\alpha}\frac{\partial T}{\partial t}$$

Let the temperature $T(x, y, t)$ be represented by

$$T(x, y, t) = T(j\Delta x, k\Delta y, n\Delta t) \equiv T^n_{j,k}$$

Write the finite-difference representation of this heat conduction equation for an internal node (j, k) using (1) an explicit method, (2) an implicit method, and (3) the Crank–Nicolson method.

12-7 By assuming that an error term can be represented in the form

$$E_{j,k,n} = e^{\gamma t} \cdot e^{i(\beta_1 x + \beta_2 y)}$$

where $t = n\Delta t, x = j\Delta x, y = k\Delta y$, show that in the Problem 12-6, the explicit finite-difference representation is stable if

$$\alpha \Delta t \left(\frac{1}{(\Delta x)^2} + \frac{1}{(\Delta y)^2} \right) \leqslant \frac{1}{2}$$

12-8 Consider the following heat conduction problem for a solid cylinder:

$$\frac{\partial^2 T}{\partial r^2} + \frac{1}{r}\frac{\partial T}{\partial r} + \frac{1}{k} g(r) = \frac{1}{\alpha}\frac{\partial T}{\partial t} \quad \text{in} \quad 0 \leqslant r < b, \quad t > 0$$

$$T = 0 \quad \text{at} \quad r = b, \quad t > 0$$

$$T = F(r) \quad \text{in} \quad 0 \leqslant r \leqslant b, \quad t = 0$$

Let the temperature $T(r, t)$ be represented by

$$T(r, t) = T(j\Delta r, n\Delta t) \equiv T^n_j$$

with $j = 0$ representing $r = 0$ and $j = N$ representing $r = b$. Write the finite-difference representation of this heat-conduction problem using (1) the explicit method and (2) the implicit method.

12-9 Repeat Problem 12-8 for a solid sphere of radius $r = b$.

12-10 Consider the heat conduction equation in the cylindrical coordinates given as

$$\frac{\partial^2 T}{\partial r^2}+\frac{1}{r}\frac{\partial T}{\partial r}+\frac{1}{r^2}\frac{\partial^2 T}{\partial \phi^2}+\frac{1}{k}g(r,\phi)=\frac{1}{\alpha}\frac{\partial T}{\partial t}$$

and let the temperature $T(r,\phi,t)$ be represented by

$$T(r,\phi,t)=T(i\Delta r, j\Delta\phi, n\Delta t)\equiv T_{i,j}^{n}$$

Write the finite difference form of this heat conduction equation for a node (i,j) using (1) the explicit scheme and (2) the implicit scheme.

12-11 Consider the heat conduction equation given as

$$\frac{\partial^2 T}{\partial r^2}+\frac{1}{r}\frac{\partial T}{\partial r}+\frac{\partial^2 T}{dz^2}+\frac{1}{k}g(r,z)=\frac{1}{\alpha}\frac{\partial T}{\partial t}$$

where $T(r,z,t)$ is represented by

$$T(r,z,t)=T(i\Delta r, k\Delta z, n\Delta t)\equiv T_{i,k}^{n}$$

Write the finite-difference form of this heat conduction equation for a node (i,k) using the Crank–Nicolson method.

12-12 Consider the following heat conduction equation

$$\frac{\partial}{\partial x}\left(k\frac{\partial T}{\partial x}\right)+g(x,t)=\rho C_p\frac{\partial T}{\partial t}$$

where $T(x,t)$ is represented by

$$T(x,t)=T(j\Delta x, n\Delta r)\equiv T_j^n$$

Write the finite-difference form of this heat conduction equation for a node (j) using (1) the explicit method and (2) the implicit method.

12-13 Consider the finite-difference form of the heat conduction equation $\partial^2 T/\partial x^2=(1/\alpha)(\partial T/\partial t)$ given in the form [see equation (12-131)]

$$\frac{T_j^{n+1}-T_j^n}{\Delta t}=\alpha\left[\theta\frac{T_{j-1}^{n+1}-2T_j^{n+1}+T_{j+1}^{n+1}}{(\Delta x)^2}+(1-\theta)\frac{T_{j-1}^n-2T_j^n+T_{j+1}^n}{(\Delta x)^2}\right]$$

Using the Fourier series method, show that for $\frac{1}{2}\leqslant\theta\leqslant 1$ the solution is unconditionally stable and for $0\leqslant\theta\leqslant\frac{1}{2}$ it is stable if

$$r=\frac{\alpha\Delta t}{(\Delta x)^2}\leqslant\frac{1}{2(1-2\theta)}$$

12-14 Consider the following one-dimensional heat conduction problem in the dimensionless form:

$$\frac{\partial^2 T}{\partial x^2} = \frac{\partial T}{\partial t} \qquad \text{in} \qquad 0 < x < 1, \quad t > 0$$

$$T = 0 \qquad \text{at} \qquad x = 0, \qquad x = 1 \quad \text{for} \quad t < 0$$

$$T = \sin \pi x \qquad \text{for} \qquad t = 0, \qquad \text{in} \quad 0 \leqslant x \leqslant 1$$

Solve this problem with finite differences using an explicit scheme by taking $\Delta x = 0.1$ and $r = \Delta t/(\Delta x)^2 = 0.1$. Compare the results with the exact solution at time $t = 0.01$ at the locations $x = 0.1$ and 0.2.

12-15 Solve Problem 12-14 using the Crank–Nicolson method and the explicit method by taking $\Delta x = 0.1$, $\Delta t = 0.005$ and compare $T(0.2, 0.01)$ with the exact solution.

12-16 Consider the following transient radial heat conduction in a solid cylinder, $0 \leqslant \eta \leqslant 1$, given in the dimensionless form as

$$\frac{\partial^2 T}{\partial \eta^2} + \frac{1}{\eta}\frac{\partial T}{\partial \eta} = \frac{\partial T}{\partial t} \qquad \text{in} \qquad 0 < \eta < 1, \quad t > 0$$

$$\frac{\partial T}{\partial \eta} = 0 \qquad \text{at} \qquad \eta = 0, \qquad t > 0$$

$$T = 0 \qquad \text{at} \qquad \eta = 1, \qquad t > 0$$

$$T = 100 J_0(\beta_1 \eta) \qquad \text{for} \qquad t = 0, \qquad 0 \leqslant \eta \leqslant 1$$

where $J_0(z)$ is the zero-order Bessel function of the first kind and β_1 is the first root of $J_0(z) = 0$. The exact analytic solution of this problem is given by

$$T(\eta, t) = 100\, J_0(\beta, \eta) e^{-\beta_1^2 t}$$

By dividing the solution domain into five equal parts and using the simple explicit scheme, solve this problem with finite-differences and compare the center temperature $T(0, t)$ with the exact solution at dimensionless times $t = 0.2, 0.4, 0.6, \ldots,$ and 1.6.

12-17 Consider the following transient heat conduction problem given in the dimensionless form as

$$\frac{\partial^2 T}{\partial x^2} = \frac{\partial T}{\partial t} \qquad \text{in} \quad 0 < x < 1, \quad t > 0$$

$$T=0 \quad \text{at} \quad x=0, \quad t>0$$

$$T=0 \quad \text{at} \quad x=1, \quad t>0$$

$$T=10\sin 2\pi x \quad \text{for} \quad t=0, \quad 0 \leqslant x \leqslant 1$$

The exact analytic solution of this problem is

$$T(x,t)=10\,e^{-4\pi^2 t}\sin 2\pi x$$

Solve this problem with finite differences using the *explicit* method, taking

(a) $\Delta x=0.1, \quad r=0.25 \quad \therefore \quad \Delta t=0.0025$

(b) $\Delta x=0.1, \quad r=0.50 \quad \therefore \quad \Delta t=0.0050$

and compare the temperature at the location $x=0.3$ with the exact results.

12-18 Perform the first three iterations of the Gauss–Seidel method for solving the following system of equations:

$$\begin{aligned} 6T_1+T_2+3T_3&=17 \\ T_1-10T_2+4T_3&=-7 \\ T_1+T_2+3T_3&=12 \end{aligned}$$

12-19 Consider the following transient heat conduction problem in a slab given in dimensionless form as

$$\frac{\partial^2 T}{\partial x^2}=\frac{\partial T}{\partial t} \quad \text{in} \quad 0<x<1, \quad t>0$$

$$\frac{\partial T}{\partial x}=0 \quad \text{at} \quad x=0, \quad t>0$$

$$T=0 \quad \text{at} \quad x=1, \quad t>0$$

$$T=100\cos\left(\frac{\pi}{2}x\right) \quad \text{for} \quad t=0, \quad 0 \leqslant x \leqslant 1$$

Solve this problem numerically with the explicit finite-difference scheme by dividing the region $0 \leqslant x \leqslant 1$ into five equal parts by using (a) a first-order accurate and (b) a second-order accurate finite differencing schemes for the boundary condition at $x=0$. The exact solution of this problem is given by

$$T(x,t) = 100\cos\left(\frac{\pi}{2}x\right)\exp\left(-\frac{\pi^2}{4}t\right)$$

Compare the temperature of the insulated surface obtained by finite-difference solution with the exact solution given above. Take $r = \frac{1}{5}$ for numerical calculations.

12-20 A 10-cm-diameter solid steel bar of thermal conductivity $k = 40$ W/(m·°C) is heated electrically by the passage of electric current that generates energy within the rod at a rate of $g = 4 \times 10^6$ W/m^3. Heat is dissipated from the surface of the rod by convection with a heat transfer coefficient $h = 400$ W/(m^2·°C) into an ambient at temperature $T_\infty = 20$ °C. By dividing the radius into five equal parts, develop the finite-difference equations for this heat conduction problem. Compare the finite-difference solution with the exact analytic solution for the cases when *first-order* and *second-order* accurate differencing are used for the convection boundary condition.

12-21 Repeat Problem 12-20 for the case of a solid sphere.

12-22 Consider the following one-dimensional, time-dependent heat conduction problem for a slab $0 \leqslant x \leqslant L$ subject to the boundary conditions of the third kind at both boundaries

$$\alpha\frac{\partial^2 T}{\partial x^2} = \frac{\partial T}{\partial t} \quad \text{in} \quad 0 < x < L, \quad t > 0$$

$$-k_1\frac{\partial T}{\partial x} + h_1 T = f_1 \quad \text{at} \quad x = 0, \quad t > 0$$

$$k_2\frac{\partial T}{\partial x} + h_2 T = f_2 \quad \text{at} \quad x = L, \quad t > 0$$

$$T = F(x) \quad \text{for} \quad t = 0, \quad \text{in} \quad 0 \leqslant x \leqslant L$$

Using the *explicit* method for the finite-differencing of the differential equation, simple *forward* and simple *backward* differences for the boundary conditions at $x = 0$ and $x = L$, respectively, write the finite-difference representation of this heat conduction problem.

12-23 Write the finite-difference representation of the heat conduction Problem 12-22 using the explicit method for the differential equation and the central differences for the boundary conditions.

12-24 Write the finite-difference representation of the heat conduction Problem 12-22 using the *Crank–Nicolson* method for the differential equation and the central differences for the boundary conditions.

13

INTEGRAL-TRANSFORM TECHNIQUE

The solution of partial-differential equations of heat conduction by the classical method of separation of variables is not always convenient when the equation and the boundary conditions involve nonhomogeneities. It is for this reason that we considered the Green's function approach for the solution of linear, nonhomogeneous boundary-value problems of heat conduction. The *integral transform technique* provides a systematic, efficient, and straightforward approach for the solution of both homogeneous and nonhomogeneous, steady-state, and time-dependent boundary-value problems of heat conduction. In this method the second partial derivatives with respect to the space variables are generally removed from the partial-differential equation of heat conduction by the application of the integral transformation. For example, in time-dependent problems, the partial derivatives with respect to the space variables are removed and the partial-differential equation is reduced to a first-order ordinary differential equation in the time variable for the transform of the temperature. The ordinary differential equation is solved subject to the transformed initial condition, and the result is inverted successively to obtain the solution for the temperature. The inversion process is straightforward, because the inversion formulas are available at the onset of the problem. The procedure is also applicable to the solution of steady-state heat conduction problems involving more than one space variable. In such cases the partial differential equation of heat conduction is reduced to an ordinary differential equation in one of the space variables. The resulting ordinary differential equation for the transformed temperature is solved, and the solution is inverted to obtain the temperature distribution.

The integral-transform technique derives its basis from the classical method of separation of variables. That is, the integral transform pairs needed for the solution of a given problem are developed by considering the representation of an arbitrary function in terms of the eigenfunctions of the corresponding eigenvalue problem. Therefore, the eigenfunctions, eigenvalues, and the normalization integrals developed in Chapters 2–4 for the solution of homogeneous problems will be utilized for the construction of the integral transform pairs.

The fundamental theory of the integral-transform technique is given in several texts [1–3] and a summary of various transform pairs and transform tables are presented in various references [4–8]. The literature on the use of the integral-transform technique for the solution of heat conduction problems is evergrowing. The reader should consult references 9–23 for the general solution of three-dimensional problems of finite regions. Its applications for the solution of specific heat conduction problems in the rectangular [24, 25], cylindrical [26–29], and spherical [30] coordinate systems are also given. Some useful convolution properties of integral transforms are discussed in references 31–34.

In this chapter a general method of analysis of three-dimensional, time-dependent heat conduction problems of finite region by the integral-transform technique is presented first. Its applications for the solution of problems of finite, semiinfinite, and infinite regions in the rectangular, cylindrical, and spherical coordinate systems are then presented systematically. More recent applications can be found in references 37–40.

13-1 THE USE OF INTEGRAL TRANSFORM IN THE SOLUTION OF HEAT CONDUCTION PROBLEMS

In this section we present the use of the integral-transform technique in the solution of three-dimensional, time-dependent, nonhomogeneous boundary-value problems of heat conduction with constant coefficients in finite regions. We consider the following heat conduction problem

$$\nabla^2 T(\mathbf{r}, t) + \frac{1}{k} g(\mathbf{r}, t) = \frac{1}{\alpha} \frac{\partial T(\mathbf{r}, t)}{\partial t} \qquad \text{in region } R, \qquad t > 0 \tag{13-1a}$$

$$k_i \frac{\partial T(\mathbf{r}_i, t)}{\partial n_i} + h_i T(\mathbf{r}_i, t) = f_i(\mathbf{r}_i, t) \qquad \text{on boundary } S_i, \quad t > 0 \tag{13-1b}$$

$$T(\mathbf{r}, t) = F(\mathbf{r}) \qquad \text{for } t = 0, \qquad \text{in region } R \tag{13-1c}$$

where, $i = 1, 2, \ldots, N$ and N is the number of continuous boundary surfaces of the region R ($s = 1$ for a semiinfinite medium, $N = 2$ for a slab, $N = 4$ for a rectangular region, etc.); $\partial/\partial n_i$ denotes the normal derivative at the boundary surface S_i in the *outward* direction; h_i and k_i are the boundary-condition coefficients at the boundary surface S_i; k is the thermal conductivity; α is the thermal diffusivity; $f_i(\mathbf{r}_i, t)$ is a specified boundary-condition function; $F(\mathbf{r})$ is a specified initial condition function; and $g(\mathbf{r}, t)$ is the heat-generation term.

The basic steps in the solution of this problem with the integral-transform technique can be summarized as follows:

1. Appropriate integral-transform pair is developed.
2. By the application of integral transformation, the partial derivatives with respect to the space variables are removed from the heat conduction equation, thus reducing it to an ordinary differential equation for the transform of temperature.
3. The resulting ordinary differential equation is solved subject to the transformed initial condition. When the transform of the temperature is inverted by the inversion formula the desired solution is obtained. The procedure is now described in detail.

1. *Development of Integral-Transform Pair.* The integral transform pair needed for the solution of the above heat conduction problem can be developed by considering the following eigenvalue problem

$$\nabla^2\psi(\mathbf{r}) + \lambda^2\psi(\mathbf{r}) = 0 \qquad \text{in region } R \tag{13-2a}$$

$$k_i\frac{\partial\psi(\mathbf{r}_i)}{\partial n_i} + h_i\psi(\mathbf{r}_i) = 0 \qquad \text{on boundary } S_i \tag{13-2b}$$

$i = 1, 2, \ldots, N$, and $k_i, h_i, \partial/\partial n_i$ are as defined previously. We note that this eigenvalue problem is obtainable by the separation of the homogeneous version of the heat-conduction problem (13-1). The eigenfunctions $\psi(\lambda_m, \mathbf{r})$ of this eigenvalue problem satisfy the following orthogonality condition (see note 1 at the end of this chapter for a proof of this orthogonality relation)

$$\int_R \psi(\lambda_m, \mathbf{r})\psi(\lambda_n, \mathbf{r})dv = \begin{cases} 0 & \text{for} \quad m \neq n \\ N(\lambda_m) & \text{for} \quad m = n \end{cases} \tag{13-3a}$$

where the *normalization integral* $N(\lambda_m)$ is defined as

$$N(\lambda_m) = \int_R [\psi(\lambda_m, \mathbf{r})]^2\, dv \tag{13-3b}$$

We now consider the representation of a function $T(\mathbf{r}, t)$, defined in the finite region R, in terms of the eigenfunctions $\psi(\lambda_m, \mathbf{r})$ in the form

$$T(\mathbf{r}, t) = \sum_{m=1}^{\infty} C_m(t)\psi(\lambda_m, \mathbf{r}) \qquad \text{in } R \tag{13-4}$$

where the summation is taken over all discrete spectrum of eigenvalues λ_m. To determine the unknown coefficients we operate on both sides of equation (13-4) by the operator

$$\int_R \psi(\lambda_n, \mathbf{r})\, dv$$

[that is, multiply by $\psi(\lambda_n, \mathbf{r})$ and integrate over the region R] then utilize the orthogonality relation (13-3) to obtain

$$C_m(t) = \frac{1}{N(\lambda_m)} \int_R \psi(\lambda_m, \mathbf{r}) T(\mathbf{r}, t)\, dv \tag{13-5}$$

This expression is introduced into equation (13-4) and the resulting representation is split up into two parts to define the *integral-transform pair* in the space variable $\mathbf{r}$ for the function $T(\mathbf{r}, t)$ as

Inversion formula: $$T(\mathbf{r}, t) = \sum_m \frac{\psi(\lambda_m, \mathbf{r})}{N(\lambda_m)} \bar{T}(\lambda_m, t) \tag{13-6a}$$

Integral transform: $$\bar{T}(\lambda_m, t) = \int_R \psi(\lambda_m, \mathbf{r}') T(\mathbf{r}', t)\, dv' \tag{13-6b}$$

where $\bar{T}(\lambda_m, t)$ is called the integral transform of the function $T(\mathbf{r}, t)$ with respect to the space variable $\mathbf{r}$. It is to be noted that in the above formal representation, the summation is actually a triple, a double, or a single summation; and the integral is a volume, a surface, or a line integral for the three-, two-, or one-dimensional regions, respectively. In the cartesian coordinate system, the eigenfunctions $\psi(\lambda_m, \mathbf{r})$ and the normalization integral $N(\lambda_m)$ are composed of the products of one-dimensional eigenfunctions and normalization integrals, respectively.

2. *Integral Transform of Heat Conduction Problem.* Having established the appropriate integral-transform pair as given above, the next step in the analysis is the removal of the partial derivatives with respect to the space variables from the differential equation (13-1a) by the application of the integral transform (13-6b). That is, both sides of equation (13-1a) is multiplied by $\psi_m(\mathbf{r})$ and integrated over the region R

$$\int_R \psi_m(\mathbf{r}) \nabla^2 T(\mathbf{r}, t)\, dv + \frac{1}{k} \int_R \psi_m(\mathbf{r}) g(\mathbf{r}, t)\, dv = \frac{1}{\alpha} \frac{\partial}{\partial t} \int_R \psi_m(\mathbf{r}) T(\mathbf{r}, t)\, dv \tag{13-7}$$

where $\psi_m(\mathbf{r}) \equiv \psi(\lambda_m, \mathbf{r})$. By utilizing the definition of the integral transform (13-6b) this expression is written as

$$\int_R \psi_m(\mathbf{r}) \nabla^2 T(\mathbf{r}, t)\, dv + \frac{1}{k} \bar{g}(\lambda_m, t) = \frac{1}{\alpha} \frac{d\bar{T}(\lambda_m, t)}{dt} \tag{13-8}$$

where $\bar{g}(\lambda_m, t)$ and $\bar{T}(\lambda_m, t)$ are the integral transforms of the function $g(\mathbf{r}, t)$ and $T(\mathbf{r}, t)$, respectively, defined as

$$\bar{g}(\lambda_m, t) = \int_R \psi_m(\mathbf{r}) g(\mathbf{r}, t)\, dv \tag{13-9a}$$

$$\bar{T}(\lambda_m, t) = \int_R \psi_m(\mathbf{r}) T(\mathbf{r}, t)\, dv \tag{13-9b}$$

The integral on the left-hand side of equation (13-8) can be evaluated by making use of the Green's theorem expressed as

$$\int_R \psi_m(\mathbf{r}) \nabla^2 T(\mathbf{r}, t)\, dv = \int_R T\, \nabla^2 \psi_m(\mathbf{r})\, dv + \sum_{i=1}^{N} \int_{S_i} \left(\psi_m \frac{\partial T}{\partial n_i} - T \frac{\partial \psi_m}{\partial n_i} \right) ds_i' \tag{13-10}$$

where $i = 1, 2, \ldots, N$ and N is the number of continuous boundary surfaces of the region R. Various terms on the right-hand side of equation (13-10) are evaluated as now described. The integral $\int_R T\, \nabla^2 \psi_m dv$ is evaluated by writing equation (13-2a) for the eigenfunction $\psi_m(\mathbf{r})$, multiplying both sides by $T(\mathbf{r}, t)$, integrating over the region R and utilizing the definition of the integral transform. We find

$$\int_R T\, \nabla^2 \psi_m\, dv = -\lambda_m^2 \int_R T \psi_m\, dv = -\lambda_m^2 \bar{T}(\lambda_m, t) \tag{13-11a}$$

The surface integral in equation (13-10) is evaluated by making use of the boundary conditions (13-1b) and (13-2b). That is, equation (13-1b) is multiplied by $\psi_m(\mathbf{r})$, equation (13-2b) is multiplied by $T(\mathbf{r}, t)$ and the results are subtracted; we obtain

$$\psi_m \frac{\partial T}{\partial n_i} - T \frac{\partial \psi_m}{\partial n_i} = \frac{\psi_m(\mathbf{r}_i)}{k_i} f_i(\mathbf{r}_i, t) \tag{13-11b}$$

Equations (13-11) are introduced into equation (13-10)

$$\int_R \psi_m(\mathbf{r}) \nabla^2 T(\mathbf{r}, t)\, dv = -\lambda_m^2 \bar{T}(\lambda_m, t) + \sum_{i=1}^{N} \int_{S_i} \frac{\psi_m(\mathbf{r}_i')}{k_i} f_i(\mathbf{r}_i', t)\, ds_i' \tag{13-12}$$

Substituting equation (13-12) into (13-8) we obtain

$$\frac{d\bar{T}(\lambda_m, t)}{dt} + \alpha \lambda_m^2 \bar{T}(\lambda_m, t) = A(\lambda_m, t) \qquad \text{for} \qquad t > 0 \tag{13-13a}$$

where

$$A(\lambda_m, t) \equiv \frac{\alpha}{k}\bar{g}(\lambda_m, t) + \alpha \sum_{i=1}^{s} \int_{S_i} \frac{\psi_m(\mathbf{r}'_i)}{k_i} f_i(\mathbf{r}', t)\, ds'_i \tag{13-13b}$$

Thus by the application of the integral-transform technique, we removed from the heat-conduction equation (13-1a) all the partial derivatives with respect to the space variables and reduced it to a first-order ordinary differential equation (13-13a) for the transform $\bar{T}(\lambda_m, t)$ of the temperature. In the process of integral transformation, we utilized the boundary conditions (13-1b); therefore the boundary conditions for the problem are incorporated in this result. The integral transform of the initial condition (13-1c) becomes

$$\bar{T}(\lambda_m, t) = \int_R \psi_m(\mathbf{r})F(\mathbf{r})\, dv \equiv \bar{F}(\lambda_m) \qquad \text{for} \qquad t = 0 \tag{13-13c}$$

3. *Solution for Transform and Inversion.* The solution of equation (13-13a) subject to the transformed initial condition (13-13c) gives the transform $\bar{T}(\lambda_m, t)$ of temperature as

$$\bar{T}(\lambda_m, t) = e^{-\alpha\lambda_m^2 t}\left[\bar{F}(\lambda_m) + \int_0^t e^{\alpha\lambda_m^2 t'} A(\lambda_m, t')\, dt'\right] \tag{13-14}$$

Introducing this integral transform into the inversion formula (13-6a), we obtain the solution of the boundary-value problem of heat conduction, equations (13-1), in the form

$$T(\mathbf{r}, t) = \sum_{m=1}^{\infty} \frac{1}{N(\lambda_m)} e^{-\alpha\lambda_m^2 t}\psi_m(\mathbf{r})\left[\bar{F}(\lambda_m) + \int_0^t e^{\alpha\lambda_m^2 t'} A(\lambda_m, t')\, dt'\right] \tag{13-15a}$$

where

$$A(\lambda_m, t') \equiv \frac{\alpha}{k}\bar{g}(\lambda_m, t') + \alpha \sum_{i=1}^{N} \int_{S_i} \frac{\psi_m(\mathbf{r}'_i)}{k_i} f_i(\mathbf{r}', t')\, ds'_i \tag{13-15b}$$

$$\bar{F}(\lambda_m) \equiv \int_R \psi_m(\mathbf{r}')F(\mathbf{r}')\, dv' \tag{13-15c}$$

$$\bar{g}(\lambda_m, t') \equiv \int_R \psi_m(\mathbf{r}')g(\mathbf{r}', t')\, dv' \tag{13-15d}$$

$$N(\lambda_m) \equiv \int_R [\psi_m(\mathbf{r}')]^2\, dv' \tag{13-15e}$$

and the summation is taken over all eigenvalues. This solution is derived formally for a boundary condition of the third kind for all boundaries. If some of the boundaries are of the second kind and some of the third kind, the general form of the solution remains unchanged. However, some modification is needed in the term $A(\lambda_m, t')$ if the problem involves boundary conditions of the first kind. Suppose the boundary condition for the surface $i = 1$ is of the first kind; this implies that the boundary-condition coefficient k_1 should be set equal to zero. This situation causes difficulty in the interpretation of the term $A(\lambda_m, t')$ given by equation (13-15b), because k_1 appears in the denominator. This difficulty can be alleviated by making the following change in equation (13-15b):

When $k_1 = 0$, replace

$$\frac{\psi_m(\mathbf{r}'_1)}{k_1}$$

by

$$-\frac{1}{h_1}\frac{\partial\psi_m(\mathbf{r}'_1)}{\partial n_1} \tag{13-15f}$$

The validity of this replacement becomes apparent if we rearrange the boundary condition (13-2b) of the eigenvalue problem for $i = 1$, in the form

$$\frac{\psi_m(\mathbf{r}_1)}{k_1} = -\frac{1}{h_1}\frac{\partial\psi_m(\mathbf{r}_1)}{\partial n_1}$$

on boundary S_1. Finally, when all boundary conditions are of the second kind, the interpretation of the general solution (13-15) requires special consideration. The reason for this, $\lambda_0 = 0$ is also an eigenvalue corresponding to the eigenfunction $\psi_0 = \text{constant} \neq 0$, for this particular case. This matter will be illustrated later in this section.

One-Dimensional Finite Region

We now consider the one-dimensional version of the heat conduction problem (13-1) in the space variable x_1 $(x, y, z, r$, etc.) for a finite region R_1. Let $\psi(\beta_m, x_1)$ be the eigenfunctions, $N(\beta_m)$ be the normalization integral, β_m the eigenvalues, and $w(x_1)$ the Sturm–Liouville weighting function of the one-dimensional version of the eigenvalue problem (13-2). As was discussed in note 1 of Chapter 2, the eigenfunctions $\psi(\beta_m, x_1)$ are orthogonal with respect to the weighting function $w(x_1)$:

$$\int_{R_1} w(x_1)\psi(\beta_m, x_1)\psi(\beta_n, x_1)\,dx_1 = \begin{cases} 0 & \text{for} \quad \lambda_m \neq \lambda_n \\ N(\beta_m) & \text{for} \quad \lambda_m = \lambda_n \end{cases} \tag{13-16a}$$

where

$$N(\beta_m) \equiv \int_{R_1} w(x_1)[\psi(\beta_m, x_1)]^2 \, dx_1 \tag{13-16b}$$

Suppose we wish to represent a function $T(x_1, t)$, defined in the finite interval R_1, in terms of the eigenfunctions $\psi(\beta_m, x_1)$. Such a representation is immediately obtained by utilizing the above orthogonality condition and the result is written as

$$T(x_1, t) = \sum_{m=1}^{\infty} \frac{\psi(\beta_m, x_1)}{N(\beta_m)} \int_{R_1} w(x'_1)\psi(\beta_m, x'_1)T(x'_1, t)\, dx'_1, \qquad \text{in region } R_1 \tag{13-16c}$$

The desired integral transform pair is constructed by splitting up this representation into two parts as

Inversion formula:
$$T(x_1, t) = \sum_{m=1}^{\infty} \frac{\psi(\beta_m, x_1)}{N(\beta_m)} \bar{T}(\beta_m, t) \tag{13-17a}$$

Integral transform:
$$\bar{T}(\beta_m, t) = \int_{R_1} w(x'_1)\psi(\beta_m, x'_1)T(x'_1, t)\, dx'_1 \tag{13-17b}$$

The solution of the one-dimensional version of the heat conduction problem (13-1) is obtained from the general solution (13-15) as

$$T(x_1, t) = \sum_{m=1}^{\infty} \frac{\psi(\beta_m, x_1)}{N(\beta_m)} e^{-\alpha\beta_m^2 t} \left[\bar{F}(\beta_m) + \int_0^t e^{\alpha\beta_m^2 t'} A(\beta_m, t')\, dt' \right] \tag{13-18a}$$

where

$$A(\beta_m, t') = \frac{\alpha}{k}\bar{g}(\beta_m, t') + \alpha \left\{ \left[w(x'_1)\frac{\psi(\beta_m, x'_1)}{k_1} \right]_{S_1} f_1(t') + \left[w(x'_1)\frac{\psi(\beta_m, x'_1)}{k_2} \right]_{S_2} f_2(t') \right\} \tag{13-18b}$$

$$\bar{F}(\beta_m) = \int_{R_1} w(x'_1)\psi(\beta_m, x'_1)F(x'_1)\, dx'_1 \tag{13-18c}$$

$$\bar{g}(\beta_m, t') = \int_{R_1} w(x'_1)\psi(\beta_m, x'_1)g(x'_1, t')\, dx'_1 \tag{13-18d}$$

$$N(\beta_m) = \int_{R_1} w(x'_1)[\psi(\beta_m, x'_1)]^2\, dx'_1 \tag{13-18e}$$

If the boundary condition is of the first kind at any of the boundaries, the following adjustment should be made in equation (13-18b):

$$\text{When } k_i = 0 \qquad \text{replace} \frac{\psi(\beta_m, x_1')}{k_i} \qquad \text{by} \qquad -\frac{1}{h_i}\frac{\partial\psi(\beta_m, x_1')}{\partial n_i} \tag{13-19}$$

where $i = 1$ or 2, and $\partial/\partial n_i$ denotes derivative along the outward-drawn normal to the boundary surface.

Boundary Condition of the Second Kind for All Boundaries

When the boundary conditions for a finite region are all of the second kind, that is all h_i values are zero in the heat conduction problem (13-1), then the eigenvalue problem (13-2) takes the form

$$\nabla^2\psi(\mathbf{r}) + \lambda^2\psi(\mathbf{r}) = 0 \qquad \text{in region } R \tag{13-20a}$$

$$\frac{\partial\psi(\mathbf{r}_i)}{\partial n_i} = 0 \qquad \text{on all boundaries } S_i \tag{13-20b}$$

For this particular case, $\lambda_0 = 0$ is also an eigenvalue corresponding to the eigenfunction $\psi_0 =$ constant $\neq 0$. The validity of this statement can be verified by integrating equation (13-20a) over the region R, applying Green's theorem to change the volume integral to the surface integral and then utilizing the boundary conditions (13-20b). We obtain

$$\lambda^2 \int_R \psi(\mathbf{r})\, dv = -\int_R \nabla^2\psi(\mathbf{r})\, dv = -\sum_{i=1}^{N}\int_{S_i}\frac{\partial\psi}{\partial n_i}\, ds_i = 0 \tag{13-21}$$

Clearly, $\lambda_0 = 0$ is also an eigenvalue corresponding to the eigenfunction $\psi_0 =$ constant $\neq 0$. We can set $\psi_0 = 1$, because ψ_0 will cancel out when it is introduced into the solution (13-15) for the eigenvalue $\lambda_0 = 0$. Then, the general solution (13-15) for the case of all boundary conditions are of the second kind [i.e., $\partial T/\partial n_i = (1/k_i) f_i(\mathbf{r}, t)$] on all S_i, takes the form

$$T(\mathbf{r}, t) = \frac{1}{N_0}\left[\bar{F}(\lambda_0) + \int_0^t A(\lambda_0, t')\, dt'\right] + \sum_{m=1}^{\infty}\frac{1}{N_m} e^{-\alpha\lambda_m^2 t}\psi(\lambda_m, \mathbf{r})\left[\bar{F}(\lambda_m) + \int_0^t e^{\alpha\lambda_m^2 t'} A(\lambda_m, t')\, dt'\right] \tag{13-22}$$

where

$$N_0 = \int_R dv' \tag{13-23a}$$

$$\bar{F}(\lambda_0) = \int_R F(\mathbf{r}')\,dv' \tag{13-23b}$$

$$A(\lambda_0, t') = \frac{\alpha}{k}\int_R g(\mathbf{r}', t')\,dv' + \alpha \sum_{i=1}^{N} \int_{S_i} \frac{f_i(\mathbf{r}'_i, t')}{k_i}\,ds'_i \tag{13-23c}$$

and the functions $A(\lambda_m, t')$, $\bar{F}(\lambda_m)$, and N_m are as defined by equations (13-15b), (13-15c), and (13-15e), respectively.

The average temperature over the region R is defined as

$$T_{av}(t) = \frac{\displaystyle\int_R T(\mathbf{r}, t)\,dv}{\displaystyle\int_V dv} \tag{13-24a}$$

and the solution (13-22) is introduced into this expression. If we take into account the following relations obtained from equation (13-21)

$$\int_R \psi(\lambda_m, \mathbf{r})\,dv = 0 \qquad \text{for} \qquad \lambda_m \neq 0 \tag{13-24b}$$

$$\int_R \psi(\lambda_0)\,dv = \int_R dv \qquad \text{since} \qquad \psi_0 = 1 \quad \text{for} \quad \lambda_0 = 0 \tag{13-24c}$$

then equation (13-24a) gives

$$T_{av}(t) = \frac{\left[\bar{F}(\lambda_0) + \displaystyle\int_0^t A(\lambda_0, t')\,dt'\right]}{\displaystyle\int_R dv} \tag{13-25a}$$

This result implies that the first term in the solution (13-22) resulting from the eigenvalue $\lambda_0 = 0$ is the average value of $T(\mathbf{r}, t)$ over the finite region R. For the special case of no heat generation and all insulated boundaries (i.e., $\partial T/\partial n_i = 0$), the quantity $A(\lambda_0, t')$ vanishes and equation (13-25a) reduces to

$$T_{av}(t) = \frac{\displaystyle\int_R F(\mathbf{r}')\,dv}{\displaystyle\int_R dv} \tag{13-25b}$$

Clearly, the expressions given by equations (13-25) are the generalization of the special cases considered in Examples 2-2 and 3-2 of problems with insulated boundaries.

13-2 APPLICATIONS IN THE RECTANGULAR COORDINATE SYSTEM

The general analysis developed in the previous section is now applied for the solution of time-dependent heat conduction problems in the rectangular coordinate system. The one-dimensional cases are considered first for the finite, semiinfinite, and infinite regions. The multidimensional problems involving any combinations of the finite, semiinfinite, and infinite regions for the $x, y,$ and z directions are then handled by the successive application of the one-dimensional integral transform.

One-Dimensional Problems of Finite Region

We consider the following heat conduction problem for a slab, $0 \leqslant x \leqslant L$:

$$\frac{\partial^2 T}{\partial x^2} + \frac{1}{k} g(x,t) = \frac{1}{\alpha}\frac{\partial T(x,t)}{\partial t} \qquad \text{in} \quad 0 < x < L, \quad t > 0 \tag{13-26a}$$

$$-k_1 \frac{\partial T}{\partial x} + h_1 T = f_1(t) \qquad \text{at} \quad x = 0, \quad t > 0 \tag{13-26b}$$

$$k_2 \frac{\partial T}{\partial x} + h_2 T = f_2(t) \qquad \text{at} \quad x = L, \quad t > 0 \tag{13-26c}$$

$$T(x,t) = F(x) \qquad \text{for} \quad t = 0, \quad \text{in} \quad 0 \leqslant x \leqslant L \tag{13-26d}$$

The eigenvalue problem associated with the solution of this problem is exactly the same as that given by equations (2-32). Clearly, this eigenvalue problem is the one-dimensional version of the general eigenvalue problem (13-2). To construct the desired integral-transform pair for the solution of the above heat conduction problem, we need the representation of an arbitrary function, defined in the interval $0 \leqslant x \leqslant L$, in terms of the eigenfunctions of the eigenvalue problem (2-32). Such a representation has already been given by equation (2-36). Then the integral-transform pair for the function $T(x, t)$ with respect to the x variable is readily obtained by splitting up the representation into two parts as

$$\text{Inversion formula:} \qquad T(x,t) = \sum_{m=1}^{\infty} \frac{X(\beta_m, x)}{N(\beta_m)} \bar{T}(\beta_m, t) \tag{13-27a}$$

$$\text{Integral transform:} \qquad \bar{T}(\beta_m, t) = \int_{x'=0}^{L} X(\beta_m, x') T(x', t)\, dx' \tag{13-27b}$$

where

$$N(\beta_m) \equiv \int_0^L [X(\beta_m, x)]^2\, dx \tag{13-27c}$$

The functions $X(\beta_m, x)$, $N(\beta_m)$, and the eigenvalues β_m are obtainable from Table 2-2 for the nine different combinations of boundary conditions at $x = 0$ and $x = L$.

To solve the heat conduction problem (13-26), we take the integral transform of equation (13-26a) by the application of the transform (13-27b). That is, we multiply both sides of equation (13-26a) by $X(\beta_m, x)$ and integrate over the region $0 \leqslant x \leqslant L$. The resulting expressions contains the term

$$\int_0^L X(\beta_m, x) \frac{\partial^2 T}{\partial x^2}\, dx$$

which is evaluated as discussed in Section 13-1 by making use of Green's theorem (or integrating it by parts twice), utilizing the eigenvalue problem (2-32), and the boundary conditions (13-26b) and (13-26c) of the above heat conduction problem. Then, the integral transform of equation (13-26a) leads to the following ordinary differential equation for the transform $\bar{T}(\beta_m, t)$ of temperature

$$\frac{d\bar{T}(\beta_m, t)}{dt} + \alpha\beta_m^2 \bar{T}(\beta_m, t) = A(\beta_m, t) \qquad \text{for} \qquad t > 0 \tag{13-28a}$$

$$\bar{T}(\beta_m, t) = \bar{F}(\beta_m) \qquad \text{for} \qquad t = 0 \tag{13-28b}$$

where $\bar{F}(\beta_m)$ is the integral transform of the initial condition function $F(x)$ and $A(\beta_m, t)$ is defined below. The solution of equations (13-28) gives the transform of temperature $\bar{T}(\beta_m, t)$; when this result is inverted by the inversion formula (13-27a) the solution of the heat conduction problem (13-26) becomes

$$T(x, t) = \sum_{m=1}^{\infty} \frac{X(\beta_m, x)}{N(\beta_m)} e^{-\alpha\beta_m^2 t} \left[\bar{F}(\beta_m) + \int_0^t e^{\alpha\beta_m^2 t'} A(\beta_m, t')\, dt' \right] \tag{13-29a}$$

where

$$A(\beta_m, t') = \frac{\alpha}{k} \bar{g}(\beta_m, t') + \alpha \left[\left. \frac{X(\beta_m, x)}{k_1} \right|_{x=0} f_1(t') + \left. \frac{X(\beta_m, x)}{k_2} \right|_{x=L} f_2(t') \right] \tag{13-29b}$$

$$\bar{F}(\beta_m) = \int_0^L X(\beta_m, x') F(x')\, dx' \tag{13-29c}$$

$$\bar{g}(\beta_m, t') = \int_0^L X(\beta_m, x') g(x', t')\, dx' \tag{13-29d}$$

$$N(\beta_m) = \int_0^L [X(\beta_m, x')]^2 \, dx' \tag{13-29e}$$

If the boundary conditions at $x = 0$ or $x = L$ or both are of the first kind, the following changes should be made in the term $A(\beta_m, t')$ defined by equation (13-29b):

$$\text{When } k_1 = 0, \quad \text{replace} \quad \left.\frac{X(\beta_m, x)}{k_1}\right|_{x=0} \quad \text{by} \quad \left.\frac{1}{h_1}\frac{dX(\beta_m, x)}{dx}\right|_{x=0} \tag{13-29f}$$

$$\text{When } k_2 = 0, \quad \text{replace} \quad \left.\frac{X(\beta_m, x)}{k_2}\right|_{x=0} \quad \text{by} \quad -\left.\frac{1}{h_2}\frac{dX(\beta_m, x)}{dx}\right|_{x=L} \tag{13-29g}$$

We also note that the solution given by equations (13-29) is also immediately obtainable from the general solution (13-18) by setting $\psi(\beta_m, x_1) = X(\beta_m, x)$, $\psi(\beta_m, x_1') = X(\beta_m, x_1')$, and $w(x_1') = 1$.

The eigenfunctions $X(\beta_m, x)$, the normalization integral $N(\beta_m)$, and the eigenvalues β_m appearing in the solution (12-39) are obtainable from Table 2-2, Chapter 2 for the nine different combinations of the boundary conditions.

Alternative Solution. In some cases it is desirable to split-up the solution, $T(x, t)$ as:

$$T(x, t) = \sum_{j=0}^{2} T_{0j}(x, t) + T_h(x, t) - \sum_{j=0}^{2} T_j(x, t) \tag{13-30}$$

where the functions $T_{0j}(x, t)$ are the solutions of the following quasi-steady-state problem

$$\frac{\partial^2 T_{0j}(x, t)}{\partial x^2} + \delta_{0j}\frac{g(x, t)}{k} = 0 \quad \text{in} \quad 0 < x < L \tag{13-31a}$$

$$-k_1\frac{\partial T_{0j}}{\partial x} + h_1 T_{0j} = \delta_{1j} f_1(t) \quad \text{at} \quad x = 0 \tag{13-31b}$$

$$k_2\frac{\partial T_{0j}}{\partial x} + h_2 T_{0j} = \delta_{2j} f_2(t) \quad \text{at} \quad x = L \tag{13-31c}$$

$$\delta_{ij} = \begin{cases} 1 & \text{for} \quad i = j \\ 0 & \text{for} \quad i \neq j \end{cases} \quad \text{and} \quad i, j = 0, 1, 2$$

The function $T_h(x, t)$ is the solution of the following homogeneous problem:

$$\frac{\partial^2 T_h(x, t)}{\partial x^2} = \frac{1}{\alpha}\frac{\partial T_h(x, t)}{\partial t} \quad \text{in} \quad 0 < x < L, \quad t > 0 \tag{13-32a}$$

$$-k_1 \frac{\partial T_h}{\partial x} + h_1 T_h = 0 \quad \text{at} \quad x = 0, \quad t > 0 \tag{13-32b}$$

$$k_2 \frac{\partial T_h}{\partial x} + h_2 T_h = 0 \quad \text{at} \quad x = L, \quad t > 0 \tag{13-32c}$$

$$T_h(x, t) = F(x) - \sum_{j=0}^{2} T_{0j}(x, 0) \quad \text{for} \quad t = 0, \quad \text{in} \quad 0 \leqslant x \leqslant L \tag{13-32d}$$

The functions $T_j(x, t)$ are related to the function $\theta_j(x, \tau, t)$ by the following relation

$$T_j(x, t) = \int_{\tau=0}^{t} \left. \frac{\partial \theta_j(x, \tau', t - \tau)}{\partial \tau'} \right|_{\tau' = \tau} d\tau \tag{13-33}$$

where $\theta_j(x, \tau, t)$ is the solution of the following homogeneous problem

$$\frac{\partial^2 \theta_j}{\partial x^2} = \frac{1}{\alpha} \frac{\partial \theta_j(x, \tau, t)}{\partial t} \quad \text{in} \quad 0 < x < L, \quad t > 0 \tag{13-34a}$$

$$-k_1 \frac{\partial \theta_j}{\partial x} + h_1 \theta_j = 0 \quad \text{at} \quad x = 0, \quad t > 0 \tag{13-34b}$$

$$k_2 \frac{\partial \theta_j}{\partial x} + h_2 \theta_j = 0 \quad \text{at} \quad x = L, \quad t > 0 \tag{13-34c}$$

$$\theta_j(x, \tau, t) = T_{0j}(x, \tau) \quad \text{for} \quad t = 0, \quad \text{in} \quad 0 \leqslant x \leqslant L \tag{13-34d}$$

where $j = 0, 1, 2$. In the following examples we consider some special cases of the one-dimensional problem (13-26).

Example 13-1

Obtain the solution of the following heat conduction problem for a slab by utilizing the general solutions given previously.

$$\frac{\partial^2 T}{\partial x^2} + \frac{1}{k} g(x, t) = \frac{1}{\alpha} \frac{\partial T(x, t)}{\partial t} \quad \text{in} \quad 0 < x < L, \quad t > 0 \tag{13-35a}$$

$$\frac{\partial T}{\partial x} = 0 \quad \text{at} \quad x = 0, \quad t > 0 \tag{13-35b}$$

$$T = 0 \quad \text{at} \quad x = L, \quad t > 0 \tag{13-35c}$$

$$T = F(x) \quad \text{for} \quad t = 0, \quad \text{in} \quad 0 \leqslant x \leqslant L \tag{13-35d}$$

Solution. The solution of this problem is immediately obtainable from the solution (13-29) as

$$T(x,t) = \sum_{m=0}^{\infty} \frac{X(\beta_m, x)}{N(\beta_m)} e^{-\alpha\beta_m^2 t} \left[\bar{F}(\beta_m) + \frac{\alpha}{k} \int_0^t e^{\alpha\beta_m^2 t'} \bar{g}(\beta_m, t')\, dt' \right] \qquad (13\text{-}36)$$

where the integral transforms $\bar{F}(\beta_m)$ and $\bar{g}(\beta_m, t')$ are defined by equations (13-29c) and (13-29d), respectively. The eigenfunctions $X(\beta_m, x)$, the normalization integral $N(\beta_m)$ and the expression defining the eigenvalues β_m are obtained from Table 2-2, case 6, as

$$X(\beta_m, x) = \cos \beta_m x, \qquad \frac{1}{N(\beta_m)} = \frac{2}{L}, \qquad \text{and} \qquad \cos \beta_m L = 0 \qquad (13\text{-}37)$$

Introducing equations (13-47) into (13-46) we find

$$T(x,t) = \frac{2}{L} \sum_{m=1}^{\infty} e^{-\alpha\beta_m^2 t} \cos \beta_m x \left\{ \int_{x'=0}^{L} F(x') \cos \beta_m x'\, dx' \right.$$
$$\left. + \frac{\alpha}{k} \int_{t'=0}^{t} e^{\alpha\beta_m^2 t'} \int_{x'=0}^{L} g(x', t') \cos \beta_m x'\, dx'\, dt' \right\} \qquad (13\text{-}38)$$

where the β_m values are the positive roots of $\cos \beta_m L = 0$ or they are given by $\beta_m = (2m-1)\pi/L, m = 1, 2, 3 \ldots.$

Example 13-2

Obtain the solution of the following heat conduction problem for a slab

$$\frac{\partial^2 T}{\partial x^2} = \frac{1}{\alpha} \frac{\partial T(x,t)}{\partial t} \qquad \text{in} \qquad 0 < x < L, \quad t > 0 \qquad (13\text{-}39a)$$

$$\frac{\partial T}{\partial x} = 0 \qquad \text{at} \qquad x = 0, \qquad t > 0 \qquad (13\text{-}39b)$$

$$T = f_2(t) \qquad \text{at} \qquad x = L, \qquad t > 0 \qquad (13\text{-}39c)$$

$$T = 0 \qquad \text{for} \qquad t = 0, \qquad \text{in} \quad 0 \leqslant x \leqslant L \qquad (13\text{-}39d)$$

Consider the case when the surface temperature is given by $f_2(t) = \gamma t$, where γ is a constant.

Solution. We solve this problem using both the solution (13-29) and its alternative form (13-30).

1. The solution of this problem is immediately obtainable from the solution (13-29) as

$$T(x,t) = -\alpha \sum_{m=1}^{\infty} \frac{X(\beta_m, x)}{N(\beta_m)} e^{-\alpha\beta_m^2 t} \int_{t'=0}^{t} e^{\alpha\beta_m^2 t'} \left[\frac{dX(\beta_m, x)}{dx} \right]_{x=L} f_2(t')dt' \quad (13\text{-}40)$$

where use is made of equation (13-29g) since the boundary condition at $x = L$ is of the first kind. The eigenfunctions $X(\beta_m, x)$, the normalization integral $N(\beta_m)$ and the eigenvalues β_m are the same as those given by equation (13-37). Then the solution (13-40) becomes

$$T(x,t) = \frac{2\alpha}{L} \sum_{m=1}^{\infty} (-1)^{m-1} e^{-\alpha\beta_m^2 t} \beta_m \cos \beta_m x \int_{t'=0}^{t} e^{\alpha\beta_m^2 t'} f_2(t')dt' \quad (13\text{-}41a)$$

since $\beta_m = (2m-1)\pi/L$ and $dX/dx|_{x=L} = -\beta_m \sin \beta_m L = -\beta_m(-1)^{m-1}$. For $f_2(t) = \gamma t$, this result reduces to

$$T(x,t) = \frac{2\alpha\gamma}{L} \sum_{m=1}^{\infty} (-1)^{m-1} e^{-\alpha\beta_m^2 t} \beta_m \cos \beta_m x \int_{t'=0}^{t} t' e^{\alpha\beta_m^2 t'} dt' \quad (13\text{-}41b)$$

The integral term is evaluated as

$$\int_0^t t' e^{\alpha\beta_m^2 t'} dt' = e^{\alpha\beta_m^2 t} \left[\frac{t}{\alpha\beta_m^2} - \frac{1}{\alpha^2\beta_m^4} \right] + \frac{1}{\alpha^2\beta_m^4} \quad (13\text{-}42)$$

Then, the solution (13-41b) takes the form

$$T(x,t) = \gamma t \frac{2}{L} \sum_{m=1}^{\infty} (-1)^{m-1} \frac{\cos \beta_m x}{\beta_m} - \frac{\gamma}{\alpha} \frac{2}{L} \sum_{m=1}^{\infty} (-1)^{m-1} \frac{\cos \beta_m x}{\beta_m^3}$$
$$+ \frac{2\gamma}{\alpha L} \sum_{m=1}^{\infty} (-1)^{m-1} e^{-\alpha\beta_m^2 t} \frac{\cos \beta_m x}{\beta_m^3} \quad (13\text{-}43)$$

Closed-form expressions for the two series are given as (see note 2 at end of this chapter for the derivation of these closed-form expressions)

$$\frac{2}{L} \sum_{m=1}^{\infty} (-1)^{m-1} \frac{\cos \beta_m x}{\beta_m} = 1 \quad (13\text{-}44a)$$

$$\frac{2}{L} \sum_{m=1}^{\infty} (-1)^{m-1} \frac{\cos \beta_m x}{\beta_m^3} = -\frac{1}{2}(x^2 - L^2) \quad (13\text{-}44b)$$

Introducing these results into equation (13-43) the solution becomes

$$T(x,t)=\gamma t+\frac{\gamma}{2\alpha}(x^2-L^2)+\frac{2\gamma}{\alpha L}\sum_{m=1}^{\infty}(-1)^{m-1}e^{-\alpha\beta_m^2 t}\frac{\cos\beta_m x}{\beta_m^3} \tag{13-45}$$

2. We now solve the problem (13-39) by utilizing the alternative form of the solution given by equations (13-30). Then we have

$$T(x,t)=T_{02}(x,t)+T_h(x,t)-T_2(x,t) \tag{13-46}$$

where the function $T_{02}(x,t)$ satisfies the following quasi-steady-state problem

$$\frac{\partial^2 T_{02}}{\partial x^2}=0 \qquad \text{in} \qquad 0<x<L \tag{13-47a}$$

$$\frac{\partial T_{02}}{\partial x}=0 \qquad \text{at} \qquad x=0 \tag{13-47b}$$

$$T_{02}=f_2(t)=\gamma t \qquad \text{at} \qquad x=L \tag{13-47c}$$

the solution of which is

$$T_{02}(x,t)=\gamma t \tag{13-48}$$

The function $T_h(x,t)$ satisfies the following homogeneous problem

$$\frac{\partial^2 T_h}{\partial x^2}=\frac{1}{\alpha}\frac{\partial T_h}{\partial t} \qquad \text{in} \qquad 0<x<L, \quad t>0 \tag{13-49a}$$

$$\frac{\partial T_h}{\partial x}=0 \qquad \text{at} \qquad x=0, \qquad t>0 \tag{13-49b}$$

$$T_h=0 \qquad \text{at} \qquad x=L, \qquad t>0 \tag{13-49c}$$

$$T_h=-T_{02}(x,0)=0 \qquad \text{for} \qquad t=0, \qquad \text{in} \quad 0\leqslant x\leqslant L \tag{13-49d}$$

which has a trivial solution; hence

$$T_h(x,t)=0 \tag{13-50}$$

Finally $T_2(x,t)$ is related to the function $\theta_2(x,t)$ by

$$T_2(x,t)=\int_{\tau=0}^{t}\left.\frac{\partial\theta_2(x,\tau',t-\tau)}{\partial\tau'}\right|_{\tau'=\tau}d\tau \tag{13-51}$$

where $\theta_2(x, \tau, t)$ is the solution of the following homogeneous problem:

$$\frac{\partial^2 \theta_2}{\partial x^2} = \frac{1}{\alpha}\frac{\partial \theta_2}{\partial t} \quad \text{in} \quad 0 < x < L, \quad t > 0 \tag{13-52a}$$

$$\frac{\partial \theta_2}{\partial x} = 0 \quad \text{at} \quad x = 0, \quad t > 0 \tag{13-52b}$$

$$\theta_2 = 0 \quad \text{at} \quad x = L, \quad t > 0 \tag{13-52c}$$

$$\theta_2 = T_{02}(x, \tau) = \gamma\tau \quad \text{for} \quad t = 0, \quad \text{in} \quad 0 \leqslant x \leqslant L \tag{13-52d}$$

When equations (13-52) are solved and $\theta_2(x, \tau, t)$ is introduced into equation (13-51) we obtain

$$T_2(x, t) = \frac{2\gamma}{\alpha L}\sum_{m=1}^{\infty}(-1)^{m-1}(1 - e^{-\alpha\beta_m^2 t})\frac{\cos\beta_m x}{\beta_m^3} \tag{13-53}$$

Introducing equations (13-48), (13-50), and (13-53) into equation (13-46), we find

$$T(x, t) = \gamma t - \frac{\gamma}{\alpha}\frac{2}{L}\sum_{m=1}^{\infty}(-1)^{m-1}\frac{\cos\beta_m x}{\beta_m^3} + \frac{2\gamma}{\alpha L}\sum_{m=1}^{\infty}(-1)^{m-1}e^{-\alpha\beta_m^2 t}\frac{\cos\beta_m x}{\beta_m^3} \tag{13-54}$$

when the closed-form expression (13-44b) is introduced, the solution (13-54) becomes

$$T(x, t) = \gamma t + \frac{\gamma}{2\alpha}(x^2 - L^2) + \frac{2\gamma}{\alpha L}\sum_{m=1}^{\infty}(-1)^{m-1}e^{-\alpha\beta_m^2 t}\frac{\cos\beta_m x}{\beta_m^3} \tag{13-55}$$

which is identical to equation (13-45).

One-Dimensional Problems of Semiinfinite and Infinite Regions

The integral-transform technique developed for the solution of heat conduction problems of finite regions is now extended for the solution of problems of semiinfinite regions. Only one of the space variables, the x variable, needs to be considered, because the same results are applicable for the solution of problems involving y- or z- variables.

Region $\mathbf{0 \leqslant x \leqslant \infty}$**.** To illustrate the basic concepts, we consider the solution of the following one-dimensional, time-dependent heat conduction problem for a semiinfinite region:

$$\frac{\partial^2 T}{\partial x^2}+\frac{1}{k}g(x,t)=\frac{1}{\alpha}\frac{\partial T(x,t)}{\partial t} \quad \text{in} \quad 0<x<\infty, \quad t>0 \tag{13-56a}$$

$$-k_1\frac{\partial T}{\partial x}+h_1 T=f_1(t) \quad \text{at} \quad x=0, \quad t>0 \tag{13-56b}$$

$$T(x,t)=F(x) \quad \text{for} \quad t=0, \quad \text{in} \quad 0\leqslant x<\infty \tag{13-56c}$$

Basic steps in the solution of this problem can be summarized as follows:

1. Develop the appropriate integral transform pair. The integral-transform pair is developed by considering the eigenvalue problem appropriate for the problem (13-56) and then representing the function $T(x,t)$, defined in the interval $0\leqslant x<\infty$, in terms of the eigenfunctions of this eigenvalue problem and then by splitting up the representation into two parts as the *inversion formula* and the *integral transform.*
2. Remove the partial derivative $\partial^2 T/\partial x^2$ from the differential equation (13-56a) by the application of the integral transform and utilizing the eigenvalue problem and the boundary conditions for the heat conduction problem.
3. Solve the resulting ordinary differential equation for the transform of temperature subject to the transformed initial condition. Invert the transform of temperature by the inversion formula to obtain the desired solution.

Step 1 is immediately obtainable from the results available in Chapter 2. That is, the eigenvalue problem is given by equations (2-48) and the representation of a function in the region $0\leqslant x<\infty$ is given by equation (2-52). Then, the integral-transform pair with respect to the x variable of the function $T(x,t)$ is immediately obtained according to equation (2-52) as

$$\text{Inversion formula:} \quad T(x,t)=\int_{\beta=0}^{\infty}\frac{X(\beta,x)}{N(\beta)}\bar{T}(\beta,t)d\beta \tag{13-57a}$$

$$\text{Integral transform:} \quad \bar{T}(\beta,t)=\int_{x'=0}^{\infty}X(\beta,x')T(x',t)dx' \tag{13-57b}$$

where the functions $X(\beta,x)$ and $N(\beta)$ are listed in Table 2-3 for three different boundary conditions at $x=0$. We note that, the eigenvalues β for a semiinfinite medium is continuous, as a result the inversion formula is an integral over β from zero to infinity instead of a summation over the discrete eigenvalues as for the finite region.

Step 2 involves taking the integral transform of equation (13-56a) by the application of the transform (13-57b); that is, we multiply both sides of equation (13-56a) by $X(\beta, x)$ and integrate with respect to x from $x = 0$ to ∞, to obtain

$$\int_0^\infty X(\beta, x)\frac{\partial^2 T}{\partial x^2}dx + \frac{1}{k}\bar{g}(\beta, t) = \frac{1}{\alpha}\frac{d\bar{T}(\beta, t)}{dt}$$

The integral on the left is performed by integrating it by parts twice and utilizing the eigenvalue problem (2-48) and the boundary condition (13-56b). (*See note 3 at end of this chapter for the details of this portion of the analysis.*) Then the resulting equation and the transform of the initial condition (13-56c), respectively, become

$$\frac{d\bar{T}(\beta, t)}{dt} + \alpha\beta^2\bar{T}(\beta, t) = \frac{\alpha}{k}\bar{g}(\beta, t) + \left.\frac{X(\beta, x)}{k_1}\right|_{x=0} \cdot f_1(t) \qquad \text{for} \quad t > 0 \tag{13-58a}$$

$$\bar{T}(\beta, t) = \bar{F}(\beta) \qquad \text{for} \quad t = 0 \tag{13-58b}$$

In step (3), equation (13-58) is solved for $\bar{T}(\beta, t)$ and the result is inverted by the inversion formula (13-57b) to obtain the solution as

$$T(x, t) = \int_{\beta=0}^\infty \frac{X(\beta, x)}{N(\beta)} e^{-\alpha\beta^2 t}\left[\bar{F}(\beta) + \int_0^t e^{\alpha\beta^2 t'} A(\beta, t')dt'\right] \tag{13-59a}$$

where

$$A(\beta, t') = \frac{\alpha}{k}\bar{g}(\beta, t') + \alpha\left.\frac{X(\beta, x')}{k_1}\right|_{x'=0} \cdot f_1(t') \tag{13-59b}$$

$$\bar{F}(\beta) = \int_0^\infty X(\beta, x')F(x')dx' \tag{13-59c}$$

$$\bar{g}(\beta, t') = \int_0^\infty X(\beta, x')g(x', t')dx' \tag{13-59d}$$

If the boundary condition at $x = 0$ is of the first kind (i.e., $k_1 = 0$) the following change should be made in the term $A(\beta, t')$:

$$\text{Replace} \quad \left.\frac{X(\beta, x')}{k_1}\right|_{x'=0} \quad \text{by} \quad \left.\frac{1}{h_1}\frac{dX(\beta, x')}{dx'}\right|_{x'=0} \tag{13-59e}$$

The functions $X(\beta, x)$ and $N(\beta)$ are obtainable from Table 2-3 for three different boundary conditions at $x = 0$.

Region $-\infty < x < \infty$. We now consider the following heat conduction problem for an infinite medium

$$\frac{\partial^2 T}{\partial x^2} + \frac{1}{k} g(x,t) = \frac{1}{\alpha} \frac{\partial T(x,t)}{\partial t} \quad \text{in} \quad -\infty < x < \infty, \quad t > 0 \tag{13-60a}$$

$$T(x,t) = F(x) \quad \text{for} \quad t = 0, \quad \text{in the region} \tag{13-60b}$$

The eigenvalue problem appropriate for the solution of this problem is given by equation (2-64d) and the representation of a function $F^*(x)$, defined in the interval $-\infty < x < \infty$, in terms of the eigenfunctions of this eigenvalue problem is given by equation (2-66d) as

$$F^*(x) = \frac{1}{\pi} \int_{\beta=0}^{\infty} \int_{x'=-\infty}^{\infty} F^*(x') \cos \beta(x' - x) dx' \, d\beta \tag{13-61}$$

This representation is expressed in the alternative form as (see note 4 at the end of this chapter for the derivation)

$$F^*(x) = \frac{1}{2\pi} \int_{\beta=-\infty}^{\infty} e^{-i\beta x} \left[\int_{x'=-\infty}^{\infty} e^{i\beta x} F^*(x') dx' \right] d\beta \tag{13-62}$$

where $i = \sqrt{-1}$.

This expression is now utilized to define the integral-transform pair for the temperature $T(x,t)$ with respect to the x variable as

$$\text{Inversion formula:} \qquad T(x,t) = \frac{1}{2\pi} \int_{\beta=-\infty}^{\infty} e^{-i\beta x} \bar{T}(\beta,t) d\beta \tag{13-63a}$$

$$\text{Integral transform:} \qquad \bar{T}(\beta,t) = \int_{-\infty}^{\infty} e^{i\beta x'} T(x',t) dx' \tag{13-63b}$$

Taking the integral transform of the heat conduction problem (13-60) according to the transform (13-63b), we obtain

$$\frac{d\bar{T}(\beta,t)}{dt} + \alpha\beta^2 \bar{T}(\beta,t) = \frac{\alpha}{k} \bar{g}(\beta,t) \quad \text{for} \quad t > 0 \tag{13-64a}$$

$$\bar{T}(\beta,t) = \bar{F}(\beta) \quad \text{for} \quad t = 0 \tag{13-64b}$$

When this equation is solved for $\bar{T}(\beta,t)$ and the result is inverted by the inversion formula (13-63a), we obtain the solution of the heat conduction problem (13-60) as

$$T(x,t)=\frac{1}{2\pi}\int_{\beta=-\infty}^{\infty} e^{-\alpha\beta^2 t-i\beta x}\left[\bar{F}(\beta)+\frac{\alpha}{k}\int_{t'=0}^{t} e^{\alpha\beta^2 t'}\bar{g}(\beta,t')dt'\right]d\beta \tag{13-65a}$$

where

$$\bar{F}(\beta)=\int_{x'=-\infty}^{\infty} e^{i\beta x'}F(x')dx' \tag{13-65b}$$

$$\bar{g}(\beta,t')=\int_{x'=-\infty}^{\infty} e^{i\beta x'}g(x',t')dx' \tag{13-65c}$$

The order of integration is changed and the result is rearranged as

$$\begin{aligned}T(x,t)=&\frac{1}{2\pi}\int_{x'=-\infty}^{\infty} F(x')dx'\int_{\beta=-\infty}^{\infty} e^{-\alpha\beta^2 t-i\beta(x-x')}d\beta\\&+\frac{1}{2\pi}\frac{\alpha}{k}\int_{t'=0}^{t} dt'\int_{x'=-\infty}^{\infty} g(x',t')dx'\int_{\beta=-\infty}^{\infty} e^{-\alpha\beta^2(t-t')-i\beta(x-x')}d\beta\end{aligned} \tag{13-66}$$

We make use of the following integral

$$\frac{1}{2\pi}\int_{\beta=-\infty}^{\infty} e^{-\alpha\beta^2 t-i\beta x}d\beta=\frac{1}{(4\pi\alpha t)^{1/2}}e^{-x^2/4\alpha t} \tag{13-67}$$

then the solution (13-66) becomes

$$\begin{aligned}T(x,t)=&\frac{1}{(4\pi\alpha t)^{1/2}}\int_{x'=-\infty}^{\infty}\exp\left[-\frac{(x-x')^2}{4\alpha t}\right]F(x')dx'\\&+\frac{\alpha}{k}\int_{t'=0}^{t}\frac{dt'}{[4\pi\alpha(t-t')]^{1/2}}\int_{x'=-\infty}^{\infty}\exp\left[-\frac{(x-x')^2}{4\alpha(t-t')}\right]g(x',t')dx'\end{aligned} \tag{13-68}$$

Example 13-3

Obtain the solution of the following heat conduction problem for a semi-infinite region

$$\frac{\partial^2 T}{\partial x^2}+\frac{1}{k}g(x,t)=\frac{1}{\alpha}\frac{\partial T(x,t)}{\partial t} \quad \text{in} \quad 0<x<\infty, \quad t>0 \tag{13-69a}$$

$$\frac{\partial T}{\partial x}=0 \quad \text{at} \quad x=0, \quad t>0 \tag{13-69b}$$

$$T=F(x) \quad \text{for} \quad t=0, \quad 0\leqslant x<\infty \tag{13-69c}$$

Solution. The solution of this problem is immediately obtainable from equations (13-59) as

$$T(x,t)=\int_{\beta=0}^{\infty}\frac{X(\beta,x)}{N(\beta)}e^{-\alpha\beta^2 t}\left[\bar{F}(\beta)+\frac{\alpha}{k}\int_{t'=0}^{t}e^{\alpha\beta^2 t'}\bar{g}(\beta,t')dt'\right]d\beta \tag{13-70a}$$

where

$$\bar{F}(\beta)=\int_{x'=0}^{\infty}X(\beta,x')F(x')dx' \tag{13-70b}$$

$$\bar{g}(\beta,t')=\int_{x'=0}^{\infty}X(\beta,x')g(x',t')dx' \tag{13-70c}$$

The functions $X(\beta,x)$ and $N(\beta)$ are determined from case 2, Table 2-3, as

$$X(\beta,x)=\cos\beta x \qquad \text{and} \qquad \frac{1}{N(\beta)}=\frac{2}{\pi} \tag{13-71}$$

Introducing equations (13-71) into (13-70) the solution becomes

$$\begin{aligned}T(x,t)=&\frac{2}{\pi}\int_{\beta=0}^{\infty}e^{-\alpha\beta^2 t}\cos\beta x\int_{x'=0}^{\infty}F(x')\cos\beta x'\,dx'\,d\beta\\&+\frac{2\alpha}{\pi k}\int_{\beta=0}^{\infty}e^{-\alpha\beta^2 t}\cos\beta x\int_{t'=0}^{t}e^{\alpha\beta^2 t'}\int_{x'=0}^{\infty}g(x',t')\cos\beta x'\,dx'\,dt'\,d\beta\end{aligned} \tag{13-72}$$

In this expression the orders of integration can be changed and the integrations with respect to β can be performed by making use of the following relation [i.e., obtained by adding equations (2-57b) and (2-57c)]:

$$\begin{aligned}&\frac{2}{\pi}\int_{\beta=0}^{\infty}e^{-\alpha\beta^2 t}\cos\beta x\cos\beta x'\,d\beta\\&=\frac{1}{(4\pi\alpha t)^{1/2}}\left[\exp\left(-\frac{(x-x')^2}{4\alpha t}\right)+\exp\left(-\frac{(x+x')^2}{4\alpha t}\right)\right]\end{aligned} \tag{13-73}$$

Then, the solution (13-82) takes the form

$$\begin{aligned}T(x,t)=&\frac{1}{(4\pi\alpha t)^{1/2}}\int_{x'=0}^{\infty}F(x')\left[\exp\left(-\frac{(x-x')^2}{4\alpha t}\right)+\exp\left(-\frac{(x+x')^2}{4\alpha t}\right)\right]dx'\\&+\frac{\alpha}{k}\int_{t'=0}^{t}\frac{dt'}{[4\pi\alpha(t-t')]^{1/2}}\int_{x'=0}^{\infty}g(x',t')\cdot\end{aligned}$$

$$\cdot\left[\exp\left(-\frac{(x-x')^2}{4\alpha(t-t')}\right)+\exp\left(-\frac{(x+x')^2}{4\alpha(t-t')}\right)\right]dx' \tag{13-74}$$

Several special cases are obtainable from this solution depending on the functional forms of the heat-generation term and the initial condition function.

Multidimensional Problems

The solution of multidimensional, time-dependent heat conduction problems by the integral-transform technique is readily handled by the successive application of one-dimensional integral transforms to remove from the equation one of the partial derivatives with respect to the space variable in each step. In the rectangular coordinate system the order of the integral transformation with respect to the space variables is immaterial. This matter is now illustrated with examples.

Example 13-4

Obtain the solution $T(x, y, t)$ of the following heat conduction problem for a semiinfinite rectangular strip, $0 \leqslant x < \infty, 0 \leqslant y \leqslant b$:

$$\frac{\partial^2 T}{\partial x^2}+\frac{\partial^2 T}{\partial y^2}+\frac{g(x,y,t)}{k}=\frac{1}{\alpha}\frac{\partial T}{\partial t} \quad \text{in} \quad 0<x<\infty, \quad 0<y<b, \quad t>0 \tag{13-75a}$$

$$T=0 \quad \text{at} \quad \text{all boundaries} \tag{13-75b}$$

$$T=0 \quad \text{for} \quad t=0, \quad \text{in the region} \tag{13-75c}$$

The integral-transform pair for $T(x, y, t)$ with respect to the x variable is defined as [see equations (13-57)]

Inversion formula:
$$T(x,y,t)=\int_{\beta=0}^{\infty}\frac{X(\beta,x)}{N(\beta)}\bar{T}(\beta,y,t)d\beta \tag{13-76a}$$

Integral transform:
$$\bar{T}(\beta,y,t)=\int_{x'=0}^{\infty}X(\beta,x')T(x',y,t)dx' \tag{13-76b}$$

and the integral transform pair for $\bar{T}(\beta, y, t)$ with respect to the y variable is defined as [see equation (13-27)]

Inversion formula:
$$\bar{T}(\beta,y,t)=\sum_{n=1}^{\infty}\frac{Y(\gamma_n,y)}{N(\gamma_n)}\bar{\bar{T}}(\beta,\gamma_n,t) \tag{13-76c}$$

Integral transform:
$$\bar{\bar{T}}(\beta,\gamma_n,t)=\int_0^b Y(\gamma_n,y')\bar{T}(\beta,y',t)dy' \tag{13-76d}$$

where the bar denotes the transform with respect to the x variable and the tilde with respect to the y variable.

We take the integral transform of the problem (13-75) first with respect to the x variable using the transform (13-76b) and then with respect to the y variable using the transform (13-76d) to obtain

$$\frac{d\tilde{\bar{T}}}{dt}+\alpha(\beta^2+\gamma_n^2)\tilde{\bar{T}}(\beta,\gamma_n,t)=\frac{\alpha}{k}\tilde{\bar{g}}(\beta,\gamma_n,t) \qquad \text{for} \qquad t>0 \tag{13-77a}$$

$$\tilde{\bar{T}}(\beta,\gamma_n,t)=0 \qquad \text{for} \qquad t=0 \tag{13-77b}$$

Equation (13-87) is solved for $\tilde{\bar{T}}$ and successively inverted by the inversion formulas (13-76c) and (13-76a) to find the solution of the problem (13-75) as

$$T(x,y,t)=\int_{\beta=0}^{\infty}\sum_{n=1}^{\infty}\frac{X(\beta,x)Y(\gamma_n,y)}{N(\beta)N(\gamma_n)}e^{-\alpha(\beta^2+\gamma_n^2)t} \cdot\frac{\alpha}{k}\int_{t'=0}^{t}e^{\alpha(\beta^2+\gamma_n^2)t'}\tilde{\bar{g}}(\beta,\gamma_n,t')dt'\,d\beta \tag{13-78a}$$

where the double transform $\tilde{\bar{g}}(\beta,\gamma_n,t)$ is defined as

$$\tilde{\bar{g}}(\beta,\gamma_n,t)=\int_{y'=0}^{b}\int_{x'=0}^{\infty}X(\beta,x')Y(\gamma_n,y')g(x',y',t')dx'\,dy' \tag{13-78b}$$

The functions $X(\beta,x)$ and $N(\beta)$ are obtained from case 3, Table 2-3, as

$$X(\beta,x)=\sin\beta x, \qquad \frac{1}{N(\beta)}=\frac{2}{\pi} \tag{13-79a}$$

and the functions $Y(\gamma_n,y)$ and $N(\gamma_n)$ from case 9, Table 2-2, as

$$Y(\gamma_n,y)=\sin\gamma_n y, \qquad \frac{1}{N(\gamma_n)}=\frac{2}{b} \tag{13-79b}$$

and the γ_n values are the positive roots of $\sin\gamma_n b=0$. Introducing the results (13-79) into equation (13-78), the solution becomes

$$T(x,y,t)=\frac{4\alpha}{\pi bk}\int_{\beta=0}^{\infty}\sum_{n=1}^{\infty}e^{-\alpha(\beta^2+\gamma_n^2)t}\sin\beta x\sin\gamma_n y \cdot\int_{t'=0}^{t}e^{\alpha(\beta^2+\gamma_n^2)t'}\int_{y'=0}^{b}\int_{x'=0}^{\infty}g(x',y',t')\sin\beta x'\sin\gamma_n y'\,dx'\,dy'\,dt'\,d\beta \tag{13-80}$$

In this solution, the integration with respect to β can be performed by making use of the following result [see equation (2-57d)]:

$$\frac{2}{\pi}\int_{\beta=0}^{\infty} e^{-\alpha\beta^2(t-t')}\sin\beta x\sin\beta x'\,d\beta$$

$$=\frac{1}{[4\pi\alpha(t-t')]^{1/2}}\left[\exp\left(-\frac{(x-x')^2}{4\alpha(t-t')}\right)-\exp\left(-\frac{(x+x')^2}{4\alpha(t-t')}\right)\right] \tag{13-81}$$

Then, the solution (13-80) takes the form

$$T(x,y,t)=\frac{2\alpha}{bk}\sum_{n=1}^{\infty} e^{-\alpha\gamma_n^2 t}\sin\gamma_n y\int_{t'=0}^{t}\frac{e^{\alpha\gamma_n^2 t'}}{[4\pi\alpha(t-t')]^{1/2}}$$

$$\cdot\int_{y'=0}^{b}\int_{x'=0}^{\infty} g(x',y',t')\sin\gamma_n y'$$

$$\cdot\left[\exp\left(-\frac{(x-x')^2}{4\alpha(t-t')}\right)-\exp\left(-\frac{(x+x')^2}{4\alpha(t-t')}\right)\right]dx'\,dy'\,dt' \tag{13-82}$$

Example 13-5

Obtain the solution $T(x,y,z,t)$ of the following heat conduction problem for a rectangular parallelepiped $0 \leqslant x \leqslant a, 0 \leqslant y \leqslant b, 0 \leqslant z \leqslant c$.

$$\frac{\partial^2 T}{\partial x^2}+\frac{\partial^2 T}{\partial y^2}+\frac{\partial^2 T}{\partial z^2}+\frac{g(x,y,z,t)}{k}=\frac{1}{\alpha}\frac{\partial T}{\partial t} \quad \text{in} \quad 0<x<a,\quad 0<y<b,\quad 0<z<c,\quad t>0 \tag{13-83a}$$

$$T=0 \quad \text{at} \quad \text{all boundaries} \tag{13-83b}$$

$$T=0 \quad \text{for} \quad t=0,\quad \text{in the region} \tag{13-83c}$$

Solution. This problem can be solved by the successive application of the one-dimensional integral transform to the x, y, and z variables, solving the resulting ordinary differential equation and then inverting the transform of temperature successively. It is also possible to write the solution immediately from the general solution (13-15) by setting

$$\psi_m(\mathbf{r}) \rightarrow X(\beta_m,x)\cdot Y(\gamma_n,y)\cdot Z(\eta_p,z)$$

$$\lambda_m^2 \rightarrow (\beta_m^2+\gamma_n^2+\eta_p^2)$$

$$\sum_m \rightarrow \sum_{m=1}^{\infty}\sum_{n=1}^{\infty}\sum_{p=1}^{\infty} \quad \text{and} \quad \int_R dv \rightarrow \int_{x'=0}^{a}\int_{y'=0}^{b}\int_{z'=0}^{c} dx'\,dy'\,dz'$$

We obtain

$$T(x,y,z,t)=\sum_{m=1}^{\infty}\sum_{n=1}^{\infty}\sum_{p=1}^{\infty}\frac{X(\beta_m,x)Y(\gamma_n,y)Z(\eta_p,z)}{N(\beta_m)N(\gamma_n)N(\eta_p)}e^{-\alpha(\beta_m^2+\gamma_n^2+\eta_p^2)t}$$

$$\cdot\frac{\alpha}{k}\int_{t'=0}^{t}e^{\alpha(\beta_m^2+\gamma_n^2+\eta_p^2)t'}\cdot\bar{\bar{\bar{g}}}(\beta_m,\gamma_n,\eta_p,t')dt' \tag{13-84a}$$

where the triple transform is defined as

$$\bar{\bar{\bar{g}}}(\beta_m,\gamma_n,\eta_p,t')=\int_{x'=0}^{a}\int_{y'=0}^{b}\int_{z'=0}^{c}X(\beta_m,x')Y(\gamma_n,y')Z(\eta_p,z')g(x',y',z',t')dx'\,dy'\,dz' \tag{13-84b}$$

The eigenfunctions, the normalization integrals and the eigenvalues are obtained from case 9, Table 2-2, as

$$X(\beta_m,x)=\sin\beta_m x,\qquad N(\beta_m)=\frac{a}{2},\qquad \sin\beta_m a=0 \tag{13-85a}$$

$$Y(\gamma_n,y)=\sin\gamma_n y,\qquad N(\gamma_n)=\frac{b}{2},\qquad \sin\gamma_n b=0 \tag{13-85b}$$

$$Z(\eta_p,z)=\sin\eta_p z,\qquad N(\eta_p)=\frac{c}{2},\qquad \sin\eta_p c=0 \tag{13-85c}$$

13-3 APPLICATIONS IN THE CYLINDRICAL COORDINATE SYSTEM

To solve the heat conduction problems in the cylindrical coordinate system with the integral-transform technique, appropriate integral-transform pairs are needed in the r, ϕ, and z variables. The integral transform pairs for the z variable depends on whether the range of z is finite, semiinfinite, or infinite as well as the boundary conditions associated with it. Since the transform pairs for the z variable are exactly the same as those discussed previously for the rectangular coordinate system, this matter is not considered here any further. Therefore, in this section we develop the integral transform pairs for the r and ϕ variables and illustrate their application to the solution of heat conduction problems involving (r,t), (r,ϕ,t), (r,z,t), and (r,ϕ,z,t) variables.

Problems in (r,t) Variables

The one dimensional, time-dependent heat conduction problems in the r variable may be confined to any one of the regions $0\leqslant r\leqslant b, a\leqslant r\leqslant b, 0\leqslant r<\infty$, and

$a \leqslant r < \infty$. The integral-transform pair for each of these cases is different. Therefore, we develop the appropriate transform pairs and illustrate the methods of solution for each of these cases.

***Problems of Region* $0 \leqslant r \leqslant b$.** We consider the following heat conduction problem for a solid cylinder of radius $r = b$:

$$\frac{\partial^2 T}{\partial r^2} + \frac{1}{r}\frac{\partial T}{\partial r} + \frac{g(r,t)}{k} = \frac{1}{\alpha}\frac{\partial T}{\partial t} \quad \text{in} \quad 0 \leqslant r < b, \quad t > 0 \tag{13-86a}$$

$$k_2 \frac{\partial T}{\partial r} + h_2 T = f_2(t) \quad \text{at} \quad r = b, \quad t > 0 \tag{13-86b}$$

$$T = F(r) \quad \text{for} \quad t = 0, \quad \text{in} \quad 0 \leqslant r \leqslant b \tag{13-86c}$$

The appropriate eigenvalue problem is given by equations (3-18) for the case $\nu = 0$ since the problem considered here possesses azimuthal symmetry. The integral-transform pair with respect to the r variable for the function $T(r, t)$ is determined according to the representation (3-23) by setting $\nu = 0$. We obtain

$$\text{Inversion formula:} \qquad T(r,t) = \sum_{m=1}^{\infty} \frac{R_0(\beta_m, r)}{N(\beta_m)} \bar{T}(\beta_m, t) \tag{13-87a}$$

$$\text{Integral transform:} \qquad \bar{T}(\beta_m, t) = \int_{r'=0}^{b} r' R_0(\beta_m, r) T(r', t)\, dr' \tag{13-87b}$$

where the functions $R_0(\beta_m, r)$, $N(\beta_m)$ and the eigenvalues β_m are obtainable from Table 3-1 for three different boundary conditions by setting $\nu = 0$.

To solve problem (13-86), we take integral transform of equation (13-86a) according to the transform (13-87b). That is we operate on both sides of equation (13-86a) by the operator $\int_0^b r R_0(\beta_m, r)\, dr$ and obtain

$$\int_0^b r R_0(\beta_m, r)\left[\frac{\partial^2 T}{\partial r^2} + \frac{1}{r}\frac{\partial T}{\partial r}\right] dr + \frac{1}{k}\bar{g}(\beta_m, t) = \frac{1}{\alpha}\frac{d\bar{T}(\beta, t)}{dt} \tag{13-88}$$

The integral on the left is evaluated either by integrating it by parts twice or by using Green's theorem and then utilizing the boundary conditions (3-18b) for $\nu = 0$ and (13-86b); we find

$$\int_0^b r R_0(\beta_m, r)\left[\frac{\partial^2 T}{\partial r^2} + \frac{1}{r}\frac{\partial T}{\partial r}\right] dr = -\beta_m^2 \bar{T}(\beta_m, r) + b \left.\frac{R_0(\beta_m, r)}{k_2}\right|_{r=b} f_2(t) \tag{13-89}$$

Introducing this expression into equation (13-88) and taking the integral transform of the initial condition (13-86c) we obtain

$$\frac{d\bar{T}(\beta_m, t)}{dt} + \alpha\beta_m^2 \bar{T}(\beta_m, t) = \frac{\alpha}{k}\bar{g}(\beta_m, t) + \alpha b \left.\frac{R_0(\beta_m, r)}{k_2}\right|_{r=b} f_2(t) \tag{13-90a}$$

$$\bar{T}(\beta_m, t) = \bar{F}(\beta) \qquad \text{for} \qquad t = 0 \tag{13-90b}$$

Equation (13-90) is solved for $\bar{T}(\beta_m, t)$ and inverted by the inversion formula (13-87a) to yield the solution of the problem (13-86) as

$$T(r, t) = \sum_{m=1}^{\infty} \frac{R_0(\beta_m, r)}{N(\beta_m)} e^{-\alpha\beta_m^2 t}\left[\bar{F}(\beta_m) + \int_0^t e^{\alpha\beta_m^2 t'} A(\beta_m, t')\, dt'\right] \tag{13-91a}$$

where

$$A(\beta_m, t') = \frac{\alpha}{k}\bar{g}(\beta_m, t') + \alpha b \left.\frac{R_0(\beta_m, r)}{k_2}\right|_{r=b} f_2(t') \tag{13-91b}$$

$$\bar{F}(\beta_m) = \int_0^b r' R_0(\beta_m, r') F(r')\, dr' \tag{13-91c}$$

$$\bar{g}(\beta_m, t') = \int_0^b r' R_0(\beta_m, r') g(r', t')\, dr' \tag{13-91d}$$

$$N(\beta_m) = \int_0^b r' [R_0(\beta_m, r')]^2\, dr' \tag{13-91e}$$

Here, $R_0(\beta_m, r)$, $N(\beta_m)$ and β_m are obtained from Table 3-1 by setting $\nu = 0$. For a boundary condition of the first kind at $r = b$, the following change should be made in equation (13-91b):

$$\text{Replace} \quad \left.\frac{R_0(\beta_m, r)}{k_2}\right|_{r=b} \quad \text{by} \quad -\frac{1}{h_2}\left.\frac{dR_0(\beta_m, r)}{dr}\right|_{r=b} \tag{13-91f}$$

Problems of Region $a \leqslant r \leqslant b$. We now consider the heat conduction problem for a hollow cylinder $a \leqslant r \leqslant b$ given as

$$\frac{\partial^2 T}{\partial r^2} + \frac{1}{r}\frac{\partial T}{\partial r} + \frac{g(r, t)}{k} = \frac{1}{\alpha}\frac{\partial T}{\partial t} \qquad \text{in} \qquad a < r < b, \quad t > 0 \tag{13-92a}$$

$$-k_1 \frac{\partial T}{\partial r} + h_1 T = f_1(t) \qquad \text{at} \qquad r = a, \qquad t > 0 \tag{13-92b}$$

$$k_2 \frac{\partial T}{\partial r} + h_2 T = f_2(t) \qquad \text{at} \qquad r = b, \qquad t > 0 \tag{13-92c}$$

$$T = F(r) \qquad \text{for} \qquad t=0, \qquad \text{in} \quad a \leqslant r \leqslant b \tag{13-92d}$$

The eigenvalue problem is given by equations (3-47) for $v=0$, and the integral-transform pair is obtained according to equation (3-51) by setting $v=0$. We find

$$\text{Inversion formula:} \qquad T(r,t) = \sum_{m=1}^{\infty} \frac{R_0(\beta_m, r)}{N(\beta_m)} \bar{T}(\beta_m, t) \tag{13-93a}$$

$$\text{Integral transform:} \qquad \bar{T}(\beta_m, t) = \int_{r'=a}^{b} r' R_0(\beta_m, r') T(r', t)\, dr' \tag{13-93b}$$

where the functions $R_0(\beta_m, r)$, $N(\beta_m)$ and the eigenvalues β_m are obtainable from Table 3-3 by setting $v=0$ for any combination of boundary conditions of the first and second kinds.

We now take the integral transform of the system (13-92) by the application of the transform (13-93b), utilize the eigenvalue problem (3-47) for $v=0$ as described previously, solve for the transform of temperature, and invert the result by the inversion formula (13-93a) to obtain the solution for the temperature as

$$T(r,t) = \sum_{m=1}^{\infty} \frac{R_0(\beta_m, r)}{N(\beta_m)} e^{-\alpha\beta_m^2 t} \left[\bar{F}(\beta_m) + \int_0^t e^{\alpha\beta_m^2 t'} A(\beta_m, t')\, dt' \right] \tag{13-94a}$$

where

$$A(\beta_m, t') = \frac{\alpha}{k} \bar{g}(\beta_m, t') + \alpha \left[a \left. \frac{R_0(\beta_m, r)}{k_1} \right|_{r=a} f_1(t') + b \left. \frac{R_0(\beta_m, r)}{k_2} \right|_{r=b} f_2(t') \right] \tag{13-94b}$$

$$\bar{F}(\beta_m) = \int_a^b r' R_0(\beta_m, r') F(r')\, dr' \tag{13-94c}$$

$$\bar{g}(\beta_m, t') = \int_a^b r' R_0(\beta_m, r') g(r', t')\, dr' \tag{13-94d}$$

$$N(\beta_m) = \int_a^b r' [R_0(\beta_m, r')]^2\, dr' \tag{13-94e}$$

Here, $R_0(\beta_m, r)$, $N(\beta_m)$, and β_m are obtainable from Table 3-2 by setting $v=0$. For a boundary condition of the first kind the following changes should be made in equation (13-94b).

$$\text{When } k_1 = 0, \quad \text{replace} \quad \left. \frac{R_0(\beta_m, r)}{k_1} \right|_{r=a} \quad \text{by} \quad \left. \frac{1}{h_1} \frac{dR_0(\beta_m, r)}{dr} \right|_{r=a} \tag{13-94f}$$

$$\text{When } k_2 = 0, \quad \text{replace} \quad \left. \frac{R_0(\beta_m, r)}{k_2} \right|_{r=b} \quad \text{by} \quad \left. -\frac{1}{h_2} \frac{dR_0(\beta_m, r)}{dr} \right|_{r=b} \tag{13-94g}$$

Problems of Region $0 \leqslant r < \infty$. We consider the following heat conduction problem for an infinite region $0 \leqslant r < \infty$:

$$\frac{\partial^2 T}{\partial r^2} + \frac{1}{r}\frac{\partial T}{\partial r} + \frac{g(r,t)}{k} = \frac{1}{\alpha}\frac{\partial T}{\partial t} \quad \text{in} \quad 0 \leqslant r < \infty, \quad t > 0 \tag{13-95a}$$

$$T = F(r) \quad \text{for} \quad t = 0, \quad \text{in} \quad 0 \leqslant r < \infty \tag{13-95b}$$

The appropriate eigenvalue problem is given by equations (3-35) for $v = 0$, and the integral-transform pair is constructed according to the representation (3-38) for $v = 0$; we obtain

$$\text{Inversion formula:} \qquad T(r,t) = \int_{\beta=0}^{\infty} \beta J_0(\beta r)\bar{T}(\beta, t)\, d\beta \tag{13-96a}$$

$$\text{Integral transform:} \qquad \bar{T}(\beta, t) = \int_{r'=0}^{\infty} r' J_0(\beta r') T(r', t)\, dr' \tag{13-96b}$$

We take the integral transform of the system (13-95) by the application of the transform (13-96b), utilize the eigenvalue problem (3-38) for $v = 0$ as discussed previously, solve for the transform of the temperature and invert the result by the inversion formula (13-96a). We obtain

$$T(r,t) = \int_{\beta=0}^{\infty} \beta J_0(\beta r) e^{-\alpha\beta^2 t}\left[\bar{F}(\beta) + \frac{\alpha}{k}\int_{t'=0}^{t} e^{\alpha\beta^2 t'}\bar{g}(\beta, t')\, dt'\right] d\beta \tag{13-97a}$$

where

$$\bar{F}(\beta) = \int_{r'=0}^{\infty} r' J_0(\beta r') F(r')\, dr' \tag{13-97b}$$

$$\bar{g}(\beta, t') = \int_{r'=0}^{\infty} r' J_0(\beta r') g(r', t')\, dr' \tag{13-97c}$$

Introducing (13-97b, c) into (13-97a) and changing the order of integrations we find

$$\begin{aligned} T(r,t) = &\int_{r'=0}^{\infty} r' F(r') \left[\int_{\beta=0}^{\infty} e^{-\alpha\beta^2 t}\beta J_0(\beta r) J_0(\beta r')\, d\beta\right] dr' \\ &+ \frac{\alpha}{k}\int_{r'=0}^{\infty}\int_{t'=0}^{t} r' g(r', t')\left[\int_{\beta=0}^{\infty} e^{-\alpha\beta^2(t-t')}\beta J_0(\beta r) J_0(\beta r')\, d\beta\right] dt'\, dr' \end{aligned} \tag{13-98}$$

We now consider the following integral [36, p. 395]

$$\int_{\beta=0}^{\infty} e^{-\alpha\beta^2 t}\beta J_\nu(\beta r)J_\nu(\beta r')\,d\beta = \frac{1}{2\alpha t}\exp\left[-\frac{r^2+r'^2}{2\alpha t}\right]I_\nu\left(\frac{rr'}{2\alpha t}\right) \tag{13-99}$$

By setting $\nu = 0$ in equation (13-99) and introducing the resulting expression into equation (13-98) we obtain

$$\begin{aligned} T(r,t) &= \frac{1}{2\alpha t}\int_{r'=0}^{\infty} r'\exp\left[-\frac{r^2+r'^2}{4\alpha t}\right]F(r')I_0\left(\frac{rr'}{2\alpha t}\right)dr' \\ &+ \frac{1}{2k}\int_{r'=0}^{\infty}\int_{t'=0}^{t}\frac{r'}{t-t'}\exp\left[-\frac{r^2+r'^2}{4\alpha(t-t')}\right]g(r',t')I_0\left(\frac{rr'}{2\alpha(t-t')}\right)dt'\,dr' \end{aligned} \tag{13-100}$$

Problems of Region $a \leqslant r < \infty$. We now consider the following heat conduction problem for a semiinfinite region $a \leqslant r < \infty$

$$\frac{\partial^2 T}{\partial r^2} + \frac{1}{r}\frac{\partial T}{\partial r} + \frac{g(r,t)}{k} = \frac{1}{\alpha}\frac{\partial T}{\partial t} \quad \text{in} \quad a < r < \infty, \quad t > 0 \tag{13-101a}$$

$$-k_1\frac{\partial T}{\partial r} + h_1 T = f_1(t) \quad \text{at} \quad r = a, \quad t > 0 \tag{13-101b}$$

$$T = F(r) \quad \text{for} \quad t = 0, \quad \text{in} \quad a \leqslant r < \infty \tag{13-101c}$$

The eigenvalue problem is given by equations (3-39) and the desired integral transform pair is obtained according to the representation (3-40) as

$$\text{Inversion formula:} \qquad T(r,t) = \int_{\beta=0}^{\infty}\frac{\beta}{N(\beta)}R_0(\beta,r)\bar{T}(\beta,t)\,d\beta \tag{13-102a}$$

$$\text{Integral transform:} \qquad \bar{T}(\beta,t) = \int_{r'=a}^{\infty} r' R_0(\beta,r')T(r',t)\,dr' \tag{13-102b}$$

where the functions $R_0(\beta, r)$ and $N(\beta)$ are available from Table 3-2 for the three different boundary conditions at $r = a$. The problem is solved by taking integral transform of the system (13-101) according to the transform (13-102b), utilizing the eigenvalue problem (3-39) as discussed previously, solving for the transform of the temperature and inverting the transform by the inversion formula (13-102a). We obtain

$$T(r,t) = \int_{\beta=0}^{\infty}\frac{\beta}{N(\beta)}R_0(\beta,r)e^{-\alpha\beta^2 t}\left[\bar{F}(\beta) + \int_{t'=0}^{\infty} e^{\alpha\beta^2 t'}A(\beta,t')\,dt'\right]d\beta \tag{13-103a}$$

where

$$A(\beta, t') = \frac{\alpha}{k}\bar{g}(\beta, t') + \alpha a \frac{R_0(\beta, r)}{k_1}\bigg|_{r=a} f_1(t') \tag{13-103b}$$

$$\bar{F}(\beta) = \int_{r'=a}^{\infty} r' R_0(\beta, r') F(r')\, dr' \tag{13-103c}$$

$$\bar{g}(\beta, t') = \int_{r'=a}^{\infty} r' R_0(\beta, r') g(r', t')\, dr' \tag{13-103d}$$

Problems in (r, ϕ, t) Variables

When the partial derivatives with respect to the r and ϕ variables are to be removed from the heat conduction equation, the order of integral transformation is important. *It should be applied first with respect to the ϕ variable and then to the r variable.* Therefore, we need the integral-transform pair that will remove from the differential equation the partial derivative with respect to the ϕ variable, that is $\partial^2 T/\partial\phi^2$. The ranges of the ϕ variable in the cylindrical coordinate system include $0 \leqslant \phi \leqslant 2\pi$ as in the case of problems of *full cylinder* and $0 \leqslant \phi \leqslant \phi_0$, for $\phi_0 < 2\pi$, as in the case of problems of a *portion of the cylinder*. The integral transform pairs for each of these two situations are different. Therefore, we first develop the integral-transform pairs with respect to the ϕ variable for these two cases and then present its application in the solution of heat conduction problems.

Transform Pair for $0 \leqslant \phi \leqslant 2\pi$. In this case since the region in the ϕ variable is a full circle, no boundary conditions are specified in ϕ except the requirement that the function should be cyclic with a period of 2π. The appropriate eigenvalue problem in ϕ is given by equation (3-52a) and the representation of a function in the interval $0 \leqslant \phi \leqslant 2\pi$ in terms of the eigenfunctions of this eigenvalue problem is given by equation (3-55b). Therefore, the integral-transform pair with respect to the ϕ variable for the function $T(r, \phi, t)$ is obtained by splitting up the representation (3-55b) into two parts as

Inversion formula:
$$T(r, \phi, t) = \frac{1}{\pi}\sum_{\nu} \bar{T}(r, \nu, t) \tag{13-104a}$$

Integral transform:
$$\bar{T}(r, \nu, t) = \int_{\phi'=0}^{2\pi} \cos \nu(\phi - \phi') T(r, \phi', t)\, d\phi' \tag{13-104b}$$

where $\nu = 0, 1, 2, 3 \ldots$ and replace π by 2π for $\nu = 0$.

Transform Pair for $0 \leqslant \phi \leqslant \phi_0 (\phi_0 < 2\pi)$. The region being a portion of a circle, boundary conditions are needed at $\phi = 0$ and $\phi = \phi_0$. Here, we consider boundary conditions of the *first* and *second* kind only. For example, for boundary conditions of the first kind at both boundaries, $\phi = 0$ and $\phi = \phi_0$, the eigenvalue problem for the ϕ variable is given by

$$\frac{d^2\Phi(\phi)}{d\phi^2} + v^2\Phi(\phi) = 0 \quad \text{in} \quad 0 < \phi < \phi_0 (< 2\pi) \tag{13-105a}$$

$$\Phi(\phi) = 0 \quad \text{at} \quad \phi = 0 \tag{13-105b}$$

$$\Phi(\phi) = 0 \quad \text{at} \quad \phi = \phi_0 \tag{13-105c}$$

which is exactly of the same form as that given by equations (2-32) for the one-dimensional finite region $0 \leqslant x \leqslant L$ in the rectangular coordinate system. Therefore, the integral-transform pair in the ϕ variable for the function $T(r, \phi, t)$, defined in the interval $0 \leqslant \phi \leqslant \phi_0$, is taken as

$$\text{Inversion formula:} \qquad T(r, \phi, t) = \sum_v \frac{\Phi(v, \phi)}{N(v)} \bar{T}(r, v, t) \tag{13-106a}$$

$$\text{Integral transform:} \qquad \bar{T}(r, v, t) = \int_{\phi=0}^{\phi_0} \Phi(v, \phi') T(r, \phi', t)\, d\phi' \tag{13-106b}$$

where

$$N(v) = \int_0^{\phi_0} [\Phi(v, \phi)]^2\, d\phi \tag{13-106c}$$

For any combination of boundary conditions of the first and second kind, the functions $\Phi(v, \phi)$, $N(v)$ and the eigenvalues v are obtainable from Table 2-2, by appropriate change in the notation.

Having established the integral-transform pairs needed for the removal of the differential operator $\partial^2 T/\partial\phi^2$ from the heat conduction equation, we now proceed to the solution of heat conduction problems involving (r, ϕ, t) variables.

Problems of Region $0 \leqslant r \leqslant b$, $0 \leqslant \phi \leqslant 2\pi$. We consider the following time-dependent heat conduction problem for a solid cylinder of radius $r = b$, in which temperature varies both r and ϕ variables:

$$\frac{\partial^2 T}{\partial r^2} + \frac{1}{r}\frac{\partial T}{\partial r} + \frac{1}{r^2}\frac{\partial^2 T}{\partial \phi^2} + \frac{g(r, \phi, t)}{k} = \frac{1}{\alpha}\frac{\partial T(r, \phi, t)}{\partial t}$$

$$\text{in} \quad 0 \leqslant r < b, \quad 0 \leqslant \phi \leqslant 2\pi, \quad t > 0 \tag{13-107a}$$

$$k_2 \frac{\partial T}{\partial r} + h_2 T = f_2(\phi, t) \qquad \text{at} \qquad r = b, \qquad t > 0 \tag{13-107b}$$

$$T(r, \phi, t) = F(r, \phi) \qquad \text{for} \qquad t = 0, \qquad \text{in the region} \tag{13-107c}$$

This problem is now solved by successive application of the integral transforms with respect to the ϕ and r variables.

The integral-transform pair in the ϕ variable for the function $T(r, \phi, t)$ is given by equation (13-104). Hence we have

$$\text{Inversion formula:} \qquad T(r, \phi, t) = \frac{1}{\pi} \sum_{\nu} \bar{T}(r, \nu, t) \tag{13-108a}$$

$$\text{Integral transform:} \qquad \bar{T}(r, \nu, t) = \int_{\phi'=0}^{2\pi} \cos \nu(\phi - \phi') T(r, \phi', t)\, d\phi' \tag{13-108b}$$

where $\nu = 0, 1, 2, 3 \ldots$ and replace π by 2π for $\nu = 0$. The integral transform of the system (13-107) by the application of the transform (13-108b) yields (see note 5 at the end of this chapter for the details)

$$\frac{\partial^2 \bar{T}}{\partial r^2} + \frac{1}{r} \frac{\partial \bar{T}}{\partial r} - \frac{\nu^2}{r^2} \bar{T} + \frac{\bar{g}(r, \nu, t)}{k} = \frac{1}{\alpha} \frac{\partial \bar{T}(r, \nu, t)}{\partial t} \qquad \text{in} \qquad 0 \leqslant r < b, \quad t > 0 \tag{13-109a}$$

$$k_2 \frac{\partial \bar{T}}{\partial r} + h_2 \bar{T} = \bar{f}_2(\nu, t) \qquad \text{at} \qquad r = b, \qquad t > 0 \tag{13-109b}$$

$$\bar{T}(r, \nu, t) = \bar{F}(r, \nu) \qquad \text{for} \qquad t = 0, \qquad \text{in } 0 \leqslant r \leqslant b \tag{13-109c}$$

where the bar denotes the integral transform with respect to the ϕ variable.

The integral-transform pair in the r variable for the function $\bar{T}(r, \nu, t)$ is obtainable according to the representation (3-22). We find

$$\text{Inversion formula:} \qquad \bar{T}(r, \nu, t) = \sum_{m=1}^{\infty} \frac{R_\nu(\beta_m, r)}{N(\beta_m)} \tilde{\bar{T}}(\beta_m, \nu, t) \tag{13-110a}$$

$$\text{Integral transform:} \qquad \tilde{\bar{T}}(\beta_m, \nu, t) = \int_0^b r' R_\nu(\beta_m, r') \bar{T}(r', \nu, t)\, dr' \tag{13-110b}$$

Here, the tilde denotes the integral transform with respect to the r variable. $R_\nu(\beta_m, r)$ and β_m are the eigenfunctions and eigenvalues associated with the eigenvalue problem given by equations (3-18). The functions $R_\nu(\beta_m, r)$, $N(\beta_m)$ and the eigenvalues β_m are obtainable from Table 3-1 for the three different boundary conditions at $r = b$.

The integral transform of the system (13-109) by the application of the transform (13-110b) yields (see note 6 at the end of this chapter for the details)

$$\frac{d\bar{\tilde{T}}}{dt}+\alpha\beta_m^2\bar{\tilde{T}}(\beta_m,\nu,t)=\frac{\alpha}{k}\bar{\tilde{g}}(\beta_m,\nu,t)+\alpha b\left.\frac{R_\nu(\beta_m,r)}{k_2}\right|_{r=b}\cdot\bar{f}_2(\nu,t) \tag{13-111a}$$

$$\bar{\tilde{T}}(\beta_m,\nu,t)=\bar{\tilde{F}}(\beta_m,\nu)\qquad\text{for}\qquad t=0 \tag{13-111b}$$

Equation (13-111) is solved for $\bar{\tilde{T}}(\beta_m,\nu,t)$ and the resulting double transform is successively inverted by the inversion formulas (13-110a) and (13-108a). Then, the solution of the problem (13-107) becomes

$$T(r,\phi,t)=\frac{1}{\pi}\sum_\nu\sum_{m=1}^{\infty}\frac{R_\nu(\beta_m,r)}{N(\beta_m)}\cdot e^{-\alpha\beta_m^2t}\cdot\left[\bar{\tilde{F}}(\beta_m,\nu)+\int_{t'=0}^{t}e^{\alpha\beta_m^2t'}A(\beta_m,\nu,t')\,dt'\right] \tag{13-112a}$$

where $\nu=0,1,2,3\ldots$ and replace π by 2π for $\nu=0$

$$A(\beta_m,\nu,t')=\frac{\alpha}{k}\bar{\tilde{g}}(\beta_m,\nu,t')+\alpha b\left.\frac{R_\nu(\beta_m,r)}{k_2}\right|_{r=b}\bar{f}_2(\nu,t') \tag{13-112b}$$

$$\bar{f}(\nu,t)=\int_{\phi'=0}^{2\pi}f_2(\phi',t)\cos\nu(\phi-\phi')\,d\phi' \tag{13-112c}$$

$$\bar{\tilde{F}}(\beta_m,\nu)=\int_{r'=0}^{b}\int_{\phi'=0}^{2\pi}r'R_\nu(\beta_m,r')\cos\nu(\phi-\phi')F(r',\phi')\,d\phi'\,dr' \tag{13-112d}$$

$$\bar{\tilde{g}}(\beta_m,\nu,t')=\int_{r'=0}^{b}\int_{\phi'=0}^{2\pi}R_\nu(\beta_m,r')\cos\nu(\phi-\phi')g(r',\phi',t')\,d\phi'\,dr' \tag{13-112e}$$

and the functions $R_\nu(\beta_m,r)$, $N(\beta_m)$ and the eigenvalues β_m are obtainable from Table 3-1.

For a boundary condition of the first kind at $r=b$, the following changes should be made in equation (3-112b).

$$\text{When } k_2=0,\text{ replace }\left.\frac{R_\nu(\beta_m,r)}{k_2}\right|_{r=b}\quad\text{by}\quad-\frac{1}{h_2}\left.\frac{dR_\nu(\beta_m,r)}{dr}\right|_{r=b} \tag{13-112f}$$

Problems of Region $a\leqslant r\leqslant b$, $0\leqslant\phi\leqslant 2\pi$. The extension of the above analysis for solid cylinder to the solution of time-dependent heat conduction problem of a hollow cylinder $a\leqslant r\leqslant b$, in which temperature varies with both r and ϕ variables is a straightforward matter. Clearly, the heat conduction problem (13-107) will involve an additional boundary condition at $r=a$. The definition of the integral-transform pair (13-108) remains the same, but that of (13-110) is modified by changing the lower limit of the integration to $r = a$; then the functions $R_\nu(\beta_m, r)$, $N(\beta_m)$ and the eigenvalues β_m are to be obtained from Table 3-3. As a result, the solution (3-112) will include an additional term in the definition of $A(\beta_m, \nu, t')$ for the effects of the boundary condition at $r = a$ and the lower limit of the integrations with respect to r' will be $r' = a$.

Problems of Region $0 \leqslant r \leqslant b, 0 \leqslant \phi \leqslant \phi_0 (< 2\pi)$. We now consider the solution by the integral-transform technique of the following time-dependent heat conduction problem for a portion of a solid cylinder of radius $r = b$, in the region $0 \leqslant \phi \leqslant \phi_0 (< 2\pi)$

$$\frac{\partial^2 T}{\partial r^2} + \frac{1}{r}\frac{\partial T}{\partial r} + \frac{1}{r^2}\frac{\partial^2 T}{\partial \phi^2} + \frac{g(r,\phi,t)}{k} = \frac{1}{\alpha}\frac{\partial T(r,\phi,t)}{\partial t}$$

$$\text{in} \quad 0 \leqslant r < b, \quad 0 < \phi < \phi_0, \quad t > 0 \tag{13-113a}$$

$$T = 0 \quad \text{at} \quad \phi = 0, \quad t > 0 \tag{13-113b}$$

$$T = 0 \quad \text{at} \quad \phi = \phi_0, \quad t > 0 \tag{13-113c}$$

$$k_4 \frac{\partial T}{\partial r} + h_4 T = f_4(\phi, t) \quad \text{at} \quad r = b, \quad t > 0 \tag{13-113d}$$

$$T(r,\phi,t) = F(r,\phi) \quad \text{for} \quad t = 0, \text{ in the region} \tag{13-113e}$$

The integral transform pair in the ϕ variable for the function $T(r, \phi, t)$ is obtained from equations (13-106) as

$$\text{Inversion formula:} \quad T(r,\phi,t) = \sum_\nu \frac{\Phi(\nu,\phi)}{N(\nu)} \bar{T}(r,\nu,t) \tag{13-114a}$$

$$\text{Integral transform:} \quad \bar{T}(r,\nu,t) = \int_0^{\phi_0} \Phi(\nu,\phi') T(r,\phi',t)\, d\phi' \tag{13-114b}$$

where the functions $\Phi(\nu, \phi)$, $N(\nu)$ and the eigenvalues ν are obtainable from Table 2-2 by appropriate change of the notation (i.e., $L \to \phi_0, \beta_m \to \nu, x \to \phi$). We note that the eigenvalues ν for this case are not integers, but are determined according to the transcendental equations given in Table 2-2.

The integral transform of the system (13-113) by the application of the transform (13-114b) yields

$$\frac{\partial^2 \bar{T}}{\partial r^2}+\frac{1}{r}\frac{\partial \bar{T}}{\partial r}-\frac{\nu^2}{r^2}\bar{T}+\frac{1}{k}\bar{g}(r,\nu,t)=\frac{1}{\alpha}\frac{\partial \bar{T}(r,\nu,t)}{\partial t}$$

$$\text{in} \quad 0 \leqslant r < b, \quad t > 0 \tag{13-115a}$$

$$k_4 \frac{\partial \bar{T}}{\partial r}+h_4 \bar{T}=\bar{f}_4(\nu,t) \quad \text{at} \quad r=b, \quad t>0 \tag{13-115b}$$

$$\bar{T}(r,\nu,t)=\bar{F}(r,\nu) \quad \text{for} \quad t=0, \quad \text{in} \quad 0 \leqslant r \leqslant b \tag{13-115c}$$

where the bar denotes the integral transform of the function with respect to the ϕ variable.

The integral-transform pair in the r variable for the function $\bar{T}(r,\nu,t)$ is immediately obtained from the transform pair (13-110) as

$$\text{Inversion formula:} \qquad \bar{T}(r,\nu,t)=\sum_{m=1}^{\infty}\frac{R_\nu(\beta_m,r)}{N(\beta_m)}\tilde{\bar{T}}(\beta_m,\nu,t) \tag{13-116a}$$

$$\text{Integral transform:} \qquad \tilde{\bar{T}}(\beta_m,\nu,t)=\int_0^b r' R_\nu(\beta_m,r')\bar{T}(r',\nu,t)\,dr' \tag{13-116b}$$

where the tilde denotes the integral transform with respect to the r variable. The functions $R_\nu(\beta_m, r)$, $N(\beta_m)$ and the eigenvalues β_m are obtainable from Table 3-1.

The integral transform of the system (13-115) by the application of the transform (13-116b) yields (i.e., the procedure is similar to that described in note 6 at the end of this chapter)

$$\frac{d\tilde{\bar{T}}}{dt}+\alpha\beta_m^2\tilde{\bar{T}}(\beta_m,\nu,t)=A(\beta_m,\nu,t) \quad \text{for} \quad t>0 \tag{13-117a}$$

$$\tilde{\bar{T}}(\beta_m,\nu,t)=\tilde{\bar{F}}(\beta_m,\nu) \quad \text{for} \quad t=0 \tag{13-117b}$$

where

$$A(\beta_m,\nu,t)\equiv\frac{\alpha}{k}\tilde{\bar{g}}(\beta_m,\nu,t)+\alpha b\left.\frac{R_\nu(\beta_m,r)}{k_4}\right|_{r=b}\cdot\bar{f}_4(\nu,t) \tag{13-118}$$

Equation (13-117) is solved for $\tilde{\bar{T}}(\beta_m, \nu, t)$, the resulting double transform of the temperature is successively inverted by the inversion formulas (13-116a) and (13-114a). Then, the solution of the problem (13-113) becomes

$$T(r, \phi, t) = \sum_{\nu} \sum_{m=1}^{\infty} \frac{\Phi(\nu, \phi) R_\nu(\beta_m, r)}{N(\nu) N(\beta_m)} e^{-\alpha\beta_m^2 t} \cdot \left[\tilde{\bar{F}}(\beta_m, \nu) + \int_{t'=0}^{t} e^{\alpha\beta_m^2 t'} A(\beta_m, \nu, t')\, dt' \right] \tag{13-119}$$

where $A(\beta_m, \nu, t')$ is defined by equations (13-118) and $\tilde{\bar{F}}, \tilde{\bar{g}}$ are the double transforms:

$$\tilde{\bar{H}}(\beta_m, \nu) = \int_{r'=0}^{b} \int_{\phi'=0}^{\phi_0} r' R(\beta_m, r') \Phi(\nu, \phi') \bar{H}(r', \phi') d\phi'\, dr', \quad \text{and} \quad H \equiv F \quad \text{or} \quad g \tag{13-120}$$

The bar denotes the integral transform with respect to the ϕ variable and the tilde the integral transform with respect to the r variable as defined by equations (13-114b) and (13-116b), respectively.

For a boundary condition of the first kind at any of these boundaries, the usual replacements should be made in the definition of $A(\beta_m, \nu, t)$ given by equation (13-118a).

Problems of Region $a \leqslant r \leqslant b, 0 \leqslant \phi \leqslant \phi_0 (\phi_0 < 2\pi)$. The extension of the above solution to the problem of time-dependent heat conduction in a hollow cylinder $a \leqslant r \leqslant b$, confined to a region $0 \leqslant \phi \leqslant \phi_0 (< 2\pi)$ is a straight-forward matter. The heat conduction problem (13-113) will include an additional boundary condition at $r = a$. The definition of the integral transform pair (13-114) remains the same, but that of given by equations (13-116) is modified by changing the lower limit of the integration to $r = a$; then, the function $R_\nu(\beta_m, r)$, $N(\beta_m)$ and the eigenvalues β_m are obtained from Table 3-3.

Problems of Region $0 \leqslant r < \infty, 0 \leqslant \phi \leqslant 2\pi$. We now consider the solution of the following time dependent heat conduction problem for an infinite medium in which temperature varies with both r and ϕ variables.

$$\frac{\partial^2 T}{\partial r^2} + \frac{1}{r}\frac{\partial T}{\partial r} + \frac{1}{r^2}\frac{\partial^2 T}{\partial \phi^2} + \frac{g(r, \phi, t)}{k} = \frac{1}{\alpha}\frac{\partial T(r, \phi, t)}{\partial t}$$

$$\text{in} \quad 0 \leqslant r < \infty, \quad 0 \leqslant \phi \leqslant 2\pi, \quad t > 0 \tag{13-121a}$$

$$T(r, \phi, t) = F(r, \phi) \quad \text{for} \quad t = 0, \quad \text{in the region} \tag{13-121b}$$

The integral-transform pair in the ϕ variable for the function $T(r, \phi, t)$ is given by equation (13-104); hence we have

$$\text{Inversion formula:} \qquad T(r, \phi, t) = \frac{1}{\pi}\sum_{v} \bar{T}(r, v, t) \tag{13-122a}$$

$$\text{Integral transform:} \qquad \bar{T}(r, v, t) = \int_{\phi'=0}^{2\pi} \cos v(\phi - \phi')T(r, \phi', t)\, d\phi' \tag{13-122b}$$

where $v = 0, 1, 2, 3\ldots$ and replace π by 2π for $v = 0$. The integral transform of the system (13-121) by the application of the transform (13-122b) yields

$$\frac{\partial^2 \bar{T}}{\partial r^2} + \frac{1}{r}\frac{\partial \bar{T}}{\partial r} - \frac{v^2}{r^2}\bar{T} + \frac{\bar{g}(r, v, t)}{k} = \frac{1}{\alpha}\frac{\partial \bar{T}(r, v, t)}{\partial t}$$

$$\text{in} \qquad 0 \leqslant r < \infty, \quad t > 0 \tag{13-123a}$$

$$\bar{T}(r, v, t) = \bar{F}(r, v) \qquad \text{for} \quad t = 0, \qquad \text{in} \quad 0 \leqslant r < \infty \tag{13-123b}$$

where the bar denotes the integral transform with respect to the ϕ variable.

The integral-transform pair in the r variable for the function $\bar{T}(r, v, t)$ is constructed according to the representation (3-38). We find

$$\text{Inversion formula:} \qquad \bar{T}(r, v, t) = \int_{\beta=0}^{\infty} \beta J_v(\beta r)\tilde{\bar{T}}(\beta, v, t)\, d\beta \tag{13-124a}$$

$$\text{Integral transform:} \qquad \tilde{\bar{T}}(\beta, v, t) = \int_{r'=0}^{\infty} r' J_v(\beta r')\bar{T}(r', v, t)\, dr' \tag{13-124b}$$

where tilde detones the integral transform with respect to the r variable and the eigenvalue problem associated with this transform pair is given by equations (3-35).

The integral transform of the system (13-123) by the application of the transform (13-124b) gives

$$\frac{d\tilde{\bar{T}}}{dt} + \alpha\beta^2\tilde{\bar{T}}(\beta, v, t) = \frac{\alpha}{k}\tilde{\bar{g}}(\beta, v, t) \qquad \text{for} \quad t > 0 \tag{13-125a}$$

$$\tilde{\bar{T}}(\beta, v, t) = \tilde{\bar{F}}(\beta, v) \qquad \text{for} \quad t = 0 \tag{13-125b}$$

Equation (13-125) is solved for $\tilde{\bar{T}}(\beta, v, t)$ and the resulting double transform is successively inverted by the inversion formulas (13-124a) and (13-122a). Then the solution of the problem (13-121) becomes

$$T(r,\phi,t)=\frac{1}{\pi}\sum_{v}\int_{\beta=0}^{\infty}\beta J_{v}(\beta r)e^{-\alpha\beta^{2}t}\left[\tilde{\bar{F}}(\beta,v)+\frac{\alpha}{k}\int_{t'=0}^{t}e^{\alpha\beta^{2}t'}\tilde{\bar{g}}(\beta,v,t')\,dt'\right]d\beta \tag{13-126a}$$

where $v=0,1,2,3\ldots$ and replace π by 2π for $v=0$

$$\tilde{\bar{F}}(\beta,v)=\int_{r'=0}^{\infty}\int_{\phi'=0}^{2\pi}r'J_{v}(\beta r')\cos v(\phi-\phi')F(r',\phi')\,d\phi' dr' \tag{13-126b}$$

$$\tilde{\bar{g}}(\beta,v,t')=\int_{r'=0}^{\infty}\int_{\phi'=0}^{2\pi}r'J_{v}(\beta r')\cos v(\phi-\phi')g(r',\phi',t')\,d\phi'\,dr' \tag{13-126c}$$

Introducing equations (13-126b,c) into equation (13-126a) and changing the order of integrations we obtain

$$\begin{aligned}T(r,\phi,t)=&\frac{1}{\pi}\sum_{v}\int_{r'=0}^{\infty}\int_{\phi'=0}^{2\pi}r'\cos v(\phi-\phi')F(r',\phi')\\&\cdot\left[\int_{\beta=0}^{\infty}e^{-\alpha\beta^{2}t}\beta J_{v}(\beta r)J_{v}(\beta r')\,d\beta\right]d\phi'\,dr'\\&+\frac{1}{\pi}\frac{\alpha}{k}\sum_{v}\int_{r'=0}^{\infty}\int_{\phi'=0}^{2\pi}r'\cos v(\phi-\phi')g(r',\phi',t')\\&\cdot\left[\int_{\beta=0}^{\infty}e^{-\alpha\beta^{2}(t-t')}\beta J_{v}(\beta r)J_{v}(\beta r')\,d\beta\right]dt' d\phi'\,dr'\end{aligned} \tag{13-127}$$

The terms inside the brackets can be evaluated by utilizing the expression (13-99). Then the solution (13-127) becomes

$$\begin{aligned}T(r,\phi,t)=&\frac{1}{2\pi\alpha t}\sum_{v}\int_{\phi'=0}^{2\pi}r'\cos v(\phi-\phi')F(r',\phi')\\&\cdot\exp\left[-\frac{r^{2}+r'^{2}}{4\alpha t}\right]I_{v}\left(\frac{rr'}{2\alpha t}\right)d\phi'\,dr'\\&+\frac{1}{2\pi k}\sum_{v}\int_{r'=0}^{\infty}\int_{\phi'=0}^{2\pi}\int_{t'=0}^{t}\frac{r'}{t-t'}\cos v(\phi-\phi')g(r',\phi',t')\\&\cdot\exp\left[-\frac{r^{2}+r'^{2}}{4\alpha(t-t')}\right]I_{v}\left(\frac{rr'}{2\alpha(t-t')}\right)dt'\,d\phi'\,dr'\end{aligned} \tag{13-128}$$

where $v=0,1,2,3\ldots$ and replace π by 2π for $v=0$. Several special cases are obtainable from this solution.

Problems in (r, z, t) Variables

The solution of time-dependent heat conduction problems in the (r, z) variables with the integral-transform technique is now a straightforward matter. The integral-transform pairs with respect to the r variable are the same as those developed in this section for the problems having azimuthal symmetry and those with respect to the z variable are the same as those for the rectangular coordinate system. Also, the order of integral transformation with respect to the r and z variables is immaterial. We illustrate this matter with the following example.

Example 13-6

Consider the solution of the following heat conduction problem for a solid cylinder of radius $r = b$ and height $z = L$:

$$\frac{\partial^2 T}{\partial r^2} + \frac{1}{r}\frac{\partial T}{\partial r} + \frac{\partial^2 T}{\partial z^2} + \frac{g(r,z,t)}{k} = \frac{1}{\alpha}\frac{\partial T(r,z,t)}{\partial t}$$

$$\text{in} \quad 0 \leqslant r < b, \quad 0 < z < L, \quad t > 0 \tag{13-129a}$$

$$T(r,z,t) = 0 \quad \text{on all boundaries,} \quad t > 0 \tag{13-129b}$$

$$T(r,z,t) = F(r,z) \quad \text{for} \quad t = 0, \quad \text{in the region} \tag{13-129c}$$

Solution. The integral transform pair for the removal of partial derivatives with respect to the r variable in the region $0 \leqslant r \leqslant b$ is the same as that given by equations (13-87). Hence the transform pair with respect to the r variable for the function $T(r, z, t)$ is

$$\text{Inversion formula:} \qquad T(r,z,t) = \sum_{m=1}^{\infty} \frac{R_0(\beta_m, r)}{N(\beta_m)} \bar{T}(\beta_m, z, t) \tag{13-130a}$$

$$\text{Integral transform:} \qquad \bar{T}(\beta_m, z, t) = \int_{r'=0}^{b} r' R_0(\beta_m, r') T(r', z, t) \tag{13-130b}$$

where $R_0(\beta_m, r)$, $N(\beta_m)$, and β_m are obtainable from Table 3-1 by setting $\nu = 0$ and the bar denotes transform with respect to the r variable. The integral transform of the system (13-129) by the application of the transform (13-130b) yields

$$-\beta_m^2 \bar{T}(\beta_m, z, t) + \frac{\partial^2 \bar{T}}{\partial z^2} + \frac{\bar{g}(\beta_m, z, t)}{k} = \frac{1}{\alpha}\frac{\partial \bar{T}(\beta_m, z, t)}{\partial t}$$

$$\text{in} \quad 0 < z < L, \quad t > 0 \tag{13-131a}$$

$$\bar{T}(\beta_m, z, t) = 0 \quad \text{at} \quad z = 0, \quad z = L \quad \text{for} \quad t > 0 \tag{13-131b}$$

$$\bar{T}(\beta_m, z, t) = \bar{F}(\beta_m, z) \quad \text{for} \quad t = 0, \quad \text{in} \quad 0 < z < L \tag{13-131c}$$

The integral-transform pair with respect to the z variable in the region $0 \leqslant z \leqslant L$ is obtained from equations (13-27) as

Inversion formula:
$$\bar{T}(\beta_m, z, t) = \sum_{p=1}^{\infty} \frac{Z(\eta_p, z)}{N(\eta_p)} \tilde{\bar{T}}(\beta_m, \eta_p, t) \tag{13-132a}$$

Integral transform:
$$\tilde{\bar{T}}(\beta_m, \eta_p, t) = \int_{z'=0}^{L} Z(\eta_p, z')\bar{T}(\beta_m, z', t)dz' \tag{13-132b}$$

where $Z(\eta_p, z)$, $N(\eta_p)$, and η_p are obtainable from Table 2-2, and the tilde denotes the integral transform with respect to the z variable. The integral transform of the system (13-131) by the application of the transform (13-132b) is

$$\frac{d\tilde{\bar{T}}}{dt} + \alpha(\beta_m^2 + \eta_p^2)\tilde{\bar{T}}(\beta_m, \eta_p, t) = \frac{\alpha}{k}\tilde{\bar{g}}(\beta_m, \eta_p, t) \qquad t > 0 \tag{13-133a}$$

$$\tilde{\bar{T}}(\beta_m, \eta_p, t) = \tilde{\bar{F}}(\beta_m, \eta_p) \qquad t = 0 \tag{13-133b}$$

Equations (13-133) are solved for $\tilde{\bar{T}}(\beta_m, \eta_p, t)$ and the resulting double transform is successively inverted by the inversion formulas (13-132a) and (13-130a). Then the solution of the problem (13-129) becomes

$$T(r, z, t) = \sum_{m=1}^{\infty} \sum_{p=1}^{\infty} \frac{R_0(\beta_m, r)Z(\eta_p, z)}{N(\beta_m)N(\eta_p)} e^{-\alpha(\beta_m^2 + \eta_p^2)t} \cdot \left[\tilde{\bar{F}}(\beta_m, \eta_p) + \frac{\alpha}{k} \int_{t'=0}^{t} e^{\alpha(\beta_m^2 + \eta_p^2)t'} \tilde{\bar{g}}(\beta_m, \eta_p, t')\, dt' \right] \tag{13-134a}$$

where the double transforms are defined as

$$\tilde{\bar{H}} = \int_{z'=0}^{L} \int_{r'=0}^{b} r' R_0(\beta_m, r')Z(\eta_p, z')H\, dr'\, dz', \quad H \equiv F \text{ or } g \tag{13-134b}$$

From Table 3-1, case 3, for $v = 0$ we have

$$R_0(\beta_m, r) = J_0(\beta_m r), \qquad \frac{1}{N(\beta_m)} = \frac{2}{b^2 J_0'^2(\beta_m b)} = \frac{2}{b^2 J_1^2(\beta_m b)}$$

and the β_m values are the roots of $J_0(\beta_m b) = 0$. From Table 2-2, case 9 we have

$$Z(\eta_p, z) = \sin \eta_p z, \qquad \frac{1}{N(\eta_p)} = \frac{2}{L}$$

and the η_p values are the roots of $\sin \eta_p L = 0$.

Problems in (r, ϕ, z, t) Variables

The solution of heat conduction problems in (r, ϕ, z, t) variables is readily handled with the integral-transform technique. The basic steps in the analysis are summarized below.

1. The partial derivative with respect to the z variable is removed by the application of transform in the z variable. The appropriate transform pairs are the same as those given for the rectangular coordinate system.
2. The partial derivative with respect to the ϕ variable is removed by the application of transform in the ϕ variable. If the range of ϕ is $0 \leqslant \phi \leqslant 2\pi$, the transform pair is given by equations (13-104). If the range of ϕ is $0 \leqslant \phi \leqslant \phi_0$, $(\phi_0 \leqslant 2\pi)$, the transform pair is given by equations (13-106) for boundary conditions of the first and the second kinds.
3. The partial derivatives with respect to the r variable are removed by the application of transform in the r variable. The transform pair to be used depends on the range of the r variable, that is, $0 \leqslant r \leqslant b, a \leqslant r \leqslant b, 0 \leqslant r < \infty$. For example, the transform pair is as given by equations (13-110) for $0 \leqslant r \leqslant b$ or given by equations (13-124) for $0 \leqslant r < \infty$.
4. The resulting ordinary differential equation with respect to the time variable is solved subject to the triple transformed initial condition. The triple transform of temperature obtained in this manner is successively inverted with respect to the r, ϕ, and z variables to obtain the solution for $T(r, \phi, z, t)$.

13-4 APPLICATIONS IN THE SPHERICAL COORDINATE SYSTEM

To solve heat conduction problems in the spherical coordinate system with the integral-transform technique, appropriate integral-transform pairs are needed in the r, μ, and ϕ variables. In this section we develop such integral-transform pairs and illustrate their application to the solution of heat conduction problems involving (r, t), (r, μ, t) and (r, μ, ϕ, t) variables.

Problems in (r, t) Variables

The time-dependent heat conduction problems involving only the r variable can be transformed into a one-dimensional, time-dependent heat conduction problem in the rectangular coordinate system by defining a new variable $U(r, t) = rT(r, t)$ as discussed in Section 4-4. The resulting heat conduction problem in the rectangular coordinate system is readily solved with the integral-transform technique as described previously. Therefore, the solution of the problems in (r, t) variables is not considered here any further.

Problems in (r, μ, t) Variables

The differential equation of heat conduction in the (r, μ, t) variables is taken in the form

$$\frac{\partial^2 T}{\partial r^2}+\frac{2}{r}\frac{\partial T}{\partial r}+\frac{1}{r^2}\frac{\partial}{\partial \mu}\left[(1-\mu^2)\frac{\partial T}{\partial \mu}\right]+\frac{g(r,\mu,t)}{k}=\frac{1}{\alpha}\frac{\partial T(r,\mu,t)}{\partial t} \tag{13-135}$$

By defining a new variable $V(r, \mu, t)$ as

$$V(r,\mu,t)=r^{1/2}T(r,\mu,t) \tag{13-136}$$

Equation (13-135) is transformed into

$$\frac{\partial^2 V}{\partial r^2}+\frac{1}{r}\frac{\partial V}{\partial r}-\frac{1}{4}\frac{V}{r^2}+\frac{1}{r^2}\frac{\partial}{\partial \mu}\left[(1-\mu^2)\frac{\partial V}{\partial \mu}\right]+\frac{r^{1/2}g}{k}=\frac{1}{\alpha}\frac{\partial V}{\partial t} \tag{13-137}$$

where $g \equiv g(r, \mu, t)$ and $V \equiv V(r, \mu, t)$.

The partial derivatives with respect to the space variables can be removed from this equation by the successive application of integral transforms with respect to the μ and r variables. *The order of transformation is important in this case; it is applied first to the μ variable and then to the r variable.* Then we need to develop the integral-transform pairs only with respect to the μ variable for the following cases: The range of μ variable is $-1 \leqslant \mu \leqslant 1$ as in the case of the *full sphere*; $0 \leqslant \mu \leqslant 1$ as in the case of the *hemisphere.*

Transform Pair for $-1 \leqslant \mu \leqslant 1$**.** This case corresponds to the full sphere. Therefore, no boundary conditions are specified in the μ variable except the requirement that the function should remain finite at $\mu = \pm 1$. The integral-transform pair in the μ variable for the function $V(r, \mu, t)$ is constructed by considering the representation of this function in a form similar to that given by equation (4-35) and then splitting up the representation into two parts. We find

Inversion formula:
$$V(r,\mu,t)=\sum_{n=0}^{\infty}\frac{2n+1}{2}P_n(\mu)\bar{V}(r,n,t) \tag{13-138a}$$

Integral transform:
$$\bar{V}(r,n,t)=\int_{\mu'=-1}^{1}P_n(\mu')V(r,\mu',t)\,d\mu' \tag{13-138b}$$

where $P_n(\mu)$ is the Legendre polynomial and $n = 0, 1, 2, 3 \ldots$.

Transform Pair for $0 \leqslant \mu \leqslant 1$**.** This case corresponds to the hemisphere. The integral transform pair is determined by splitting up the expansion (4-51) as

Inversion formula:
$$V(r,\mu,t)=\sum_{n}(2n+1)P_n(m)\bar{V}(r,n,t) \tag{13-139a}$$

Integral transform:
$$\bar{V}(r,n,t)=\int_{\mu'=0}^{1}P_n(\mu')V(r,\mu',t)\,d\mu' \tag{13-139b}$$

where the values of n are $n = 1, 3, 5, \ldots$ (i.e., odd integers) for boundary condition of the first kind at $\mu = 0$, and $n = 0, 2, 4, \ldots$ (i.e., even integers) for boundary condition of the second kind at $\mu = 0$.

Example 13-7

We consider the solution of the following time-dependent heat conduction problem for a solid sphere of radius $r = b$:

$$\frac{\partial^2 T}{\partial r^2} + \frac{2}{r}\frac{\partial T}{\partial r} + \frac{1}{r^2}\frac{\partial}{\partial \mu}\left[(1-\mu^2)\frac{\partial T}{\partial \mu}\right] + \frac{g(r,\mu,t)}{k} = \frac{1}{\alpha}\frac{\partial T(r,\mu,t)}{\partial t}$$

$$\text{in} \quad 0 \leqslant r < b, \quad -1 \leqslant \mu \leqslant 1 \quad t > 0 \tag{13-140}$$

$$T(r,\mu,t) = 0 \quad \text{at} \quad r = b, \quad t > 0 \tag{13-141a}$$

$$T(r,\mu,t) = F(r,\mu) \quad \text{for} \quad t = 0, \quad \text{in the region} \tag{13-141b}$$

Solution. Here we considered a homogeneous boundary condition of the first kind for simplicity in the analysis; the analysis for boundary condition of the third kind is performed in a similar manner.

A new dependent variable $V(r,\mu,t)$ is defined as

$$V(r,\mu,t) = r^{1/2}T(r,\mu,t) \tag{13-142}$$

Then, the problem (13-140)–(13-141) is transformed into

$$\frac{\partial^2 V}{\partial r^2} + \frac{1}{r}\frac{\partial V}{\partial r} - \frac{1}{4}\frac{V}{r^2} + \frac{1}{r^2}\frac{\partial}{\partial \mu}\left[(1-\mu^2)\frac{\partial V}{\partial \mu}\right] + \frac{r^{1/2}g(r,\mu,t)}{k} = \frac{1}{\alpha}\frac{\partial V(r,\mu,t)}{\partial t}$$

$$\text{in} \quad 0 \leqslant r < b, \quad -1 \leqslant \mu \leqslant 1 \quad t > 0 \tag{13-143a}$$

$$V(r,\mu,t) = 0 \quad \text{at} \quad r = b, \quad t > 0 \tag{13-143b}$$

$$V(r,\mu,t) = r^{1/2}F(r,\mu) \quad \text{for} \quad t = 0, \quad \text{in the region} \tag{13-143c}$$

The integral-transform pair with respect to the μ variable for $-1 \leqslant \mu \leqslant 1$ is obtained from equations (13-138) as

$$\text{Inversion formula:} \qquad V(r,\mu,t) = \sum_{n=0}^{\infty} \frac{2n+1}{2} P_n(\mu)\bar{V}(r,n,t) \tag{13-144a}$$

$$\text{Integral transform:} \qquad \bar{V}(r,n,t) = \int_{\mu=-1}^{1} P_n(\mu')V(r,\mu',t)\,d\mu' \tag{13-144b}$$

where $n = 0, 1, 2, 3 \ldots$.

We take the integral transform of the system (13-143) by the application of the transform (13-144b) to obtain (see note 7 at the end of this chapter for details)

$$\frac{\partial^2 \bar{V}}{\partial r^2}+\frac{1}{r}\frac{\partial \bar{V}}{\partial r}-\frac{(n+\frac{1}{2})^2}{r^2}\bar{V}+\frac{\bar{g}^*(r,n,t)}{k}=\frac{1}{\alpha}\frac{\partial \bar{V}(r,n,t)}{\partial t}$$

$$\text{in} \quad 0\leqslant r<b, \quad t>0 \tag{13-145a}$$

$$\bar{V}(r,n,t)=0 \quad \text{at} \quad r=b, \quad t>0 \tag{13-145b}$$

$$\bar{V}(r,n,t)=\bar{F}^*(r,n) \quad \text{for} \quad t=0, \quad \text{in} \quad 0\leqslant r\leqslant b \tag{13-145c}$$

where

$$g^*(r,\mu,t)=r^{1/2}g(r,\mu,t), \quad F^*(r,\mu)=r^{1/2}F(r,\mu) \tag{13-145d}$$

and the bar denotes the integral transform with respect to the μ variable according to the transform (13-144b).

Equation (13-145a) is similar in form to equation (13-109a) in the cylindrical coordinate system. Therefore, the integral-transform pair needed to remove the partial derivatives with respect to the r variable is immediately obtainable from equation (13-110) or (13-116) by setting $v\equiv n+\frac{1}{2}$. We find

Inversion formula:
$$\bar{V}(r,n,t)=\sum_{p=1}^{\infty}\frac{R_{n+1/2}(\lambda_{np},r)}{N(\lambda_{np})}\tilde{\bar{V}}(\lambda_{np},n,t) \tag{13-146a}$$

Integral transform:
$$\tilde{\bar{V}}(\lambda_{np},n,t)=\int_0^b R_{n+1/2}(\lambda_{np},r)\bar{V}(r',n,t)\,dr' \tag{13-146b}$$

where the tilde denotes the integral transform with respect to the r variable. The functions $R_{n+1/2}(\lambda_{np},r)$, $N(\lambda_{np})$ and the eigenvalues λ_{np} for the boundary condition of the first kind considered in the problem (13-145) are obtainable from Table 3-1, case 3, as

$$R_{n+1/2}(\lambda_{np},r)=J_{n+1/2}(\lambda_{np}r), \quad \frac{1}{N(\lambda_{np})}=\frac{2}{b^2[J'_{n+1/2}(\lambda_{np}b)]^2} \tag{13-146c}$$

and the λ_{np} values are the positive roots of

$$J_{n+1/2}(\lambda_{np}b)=0 \tag{13-146d}$$

Taking the integral transform of the system (13-145) by the application of the transform (13-146b) we obtain

$$\frac{d\tilde{\bar{V}}}{dt}+\alpha\lambda_{np}^2\tilde{\bar{V}}(\lambda_{np},n,t)=\frac{\alpha}{k}\tilde{\bar{g}}^* \quad \text{for} \quad t>0 \tag{13-147a}$$

$$\bar{\tilde{V}}(\lambda_{np}, n, t) = \bar{\tilde{F}}^* \qquad \text{for} \qquad t = 0 \tag{13-147b}$$

Equation (13-147) is solved for $\bar{\tilde{V}}(\lambda_{np}, n, t)$, the resulting double transform is inverted successively by the inversion formulas (13-146a) and (13-144a) to obtain $V(r, \mu, t)$. When $V(r, \mu, t)$ is transformed by the expression (13-142d) into $T(r, \mu, t)$, the solution of the problem (13-141) is obtained as

$$T(r, \mu, t) = \sum_{n=0}^{\infty} \sum_{p=1}^{\infty} \frac{(2n+1) r^{-1/2} J_{n+1/2}(\lambda_{np} r) P_n(\mu)}{b^2 [J'_{n+1/2}(\lambda_{np} b)]^2} e^{-\alpha \lambda_{np}^2 t} \cdot \left[\bar{\tilde{F}}^* + \frac{\alpha}{k} \int_0^t e^{\alpha \lambda_{np}^2 t'} \bar{\tilde{g}}^* dt' \right] \tag{13-148a}$$

where

$$\bar{\tilde{F}}^* = \int_{r'=0}^{b} \int_{\mu'=-1}^{1} r'^{3/2} J_{n+1/2}(\lambda_{np} r') P_n(\mu') F(r', \mu')\, d\mu'\, dr' \tag{13-148b}$$

$$\bar{\tilde{g}}^* = \int_{r'=0}^{b} \int_{\mu'=-1}^{1} r'^{3/2} J_{n+1/2}(\lambda_{np} r') P_n(\mu') g(r', \mu', t')\, d\mu'\, dt' \tag{13-148c}$$

where $n = 0, 1, 2 \ldots$ and the λ_{np} values are the roots of equation (13-146d). For the case of no heat generation, this solution reduces to that given by equation (4-88).

Example 13-8

We consider the following time-dependent heat conduction equation for a hemisphere of radius $r = b$:

$$\frac{\partial^2 T}{\partial r^2} + \frac{2}{r}\frac{\partial T}{\partial r} + \frac{1}{r^2}\frac{\partial}{\partial \mu}\left[(1-\mu^2)\frac{\partial T}{\partial \mu}\right] + \frac{g(r, \mu, t)}{k} = \frac{1}{\alpha}\frac{\partial T(r, \mu, t)}{\partial t}$$

$$\text{in} \qquad 0 \leqslant r < b, \quad 0 \leqslant \mu \leqslant 1, \qquad t > 0 \tag{13-149a}$$

$$T(r, \mu, t) = 0 \qquad \text{at} \qquad r = b, \qquad \mu = 0, \quad \text{for} \quad t > 0 \tag{13-149b}$$

$$T(r, \mu, t) = F(r, \mu) \qquad \text{for} \qquad t = 0, \qquad \text{in the region} \tag{13-149c}$$

Solution. The basic steps for the solution of this problem are exactly the same as those described above for the solution of problem (13-140). The only difference is that, the range of μ being $0 \leqslant \mu \leqslant 1$, the integral-transform pair with respect to the μ variable is determined according to equations (13-139) as

$$\text{Inversion formula:} \qquad V(r, \mu, t) = \sum_n (2n+1) P_n(\mu) \bar{V}(r, n, t) \tag{13-150a}$$

$$\text{Integral transform:} \qquad \bar{V}(r, n, t) = \int_{\mu'=0}^{1} P_n(\mu') V(r, \mu', t)\, d\mu' \tag{13-150b}$$

where the values of n are $n = 1, 3, 5, \ldots$ (i.e., odd integers) since the boundary condition at $\mu = 0$ is of the first kind.

The integral-transform pair with respect to the r variable is taken the same as that given by equations (13-146). The system (13-149) is transformed from the $T(r, \mu, t)$ variable into the $V(r, \mu, t)$ variable. Then, by the application of the integral transform with respect to the μ and r variables an ordinary differential equation is obtained for the double transform $\tilde{\bar{V}}(\lambda_{np}, n, t)$. The resulting ordinary differential equation is solved and the double transform is successively inverted by the inversion formulas (13-146a) and (13-150a) to obtain $V(r, \mu, t)$. When $V(r, \mu, t)$ is transformed by the transformation (13-142), the solution $T(r, \mu, t)$ of the problem (13-149) is obtained as

$$T(r, \mu, t) = \sum_{n=1,3,5\ldots}^{\infty} \sum_{p=1}^{\infty} \frac{2(2n+1) r^{-1/2} J_{n+1/2}(\lambda_{np} r) P_n(\mu)}{b^2 [J'_{n+1/2}(\lambda_{np} b)]^2} e^{-\alpha \lambda_{np}^2 t} \cdot \left[\tilde{\bar{F}}^* + \frac{\alpha}{k} \int_0^t e^{\alpha \lambda_{np}^2 t'} \tilde{\bar{g}}^* \, dt' \right] \tag{13-151a}$$

where

$$\tilde{\bar{F}}^* = \int_{r'=0}^{b} \int_{\mu'=0}^{1} r'^{3/2} J_{n+1/2}(\lambda_{np} r') P_n(\mu') F(r', \mu') \, d\mu' \, dr' \tag{13-151b}$$

$$\tilde{\bar{g}}^* = \int_{r'=0}^{b} \int_{\mu'=0}^{1} r'^{3/2} J_{n+1/2}(\lambda_{np} r') P_n(\mu') g(r', \mu', t') \, d\mu' \, dr' \tag{13-151c}$$

and the λ_{np} values are the roots of

$$J_{n+1/2}(\lambda_{np} b) = 0. \tag{13-151d}$$

We note that for the case of no heat generation, the solution (13-151) reduces to that given by equations (4-98).

Problems in (r, μ, ϕ, t) Variables

The differential equation of heat conduction in the (r, μ, ϕ, t) variables is taken in the form

$$\frac{\partial^2 T}{\partial r^2} + \frac{2}{r} \frac{\partial T}{\partial r} + \frac{1}{r^2} \frac{\partial}{\partial \mu} \left[(1 - \mu^2) \frac{\partial T}{\partial \mu} \right] + \frac{1}{r^2 (1 - \mu^2)} \frac{\partial^2 T}{\partial \phi^2} + \frac{g(r, \mu, \phi, t)}{k} = \frac{1}{\alpha} \frac{\partial T(r, \mu, \phi, t)}{\partial t} \tag{13-152}$$

We consider the problem of *full sphere*, hence choose the ranges of μ and ϕ variables as $0 \leqslant \phi \leqslant 2\pi$ and $-1 \leqslant \mu \leqslant 1$. The range of the r variable may be finite or infinite.

Now a new variable $V(r, \mu, \phi, t)$ is defined as

$$V(r, \mu, \phi, t) = r^{1/2} T(r, \mu, \phi, t) \tag{13-153}$$

Then equation (13-152) is transformed into

$$\frac{\partial^2 V}{\partial r^2} + \frac{1}{r}\frac{\partial V}{\partial r} - \frac{1}{4}\frac{V}{r^2} + \frac{1}{r^2}\frac{\partial}{\partial \mu}\left[(1-\mu^2)\frac{\partial V}{\partial \mu}\right] + \frac{1}{r^2(1-\mu^2)}\frac{\partial^2 V}{\partial \phi^2} + \frac{r^{1/2} g}{k} = \frac{1}{\alpha}\frac{\partial V}{\partial t} \tag{13-154}$$

where $0 \leqslant \phi \leqslant 2\pi$, $-1 \leqslant \mu \leqslant 1$ and the range of r is finite or infinite.

The partial derivatives with respect to the space variables can be removed from this equation by the application of integral transform with respect to the $\phi, \mu,$ and r variable. For this particular case the order of transformation is important. *That is, the transformation should be applied first with respect to the ϕ variable, then to the μ variable and finally to the r variable.* The procedure is as follows.

Removal of the Derivative in the ϕ Variable. The range of the ϕ variable being in $0 \leqslant \phi \leqslant 2\pi$, the transform pair with respect to the ϕ variable is obtained from equations (13-104) as

$$\text{Inversion formula:} \quad V(r, \mu, \phi, t) = \frac{1}{\pi}\sum_{m=0}^{\infty} \bar{V}(r, \mu, m, t) \tag{13-155a}$$

$$\text{Integral transform:} \quad \bar{V}(r, \mu, m, t) = \int_{\phi'=0}^{2\pi} \cos m(\phi - \phi') V(r, \mu, \phi', t)\, d\phi' \tag{13-155b}$$

where $m = 0, 1, 2, 3 \ldots$ and replace π by 2π for $m = 0$ and the eigenvalue problem associated with this transform pair is the same as that given by equations (3-116a).

The integral transform of equation (13-154) by the application of the transform (13-155b) is

$$\frac{\partial^2 \bar{V}}{\partial r^2} + \frac{1}{r}\frac{\partial \bar{V}}{\partial r} - \frac{1}{4}\frac{\bar{V}}{r^2} + \frac{1}{r^2}\left\{\frac{\partial}{\partial \mu}\left[(1-\mu^2)\frac{\partial \bar{V}}{\partial \mu}\right] - \frac{m^2}{1-\mu^2}\bar{V}\right\} + \frac{\bar{g}^*}{k} = \frac{1}{\alpha}\frac{\partial \bar{V}}{\partial t} \tag{13-156}$$

where $g^* \equiv r^{1/2} g(r, \mu, \phi, t)$, $\bar{V} \equiv \bar{V}(r, \mu, m, t)$ and the bar denotes the integral transform with respect to the ϕ variable. Thus, by the application of the transform (13-155b), we removed from the differential equation the partial derivative with respect to the ϕ variable, that is, $\partial^2 V/\partial \phi^2$.

The Removal of the Derivative in the μ Variable. In equation (13-156), the differential operator with respect to the μ variable is in the form

$$\frac{\partial}{\partial\mu}\left[(1-\mu^2)\frac{\partial\bar{V}}{\partial\mu}\right]-\frac{m^2}{1-\mu^2}\bar{V},$$

and the range of μ is $-1\leqslant\mu\leqslant 1$. The integral transform pair to remove this differential operator can be constructed by considering the representation of a function, defined in the interval $-1\leqslant\mu\leqslant 1$, in terms of the eigenfunctions of Legendre's associated differential equation [see equation (4-13c) or (4-22)]

$$\frac{d}{d\mu}\left[(1-\mu^2)\frac{dM}{d\mu}\right]+\left[n(n+1)-\frac{m^2}{1-\mu^2}\right]M=0 \tag{13-157}$$

and then splitting up the representation into two parts. The resulting integral transform pair with respect to the μ variable for the function $\bar{V}(r,\mu,m,t)$ is given as

$$\text{Inversion formula:}\quad \bar{V}(r,\mu,m,t)=\sum_{m=0}^{n}\frac{P_n^m(\mu)}{N(m,n)}\tilde{\bar{V}}(r,n,m,t) \tag{13-158a}$$

$$\text{Integral transform:}\quad \tilde{\bar{V}}(r,n,m,t)=\int_{-1}^{1}P_n^m(\mu')\bar{V}(r,\mu',m,t)\,d\mu' \tag{13-158b}$$

where

$$\frac{1}{N(m,n)}=\frac{2n+1}{2}\frac{(n-m)!}{(n+m)!}\qquad n\geqslant m \tag{13-158c}$$

n and m are integers, $P_n^m(\mu)$ is the associated Legendre function of degree n, order m, of the first kind; and the tilde denotes the integral transform with respect to the μ variable. It is to be noted that, when the integral-transform pairs (13-155) and (13-158) are combined, the expansion given by equation (4-43) in Chapter 4 is obtained.

Taking the integral transform of equation (13-156) by the application of the transform (13-158b) and utilizing equation (13-157) we obtain (see note 8 at the end of this chapter for details)

$$\frac{\partial^2\tilde{\bar{V}}}{\partial r^2}+\frac{1}{r}\frac{\partial\tilde{\bar{V}}}{\partial r}-\frac{(n+\frac{1}{2})^2}{r^2}\tilde{\bar{V}}+\frac{\tilde{\bar{g}}^*}{k}=\frac{1}{\alpha}\frac{\partial\tilde{\bar{V}}(r,n,m,t)}{\partial t} \tag{13-159}$$

where the tilde denotes the transform with respect to the μ variable.

The Removal of the Derivative in the r Variable. The differential operator with respect to the r variable can readily be removed from equation (13-159) by the application of an appropriate transform in the r variable developed previously for the solution of problems in the cylindrical coordinate system. The form of the transform pair depends on the range of r, whether it is finite or infinite. We now illustrate the application with an example given below.

Example 13-9

Solve the following time-dependent, three-dimensional heat conduction problem for a solid sphere of radius $r = b$:

$$\frac{\partial^2 T}{\partial r^2}+\frac{2}{r}\frac{\partial T}{\partial r}+\frac{1}{r^2}\frac{\partial}{\partial \mu}\left[(1-\mu^2)\frac{\partial T}{\partial \mu}\right]+\frac{1}{r^2(1-\mu^2)}\frac{\partial^2 T}{\partial \phi^2}+\frac{g}{k}=\frac{1}{\alpha}\frac{\partial T}{\partial t}$$

$$\text{in} \quad 0 \leqslant r < b, \quad -1 \leqslant \mu \leqslant 1, \quad 0 \leqslant \phi \leqslant 2\pi, \quad t > 0 \tag{13-160a}$$

$$T = 0 \quad \text{at} \quad r = b, \quad t > 0 \tag{13-160b}$$

$$T = F(r, \mu, \phi) \quad \text{for} \quad t = 0, \quad \text{in the region} \tag{13-160c}$$

where $g \equiv g(r, \mu, \phi, t)$ and $T \equiv T(r, \mu, \phi, t)$.

Solution. By defining a new dependent variable as

$$V(r, \mu, \phi, t) = r^{1/2} T(r, \mu, \phi, t) \tag{13-161}$$

the system (13-160) is transformed into

$$\frac{\partial^2 V}{\partial r^2}+\frac{1}{r}\frac{\partial V}{\partial r}-\frac{1}{4}\frac{V}{r^2}+\frac{1}{r^2}\frac{\partial}{\partial \mu}\left[(1-\mu^2)\frac{\partial V}{\partial \mu}\right]+\frac{1}{r^2(1-\mu^2)}\frac{\partial^2 V}{\partial \phi^2}+\frac{r^{1/2}g}{k}=\frac{1}{\alpha}\frac{\partial V}{\partial t}$$

$$\text{in} \quad 0 \leqslant r < b, \quad -1 \leqslant \mu \leqslant 1, \quad 0 \leqslant \phi \leqslant 2\pi, \quad t > 0 \tag{13-162a}$$

$$V = 0 \quad \text{at} \quad r = b, \quad t > 0 \tag{13-162b}$$

$$V = r^{1/2} F(r, \mu, \phi) \quad \text{for} \quad t = 0, \quad \text{in the region} \tag{13-162c}$$

The integral transform of this system with respect to the ϕ variable by the application of the transform (13-155b) yields

$$\frac{\partial^2 \bar{V}}{\partial r^2}+\frac{1}{r}\frac{\partial \bar{V}}{\partial r}-\frac{1}{4}\frac{\bar{V}}{r^2}+\frac{1}{r^2}\left\{\frac{\partial}{\partial \mu}\left[(1-\mu^2)\frac{\partial \bar{V}}{\partial \mu}\right]-\frac{m^2}{1-\mu^2}\bar{V}\right\}+\frac{\bar{g}^*}{k}=\frac{1}{\alpha}\frac{\partial \bar{V}}{\partial t}$$

$$\text{in} \quad 0 \leqslant r < b, \quad -1 \leqslant \mu \leqslant 1, \quad t > 0 \tag{13-163a}$$

$$\bar{V}=0 \qquad \text{at} \qquad r=b, \quad t>0 \tag{13-163b}$$

$$\bar{V}=\bar{F}^*(r,\mu,m) \qquad \text{for} \qquad t=0, \quad \text{in} \quad 0\leqslant r\leqslant b, \quad -1\leqslant\mu\leqslant 1 \tag{13-163c}$$

where $g^* \equiv r^{1/2}g(r,\mu,\phi,t)\, F^* \equiv r^{1/2}F(r,\mu,\phi)$, $\bar{V}\equiv\bar{V}(r,\mu,m,t)$, and the bar denotes the transform with respect to the ϕ variable. Now, the integral transform of the system (13-163) with respect to the μ variable, by the application of the transform (13-158b) gives

$$\frac{\partial^2\tilde{\bar{V}}}{\partial r^2}+\frac{1}{r}\frac{\partial\tilde{\bar{V}}}{\partial r}-\frac{(n+\frac{1}{2})^2}{r^2}\tilde{\bar{V}}+\frac{\tilde{\bar{g}}^*}{k}=\frac{1}{\alpha}\frac{\partial\tilde{\bar{V}}}{\partial t} \qquad \text{in} \quad 0\leqslant r<b, \quad t>0 \tag{13-164a}$$

$$\tilde{\bar{V}}=0 \qquad \text{at} \quad r=b, \qquad t>0 \tag{13-164b}$$

$$\tilde{\bar{V}}=\tilde{\bar{F}}^*(r,n,m) \qquad \text{for} \quad t=0, \qquad 0\leqslant r\leqslant b \tag{13-164c}$$

where $\tilde{\bar{V}}\equiv V(r,n,m,t)$ and the tilde denotes transform with respect to the μ variable. This system is now exactly of the same form as that given by equations (13-145). To remove the differential operator with respect to the r variable the appropriate integral transform pair is exactly the same as that given by equations (13-146). Therefore, taking the integral transform of the system (13-164) by the application of the transform (13-146b) we find

$$\frac{d\overset{x}{\tilde{\bar{V}}}}{dt}+\alpha\lambda_{np}^2\overset{x}{\tilde{\bar{V}}}(\lambda_{np},n,m,t)=\frac{\alpha}{k}\overset{x}{\tilde{\bar{g}}}{}^* \qquad \text{for} \qquad t>0 \tag{13-165a}$$

$$\overset{x}{\tilde{\bar{V}}}=\overset{x}{\tilde{\bar{F}}}{}^*(\lambda_{np},n,m) \qquad \text{for} \qquad t=0 \tag{13-165b}$$

where superscript x denotes the integral transform with respect to the r variable. Equation (13-165) is solved for

$$\overset{x}{\tilde{\bar{V}}}(\lambda_{np},n,m,t)$$

The resulting triple transform is successively inverted by the inversion formulas (13-146a), (13-158a), and (13-155a) to obtain $V(r,\mu,\phi,t)$. When the function $V(r,\mu,\phi,t)$ is transformed by the expression (13-161), we obtain the solution $T(r,\mu,\phi,t)$ of the problem (13-160) as

$$T(r,\mu,\phi,t)=\frac{1}{\pi}\sum_{n=0}^{\infty}\sum_{m=0}^{n}\sum_{p=1}^{\infty}\frac{r^{-1/2}J_{n+1/2}(\lambda_{np}r)P_n^m(\mu)}{N(m,n)N(\lambda_{np})}e^{-\alpha\lambda_{np}^2t} \cdot\left[\overset{x}{\tilde{\bar{F}}}+\frac{\alpha}{k}\int_0^t e^{-\alpha\lambda_{np}^2t'}\overset{x}{\tilde{\bar{g}}}{}^*\,dt'\right] \tag{13-166a}$$

and replace π by 2π for $m=0$. Various quantities are defined as

$$\tilde{\bar{F}} = \int_{r'=0}^{b}\int_{\mu'=-1}^{1}\int_{\phi'=0}^{2\pi} r'^{3/2} J_{n+1/2}(\lambda_{np} r') P_n^m(\mu') \cdot \cos m(\phi-\phi') F(r',\mu',\phi')\, d\phi'\, d\mu'\, dr' \tag{13-166b}$$

$$\tilde{\bar{g}}^* = \int_{r'=0}^{b}\int_{\mu'=-1}^{1}\int_{\phi'=0}^{2\pi} r'^{3/2} J_{n+1/2}(\lambda_{np} r') P_n^m(\mu') \cdot \cos m(\phi-\phi') g(r',\mu',\phi',t')\, d\phi'\, d\mu'\, dr' \tag{13-166c}$$

$$\frac{1}{N(m,n)} = \frac{2n+1}{2}\frac{(n-m)!}{(n+m)!} \tag{13-166d}$$

$$\frac{1}{N(\lambda_{np})} = \frac{2}{b^2[J'_{n+1/2}(\lambda_{np} b)]^2} \tag{12-166e}$$

and the λ_{np} values are the positive roots of

$$J_{n+1/2}(\lambda_{np} b) = 0 \tag{13-166f}$$

For the case of no heat generation, this solution reduces to that given by equations (4-118).

13-5 APPLICATIONS IN THE SOLUTION OF STEADY-STATE PROBLEMS

The integral-transform technique is also very effective in the solution of multi-dimensional, steady-state heat conduction problems, because, by the successive application of the integral transform, the partial-differential equation is reduced to an ordinary differential equation in one of the space variables. The resulting ordinary differential equation is solved for the transform of the temperature, which is then inverted successively to obtain the desired solution. This procedure is now illustrated with examples.

Example 13-10

Solve the following steady-state heat conduction problem for a rectangular region $0 \leqslant x \leqslant a$, $0 \leqslant y \leqslant b$

$$\frac{\partial^2 T}{\partial x^2} + \frac{\partial^2 T}{\partial y^2} = 0 \quad \text{in} \quad 0<x<a, \quad 0<y<b \tag{13-167a}$$

$$T = 0 \quad \text{at} \quad x=0, \quad x=a \quad \text{and} \quad y=b \tag{13-167b}$$

$$T = f(x) \quad \text{at} \quad y=0 \tag{13-167c}$$

Solution. In this example we prefer to take the integral transform with respect to the x variable, because in the resulting ordinary differential equation for the transform of T the boundary condition at $y = b$ becomes a constant, hence its integration is readily performed. The integral-transform pair with respect to the x variable, for $0 \leqslant x \leqslant a$, of function $T(x, y)$ is defined as

$$\text{Inversion formula:} \qquad T(x, y) = \sum_{m=1}^{\infty} \frac{X(\beta_m, x)}{N(\beta_m)} \bar{T}(\beta_m, y) \tag{13-168a}$$

$$\text{Integral transform:} \quad \bar{T}(\beta_m, y) = \int_{x'=0}^{b} X(\beta_m, x') T(x', y)\, dx' \tag{13-168b}$$

where $X(\beta_m, x)$, $N(\beta_m)$, and β_m are obtained from Table 2-2, case 9 as

$$X(\beta_m, x) = \sin \beta_m x, \frac{1}{N(\beta_m)} = \frac{2}{a} \qquad \text{and} \qquad \sin \beta_m a = 0 \tag{13-168c}$$

The integral transform of the system (13-167) by the application of the transform (13-168b) yields

$$\frac{d^2 \bar{T}}{dy^2} - \beta_m^2 \bar{T}(\beta_m, y) = 0 \qquad \text{in} \qquad 0 < y < b \tag{13-169a}$$

$$\bar{T} = \bar{f}(\beta_m) \qquad \text{at} \qquad y = 0 \tag{13-169b}$$

$$\bar{T} = 0 \qquad \text{at} \qquad y = b \tag{13-169c}$$

The solution of equations (13-179) is

$$\bar{T}(\beta_m, y) = \bar{f}(\beta_m) \frac{\sinh \beta_m (b - y)}{\sinh \beta_m b} \tag{13-170}$$

The inversion of this result with the inversion formula (13-168a) gives

$$T(x, y) = \frac{2}{a} \sum_{m=1}^{\infty} \sin \beta_m x \frac{\sinh \beta_m (b - y)}{\sinh \beta_m b} \int_{x'=0}^{a} \sin \beta_m x' f(x')\, dx' \tag{13-171a}$$

where

$$\beta_m = m\pi/a, \quad m = 1, 2, 3 \ldots \tag{13-171b}$$

Example 13-11

Solve the following steady-state heat conduction problem for a long solid cylinder:

$$\frac{\partial^2 T}{\partial r^2}+\frac{1}{r}\frac{\partial T}{\partial r}+\frac{1}{r^2}\frac{\partial^2 T}{\partial \phi^2}=0 \quad \text{in} \quad 0 \leqslant r < b, \quad 0 \leqslant \phi \leqslant 2\pi \tag{13-172a}$$

$$k_2\frac{\partial T}{\partial r}+h_2 T=f_2(\phi) \quad \text{at} \quad r=b \tag{13-172b}$$

Solution. The integral-transform pair with respect to the ϕ variable over the range $0 \leqslant \phi \leqslant 2\pi$ is obtained from equations (13-104) as

$$\text{Inversion formula:} \quad T(r,\phi)=\frac{1}{\pi}\sum_{v}\bar{T}(r,v) \tag{13-173a}$$

$$\text{Integral transform:} \quad \bar{T}(r,v)=\int_{\phi'=0}^{2\pi}\cos v(\phi-\phi')T(r,\phi')\,d\phi' \tag{13-173b}$$

where $v=0,1,2,3\ldots$ and replace π by 2π for $v=0$. The integral transform of the system (13-172) by the application of the transform (13-173b) yields

$$\frac{d^2\bar{T}}{dr^2}+\frac{1}{r}\frac{d\bar{T}}{dr}-\frac{v^2}{r^2}\bar{T}(r,v)=0 \quad \text{in} \quad 0 \leqslant r < b \tag{13-174a}$$

$$k_2\frac{d\bar{T}}{dr}+h_2\bar{T}=\bar{f}_2(v) \quad \text{at} \quad r=b \tag{13-174b}$$

The solution of equations (13-174) is

$$\bar{T}(r,v)=b\left(\frac{r}{b}\right)^{v}\frac{\bar{f}_2(v)}{k_2 v+h_2 b} \tag{13-175}$$

The inversion of this result by the inversion formula (13-173a) gives the temperature distribution as

$$T(r,\phi)=\frac{1}{\pi}\sum_{v} b\left(\frac{r}{b}\right)^{v}\frac{\bar{f}_2(v)}{k_2 v+h_2 b} \tag{13-176a}$$

where

$$\bar{f}_2(v)=\int_{\phi'=0}^{2\pi}\cos v(\phi-\phi')f_2(\phi')\,d\phi' \tag{13-176b}$$

$v=0,1,2,3\ldots$ and replace π by 2π for $v=0$.

Example 13-12

Solve the following steady-state heat conduction problem for a solid hemisphere of radius $r = b$.

$$\frac{\partial}{\partial r}\left(r^2 \frac{\partial T}{\partial r}\right) + \frac{\partial}{\partial \mu}\left[(1-\mu^2)\frac{\partial T}{\partial \mu}\right] = 0 \qquad 0 \leqslant r < b, \quad 0 < \mu \leqslant 1 \qquad \text{(13-177a)}$$

$$\frac{\partial T}{\partial \mu} = 0 \qquad \text{at} \quad \mu = 0 \qquad \text{(13-177b)}$$

$$T(r, \mu) = f(\mu) \qquad \text{at} \quad r = b \qquad \text{(13-177c)}$$

Solution. This problem is the same as that considered in system (4-128). The integral-transform pair with respect to the μ variable for $0 \leqslant \mu \leqslant 1$ and the boundary condition of the second kind at $\mu = 0$ is obtained from equations (13-150) as

$$\text{Inversion formula:} \qquad T(r, \mu) = \sum_n (2n+1) P_n(\mu) \bar{T}(r, n) \qquad \text{(13-178a)}$$

$$\text{Integral transform:} \qquad \bar{T}(r, n) = \int_{\mu'=0}^{1} P_n(\mu') T(r, \mu')\, d\mu' \qquad \text{(13-178b)}$$

where $n = 0, 2, 4, 6 \ldots$ (even integers). The integral transform of the system (13-177) by the application of the transform (13-178b) yields

$$\frac{d}{dr}\left(r^2 \frac{d\bar{T}}{dr}\right) - n(n+1)\bar{T}(r, n) = 0 \qquad 0 \leqslant r < b \qquad \text{(13-179a)}$$

$$\bar{T}(r, n) = \bar{f}(n) \qquad \text{at} \quad r = b \qquad \text{(13-179b)}$$

The solution of equations (13-179) is

$$\bar{T}(r, n) = \left(\frac{r}{b}\right)^n \bar{f}(n) \qquad \text{(13-180)}$$

The inversion of this result by the inversion formula (13-178a) gives the solution for the temperature as

$$T(r, \mu) = \sum_{n=0,2,4\ldots}^{\infty} (2n+1) P_n(\mu) \left(\frac{r}{b}\right)^n \int_{\mu'=0}^{1} P_n(\mu') f(\mu')\, d\mu' \qquad \text{(13-181)}$$

REFERENCES

1. M. D. Mikhailov and M. N. Özışık, *Unified Analysis and Solutions of Heat and Mass Diffusion*, Wiley, New York, 1984.
2. Ian N. Sneddon, *The Use of Integral Transforms*, McGraw-Hill, New York, 1972.
3. E. C. Titchmarsh, *Fourier Integrals*, 2nd ed., Clarendon Press, London, 1948.
4. A. Erdelyi, W. Magnus, F. Oberhettinger, and F. G. Tricomi, *Tables of Integrals Transforms*, McGraw-Hill, New York, 1954.
5. C. J. Tranter, *Integral Transforms in Mathematical Physics*, Wiley, New York, 1962.
6. V. A. Ditkin and A. P. Produnikov, *Integral Transforms and Operational Calculus*, Pergamon Press, New York, 1965.
7. M. N. Özışık, Integral Transform in the Solution of Heat-Conduction Equation in the Rectangular coordinate System, ASME Paper 67-WA/HT-46, 1967.
8. M. N. Özışık, *Boundary Value Problems of Heat Conduction*, International Textbook Co., Scranton, Pa., 1968.
9. A. C. Eringen, *Q. J. Math. Oxford* **5** (2), 120–129, 1954.
10. R. V. Churchill, *Mich. Math. J.* **3**, 85, 1955–1956.
11. A. McD. Mercer, *Q. J. Math. Oxford.* **14**, 9–15, 1963.
12. N. Y. Ölçer, *Österr. Ing.-Arch* **18**, 104–113, 1964.
13. N. Y. Ölçer, *Int. J. Heat Mass Transfer* **7**, 307–314, 1964.
14. N. Y. Ölçer, *Int. J. Heat Mass Transfer* **8**, 529–556, 1965.
15. N. Y. Ölçer, *J. Math Phys.* **46**, 99–106, 1967.
16. M. D. Mikhailov, *Int. J. Eng. Sci.* **10**, 577–591, 1972.
17. M. D. Mikhailov, *Int. J. Heat Mass Transfer* **16**, 2155–2164, 1973.
18. M. D. Mikhailov, *Int. J. Heat Mass Transfer* **17**, 1475–1478, 1974.
19. M. D. Mikhailov, *Int. J. Eng. Sci.* **11**, 235–241, 1973.
20. M. N. Özışık and R. L. Murray, *J. Heat Transfer* **96c**, 48–51, 1974.
21. Y. Yener and M. N. Özışık, On the Solution of Unsteady Heat Conduction in Multi-Region Media with Time Dependent Heat Transfer Coefficient, *Proc. 5th. Int. Heat Trans. Conference*, Cu 2.5, pp. 188–192, Tokyo, Sept. 1974.
22. M. D. Mikhailov, *Int. J. Heat Mass Transfer* **18**, 344–345, 1975.
23. M. D. Mikhailov, *Int. J. Heat Mass Transfer* **20**, 1409–1415, 1977.
24. N. Y. Ölçer, *Int. J. Heat Mass Transfer* **12**, 393–411, 1969.
25. K. Kobayashi, N. Ohtani, and J. Jung, *Nucl. Sci. Eng.* **55**, 320–328, 1974.
26. I. N. Sneddon, *Phil. Mag.* **37**, 17, 1946.
27. G. Cinelli, *Int. J. Engl. Sci.* **3**, 539–559, 1965.
28. N. Y. Ölçer, *Brit. J. Appl. Phys.* **18**, 89–105, 1967.
29. N. Y. Ölçer, *Proc. Comb. Phil. Soc.* **64**, 193–202, 1968.
30. N. Y. Ölçer, *J. Heat Transfer* **91c**, 45–50, 1969.
31. A. W. Jacobson, *Quart. Appl. Math.* **7**, 293–302, 1949.
32. C. J. Tranter, *Quart. J. Math. Oxford* **1**, 1–8, 1950.
33. R. V. Churchill and C. L. Dolph, *Proc. Amer. Math. Soc.* **5**, 93, 1954.
34. R. V. Churchill, *J. Math. Phys.* **33**, 165–178, 1954–1955.

35. R. Courant and D. Hilbert, *Methods of Mathematical Physics*, Vol. 1, Interscience Publishers, New York, 1953, p. 106.
36. G. N. Watson, *A Treatise on the Theory of Bessel Functions*, Cambridge University Press, London, 1966.
37. M. D. Mikhailov, M. N. Özışık, and N. L. Vulchanov, *Int. J. Heat Mass Transfer* **26**, 1131–1141, 1983.
38. M. D. Mikhailov and M. N. Özışık, *Int. J. Heat Mass Transfer* **28**, 1039–1045, 1985.
39. M. D. Mikhailov and M. N. Özışık, *J. Franklin Institute* **321**, 299–307, 1986.
40. R. M. Cotta and M. N. Özışık, *Can. J. Chem. Eng.*, **64**, 734–742, 1986.

PROBLEMS

13-1 Solve the one-dimensional, time-dependent heat conduction problem for a slab $0 \leqslant x \leqslant L$, which is initially at zero temperature and for times $t > 0$ the boundaries at $x = 0$ and $x = L$ are kept at temperatures zero and $f_2(t)$, respectively. Consider the case when the surface temperature is given by $f_2(t) = \gamma t$, where γ is a constant.

13-2 Solve the one-dimensional, time-dependent heat conduction problem for a slab $0 \leqslant x \leqslant L$, which is initially at zero temperature, and for times $t > 0$ heat is generated in the medium at a rate of $g(x, t)$, W/m^3, while the boundary surface at $x = 0$ is kept insulated and the boundary surface at $x = L$ is kept at zero temperature. Consider the case when the heat source is an instantaneous plane heat source of total strength g_s^i Ws/m^2, situated at $x = b$ and release its heat spontaneously at time $t = 0$, that is, $g(x, t) = g_s^i \delta(x - b)\delta(t)$.

13-3 A semiinfinite medium, $0 \leqslant x < \infty$, is initially at zero temperature. For times $t > 0$ the boundary at $x = 0$ is kept at zero temperature, while heat is generated in the medium at a rate of $g(x, t)$ W/m^3. Obtain an expression for the temperature distribution $T(x, t)$ in the medium. Consider the cases (1) the heat source is a continuous plane-surface heat source of strength $g_s^c(t)$ W/m^2, which is situated at $x = b$, that is $g(x, t) = g_s^c(t)\delta(x - b)$, (2) the heat source is a constant heat source, that is, $g(x, t) = g_0 = \text{constant}$ W/m^3.

13-4 An infinite medium $-\infty < x < \infty$ is initially at zero temperature. A plane-surface heat source of strength $g_s^c(t)$ W/m^2, situated at $x = 0$, releases heat continuously for times $t > 0$. Obtain an expression for the temperature distribution $T(x, t)$ in the medium for times $t > 0$ [i.e., $g(x, t) = g_s(t)\delta(x)$].

13-5 A rectangular region $0 \leqslant x \leqslant a$, $0 \leqslant y \leqslant b$ is initially at zero temperature. For times $t > 0$ heat is generated in the medium at a rate of $g(x, y, t)$ W/m^3, while the boundaries are kept at zero temperature. Obtain an expression for the temperature distribution $T(x, y, t)$ in the region. Also consider the special case, when the heat source is an instantaneous line heat source g_{Li} of strength Ws/m, situated at (x_1, y_1) within the region and releases its heat spontaneously at time $t = 0$, that is, $g(x, y, t) = g_L^i \delta(x - x_1)\delta(y - y_1)\delta(t)$.

13-6 A three-dimensional infinite medium $-\infty < x < \infty$, $-\infty < y < \infty$, $-\infty < z < \infty$ is initially at zero temperature. For times $t > 0$, heat is generated in the medium at a rate of $g(x, y, z, t)$ W/m^3. Obtain an expression for the temperature distribution $T(x, y, z, t)$ in the medium. Also consider the special case when the heat source is an instantaneous point heat source of strength g_s^i Ws, situated at $x = 0$, $y = 0$, $z = 0$ and releasing its heat spontaneously at time $t = 0$, that is, $g(x, y, z, t) = g_p^i \delta(x)\delta(y)\delta(z)\delta(t)$.

13-7 A solid cylinder $0 \leqslant r \leqslant b$, is initially at zero temperature. For times $t > 0$, heat is generated within the region at a rate of $g(r, t)$ W/m^3, while the boundary at $r = b$ is kept at zero temperature. Obtain an expression for the temperature distribution $T(r, t)$ in the cylinder. Consider the special cases (1) the heat is generated at a constant rate g_0 W/m^3, in the region, (2) the heat source is a line heat source of strength $g_L(t)$ W/m, situated along the axis of the cylinder, that is, $g(r, t) = \dfrac{1}{2\pi r} g_L(t)\delta(r)$.

13-8 A long solid cylinder, $0 \leqslant r \leqslant b$, is initially at temperature $F(r)$. For times $t > 0$ the boundary at $r = b$ is kept insulated. Obtain an expression for the temperature distribution $T(r, t)$ in the cylinder.

13-9 A long hollow cylinder, $a \leqslant r \leqslant b$, is initially at temperature $F(r)$. For times $t > 0$ the boundaries at $r = a$ and $r = b$ are kept insulated. Obtain an expression for the temperature distribution $T(r, t)$ in the region.

13-10 A long hollow cylinder, $a \leqslant r \leqslant b$, is initially at zero temperature. For times $t > 0$ heat is generated in the medium at a rate of $g(r, t)$ W/m^3, while the boundaries at $r = a$ and $r = b$ are kept at zero temperature. Obtain an expression for the temperature distribution $T(r, t)$ in the cylinder. Consider the special cases (1) the heat-generation rate is constant, that is, $g_0 =$ constant, and (2) the heat source is an instantaneous cylindrical heat source of radius $r = r_1$ (i.e., $a < r_1 < b$) of strength g_s^i Ws/m, per linear length of the cylinder, which is situated inside the cylinder coaxially and releases its heat spontaneously at time $t = 0$, that is, $g(r, t) = \dfrac{1}{2\pi r} g_s^i \delta(r - r_1)\delta(t)$.

13-11 An infinite region, $0 \leqslant r < \infty$, is initially at zero temperature. For times $t > 0$ heat is generated in the medium at a rate of $g(r, t)$ W/m^3. Obtain an expression for the temperature distribution $T(r, t)$ in the medium for times $t > 0$. Consider the special cases (1) the heat source is of constant strength, that is, $g(r, t) = g_0 =$ constant, (2) the heat source is an instantaneous line-heat source of strength g_L^i Ws/m, situated along the z axis in the medium and releases its heat spontaneously at time $t=0$, that is, $g(r,t)=(1/2\pi r)g_L^i\delta(r)\delta(t)$.

13-12 The region $a \leqslant r < \infty$ is initially at zero temperature. For times $t>0$, heat is generated in the medium at a rate of $g(r,t)$ W/m^3, while the boundary surface at $r=a$ is kept at zero temperature. Obtain an expression for the temperature distribution $T(r,t)$ in the medium for times $t>0$. Consider the special case of constant heat generation in the medium.

13-13 The cylindrical region $0 \leqslant r \leqslant b$, $0 \leqslant \phi \leqslant 2\pi$ is initially at temperature $F(r,\phi)$. For times $t>0$ the boundary surface at $r=b$ is kept insulated. Obtain an expression for the temperature distribution $T(r,\phi,t)$ in the region for times $t>0$.

13-14 The cylindrical region $0 \leqslant r \leqslant b, 0 \leqslant \phi \leqslant 2\pi$ is initially at zero temperature. For times $t>0$, heat is generated in the medium at a rate of $g(r,\phi,t)$ W/m^3, while the boundary at $r=b$ is kept at zero temperature. Obtain an expression for the temperature distribution $T(r,\phi,t)$ in the region for times $t>0$.

13-15 The cylindrical region $a \leqslant r \leqslant b, 0 \leqslant \phi \leqslant 2\pi$ is initially at temperature $F(r,\phi)$. For times $t>0$, the boundaries at $r=a$ and $r=b$ are kept at zero temperatures. Obtain an expression for the temperature distribution $T(r,\phi,t)$ in the region for times $t>0$.

13-16 The cylindrical region consisting of a portion of a cylinder, $0 \leqslant r \leqslant b$, $0 \leqslant \phi \leqslant \phi_0$ (where $\phi_0 < 2\pi$) is initially at zero temperature. For times $t>0$ heat is generated in the medium at a rate of $g(r,\phi,t)$ W/m^3, while all boundary surfaces are kept at zero temperature. Obtain an expression for the temperature distribution $T(r,\phi,t)$ in the region for times $t>0$. Also consider the special case of $g(r,\phi,t)=g_0=\text{constant}$.

13-17 The cylindrical region consisting of a portion of a cylinder, $a \leqslant r \leqslant b$, $0 \leqslant \phi \leqslant \phi_0$ (where $\phi_0 < 2\pi$) is initially at temperature $F(r,\phi)$. For times $t>0$ all boundary surfaces are kept at zero temperature. Obtain an expression for the temperature distribution $T(r,\phi,t)$ in the region for times $t>0$. Also consider the special case of uniform initial temperature distribution, that is, $F(r,\phi)=T_0=\text{constant}$.

13-18 The cylindrical region $a \leqslant r \leqslant b$, $0 \leqslant \phi \leqslant \phi_0$ (where $\phi_0 < 2\pi$) is initially at zero temperature. For times $t>0$ heat is generated in the medium at a rate of $g(r,\phi,t)$ W/m^3, while the boundaries are kept at zero temperature. Obtain an expression for the temperature distribution $T(r,\phi,t)$ in the medium for times $t>0$.

13-19 A cylindrical region $0 \leqslant r \leqslant b$, $0 \leqslant z \leqslant L$ is initially at zero temperature. For times $t>0$ heat is generated in the medium at a rate of $g(r,z,t)$ W/m^3, while the boundary surface at $z=0$ is kept insulated and all the remaining boundaries are kept at zero temperature. Obtain an expression for the temperature distribution $T(r, z, t)$ in the region.

13-20 A cylindrical region $0 \leqslant r \leqslant b, 0 \leqslant z < \infty$ is initially at temperature $F(r, z)$. For times $t > 0$ all the boundary surfaces are kept at zero temperature. Obtain an expression for the temperature distribution $T(r, z, t)$ in the region for times $t > 0$.

13-21 A hemispherical region $0 \leqslant r \leqslant b$, $0 \leqslant \mu \leqslant 1$ is initially at temperature $F(r, \mu)$. For times $t > 0$ the boundary surface at $\mu = 0$ is kept insulated and the boundary surface at $r = b$ is kept at zero temperature. Obtain an expression for the temperature distribution $T(r, \mu, t)$ in the hemisphere for times $t > 0$.

13-22 A hemispherical region $0 \leqslant r \leqslant b, 0 \leqslant \mu \leqslant 1$ is initially at zero temperature. For times $t > 0$ heat is generated in the medium at a rate of $g(r, \mu, t)$ $\mathrm{W/m^3}$, while the boundary surface at $\mu = 0$ is kept insulated and the boundary at $r = b$ is kept at zero temperature. Obtain an expression for the temperature distribution $T(r, \mu, t)$ in the region for times $t > 0$.

13-23 A hollow hemispherical region $a \leqslant r \leqslant b$, $0 \leqslant \mu \leqslant 1$ is initially at zero temperature. For times $t > 0$ heat is generated in the medium at a rate of $g(r, \mu, t)$ $\mathrm{W/m^3}$, while the boundaries are kept at zero temperature. Obtain an expression for the temperature distribution $T(r, \mu, t)$ in the region.

13-24 A solid sphere of radius $r = b$ is initially at temperature $F(r, \mu, \phi)$. For times $t > 0$ the boundary surface at $r = b$ is kept insulated. Obtain an expression for the temperature distribution $T(r, \mu, \phi, t)$ in the sphere for times $t > 0$.

13-25 Solve for the steady-state temperature distribution $T(x, y)$ in a rectangular strip $0 \leqslant y \leqslant b$, $0 \leqslant x < \infty$ subject to the boundary conditions $T = f(y)$ at $x = 0$ and $T = 0$ at $y = 0$ and $y = b$.

13-26 Solve for the steady-state temperature distribution $T(r, \mu, \phi)$ in a solid sphere of radius $r = b$ subject to the boundary condition $T = f(\mu, \phi)$ at the boundary surface $r = b$.

NOTES

1. To prove the orthogonality relation given by equations (13-3), equation (13-2a) is written for two different eigenfunctions $\psi_m(\mathbf{r})$ and $\psi_n(\mathbf{r})$, corresponding to two different eigenvalues λ_m and λ_n as

$$\nabla^2 \psi_m(\mathbf{r}) + \lambda_m^2 \psi_m(\mathbf{r}) = 0 \quad \text{in} \quad R \tag{1a}$$

$$\nabla^2 \psi_n(\mathbf{r}) + \lambda_n^2 \psi_n(\mathbf{r}) = 0 \quad \text{in} \quad R \tag{1b}$$

The first equation is multiplied by $\psi_n(\mathbf{r})$, the second by $\psi_m(\mathbf{r})$, the results are subtracted and integrated over the region R

$$(\lambda_m^2 - \lambda_n^2) \int_R \psi_m(\mathbf{r})\psi_n(\mathbf{r})\, dv = \int_R [\psi_m \nabla^2 \psi_n - \psi_n \nabla^2 \psi_m]\, dv \tag{2}$$

The volume integral on the right is changed to surface integral by Green's theorem as discussed in note 2 at the end of Chapter 6. We find

$$\begin{aligned}(\lambda_m^2 - \lambda_n^2) \int_R \psi_m(\mathbf{r})\psi_n(\mathbf{r})\, dv &= \int_S \left(\psi_m \frac{\partial \psi_n}{\partial n} - \psi_n \frac{\partial \psi_m}{\partial n} \right) ds \\ &= \sum_{i=1}^{N} \int_{S_i} \left(\psi_m \frac{\partial \psi_n}{\partial n_i} - \psi_n \frac{\partial \psi_m}{\partial n_i} \right) ds_i \end{aligned} \tag{3}$$

The boundary condition (13-2b) is written for two different eigenfunctions $\psi_m(\mathbf{r})$ and $\psi_n(\mathbf{r})$

$$k_i \frac{\partial \psi_m}{\partial n_i} + h_i \psi_m = 0 \tag{4a}$$

$$k_i \frac{\partial \psi_n}{\partial n_i} + h_i \psi_n = 0 \tag{4b}$$

The first is multiplied by ψ_n, the second by ψ_n, the results are subtracted

$$\psi_m \frac{\partial \psi_n}{\partial n_i} - \psi_n \frac{\partial \psi_m}{\partial n_i} = 0 \tag{5}$$

when this result is introduced into equation (3) we obtain

$$(\lambda_m^2 - \lambda_n^2) \int_R \psi_m(\mathbf{r})\psi_n(\mathbf{r})\, dv = 0 \tag{6}$$

Thus

$$\int_R \psi_m(\mathbf{r})\psi_n(\mathbf{r})\, dv = 0 \qquad \text{for} \qquad m \neq n \tag{7}$$

2. The closed-form expressions given by equations (13-44) can be derived as now described. Consider the problem

$$\frac{\partial^2 \theta}{\partial x^2} = \frac{1}{\alpha} \frac{\partial \theta}{\partial t} \qquad \text{in} \qquad 0 < x < L, \quad t > 0 \tag{1a}$$

$$\frac{\partial \theta}{\partial x} = 0 \qquad \text{at} \qquad x = 0, \qquad t > 0 \tag{1b}$$

$$\theta = 0 \qquad \text{at} \qquad x = L, \qquad t > 0 \tag{1c}$$

$$\theta = 1 \qquad \text{for} \qquad t = 0, \qquad \text{in} \quad 0 \leqslant x \leqslant L \tag{1d}$$

The solution of this problem is given as

$$\begin{aligned}\theta(x,t) &= \frac{2}{L}\sum_{m=1}^{\infty} e^{-\alpha\beta_m^2 t}\cos\beta_m x \int_0^L 1\cdot\cos\beta_m x'\,dx' \\ &= \frac{2}{L}\sum_{m=1}^{\infty} e^{-\alpha\beta_m^2 t}\frac{\cos\beta_m x}{\beta_m}(-1)^{m-1}\end{aligned} \tag{2}$$

where $\beta_m = (2m-1)\pi/2L$. For $t = 0$ equation (2) should be equal to the initial condition (1d); hence

$$\frac{2}{L}\sum_{m=1}^{\infty}(-1)^{m-1}\frac{\cos\beta_m x}{\beta_m} = 1 \tag{3}$$

which is the result given by equation (13-44a).
We now consider the problem given by equations (1) for an initial condition $(x^2 - L^2)$. The solution becomes

$$\theta(x,t) = \frac{2}{L}\sum_{m=1}^{\infty} e^{-\alpha\beta_m^2 t}\cos\beta_m x \int_0^L (x'^2 - L^2)\cos\beta_m x'\,dx' \tag{4}$$

After performing the integration we obtain

$$\theta(x,t) = -\frac{4}{L}\sum_{m=1}^{\infty} e^{-\alpha\beta_m^2 t}(-1)^{m-1}\frac{\cos\beta_m x}{\beta_m^3} \tag{5}$$

For $t = 0$, we have $\theta(x,t) = x^2 - L^2$; then

$$x^2 - L^2 = -\frac{4}{L}\sum_{m=1}^{\infty}(-1)^{m-1}\frac{\cos\beta_m x}{\beta_m^3} \tag{6}$$

which is the result given by equation (13-44b)

3. The integral transform of equation (13-56a) according to the definition of the integral transform (13-57b) is

$$\int_{x=0}^{\infty} X(\beta,x)\frac{\partial^2 T}{\partial x^2}dx + \frac{1}{k}\bar{g}(\beta,t) = \frac{1}{\alpha}\frac{d\bar{T}(\beta,t)}{dt} \tag{1}$$

The first term on the left is evaluated by integrating it by parts twice.

$$\begin{aligned}\int_{x=0}^{\infty} X(\beta,x)\frac{\partial^2 T}{\partial x^2}dx &= \left[X\frac{\partial T}{\partial x}\right]_0^{\infty} - \int_0^{\infty}\frac{dX}{dx}\frac{\partial T}{\partial x}dx \\ &= \left[X\frac{\partial T}{\partial x} - T\frac{dX}{dx}\right]_0^{\infty} + \int_0^{\infty} T\frac{d^2X}{dx^2}dx\end{aligned} \tag{2}$$

The term inside the bracket vanishes at the upper limit; it is evaluated at the lower limit by utilizing the boundary conditions (13-56b) and (2-48b); we obtain

$$\left[X\frac{\partial T}{\partial x}-T\frac{dX}{dx}\right]_0^\infty=\left.\frac{X(\beta,x)}{k_1}\right|_{x=0}\cdot f_1(t) \tag{3}$$

The second term on the right in equation (2) is evaluated by multiplying equation (2-48a) by $T(x,t)$ and utilizing the definition of the integral transform (13-57b). We obtain

$$\int_0^\infty T\frac{\partial^2 X}{dx^2}dx=-\beta^2\int_0^\infty TX\,dx=-\beta^2\bar{T}(\beta,t) \tag{4}$$

Introducing equations (2) to (4) into (1) we find

$$\frac{d\bar{T}(\beta,t)}{dt}+\alpha\beta^2\bar{T}(\beta,t)=\frac{\alpha}{k}\bar{g}(\beta,t)+\left.\frac{X(\beta,x)}{k_1}\right|_{x=0}\cdot f_1(t) \tag{5}$$

which is the result given by equation (13-58a).

4. We consider the representation

$$F^*(x)=\frac{1}{\pi}\int_{\beta=0}^\infty\int_{x'=-\infty}^\infty F^*(x')\cos\beta(x-x')dx'\,d\beta \tag{1}$$

The integral of the cosine term with respect to β is expressed as a complex integral in the form

$$\int_0^L\cos\beta(x'-x)d\beta=\frac{1}{2}\int_{-L}^L\cos\beta(x'-x)d\beta=\frac{1}{2}\int_{-L}^L e^{i\beta(x'-x)}d\beta \tag{2a}$$

since

$$\int_{-L}^L\sin\beta(x'-x)d\beta=0 \tag{2b}$$

Then, in the limit $L\to\infty$, equation (1) can be written as

$$F^*(x)=\frac{1}{2\pi}\int_{-\infty}^\infty e^{-i\beta x}\int_{-\infty}^\infty e^{i\beta x'}F^*(x')dx'\,d\beta \tag{3}$$

which is the result given by equation (13-62).

5. The integral transform of equation (13-107a) with respect to the ϕ variable, by the application of the transform (13-108b) is

$$\frac{\partial^2\bar{T}}{\partial r^2}+\frac{1}{r}\frac{\partial\bar{T}}{\partial r}+\frac{1}{r^2}\int_{\phi=0}^{2\pi}\cos\nu(\phi-\phi')\frac{\partial^2 T}{\partial\phi^2}d\phi+\frac{\bar{g}}{k}=\frac{1}{\alpha}\frac{\partial\bar{T}}{\partial t} \tag{1}$$

The integral term is evaluated as follows. Let $\Phi(\phi) \equiv \cos \nu(\phi - \phi')$. Then, the integral term becomes

$$\int_0^{2\pi} \Phi(\phi) \frac{\partial^2 T}{\partial \phi^2} d\phi = \left[\Phi \frac{\partial T}{\partial \phi} \right]_0^{2\pi} - \int_0^{2\pi} \frac{d\Phi}{d\phi} \frac{\partial T}{\partial \phi} d\phi$$

$$= \left[\Phi \frac{\partial T}{\partial \phi} - T \frac{d\Phi}{d\phi} \right]_0^{2\pi} + \int_0^{2\pi} T \frac{d^2 \Phi}{d\phi^2} d\phi$$

$$= \int_0^{2\pi} T \frac{d^2 \Phi}{d\phi^2} d\phi \tag{2}$$

since the terms in the bracket vanish because the functions are cyclic with a period of 2π. To evaluate this integral, we consider the eigenvalue problem given as

$$\frac{d^2 \Phi(\phi)}{d\phi^2} + \nu^2 \Phi(\phi) = 0 \qquad \text{in} \qquad 0 \leqslant \phi \leqslant 2\pi \tag{3}$$

The function $\Phi(\phi)$ is cyclic with a period of 2π. We multiply this equation by T, integrate with respect to ϕ from 0 to 2π and utilize the definition of the integral transform (13-108b) to obtain

$$\int_0^{2\pi} T \frac{d^2 \Phi(\phi)}{d\phi^2} d\phi = -\nu^2 \int_0^{2\pi} \Phi(\phi) T \, d\phi = -\nu^2 \bar{T} \tag{4}$$

Introducing equations (2) and (4) into (1) we find

$$\frac{\partial^2 \bar{T}}{\partial r^2} + \frac{1}{r} \frac{\partial \bar{T}}{\partial r} - \frac{\nu^2}{r^2} \bar{T} + \frac{\bar{g}}{k} = \frac{1}{\alpha} \frac{\partial \bar{T}}{\partial t} \tag{5}$$

which is the result given by equation (13-109a).

6. The integral transform of equation (13-109a) by the application of the transform (13-110b) is

$$\int_0^b r R_\nu(\beta_m, r) \left[\frac{\partial^2 \bar{T}}{\partial r^2} + \frac{1}{r} \frac{\partial \bar{T}}{\partial r} - \frac{\nu^2}{r^2} \bar{T} \right] dr + \frac{\tilde{\bar{g}}}{k} = \frac{1}{\alpha} \frac{d \tilde{\bar{T}}}{dt} \tag{1}$$

The integral term is evaluated by integrating it by parts twice, utilizing the eigenvalue problem for the R_ν function and the boundary conditions as

$$I \equiv \int_0^b r R_\nu \left[\frac{\partial^2 \bar{T}}{\partial r^2} + \frac{1}{r} \frac{\partial \bar{T}}{\partial r} - \frac{\nu^2}{r^2} \bar{T} \right] dr$$

$$= \left[r \left(R_\nu \frac{\partial \bar{T}}{\partial r} - \bar{T} \frac{dR_\nu}{dr} \right) \right]_0^b + \int_0^b r \left(\frac{d^2 R_\nu}{dr^2} + \frac{1}{r} \frac{dR_\nu}{dr} - \frac{\nu^2}{r^2} R_\nu \right) \bar{T} \, dr$$

$$= b \left(R_\nu \frac{\partial \bar{T}}{\partial r} - \bar{T} \frac{dR_\nu}{\partial r} \right) \bigg|_{r=b} + \int_0^b r \left(\frac{d^2 R_\nu}{dr^2} + \frac{1}{r} \frac{dR_\nu}{dr} - \frac{\nu^2}{r^2} R_\nu \right) \bar{T} \, dr \tag{2}$$

since the term inside the bracket vanishes at the lower limit. From the eigenvalue problem [i.e., equation (3-18a)] we have

$$\frac{d^2R_\nu}{dr^2}+\frac{1}{r}\frac{dR_\nu}{dr}+\left(\beta_m^2-\frac{\nu^2}{r^2}\right)R_\nu=0 \qquad \text{in} \qquad 0\leqslant r<b \tag{3}$$

Multiplying equation (3) by $r\bar{T}$, integrating it with respect to r from $r=0$ to $r=b$, and utilizing the definition of the integral transform (13-110b) we obtain

$$\int_0^b r\left(\frac{d^2R_\nu}{dr^2}+\frac{1}{r}\frac{dR_\nu}{dr}-\frac{\nu^2}{r^2}R_\nu\right)\bar{T}\,dr=-\beta_m^2\int_0^b rR_\nu\bar{T}\,dr=-\beta_m^2\bar{\bar{T}} \tag{4}$$

From the boundary conditions (13-109b) and (13-18b) we have

$$k_2\frac{\partial\bar{T}}{\partial r}+h_2\bar{T}=\bar{f}_2(\nu,t) \qquad \text{at} \qquad r=b \tag{5a}$$

$$k_2\frac{dR_\nu}{dr}+h_2R_\nu=0 \qquad \text{at} \qquad r=b \tag{5b}$$

Multiplying (5a) by R_ν and (5b) by $\bar{T}$ and subtracting the results we obtain

$$\left[R_\nu\frac{\partial\bar{T}}{\partial r}-\bar{T}\frac{dR_\nu}{dr}\right]_{r=b}=b\left.\frac{R_\nu}{k_2}\right|_{r=b}\cdot\bar{f}_2(\nu,t) \tag{6}$$

Introducing equations (6), (4), and (2) into equation (1) we find

$$-\beta_m^2\bar{\bar{T}}(\beta_m,\nu,t)+b\left.\frac{R_\nu(\beta_m,r)}{k_2}\right|_{r=b}\cdot\bar{f}_2(\nu,t)+\frac{\bar{\bar{g}}(\beta_m,\nu,t)}{k}=\frac{1}{\alpha}\frac{d\bar{\bar{T}}}{dt}$$

or

$$\frac{d\bar{\bar{T}}(\beta_m,\nu,t)}{dt}+\alpha\beta_m^2\bar{\bar{T}}(\beta_m,\nu,t)=\frac{\alpha}{k}\bar{\bar{g}}(\beta_m,\nu,t)+\alpha b\left.\frac{R_\nu(\beta_m,r)}{k_2}\right|_{r=b}\cdot\bar{f}_2(\nu,t) \tag{7}$$

which is the result given by equation (13-111a).

7. When taking the integral transform of the differential equation (13-143a) by the application of the transform (13-144b), we need to consider only the removal of the following differential operator

$$\nabla^2V\equiv-\frac{V}{4}+\frac{\partial}{\partial\mu}\left[(1-\mu^2)\frac{\partial V}{\partial\mu}\right] \qquad \text{in} \qquad -1\leqslant\mu\leqslant1 \tag{1}$$

since the integral transform of the remaining terms is straightforward. The integral transform of this operator under the transform (13-144b) is

$$\overline{\nabla^2V}=-\frac{1}{4}\bar{V}+\int_{-1}^1 P_n(\mu)\frac{\partial}{\partial\mu}\left[(1-\mu^2)\frac{\partial V}{\partial\mu}\right]d\mu \tag{2}$$

The integration is performed by integrating it by parts twice:

$$\overline{\nabla^2 V} \equiv -\frac{1}{4}\bar{V} + \left[(1-\mu^2)\left(P_n\frac{\partial V}{\partial \mu} - V\frac{dP_n}{d\mu}\right)\right]_{-1}^{1} + \int_{-1}^{1} V\frac{d}{d\mu}\left[(1-\mu^2)\frac{dP_n}{d\mu}\right]d\mu \tag{3}$$

The terms inside the bracket vanishes at both limits. The integral terms is evaluated by noting that $P_n(\mu)$ function satisfies the Legendre equation

$$\frac{d}{d\mu}\left[(1-\mu^2)\frac{dP_n}{d\mu}\right] + n(n+1)P_n(\mu) = 0 \tag{4}$$

Multiplying this equation by V and integrating from $\mu = -1$ to 1 we find

$$\int_{-1}^{1} V\frac{d}{d\mu}\left[(1-\mu^2)\frac{dP_n}{d\mu}\right]d\mu = -n(n+1)\int_{-1}^{1} P_n(\mu)V\,d\mu = -n(n+1)\bar{V} \tag{5}$$

Introducing equation (5) into (3) we obtain

$$\overline{\nabla^2 V} = -\tfrac{1}{4}\bar{V} - n(n+1)\bar{V} = -(n+\tfrac{1}{2})^2\bar{V} \tag{6}$$

which is the term that appears in equation (13-145a).

8. When taking the integral transform of the equation (13-156) by the application of the transform (13-158b) we need to consider only the removal of the following differential operator:

$$\nabla^2\bar{V} \equiv -\frac{1}{4}\bar{V} + \left\{\frac{\partial}{\partial\mu}\left[(1-\mu^2)\frac{\partial\bar{V}}{\partial\mu} - \frac{m^2\bar{V}}{1-\mu^2}\right]\right\} \quad \text{in} \quad -1 \leqslant \mu \leqslant 1 \tag{1}$$

The integral transform of this operator under the transform (13-158b) is

$$\widetilde{\nabla^2\bar{V}} = -\frac{1}{4}\tilde{\bar{V}} + \int_{-1}^{1} P_n^m\left\{\frac{\partial}{\partial\mu}\left[(1-\mu^2)\frac{\partial\bar{V}}{\partial\mu}\right]\right\}d\mu - \int_{-1}^{1}\frac{m^2}{1-\mu^2}P_n^m\bar{V}\,d\mu \tag{2}$$

The first integral is performed by integrating by parts twice

$$\begin{aligned}\widetilde{\nabla^2\bar{V}} \equiv & -\frac{1}{4}\tilde{\bar{V}} + \left[(1-\mu^2)\left(P_n^m\frac{\partial\bar{V}}{\partial\mu} - \bar{V}\frac{dP_n^m}{d\mu}\right)\right]_{-1}^{1} \\ & + \int_{-1}^{1}\bar{V}\left\{\frac{\partial}{\partial\mu}\left[(1-\mu^2)\frac{dP_n^m}{d\mu}\right] - \frac{m^2}{1-\mu^2}\bar{V}\right\}d\mu\end{aligned} \tag{3}$$

The second term in the bracket vanishes at both limits. The integral term is evaluated by noting that $P_n^m(\mu)$ function satisfies Legendre's associated differential equation (13-157):

$$\frac{d}{d\mu}\left[(1-\mu^2)\frac{dP_n^m}{d\mu}\right] + \left[n(n+1) - \frac{m^2}{1-\mu^2}\right]P_n^m = 0 \tag{4}$$

Multiplying this equation by $\bar{V}$ and integrating from $\mu = -1$ to 1, we find

$$\int_{-1}^{1} \bar{V}\left\{\frac{d}{d\mu}\left[(1-\mu^2)\frac{dP_n^m}{d\mu}\right] - \frac{m^2}{1-\mu^2}P_n^m\right\}d\mu = -n(n+1)\int_{-1}^{1} P_n^m \bar{V}\,d\mu$$
$$= -n(n+1)\tilde{\bar{V}} \tag{5}$$

Introducing equation (5) into (3) we obtain

$$\widetilde{\nabla^2 \bar{V}} \equiv -\tfrac{1}{4}\tilde{\bar{V}} - n(n+1)\tilde{\bar{V}} = -(n+\tfrac{1}{2})^2\tilde{\bar{V}} \tag{6}$$

which is the term that appears in equation (13-159).

14

INVERSE HEAT CONDUCTION PROBLEMS (IHCP)

Inverse problems are encountered in various branches of science and engineering. Mechanical, aerospace and chemical engineers; mathematicians, astrophysicists, geophysicists, statisticians and specialists of many other disciplines are all interested in inverse problems, each with different applications in mind. In the field of heat transfer, the use of inverse analysis for the estimation of surface conditions such as temperature and heat flux, or the determination of thermal properties such as thermal conductivity and heat capacity of solids by utilizing the transient temperature measurements taken within the medium has numerous practical applications. For example, the direct measurement of heat flux at the surface of a wall subjected to fire, at the outer surface of a reentry vehicle, or at the inside surface of a combustion chamber is extremely difficult. In such situations, the inverse method of analysis, using transient temperature measurements taken within the medium can be applied for the estimation of such quantities. However, difficulties associated with the implementation of inverse analysis should be also recognized. The main difficulty comes from the fact that inverse solutions are very sensitive to changes in input data resulting from measurement and modelling errors, hence may not be unique. Mathematically, the inverse problems belong to the class of problems called the *ill-posed* problems; that is, their solution does not satisfy the general requirement of *existence*, *uniqueness*, and *stability* under small changes to the input data. To overcome such difficulties, a variety of techniques for solving inverse heat conduction problems have been proposed. In this chapter an introductory treatment is presented to the theory and application of inverse heat conduction problems.

14-1 AN OVERVIEW OF IHCP

The standard heat conduction problems are concerned with the determination of temperature-distribution in the interior of the solid when the boundary and initial conditions, the energy generation rate and thermophysical properties of the medium are specified. In contrast, *inverse heat conduction problems* (IHCPs) are concerned with the determination of boundary condition, energy-generation rate, or thermophysical properties by utilizing the measured temperature history at one or more locations in the solid. The most common usage of IHCP is concerned with the estimation of the surface heat flux or temperature history from temperature measurement taken at an interior location at different times. To illustrate this matter, we consider the following one-dimensional transient heat conduction problem in a slab

$$\frac{\partial}{\partial x}\left(k\frac{\partial T}{\partial x}\right)=\rho C_p\frac{\partial T}{\partial t} \qquad \text{in} \qquad 0<x<L, \quad t>0 \tag{14-1a}$$

$$-k\frac{\partial T}{\partial x}=f(t) \qquad \text{at} \qquad x=0, \qquad t>0 \tag{14-1b}$$

$$T=T_L \qquad \text{at} \qquad x=L, \qquad t>0 \tag{14-1c}$$

$$T=F(x) \qquad \text{for} \qquad t=0, \qquad \text{in} \quad 0\leqslant x\leqslant L \tag{14-1d}$$

where $f(t), T_L, F(x), \rho C_p$, and k are all considered known. Then the problem is concerned with the determination of temperature distribution $T(x,t)$ in the interior region of the solid as a function of time and position. We shall refer to such traditional problems as the *direct heat conduction* problems.

We now consider a problem similar to that given by equations (14-1), but the boundary condition function $f(t)$ at one of the boundary surfaces is unknown. To compensate for the lack of information on the boundary condition, measured temperatures $T(x_1,t_j)\equiv Y_j$ are given at an interior point x_1 at different times t_j $(j=1,2,\ldots,N)$ over a specified time interval $0<t\leqslant t_f$, where t_f is the final time. This is an IHCP because it is concerned with the estimation of the unknown surface condition $f(t)$. The mathematical formulation of this problem is given by

$$\frac{\partial}{\partial x}\left(k\frac{\partial T}{\partial x}\right)=\rho C_p\frac{\partial T}{\partial x} \qquad \text{in} \qquad 0<x<L, \quad 0<t\leqslant t_f \tag{14-2a}$$

$$-k\frac{\partial T}{\partial x}=f(t)=?\,(\text{unknown}) \qquad \text{at} \qquad x=0, \qquad 0<t\leqslant t_f \tag{14-2b}$$

$$T=T_L \qquad \text{at} \qquad x=L, \qquad 0<t\leqslant t_f \tag{14-2c}$$

$$T=F(x) \qquad \text{for} \qquad t=0, \qquad \text{in} \quad 0\leqslant x\leqslant L \tag{14-2d}$$

and temperature measurements at an interior location x_1, at different times t_j are given by

$$T(x_1, t_j) \equiv Y_j \quad \text{at} \quad x = x_1, \quad \text{for} \quad t = t_j \quad (j = 1, 2, \ldots, N) \tag{14-3}$$

This is an IHCP of *unknown surface condition.* Analogously, one envisions IHCP of unknown thermophysical properties, energy generation rate, initial condition, and so forth. In addition, one can also envision inverse problems of convection, radiation, phase-change, and other factors. The principal difficulty in the solution of inverse problems is that they are ill-posed; as a result the solution becomes unstable under small changes to the input data.

Applications

The space program played a significant role in the advancement of solution techniques for inverse heat conduction problems in late 1950s. An excellent discussion of this subject and various historical developments are given in references 1–3. For example, the aerodynamic heating of space vehicles during reentry is so high that the surface temperature cannot be measured directly with the sensors. Therefore, the sensors are placed beneath the surface of the radiation shield and the temperature of the hot surface is estimated by inverse analysis. The problems of this type belong to the inverse problem of unknown surface condition and have been the subject of numerous investigations [4–18]. Other applications include, among many others, the determination of physical properties k and ρC_p [19–31], temperature in internal-combustion engines [32, 33], interface conductance between periodically contacting surfaces [16, 17], the bulk radiation properties such as absorption and scattering coefficients [34, 35] and wall heat flux in forced convection inside ducts [36, 37].

Difficulties in Inverse Solution

The principal difficulty in the solution of inverse problems arises from the fact that they are *ill-posed* [38–41]. The standard heat conduction problems are well-posed, because (1) The solution exists, (2) is unique, and (3) is stable under small changes to the input data. The problems that fail to satisfy any one of these conditions are called *ill-posed.*

To illustrate the sensitivity of inverse solution to small changes in the measured input data, we consider one-dimensional quasi-stationary temperature fields in a semiinfinite solid subjected to periodically varying heat flux at the boundary surfacc $x = 0$. The physical problem is stated as follows. A semiinfinite solid confined to the domain $0 < x < \infty$ is initially at zero temperature. For times $t > 0$, the boundary surface at $x = 0$ is subjected to a periodically varying heat flux given in the form $q_0 \cos \omega t$, where q_0 is a constant heat flux and ω is the frequency of oscillations. After the transients have passed, the quasi-stationary temperature distribution in the solid at anytime, at a position x is given by [42, p. 114]

$$T(x,t) = \frac{q_0}{k}\sqrt{\frac{\alpha}{\omega}}\exp\left(-x\sqrt{\frac{\omega}{2\alpha}}\right)\cos\left(\omega t - x\sqrt{\frac{\omega}{2\alpha}} - \frac{\pi}{4}\right) \tag{14-4a}$$

The maximum amplitude at any location is obtained by setting $\cos(\cdot) = 1$:

$$[T(x,t)]_{\max} = \frac{q_0}{k}\sqrt{\frac{\alpha}{\omega}}\exp\left(-x\sqrt{\frac{\omega}{2\alpha}}\right) \tag{14-4b}$$

Equation (14-4b) shows that the maximum temperature in the interior region attenuates exponentially with increasing distance x from the surface and the square root of the frequency. However, if the surface temperature is to be determined by utilizing the measured temperatures at an interior points, any measurement error will be magnified exponentially with the distance x and the square root of the frequency ω. Therefore, depending on the location of the sensor and the frequency of oscillations, the solution of the inverse problem may become very sensitive to measurement errors in the input data.

Methods of Solution of IHCP

A variety of analytic and numerical approaches have been proposed for the solution of IHCP. Stoltz [43] was one of the earliest investigators who developed an analytic solution for a linear inverse heat conduction problem by using Duhamel's method, but the solution was found to be unstable for small time steps. This shortcoming was amended [6, 44, 45] through the use of future data concept; as a result, the improved solution permitted the use of much smaller time steps than that used in reference 43.

The analytic solutions developed by using integral or Laplace transform techniques [46–48] required continuously differentiable data; as a result, they were not so useful for practical applications; however, they provided a good insight into the nature of IHCP.

In order to cast IHCP as a well-posed problem, the traditional heat conduction equation was replaced by a hyperbolic heat conduction equation and the well established techniques were used to solve the resulting IHCP [49].

The analytic solutions are strictly applicable to linear problems. To extend the technique to nonlinear problems, the numerical methods such as FDM [48, 50–57] and FEM [58–60] have been used in the solution of IHCP.

The terminology, *function estimation*, and *parameter estimation*, frequently used in the study of inverse analysis, needs some clarification. If the problem involves the determination of an unknown function such as the timewise variation of the surface heat flux with no prior knowledge of the functional form of the unknown quantity, the problem is referred to as a problem of *function estimation*. Thus, the function estimation requires the determination of a large number of surface heat flux components q_i ($i = 1, 2, \ldots, N$); hence, it is referred to as an *infinite dimensional* minimization problem. On the other hand, if some prior knowledge is available on the functional form of $q(t)$, it can be parameterized and the inverse problem is called a problem of *parameter estimation* because only a limited number of parameters are to be estimated. Such problems are referred to as *finite dimensional* minimization problem. However, if the number of parameters to be estimated is increased, it may not be possible to make a clear distinction between the parameter and the function estimation problems.

Let us consider an inverse problem of *parameter estimation*, involving M unknown parameters, by utilizing the measured data. In the ideal case, if we have a perfect sensor with no measurement error, it would be sufficient to have the number of temperature data equal to the number of unknown parameters M to solve the inverse problem. However, as the temperature data always contain measurement errors, more measurements are needed than the number of unknowns; then, the system of equations to be solved becomes *overdetermined*. One way of solving a system of overdetermined equations is the use of the traditional *least-squares approach* coupled to an appropriate minimization procedure.

In the case of *function estimation*, the *sequential estimation* technique [6, 52, 57, 63], and the least squares method modified by the addition of *regularization* term [2, 63, 64] have been used. The *conjugate gradient method* with adjoint equation [65–70] has the advantage that the regularization is implicitly build in the computational procedure; some mathematical aspects of this method have been discussed [71].

Inverse radiation problems involving the determination of bulk radiation properties of semitransparent materials from the knowledge of the exit distribution of radiation intensity have also been the subject to numerous investigations [72–76].

The study of inverse problems requires some background in statistical analysis. Therefore, we first present some background material in statistical analysis and then introduce the methods of solving IHCP with some representative examples such as the determination of unknown surface heat flux and the thermophysical properties of materials.

14-2 BACKGROUND STATISTICAL MATERIAL

The purpose of this section is to present some basic statistical material needed in the analysis and solution of IHCP that is not generally covered in the customary courses in engineering. Therefore, a survey of pertinent statistical terminologies, statistical description of errors and some standard assumptions regarding the temperature measurement will be reviewed. Readers should consult references 77–81 for a more in-depth discussion of such matters.

Random Variable

A function whose value is a real number determined by each element in the sample space is called a *random variable*. Here, the sample space refers to the set of all possible outcomes from a given experiment.

Probability Distribution Function

Let the capital letter X denote a random variable. It is called a *discrete random variable* if it is a vector of a set of discrete numbers $x_n, n = 1, 2, \ldots, N$, and a *continuous random variable* if it contains an infinite number of points x. Depending on whether X is discrete or continuous, the *probability distribution function* $f(x)$ is defined as

$$\sum_{n=1}^{N} f(x_n) = 1 \qquad \text{when } X \text{ is discrete} \tag{14-5a}$$

$$\int_{-\infty}^{\infty} f(x)dx = 1 \qquad \text{when } X \text{ is continuous} \tag{14-5b}$$

Expected Value of X

Let X be a random variable, discrete or continuous, with the corresponding probability distribution functions $f(x_n)$ or $f(x)$, respectively. The expected value of X, denoted by $E(X)$, is defined as

$$E(X) = \begin{cases} \displaystyle\sum_{n=1}^{N} x_n f(x_n) & \text{when } X \text{ is discrete} \quad \text{(14-6a)} \\ \displaystyle\int_{-\infty}^{\infty} x f(x)dx & \text{when } X \text{ is continuous} \quad \text{(14-6b)} \end{cases}$$

The expected value of any random variable X is obtained by multiplying each value of a random variable by its corresponding probability distribution function and then summing up the results if X is discrete and integrating the results if X is continuous. Clearly, the expected value of X is a *weighted* mean of all possible values with the weight factor $f(x)$. If the weights are equal, that is, $f(x) = 1$, then the expected value becomes an *arithmetic mean* of X.

Expected Value of a Function $g(X)$

Consider a random variable X and the probability distribution function $f(x)$ associated with it. The expected value of the function $g(X)$, denoted by $E[g(X)]$, is given by

$$E[g(X)] = \begin{cases} \displaystyle\sum_{n=1}^{N} g(x_n) f(x_n) & \text{when } X \text{ is discrete and } x_n \text{ being one of its } N \text{ values} \quad \text{(14-7a)} \\ \displaystyle\int_{-\infty}^{\infty} g(x) f(x)dx & \text{when } x \text{ is continuous} \quad \text{(14-7b)} \end{cases}$$

Example 14-1

The elements x_n of a discrete random variable X and the probability distribution $f(x_n)$ associated with each value of x_n are listed below:

x_n	0	1	2	3	4	5
$f(x_n)$	0.1	0.2	0.3	0.2	0.1	0.1

Calculate (1) the expected value $E(X)$ of the random variable X and (2) the expected value $E[g(X)]$ of the function $g(X) = (2x - 1)^2$.

Solution

1. The expected value $E(X)$ of the random variable X is calculated according to equation (14-6a). We obtain

$$\begin{aligned} E(X) &= \sum_{n=1}^{N} x_n f(x_n) \\ &= 0 \times f(0) + 1 \times f(1) + 2 \times f(2) + 3 \times f(3) + 4 \times f(4) + 5 \times f(5) \\ &= 0 \times 0.1 + 1 \times 0.2 + 2 \times 0.3 + 3 \times 0.2 + 4 \times 0.1 + 5 \times 0.1 \\ &= 0 + 0.2 + 0.6 + 0.6 + 0.4 + 0.5 \\ &= 2.3 \end{aligned}$$

2. The expected value $E[g(X)]$ of the function $g(X)$ is calculated according to equation (14-7a) as

$$\begin{aligned} E[g(X)] &= \sum_{n=1}^{N} g(x_n) f(x_n) = \sum_{n=1}^{N} (2x_n - 1)^2 f(x_n) \\ &= (-1)^2 f(0) + (1)^2 f(1) + (3)^2 f(2) + (5)^2 f(3) + (7)^2 f(4) + (9)^2 f(5) \\ &= 1 \times 0.1 + 1 \times 0.2 + 9 \times 0.3 + 25 \times 0.2 + 49 \times 0.1 + 81 \times 0.1 \\ &= 0.1 + 0.2 + 2.7 + 5 + 4.9 + 8.1 \\ &= 21.0 \end{aligned}$$

Variance of a Random Variable X

The variance of a random variable X, denoted by σ_x^2 or simply σ^2, is defined by

$$\sigma_x^2 \equiv \sigma^2 = E[(X - \mu)^2] \qquad \text{where } \mu = E(x) \tag{14-8a}$$

or an alternative form is obtained by expanding this expression

$$\sigma^2 = E(X^2) - \mu^2 \tag{14-8b}$$

since $E(\mu^2) = \mu^2$.

The positive square root σ of the variance, σ^2, is called the *standard deviation.*

Example 14-2

A discrete random variable X and the corresponding probability distribution function $f(x)$ are listed

x_n	0	1	2	3
$f(x_n)$	$\frac{1}{40}$	$\frac{10}{40}$	$\frac{25}{40}$	$\frac{4}{40}$

calculate the variance σ^2 of the random variable X.

Solution. We use equation (14-8b) to calculate the variance of X, that is,

$$\sigma^2 = E(X^2) - \mu^2$$

where

$$\mu = E(X)$$

$$E(X) = 0 \times \frac{1}{40} + 1 \times \frac{10}{40} + 2 \times \frac{25}{40} + 3 \times \frac{4}{40} = \frac{72}{40} = 1.8 \equiv \mu$$

$$E(X^2) = 0^2 \times \frac{1}{40} + 1^2 \times \frac{10}{40} + 2^2 \times \frac{25}{40} + 3^2 \times \frac{4}{40} = \frac{146}{40} = 3.65$$

Hence, $\sigma^2 = 3.65 - (1.8)^2 = 0.41$.

Convariance of Two Random Variables X, Y

The *covariance* of two random variables X, Y, denoted by σ_{XY}, is given as

$$\text{cov}(X, Y) \equiv \sigma_{XY} = E[(X - \mu_X)(Y - \mu_Y)] \tag{14-9a}$$

and an alternative form of this expression is obtained by expanding this result as

$$\text{cov}(X, Y) \equiv \sigma_{XY} = E(\mu_X \mu_Y) - \mu_X \mu_Y \tag{14-9b}$$

where $\mu_X = E(X), \mu_Y = E(Y)$ and $E(XY) = \mu_X \mu_Y$.

Some Properties of Variance

We list below some useful properties of the variance and covariance.

1. If X is a random variable and a is a constant, then

$$\sigma^2_{X+a} = \sigma^2_X \equiv \sigma^2 \tag{14-10}$$

This result states that the variance remain unchanged if a constant is added to or subtracted from a random variable. The following expression

$$\sigma_{aX}^2 = a^2\sigma_X^2 \equiv a^2\sigma^2 \tag{14-11}$$

shows that, if a random variable is multiplied or divided by a constant, the variance is multiplied or divided by the square of the same constant.

2. If X and Y are *independent variables* and a and b are constants, then

$$\sigma_{aX+bY}^2 = a^2\sigma_X^2 + b^2\sigma_Y^2 \tag{14-12a}$$

and replacing b by $-b$ the right hand side remains invariant:

$$\sigma_{aX-bY}^2 = a^2\sigma_X^2 + b^2\sigma_Y^2 \tag{14-12b}$$

3. The covariance of the *independent variables X and Y* is zero:

$$\sigma_{XY} = E(XY) - \mu_X\mu_Y = 0 \tag{14-13}$$

since $E(XY) = E(X)E(Y)$ and $\mu_x = E(X), \mu_y = E(Y)$.

4. If a and b are constants, then

$$E(b) = b \tag{14-14a}$$

$$E(ax) = aE(x) \tag{14-14b}$$

$$E(ax + b) = aE(x) + b \tag{14-14c}$$

Normal Distribution

The most frequently used continuous probability distribution function is the normal (or Gaussian) distribution which has a bell shaped curve about the mean value. The normal (Gaussian) probability distribution function with a mean μ and variance σ^2 is given by

$$f(x) = \frac{1}{\sigma\sqrt{2\pi}} \exp\left[-\frac{1}{2}\left(\frac{x-\mu}{\sigma}\right)^2\right] \tag{14-15}$$

The area under the integration of this function from $x = -\infty$ to x represents the probability $P(-\infty < X < x)$ that a random variable X with a mean μ and variance σ^2 assumes a value between $x = -\infty$ and x. That is, $P(-\infty < X \leqslant x)$ is defined by

$$P(-\infty < X \leqslant x) = \frac{1}{\sigma\sqrt{2\pi}} \int_{-\infty}^{x} \exp\left[-\frac{1}{2}\left(\frac{x'-\mu}{\sigma}\right)^2\right] dx' \tag{14-16}$$

To alleviate the difficulty in the calculation of this integral for each given set of values of σ, μ and x, a new independent variable Z is defined as

$$Z = \frac{X - \mu}{\sigma} \qquad \text{or} \qquad z = \frac{x - \mu}{\sigma} \tag{14-17}$$

Then the integral (14-16) becomes

$$P(-\infty < Z \leqslant z) = \frac{1}{\sqrt{2\pi}} \int_{-\infty}^{z} e^{-z'^2/2} dz' \tag{14-18}$$

which has only one independent variable z, hence the results of integration can be tabulated as given in Table 14-1. Readers should consult reference 77 for more comprehensive tabulation of the probability function $P(-\infty < Z < z)$.

Table 14-1 can be used to determine the normal probability, $P(x_1 < X < x_2)$, of a random variable X having a mean μ, variance σ^2, assuming a value between $x_1 < X < x_2$:

$$P(x_1 < X < x_2) = P\left(\frac{x_1 - \mu}{\sigma} < Z < \frac{x_2 - \mu}{\sigma}\right) \tag{14-18a}$$

$$= P(z_1 < Z < z_2) \tag{14-18b}$$

$$= P(-\infty < Z < z_2) - P(-\infty < Z < z_1) \tag{14-18c}$$

where $P(-\infty < Z < z_2)$ and $P(-\infty < Z < z_1)$ are determined from Table 14-1.

Example 14-3

The average life of a certain type of watch is $\mu = 10$ yr with a standard deviation of $\sigma = 1$ yr. Assuming that the life expectancy of such watches has a normal (Gaussian) distribution, determine the probability that a given watch will last less than $x = 8.5$ yr.

Solution. The shaded area under the probability curve in the accompanying figure represents the desired probability for this example.

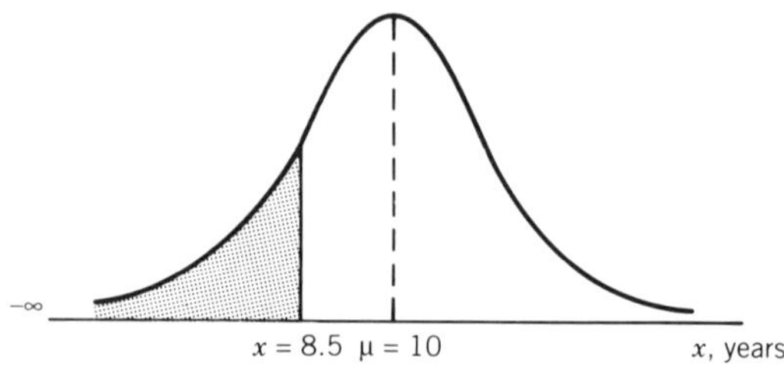

Example 14-3

TABLE 14-1 Normal Probability Distribution Function $P(-\infty < Z < z)$ Defined by Equation (14-18)

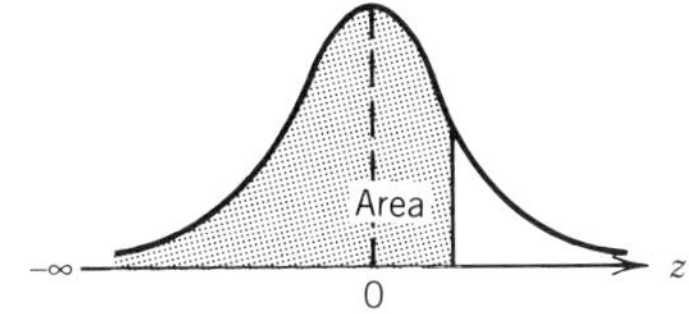

z	$P(-\infty < Z < z)$	z	$P(-\infty < Z < z)$
−2.9	0.0019	0.0	0.5000
−2.8	0.0026	0.1	0.5398
−2.7	0.0035	0.2	0.5793
−2.6	0.0047	0.3	0.6179
−2.5	0.0062	0.4	0.6554
−2.4	0.0082	0.5	0.6915
−2.3	0.0107	0.6	0.7257
−2.2	0.0139	0.7	0.7580
−2.1	0.0179	0.8	0.7881
−2.0	0.0228	0.9	0.8159
−1.9	0.0287	1.0	0.8413
−1.8	0.0359	1.1	0.8643
−1.7	0.0446	1.2	0.8849
−1.6	0.0548	1.3	0.9032
−1.5	0.0668	1.4	0.9192
−1.4	0.0808	1.5	0.9332
−1.3	0.0968	1.6	0.9452
−1.2	0.1151	1.7	0.9554
−1.1	0.1357	1.8	0.9641
−1.0	0.1587	1.9	0.9713
−0.9	0.1841	2.0	0.9772
−0.8	0.2119	2.1	0.9821
−0.7	0.2420	2.2	0.9861
−0.6	0.2743	2.3	0.9893
−0.5	0.3085	2.4	0.9918
−0.4	0.3446	2.5	0.9938
−0.3	0.3821	2.6	0.9953
−0.2	0.4207	2.7	0.9965
−0.1	0.4602	2.8	0.9974
−0.0	0.5000	2.9	0.9981

We have

$$z = \frac{x - \mu}{\sigma} = \frac{8.5 - 10}{1} = -1.5$$

Then, using Table 14-1, we obtain

$$P(-\infty < X < 8.5) = P(-\infty < Z < -1.5) = 0.0668$$

Thus the probability that a watch will last less than 8.5 yrs is 0.0668.

Example 14-4

In the Example 14-3, what is the probability that a given watch will last more than $x = 11.5$ years?

Solution. The shaded area under the probability curve in the accompanying figure represents the probability for this example.

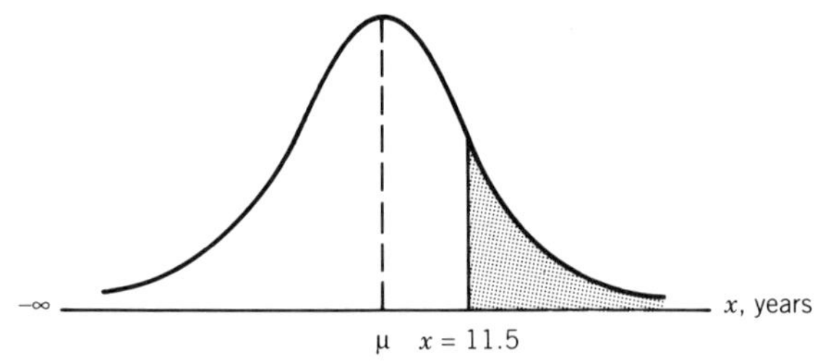

Example 14-4

we have

$$z = \frac{x - \mu}{\sigma} = \frac{11.5 - 10}{1} = 1.5$$

$$P(X > 11.5) = P(Z > 1.5)$$

and using Table 14-1 we obtain

$$P(Z > 1.5) = 1 - P(-\infty < Z < 1.5) = 1 - 0.9332 = 0.0668$$

As expected, the result is the same as that in Example 14-3.

Example 14-5

Determine the values of $\pm z$ such that the shaded area under the normal probability curve shown in the accompanying figure will be equal to 99%.

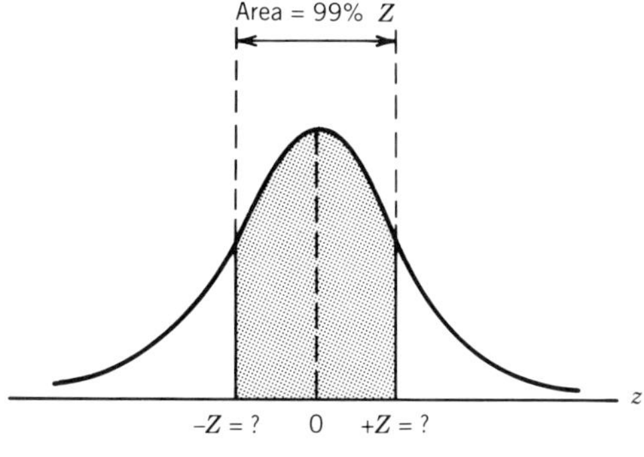

Example 14-5

Solution. We need to determine the z value such that

$$P(-\infty < Z \leqslant +z) - P(-\infty < Z \leqslant -z) = 0.99$$

By using the more extensive table of normal probability distribution it can be verified that, for $z = 2.576$ we have

$$P(-\infty < Z \leqslant +2.576) = 0.995$$
$$P(-\infty < Z \leqslant -2.576) = 0.005$$

Hence

$$P(-\infty < Z < 2.576) - P(-\infty < Z < +2.576) = 0.99$$

Assumptions in Errors

The random measurement errors are generally the major source of errors in estimates made by the inverse analysis. A statistical description of errors is useful in the analysis of random errors. Eight standard assumptions regarding the temperature measurements described in references 1, 78 are listed below.

1. The errors are additive, that is

$$Y_i = T_i + \epsilon_i \tag{14-19a}$$

   where Y_i is the measured temperature, T_i is the "true" temperature and ϵ_i is the random error at time t_i.

2. The temperature errors, ϵ_i, have a zero mean, that is,

$$E(\epsilon_i) = 0 \tag{14-19b}$$

   where $E(\cdot)$ is the expected value operator.

3. The errors have constant variance, that is,

$$E\{[y_i - E(y_i)]^2\} = \sigma^2 = \text{constant} \tag{14-19c}$$

which means that the variance of y_i is independent of time.

4. Two measurement errors ϵ_i and ϵ_j, where $i \neq j$, are uncorrelated if the covariance of ϵ_i and ϵ_j is zero:

$$\text{cov}(\epsilon_i, \epsilon_j) \equiv E\{[\epsilon_i - E(\epsilon_i)][\epsilon_j - E(\epsilon_j)]\} = 0 \qquad \text{for} \qquad i \neq j \tag{14-19d}$$

The different errors ϵ_i and ϵ_j are called uncorrelated if each has no effect on or relationship to the other.

5. Measurement errors have a normal (Gaussian) distribution, that is, the probability distribution function of ϵ_i is given by

$$f(\epsilon_i) = \frac{1}{\sigma\sqrt{2\pi}} \exp\left(\frac{-\epsilon_i^2}{2\sigma^2}\right) \tag{14-19e}$$

6. The statistical parameters such as σ^2 describing ϵ_i are known.
7. The measurement times $t_1, t_2, \ldots, t_n$, the measurement positions $x_1, x_2, \ldots, x_j$, the dimensions of the specimens, and thermal properties are all accurately known.
8. There is no prior information regarding the distribution of the function to be estimated. If such information exists, it can be utilized to obtain estimates.

14-3 IHCP OF ESTIMATING UNKNOWN SURFACE HEAT FLUX

We consider a flat plate of thickness L initially at zero temperature. For times $t > 0$, a timewise varying unknown heat flux $q(t)$ is applied at the boundary surface $x = 0$, while the boundary surface at $x = L$ is kept insulated. It is assumed that $q(t)$ is a single-valued continuous function of time that may rise or fall arbitrarily, but no other prior information is available on the functional form. In order to estimate $q(t)$, temperature measurements are taken with a sensor positioned at the insulated surface $x = L$, over a time period $0 < t < t_f$, at successive time intervals t_j $(j = 1, 2, \ldots, M)$. Here, t_f denotes the final time for temperature measurements. Figure 14-1 illustrates the geometry and the coordinates. Our objective is, by utilizing the measured temperature data, to estimate the unknown surface heat flux function $q(t)$, over the whole time domain $0 < t \leqslant t_f$. Since no prior information is available regarding the functional form of $q(t)$, we need to compute $q_i = q(t_i)$ at a sufficiently large number of times $t_i (i = 1, 2, 3, \ldots, N)$ in order to make a reasonably accurate estimation of the function $q(t)$.

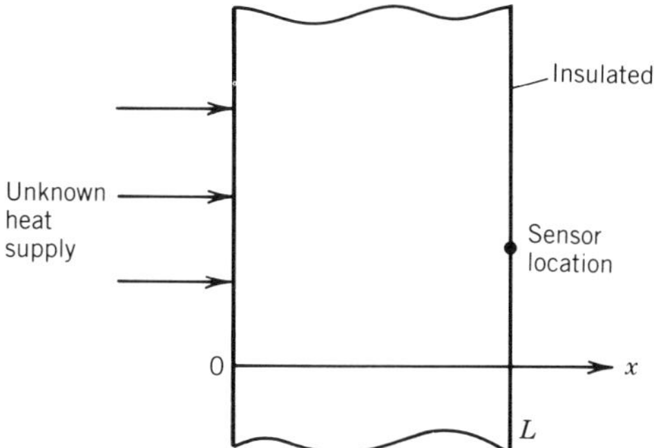

Fig. 14-1 Inverse problem of unknown surface heat flux.

Therefore, the problem may be regarded as a problem of *function estimation.* The analysis and solution of this inverse problem are presented in the following basic steps:

1. The formulation of direct and inverse problems
2. The transformation of the inverse problem into a system of least-squares equations
3. Physical significance of sensitivity coefficients
4. The solution of the least-squares equations
5. The determination of the sensitivity coefficients
6. Numerical results

1. The Direct and Inverse Problem Basic to the solution of this inverse problem is the solution of the related direct problem in which the surface heat flux $q(t)$ is considered known. The mathematical formulation of the *direct problem* is given by

$$k\frac{\partial^2 T}{\partial x^2} = \rho C_p \frac{\partial T}{\partial t} \quad \text{in} \quad 0 < x < L, \quad t > 0 \tag{14-20a}$$

$$-k\frac{\partial T}{\partial x} = q(t) \quad \text{at} \quad x = 0, \quad t > 0 \tag{14-20b}$$

$$\frac{\partial T}{\partial x} = 0 \quad \text{at} \quad x = L, \quad t > 0 \tag{14-20c}$$

$$T = 0 \quad \text{for} \quad t = 0, \quad \text{in } 0 \leqslant x \leqslant L \tag{14-20d}$$

This direct problem can readily be solved by classical solution techniques; we prefer to use Duhamel's theorem.

The mathematical formulation of the *inverse problem* is given by

$$k\frac{\partial^2 T}{\partial x^2}=\rho C_p\frac{\partial T}{\partial t} \qquad \text{in} \qquad 0<x<L, \qquad \text{for} \quad 0<t<t_f \tag{14-21a}$$

$$-k\frac{\partial T}{\partial x}=q(t)=? \qquad \text{at} \qquad x=0 \tag{14-21b}$$

$$\frac{\partial T}{\partial x}=0 \qquad \text{at} \qquad x=L \tag{14-21c}$$

$$T=0 \qquad \text{for} \qquad t=0, \qquad \text{in} \quad 0\leqslant x\leqslant L \tag{14-21d}$$

where the applied surface heat flux $q(t)$ is unknown; instead, temperature measurements

$$T(L,t_j)\equiv Y_j, \qquad j=1,2,\ldots,M$$

taken with a single sensor placed at $x=L$ at times t_j are given over the whole time domain $0<t\leqslant t_f$, where t_f is the final measurement time.

Then the inverse problem can be stated as follows: By utilizing the M measured data $Y_j(j=1,2,\ldots,M)$, estimate the M heat flux components $q(t_j)\equiv q_j(j=1, 2,\ldots,M)$.

2. *Transformation to a Least Squares Problem* The inverse problem as stated above is mathematically ill-posed in a sense that its *existence*, *uniqueness*, and/or *stability* is not yet ensured. A successful solution of an inverse problem generally involves the transformation of the inverse problem into a well posed-approximate solution.

The *existence* of an inverse solution is guaranteed by requiring that the inverse solution minimize the *least squares norm* rather than make it necessarily zero. Since the computation of the direct solution is implicit in the least squares method, the establishment of the direct solution given by equations (14-20) is obviously basic to the procedure.

To solve the inverse problem on such a basis, we require that the estimated temperature $\hat{T}_j(\hat{q}_i), j=1,2,\ldots,M$, computed from the solution of the direct problem by using the estimated values of the heat flux components $\hat{q}_i, i=1,2,\ldots,M$, should match the measured temperatures $Y_j, j=1,2,\ldots,M$ as closely as possible over a specified time domain $0<t<t_f$. Here, the superscript ^ over T or q denotes the estimated values. One way to realize such a matching is to require that the traditional least squares norm is minimized with respect to each of the unknown heat flux components $q_i, i=1,2,\ldots,M$. Here we consider the least squares norm modified by the addition of a zeroth-order regularization term

$$S(\hat{\mathbf{q}})=\sum_{j=1}^{M}[Y_j-\hat{T}_j(\hat{\mathbf{q}})]^2+\alpha^*\sum_{j=1}^{M}\hat{q}_j^2 \tag{14-22}$$

where $\hat{\mathbf{q}} \equiv [\hat{q}_i \text{ for } i = 1, 2, \ldots, M]$ and the superscript ^ denotes the estimated values and other quantities defined by

$S(\hat{\mathbf{q}})$ = sum of squares
$\hat{q}_j \equiv \hat{q}(t_j)$ = estimated surface heat flux at the boundary
$Y_j \equiv Y(t_j)$ = measured temperature at surface $x = L$, at times t_j
$\hat{T}_j(\hat{\mathbf{q}})$ = estimated temperature at the surface $x = 0$ at times $t = t_j$ computed by using estimated heat flux, $[\hat{q}_i, i = 1, 2, \ldots, M] = \hat{\mathbf{q}}$
α^* = the regularization parameter > 0

In equation (14-22), the first summation term on the right-hand side is the traditional least squares. The second summation is the zero-order regularization term which is added in order to reduce *instability* or *oscillations* that are inherent in the solution of ill-posed problems when a large number of parameters are to be estimated. Readers should consult references 1 and 63 for a discussion of the first- and second-order regularization terms. The concept of regularization was originally introduced in reference 41. The procedure is essentially related to the damped least squares or the ridge method discussed in references 80–85.

The coefficient α^* is called the *regularization parameter*. As $\alpha^* \to 0$, the solution exhibits oscillatory behavior and becomes unstable if a large number of parameters are to be estimated (i.e., function estimation). For large values of α^*, the solution is damped and deviates from the exact results. By proper selection of α^*, instability can be alleviated. A discussion of the ways to select value for α^* can be found in references 2, 63, and 64. Fortunately, a relatively wide range of values of α^* can be used. For example, the values for α^* ranged from 10^{-1} to 10^{-4} [63]. A cross-validation approach discussed in references 87–90 has been used [87] to calculate the optimum value of α^*. Depending on the value of the standard deviation σ of the measurement errors, the optimum value of α^* ranged from 10^{-2} to 10^{-4}.

The *next step* in the analysis is the *minimization of the least-squares equation* (14-22) by differentiating it with respect to each of the unknown heat flux components $\hat{q}_i, i = 1, 2, \ldots, M$, and then setting the resulting expression equal to zero. We find

$$\frac{\partial S(\hat{\mathbf{q}})}{\partial \hat{q}_i} = 2 \sum_{j=1}^{M} \frac{\partial \hat{T}_j(\hat{\mathbf{q}})}{\partial \hat{q}_i} [\hat{T}_j(\hat{\mathbf{q}}) - Y_j] + 2\alpha^* \sum_{j=1}^{M} \hat{q}_j \frac{\partial \hat{q}_j}{\partial \hat{q}_i} = 0 \tag{14-23a}$$

where $i = 1, 2, \ldots, M$, and

$$\frac{\partial \hat{q}_j}{\partial \hat{q}_i} = \begin{cases} 0 & \text{for} \quad i \neq j \\ 1 & \text{for} \quad i = j \end{cases} \tag{14-23b}$$

The total number of temperature measurements should be at least equal to or more than the number of the parameters to be estimated. Equation (14-23a) is rearranged in the form

$$\sum_{j=1}^{M} \frac{\partial \hat{T}_j(\hat{\mathbf{q}})}{\partial \hat{q}_i}[Y_j - \hat{T}_j(\hat{\mathbf{q}})] = \alpha^* \sum_{j=1}^{M} \hat{q}_j \frac{\partial \hat{q}_j}{\partial \hat{q}_i} \tag{14-23c}$$

where $i = 1, 2, \ldots, M$ and

$$\frac{\partial \hat{T}_j(\hat{\mathbf{q}})}{\partial \hat{q}_i} = \frac{\partial \hat{T}_j(\hat{q}_1, \hat{q}_2, \ldots, \hat{q}_M)}{\partial \hat{q}_i} \equiv X_{ji} = \begin{array}{l}\text{sensitivity coefficients}\\ \text{with respect to } \hat{q}_i\end{array} \tag{14-23d}$$

As discussed previously, it is desirable to have large uncorrelated sensitivity coefficients X_{ij}. If the sensitivity coefficients are linearly dependent, the minimization procedure defined by equation (14-23) will not have unique solution with $\alpha^* = 0$. Equation (14-23c) is written in the matrix form as

$$\mathbf{X}^T(\mathbf{Y} - \mathbf{T}) = \alpha^* \mathbf{q} \tag{14-24a}$$

where various vectors are given by

$$\mathbf{T} = \begin{bmatrix} \hat{T}_1 \\ \hat{T}_2 \\ \vdots \\ \hat{T}_M \end{bmatrix}, \quad \mathbf{Y} = \begin{bmatrix} Y_1 \\ Y_2 \\ \vdots \\ Y_M \end{bmatrix}, \quad \mathbf{q} = \begin{bmatrix} \hat{q}_1 \\ \hat{q}_2 \\ \vdots \\ \hat{q}_M \end{bmatrix} \tag{14-24b,c,d}$$

and the sensitivity matrix $\mathbf{X}$ with respect to $\mathbf{q}$ is written in the explicit form as

$$\mathbf{X} \equiv \frac{\partial \mathbf{T}}{\partial \mathbf{q}^T} = \begin{bmatrix} \frac{\partial T_1}{\partial q_1} & \frac{\partial T_1}{\partial q_2} & \cdots & \frac{\partial T_1}{\partial q_M} \\ \frac{\partial T_2}{\partial q_1} & \frac{\partial T_2}{\partial q_2} & & \frac{\partial T_2}{\partial q_M} \\ \vdots & & & \\ \frac{\partial T_M}{\partial q_1} & \frac{\partial T_M}{\partial q_2} & \cdots & \frac{\partial T_M}{\partial q_M} \end{bmatrix} \tag{14-24e}$$

We note that, in this sensitivity matrix, *the terms above the main diagonal must be zero, because the temperatures $\hat{T}_i$ calculated at any instant of time t_i must be independent of the future heat fluxes, $q_j, j > i$.*

3. Physical Significance of Sensitivity Coefficients The sensitivity coefficient X_{ji} defined by equation (14-23d) or (14-24e) is the first derivative of the *dependent variable* (i.e., temperature) with respect to the *unknown parameter* (i.e., heat flux components). It represents the changes in T_i with respect to the changes in the unknown parameters q_i. It is preferable to have large, uncorrelated values of the sensitivity coefficients X_{ij}. Therefore, sensors should be placed at such locations where the readings are most sensitive to changes in the unknown parameters. A small value of X_{ji} indicates insensitivity of the dependent variable to changes in the value of the unknown parameter; for such cases the inverse analysis become very sensitive to measurement errors and the estimation process become difficult. To establish the best sensor locations, measurement times, and so on, it is desirable to examine the effects of measurements locations, measurement times, and similar parameters on the relative values of the sensitivity coefficients.

If the sensitivity coefficients are functions of the parameters to be estimated, then the resulting estimation problem is *nonlinear*; conversely, if they are *not* functions of the parameters, the estimation problem is *linear*.

Consider, for example, transient temperature distribution in a semiinfinite medium, initially at zero temperature, for times $t > 0$, where a constant heat flux q_0 W/m^2 is applied at the boundary surface $x = 0$. The temperature at the surface $x = 0$ is a function of time given by

$$T(0, t) = \frac{2q_0}{k}\left(\frac{\alpha t}{\pi}\right)^{1/2} \tag{14-25a}$$

where α is the thermal diffusivity and k is the thermal conductivity. The sensitivity coefficient with respect to the surface heat flux q_0 given by

$$X_{q_0}(0, t) \equiv \frac{\partial T(0, t)}{\partial q_0} = \frac{2}{k}\left(\frac{\alpha t}{\pi}\right)^{1/2} \tag{14-25b}$$

is independent of q_0, whereas the sensitivity coefficient with respect to the thermal diffusivity

$$X_\alpha(0, t) \equiv \frac{\partial T(0, t)}{\partial \alpha} = \frac{q_0}{k}\left(\frac{t}{\pi \alpha}\right)^{1/2} \tag{14-25c}$$

depends on α; hence the inverse problem of estimation of α is nonlinear.

The final step in the analysis is the solution of the above system of least squares equations (14-24) in order to calculate the heat flux components $\hat{q}_j, j = 1, 2, \ldots, M$ as discussed next.

4. Solution of the Least-Squares Equations Now the IHCP is reduced to that of solving the system of least-squares equations (14-23) or (14-24) by a suitable solution algorithm. It is desirable to express equation (14-23) in a more convenient form for the calculation of $\hat{q}_i$; this can be achieved by expanding $\hat{T}_i(\hat{\mathbf{q}})$ in a Taylor series with respect to an arbitrary value of the heat flux as

$$\hat{T}_j = \hat{T}_{0j} + \sum_{k=1}^{M} \frac{\partial \hat{T}_j}{\partial \hat{q}_k}(\hat{q}_k - \hat{q}_0) \tag{14-26a}$$

This result is expressed in the matrix form as

$$\mathbf{T} = \mathbf{T}_0 + \frac{\partial \mathbf{T}}{\partial \mathbf{q}^T}(\mathbf{q} - \mathbf{q}_0) \tag{14-26b}$$

If we choose $\mathbf{T}_0 = 0$ and $\mathbf{q}_0 = 0$, equations (14-26a) and (14-26b), respectively, reduce to

$$\hat{T}_j = \sum_{k=1}^{M} \frac{\partial \hat{T}_j}{\partial q_k} \hat{q}_k \tag{14-27a}$$

and

$$\mathbf{T} = \frac{\partial \mathbf{T}}{\partial \mathbf{q}^T} \mathbf{q} \equiv \mathbf{X}\mathbf{q} \tag{14-27b}$$

Since $\partial \mathbf{T}/\partial \mathbf{q}^T \equiv \mathbf{X}$.

We now substitute equation (14-27a) into equation (14-23c)

$$\sum_{j=1}^{M} \frac{\partial \hat{T}_j}{\partial \hat{q}_i}\left(Y_j - \sum_{k=1}^{M} \frac{\partial \hat{T}_j}{\partial \hat{q}_k} \hat{q}_k \right) = \alpha^* \sum_{j=1}^{M} \hat{q}_j \frac{\partial \hat{q}_j}{\partial \hat{q}_i}, \qquad i = 1, 2, \ldots, M \tag{14-28a}$$

The matrix form of this equation is obtained by introducing equation (14-27b) into equation (14-24a)

$$\mathbf{X}^T(\mathbf{Y} - \mathbf{X}\mathbf{q}) = \alpha^* \mathbf{q} \tag{14-28b}$$

The equivalence of equations (14-28a) and (14-28b) can be verified by expanding equation (14-28b). The solution of equations (14-28a) or (14-28b) gives the estimated values of the heat flux components $\hat{q}_i$ at each time $t_i (i = 1, 2, \ldots, M)$. It is convenient to express the solution for the heat flux $\mathbf{q}$ in the matrix form as

$$\mathbf{q} = (\mathbf{X}^T\mathbf{X} + \alpha^* \mathbf{I})^{-1} \mathbf{X}^T \mathbf{Y} \tag{14-29}$$

Equation (14-29) is the formal solution of the inverse heat conduction problem considered here for the unknown surface heat flux $\mathbf{q}$ over the period $0 < t < t_f$. Once the sensitivity coefficients $\mathbf{X}$, the regularization parameter α^*, and the measured temperature data $\mathbf{Y}$ are available, the surface heat flux $\mathbf{q}$ is computed from equation (14-29). The sensitivity coefficients $\mathbf{X}$ are determined as described next.

5. Determination of Sensitivity Coefficients Because the direct problem associated with this inverse problem is linear, analytic expression can readily be developed for the sensitivity coefficients by the application of Duhamel's theorem as follows.

By definition, the sensitivity coefficients are given by $X_{ij} \equiv \partial \hat{T}_i / \partial \hat{q}_j$. We consider the solution $T(x, t)$ of the direct problem (14-20) determined by Duhamel's theorem [see equation 5-8]

$$T(x,t) = \int_{\lambda=0}^{t} q(\lambda) \frac{\partial \phi(x, t-\lambda)}{\partial t} d\lambda \tag{14-30}$$

where $\phi(x, t)$ is the solution of the following auxiliary problem

$$k \frac{\partial^2 \phi}{\partial x^2} = \rho C_p \frac{\partial \phi}{\partial t} \qquad \text{in} \qquad 0 < x < L \tag{14-31a}$$

$$k \frac{\partial \phi}{\partial x} = 1 \qquad \text{at} \qquad x = 0 \tag{14-31b}$$

$$\frac{\partial \phi}{\partial x} = 0 \qquad \text{at} \qquad x = L \tag{14-31c}$$

$$\phi = 0 \qquad \text{for} \qquad t = 0 \tag{14-31d}$$

Duhamel's theorem given by equation (14-30) is written in the alternative form as

$$T(x,t) = \int_{\lambda=0}^{t} q(\lambda) \frac{\partial \phi(x, t-\lambda)}{\partial \lambda} d\lambda \tag{14-32a}$$

since

$$\frac{\partial \phi(x, t-\lambda)}{\partial t} = -\frac{\partial \phi(x, t-\lambda)}{\Delta \lambda} \tag{14-32b}$$

The integral in equation (14-32a) is discretized as

$$T(x, t_M) = \sum_{n=1}^{M} q_n \frac{\phi(x, t_M - \lambda_{n-1}) - \phi(x, t_M - \lambda_n)}{\Delta \lambda} \Delta \lambda \tag{14-33a}$$

$$= \sum_{n=1}^{M} q_n [\phi(x, t_{M-(n-1)}) - \phi(x, t_{M-n})] \tag{14-33b}$$

which is written more compactly as

$$T_M = \sum_{n=1}^{M} q_n \Delta \phi_{M-n} \qquad n = 1, 2, \ldots, M \tag{14-34}$$

where q_n is evaluated at time $(n - \frac{1}{2})\Delta t$. Equations (14-34) can be expressed in the matrix form as

$$\begin{bmatrix} T_1 \\ T_2 \\ \vdots \\ \vdots \\ T_M \end{bmatrix} = \begin{bmatrix} \Delta\phi_0 & & & & 0 \\ \Delta\phi_1 & \Delta\phi_0 & & & \\ \vdots & & \ddots & & \\ \vdots & & & \Delta\phi_0 & \\ \Delta\phi_{M-1} & \cdots & \cdots & \Delta\phi_1 & \Delta\phi_0 \end{bmatrix} \begin{bmatrix} q_1 \\ q_2 \\ \vdots \\ \vdots \\ q_M \end{bmatrix} \tag{14-35}$$

where $\Delta\phi_i \equiv \phi_{i+1} - \phi_i$ and $\phi_i \equiv \phi(x, t_i)$. Therefore, ϕ_i represents the temperature rise in the solid for a unit step increase in the surface heat flux as computed from the solution of the problem (14-31). Recalling the definition of $\mathbf{X}$ given by equation (14-24e),

$$\frac{\partial \mathbf{T}}{\partial \mathbf{q}^T} = \mathbf{X} \tag{14-36}$$

we conclude that the coefficient matrix in equation (14-35) must be the sensitivity matrix $\mathbf{X}$. Hence we write

$$\mathbf{X} \equiv \mathbf{X}_{ij} = \begin{bmatrix} \Delta\phi_0 & & & & \\ \Delta\phi_1 & \Delta\phi_0 & & & \\ \Delta\phi_2 & \Delta\phi_1 & \Delta\phi_0 & & \\ \vdots & \cdots & \cdots & \ddots & \\ \Delta\phi_{M-1} & \cdots & \cdots & \Delta\phi_1 & \Delta\phi_0 \end{bmatrix} \tag{14-37}$$

Thus the sensitivity coefficients X_{ij} are evaluated exactly, since $\Delta\phi_i$ are obtainable exactly from the solution of the auxiliary problem (14-31).

Knowing $\mathbf{X}$, $\mathbf{Y}$, and α^*, we can determine the unknown surface heat flux vector $\mathbf{q}$ from equation (14-29).

6. Numerical Results To examine the accuracy of predictions by the inverse analysis, we consider the inverse problem of estimating the unknown boundary surface heat flux $q(t)$ in the problem defined by equation (14-21). The problem is expressed in the dimensionless form as

$$\frac{\partial^2\theta}{\partial X^2} = \frac{\partial\theta(x,\tau)}{\partial\tau} \quad \text{in} \quad 0 < X < 1, \quad \tau > 0 \tag{14-38}$$

$$-\frac{\partial\theta}{\partial X} = Q(\tau) \quad \text{at} \quad X = 0, \quad \tau > 0 \tag{14-39}$$

$$\frac{\partial \theta}{\partial X} = 0 \qquad \text{at} \qquad X = 1, \qquad \tau > 0 \tag{14-40}$$

$$\theta = 0 \qquad \text{for} \qquad \tau = 0, \qquad 0 \leqslant x \leqslant 1 \tag{14-41}$$

and the temperature measurements

$$\theta(1, \tau_j) = Y_j^*, \qquad j = 1, 2, \ldots, M \tag{14-42}$$

taken at the surface $X = 1$ are available over the whole time domain $0 \leqslant \tau \leqslant \tau_f$. Various dimensionless quantities are defined as

$$\theta(X, \tau) = \frac{T}{q_R L/k}, \qquad X = \frac{x}{L}, \qquad \tau = \frac{\alpha t}{L^2}$$

$$Q(\tau) = \frac{q(t)}{q_R} = \text{unknown dimensionless applied surface heat flux} \tag{14-43}$$

$q_R =$ the reference heat flux that may be chosen as the maximum value of the heat flux over the considered range

We assume that no prior information is available on the functional form of $Q(\tau)$. In order to make a strict test for the accuracy of the inverse analysis in estimating the unknown surface heat flux $Q(\tau)$, we consider a flux having a triangular shape as illustrated with the solid line in Fig. 14-2 over the whole time domain $0 \leqslant \tau \leqslant 1.8$. The measured data Y_j^* are simulated in the following manner.

Temperature readings are considered taken at the insulated surface, $X = 1$, with small dimensionless time steps $\Delta\tau = 0.03$ over the whole time domain

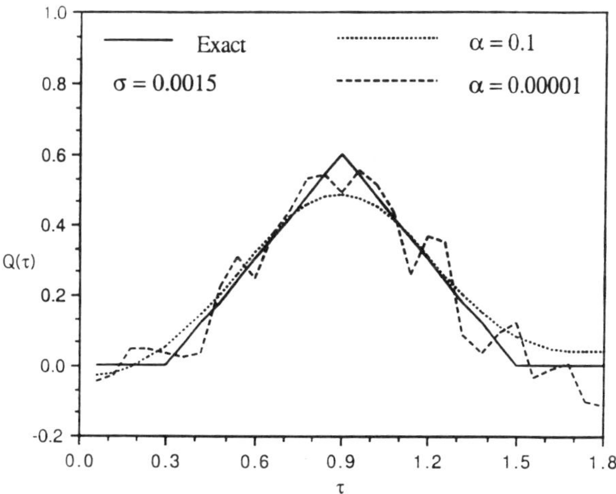

Fig. 14-2 Effect of regularization parameter α^* on the stability of inverse solution.

$0 < \tau \leqslant 1.8$. Thus we have 60 measurements. Then the simulated dimensionless measured data, $Y^*_{\text{meas.}}$, is generated by introducing random error $\omega\sigma$ to the exact dimensionless temperature θ_{exact} as

$$Y^*_{\text{meas.}} = \theta_{\text{exact}} + \omega\sigma \tag{14-44}$$

where θ_{exact} is computed from the solution of the problem given by equations (14-38 to 14-42), by using the exact value of the surface heat flux, $Q(\tau)$.

For normally distributed errors, there is a 99% probability of a value of ω lying in the range

$$-2.576 < \omega < 2.576 \tag{14-45}$$

A random number generator, such as the subroutine DRNNOR [101] of the IMSL library, can be used to generate values of ω.

Once the values of **Y**, **X** and α^* are available, the heat flux vector **q** is estimated from equation (14-29). Figure 14-2 shows the results of such calculations for two different values of the regularization parameter (i.e., $\alpha^* = 10^{-1}$ and 10^{-5}) for a standard deviation of measurement errors $\sigma = 0.0015$. We note that with a very small value of $\alpha^* = 10^{-5}$, the solution exhibits oscillatory behaviour, whereas for $\alpha^* = 10^{-1}$ the solution is damped and deviates from the exact result. A CPU time of about 50 s is required on the VAX-785 computer to perform the calculations needed for the construction of Fig. 14-2.

14-4 IHCP OF ESTIMATING SPATIALLY VARYING THERMAL CONDUCTIVITY AND HEAT CAPACITY

In this section we present the formulation and solution of an inverse heat conduction problem of simultaneously estimating the unknown thermal conductivity $k(x)$ and heat capacity $C(x) \equiv \rho C_p(x)$ which are assumed to vary linearly and

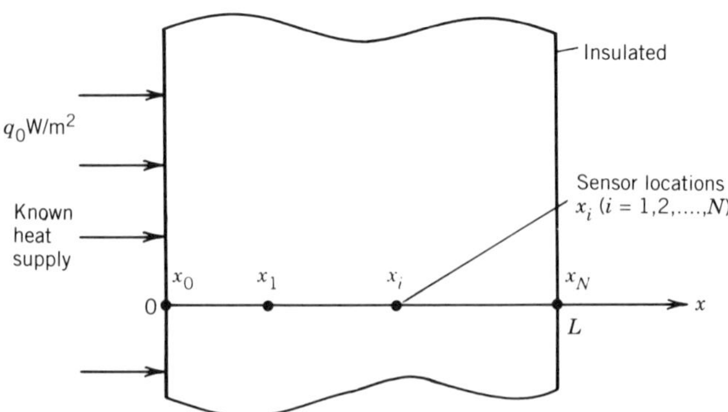

Fig. 14-3 The geometry and coordinates.

continuously in the direction normal to the surface of the sample plate. Even though the spatial variation of $k(x)$ and $C(x)$ is not encountered in common applications, this example is included in order to illustrate the capabilities of the inverse analysis. Of course as a special case one obtains k and C as constants. Figure 14-3 shows the geometry and coordinates for the plate of thickness L which is initially at zero temperature. For times $t > 0$, a known constant heat flux q_0 W/m^2 is applied at the surface $x = 0$ while the boundary surface at $x = L$ is kept insulated. Temperature sensors are installed at multiple space locations $x_i (i = 0, 1, 2, \ldots, N)$ and temperature readings are taken at times $t_j (j = 0, 1, \ldots, M)$. Thus a total of $(M + 1)(N + 1) = \Omega$ temperature measurements are available. The first step in the analysis is the identification of the related *direct problem*, which is given

$$\frac{\partial}{\partial x}\left[k(x)\frac{\partial T}{\partial x}\right] = C(x)\frac{\partial T(x,t)}{\partial t} \quad \text{in} \quad 0 < X < L, \quad t > 0 \tag{14-46a}$$

$$-k(0)\frac{\partial T(0,t)}{\partial x} = q_0 \quad \text{at} \quad x = 0, \quad t > 0 \tag{14-46b}$$

$$\frac{\partial T(L,t)}{\partial x} = 0 \quad \text{at} \quad x = L, \quad t > 0 \tag{14-46c}$$

$$T(x,0) = 0 \quad \text{for} \quad t = 0, \quad \text{in} \quad 0 \leqslant x \leqslant L \tag{14-46d}$$

where, $k(x)$, $C(x)$ and q_0 are regarded as known.

The mathematical formulation of the *inverse problem* is the same as that given by equations (14-46), except the thermophysical properties $k(x)$ and $C(x)$ are unknown; instead $(M + 1)(N + 1) = \Omega$ temperature measurements Y_k, $(k = 1, 2, \ldots, \Omega)$ are considered available.

Method of Analysis

This problem can be regarded as a problem of parameter estimation because the functional forms of $k(x)$ and $C(x)$ are being specified to vary linearly with x and they can be parameterized such that the inverse problem becomes one estimating of only four unknown parameters. For such a case there is no need for regularization, since only four parameters are to be estimated.

We consider four sensors placed in the plate such that two are placed at the locations x_1 and x_2 inside the plate and the remaining two are placed at the boundary surfaces $x_0 = 0$ and $x_3 = L$ as illustrated in Fig. 14-4. Since $k(x)$ and

Fig. 14-4 Sensor locations.

$C(x)$ vary linearly with x, we choose the boundary values K_0, K_3 of $k(x)$ and C_0, C_3 of $C(x)$ as the four unknown parameters. Then $k(x)$ and $C(x)$ are expressed in the form

$$k(x) = \frac{K_3 - K_0}{x_3 - x_0}(x - x_0) + K_0 \tag{14-47a}$$

$$C(x) = \frac{C_3 - C_0}{x_3 - x_0}(x - x_0) + C_0 \tag{14-47b}$$

Thus, by parameterizing the unknown thermal property functions $k(x)$ and $C(x)$, we reduce the problem from an infinite-dimension (i.e., large number of parameters) function estimation problem to a finite-dimension (i.e., few parameters) parameter estimation problem.

The *existence* of the inverse solution is guaranteed by requiring that the inverse solution minimizes the least squares norm. Since the number of unknown parameters are few (i.e., four only) the regularization term is not needed, we write

$$S(\hat{\mathbf{p}}) = \sum_{i=1}^{\Omega} [Y_i - \hat{T}_i(\hat{p}_j)]^2, \qquad j = 1, 2, 3, 4 \tag{14-48}$$

where

$Y_i =$ measured temperature

$\hat{T}_i(\hat{p}_j) =$ estimated temperature obtained from the solution of the direct problem (14-46) by using the estimated values of the unknown parameters.

$\hat{p}_j =$ elements of the unknown parameter vector $\hat{\mathbf{p}} \equiv \{K_0, K_3, C_0, C_3\}^T$

Equation (14-48) is minimized by differentiating it with respect to each of the unknown parameters $\hat{p}_j$ $(j = 1, 2, 3, 4)$ and then setting the resulting expression equal to zero.

$$\frac{\partial S}{\partial \hat{p}_j} = 2 \sum_{i=1}^{\Omega} \left(\frac{\partial \hat{T}_i}{\partial \hat{p}_j}\right)(\hat{T}_i(\hat{\mathbf{p}}) - Y_i) = 0, \qquad j = 1, 2, 3, 4 \tag{14-49}$$

Here, the total number of measurements Ω should be larger than the number of unknowns. In addition, the number of spatial measurement locations should also ensure uniqueness of the estimated thermal property parameters. In this example, for linear variation of the thermal properties, the use of four sensor locations within the medium ensures uniqueness. No general spatial uniqueness conditions applicable to the problem of estimating $k(x)$ and $C(x)$ for more general functional forms could be located in the literature. However, the numerical experimentation conducted in reference 29 suggest that the number of spatial measurement positions must equal or exceed the number of parameters. Equations (14-49) are written in the matrix form as

$$\frac{\partial S}{\partial \hat{\mathbf{p}}} = 2\mathbf{X}^T(\mathbf{T} - \mathbf{Y}) = 0 \tag{14-50a}$$

where

$$\mathbf{T} = \begin{bmatrix} T_1 \\ T_2 \\ \vdots \\ T_\Omega \end{bmatrix}, \qquad \mathbf{Y} = \begin{bmatrix} Y_1 \\ Y_2 \\ \vdots \\ Y_\Omega \end{bmatrix}, \qquad \hat{\mathbf{p}} = \begin{bmatrix} \hat{p}_1 \\ \hat{p}_2 \\ \vdots \\ \hat{p}_\Omega \end{bmatrix} \tag{14-50b,c,d}$$

$$\mathbf{X} = \frac{\partial \mathbf{T}}{\partial \hat{\mathbf{p}}^T} = \begin{bmatrix} \dfrac{\partial \hat{T}_1}{\partial \hat{p}_1} & \dfrac{\partial \hat{T}_1}{\partial \hat{p}_2} & \cdots & \dfrac{\partial \hat{T}_1}{\partial \hat{p}_4} \\ \dfrac{\partial \hat{T}_2}{\partial \hat{p}_1} & \dfrac{\partial \hat{T}_2}{\partial \hat{p}_2} & \cdots & \dfrac{\partial \hat{T}_2}{\partial \hat{p}_4} \\ \vdots & & & \\ \dfrac{\partial \hat{T}_\Omega}{\partial \hat{p}_1} & \dfrac{\partial \hat{T}_\Omega}{\partial \hat{p}_2} & \cdots & \dfrac{\partial \hat{T}_\Omega}{\partial \hat{p}_4} \end{bmatrix} = \begin{array}{l}\text{sensitivity coefficient} \\ \text{matrix with respect to } \mathbf{p}\end{array} \tag{14-50e}$$

and the elements of this matrix

$$X_{ij} \equiv \frac{\partial \hat{T}_i}{\partial \hat{p}_j}, \qquad i = 1, 2, \ldots, \Omega \quad \text{and} \quad j = 1, 2, 3, 4 \tag{14-50f}$$

are the sensitivity coefficients.

Method of Solution

Because the system of equations (14-49) are nonlinear, an iterative technique is necessary for its solution. The modified *Levenberg–Marquardt* algorithm, available as the subroutine DBCLSJ in the IMSL library edition 10.0, is used to solve the nonlinear least-squares equations (14-49) by iteration. This algorithm is a combination of the *Newton method* which converges fast but requires a good initial guess, and the *steepest descent* method which converges slowly but does not require a good initial guess. The Levenberg–Marquardt algorithm is given by

$$\hat{\mathbf{p}}^{k+1} = \hat{\mathbf{p}}^k + (\mathbf{J}^T\mathbf{J} + \mu_k \mathbf{I})^{-1}\mathbf{J}^T(\mathbf{Y} - \mathbf{T}), \qquad k = 1, 2, 3, \ldots \tag{14-51}$$

where

$\mathbf{p} = \{K_0, K_3, C_0, C_3\}^T$ = estimated parameters vector
$\mathbf{Y}$ = measured temperature vector
μ_k = damping parameters

and the Jacobian matrix $\mathbf{J}$ is the sensitivity coefficient matrix $\mathbf{X}$ defined by equation (14-50e):

$$\mathbf{J} \equiv \partial \mathbf{T}/\partial \hat{\mathbf{p}}^T = X \tag{14-52}$$

Clearly, for $\mu_k \to 0$, equation (14-51) reduces to Newton's method and for $\mu_k \to \infty$ it becomes the steepest descent method. Calculations are started with large values of μ_k and its value is gradually reduced as the solution approaches the converged result.

The *solution algorithm* with the Levenberg–Marquardt method is as follows. Suppose $\hat{\mathbf{p}}^k$ at the kth iteration are available.

1. Solve the direct problem (14-46) with finite-differences by using the estimated values of the parameters $\hat{\mathbf{p}} \equiv \{K_0, K_3, C_0, C_3\}^T$ at the kth iteration and compute $\mathbf{T}$.
2. Since the problem involves four known parameters, solve the direct problem (14-46) four more times, each time perturbing only one of the parameters by a small amount and compute

$$\begin{gathered} T(K_0 + \Delta K_0, K_3, C_0, C_3; t) \\ T(K_0, K_3 + \Delta K_3, C_0, C_3; t) \\ T(K_0, K_3, C_0 + \Delta C_0, C_3; t) \\ T(K_0, K_3, C_0, C_3 + \Delta C_3; t) \end{gathered}$$

For each case $T_i \equiv T(t_i)$, for $i = 1, 2, \ldots, \Omega$ are readily available.

3. Compute the sensitivity coefficients defined by equation (14-50f) for each parameter K_0, K_3, C_0, C_3. For example, with respect to K_0 we have

$$\frac{\partial T_i}{\partial K_0} = \frac{T(K_0 + \Delta K_0, K_3, C_0, C_3; t_i) - T(K_0, K_3, C_0, C_3; t_i)}{\Delta K_0} \tag{14-53}$$

for $i = 1, 2, 3, \ldots, \Omega$ and determine the Jacobian matrix $\mathbf{J}$ defined by equation (14-52).

4. Compute $(\mathbf{J}^T\mathbf{J} + \mu_k \mathbf{I})^{-1}\mathbf{J}^T(\mathbf{Y} - \mathbf{T})$.
5. Compute $\mathbf{p}^{k+1}$ according to equation (14-51).
6. Repeat the calculations until any one of the following convergence criteria is satisfied.

$$\text{(i)} \quad S^{k+1} < \epsilon_1$$

$$\text{(ii)} \quad \frac{|S^{k+1} - S^k|}{S^{k+1}} < \epsilon_2$$

$$\text{(iii)} \quad |\mathbf{p}^{k+1} - \mathbf{p}^k| < \epsilon_3$$

Calculations are performed with $\epsilon_1 = \epsilon_2 = \epsilon_3 = 10^{-5}$.

Numerical Results

To illustrate the application we consider linear variation of $k(x)$ and $C(x)$ with x, according to equations (14-47) as

$$k(x) = (K_3 - K_0)\frac{x}{L} + K_0 \tag{14-54}$$

$$C(x) = (C_3 - C_0)\frac{x}{L} + C_0 \tag{14-55}$$

where K_0, K_3, C_0 and C_3 are the four unknown parameters to be estimated by the inverse analysis.

Two different samples considered included a metal-like iron and an insulating material such as fiberglass, covering a broad range of property variation. Table 14-2 shows the values selected for each of these four coefficients and the applied surface heat flux q_0 at the boundary surface $x = 0$. This problem has been studied in reference 30.

We consider a test specimen of thickness $L = 0.03$ m, initially at zero temperature and for times $t > 0$ the boundary at $x = 0$ is subjected to a constant heat flux q_0 W/m^2 while the boundary surface at $x = L$ is kept insulated. Temporal temperature readings are taken with sensors at four locations (i.e., $x_0 = 0.0$, $x_1 = 0.01$, $x_2 = 0.02$ and $x_3 = L = 0.03$ m) over a period of $0 < t \leqslant 300$ seconds with $\Delta t = 20$ s time interval. This corresponds to 15 temperature readings per thermocouple or a total of 60 readings for the four sensors. The problem is overdetermined because we have 60 readings, only four unknown coefficients.

To simulate the measured temperatures that contain measurement errors, random errors $\omega\sigma$ are introduced to the exact temperatures as

$$Y = T_{\text{exact}} + \omega\sigma \tag{14-56}$$

where the exact temperature T_{exact} is determined from the solution of the direct problem (14-46) by using the exact values of the coefficients listed in Table 14-2. For normally distributed errors, there is a 99% probability of a value of ω lying

TABLE 14-2 Exact Values of the Coefficients and Heat Fluxes

Material	q_0 (w/m^2)	K_0 (w/m°C)	K_3 (w/m°C)	C_0 (kJ/m^3°C)	C_3 (kJ/m^3°C)
Metal	25,000	50.0	59.0	3,600	4,500
Insulating material	100	0.04	0.07	13	14

in the range $-2.576 < \omega < 2.576$. A random-number generator, such as, the subroutine DRNNOR of the IMSL library, can be used to generate values for ω.

Knowing the simulated measured temperatures $\mathbf{Y}$, computing the sensitivity coefficients $\partial T_i/\partial p_j$ (i.e., X) with finite differences according to equation (14-53), the Levenberg–Marquardt algorithm given by equation (14-51) is used to estimate the four unknown parameters $\hat{\mathbf{p}} \equiv \{K_0, K_3, C_0, C_3\}^{\mathrm{T}}$.

Figure 14-5 shows the exact and estimated values of $k(x)$ and $C(x)$ for a metal like iron having thermal conductivity of the order of 50 W(m°C), plotted as a function of position for a standard deviation $\sigma = 0.1$ for the measurement errors. Figure 14-6 show similar graphs for an insulating material like fiberglass having a thermal conductivity of the order of 0.04 W/m°C. These results cover a broad

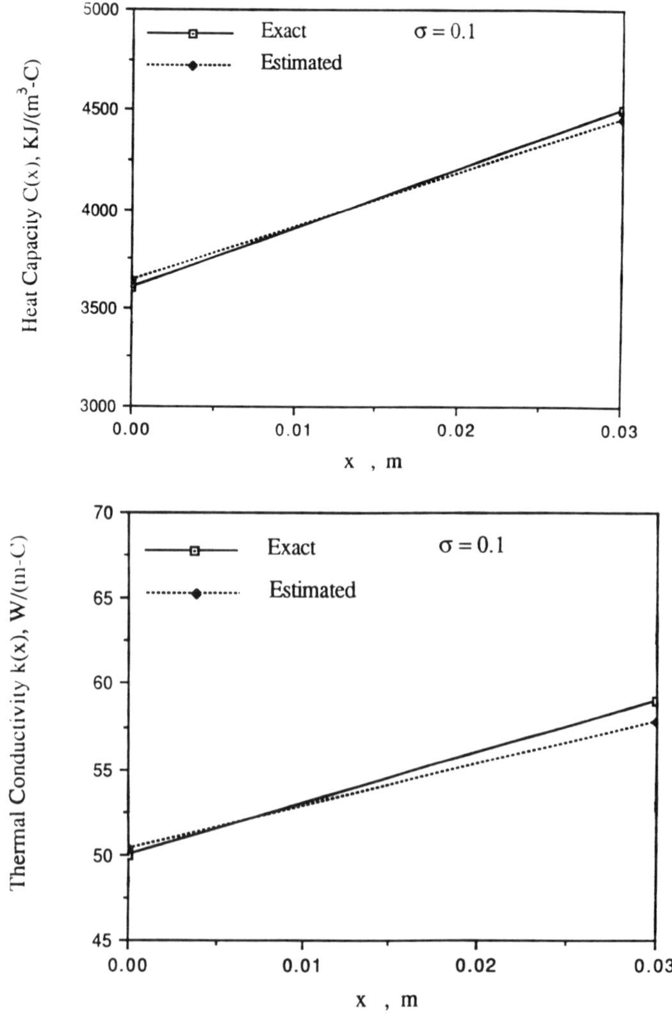

Fig. 14-5 Estimated thermal conductivity $k(x)$, heat capacitance $c(x)$ for a metal like iron.

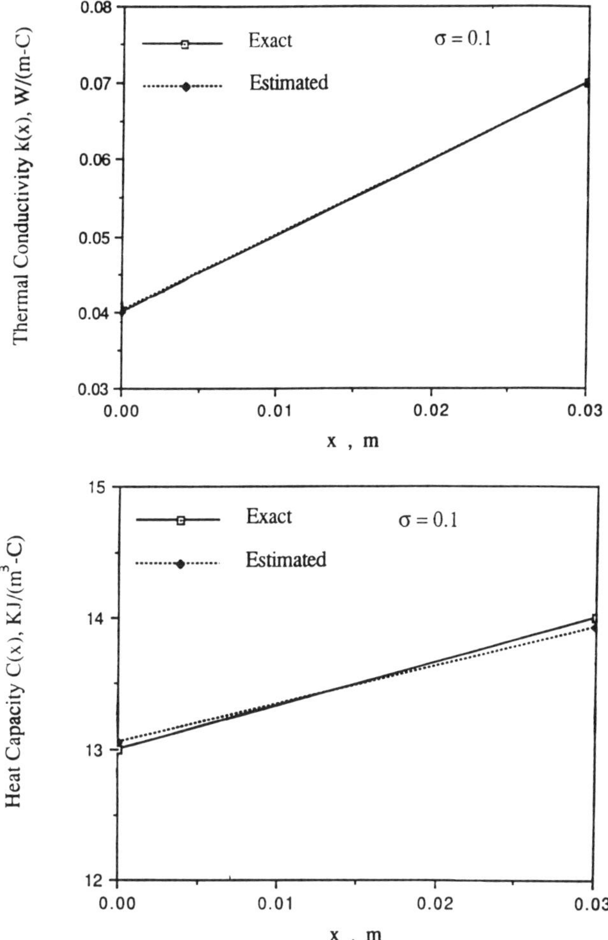

Fig. 14-6 Estimated thermal conductivity $k(x)$, heat capacitance $c(x)$ for insulating materials such as fiberglass.

range of property variation. The agreement between the exact and estimated properties is very good.

Solutions would converge within about 6 min of CPU time on the VAX-785 computer with initial guesses deviating from the exact values by a factor of 2 or $\frac{1}{2}$.

14-5 CONJUGATE GRADIENT METHOD WITH ADJOINT EQUATION FOR SOLVING IHCP AS A FUNCTION ESTIMATION PROBLEM

In this section we present a powerful *conjugate gradient method* of minimization, utilizing an adjoint problem [65-68] for solving IHCP as a *function estimation*

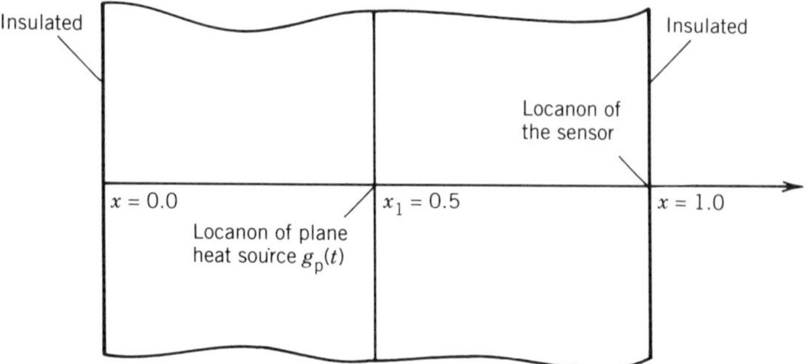

Fig. 14-7 Locations of the plane surface heat source and sensor.

problem with no prior information on the functional form of the unknown. The advantage of the method is that the regularization is implicitly built into the computational procedure. Recently, the method has been successfully used by several investigators [37, 69, 70]. To illustrate the basic steps in the application of this technique, we consider the problem involving the determination of the functional form of the timewise variation of the unknown strength of a plane surface heat source at a specified location inside a flat plate, as described below.

Consider a plate heated by a plane surface heat source of unknown strength $g_p(t)$ located at a specified position $x = x_1$, inside the plate releasing its heat continuously for times $t > 0$. Both boundaries of the plate are insulated. The IHCP is concerned with the determination of the unknown timewise varying strength $g_p(t)$ of the heat source from the temperature measurements taken with a sensor located at the surface $x = 1$ of the plate at times t_j, $j = 1, 2, \ldots, M$, thus providing a total of M temperature measurements. Figure 14-7 shows the geometry and coordinate. The steps in the solution of this IHCP problem are:

1. The direct problem
2. The sensitivity problem
3. The adjoint problem
4. The gradient equation
5. The conjugate gradient method for minimization
6. The stopping criteria
7. The computational algorithm

We now present the procedure in each of these distinct steps.

1. The Direct Problem The mathematical formulation of the direct problem is given in the dimensionless form as

$$\frac{\partial^2 T(x,t)}{\partial x^2} + g_p(t)\delta(x - x_1) = \frac{\partial T(x,t)}{\partial t}, \quad \text{in} \quad 0 < x < 1 \tag{14-57a}$$

$$\frac{\partial T(0,t)}{\partial x} = 0 \qquad \text{at} \qquad x = 0 \tag{14-57b}$$

$$\frac{\partial T(1,t)}{\partial x} = 0 \qquad \text{at} \qquad x = 1 \tag{14-57c}$$

$$T(x,0) = 0 \qquad \text{at} \qquad t = 0 \tag{14-57d}$$

Here $\delta(\cdot)$ is the Dirac delta function and the plane surface heat source of strength $g_p(t)$, located at $x_1 = 0.5$, releases its energy continuously over the time domain.

In the inverse problem considered here, the source term $g_p(t)$ is an unknown function of time, but the measured transient temperature data $Y(1,t)$, taken at the location $x = 1$, are available over the whole time domain $0 < t < t_f$, where t_f is the final measurement time. It will soon be apparent that, to solve this inverse problem with the conjugate gradient method, two auxiliary functions, called the *sensitivity function* $\Delta T(x,t)$ and the *Lagrange multiplier* $\lambda(x,t)$, will appear in the variation ΔJ of the functional J [(see equation (14-61)] and will be needed in the minimization procedure. Therefore, we need to develop the auxiliary problems, called the *sensitivity problem* and the *adjoint problem*, in order to determine $\Delta T(x,t)$ and $\lambda(x,t)$, respectively.

2. The Sensitivity Problem It is assumed that when $g_p(t)$ undergoes an increment $\Delta g_p(t)$, the temperature $T(x,t)$ changes by an amount $\Delta T(x,t)$. Then, to construct the sensitivity problem satisfying the function $\Delta T(x,t)$, we replace $T(x,t)$ by $T(x,t) + \Delta T(x,t)$, and $g_p(t)$ by $g_p(t) + \Delta g_p(t)$ in the direct problem (14-57) and subtract from it the original problem (14-57). The following sensitivity problem is obtained

$$\frac{\partial^2 \Delta T(x,t)}{\partial x^2} + \Delta g_p(t)\delta(x - 0.5) = \frac{\partial \Delta T(x,t)}{\partial t}, \qquad 0 < x < 1 \tag{14-58a}$$

$$\frac{\partial \Delta T(0,t)}{\partial x} = 0 \qquad \text{at} \qquad x = 0 \tag{14-58b}$$

$$\frac{\partial \Delta T(1,t)}{\partial x} = 0 \qquad \text{at} \qquad x = 1 \tag{14-58c}$$

$$\Delta T(x,0) = 0 \qquad \text{for} \qquad t = 0 \tag{14-58d}$$

Clearly, $\Delta T(x,t)$ represents changes in $T(x,t)$ with respect to the changes in the unknown function $g_p(t)$: hence it is a sensitivity function. In the sensitivity problem (14-58), $\Delta g_p(t)$ is the only forcing function needed for the solution of this problem. The choice of $\Delta g_p(t)$ will be described later in the analysis.

3. The Adjoint Problem The IHCP is recast as an optimum control problem of finding the unknown control function $g_p(t)$ such that it minimizes the following functional, $J(g_p)$:

$$J(g_p) = \int_{t=0}^{t_f} [T(1,t;g_p) - Y(1,t)]^2 \, dt \tag{14-59}$$

where $T(1,t;g_p)$ is the temperature at $x = 1$ computed from the solution of the direct problem (14-57) by using the estimated values for $g_p(t)$ and $Y(1,t)$ is the measured temperature at $x = 1$.

To develop the *adjoint problem* we introduce a new function $\lambda(x,t)$, called the *Lagrange multiplier*. We multiply equation (14-57a) with $\lambda(x,t)$, integrate the resulting expression over the spatial domain from $x = 0$ to $x = 1$, and then over the time domain from $t = 0$ to $t = t_f$ where t_f is the dimensionless final time. The result obtained in this manner is added to the functional $J[g_p(t)]$ given by equation (14-59). We obtain

$$\begin{aligned} J(g_p) = & \int_{t=0}^{t_f} [T(1,t;g_p) - Y(1,t)]^2 \, dt \\ & + \int_{t=0}^{t_f} \int_{x=0}^{1} \lambda(x,t) \left[\frac{\partial^2 T(x,t)}{\partial x^2} + g_p(t)\delta(x-0.5) - \frac{\partial T(x,t)}{\partial t} \right] dx \, dt \end{aligned} \tag{14-60}$$

Clearly, when $T(x,t)$ is the exact solution to the problem, the terms inside the bracket vanish and equation (14-60) reduces to equation (14-59). An expression for the variation $\Delta J(g_p)$ of the functional $J(g_p)$ is developed by perturbing $T(x,t)$ by $\Delta T(x,t)$ and $g_p(t)$ by $\Delta g_p(t)$ in equation (14-60) and then by subtracting from the resulting expression the original equation (14-60). We find

$$\begin{aligned} \Delta J = & \int_{t=0}^{t_f} 2[T(1,t;g_p) - Y(1,t)] \Delta T(1,t) \, dt \\ & + \int_{t=0}^{t_f} \int_{x=0}^{1} \lambda \left[\frac{\partial^2 \Delta T(x,t)}{\partial x^2} + \Delta g_p(t)\, \delta(x-0.5) - \frac{\partial \Delta T(x,t)}{\partial t} \right] dx \, dt \end{aligned} \tag{14-61}$$

The second term of the right-hand side of this equation is simplified by integration by parts; after rearrangement equation (14-61) takes the form

$$\Delta J = \int_{t=0}^{t_f} \int_{x=0}^{1} \left(\frac{\partial^2 \lambda(x,t)}{\partial x^2} + \frac{\partial \lambda(x,t)}{\partial t} \right) \Delta T(x,t) dx \, dt + \int_{t=0}^{t_f} \frac{\partial \lambda(0,t)}{\partial x} \Delta T(0,t) dt$$

$$+\int_{t=0}^{t_f}\left[2[T(1,t;g_p)-Y(1,t)]-\frac{\partial\lambda(1,t)}{\partial x}\right]\Delta T(1,t)dt$$
$$-\int_{x=0}^{1}\lambda(x,t_f)\Delta T(x,t_f)dx+\int_{t=0}^{t_f}\lambda(0.5,t)\Delta g_p(t)dt \qquad (14\text{-}62)$$

The boundary value problem satisfying the function $\lambda(x,t)$ is obtained by letting the first four integral terms containing ΔT on the right-hand side of equation (14-62) to vanish. This requirement leads to the following *adjoint problem*

$$\frac{\partial^2\lambda(x,t)}{\partial x^2}+\frac{\partial\lambda(x,t)}{\partial t}=0 \qquad \text{in} \qquad 0<x<1 \qquad (14\text{-}63a)$$

$$\frac{\partial\lambda(0,t)}{\partial x}=0 \qquad \text{at} \qquad x=0 \qquad (14\text{-}63b)$$

$$\frac{\partial\lambda(1,t)}{\partial x}=2[T(1,t)-Y(1,t)] \qquad \text{at} \qquad x=1 \qquad (14\text{-}63c)$$

$$\lambda(x,t_f)=0 \qquad \text{at} \qquad t=t_f \qquad (14\text{-}63d)$$

We note that, in this adjoint problem, the condition (14-63d) is the value of the function $\lambda(x,t)$ at the final time $t=t_f$. In the conventional initial value problem, the value of the function is specified at time $t=0$. However, the final value problem (14-63) can be transformed to an initial value problem by defining a new time variable given by $\tau=t_f-t$.

4. The Gradient Equation Finally, the only integral term left on the right-hand side of equation (14-62) is

$$\Delta J=\int_{t=0}^{t_f}\lambda(0.5,t)\Delta g_p(t)dt \qquad (14\text{-}64)$$

We note that ΔJ, by definition, is given by [91]

$$\Delta J=\int_{t=0}^{t_f}J'(t)\Delta g_p(t)dt \qquad (14\text{-}65)$$

where $J'(t)$ is the gradient of the functional $J(g)$.

From the comparison of equations (14-64) and (14-65) we conclude that

$$J'(t)=\lambda(0.5,t) \qquad (14\text{-}66)$$

which is the *gradient equation* for the functional, $J(g)$.

5. The Conjugate Gradient Method for Minimization The mathematical development given above provides three distinct problems defined by equations (14-57), (14-58) and (14-63) called the direct, sensitivity and adjoint problems for the unknown functions $T(x,t), \Delta T(x,t)$ and $\lambda(x,t)$, respectively. The measured data $Y(1,t)$ are considered available and the gradient $J'(t)$ is related to λ by equation (14-66).

The unknown generation function $g_p(t)$ can be determined by a procedure based on the minimization of the functional $J[g_p(t)]$ with an iterative approach by proper selection of the direction of descent and the step size in going from step k to $k+1$. In the conjugate gradient method of minimization [91,92], we consider the following iterative scheme for the determination of $g_p(t)$.

$$g_p^{k+1} = g_p^k - \beta^k P^k, \qquad k = 0, 1, 2, \ldots \tag{14-67}$$

where β^k is the *step size* in going from step k to step $k+1$ and P^k is the *direction of descent*, defined as

$$P^0 = J'^0 \tag{14-68a}$$

$$P^k = J'^k + \gamma^k P^{k-1} \qquad \text{with} \qquad \gamma^0 = 0 \quad \text{and} \quad k = 1, 2, \ldots \tag{14-68b}$$

Different definitions of the *conjugate coefficient* γ^k can be found in the standard texts on mathematics; here we choose the form [93,95]

$$\gamma^k = \frac{\displaystyle\int_{t=0}^{t_f} [J'^k(t)]^2\, dt}{\displaystyle\int_{t=0}^{t_f} [J'^{k-1}(t)]^2\, dt}, \qquad k = 1, 2, \ldots \tag{14-69}$$

The step size β^k is determined by minimizing the functional $J[g_p(t)]$ given by equation (14-59) with respect to β, that is

$$\min_\beta J(g^{k+1}) = \min_\beta \int_{t=0}^{t_f} [T(g_p^k - \beta^k P^k) - Y(1,t)]^2\, dt \tag{14-70a}$$

and by a Taylor series expansion we obtain

$$\min_\beta J(g_p^{k+1}) = \min_\beta \int_{t=0}^{t_f} [T(g_p^k) - \beta^k \Delta T(P^k) - Y(1,t)]^2\, dt \tag{14-70b}$$

To minimize equation (14-70b), we differentiate it with respect to β^k and set the resulting expression equal to zero. After rearrangement, the following expression is obtained for determining the step size β^k

$$\beta^k = \frac{\displaystyle\int_{t=0}^{t_f} \Delta T(P^k)[T(g_p^k) - Y]\, dt}{\displaystyle\int_{t=0}^{t_f} [\Delta T(P^k)]^2\, dt} \tag{14-71}$$

An initial estimate can be chosen for $g_p(t)$ to start the iterations. However, a special stopping criterion is needed to terminate the iterations because the measurement data contains error. The stopping criteria based on the discrepancy principle is described next.

6. The Stopping Criterion If the problem contains no measurement errors, the traditional check condition specified as [96]

$$J(g_p^{k+1}) < \epsilon^* \tag{14-72}$$

where ϵ^* is a small specified number, could be used. However, the observed temperature data contains measurement errors; as a result, the inverse solution will tend to approach the perturbed input data and the solution will exhibit oscillatory behavior as the number of iteration is increased [91]. The computational experience shows that it is advisable to use the *discrepancy principle* [97–99] for terminating the iteration process. Assuming $T(1, t; g_p) - Y(1, t) \cong \sigma =$ constant, the discrepancy principle that establishes the value of the stopping criteria is obtained from equation (14-59) as

$$\int_0^{t_f} \sigma^2\, dt = \sigma^2 t_f \equiv \epsilon^2 \tag{14-73a}$$

where σ is the standard deviation of the measurement error. Then the stopping criterion is taken as

$$J(g_p^{k+1}) < \epsilon^2 \tag{14-73b}$$

here ϵ^2 is determined from equation (14-73a).

7. Computational Algorithm The iterative computational procedure for the conjugate gradient method can be summarized as follow. To start the iterations an initial estimate is made for the function $g_p(t)$, which may be chosen as constant, say, zero. To solve the sensitivity problem (14-58), a value is needed for the source term Δg_p. However, from equation (14-67), we have $\Delta g_p = \beta^k P^k(t)$, which is equivalent to dividing the sensitivity coefficients by the constant β^k and then setting $\Delta g_p = P^k(t)$ in the sensitivity problem (14-58).

Suppose $g_p^k(t)$ is available at the kth iteration. The computational procedure is as follows:

Step 1. Solve the direct problem (14-57) and compute $T(x, t)$, based on $g_p^k(t)$.

Step 2. Knowing $T(x, t)$ and measured temperature $Y(x_1, t)$, solve the adjoint problem (14-63) and compute $\lambda(x_1, t)$; $x_1 = 0.5$.

Step 3. Knowing $\lambda(0.5, t)$, compute $J'(t)$ from equation (14-66).

Step 4. Knowing the gradient $J'(t)$, compute γ^k from equation (14-69), then the direction of descent P^k from equation (14-68).

Step 5. Setting $\Delta g_p^k(t) = P^k$, solve the sensitivity problem (14-58) and obtain $\Delta T(P^k)$.

Step 6. Knowing $\Delta T(P^k)$, compute step size β^k from equation (14-71).

Step 7. Knowing step size β^k, compute new $g^{k+1}(t)$ from equation (14-67), new $T(x, t)$ from the solution of the direct problem (14-57) and new $J[g_p^{k+1}]$ from equation (14-59).

Step 8. Check if the stopping criterion (14-73b) is satisfied; if not repeat the above calculational procedure until the discrepancy principle defined by equations (14-73) is satisfied.

In the above algorithm, the direct problem (14-57) appearing in Step 1, the adjoint problem (14-63) appearing in Step 2, and the sensitivity problem (14-58) appearing in Step 5 can readily be solved with finite differences. Analytic approaches can also be used for their solution, if possible.

RESULTS. To illustrate the accuracy of the method we refer to the following example [100].

Consider a slab of dimensionless thickness $L = 1$. The final dimensionless time is taken as $t_f = 1.8$ and the timewise variation of the strength of the internal plane heat source $g_p(t)$ located at $x = 0.5$ is defined as

$$g_p(t) = \begin{cases} 0.3 + \frac{7}{9}t & 0 \leqslant t < 0.9 \\ 1.5 - \frac{5}{9}t & 0.9 \leqslant t \leqslant 1.8 \end{cases} \tag{14-74}$$

which represents a triangular variation over time.

Finite differencing, with space increment $dx = 0.02$, time increment $dt = 0.03$ is used and all properties are taken as unity. Random noise levels of $\sigma = 0.001$ and 0.005 were added to the simulated exact temperature to generate the measured temperature data, i.e.,

$$Y_{\text{measured}} = Y_{\text{exact}} + \omega\sigma \tag{14-75}$$

where σ is the standard deviation of measurement errors and the values of ω are calculated randomly by the IMSL [101] subroutine DRNNOR. In the present calculation the value of ω is chosen over the range $-2.576 \leqslant \omega \leqslant 2.576$ which represents the 99% confidence for the measured temperature data.

The results of inverse solution are presented in figure 14-8 for $\sigma = 0.001$ and $\sigma = 0.005$. For a temperature of unity and a 99% confidence, these standard deviations correspond to measurement errors of 0.26% and 1.3%, respectively. The prediction is in excellent agreement with the exact results for both cases except for small deviation near the final time $t = t_f = 1.8$.

The reason for the inaccuracy of the estimation at the final time $t = t_f$ is due to the fact that, with the conjugate gradient method of minimization described

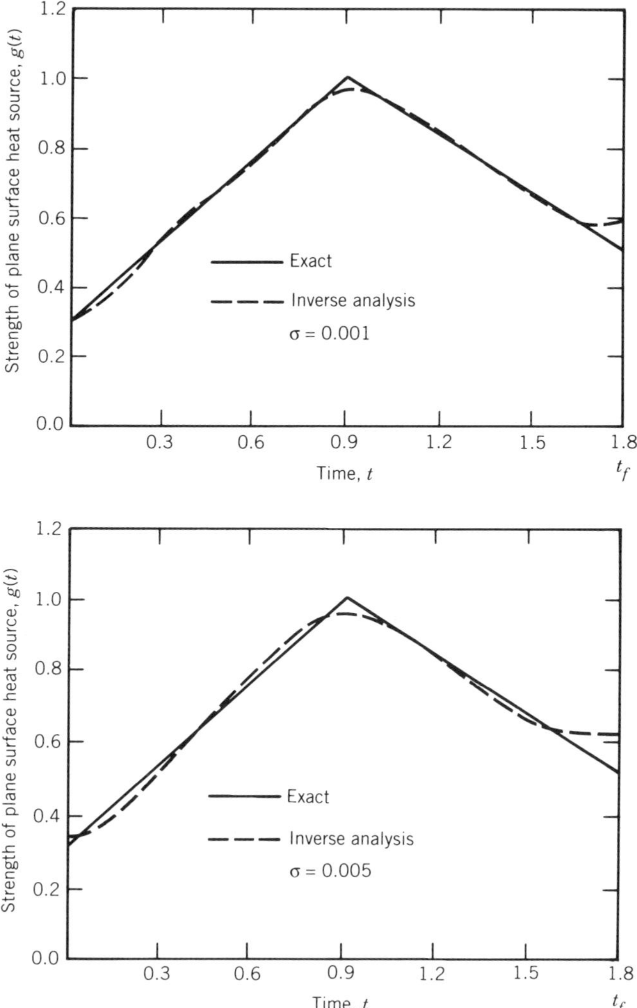

Fig 14-8 Prediction of the timewise variation of the strength of a plane surface heat source, with noise levels $\sigma = 0.001$ and 0.005 (From Ref. [100]).

here, the estimated value of $g_p^{k+1}(t_f)$ will always be equal to the initial guess $g_p^0(t_f)$ [102].

This difficulty at the final time t_f can be alleviated by repeating the calculations with initial guesses for g_p taken at a time few time steps before the final time t_f and omitting the results for the last few time steps.

An examination of the exact and the estimated values of $g_p(t)$ shown in Fig. 14-8 reveals that the estimations are accurate up to times very close to the final time t_f.

All the calculations in this work are performed on VAX-785, and the computation times for each of the figures required about 20 s of CPU time. The number of iterations depended on the measurement error, but always remained in the range $5 < k < 10$.

REFERENCES

1. V. J. Beck, B. Blackwell, and C. A. St. Clair, *Inverse Heat Conduction*, Wiley, New York, 1985.
2. E. Hensel, *Inverse Theory and Applications for Engineers*, Prentice Hall, Englewood Cliffs, N.J., 1991.
3. O. M. Alifanov, *An Introduction to the Theory of Inverse Heat Transfer Problems*, Mashinestroenie Publishing Agency, Moscow, 1991.
4. O. M. Alifanov and N. V. Kerov, *J. Eng. Phys.* **41**, 1049–1053, 1981.
5. O. M. Alifanov and V. V. Mikhailov, *J. Eng. Phys.* **45**, 724–730, 1983.
6. J. V. Beck, *Nucl. Eng. Des.* **7**, 170–178, 1968.
7. J. V. Beck, *Nucl. Eng. Des.* **7**, 9–12, 1968.
8. W. H. Giedt, *Jet Propulsion* **25**, 158–162, 1955.
9. T. Kuroyanagi, *Bulletin of JSME* **29**, 2961–2969, 1986.
10. D. A. Murio, *Computational and Applied Mathematics* **8**, 111–119, 1982.
11. G. A. Surkov, F. B. Yurevich, A. N. Filomenko, and V. V. Chupnason, *J. Eng. Phys.* **45**, 1109–1111, 1983.
12. N. M. Alnajem and M. N. Özişik, *J. Heat Transfer* **107**, 700–703, 1985.
13. N. M. Alnajem and M. N. Özişik, *Int. J. Heat Mass Transfer* **28**, 2121–2128, 1985.
14. N. M. Alnajem and M. N. Özişik, *Wärme-und Stoffübertragung* **20**, 89–96, 1986.
15. G. P. Flach and M. N. Özişik, *Annals of Nuclear Energy* **13**, 325–336, 1986.
16. G. P. Flach and M. N. Özişik, *Int. J. Heat Mass Transfer* **30**, 869–880, 1987.
17. G. P. Flach and M. N. Özişik, *J. Heat Transfer* **110**, 821–829, 1988.
18. G. P. Flach and M. N. Özişik, *J. Heat Transfer* **114**, 5–13, 1992.
19. O. M. Alifanov and V. V. Mikhailov, *High Temp.* **35**, 1501–1506, 1978.
20. O. M. Alifanov and M. V. Klibanov, *J. Eng. Phys.* **48**, 730–735, 1985.
21. E. A. Artyukhin, *J. Eng. Phys.* **41**, 1054–1058, 1981.
22. E. A. Artyukhin and A. S. Okhapkin, *J. Eng. Phys.* **45**, 1275–1281, 1983.
23. Y. M. Chen and J. Q. Liu, *J. Computational Phys.* **43**, 315–326, 1981.
24. W. H. Chen and J. H. Seinfield, *Int. J. Control* **15**, 487–495, 1972.

25. L. Carotenuto, et al., *Proc. Fourth IFAC Symposium*, Tiblisi, pp. 1539–1556, North-Holland, Amsterdam, 1978.
26. C. C. Travis and L. W. White, *Math. Biosci.* **77**, 341–352, 1985.
27. L. Carotenuto and G. Raiconi, *Int. J. Syst. Sci.* **11**, 1035–1049, 1980.
28. X. Y. Liu and Y. M. Chen, *SIAM J. Sci. Stat. Comp.* **8**, 436–445, 1987.
29. G. P. Flach and M. N. Özişik, *Num. Heat Transfer* **16**, 249–266, 1989.
30. C. H. Huang and M. N. Özişik, *Int. J. Heat and Fluid Flow* **11**, 262–268, 1990.
31. C. H. Huang and M. N. Özişik, *Num. Heat Transfer* (in press).
32. A. L. Alkidas, *J. Heat Transfer* **102**, 189–193, 1980.
33. V. D. Overbye, et al., *SAE Trans* **69**, 461–494, 1961.
34. C. H. Ho and M. N. Özişik, *Int. J. Heat Mass Transfer* **32**, 335–341, 1989.
35. C. H. Ho and M. N. Özişik, *JQSRT* **40**, 553–560, 1988.
36. A. Moutsoglu, *J. Heat Transfer* **111**, 37–43, 1989.
37. C. H. Huang and M. N. Özişik, *Num. Heat Transfer*, Part A, **21**, 101–116, 1992.
38. O. M. Alifanov, *High Temp.* **15**, 498–504, 1977.
39. J. V. Beck, *Int. J. Heat Mass Transfer* **13**, 703–716, 1970.
40. C. F. Weber, *Int. J. Heat Mass Transfer* **24**, 1783–1792, 1981.
41. A. N. Tikhonov and V. Y. Arsenin, *Solutions of Ill-Posed Problems*, Winston & Sons, Washington, DC, 1977.
42. M. N. Özişik, *Boundary Value Problems of Heat Conduction*, Dover, New York, 1989.
43. G. Stoltz, Jr., *J. Heat Transfer* **82**, 20–26, 1960.
44. J. V. Beck, *ASME Paper*, No. 75-WA/HT, 1982.
45. J. V. Beck and H. Wolf, *ASME Paper*, No. 65-HT-40, 1965.
46. O. R. Burggraf, *J. Heat Transfer* **86**, 373–382, 1964.
47. M. Imber and J. Khan, *AIAA Journal* **10**, 784–789, 1972.
48. D. Langford, *Q. Appl. Math.* **24**, 315–322, 1967.
49. C. F. Weber, *Int. J. Heat Mass Transfer* **24**, 1783–1792, 1981.
50. W. B. Powell and T. W. Price, *ISA Trans* **3**, 246–254, 1964.
51. T. P. Fidelle and G. E. Zinsmeister, *ASME Paper*, No. 68-WA/HT-26, 1968.
52. J. V. Beck, *Int. J. Heat Mass Transfer* **13**, 703–716, 1970.
53. N. D'Souza, *ASME Paper*, No. 75-WA/HT-81, 1975.
54. L. Garifo, V. E. Schrock, and E. Spedicato, *Energia Nucleare* **22**, 452–464, 1975.
55. J. D. Randal, Technical Report, John Hopkins University, Laural, MD, 1976.
56. J. V. Beck, B. Litkovhi, and C. R. St. Clair, *Num. Heat Transfer* **5**, 275–286, 1982.
57. B. F. Blackwell, M., *Num. Heat Transfer* **4**, 229–239, 1981.
58. B. R. Bass, *J. Eng. Ind.* **102**, 168–176, 1980.
59. P. S. Hore, G. W. Krutz, and R. J. Schoenhals, *ASME Paper*, No. 77-WA/TM-4, 1977.
60. B. R. f. Bass and L. J. Ott, *Adv. Comp. Tech.* **2**, 238–248, 1980.
61. I. Frank, *J. Heat Transfer* **85**, 378–379, 1963.
62. J. M. Davies, *J. Heat Transfer* **88**, 154–160, 1966.
63. E. Scott and J. V. Beck, *ASME Paper*, No. 85-WA/HT-43, 1985.

64. A. N. Tikhonov and Y. V. Arsenin, *Solutions of Ill-Posed Problems*, Winston & Sons, Washington, DC, 1977.
65. O. M. Alifanov and S. V. Rumyatsev, *J. of Eng. Phys.* **34**, 223–226, 1978.
66. S. V. Rumyantsev, *J. Eng. Phys.* **49**, 1418–1421, 1986.
67. O. M. Alifanov and S. V. Rumyatsev, *J. Eng. Phys.* **53**, 1335–1342, 1987.
68. Y. Jarny, M. N. Özişik, and J. P. Bardon, *Int. J. Heat Mass Transfer* **34**, 2911–2919, 1991.
69. Y. Jarny, D. Delaunay and J. Bransier, *8th Proc. Int. Heat Transfer Conf.* **4**, 1811–1816, 1986.
70. C. H. Huang and M. N. Özişik, *Numerical Heat Transfer* **21**, 101–116, 1992.
71. P. K. Lamm, *SIAM J. Control* **25**, 18–37, 1987.
72. C. E. Siewert, *Nucl. Sci. Eng.* **67**, 259–260, 1978.
73. C. E. Siewert, *J. Math. Phys.* **19**, 2619–2621, 1978.
74. H. Y. Li and M. N. Özişik, *J. Heat Transfer* (in press).
75. N. J. McCormick, *JQSRT* **42**, 303–309, 1989.
76. N. J. McCormick and G. E. Rinaldi, *Appl. Opt.* **28**, 2605–2613, 1989.
77. R. E. Walpole and R. H. Myers, *Probability and Statistics for Engineers and Scientists*, Macmillan, New York, 1978.
78. J. V. Beck and K. J. Arnold, *Parameter Estimation in Engineering and Science*, Wiley, New York, 1977.
79. L. E. Scales, *Introduction to Non-Linear Optimization*, Springer-Verlag, New York, 1985.
80. G. V. Reclaitis, A. Ravindran, and K. M. Ragsdell, *Engineering Optimization*, Wiley, New York, 1983.
81. R. W. Farebrother, *Linear Least Squares Computations*, Marcel Dekker, New York, 1988.
82. N. R. Draper and R. C. Van Nostrand, *Ridge Regression*, Univ. Wisconsin, Dept. of Statistics, Technical Report 501, March 1978.
83. K. A. Levenberg, *Q. Appl. Math.* **2**, 164–168, 1944.
84. D. W. Marquardt, *J. Soc. Ind. Appl. Math.* **11**, 431–441, 1963.
85. D. W. Marquardt, *Technometrics* **12**, 591–612, 1970.
86. H. S. Carslaw and J. C. Jaeger, *Conduction of Heat in Solids*, 2nd ed., Oxford University Press, London, 1959.
87. C. H. Huang and M. N. Özişik, *Int. J. Heat Fluid Flow* **12**, 173–178, 1991.
88. P. Craven and G. Wahba, *Numerische Mathematic* **31**, 377–403, 1979.
89. G. H. Golub, *Technometrics* **21**, No. 2, 1979.
90. H. R. Busby and D. M. Trujillo, *Int. J. Num. Meth. Eng.* **21**, 349–359, 1985.
91. O. M. Alifanov, *J. Eng. Phys.* **26**, No. (4), 471–476, 1974.
92. F. S. Beckman, "The Solution of Linear Equation by Conjugate Gradient Method," *Mathematical Method for Digital Computer*, A. Ralston and H. S. Wilf (Eds), Chapter 4, Wiley, New York, 1960.
93. R. Fletcher and C. M. Reeves, *Computer J.* **7**, 149–154, July 1964.
94. L. S. Lasdon, S. K. Mitter, and A. D. Waren, *IEEE Trans. Automatic Control* **12**, No. (2), 132–154, 1967.

95. W. J. Kammerer and M. Z. Nashed, *SIAM J. Num. Anal.* **9**, No. 1, 165–181, 1972.
96. O. M. Alifanov and V. V. Mikhailov, *J. Eng. Phys.* **35**, 1501–1506, 1978.
97. O. M. Alifanov and E. A. Artyukhin, *J. Eng. Phys.* **29**, 934–938, 1975.
98. O. M. Alifanov, *J. Eng. Phys.* **23**, 1566–1571, 1972.
99. H. W. Engl, *J. Optimization Theory and Appl.* **52**, No. 2, 209–215, 1987.
100. C. H. Huang and M. N. Özişik, *J. Franklin Institute*, (in press).
101. IMSL Library, Edition 10, 1987, NBC Building, 7500 Ballaire Blvd, Houston, TX.
102. A. J. Silva Neto and M. N. Özişik, *J. Appl. Phys.* **71**, No. (11), 1–8, 1992.

PROBLEMS

14-1 Elements of a discrete random variable X and the probability distribution function $f(x)$ associated with it are listed below. Calculate the expected value $E(X)$ of the random variable.

x_n	0	2	4	6	8	10	9
$f(x_n)$	$\frac{1}{15}$	$\frac{4}{15}$	$\frac{3}{15}$	$\frac{1}{15}$	$\frac{2}{25}$	$\frac{3}{15}$	$\frac{1}{15}$

14-2 Element of a discrete random variable X and the probability distribution function $f(x)$ associated with it are listed below. Calculate the variance σ^2 of the random variable.

x_n	4	3	2	1	0
$f(x_n)$	0.2	0.3	0.1	0.3	0.1

14-3 Consider the function $g(X) = X - 1$ of the continuous random variable X. Given the probability distribution function

$$f(x) = \frac{x}{5} \qquad \text{for} \qquad -2 < x < 2$$

calculate the expected value of $g(X)$.

14-4 Find the expected value of the continuous random variable X having a probability distribution function $f(x)$ given by

$$f(x) = \begin{cases} 3 - x & \text{for} \quad 0 < x < 2 \\ 0 & \text{elsewhere} \end{cases}$$

14-5 Probabilities of number of defectives in a sample of six are given by

x	0	1	2	3	4	5
$f(x)$	0.80	0.10	0.05	0.03	0.02	0.00

Calculate the variance $\sigma^2 = E(x^2) - \mu$.

14-6 The probability distribution function $f(x)$ for continuous random variable X is given by

$$f(x) = \begin{cases} x - 1 & \text{for} \quad 1 < x < 3 \\ 0 & \text{elsewhere} \end{cases}$$

Calculate the mean and variance of X.

14-7 The average life of a certain type of engine is 20 years with a standard deviation of $\sigma = 2$ years. Assuming the life expectancy of such engines to have a normal distribution, determine the probability that a given engine will last 18 years.

14-8 Determine the value of $\pm z$ such that the shaded area under the normal probability curve shown below should be equal to 95%.

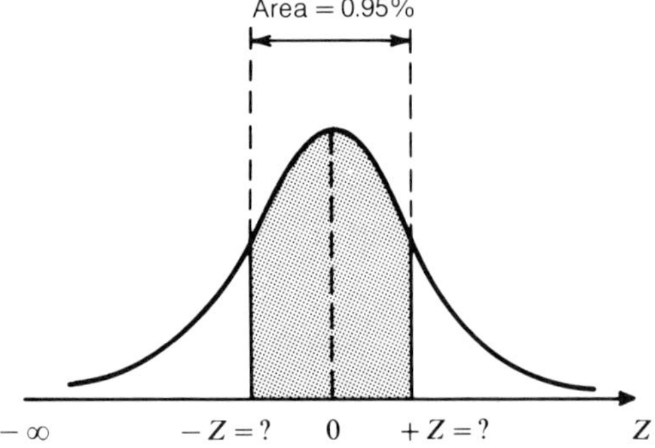

14-9 Consider a semiinfinite solid ($x \geqslant 0$) initially at zero temperature. For times $t > 0$, a constant heat flux q_0 W/m^2 is applied to the surface at $x = 0$. The temperature of the boundary surface $x = 0$ is given by [see: equation 14-25a].

$$T(0, t) = \frac{2q_0}{k}\left(\frac{\alpha t}{\pi}\right)^{1/2}$$

(a) Determine the sensitivity coefficient $X_{q_0}(0, t)$ with respect to the surface heat flux q_0.

(b) Determine the sensitivity coefficient $X_{\alpha}(0, t)$ with respect to the thermal diffusivity α.

(c) Let the dimensionless sensitivity coefficient X^* and the dimensionless surface temperature T^* be defined as

$$X^* = \frac{X_{q_0}(0, t)k}{L}, \qquad T^* = \frac{T(0, t)k}{q_0 L}$$

Then show that $X^* = T^*$.

14-10 To determine the unknown constant thermal conductivity k, W/(m·°C) and heat capacity $\rho C_p \equiv C, kJ/(m^3 \cdot °C)$ of a solid, the following transient heat conduction problem is considered.

A plate of thickness L, initially at a uniform temperature, is suddenly subjected to a constant heat flux q_0, W/m^2, at the surface $x = 0$ while the surface at $x = L$ is kept insulated. Temperature recordings are taken with a sensor located at the insulated surface as a function of time. The following measured data are given:

$$L = 0.03 \text{ m}, \qquad q_0 = 100 \text{ W/m}^2$$

The standard deviation of measurement $\equiv \sigma = 0.1$

Time (s)	Measured T, (°C)	Time (s)	Measured T, (°C)
0	0.000	160	28.45
20	0.282	180	33.71
40	1.892	200	38.71
60	5.110	220	43.96
80	9.124	240	49.20
100	13.80	260	54.12
120	18.47	280	59.32
140	23.57	300	64.41

Utilizing the above measured data and using the least-squares approach with the Levenberg–Marquardt algorithm, estimate the thermal conductivity k and heat capacity C of the solid.

14-11 Repeat problem 14-10 by utilizing the following measured data:

$$L = 0.03 \text{ m}, \qquad q_0 = 25{,}000 \text{ W/m}^2$$

Standard deviation of temperature measurements $= \sigma = 0.1$

Time (s)	Measured T, (°C)	Time (s)	Measured T, (°C)
0	0.000	160	34.37
20	2.305	180	39.18
40	6.740	200	43.70
60	11.41	220	48.46
80	15.97	240	53.21
100	20.70	260	57.64
120	25.16	280	62.34
140	29.91	300	66.93

14-12 Consider the inverse heat conduction problem given by

$$\frac{\partial^2 T}{\partial x^2} = \frac{1}{\alpha}\frac{\partial T}{\partial t} \qquad \text{in} \quad 0 < x < L, \quad 0 \leqslant t \leqslant t_f$$

$$-k\frac{\partial T}{\partial x} = q(t) = \text{unknown} \qquad \text{at} \quad x = 0, \qquad 0 \leqslant t \leqslant t_f$$

$$-k\frac{\partial T}{\partial x} = 0 \qquad \text{at} \quad x = L, \qquad 0 \leqslant t \leqslant t_f$$

$$T = F(x) \qquad \text{in} \quad 0 \leqslant x \leqslant L, \quad \text{for} \quad t = 0$$

Here, the unknown boundary surface heat flux $q(t)$ is to be determined by solving this inverse problem with the conjugate gradient method with adjoint equation. Develop the sensitivity and adjoint problems and the gradient equation needed for the solution and write the solution algorithm.

15

HEAT CONDUCTION IN ANISOTROPIC SOLIDS

In the previous chapter we considered heat conduction in solids that are said to be *isotropic*; that is, the thermal conductivity does not depend on direction. There are many natural and synthetic materials in which the thermal conductivity varies with direction; they are called *anisotropic* materials. For example, crystals, wood, sedimentary rocks, metals that have undergone heavy cold pressing, laminated sheets, cables, heat shielding materials for space vehicles, fiber-reinforced composite structures, and many others are anisotropic materials. In wood, the thermal conductivity is different along the grain, across the grain, and circumferentially. In laminated sheets the thermal conductivity is not the same along and across the laminations. Therefore, heat conduction in anisotropic materials has numerous important applications in various branches of science and engineering.

Most of the earlier work have been limited to the problems of one-dimensional heat flow in crystal physics [1, 2]. The differential equation of heat conduction for anisotropic solids involves cross-derivatives of the space variables; therefore, the general analysis of heat conduction in anisotropic solids is complicated. When the cross-derivatives are absent from the heat conduction equation, as in the case of *orthotropic* solids, the analysis of heat transfer is significantly simplified and has been considered in several references [3–12]. In recent years several works have appeard in the literature on the solution of heat conduction in anisotropic media [13–30]. Experimental work on heat diffusion in anisotropic solids is very limited; the available work [2, 5, 17] deals with either the one-dimensional situation or the orthotropic materials.

In this chapter we present the differential equation of heat conduction and the boundary conditions for anisotropic solids, discuss the thermal conductivity coefficients for crystal structures, and illustrate the solution of the steady-state and time-dependent heat conduction problems in anisotropic solids with representative examples.

15-1 HEAT FLUX FOR ANISOTROPIC SOLIDS

The heat flux in isotropic solids, as discussed in Chapter 1, obeys the Fourier law

$$\mathbf{q} = -k\nabla T \tag{15-1}$$

where the thermal conductivity is independent of direction and the heat flux vector $\mathbf{q}$ is normal to the isothermal surface passing through the spacial position considered.

In the case of anisotropic solids, the component of the heat flux, say, q_1, along $0x_1$, depends in general on a linear combination of the temperature gradients along the $0x_1$, $0x_2$, and $0x_3$ directions. With this consideration, the general expressions for the three components of the heat flux q_1, q_2, and q_3 along the $0x_1$, $0x_2$, and $0x_3$ directions in the rectangular coordinate system are given, respectively, as [31]

$$-q_1 = k_{11}\frac{\partial T}{\partial x_1} + k_{12}\frac{\partial T}{\partial x_2} + k_{13}\frac{\partial T}{\partial x_3} \tag{15-2a}$$

$$-q_2 = k_{21}\frac{\partial T}{\partial x_1} + k_{22}\frac{\partial T}{\partial x_2} + k_{23}\frac{\partial T}{\partial x_3} \tag{15-2b}$$

$$-q_3 = k_{31}\frac{\partial T}{\partial x_1} + k_{32}\frac{\partial T}{\partial x_2} + k_{33}\frac{\partial T}{\partial x_3} \tag{15-2c}$$

which can be written more compactly in the form

$$q_i = -\sum_{j=1}^{3} k_{ij}\frac{\partial T}{\partial x_j} \qquad i = 1, 2, 3 \tag{15-3}$$

Therefore, for an anisotropic solid the heat flux vector $\mathbf{q}$ is not necessarily normal to the isothermal surface passing through the point considered. The thermal conductivity of an anisotropic solid involves nine components, k_{ij}, called the *conductivity coefficients*, that are considered to be the components of a second-order tensor $\bar{\bar{k}}$

$$\bar{\bar{k}} \equiv \left| \begin{matrix} k_{11} & k_{12} & k_{13} \\ k_{21} & k_{22} & k_{23} \\ k_{31} & k_{32} & k_{33} \end{matrix} \right| \tag{15-4a}$$

From Onsagar's [31] principles of thermodynamics of irreversible processes it is shown that when the fluxes (i.e., q_i) and the forces (i.e., $\partial T/\partial x_i$) are related to each other linearly as given by equations (15-2), the phenomenologic coefficients obey the reciprocity relation. A discussion of the application of Onsagar's reciprocity relation for the thermal conductivity coefficients associated with heat conduction in anisotropic solids is given by Casimir [32]. Therefore, the conductivity coefficients k_{ij} can be considered to obey the reciprocity relation

$$k_{ij} = k_{ji} \qquad i, j = 1, 2, 3 \tag{15-4b}$$

Furthermore, as discussed in reference [33], according to irreversible thermodynamics, the coefficients k_{11}, k_{22}, and k_{33} are positive, that is,

$$k_{ii} > 0 \tag{15-4c}$$

and the magnitude of the coefficients k_{ij}, for $i \neq j$, is limited by the requirement [31]

$$k_{ii} k_{jj} > k_{ij}^2 \qquad \text{for} \qquad i \neq j \tag{15-4d}$$

The expression for the heat flux components, given by equation (15-3) for the rectangular coordinate system, can readily be generalized for the orthogonal curvilinear coordinate system (u_1, u_2, u_3) as

$$q_i = -\sum_{j=1}^{3} \frac{1}{a_j} k_{ij} \frac{\partial T}{\partial u_j} \qquad i = 1, 2, 3 \tag{15-5}$$

where a_j, are the scale factors discussed in Chapter 1.

For the (x_1, x_2, x_3) *rectangular coordinate* system equation (15-5) reduces to equations (15-2).

For the (r, ϕ, z) *cylindrical coordinate* system we set $u_1 = r$, $u_2 = \phi$, $u_3 = z$ and $a_1 = 1$, $a_2 = r$, $a_3 = 1$; then equation (15-5) gives

$$-q_r = k_{11} \frac{\partial T}{\partial r} + k_{12} \frac{1}{r} \frac{\partial T}{\partial \phi} + k_{13} \frac{\partial T}{\partial z} \tag{15-6a}$$

$$-q_\phi = k_{21} \frac{\partial T}{\partial r} + k_{22} \frac{1}{r} \frac{\partial T}{\partial \phi} + k_{23} \frac{\partial T}{\partial z} \tag{15-6b}$$

$$-q_z = k_{31} \frac{\partial T}{\partial r} + k_{32} \frac{1}{r} \frac{\partial T}{\partial \phi} + k_{33} \frac{\partial T}{\partial z} \tag{15-6c}$$

For the (r, ϕ, θ) *spherical coordinate system* we set $u_1 = r$, $u_2 = \phi$, $u_3 = \theta$ and $a_1 = 1$, $a_2 = r\sin\theta$, $a_3 = r$ and obtain

$$-q_r = k_{11}\frac{\partial T}{\partial r} + k_{12}\frac{1}{r\sin\theta}\frac{\partial T}{\partial \phi} + k_{13}\frac{1}{r}\frac{\partial T}{\partial \theta} \tag{15-7a}$$

$$-q_\phi = k_{21}\frac{\partial T}{\partial r} + k_{22}\frac{1}{r\sin\theta}\frac{\partial T}{\partial \phi} + k_{23}\frac{1}{r}\frac{\partial T}{\partial \theta} \tag{15-7b}$$

$$-q_\theta = k_{31}\frac{\partial T}{\partial r} + k_{32}\frac{1}{r\sin\theta}\frac{\partial T}{\partial \phi} + k_{33}\frac{1}{r}\frac{\partial T}{\partial \theta} \tag{15-7c}$$

15-2 HEAT CONDUCTION EQUATION FOR ANISOTROPIC SOLIDS

The differential equation of heat conduction for an anisotropic solid in the orthogonal curvilinear coordinate system (u_1, u_2, u_3) is given as

$$-\frac{1}{a_1 a_2 a_3}\left[\frac{\partial}{\partial u_1}(a_2 a_3 q_1) + \frac{\partial}{\partial u_2}(a_1 a_3 q_2) + \frac{\partial}{\partial u_3}(a_1 a_2 q_3)\right] + g = \rho C_p \frac{\partial T}{\partial t} \tag{15-8}$$

where q_1, q_2, and q_3 are the three components of the heat flux vector defined by equation (15-5), g is the heat-generation term, and the other quantities are as defined previously.

We now present explicit form of the heat conduction equation (15-8) for the rectangular, cylindrical, and spherical coordinates for the case of constant conductivity coefficients.

Rectangular Coordinate System

For the (x, y, z) rectangular coordinate system we set $u_1 = x$, $u_2 = y$, $u_3 = z$, and $a_1 = a_2 = a_3 = 1$; then equation (15-8), with q_i, given by equation (15-5), yields

$$k_{11}\frac{\partial^2 T}{\partial x^2} + k_{22}\frac{\partial^2 T}{\partial y^2} + k_{33}\frac{\partial^2 T}{\partial z^2} + (k_{12} + k_{21})\frac{\partial^2 T}{\partial x \partial y} + (k_{13} + k_{31})\frac{\partial^2 T}{\partial x\, \partial z}$$
$$+ (k_{23} + k_{32})\frac{\partial^2 T}{\partial y\, \partial z} + g(x, y, z, t) = \rho C_p \frac{\partial T(x, y, z, t)}{\partial t} \tag{15-9}$$

where $k_{12} = k_{21}$, $k_{13} = k_{31}$, and $k_{23} = k_{32}$ by the reciprocity relation.

Cylindrical Coordinate System

For the (r, ϕ, z) cylindrical coordinate system we set $u_1 = r$, $u_2 = \phi$, $u_3 = z$, and $a_1 = 1$, $a_2 = r$, $a_3 = 1$. Then, from equations (15-8) and (15-5) we obtain

$$k_{11}\frac{1}{r}\frac{\partial}{\partial r}\left(r\frac{\partial T}{\partial r}\right) + k_{22}\frac{1}{r^2}\frac{\partial^2 T}{\partial \phi^2} + k_{33}\frac{\partial^2 T}{\partial z^2} + (k_{12} + k_{21})\frac{1}{r}\frac{\partial^2 T}{\partial \phi\, \partial z}$$

$$+ (k_{13} + k_{31})\frac{\partial^2 T}{\partial r\, \partial z} + \frac{k_{13}}{r}\frac{\partial T}{\partial z} + (k_{23} + k_{32})\frac{1}{r}\frac{\partial^2 T}{\partial \phi\, \partial z} + g(r, \phi, z, t)$$

$$= \rho C_p \frac{\partial T(r, \phi, z, t)}{\partial t} \tag{15-10}$$

where $k_{ij} = k_{ji}$, $i \neq j$.

Spherical Coordinate System

For the (r, ϕ, θ) spherical coordinate system we set $u_1 = r$, $u_2 = \phi$, $u_3 = \theta$, and $a_1 = 1$, $a_2 = r \sin\theta$, $a_3 = r$; then from equations (15-8) and (15-5) obtain

$$k_{11}\frac{1}{r^2}\frac{\partial}{\partial r}\left(r^2\frac{\partial T}{\partial r}\right) + k_{22}\frac{1}{r^2 \sin^2\theta}\frac{\partial^2 T}{\partial \phi^2} + k_{33}\frac{1}{r^2 \sin\theta}\frac{\partial}{\partial \theta}\left(\sin\theta\frac{\partial T}{\partial \theta}\right)$$

$$+ \frac{(k_{12} + k_{21})}{r \sin\theta}\frac{\partial^2 T}{\partial r\, \partial \phi} + k_{12}\frac{1}{r^2 \sin\theta}\frac{\partial T}{\partial \phi} + \frac{(k_{13} + k_{31})}{r}\frac{\partial^2 T}{\partial r\, \partial \theta} + k_{13}\frac{1}{r^2}\frac{\partial T}{\partial \theta}$$

$$+ (k_{23} + k_{32})\frac{1}{r^2 \sin\theta}\frac{\partial^2 T}{\partial \theta\, \partial \phi} + k_{31}\frac{\cos\theta}{r \sin\theta}\frac{\partial T}{\partial r} + g(r, \phi, \theta, t)$$

$$= \rho C_p \frac{\partial T(r, \phi, \theta, t)}{\partial t} \tag{15-11}$$

where $k_{ij} = k_{ji}$, $i \neq j$.

15-3 BOUNDARY CONDITIONS

The boundary conditions for the heat conduction equation for an anisotropic medium may be of the first, second, or third kind. We consider a boundary surface S_i normal to the coordinate axis u_i. The boundary condition of the third kind can be written as

$$\mp \delta_i k_{\text{ref}} \cdot \frac{\partial T}{\partial n^*} + h_i T = f_i \qquad \text{on boundary } S_i \tag{15-12}$$

where

$$\frac{\partial T}{\partial n^*} \equiv \sum_{j=1}^{3} \frac{1}{a_j}\frac{k_{ij}}{k_{\text{ref}}}\frac{\partial T}{\partial u_j} \tag{15-13}$$

where

a_j = scale factor

k_{ref} = a reference conductivity that may be chosen as k_{11}, k_{22}, or k_{33}.

δ_i = zero or unity; that is, by setting $\delta_i = 0$, the boundary condition of the first kind is obtained

Here $\partial/\partial n^*$ is the derivative as defined by equation (15-13). The choice of *plus* or *minus* sign in equation (15-12) depends on whether the outward drawn normal to the boundary surface S_i is pointing in the *positive* or *negative* u_i direction respectively.

We illustrate the boundary conditions for the anisotropic medium with specific examples given below.

Example 15-1

Write the boundary conditions of the third kind for an anisotropic slab at the boundary surfaces $x = 0$ and $x = L$.

Solution. For the (x, y, z) rectangular coordinate system we write

$$-\left(k_{11}\frac{\partial T}{\partial x} + k_{12}\frac{\partial T}{\partial y} + k_{13}\frac{\partial T}{\partial z}\right) + h_1 T = f_1 \qquad \text{at} \qquad x = 0 \qquad (15\text{-}14\text{a})$$

$$+\left(k_{11}\frac{\partial T}{\partial x} + k_{12}\frac{\partial T}{\partial y} + k_{13}\frac{\partial T}{\partial z}\right) + h_2 T = f_2 \qquad \text{at} \qquad x = L \qquad (15\text{-}14\text{b})$$

Equations (15-14) can be written more compactly in the form given by equation (15-12) by setting $k_{\text{ref}} \equiv k_{11}$:

$$-k_{11}\frac{\partial T}{\partial n^*} + h_1 T = f_1 \qquad \text{at} \qquad x = 0 \qquad (15\text{-}14\text{c})$$

$$k_{11}\frac{\partial T}{\partial n^*} + h_2 T = f_2 \qquad \text{at} \qquad x = L \qquad (15\text{-}14\text{d})$$

where

$$\frac{\partial}{\partial n^*} \equiv \frac{\partial}{\partial x} + \epsilon_{12}\frac{\partial}{\partial y} + \epsilon_{13}\frac{\partial}{\partial z} \qquad (15\text{-}14\text{e})$$

$$\epsilon_{ij} \equiv k_{ij}/k_{11} \qquad (15\text{-}14\text{f})$$

Example 15-2

Write the boundary conditions of the third kind for an anisotropic hollow cylinder at the boundary surfaces $r = a$ and $r = b$.

Solution. For the (r, ϕ, z) cylindrical coordinate system we take the scale factors as $a_1 = 1$, $a_2 = r$, and $a_3 = 1$ and write

$$-\left(k_{11}\frac{\partial T}{\partial r} + k_{12}\frac{1}{r}\frac{\partial T}{\partial \phi} + k_{13}\frac{\partial T}{\partial z}\right) + h_1 T = f_1 \qquad \text{at} \qquad r = a \tag{15-15a}$$

$$+\left(k_{11}\frac{\partial T}{\partial r} + k_{12}\frac{1}{r}\frac{\partial T}{\partial \phi} + k_{13}\frac{\partial T}{\partial z}\right) + h_2 T = f_2 \qquad \text{at} \qquad r = b \tag{15-15b}$$

Equations (15-15) can be written more compactly in the form given by equation (15-12) by setting $k_{\text{ref}} \equiv k_{11}$:

$$-k_{11}\frac{\partial T}{\partial n^*} + h_1 T = f_1 \qquad \text{at} \qquad r = a \tag{15-15c}$$

$$k_{11}\frac{\partial T}{\partial n^*} + h_2 T = f_2 \qquad \text{at} \qquad r = b \tag{15-15d}$$

where

$$\frac{\partial}{\partial n^*} \equiv \frac{\partial}{\partial r} + \epsilon_{12}\frac{1}{r}\frac{\partial}{\partial \phi} + \epsilon_{13}\frac{\partial}{\partial z} \tag{15-15e}$$

$$\epsilon_{ij} = \frac{k_{ij}}{k_{11}} \tag{15-15f}$$

15-4 THERMAL-RESISTIVITY COEFFICIENTS

In the previous sections we expressed each component of the heat flux vector as a linear sum of temperature gradients along the $0x_1$, $0x_2$, and $0x_3$ axes as given by equation (15-2). Sometimes it is desirable to express the temperature gradient in a given direction as linear combination of the heat flux components in the $0x_1$, $0x_2$, and $0x_3$ directions. To obtain such a relationship in the (x_1, x_2, x_3) rectangular coordinate system we write equations (15-2) in matrix notation as

$$-[k_{ij}]\left[\frac{\partial T}{\partial x_i}\right] = [q_i] \tag{15-16a}$$

or

$$-\left[\frac{\partial T}{\partial x_i}\right] = [k_{ij}]^{-1}[q_i] \tag{15-16b}$$

Let r_{ij} be the elements of the inverse matrix $[k_{ij}]^{-1}$; then equation (15-16b) is written explicitly as

$$-\frac{\partial T}{\partial x_1}=r_{11}q_1+r_{12}q_2+r_{13}q_3 \tag{15-17a}$$

$$-\frac{\partial T}{\partial x_2}=r_{21}q_1+r_{22}q_2+r_{23}q_3 \tag{15-17b}$$

$$-\frac{\partial T}{\partial x_3}=r_{31}q_1+r_{32}q_2+r_{33}q_3 \tag{15-17c}$$

where the coefficients r_{ij} are called the *thermal resistivity coefficients.* The coefficients r_{ij} can be determined in terms of k_{ij} by the matrix inversion procedure. Since $k_{ij}=k_{ji}$, it can be shown that r_{ij}'s are given by

$$r_{ij}=(-1)^{i+j}\frac{a_{ij}}{\Delta} \tag{15-18a}$$

where Δ is the symmetrical thermal conductivity tensor given by

$$\Delta\equiv\begin{vmatrix} k_{11} & k_{12} & k_{13}\\ k_{21} & k_{22} & k_{23}\\ k_{31} & k_{32} & k_{33}\end{vmatrix} \tag{15-18b}$$

a_{ij} is the cofactor obtained from Δ by omitting the ith row and the jth column. As in the case of thermal conductivity coefficients, k_{ij}, the thermal resistivity coefficients, r_{ij}, obey the reciprocity relation, $r_{ij}=r_{ji}$.

To illustrate the application of equation (15-18), we write below the thermal-resistivity coefficient r_{12} and r_{11} in terms of the thermal conductivity coefficients as

$$r_{12}=(-1)^3\frac{\begin{vmatrix} k_{21} & k_{23}\\ k_{31} & k_{33}\end{vmatrix}}{\Delta}=\frac{k_{23}k_{31}-k_{21}k_{33}}{\Delta} \tag{15-19a}$$

$$r_{11}=(-1)^2\frac{\begin{vmatrix} k_{22} & k_{23}\\ k_{32} & k_{33}\end{vmatrix}}{\Delta}=\frac{k_{22}k_{33}-k_{23}k_{32}}{\Delta} \tag{15-19b}$$

15-5 DETERMINATION OF PRINCIPAL CONDUCTIVITIES AND PRINCIPAL AXES

We consider the heat conduction equation for an anisotropic solid in the x_1, x_2, x_3 rectangular coordinate system written as

$$k_{11}\frac{\partial^2 T}{\partial x_1^2}+k_{22}\frac{\partial^2 T}{\partial x_2^2}+k_{33}\frac{\partial^2 T}{\partial x_3^2}+(k_{12}+k_{21})\frac{\partial^2 T}{\partial x_1\,\partial x_2}+(k_{13}+k_{31})\frac{\partial^2 T}{\partial x_1\,\partial x_3}$$
$$+(k_{23}+k_{32})\frac{\partial^2 T}{\partial x_2\,\partial x_3}+g=\sigma C_p\frac{\partial T}{\partial t} \tag{15-20}$$

where $k_{ij}=k_{ji}$. When the conductivity matrix given by equation (15-4a) is symmetric, it is possible to find a new system of rectangular coordinates ξ_1, ξ_2, and ξ_3 that can transform it to a diagonal form as

$$\begin{vmatrix} k_1 & 0 & 0 \\ 0 & k_2 & 0 \\ 0 & 0 & k_3 \end{vmatrix} \tag{15-21}$$

where k_1, k_2, and k_3 are called the *principal conductivities* along the principal coordinate axes ξ_1, ξ_2 and ξ_3, respectively. Then the heat conduction equation (15-20), in terms of the principal coordinates, becomes

$$\frac{\partial^2 T}{\partial \xi_1^2}+\frac{\partial^2 T}{\partial \xi_2^2}+\frac{\partial^2 T}{\partial \xi_3^2}+g=\rho C_p\frac{\partial T}{\partial t} \tag{15-22}$$

The principal conductivities k_1, k_2, and k_3 are determined in the following manner. Let the therrmal conductivity matrix be denoted by

$$\bar{\bar{k}} \equiv \begin{vmatrix} k_{11} & k_{12} & k_{13} \\ k_{21} & k_{22} & k_{23} \\ k_{31} & k_{32} & k_{33} \end{vmatrix} \tag{15-23}$$

Then the principal conductivities k_1, k_2, and k_3 are the eigenvalues of the following equation:

$$\begin{vmatrix} k_{11}-\lambda & k_{12} & k_{13} \\ k_{21} & k_{22}-\lambda & k_{23} \\ k_{31} & k_{32} & k_{33}-\lambda \end{vmatrix} = 0 \tag{15-24}$$

This is a cubic equation in λ and has three roots. Each of these roots is a real number because the conductivity coefficients k_{ij} are real numbers (see reference 34 for proof) and each corresponds to a principal conductivity, that is, $\lambda_1=k_1$, $\lambda_2=k_2$, and $\lambda_3=k_3$ along the principal axes ξ_1, ξ_2, and ξ_3, respectively.

The principal axes ξ_1, ξ_2, and ξ_3 are determined in the following manner:

Let l_1, l_2, l_3 be the direction cosines of the principal axis $0\xi_1$ with respect to the axes $0x_1$, $0x_2$, $0x_3$, and $\lambda_1=k_1$ be the principal conductivity along the direction $0\xi_1$. Then l_1, l_2, l_3 satisfy the relation

$$\begin{vmatrix} k_{11}-\lambda_1 & k_{12} & k_{13} \\ k_{21} & k_{22}-\lambda_1 & k_{23} \\ k_{31} & k_{32} & k_{33}-\lambda_1 \end{vmatrix} \begin{vmatrix} l_1 \\ l_2 \\ l_3 \end{vmatrix} = 0 \tag{15-25a}$$

which provides three homogeneous equations for the three unknowns l_1, l_2, l_3; only two of these equations are linearly independent. An additional relation is obtained from the requirement that the direction cosines satisfy

$$l_1^2 + l_2^2 + l_3^2 = 1 \tag{15-25b}$$

Thus the three direction cosines of the principal axis $0\xi_1$ are determined from equations (15-25).

The procedure is repeated with $\lambda_2 = k_2$ for the determination of m_1, m_2, m_3 of the principal axis $0\xi_2$ and with $\lambda_3 = k_3$ for n_1, n_2, n_3 of the principal axes $0\xi_3$.

15-6 CONDUCTIVITY MATRIX FOR CRYSTAL SYSTEMS

With symmetry considerations, crystals can be grouped into seven distinct systems identified as triclinic, monoclinic, orthorhombic hexagonal, tetragonal, trigonal, and cubic systems. Readers should consult references 1 and 2 for an in-depth discussion of this matter. Here we are concerned with the thermal conductivity tensors associated with such systems and summarize the results as follows:

1. *Triclinic.* In this system there are no limitations imposed on the conductivity coefficients by symmetry considerations; hence all nine components of k_{ij} can be nonzero, and we have

$$\bar{\bar{k}} \equiv \begin{vmatrix} k_{11} & k_{12} & k_{13} \\ k_{21} & k_{22} & k_{23} \\ k_{31} & k_{32} & k_{33} \end{vmatrix} \tag{15-26}$$

2. *Monoclinic.* Some of the components become zero with symmetry considerations, hence we have

$$\bar{\bar{k}} \equiv \begin{vmatrix} k_{11} & k_{12} & 0 \\ k_{21} & k_{22} & 0 \\ 0 & 0 & k_{23} \end{vmatrix} \tag{15-27}$$

3. *Orthorhombic.* The conductivity coefficients are given by

$$\bar{\bar{k}} \equiv \begin{vmatrix} k_{11} & 0 & 0 \\ 0 & k_{22} & 0 \\ 0 & 0 & k_{33} \end{vmatrix} \tag{15-28}$$

TABLE 15-1 Values of Principal Conductivities for Some Crystals at 30°C, in W/(mK)

Crystal	System	k_1, k_2	k_3
Quartz	trigonal	6.50	11.30
Calcite	trigonal	4.18	4.98
Bismuth	trigonal	9.24	6.65
Graphite	hexagonal	355.00	89.00

Note: For crystals listed here $k_1 = k_2$. From International Critical Tables (1929), Vol. 5, p. 231

4. *Cubic.* In this system we have $k_{11} = k_{22} = k_{33}$; hence we write

$$\bar{\bar{k}} \equiv \begin{vmatrix} k_{11} & 0 & 0 \\ 0 & k_{11} & 0 \\ 0 & 0 & k_{11} \end{vmatrix} \tag{15-29}$$

5. *Hexagonal, Tetragonal and Trigonal.*

$$\bar{\bar{k}} \equiv \begin{vmatrix} k_{11} & k_{12} & 0 \\ -k_{12} & k_{11} & 0 \\ 0 & 0 & k_{33} \end{vmatrix} \tag{15-30}$$

It was previously stated that, whenever the heat flux law of the form given by equations (15-2) holds, the classical thermodynamic considerations lead to the reciprocity relationship given by equation (15-4b). In the case of crystals, the results on the conductivity coefficients given above have been derived from the considerations of macroscopic symmetry. Since no general proof is available to show that the coefficients are symmetric, it has been necessary to rely on experiments. If the relation given by equation (15-4b) should apply, then it implies that $k_{12} = 0$ in equation (15-30) and $k_{21} = k_{12}$ in equation (15-27). Experimentally, principal conductivities are always found to be positive. Table 15-1 lists the values of principal conductivities for some crystals.

15-7 TRANSFORMATION OF HEAT CONDUCTION EQUATION FOR ORTHOTROPIC MEDIUM

The heat conduction equation for an orthotropic medium can be transformed to a standard heat conduction equation for an isotropic solid as described below.

We consider the heat conduction equation for an orthotropic medium in the rectangular coordinate system given by

$$k_1 \frac{\partial^2 T}{\partial x^2} + k_2 \frac{\partial^2 T}{\partial y^2} + k_3 \frac{\partial^2 T}{\partial z^2} + g = \rho C_p \frac{\partial T}{\partial t} \tag{15-31}$$

New independent variables X, Y, and Z are defined as

$$X = \left(\frac{k}{k_1}\right)^{1/2} x, \qquad Y = \left(\frac{k}{k_2}\right)^{1/2} y, \qquad Z = \left(\frac{k}{k_3}\right)^{1/2} z \tag{15-32}$$

where k is a reference conductivity. Equation (15-31) becomes

$$k\left(\frac{\partial^2 T}{\partial X^2} + \frac{\partial^2 T}{\partial Y^2} + \frac{\partial^2 T}{\partial Z^2}\right) + g = \rho C_p \frac{\partial T}{\partial t} \tag{15-33}$$

which looks like the standard heat conduction equation for an isotropic solid. However, the choice of the reference thermal conductivity is not arbitrary. The reason for this is that a volume element in the original space "$dx\,dy\,dz$" transforms, under the transformation (15-32), into

$$\frac{(k_1 k_2 k_3)^{1/2}}{k^{3/2}} dX\,dY\,dZ \tag{15-34}$$

If the quantities ρC_p and the generation term g defined on the basis of unit volume should have the same physical significance, we should have

$$\frac{(k_1 k_2 k_3)^{1/2}}{k^{3/2}} = 1 \qquad \text{or} \qquad k = (k_1 k_2 k_3)^{1/3} \tag{15-35a, b}$$

Then the heat conduction equation (15-33) takes the form

$$(k_1 k_2 k_3)^{1/3}\left(\frac{\partial^2 T}{\partial X^2} + \frac{\partial^2 T}{\partial Y^2} + \frac{\partial^2 T}{\partial Z^2}\right) + g = \rho C_p \frac{\partial T}{\partial t} \tag{15-36}$$

where X, Y, and Z are as defined by equation (15-32). This implies an isotropic medium of thermal conductivity $(k_1 k_2 k_3)^{1/3}$.

Several other ways of arriving at the result given by equation (15-35) are discussed in reference 8. Similar transformations are applicable to transform the equation into the standard form for the cylindrical and spherical coordinate systems.

Under the transformation discussed above, the solution of the resulting heat conduction equation is a straightforward matter, but the transformation of the solution to the original physical space requires additional commutations according to the transformation used. That is, the corresponding isotropic heat conduction problem of thermal conductivity $(k_1 k_2 k_3)^{1/3}$ is readily solved. The region is then distorted according to the transformation (15-32).

15-8 SOME SPECIAL CASES

We now examine some special situations that may give some insight to the physical significance of heat flow in an anisotropic medium.

Temperature Depending on x_1 and x_2 Only

For such a case we have $(\partial T/\partial x_3) = 0$; then equations (15-2) for the heat flux components reduce to

$$-q_1 = k_{11}\frac{\partial T}{\partial x_1} + k_{12}\frac{\partial T}{\partial x_2} \tag{15-37a}$$

$$-q_2 = k_{21}\frac{\partial T}{\partial x_1} + k_{22}\frac{\partial T}{\partial x_2} \tag{15-37b}$$

$$-q_3 = k_{31}\frac{\partial T}{\partial x_1} + k_{32}\frac{\partial T}{\partial x_2} \tag{15-37c}$$

This result implies that there is still a heat flux component q_3 in the x_3 direction even though there is no temperature gradient in that direction.

The heat conduction equation (15-9) simplifies to

$$k_{11}\frac{\partial^2 T}{\partial x_1^2} + k_{22}\frac{\partial^2 T}{\partial x_2^2} + (k_{12} + k_{21})\frac{\partial^2 T}{\partial x_1 \partial x_2} = \rho C_p \frac{\partial T}{\partial t} \tag{15-38}$$

where we assumed no energy generation in the medium.

Temperature Depending on x_1 Only

For such a case we have $(\partial T \partial z_2) = 0, (\partial T/\partial x_3) = 0$; then equation (15-21) for the heat flux components reduces to

$$-q_1 = k_{11}\frac{\partial T}{\partial x_1}, \quad -q_2 = k_{21}\frac{\partial T}{\partial x_1}, \quad -q_3 = k_{31}\frac{\partial T}{\partial x_1} \tag{15-39a,b,c}$$

This result implies that there is heat flow in the x_2 and x_3 directions even though temperature gradients are assumed to be zero in those directions.

The heat conduction equation (15-9) reduces to

$$k_{11}\frac{\partial^2 T}{\partial x_1^2} = \rho C_p \frac{\partial T}{\partial t} \tag{15-40}$$

where we assumed no energy generation. This equation is similar to the one-dimensional heat conduction equation for an isotropic medium.

A physical situation simulating one-dimensional heat flow through an isotropic solid can be realized as follows.

Consider a large, thin plate of crystal placed between two highly conducting materials maintained at constant uniform temperatures T_1 and T_2 as illustrated in Fig. 15-1. Since the crystal is thin and large, the isothermal surfaces are parallel

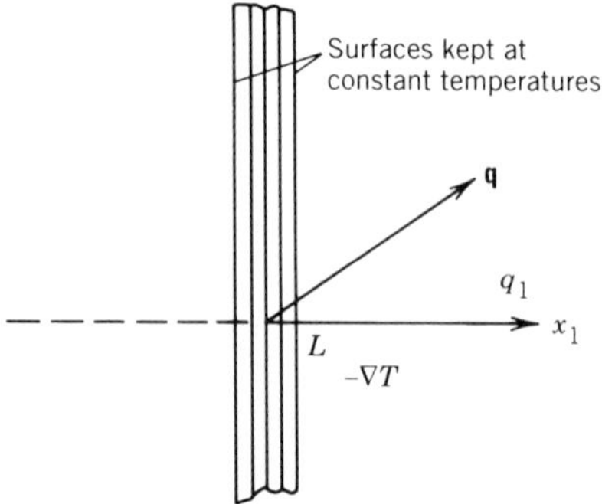

Fig. 15-1 Heat flow across a large thin crystal plate.

to the large faces of the crystal except in the region near the edges. If the plate thickness is small compared to the lateral dimensions, the edge effects become negligible. We note that the temperature gradient vector ∇T is along the $0x_1$ axis, but the heat flux vector $\mathbf{q}$ is not parallel to ∇T. The total heat flux flowing normal to the plate is q_1, since the heat flux components q_2 and q_3 do not carry heat in that direction. Then the quantities that can readily be measured with experiments are $\partial T/\partial x_1$ and q_1; hence under steady-state condition, equation (15-39a), that is

$$-q_1 = k_{11}\frac{\partial T}{\partial x_1} \tag{15-41a}$$

can be used to determine the conductivity coefficient k_{11}. The variation of k_{11} with the orientation of the $0x_1$ axis with reference to the principal axis is given by the relation [1–4]

$$k_{11} = l_1^2 k_1 + l_2^2 k_2 + l_3^2 k_3 \tag{15-41b}$$

where k_1, k_2, and k_3 are the principal conductivities and l_1, l_2, and l_3 are the direction cosines of the $0x_1$ axis relative to the principal axes $0\xi_1, 0\xi_2$, and $0\xi_3$, respectively.

Heat Flow in the x_1 Direction

The physical situation simulating such a condition can be realized by considering a long, thin crystal rod with two ends kept at different constant temperatures and the lateral surfaces insulated as illustrated in Fig. 15-2.

The heat flow is along the $0x_1$ direction only, since the lateral surface of the rod is insulated. Then we have

$$q_2 = q_3 = 0 \tag{15-41c}$$

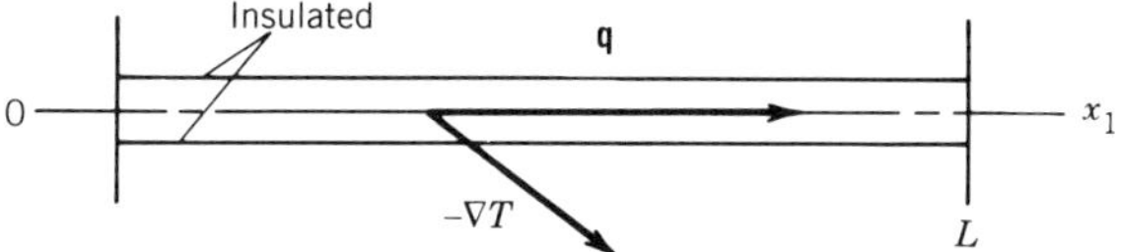

Fig. 15-2 Heat flow along a thin, long rod.

which implies that the heat flux vector **q** is along the $0x_1$ axis. When the results given by equations (15-41c) are introduced into equations (15-17), the three components of the temperature gradient vector become

$$-\frac{\partial T}{\partial x_1}=r_{11}q_1, \quad -\frac{\partial T}{\partial x_2}=r_{21}q_1, \quad -\frac{\partial T}{\partial x_3}=r_{31}q_1 \tag{15-42a,b,c}$$

Here, the temperature gradient $\partial T/\partial x_1$ and the heat flux q_1 in the x_1 direction along the rod are the measurable quantities. Then equation (15-42a) can be used to determine the resistivity coefficient r_{11}. The variation of r_{11} with the orientation of the $0x_1$ axis with reference to the principal axes is given by the relation [1–4]

$$r_{11}=l_1^2 r_1+l_2^2 r_2+l_3^2 r_3 \tag{15-42d}$$

where r_1, r_2, and r_3 are the principal resistivities and l_1, l_2, and l_3 are the direction cosines of the $0x_1$ axis relative to the principal axes $0\xi_1, 0\xi_2$, and $0\xi_3$, respectively.

15-9 HEAT CONDUCTION IN AN ORTHOTROPIC MEDIUM

In the case of noncrystalline anisotropic solids, such as wood, the thermal conductivities k_1, k_2, and k_3 are in the mutually perpendicular directions. Then the three components of the heat flux (q_1, q_2, q_3) are given in the (u_1, u_2, u_3) orthogonal curvilinear coordinate system as

$$q_1=-\frac{k_1}{a_1}\frac{\partial T}{\partial u_1}, \quad q_2=-\frac{k_2}{a_2}\frac{\partial T}{\partial u_2}, \quad q_3=-\frac{k_3}{a_3}\frac{\partial T}{\partial u_3} \tag{15-43a}$$

where a_1, a_2, a_3 are the scale factors.

Introducing equation (15-43a) into the energy equation (15-8), the heat-conduction equation for an orthotropic solid becomes

$$\frac{1}{a_1 a_2 a_3}\left[\frac{\partial}{\partial u_1}\left(\frac{a_2 a_3}{a_1}k_1\frac{\partial T}{\partial u_1}\right)+\frac{\partial}{\partial u_2}\left(\frac{a_1 a_3}{a_2}k_2\frac{\partial T}{\partial u_2}\right)+\frac{\partial}{\partial u_3}\left(\frac{a_1 a_2}{a_3}k_3\frac{\partial T}{\partial u_3}\right)\right]$$
$$+g=\rho C_p\frac{\partial T}{\partial t} \tag{15-43b}$$

Assuming k_1, k_2, k_3 constant, equation (15-43) for the rectangular, cylindrical, and spherical coordinates takes the following forms.

Rectangular coordinate system (x, y, z):

$$k_1 \frac{\partial^2 T}{\partial x^2} + k_2 \frac{\partial^2 T}{\partial y^2} + k_3 \frac{\partial^2 T}{\partial z^2} + g = \rho C_p \frac{\partial T}{\partial t} \tag{15-44}$$

Cylindrical coordinate system (r, ϕ, z):

$$k_1 \frac{1}{r}\frac{\partial}{\partial r}\left(r \frac{\partial T}{\partial r} \right) + k_2 \frac{1}{r^2}\frac{\partial^2 T}{\partial \phi^2} + k_3 \frac{\partial^2 T}{\partial z^2} + g = \rho C_p \frac{\partial T}{\partial t} \tag{15-45}$$

Spherical coordinate system (r, ϕ, θ):

$$k_1 \frac{1}{r^2}\frac{\partial}{\partial r}\left(r^2 \frac{\partial T}{\partial r} \right) + k_2 \frac{1}{r^2 \sin^2\theta}\frac{\partial^2 T}{\partial \phi^2} + k_3 \frac{1}{r^2 \sin\theta}\frac{\partial}{\partial \theta}\left(\sin\theta \frac{\partial T}{\partial \theta} \right) + g = \rho C_p \frac{\partial T}{\partial t} \tag{15-46}$$

We illustrate below with examples the solution of heat conduction in orthotropic medium for both the steady-state and time-dependent situations.

Example 15-3

Consider a point source of strength Q watts, located at the origin of the rectangular coordinate system, releasing its heat continuously over time at a constant rate in an orthotropic medium. In the regions away from the source the region is at a temperature T_∞. Develop an expression for the steady-state temperature distribution in the solid.

Solution. We consider the transformed equation (15-33). For the steady state problem in the region outside the origin where there is no energy generation we write

$$\frac{\partial^2 T}{\partial X^2} + \frac{\partial^2 T}{\partial Y^2} + \frac{\partial^2 T}{\partial Z^2} = 0 \quad \text{in} \quad 0 < X < \infty, \quad 0 < Y < \infty, \quad 0 < Z < \infty \tag{15-47}$$

where

$$X = \left(\frac{k}{k_1} \right)^{1/2} x, \quad Y = \left(\frac{k}{k_2} \right)^{1/2} y, \quad Z = \left(\frac{k}{k_3} \right)^{1/2} z \tag{15-48a,b,c}$$

$$k = (k_1 k_2 k_3)^{1/3} \tag{15-48d}$$

The boundary condition at the origin is obtained by drawing a small sphere of radius R around the point source and equating the rate of energy released by the source to the heat conducted into the medium:

$$(4\pi R^2)\left(-k\frac{\partial T}{\partial R}\right)=Q \qquad \text{as} \qquad R\to 0 \tag{15-49}$$

where $R=(X^2+Y^2+Z^2)^{1/2}$. The boundary condition at infinity is

$$T=T_\infty \qquad \text{as} \qquad R\to\infty \tag{15-50}$$

Equation (15-47) is Laplace's equation and its solution satisfying the boundary condition (15-50) is written as

$$T(R)=\frac{C}{R}+T_\infty \tag{15-51}$$

where the unknown constant C is determined by the application of the boundary condition (15-49) as

$$(4\pi R^2)\left(k\frac{C}{R^2}\right)=Q \tag{15-52}$$

or

$$C=\frac{1}{k}\frac{Q}{4\pi} \tag{15-53}$$

After equation (15-53) is introduced into (15-51), the solution becomes

$$T(R)-T_\infty=\frac{Q}{4\pi}\frac{1}{kR} \tag{15-54a}$$

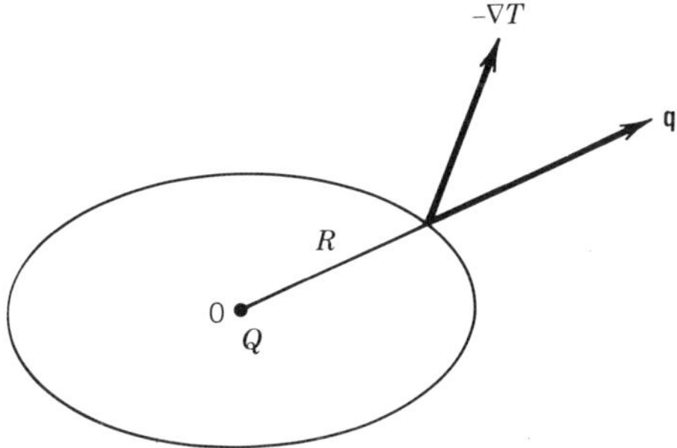

Fig. 15-3 Ellipsoidal isothermal surfaces around a point source, Q.

or

$$T(x, y, z) - T_\infty = \frac{Q}{4\pi}(k_1 k_2 k_3)^{-1/2}\left(\frac{x^2}{k_1} + \frac{y^2}{k_2} + \frac{z^2}{k_3}\right)^{-1/2} \tag{15-54b}$$

Clearly, $T(R) - T_\infty$ decreases with increasing distance R from the origin. Figure 15-3 shows that the heat flux vector $\mathbf{q}$ is along the R coordinate lines and the maximum temperature gradient ∇T is normal to the ellipsoidal isothermal surfaces. We note that the vectors q and ∇T are not necessarily parallel to each other.

Example 15-4

Consider the steady-state heat conduction problem for an orthotropic rectangular region $0 \leqslant x \leqslant a$, $0 \leqslant y \leqslant b$ in which heat is generated at a constant rate of g_0 W/m^3. Boundaries at $x = 0$ and $y = 0$ are kept insulated and those at $x = a$ and $y = b$ are dissipating heat by convection into an environment at zero temperature. The orthotropic thermal conductivities in the $0x$ and $0y$ directions are, respectively, k_1 and k_2. Obtain an expression for the steady-state temperature distribution in the region.

Solution. The mathematical formulation of this problem is given as

$$\frac{\partial^2 T}{\partial x^2} + \frac{1}{\epsilon^2}\frac{\partial^2 T}{\partial y^2} = -\frac{g_0}{k_1} \quad \text{in} \quad 0 < x < a, \quad 0 < y < b \tag{15-55}$$

$$\frac{\partial T}{\partial x} = 0 \quad \text{at} \quad x = 0 \tag{15-56}$$

$$\frac{\partial T}{\partial x} + H_1 T = 0 \quad \text{at} \quad x = a \tag{15-57}$$

$$\frac{\partial T}{\partial y} = 0 \quad \text{at} \quad y = 0 \tag{15-58}$$

$$\frac{\partial T}{\partial y} + H_2 T = 0 \quad \text{at} \quad y = b \tag{15-59}$$

where

$$\epsilon^2 \equiv \frac{k_1}{k_2}, \quad H_1 = \frac{h_1}{k_1}, \quad H_2 = \frac{h_2}{k_2}$$

We define the integral transform pair with respect to the x variable as

$$\text{Transform:} \quad \bar{T}(\beta_m, y) = \int_0^a X(\beta_m, x')T(x', y)\,dx' \tag{15-60a}$$

Inversion: $$T(x, y) = \sum_{m=1}^{\infty} \frac{1}{N(\beta_m)} X(\beta_m, x) \bar{T}(\beta_m, y) \tag{15-60b}$$

where $X(\beta_m, x)$, $N(\beta_m)$, and β_m are obtained from Table 2-2, case 4 as

$$X(\beta_m, x) = \cos \beta_m x, \qquad \frac{1}{N(\beta_m)} = 2\frac{\beta_m^2 + H_1^2}{a(\beta_m^2 + H_1^2) + H_1} \tag{15-60c}$$

and the β_m values are the roots of

$$\beta_m \tan \beta_m a = H_1 \tag{15-60d}$$

Taking the integral transform of system (15-55)–(15-59) by the application of the transform (15-60a) we obtain

$$\frac{d^2\bar{T}}{dy^2} - \beta_m^2 \epsilon^2 \bar{T}(\beta_m, y) = -\frac{\epsilon^2}{k_1} \bar{g}_0 \qquad \text{in} \qquad 0 < y < b \tag{15-61a}$$

$$\frac{d\bar{T}}{dy} = 0 \qquad \text{at} \qquad y = 0 \tag{15-61b}$$

$$\frac{d\bar{T}}{dy} + H_2 \bar{T} = 0 \qquad \text{at} \qquad y = b \tag{15-61c}$$

The solution of the system (15-61) is

$$\bar{T}(\beta_m, y) = \frac{1}{k_1 \beta_m^2} \bar{g}_0 - \frac{1}{k_1 \beta_m^2} \bar{g}_0 \frac{\cosh \beta_m \epsilon y}{\dfrac{\beta_m \epsilon}{H_2} \sinh \beta_m \epsilon b + \cosh \beta_m \epsilon b} \tag{15-62a}$$

where

$$\bar{g} = \int_0^a g_0 \cos \beta_m x \, dx = \frac{\sin \beta_m a}{\beta_m} g_0 \tag{15-62b}$$

The inversion of (15-62) by the inversion formula (15-60b) yields

$$\begin{aligned} T(x, y) = {} & \frac{g_0}{k_1} \sum_{m=1}^{\infty} \frac{1}{N(\beta_m)} \frac{\cos \beta_m x \sin \beta_m a}{\beta_m^3} \\ & - \frac{g_0}{k_1} \sum_{m=1}^{\infty} \frac{1}{\beta_m^3 N(\beta_m)} \frac{\cos \beta_m x \sin \beta_m a \cosh \beta_m \epsilon y}{\dfrac{\beta_m \epsilon}{H_2} \sinh \beta_m \epsilon b + \cosh \epsilon b} \end{aligned} \tag{15-63}$$

A closed-form expression for the first summation on the right is determined as (see note 1 at end of this chapter)

$$\sum_{m=1}^{\infty} \frac{1}{N(\beta_m)} \frac{\cos \beta_m x \sin \beta_m a}{\beta_m^3} = \frac{a}{H_1} + \frac{1}{2}(a^2 - x^2) \tag{15-64}$$

Then the solution (15-73) takes the form

$$T(x, y) = \frac{g_0 a}{k_1 H_1} + \frac{g_0}{2k_1}(a^2 - x^2) - \frac{2g_0}{k_1} \sum_{m=1}^{\infty} \frac{1}{\beta_m^3} \frac{\beta_m^2 + H_1^2}{a(\beta_m^2 + H_1^2) + H_1} \cdot \frac{\cos \beta_m x \sin \beta_m a \cosh \beta_m \epsilon y}{\dfrac{\beta_m \epsilon}{H_2} \sinh \beta_m \epsilon b + \cosh \beta_m \epsilon b} \tag{15-65a}$$

where the β_m values are the positive roots of

$$\beta_m \tan \beta_m a = H_1 \tag{15-65b}$$

Example 15-5

Consider the time-dependent heat conduction problem for an orthotropic rectangular region $0 \leqslant x \leqslant a$, $0 \leqslant y \leqslant b$. Initially the region is at a uniform temperature T_0. For times $t > 0$, the boundaries at $x = 0$ and $y = 0$ are kept insulated and those at $x = a$ and $y = b$ are dissipating heat by convection into an environment at zero temperature, while heat is generated in the region at a constant rate of g_0 W/m^3. The orthotropic thermal conductivities in the $0x$ and $0y$ directions are, respectively, k_1 and k_2. Obtain an expression for the time-dependent temperature distribution $T(x, y, t)$ in the region for times $t > 0$.

Solution. The mathematical formulation of this problem is given as

$$\frac{\partial^2 T}{\partial x^2} + \frac{1}{\epsilon^2} \frac{\partial^2 T}{\partial y^2} + \frac{g_0}{k_1} = \frac{1}{\alpha_1} \frac{\partial T}{\partial t} \quad \text{in} \quad 0 < x < a, \quad 0 < y < b, \quad t > 0 \tag{15-66a}$$

$$\frac{\partial T}{\partial x} = 0 \quad \text{at} \quad x = 0, \quad t > 0 \tag{15-66b}$$

$$\frac{\partial T}{\partial x} + H_1 T = 0 \quad \text{at} \quad x = a, \quad t > 0 \tag{15-66c}$$

$$\frac{\partial T}{\partial y} = 0 \quad \text{at} \quad y = 0, \quad t > 0 \tag{15-66d}$$

$$\frac{\partial T}{\partial y}+H_2 T=0 \qquad \text{at} \quad y=b, \qquad t>0 \qquad (15\text{-}66e)$$

$$T=T_0 \qquad \text{for} \quad t=0, \quad \text{in the region} \qquad (15\text{-}66f)$$

where

$$\epsilon^2 \equiv \frac{k_1}{k_2}, \quad H_1=\frac{h_1}{k_1}, \quad H_2=\frac{h_2}{k_2}, \quad \alpha_1=\frac{k_1}{\rho C_p}$$

It is convenient to split up this problem into two simpler problems as

$$T(x, y, t)=T_s(x, y)+T_h(x, y, t) \qquad (15\text{-}67)$$

Where the steady-state temperature $T_s(x, y)$ is the solution of the following problem

$$\frac{\partial^2 T_s}{\partial x^2}+\frac{1}{\epsilon^2}\frac{\partial^2 T_s}{\partial y^2}+\frac{g_0}{k_1}=0 \qquad \text{in} \quad 0<x<a, \quad 0<y<b \qquad (15\text{-}68a)$$

$$\frac{\partial T_s}{\partial x}=0 \qquad \text{at} \quad x=0 \qquad (15\text{-}68b)$$

$$\frac{\partial T_s}{\partial x}+H_1 T_s=0 \qquad \text{at} \quad x=a \qquad (15\text{-}68c)$$

$$\frac{\partial T_s}{\partial y}=0 \qquad \text{at} \quad y=0 \qquad (15\text{-}68d)$$

$$\frac{\partial T_s}{\partial y}+H_2 T_s=0 \qquad \text{at} \quad y=b \qquad (15\text{-}68e)$$

and the transient temperature $T_h(x, y, t)$ is the solution of the following homogeneous problem:

$$\frac{\partial^2 T_h}{\partial x^2}+\frac{1}{\epsilon^2}\frac{\partial^2 T_h}{\partial y^2}=\frac{1}{\alpha_1}\frac{\partial T_h}{\partial t} \qquad \text{in} \quad 0<x<a, \quad 0<y<b, \quad t>0 \qquad (15\text{-}69a)$$

$$\frac{\partial T_h}{\partial x}=0 \qquad \text{at} \quad x=0, \qquad t>0 \qquad (15\text{-}69b)$$

$$\frac{\partial T_h}{\partial x}+H_1 T_h=0 \qquad \text{at} \quad x=a, \qquad t>0 \qquad (15\text{-}69c)$$

$$\frac{\partial T_h}{\partial y} = 0 \qquad \text{at} \qquad y = 0, \qquad t > 0 \tag{15-69d}$$

$$\frac{\partial T_h}{\partial y} + H_2 T_h = 0 \qquad \text{at} \qquad y = b, \qquad t > 0 \tag{15-69e}$$

$$T_h = T_0 - T_s(x, y) \equiv F(x, y) \qquad \text{for} \qquad t = 0, \quad \text{in the region} \tag{15-69f}$$

The steady-state problem (15-68) is exactly the same as that considered in Example 15-4; therefore its solution is immediately obtainable from equation (15-65). The homogeneous problem defined by equations (15-69) can readily be solved by the integral-transform technique as now described. We define the integral-transform pair with respect to the x variable as

$$\text{Transform:} \quad \bar{T}(\beta_m, y, t) = \int_0^a X(\beta_m, x') T(x', y, t)\, dx' \tag{15-70a}$$

$$\text{Inversion:} \quad T(x, y, t) = \sum_{m=1}^{\infty} \frac{1}{N(\beta_m)} X(\beta_m, x) \bar{T}(\beta_m, y, t) \tag{15-70b}$$

where

$$X(\beta_m, x) = \cos \beta_m x, \qquad \frac{1}{N(\beta_m)} = 2 \frac{\beta_m^2 + H_1^2}{a(\beta_m^2 + H_1^2) + H_1} \tag{15-70c}$$

and the β_m values are the positive roots of

$$\beta_m \tan \beta_m a = H_1 \tag{15-70d}$$

The integral transform pair with respect to the y variable is defined as

$$\text{Transform:} \quad \tilde{\bar{T}}(\beta_m, \gamma_n, t) = \int_0^b Y(\gamma_n, y') \bar{T}(\beta_m, y', t)\, dy' \tag{15-71a}$$

$$\text{Inversion:} \quad \bar{T}(\beta_m, y, t) = \sum_{n=1}^{\infty} \frac{Y(\gamma_n, y)}{N(\gamma_n)} \tilde{\bar{T}}(\beta_m, \gamma_n, t) \tag{15-71b}$$

where

$$Y(\gamma_n, y) = \cos \gamma_n y, \qquad \frac{1}{N(\gamma_n)} = 2 \frac{\gamma_n^2 + H_2^2}{b(\gamma_n^2 + H_2^2) + H_2} \tag{15-71c}$$

and the γ_n values are the positive roots of

$$\gamma_n \tan \gamma_n b = H_2 \tag{15-71d}$$

The integral transform of the system (15-69) with respect to the x variable by the application of the transform (15-70a) is

$$-\beta_m^2 \bar{T}_h(\beta_m, y, t) + \frac{1}{\epsilon^2}\frac{\partial^2 \bar{T}_h}{\partial y^2} = \frac{1}{\alpha_1}\frac{\partial \bar{T}}{\partial t} \quad \text{in} \quad 0 < y < b, \quad t > 0 \tag{15-72a}$$

$$\frac{\partial \bar{T}_h}{\partial y} = 0 \quad \text{at} \quad y = 0, \quad t > 0 \tag{15-72b}$$

$$\frac{\partial \bar{T}_h}{\partial y} + H_2 \bar{T}_h = 0 \quad \text{at} \quad y = b, \quad t > 0 \tag{15-72c}$$

$$\bar{T}_h = \bar{F}(\beta_m, y) \quad \text{for} \quad t = 0, \quad \text{in} \quad 0 \leqslant y \leqslant b \tag{15-72d}$$

The integral transform of the system (15-72) with respect to the y variable by the application of the transform (15-71a) gives

$$-\beta_m^2 \bar{\bar{T}}_h(\beta_m, \gamma_n, t) - \frac{1}{\epsilon^2}\gamma_n^2 \bar{\bar{T}}_n = \frac{1}{\alpha_1}\frac{d\bar{\bar{T}}}{dt}$$

or

$$\frac{d\bar{\bar{T}}_h}{dt} + \alpha_1 \lambda_{mn}^2 \bar{\bar{T}}_h = 0 \quad \text{for} \quad t > 0 \tag{15-73a}$$

$$\bar{\bar{T}}_h(\beta_m, \gamma_n, t) = \bar{\bar{F}}(\beta_m, \gamma_n) \tag{15-73b}$$

where

$$\lambda_{mn}^2 = \beta_m^2 + \frac{1}{\epsilon^2}\gamma_n^2 \tag{15-73c}$$

The solution of equation (15-73) is

$$\bar{\bar{T}}_h(\beta_m, \gamma_n, t) = e^{-\alpha_1 \lambda_{mn}^2 t}\, \bar{\bar{F}}(\beta_m, \gamma_n) \tag{15-74}$$

The inversion of equation (15-74) successively by the inversion formulas (15-71b) and (15-70b) gives the solution for $T_h(x, y, t)$ as

$$T_h(x, y, t) = \sum_{m=1}^{\infty} \sum_{n=1}^{\infty} \frac{e^{-\alpha_1 \lambda_{mn}^2 t}}{N(\beta_m) N(\gamma_n)} \cos \beta_m x \cos \gamma_n y \cdot \int_{x'=0}^{a} \int_{y'=0}^{b} \cos \beta_m x' \cos \gamma_n y' F(x', y')\, dx'\, dy' \tag{15-75}$$

where $N(\beta_m)$ and $N(\gamma_n)$ are defined by equations (15-70c) and (15-71c), respectively; β_m and γ_n are the roots of the transcendental equations (15-70d) and (15-71d), respectively; and λ_{mn}^2 is defined by equation (15-73c). The function $F(x', y')$ being specified according to equations (15-69f) and (15-65), the integral with respect to the space variables in equation (15-65) can be evaluated analytically or numerically.

15-10 MULTIDIMENTIONAL HEAT CONDUCTION IN AN ANISOTROPIC MEDIUM

The multidimensional heat conduction equation for the case of general anisotropy involves cross-derivatives, whereas the boundary conditions may contain various partial derivatives with respect to the space variables. As a result, the analytic solution of the multidimensional heat conduction problem for the general anisotropic case is difficult to obtain, especially for finite regions. However, the solutions can be obtained for special situations involving semiinfinite or infinite regions as illustrated in the following examples.

Example 15-6

We consider a two-dimensional, time-dependent heat conduction problem for an anisotropic region $0 \leqslant x \leqslant \infty$, $-\infty < y < \infty$ in the rectangular coordinate system. The medium is initially at temperature $F(x, y)$ and for times $t > 0$ the boundary surface at $x = 0$ is kept at zero temperature. Obtain an expression for the temperature distribution $T(x, y, t)$ in the region for times $t > 0$.

Solution. Since no temperature variation is considered in the z direction, we have $\partial T/\partial z = 0$. Then, the heat-conduction equation (15-9) reduces to

$$\frac{\partial^2 T}{\partial x^2} + \epsilon_{22}\frac{\partial^2 T}{\partial y^2} + 2\epsilon_{12}\frac{\partial^2 T}{\partial x \partial y} = \frac{1}{\alpha_{11}}\frac{\partial T}{\partial t} \quad \text{in} \quad 0 < x < \infty, \quad -\infty < y < \infty, \quad t > 0 \tag{15-76a}$$

with the boundary and initial conditions

$$T = 0 \quad \text{at} \quad x = 0, \quad \text{for} \quad t > 0 \tag{15-76b}$$

$$T = F(x, y) \quad \text{for} \quad t = 0, \quad \text{in} \quad 0 \leqslant x < \infty, \quad -\infty < y < \infty \tag{15-76c}$$

where we defined

$$\epsilon_{ij} = \frac{k_{ij}}{k_{11}}, \quad k_{ij} = k_{ji}, \quad \text{and} \quad \alpha_{11} = \frac{k_{11}}{\rho C_p} \tag{15-76d}$$

We note that the differential equation involves one cross-derivative and the region in the y direction is infinite in extend. Therefore, the integral transform with respect to the y variable can be applied to remove from this equations the first and second partial derivatives with respect to the y variable. The integral transform pair with respect to the y variable is defined as [see equation (13-63)]

$$\text{Inversion:} \qquad T(x,y,t) = \frac{1}{2\pi}\int_{-\infty}^{\infty} e^{-i\gamma y}\bar{T}(x,\gamma,t)\,d\gamma \tag{15-77a}$$

$$\text{Transform:} \qquad \bar{T}(x,\gamma,t) = \int_{y'=-\infty}^{\infty} e^{i\gamma y'}T(x,y',t)\,dy' \tag{15-77b}$$

where the bar denotes the integral transform with respect to the y variable.

The integral transform of the system (15-76) by the application of the transform (15-77b) yields (see note 2 at the end of this chapter for the transform of the second and the first derivatives with respect to the y variable)

$$\frac{\partial^2 \bar{T}}{\partial x^2} - \gamma^2\epsilon_{22}\bar{T} - 2i\gamma\epsilon_{12}\frac{\partial \bar{T}}{\partial x} = \frac{1}{\alpha_{11}}\frac{\partial \bar{T}}{\partial t} \quad \text{in} \quad 0 < x < \infty, \quad t > 0 \tag{15-78a}$$

$$\bar{T} = 0 \quad \text{at} \quad x = 0, \quad t > 0 \tag{15-78b}$$

$$\bar{T} = \bar{F}(x,\gamma) \quad \text{for} \quad t = 0, \quad \text{in} \quad 0 \leqslant x < \infty \tag{15-78c}$$

where $\bar{T} \equiv \bar{T}(x,\gamma,t)$. The partial derivative $\partial\bar{T}/\partial x$ can be removed from this equation by defining a new variable $\bar{w}(x,\gamma,t)$ as

$$\bar{T}(x,\gamma,t) = \bar{w}(x,\gamma,t)e^{i\gamma\epsilon_{12}x} \tag{15-79}$$

Then, the system (15-78) is transformed to

$$\frac{\partial^2 \bar{w}}{\partial x^2} - \gamma^2(\epsilon_{22} - \epsilon_{12}^2)\bar{w} = \frac{1}{\alpha_{11}}\frac{\partial \bar{w}}{\partial t} \quad \text{in} \quad 0 < x < \infty, \quad t > 0 \tag{15-80a}$$

$$\bar{w} = 0 \quad \text{at} \quad x = 0, \quad t > 0 \tag{15-80b}$$

$$\bar{w} = e^{-i\gamma\epsilon_{12}x}\bar{F}(x,\gamma) \quad \text{for} \quad t = 0, \quad \text{in} \quad 0 \leqslant x < \infty \tag{15-80c}$$

To remove the partial derivative with respect to the x variable from this system, the integral-transform pair with respect to the x variable for the region

$0 < x < \infty$ is defined as [see equations (13-57) and Table 2-3, case 3]

$$\text{Inversion:} \qquad \bar{w}(x, \gamma, t) = \frac{2}{\pi} \int_{\beta=0}^{\infty} \sin \beta x \, \tilde{\bar{w}}(\beta, \gamma, t) \, d\beta \tag{15-81a}$$

$$\text{Transform:} \qquad \tilde{\bar{w}}(\beta, \gamma, t) = \int_{x'=0}^{\infty} \sin \beta x' \, \bar{w}(x', \gamma, t) \, dx' \tag{15-81b}$$

where the tilde denotes the transform with respect to the x variable. The integral transform of the system (15-80) by the application of the transform (15-81b) gives

$$\frac{d\tilde{\bar{w}}}{dt} + \alpha_{11}\lambda^2 \tilde{\bar{w}}(\beta, \gamma, t) = 0 \qquad \text{for} \qquad t > 0 \tag{15-82a}$$

$$\tilde{\bar{w}} = \tilde{\bar{H}}(\beta, \gamma) \qquad \text{for} \qquad t = 0 \tag{15-82b}$$

where

$$\lambda^2 \equiv \beta^2 + \gamma^2(\epsilon_{22} - \epsilon_{12}^2) \tag{15-83a}$$

$$\epsilon_{22} - \epsilon_{12}^2 > 0 \qquad \text{according to equation (15-4d)} \tag{15-83b}$$

$$\tilde{\bar{H}}(\beta, \gamma) = \int_{x'=0}^{\infty} e^{-i\gamma\epsilon_{12}x'} \bar{F}(x', \gamma) \sin \beta x' \, dx' \tag{15-83c}$$

$$\bar{F}(x', \gamma) = \int_{y'=-\infty}^{\infty} e^{i\gamma y'} F(x', y') \, dy' \tag{15-83d}$$

The solution of equation (15-82) is

$$\tilde{\bar{w}}(\beta, \gamma, t) = \tilde{\bar{H}}(\beta, \gamma) e^{-\alpha_{11}\lambda^2 t} \tag{15-84}$$

The inversion of equation (15-84) by the inversion formula (15-81a) and then the application of equation (15-79) yields

$$\bar{T}(x, \gamma, t) = \frac{2}{\pi} \int_{\beta=0}^{\infty} \sin \beta x \tilde{\bar{H}}(\beta, \gamma) e^{-\alpha_{11}\lambda^2 t + i\gamma\epsilon_{12}x} d\beta \tag{15-85}$$

This result is inverted by the inversion formula (15-77a), the explicit form of $\tilde{\bar{H}}(\beta, \gamma)$ is introduced and the order of integrations is rearranged:

$$T(x, y, t) = \int_{x'=0}^{\infty} \int_{y'=-\infty}^{\infty} F(x', y')$$

$$\cdot\left[\frac{1}{2\pi}\int_{\gamma=-\infty}^{\infty} e^{-\alpha_{11}(\epsilon_{22}-\epsilon_{12}^2)\gamma^2 t - i\gamma[(y-y')-\epsilon_{12}(x-x')]}d\gamma\right]$$

$$\cdot\left[\frac{2}{\pi}\int_{\beta=0}^{\infty} e^{-\alpha_{11}\beta^2 t}\sin\beta x\sin\beta x'\,d\beta\right]dy'\,dx' \tag{15-86}$$

In this result the integrals with respect to the variables γ and β can be evaluated by making use of the integrals given by equations (13-67) and (13-81), respectively, that is

$$\frac{1}{2\pi}\int_{\gamma=-\infty}^{\infty} e^{-\gamma^2 At - i\gamma z}d\gamma = \frac{1}{(4\pi At)^{1/2}}e^{-z^2/4At} \tag{15-87a}$$

$$\frac{2}{\pi}\int_{\beta=0}^{\infty} e^{-\beta^2\alpha t}\sin\beta x\sin\beta x'\,d\beta$$
$$= \frac{1}{(4\pi\alpha t)^{1/2}}\left[\exp\left(-\frac{(x-x')^2}{4\alpha t}\right)-\exp\left(-\frac{(x+x')^2}{4\alpha t}\right)\right] \tag{15-87b}$$

Then, the solution (15-86) takes the form

$$T(x,y,t) = \frac{1}{[4\pi\alpha_{11}(\epsilon_{22}-\epsilon_{12}^2)t]^{1/2}[4\pi\alpha_{11}t]^{1/2}}\int_{x'=0}^{\infty}\int_{y'=-\infty}^{\infty} F(x',y')$$
$$\cdot\exp\left(-\frac{[(y-y')^2-\epsilon_{12}(x-x')^2]^2}{4\pi\alpha_{11}(\epsilon_{22}-\epsilon_{12}^2)t}\right)$$
$$\cdot\left[\exp\left(-\frac{(x-x')^2}{4\alpha_{11}t}\right)-\exp\left(-\frac{(x+x')^2}{4\alpha_{11}t}\right)\right]dy'\,dx' \tag{15-88}$$

Example 15-7

An anisotropic medium $0 \leqslant x < \infty$, $-\infty < y < \infty$ is initially at zero temperature. For times $t > 0$, heat is generated in the medium at a rate of $g(x, y, t)$ W/m^3 while the boundary surface at $x = 0$ is kept at zero temperature. Obtain an expression for the temperature distribution $T(x, y, t)$ in the region for times $t > 0$.

Solution. The mathematical formulation of the heat-conduction problem is given as

$$\frac{\partial^2 T}{\partial x^2} + \epsilon_{22}\frac{\partial^2 T}{\partial y^2} + 2\epsilon_{12}\frac{\partial^2 T}{\partial x\,\partial y} + \frac{1}{k_{11}}g(x,y,t) = \frac{1}{\alpha_{11}}\frac{\partial T}{\partial t}$$
$$\text{in}\quad 0<x<\infty,\quad -\infty<y<\infty,\quad t>0 \tag{15-89a}$$

$$T = 0 \quad \text{at} \quad x = 0, \quad t > 0 \tag{15-89b}$$

$$T = 0 \quad \text{for} \quad t = 0, \quad \text{in} \quad 0 \leqslant x < \infty, \quad -\infty < y < \infty \tag{15-89c}$$

where we defined

$$\epsilon_{ij} = \frac{k_{ij}}{k_{11}}, \quad k_{ij} = k_{ji}, \quad \text{and} \quad \alpha_{11} = \frac{k_{11}}{\rho C_p} \tag{15-89d}$$

This problem is similar to that considered in Example 15-6, except for the heat generation and the zero initial condition. Therefore, the integral transform pairs defined in the previous example are applicable for the solution of this problem. The integral transform of the system (15-89) by the application of the transform (15-77b) gives

$$\frac{\partial^2 \bar{T}}{\partial x^2} - \gamma^2 \epsilon_{22} \bar{T} - 2i\gamma\epsilon_{12} \frac{\partial \bar{T}}{\partial x} + \frac{1}{k_{11}} \bar{g}(x, \gamma, t) = \frac{1}{\alpha_{11}} \frac{\partial \bar{T}}{\partial t}$$

$$\text{in} \quad 0 < x < \infty, \quad t > 0 \tag{15-90a}$$

$$\bar{T} = 0 \quad \text{at} \quad x = 0, \quad t > 0 \tag{15-90b}$$

$$\bar{T} = 0 \quad \text{for} \quad t = 0, \quad \text{in} \quad 0 \leqslant x < \infty \tag{15-90c}$$

where $\bar{T} = \bar{T}(x, \gamma, t)$. The partial derivative $\partial \bar{T}/\partial x$ can be removed from equation (15-90a) by the application of the transform (15-79). Then the system (15-90) is transformed to

$$\frac{\partial^2 \bar{w}}{\partial x^2} - \gamma^2(\epsilon_{22} - \epsilon_{12}^2)\bar{w} + \frac{1}{k_{11}} e^{-i\gamma\epsilon_{12}x} \bar{g}(x, \gamma, t) = \frac{1}{\alpha_{11}} \frac{\partial \bar{w}}{\partial t}$$

$$\text{in} \quad 0 < x < \infty, \quad t > 0 \tag{15-91a}$$

$$\bar{w} = 0 \quad \text{at} \quad x = 0, \quad t > 0 \tag{15-91b}$$

$$\bar{w} = 0 \quad \text{for} \quad t = 0, \quad \text{in} \quad 0 \leqslant x < \infty \tag{15-91c}$$

The partial derivative with respect to the x variable is removed from equation (15-91a) by the application of the transform (15-81b). Then, the system (15-91) is reduced to the following ordinary differential equation

$$\frac{d\tilde{\bar{w}}}{\partial t} + \alpha_{11}\lambda^2 \tilde{\bar{w}}(\beta, \gamma, t) = \frac{\alpha_{11}}{k_{11}} \tilde{\bar{G}}(\beta, \gamma, t) \quad \text{for} \quad t > 0 \tag{15-92a}$$

$$\tilde{\bar{w}}(\beta, \gamma, t) = 0 \quad \text{for} \quad t = 0 \tag{15-92b}$$

where

$$\lambda^2 \equiv \beta^2 + \gamma^2(\epsilon_{22} - \epsilon_{12}^2) \tag{15-93a}$$

$$\epsilon_{22} - \epsilon_{12}^2 > 0 \tag{15-93b}$$

$$\bar{\bar{G}}(\beta, \gamma, t) = \int_{x'=0}^{\infty} e^{-i\gamma\epsilon_{12}x'} \bar{g}(x', \gamma, t) \sin \beta x' \, dx' \tag{15-93c}$$

$$\bar{g}(x', \gamma, t) = \int_{y'=-\infty}^{\infty} e^{i\gamma y'} g(x', y', t) dy' \tag{15-93d}$$

The solution of equations (15-92) is

$$\tilde{\bar{w}}(\beta, \gamma, t) = e^{-\alpha_{11}\lambda^2 t} \int_0^t \frac{\alpha_{11}}{k_{11}} \bar{\bar{G}}(\beta, \gamma, t) e^{\alpha_{11}\lambda^2 t'} dt' \tag{15-94}$$

The inversion of this result by the inversion formula (15-81a) and then the application of equation (15-79) yields

$$T(x, \gamma, t) = \frac{2}{\pi} \int_{\beta=0}^{\infty} \int_{t'=0}^{t} \frac{\alpha_{11}}{k_{11}} \sin \beta x \bar{\bar{G}}(\beta, \gamma, t) e^{-\alpha_{11}\lambda^2(t-t') + i\gamma\epsilon_{12}x} dt' \, d\beta \tag{15-95}$$

This result is inverted by the inversion formula (15-77a), the explicit form of $\bar{\bar{G}}(\beta, \gamma, t)$ defined by equation (15-93) is introduced and the order of the integrations is rearranged:

$$\begin{aligned} T(x, y, t) = & \int_{x'=0}^{\infty} \int_{y'=-\infty}^{\infty} \int_{t'=0}^{t} g(x', y', t') \\ & \cdot \left\{ \frac{1}{2\pi} \int_{\gamma=-\infty}^{\infty} e^{-\alpha_{11}(\varepsilon_{22} - \epsilon_{12}^2)\gamma^2(t-t') - i\gamma[(y-y') - \epsilon_{12}(x-x')]} d\gamma \right\} \\ & \cdot \left\{ \frac{2}{\pi} \int_{\beta=0}^{\infty} e^{-\alpha_{11}\beta^2(t-t')} \sin \beta x \sin \beta x' \, d\beta \right\} dt' \, dy' \, dx' \end{aligned} \tag{15-96}$$

The integrals with respect to the variables γ and β can be evaluated by making use of the integrals (15-87a) and (15-87b); then the solution (15-96) takes the form

$$\begin{aligned} T(x, y, t) = & \frac{1}{[4\pi\alpha_{11}(\epsilon_{22} - \epsilon_{12}^2)(t-t')]^{1/2}[4\pi\alpha_{11}(t-t')]^{1/2}} \\ & \cdot \int_{x'=0}^{\infty} \int_{y'=-\infty}^{\infty} \int_{t'=0}^{t} g(x', y', t') \cdot \exp\left(- \frac{[(y-y') - \epsilon_{12}(x-x')]^2}{4\pi\alpha_{11}(\alpha_{22} - \epsilon_{12}^2)(t-t')} \right) \\ & \cdot \left[\exp\left(- \frac{(x-x')^2}{\alpha_{11}(t-t')} \right) - \exp\left(- \frac{(x+x')^2}{4\alpha_{11}(t-t')} \right) \right] dt' \, dy' \, dx' \end{aligned} \tag{15-97}$$

Example 15-8

An anisotropic cylindrical region $0 \leqslant r \leqslant b, -\infty < z < \infty$ is initially at temperature $F(r, z)$. For times $t > 0$ the boundary surface at $r = b$ is kept at zero temperature. Obtain an expression for the temperature distribution $T(r, z, t)$ in the cylinder for times $t > 0$.

Solution. Since there is no azimuthal variation of temperature, we have $\partial T/\partial\phi = 0$. Then the heat conduction equation (15-10) becomes

$$\frac{1}{r}\frac{\partial}{\partial r}\left(r\frac{\partial T}{\partial r}\right) + \epsilon_{33}\frac{\partial^2 T}{\partial z^2} + 2\epsilon_{13}\frac{\partial^2 T}{\partial r \partial z} + \epsilon_{13}\frac{1}{r}\frac{\partial T}{\partial z} = \frac{1}{\alpha_{11}}\frac{\partial T}{\partial t}$$

$$\text{in} \quad 0 \leqslant r < b, \quad -\infty < z < \infty, \quad t > 0 \tag{15-98a}$$

with the boundary and initial conditions

$$T = 0 \qquad \text{at} \qquad t = b, \quad t > 0 \tag{15-98b}$$

$$T = F(r, z) \qquad \text{for} \qquad t = 0, \quad \text{in} \quad 0 \leqslant r < b, \quad -\infty < z < \infty \tag{15-98c}$$

where we defined

$$\epsilon_{ij} = \frac{k_{ij}}{k_{11}}, \qquad k_{ij} = k_{ji}, \qquad \alpha_{11} = \frac{k_{11}}{\rho C_p} \tag{15-98d}$$

This problem is now solved by the application of integral-transform technique as now described. The integral-transform pair with respect to the z variable for $-\infty < z < \infty$ is defined as

$$\text{Inversion:} \qquad T(r, z, t) = \frac{1}{2\pi}\int_{-\infty}^{\infty} e^{-i\gamma z}\bar{T}(r, \gamma, t)\, d\gamma \tag{15-99a}$$

$$\text{Transform:} \qquad \bar{T}(r, \gamma, t) = \int_{z'=-\infty}^{\infty} e^{i\gamma z'}T(r, z', t)\, dz' \tag{15-99b}$$

The integral transform of the system (15-98) by the application of the transform (15-99b) yields

$$\frac{\partial^2 \bar{T}}{\partial r^2} + \frac{1}{r}\frac{\partial \bar{T}}{\partial r} - \gamma^2\epsilon_{33}\bar{T} - 2i\gamma\epsilon_{13}\frac{\partial \bar{T}}{\partial r} - i\gamma\frac{\epsilon_{13}}{r}\bar{T} = \frac{1}{\alpha_{11}}\frac{\partial \bar{T}}{\partial t}$$

or

$$\frac{\partial^2 \bar{T}}{\partial r^2} + \left(\frac{1}{r} - 2i\gamma\epsilon_{13}\right)\frac{\partial \bar{T}}{\partial r} - \left(\frac{i\gamma\epsilon_{13}}{r} + \gamma^2\varepsilon_{33}\right)\bar{T} = \frac{1}{\alpha_{11}}\frac{\partial \bar{T}}{\partial t}$$

$$\text{in} \quad 0 \leqslant r < b, \quad t > 0 \tag{15-100a}$$

$$\bar{T} = 0 \qquad \text{at} \qquad r = b, \quad t > 0 \tag{15-100b}$$

$$\bar{T} = \bar{F}(r, \gamma) \qquad \text{for} \qquad t = 0, \quad \text{in} \quad 0 \leqslant r < b \tag{15-100c}$$

where $\bar{T} \equiv \bar{T}(r, \gamma, t)$. A new variable $\bar{w}(r, \gamma, t)$ is defined as

$$\bar{T}(r, \gamma, t) = \bar{w}(r, \gamma, t) \cdot e^{i\gamma\epsilon_{13}r} \tag{15-101}$$

Then the system (15-100) is transformed to

$$\frac{\partial^2 \bar{w}}{\partial r^2} + \frac{1}{r}\frac{\partial \bar{w}}{\partial r} - (\epsilon_{33} - \epsilon_{13}^2)\gamma^2 \bar{w} = \frac{1}{\alpha_{11}}\frac{\partial \bar{w}}{\partial t} \qquad \text{in} \qquad 0 \leqslant r < b, \qquad t > 0 \tag{15-102a}$$

$$\bar{w} = 0 \qquad \text{at} \qquad r = b, \qquad t > 0 \tag{15-102b}$$

$$\bar{w} = e^{-i\gamma\epsilon_{13}r}\bar{F}(r, \gamma) \qquad \text{for} \qquad t = 0, \qquad \text{in} \qquad 0 \leqslant r \leqslant b \tag{15-102c}$$

To remove the partial derivative with respect to the r variable, the integral-transform pair is defined as [see equations (13-87) and Table 3-1, case 3]

$$\text{Inversion:} \qquad \bar{w}(r, \gamma, t) = \sum_{m=1}^{\infty} \frac{1}{N(\beta_m)} J_0(\beta_m r)\tilde{\bar{w}}(\beta_m, \gamma, t) \tag{15-103a}$$

$$\text{Transform:} \qquad \tilde{\bar{w}}(\beta_m, \gamma, t) = \int_{r'=0}^{b} r' J_0(\beta_m r')\bar{w}(r', \gamma, t)\, dr' \tag{15-103b}$$

where

$$\frac{1}{N(\beta_m)} = \frac{2}{b^2} \frac{1}{J_0'^2(\beta_m b)} = \frac{2}{b^2 J_1^2(\beta_m b)} \tag{15-103c}$$

and the β_m values are the roots of

$$J_0(\beta_m b) = 0 \tag{15-103d}$$

The integral transform of the system (15-102) by the application of transform (15-103b) is

$$\frac{d\tilde{\bar{w}}}{dt} + \alpha_{11}\lambda^2 \tilde{\bar{w}}(\beta_m, \gamma, t) = 0 \qquad \text{for} \qquad t > 0 \tag{15-104a}$$

$$\tilde{\bar{w}}(\beta_m, \gamma, t) = \tilde{\bar{H}}(\beta_m, \gamma) \qquad \text{for} \qquad t = 0 \tag{15-104b}$$

where

$$\lambda^2 \equiv \beta_m^2 + \gamma^2(\epsilon_{33} - \epsilon_{13}^2) \tag{15-105a}$$

$$\epsilon_{33} - \epsilon_{13}^2 > 0 \tag{15-105b}$$

$$\tilde{\bar{H}}(\beta_m, \gamma) = \int_{r'=0}^{b} r' J_0(\beta_m, r') e^{-i\gamma\epsilon_{13}r'} \bar{F}(r', \gamma)\, dr' \tag{15-105c}$$

$$\bar{F}(r', \gamma) = \int_{z'=-\infty}^{\infty} e^{i\gamma z'} F(r', z')\, dz' \tag{15-105d}$$

The solution of equation (15-104) is

$$\tilde{\bar{w}}(\beta_m, \gamma, t) = e^{-\alpha_{11}\lambda^2 t} \tilde{\bar{H}}(\beta_m, \gamma) \tag{15-106}$$

The inversion of (15-106) by the inversion formula (15-103a) gives

$$\bar{w}(r, \gamma, t) = \sum_{m=1}^{\infty} \frac{1}{N(\beta_m)} J_0(\beta_m r) e^{-\alpha_{11}\lambda^2 t} \tilde{\bar{H}}(\beta_m, \gamma) \tag{15-107}$$

This result is introduced into equation (15-101) to obtain

$$\bar{T}(r, \gamma, t) = \sum_{m=1}^{\infty} \frac{1}{N(\beta_m)} J_0(\beta_m r) \cdot e^{-\alpha_{11}\lambda^2 t + i\gamma\epsilon_{13}r} \tilde{\bar{H}}(\beta_m, \gamma) \tag{15-108}$$

The inversion of equation (15-108) by the inversion formula (15-99a) gives

$$T(r, z, t) = \frac{1}{2\pi} \sum_{m=1}^{\infty} \int_{\gamma=-\infty}^{\infty} \frac{1}{N(\beta_m)} \cdot J_0(\beta_m r) \tilde{\bar{H}}(\beta_m, \gamma) e^{-\alpha_{11}\lambda^2 t - i\gamma(z - \epsilon_{13}r)}\, d\gamma \tag{15-109}$$

where

$$\lambda^2 \equiv \beta_m^2 + \gamma^2(\epsilon_{33} - \epsilon_{13}^2)$$

The term $\tilde{\bar{H}}(\beta_m, \gamma)$ defined by equations (15-105) is introduced into equation (15-109) and the order of integrations is rearranged:

$$T(r, z, t) = \sum_{m=1}^{\infty} e^{-\alpha_{11}\beta_m^2 t} \frac{1}{N(\beta_m)} J_0(\beta_m, r) \int_{z'=-\infty}^{\infty} \int_{r'=0}^{b} r' J_0(\beta_m r') F(r', z')\, dr'\, dz' \cdot \left\{ \frac{1}{2\pi} \int_{\gamma=-\infty}^{\infty} e^{-\gamma^2\alpha_{11}(\epsilon_{33} - \epsilon_{13})t - i\gamma[(z - z') + \epsilon_{13}(r' - r)]}\, d\gamma \right\} \tag{15-110}$$

The integral with respect to γ can be evaluated according to equation (15-87a). Then, the solution becomes

$$T(r,z,t)=\frac{1}{\sqrt{4\pi\alpha_{11}(\epsilon_{33}-\epsilon_{13}^2)t}}\sum_{m=1}^{\infty}e^{-\alpha_{11}\beta_m^2 t}\cdot\frac{1}{N(\beta_m)}J_0(\beta_m r)$$
$$\cdot\int_{z'=-\infty}^{\infty}\int_{r'=0}^{b}r'J_0(\beta_m r')\cdot F(r',z')$$
$$\cdot\exp\left\{-\frac{[(z-z')+\epsilon_{13}(r'-r)]^2}{4\alpha_{11}(\epsilon_{33}-\epsilon_{13}^2)t}\right\}dr'\,dz' \qquad (15\text{-}111)$$

where $N(\beta_m)$ is given by equation (15-103c) and the β_m values are the roots of equation (15-103d).

REFERENCES

1. W. A. Wooster, *A Textbook in Crystal Physics*, Cambridge University Press, London, 1938.
2. J. F. Nye, *Physical Properties of Crystals*, Clarendon Press, London, 1957.
3. H. S. Carslaw and J. C. Jeager, *Conduction of Heat in Solids*, Clarendon Press, London, 1959.
4. M. N. Özisik, *Boundary Value Problems of Heat Conduction*, International Textbook, Scranton, Pa., 1968; Dover Publications, New York, 1989.
5. W. H. Giedt and D. R. Hornbaker, *ARS J.* **32**, 1902–1909, 1962.
6. K. J. Touryan, *AIAA J.* **2**, 124–126, 1964.
7. B. Venkatraman, S. A. Patel, and F. V. Pohle, *J. Aerospace Sci.* **29**, 628–629, 1962.
8. B. T. Chao, *Appl. Sci. Res.* **A12**, 134–138, 1963.
9. H. F. Cooper, Joulean Heating of an Infinite Rectangular Rod with Orthotropic Thermal Properties, ASME Paper No. 66-WA/HT-14, 1966.
10. H. F. Cooper, Transient and Steady State Temperature Distribution in Foil Wound Solenoids and Other Electric Apparatus of Rectangular Cross-Section, 1965 *IEEE International Convention Record*, Part 10, March 1965, pp. 67–75.
11. N. Vutz and S. W. Angrist, Thermal Contact Resistance of Anisotropic Materials, ASME Paper 69-HT-47, 1969.
12. R. C. Pfahl, *Int. J. Heat Mass Transfer* **18**, 191–204, 1975.
13. B. F. Blackwell, *An Introduction to Heat Conduction in an Anisotropic Medium*, SC-RR-69-542, Oct. 1969, Sandia Lab., Albuquerque, N.M.
14. Y. P. Chang, C. S. Kang, and D. J. Chen, *Int. J. Heat Mass Transfer* **16**, 1905–1918, 1973.
15. J. Padovan, *J. Heat Transfer* **96c**, 428–431, 1974.
16. J. Padovan, *AIAA J.* **10**, 60–64, 1972.
17. K. Katayama, Transient Heat Conduction in Anisotropic Solids, *Proceedings of the 5th International Heat Transfer Conference*, Cu 1.4, pp. 137–141, Tokyo, Sept. 3974.

18. M. H. Cobble, *Int. J. Heat Mass Transfer* **17**, 379–380, 1974.
19. G. P. Mulholland and B. P. Gupta, *J. Heat Transfer* **99c**, 135–137, 1977.
20. Y. P. Chang and C. H. Tsou, *J. Heat Transfer* **99c**, 132–134, 1977.
21. Y. P. Chang and C. H. Tsou, *J. Heat Transfer* **99c**, 41–47, 1977.
22. Y. P. Chang, *Int. J. Heat Mass Transfer* **20**, 1019–1025, 1977.
23. M. N. Özişik and S. M. Shouman, Transient Heat Conduction in an Anisotropic Medium in Cylindrical Coordinates, *J. Franklin Institute* **309**, 457–472, 1980.
24. J. Padovan, *J. Heat Transfer* **96c**, 313–318, 1974.
25. G. P. Mulholland, Diffusion Through Laminated Orthotropic Cylinders, *Proceedings of the 5th International Heat Transfer Conference*, Cu 4.3, pp. 250–254, Tokyo, 1974.
26. M. H. Sadd and I. Miskioglu, *J. Heat Transfer* 100, 553–555, 1978.
27. K. C. Poon, R. C. H. Tsou and Y. P. Chang, *J. Heat Transfer* **101**, 340–345, 1979.
28. M. D. Mikhailov and M. N. Özişik, *Lett. Heat Mass Transfer* **8**, 329–335, 1981.
29. S. C. Huang and Y. P. Chang, *J. Heat Transfer* **106**, 646–648, 1984.
30. W. S. Wang and T. W. Chou, *J. Composite Materials* **19**, 424–442, 1985.
31. L. Onsagar, *Phys Rev.* **37**, 405–426, 1931; **38**, 2265–2279, 1931.
32. H. B. G. Casimir, *Rev. Mod. Phys.* **17**, 343–350, 1945.
33. I. Prigogine, *Thermodynamics of Irreversible Processes*, Wiley-Interscience, New York 1961.
34. L. P. Eisenhart, *Coordinate Geometry*, Dover Publications, New York, 1962.
35. A. R. Amir-Moe'z and A. L. Fass, *Elements of Linear Spaces*, Pergamon Press, New York, 1962.
36. S. H. Maron and C. F. Prutton, *Principles of Physical Chemistry*, Macmillan, New York, 1958.

PROBLEMS

15-1 Write the expressions for the three components of the heat flux, q_i, $i = 1, 2, 3$, for an anisotropic medium in the following orthogonal coordinate systems: (1) prolate spheroid; (2) oblate spheroid.

15-2 Write the time dependent heat conduction equation for an anisotropic medium with constant conductivity coefficients for the following cases:

1. In the cylindrical coordinate system when temperature is a function of r, ϕ variables.
2. In the spherical coordinate system when temperature is a function of r, ϕ variables.

15-3 Write the boundary conditions of the third kind for an anisotropic solid at the following boundary surfaces.

1. At the boundary surfaces $z = 0$, $z = L$, and $r = b$ of a solid cylinder of radius b, height L.
2. At the surface $r = b$ of a solid sphere.

15-4 Write the thermal resistivity coefficients r_{11}, r_{13}, and r_{23} in terms of the thermal conductivity coefficients k_{ij}.

15-5 Consider two-dimensional steady state heat conduction in an orthotropic rectangular solid in the region $0 \leqslant x \leqslant a, 0 \leqslant y \leqslant b$ with thermal conductivities k_1 and k_2 in the x- and y-directions, respectively. The boundaries at $x = 0, x = a$ and $y = b$ are kept at zero temperature, while the boundary at $y = 0$ is maintained at a temperature $T = f(x)$. Develop an expression for the steady state temperature distribution $T(x, y)$ in the solid.

15-6 Consider steady state heat conduction in an orthotropic solid cylinder $0 \leqslant r \leqslant b$, $0 \leqslant z \leqslant L$ in which heat is generated at a uniform rate of g_0 W/m^3 while the boundaries are kept at zero temperature. The thermal conductivity coefficients in the r and z directions are k_1 and k_2, respectively. Obtain an expression for the steady state temperature distribution $T(r, z)$ in the cylinder.

15-7 Consider an orthotropic region $0 \leqslant x < \infty, 0 \leqslant y < \infty$, which is initially at temperature $F(x, y)$ and for times $t > 0$ the boundaries at $x = 0$ and $y = 0$ are kept at zero temperature. The thermal conductivity coefficients for the x and y directions are k_1 and k_2, respectively. Obtain an expression for the temperature distribution $T(x, y, t)$ in the medium for times $t > 0$.

15-8 An orthotropic solid cylinder $0 \leqslant r \leqslant b, 0 \leqslant z \leqslant L$ is initially at temperature $F(r, z)$. For times $t > 0$ the boundaries are kept at zero temperature. The thermal conductivity coefficients for the r and z directions and k_1 and k_2, respectively. Obtain an expression for the temperature distribution $T(r, z, t)$ in the solid for times $t > 0$.

15-9 Consider two-dimensional steady state heat conduction in an orthotropic solid cylinder of radius $r = b$ and height $z = L$ with thermal conductivities k_1 and k_2 in the r- and z-directions, respectively. The boundary surfaces at $r = b$ and $z = L$ are kept at zero temperatures, while the boundary surface at $z = 0$ is kept at temperature $T = f(r)$. Develop an expression for the steady-state temperature $T(r, z)$.

15-10 Consider time-dependent, two-dimensional heat conduction problem for an anisotropic medium $0 \leqslant x < \infty, \; -\infty < y < \infty$ which is initially at temperature $F(x, y)$ and for times $t > 0$ the boundary surface at $x = 0$ is kept insulated. Obtain an expression for the temperature distribution $T(x, y, t)$ in the medium for times $t > 0$.

15-11 Consider time dependent, two-dimensional heat conduction problem for an anisotropic region $0 \leqslant x < \infty, \; -\infty < y < \infty$ that is initially at zero temperature. For times $t > 0$, heat is generated in the medium at a rate of $g(x, y, t)$ W/m^3, while the boundary at $x = 0$ is kept insulated. Obtain an expression for the temperature distribution $T(x, y, t)$ in the medium for times $t > 0$.

15-12 Consider time dependent, two dimensional heat conduction in an anisotropic hollow cylinder $a \leqslant r \leqslant b$, $-\infty < z < \infty$, which is initially at temperature $F(r, z)$. For times $t > 0$, the boundaries at $r = a$ and $r = b$ are kept at zero temperature. Obtain an expression for the temperature distribution $T(r, z, t)$ in the medium for times $t > 0$.

15-13 Transform the heat conduction equation

$$k_{11}\frac{\partial^2 T}{\partial x^2} + k_{22}\frac{\partial^2 T}{\partial y^2} + g = \pi C_p \frac{\partial T}{\partial t}$$

into a one similar to that for the isotropic medium.

NOTES

1. The closed-form expression given by equation (15-74) is determined as now described. We consider the following heat conduction problem:

$$\frac{d^2 T}{dx^2} + \frac{g_0}{k_1} = 0 \quad \text{in} \quad 0 < x < a \tag{1a}$$

$$\frac{dT}{dx} = 0 \quad \text{at} \quad x = 0 \tag{1b}$$

$$\frac{dT}{dx} + H_1 T = 0 \quad \text{at} \quad x = a \tag{1c}$$

This problem is solved both by direct integration and using the integral transform technique as given below.

a. When it is solved by direct integration we obtain

$$T = \frac{g_0 a}{k_1 H_1} + \frac{g_0}{2k_1}(a^2 - x^2) \tag{2}$$

b. To solve the system, equation (1), by the integral transform technique, we take its transform by the application of transform (15-70a) and obtain

$$\bar{T} = \frac{1}{k_1 \beta^2}\bar{g}_0 \tag{3a}$$

where

$$\bar{g}_0 = \int_0^a g_0 \cos \beta_m x \, dx = \frac{\sin \beta_m a}{\beta_m} g_0 \tag{3b}$$

Introducing the transform (3) into the inversion formula (15-60b), we obtain the solution as

$$T = \frac{g_0}{k_1} \sum_{n=0}^{\infty} \frac{1}{N(\beta_m)} \cdot \frac{\cos \beta_m x \sin \beta_m a}{\beta_m^3} \tag{4}$$

Since equations (2) and (4) are the solution of the same problem, by equating them we obtain

$$\sum_{m=1}^{\infty} \frac{1}{N(\beta_m)} \frac{\cos \beta_m x \sin \beta_m a}{\beta_m^3} = \frac{a}{H_1} + \frac{1}{2}(a^2 - x^2) \tag{5}$$

which is the result given by equation (15-64).

2. The integral transform of $\partial^2 T/\partial y^2$ by the application of the transform (15-77b) is determined as

$$\int_{-\infty}^{\infty} e^{i\gamma y} \frac{\partial^2 T}{\partial y^2} dy = \left[\frac{\partial T}{\partial y} e^{i\gamma y} - i\gamma T e^{i\gamma y} \right]_{y=-\infty}^{\infty} - \gamma^2 \int_{-\infty}^{\infty} e^{i\gamma y} T \, dy$$

$$= -\gamma^2 \int_{-\infty}^{\infty} e^{i\gamma y} T dy = -\gamma^2 \bar{T} \tag{1}$$

To obtain this result we integrated by parts twice, assumed that T and $\partial T/\partial y$ both vanish as $y \to \pm\infty$, and utilized the definition of the transform (15-77b). The integral transform of $\partial^2 T/\partial x \partial y$ is determined as

$$\int_{-\infty}^{\infty} e^{i\gamma y} \frac{\partial^2 T}{\partial x \partial y} dy = \left[\frac{\partial T}{\partial x} e^{i\gamma y} \right]_{-\infty}^{\infty} - i\gamma \frac{\partial}{\partial x} \int_{-\infty}^{\infty} e^{i\gamma y} T \, dy$$

$$= -i\gamma \frac{\partial}{\partial x} \int_{-\infty}^{\infty} e^{i\gamma y} T dy = -i\gamma \frac{\partial \bar{T}}{\partial x} \tag{2}$$

where we assumed that $\partial T/\partial x$ vanish at $y \to \pm\infty$.

APPENDIXES

APPENDIX I

PHYSICAL PROPERTIES

TABLE I-1 Physical Properties of Metals

Metal	Melting Point °C	ρ, $\frac{kg}{m^3}$	C_p, $\frac{kJ}{kg \cdot °C}$	k, $\frac{W}{m \cdot °C}$	α, $\frac{m^2}{s} \times 10^5$
		Properties at 20°C			
Aluminum					
Pure	660	2,707	0.896	204	8.418
Al-Cu (Duralumin), 94–96% Al, 3–5% Cu, trace Mg		2,787	0.883	164	6.676
Beryllium	1277	1,850	1.825	200	5.92
Bismuth	272	9,780	0.122	7.86	0.66
Cadmium	321	8,650	0.231	96.8	4.84
Copper					
Pure	1085	8,954	0.3831	386	11.234
Aluminum bronze 95% Cu, 5% Al		8,666	0.410	83	2.330
Constantan		8,922	0.410	22.7	0.612
60% Cu, 40% Ni					
Iron					
Pure	1537	7,897	0.452	73	2.034
Wrought iron, 0.5% C		7,849	0.46	59	1.626
Carbon steel					
C ≈ 0.5%		7,833	0.465	54	1.474
1.0%		7,801	0.473	43	1.172

TABLE I-1 (*Continued*)

Metal	Melting Point °C	Properties at 20°C			
		ρ, $\frac{kg}{m^3}$	C_p, $\frac{kJ}{kg \cdot °C}$	k, $\frac{W}{m \cdot °C}$	α, $\frac{m^2}{s} \times 10^5$
Chrome steel					
Cr = 0%		7,897	0.452	73	2.026
1%		7,865	0.46	61	1.665
5%		7,833	0.46	40	1.110
Nickel steel					
Ni ≃ 0%		7,897	0.452	73	2.026
20%		7,933	0.46	19	0.526
Lead	328	11,373	0.130	35	2.343
Magnesium					
Pure	650	1,746	1.013	171	9.708
Mg-Al (electrolytic)					
6 – 8% Al, 1–2% Zn		1,810	1.00	66	3.605
Molybdenum	2,621	10,220	0.251	123	4.790
Nickel					
Pure (99.9%)	1,455	8,906	0.4459	90	2.266
Ni-Cr					
90% Ni, 10% Cr		8,666	0.444	17	0.444
80% Ni, 20% Cr		8,314	0.444	12.6	0.343
Silver					
Purest	962	10,524	0.2340	419	17.004
Pure (99.9%)		10,525	0.2340	407	16.563
Tin, pure	232	7,304	0.2265	64	3.884
Tungsten	3,387	19,350	0.1344	163	6.271
Uranium	1,133	19,070	0.116	27.6	1.25
Zinc, pure	420	7,144	0.3843	112.2	4.106

TABLE I-2 Physical Properties of Nonmetals

Material	T, °C	k, $\frac{W}{m \cdot °C}$	ρ, $\frac{kg}{m^3}$	C_p, $\frac{kJ}{kg \cdot °C}$	α, $\frac{m^2}{s} \times 10^7$
Asphalt	20–55	0.74–0.76			
Brick					
Building brick,	20	0.69	1600	0.84	5.2
common face		1.32	2000		
Carborundum	600	18.5			
brick	1400	11.1			
Chrome brick	200	2.32	3000	0.84	9.2
	900	1.99			7.9
Diatomaceous					
earth, molded	200	0.24			
and fired	870	0.31			
Fireclay brick,	500	1.04	2000	0.96	5.4
burned 1330°C	800	1.07			
Clay	30	1.3	1460	0.88	
Cement, portland	23	0.29	1500		
Coal. anthracite	30	0.26	1200–1500	1.26	
Concrete, cinder	23	0.76			
Stone 1-2-4 mix	20	1.37	1900–2300	0.88	8.2–6.8
Cotton	20	0.06	80	1.30	
Glass, window	20	0.78 (avg)	2700	0.84	3.4
Pyrex	30	1.4	2225	0.835	
Paper	30	0.011	930	1.340	
Paraffin	30	0.020	900	2.890	
Plaster, gypsum	20	0.48	1440	0.84	4.0
Rubber, vulcanized					
Soft	30	0.012	1100	2.010	
Hard	30	0.013	1190	—	
Sand	30	0.027	1515	0.800	
Stone					
Granite		1.73–3.98	2640	0.82	8–18
Limestone	100–300	1.26–1.33	2500	0.90	5.6–5.9
Marble		2.07–2.94	2500–2700	0.80	10–13.6
Sandstone	40	1.83	2160–2300	0.71	11.2–11.9
Teflon	30	0.35	2200	—	
Tissue, human skin	30	0.37	—	—	
Wood (across grain)					
Balsa	30	0.055	140		
Cypress	30	0.097	460		
Fir	23	0.11	420	2.72	0.96
Maple or oak	30	0.166	540	2.4	1.28
Yellow pine	23	0.147	640	2.8	0.82
White pine	30	0.112	430		

TABLE I-3 Physical Properties of Insulating Materials

Material	T, °C	k, $\frac{W}{m \cdot °C}$	ρ, $\frac{kg}{m^3}$	C_p, $\frac{kJ}{kg \cdot °C}$	α, $\frac{m^2}{s} \times 10^7$
Asbestos					
Loosely packed	0	0.154	470–570	0.816	3.3–4
	100	0.161			
Asbestos-cement boards	20	0.74			
Sheets	51	0.166			
Felt, 40 laminations in	38	0.057			
Balsam wool	32	0.04	35		
Board and slab					
Cellular glass	30	0.058	145	1.000	
Glass fiber, organic bonded	30	0.036	105	0.795	
Polystyrene, expanded extruded (R-12)	30	0.027	55	1.210	
Mineral fiberboard; roofing material	30	0.049	265		
Cardboard, corrugated	—	0.064			
Celotex	32	0.048			
Corkboard	30	0.043	160		
Diatomaceous earth (Sil-o-cel)	0	0.061	320		
Felt, hair	30	0.036	130–200		
Wool	30	0.052	330		
Fiber, insulating board	20	0.048	240		
Glass wool	23	0.038	24	0.7	22.6
Loose fill					
Cork, granulated	30	0.045	160		
Glass fiber, poured or blown	30	0.043	16	0.835	
Vermiculite, flakes	30	0.068	80	0.835	
Magnesia, 85%	38	0.067	270		
	150	0.074			
	204	0.080			
Rock wool, 10 lb/ft^3	32	0.040	160		
Loosely packed	150	0.067	64		
	260	0.087			
Sawdust	23	0.059			
Silica aerogel	32	0.024	140		
Wood shavings	23	0.059			

APPENDIX II

ROOTS OF TRANSCENDENTAL EQUATIONS

First Six Roots β_n of $\beta \tan \beta = c$

c	β_1	β_2	β_3	β_4	β_5	β_6
0	0	3.1416	6.2832	9.4248	12.5664	15.7080
0.001	0.0316	3.1419	6.2833	9.4249	12.5665	15.7080
0.002	0.0447	3.1422	6.2835	9.4250	12.5665	15.7081
0.004	0.0632	3.1429	6.2838	9.4252	12.5667	15.7082
0.006	0.0774	3.1435	6.2841	9.4254	12.5668	15.7083
0.008	0.0893	3.1441	6.2845	9.4256	12.5670	15.7085
0.01	0.0998	3.1448	6.2848	9.4258	12.5672	15.7086
0.02	0.1410	3.1479	6.2864	9.4269	12.5680	15.7092
0.04	0.1987	3.1543	6.2895	9.4290	12.5696	15.7105
0.06	0.2425	3.1606	6.2927	9.4311	12.5711	15.7118
0.08	0.2791	3.1668	6.2959	9.4333	12.5727	15.7131
0.1	0.3111	3.1731	6.2991	9.4354	12.5743	15.7143
0.2	0.4328	3.2039	6.3148	9.4459	12.5823	15.7207
0.3	0.5218	3.2341	6.3305	9.4565	12.5902	15.7270
0.4	0.5932	3.2636	6.3461	9.4670	12.5981	15.7334
0.5	0.6533	3.2923	6.3616	9.4775	12.6060	15.7397
0.6	0.7051	3.3204	6.3770	9.4879	12.6139	15.7460
0.7	0.7506	3.3477	6.3923	9.4983	12.6218	15.7524
0.8	0.7910	3.3744	6.4074	9.5087	12.6296	15.7587
0.9	0.8274	3.4003	6.4224	9.5190	12.6375	15.7650
1.0	0.8603	3.4256	6.4373	9.5293	12.6453	15.7713
1.5	0.9882	3.5422	6.5097	9.5801	12.6841	15.8026
2.0	1.0769	3.6436	6.5783	9.6296	12.7223	15.8336
3.0	1.1925	3.8088	6.7040	9.7240	12.7966	15.8945

First Six Roots β_n of $\beta \tan \beta = c$ *(Continued)*

c	β_1	β_2	β_3	β_4	β_5	β_6
4.0	1.2646	3.9352	6.8140	9.8119	12.8678	15.9536
5.0	1.3138	4.0336	6.9096	9.8928	12.9352	16.0107
6.0	1.3496	4.1116	6.9924	9.9667	12.9988	16.0654
7.0	1.3766	4.1746	7.0640	10.0339	13.0584	16.1177
8.0	1.3978	4.2264	7.1263	10.0949	13.1141	16.1675
9.0	1.4149	4.2694	7.1806	10.1502	13.1660	16.2147
10.0	1.4289	4.3058	7.2281	10.2003	13.2142	16.2594
15.0	1.4729	4.4255	7.3959	10.3898	13.4078	16.4474
20.0	1.4961	4.4915	7.4954	10.5117	13.5420	16.5864
30.0	1.5202	4.5615	7.6057	10.6543	13.7085	16.7691
40.0	1.5325	4.5979	7.6647	10.7334	13.8048	16.8794
50.0	1.5400	4.6202	7.7012	10.7832	13.8666	16.9519
60.0	1.5451	4.6353	7.7259	10.8172	13.9094	17.0026
80.0	1.5514	4.6543	7.7573	10.8606	13.9644	17.0686
100.0	1.5552	4.6658	7.7764	10.8871	13.9981	17.1093
∞	1.5708	4.7124	7.8540	10.9956	14.1372	17.2788

Roots are all real if $c > 0$.

First Six Roots β_n of $\beta \cot \beta = -c$

c	β_1	β_2	β_3	β_4	β_5	β_6
−1.0	0	4.4934	7.7253	10.9041	14.0662	17.2208
−0.995	0.1224	4.4945	7.7259	10.9046	14.0666	17.2210
−0.99	0.1730	4.4956	7.7265	10.9050	14.0669	17.2213
−0.98	0.2445	4.4979	7.7278	10.9060	14.0676	17.2219
−0.97	0.2991	4.5001	7.7291	10.9069	14.0683	17.2225
−0.96	0.3450	4.5023	7.7304	10.9078	14.0690	17.2231
−0.95	0.3854	4.5045	7.7317	10.9087	14.0697	17.2237
−0.94	0.4217	4.5068	7.7330	10.9096	14.0705	17.2242
−0.93	0.4551	4.5090	7.7343	10.9105	14.0712	17.2248
−0.92	0.4860	4.5112	7.7356	10.9115	14.0719	17.2254
−0.91	0.5150	4.5134	7.7369	10.9124	14.0726	17.2260
−0.90	0.5423	4.5157	7.7382	10.9133	14.0733	17.2266
−0.85	0.6609	4.5268	7.7447	10.9179	14.0769	17.2295
−0.8	0.7593	4.5379	7.7511	10.9225	14.0804	17.2324
−0.7	0.9208	4.5601	7.7641	10.9316	14.0875	17.2382
−0.6	1.0528	4.5822	7.7770	10.9408	14.0946	17.2440
−0.5	1.1656	4.6042	7.7899	10.9499	14.1017	17.2498
−0.4	1.2644	4.6261	7.8028	10.9591	14.1088	17.2556
−0.3	1.3525	4.6479	7.8156	10.9682	14.1159	17.2614
−0.2	1.4320	4.6696	7.8284	10.9774	14.1230	17.2672
−0.1	1.5044	4.6911	7.8412	10.9865	14.1301	17.2730
0	1.5708	4.7124	7.8540	10.9956	14.1372	17.2788

First Six Roots β_n of $\beta \cot \beta = -c$ *(Continued)*

c	β_1	β_2	β_3	β_4	β_5	β_6
0.1	1.6320	4.7335	7.8667	11.0047	14.1443	17.2845
0.2	1.6887	4.7544	7.8794	11.0137	14.1513	17.2903
0.3	1.7414	4.7751	7.8920	11.0228	14.1584	17.2961
0.4	1.7906	4.7956	7.9046	11.0318	14.1654	17.3019
0.5	1.8366	4.8158	7.9171	11.0409	14.1724	17.3076
0.6	1.8798	4.8358	7.9295	11.0498	14.1795	17.3134
0.7	1.9203	4.8556	7.9419	11.0588	14.1865	17.3192
0.8	1.9586	4.8751	7.9542	11.0677	14.1935	17.3249
0.9	1.9947	4.8943	7.9665	11.0767	14.2005	17.3306
1.0	2.0288	4.9132	7.9787	11.0856	14.2075	17.3364
1.5	2.1746	5.0037	8.0385	11.1296	14.2421	17.3649
2.0	2.2889	5.0870	8.0962	11.1727	14.2764	17.3932
3.0	2.4557	5.2329	8.2045	11.2560	14.3434	17.4490
4.0	2.5704	5.3540	8.3029	11.3349	14.4080	17.5034
5.0	2.6537	5.4544	8.3914	11.4086	14.4699	17.5562
6.0	2.7165	5.5378	8.4703	11.4773	14.5288	17.6072
7.0	2.7654	5.6078	8.5406	11.5408	14.5847	17.6562
8.0	2.8044	5.6669	8.6031	11.5994	14.6374	17.7032
9.0	2.8363	5.7172	8.6587	11.6532	14.6870	17.7481
10.0	2.8628	5.7606	8.7083	11.7027	14.7335	17.7908
15.0	2.9476	5.9080	8.8898	11.8959	14.9251	17.9742
20.0	2.9930	5.9921	9.0019	12.0250	15.0625	18.1136
30.0	3.0406	6.0831	9.1294	12.1807	15.2380	18.3018
40.0	3.0651	6.1311	9.1987	12.2688	15.3417	18.4180
50.0	3.0801	6.1606	9.2420	12.3247	15.4090	18.4953
60.0	3.0901	6.1805	9.2715	12.3632	15.4559	18.5497
80.0	3.1028	6.2058	9.3089	12.4124	15.5164	18.6209
100.0	3.1105	6.2211	9.3317	12.4426	15.5537	18.6650
∞	3.1416	6.2832	9.4248	12.5664	15.7080	18.8496

Roots are all real if $c > -1$.

APPENDIX III

ERROR FUNCTIONS

Numerical Values of Error Function $\mathrm{erf}(z) = \dfrac{2}{\sqrt{\pi}} \displaystyle\int_0^z e^{-\xi^2}\, d\xi$

z	erf z	z	erf z	z	erf z	z	erf z	z	erf z
0.00	0.00000	0.50	0.52049	1.00	0.84270	1.50	0.96610	2.00	0.99532
0.01	0.01128	0.51	0.52924	1.01	0.84681	1.51	0.96727	2.20	0.99814
0.02	0.02256	0.52	0.53789	1.02	0.85083	1.52	0.96841	2.40	0.99931
0.03	0.03384	0.53	0.54646	1.03	0.85478	1.53	0.96951	2.60	0.99976
0.04	0.04511	0.54	0.55493	1.04	0.85864	1.54	0.97058	2.80	0.99992
0.05	0.05637	0.55	0.56332	1.05	0.86243	1.55	0.97162	3.00	0.99998
0.06	0.06762	0.56	0.57161	1.06	0.86614	1.56	0.97262		
0.07	0.07885	0.57	0.57981	1.07	0.86977	1.57	0.97360		
0.08	0.09007	0.58	0.58792	1.08	0.87332	1.58	0.97454		
0.09	0.10128	0.59	0.59593	1.09	0.87680	1.59	0.97546		
0.10	0.11246	0.60	0.60385	1.10	0.88020	1.60	0.97634		
0.11	0.12362	0.61	0.61168	1.11	0.88353	1.61	0.97720		
0.12	0.13475	0.62	0.61941	1.12	0.88678	1.62	0.97803		
0.13	0.14586	0.63	0.62704	1.13	0.88997	1.63	0.97884		
0.14	0.15694	0.64	0.63458	1.14	0.89308	1.64	0.97962		
0.15	0.16799	0.65	0.64202	1.15	0.89612	1.65	0.98037		
0.16	0.17901	0.66	0.64937	1.16	0.89909	1.66	0.98110		
0.17	0.18999	0.67	0.65662	1.17	0.90200	1.67	0.98181		
0.18	0.20093	0.68	0.66378	1.18	0.90483	1.68	0.98249		
0.19	0.21183	0.69	0.67084	1.19	0.90760	1.69	0.98315		
0.20	0.22270	0.70	0.67780	1.20	0.91031	1.70	0.98379		

Numerical Values of Error Function $\mathbf{erf}(z) = \dfrac{2}{\sqrt{\pi}} \int_0^z e^{-\xi^2}\, d\xi$

z	erf z	z	erf z	z	erf z	z	erf z
0.21	0.23352	0.71	0.68466	1.21	0.91295	1.71	0.98440
0.22	0.24429	0.72	0.69143	1.22	0.91553	1.72	0.98500
0.23	0.25502	0.73	0.69810	1.23	0.91805	1.73	0.98557
0.24	0.26570	0.74	0.70467	1.24	0.92050	1.74	0.98613
0.25	0.27632	0.75	0.71115	1.25	0.92290	1.75	0.98667
0.26	0.28689	0.76	0.71753	1.26	0.92523	1.76	0.98719
0.27	0.29741	0.77	0.72382	1.27	0.92751	1.77	0.98769
0.28	0.30788	0.78	0.73001	1.28	0.92973	1.78	0.98817
0.29	0.31828	0.79	0.73610	1.29	0.93189	1.79	0.98864
0.30	0.32862	0.80	0.74210	1.30	0.93400	1.80	0.98909
0.31	0.33890	0.81	0.74800	1.31	0.93606	1.81	0.98952
0.32	0.34912	0.82	0.75381	1.32	0.93806	1.82	0.98994
0.33	0.35927	0.83	0.75952	1.33	0.94001	1.83	0.99034
0.34	0.36936	0.84	0.76514	1.34	0.94191	1.84	0.99073
0.35	0.37938	0.85	0.77066	1.35	0.94376	1.85	0.99111
0.36	0.38932	0.86	0.77610	1.36	0.94556	1.86	0.99147
0.37	0.39920	0.87	0.78143	1.37	0.94731	1.87	0.99182
0.38	0.40900	0.88	0.78668	1.38	0.94901	1.88	0.99215
0.39	0.41873	0.89	0.79184	1.39	0.95067	1.89	0.99247
0.40	0.42839	0.90	0.79690	1.40	0.95228	1.90	0.99279
0.41	0.43796	0.91	0.80188	1.41	0.95385	1.91	0.99308
0.42	0.44746	0.92	0.80676	1.42	0.95537	1.92	0.99337
0.43	0.45688	0.93	0.81156	1.43	0.95685	1.93	0.99365
0.44	0.46622	0.94	0.81627	1.44	0.95829	1.94	0.99392
0.45	0.47548	0.95	0.82089	1.45	0.95969	1.95	0.99417
0.46	0.48465	0.96	0.82542	1.46	0.96105	1.96	0.99442
0.47	0.49374	0.97	0.82987	1.47	0.96237	1.97	0.99466
0.48	0.50274	0.94	0.83423	1.48	0.96365	1.98	0.99489
0.49	0.51166	0.99	0.83850	1.49	0.96489	1.99	0.99511

The error function of argument x is defined as

$$\operatorname{erf}(x) = \frac{2}{\sqrt{\pi}} \int_0^x e^{-\eta^2}\, d\eta \tag{1}$$

and we have

$$\operatorname{erf}(\infty) = 1 \qquad \text{and} \qquad \operatorname{erf}(-x) = -\operatorname{erf}(x) \tag{2}$$

The complimentary error function, erfc(x), is defined as

$$\operatorname{erfc}(x) = 1 - \operatorname{erf}(x) = \frac{2}{\sqrt{\pi}} \int_x^{\infty} e^{-\eta^2} d\eta \tag{3}$$

The derivatives of error function are given as

$$\frac{d}{dx}\operatorname{erf}(x) = \frac{2}{\sqrt{\pi}} e^{-x^2}, \qquad \frac{d^2}{dx^2}\operatorname{erf}(x) = -\frac{4}{\sqrt{\pi}} x e^{-x^2}, \text{etc.} \tag{4}$$

The repeated integrals of error function are defined as

$$i^n \operatorname{erfc}(x) = \int_x^{\infty} i^{n-1} \operatorname{erfc} \eta \, d\eta, \qquad n = 0, 1, 2 \ldots \tag{5a}$$

with

$$i^{-1} \operatorname{erfc}(x) = \frac{2}{\sqrt{\pi}} e^{-x^2}, \qquad i^0 \operatorname{erfc} x = \operatorname{erfc} x \tag{5b}$$

Then we have

$$i \operatorname{erfc}(x) = \frac{1}{\sqrt{\pi}} e^{-x^2} - x \operatorname{erfc} x \tag{6}$$

$$i^2 \operatorname{erfc}(x) = \frac{1}{4}\left[(1 + 2x^2) \operatorname{erfc} x - \frac{2}{\sqrt{\pi}} x e^{-x^2} \right] \tag{7}$$

Series expansion for error function is given as

$$\operatorname{erf}(x) = \frac{2}{\sqrt{\pi}} \sum_{n=0}^{\infty} (-1)^n \frac{x^{2n+1}}{n!(2n+1)} \tag{8}$$

For large values of x, its asymptotic expansion is

$$\operatorname{erfc}(x) = 1 - \operatorname{erf}(x) \cong \frac{e^{-x^2}}{\sqrt{\pi} x} \left[1 + \sum_{n=1}^{\infty} (-1)^n \frac{1.3 \cdots (2n-1)}{(2x^2)^n} \right] \tag{9}$$

The error function, its derivatives, and its integrals have been tabulated [1, 2].

REFERENCES

1. M. Abramowitz and I. A. Stegun, *Handbook of Mathematical Functions*, National Bureau of Standards, Applied Mathematic Series 55, U.S. Government Printing Office, Washington, D.C., 1964.
2. E. Jahnke and F. Emde, *Tables of Functions*, 2nd ed., Dover Publications, New York, 1945.

APPENDIX IV

BESSEL FUNCTIONS

The differential equation

$$\frac{d^2R}{dz^2}+\frac{1}{z}\frac{dR}{dz}+\left(1-\frac{v^2}{z^2}\right)R=0 \tag{1}$$

is called *Bessels's differential equation of order v*. Two linearly independent solutions of this equation for all values of v are $J_v(z)$, the Bessel function of the first kind of order v and $Y_v(z)$, the Bessel function of the second kind of order v. Thus, the general solution of equation (1) is written as [1, 2, 3]

$$R(z)=c_1 J_v(z)+c_2 Y_v(z) \tag{2}$$

The Bessel function $J_v(z)$ in series form is defined as

$$J_v(z)=(\tfrac{1}{2}z)^v\sum_{k=0}^{\infty}(-1)^k\frac{(\frac{1}{2}z)^{2k}}{k!\Gamma(v+k+1)} \tag{3}$$

where $\Gamma(x)$ is the gamma function.

The differential equation

$$\frac{d^2R}{dz^2}+\frac{1}{z}\frac{dR}{dz}-\left(1+\frac{v^2}{z^2}\right)R=0 \tag{4}$$

is called *Bessel's modified differential equation of order v*. Two linearly independent solutions of this equation for all values of v are $I_v(z)$, the modified Bessel function of the first kind of order v and $K_v(z)$, the modified Bessel function of the second kind of order ν. Thus, the general solution of equation (4) is written as

$$R(z) = c_1 I_\nu(z) + c_2 K_\nu(z) \tag{5}$$

$I_\nu(z)$ and $K_\nu(z)$ are real and positive when $\nu > -1$ and $z > 0$. The Bessel function $I_\nu(z)$ in series form is given by

$$I_\nu(z) = (\tfrac{1}{2}z)^\nu \sum_{k=0}^{\infty} \frac{(\tfrac{1}{2}z)^{2k}}{k!\Gamma(\nu + k + 1)} \tag{6}$$

When ν *is not zero or not a positive integer*, the general solutions (2) and (5) can be taken, respectively, in the form

$$R(z) = c_1 J_\nu(z) + c_2 J_{-\nu}(z) \tag{7a}$$

$$R(z) = c_1 I_\nu(z) + c_2 I_{-\nu}(z) \tag{7b}$$

When $\nu = n$ is a positive integer, the solutions $J_n(z)$ and $J_{-n}(z)$ are not independent; they are related by

$$J_n(z) = (-1)^n J_{-n}(z) \qquad \text{and} \qquad J_{-n}(z) = J_n(-z)(n = \text{integer}) \tag{8}$$

similarly, when $\nu = n$ is a positive integer, the solutions $I_n(z)$ and $I_{-n}(z)$ are not independent.

We summarize various forms of solutions of equation (1) as [2]

$$R(z) = c_1 J_\nu(z) + c_2 Y_\nu(z) \qquad \text{always} \tag{9a}$$

$$R(z) = c_1 J_\nu(z) + c_2 J_{-\nu}(z) \qquad \nu \text{ is not zero or a positive integer} \tag{9b}$$

and the solutions of equation (4) as [2]

$$R(z) = c_1 I_\nu(z) + c_2 K_\nu(z) \qquad \text{always} \tag{10a}$$

$$R(z) = c_1 I_\nu(z) + c_2 I_{-\nu}(z) \qquad \nu \text{ is not zero or positive integer} \tag{10b}$$

GENERALIZED BESSEL EQUATION

Sometimes a given differential equation, after suitable transformation of the independent variable, yields a solution that is a linear combination of Bessel functions. A convenient way of finding out whether a given differential equation possesses a solution in terms of Bessel functions is to compare it with the *generalized Bessel equation* developed by Douglas [in Ref. 4, p. 210]

$$\frac{d^2R}{dx^2}+\left[\frac{1-2m}{x}-2\alpha\right]\cdot\frac{dR}{dx}+\left[p^2a^2x^{2p-2}+\alpha^2+\frac{\alpha(2m-1)}{x}+\frac{m^2-p^2\nu^2}{x^2}\right]R=0 \tag{11a}$$

and the corresponding solution of which is

$$R=x^m\cdot e^{\alpha x}[c_1J_\nu(ax^p)+c_2Y_\nu(ax^p)] \tag{11b}$$

where c_1 and c_2 are arbitrary constants.

For example, by comparing the differential equation

$$\frac{d^2R}{dx^2}+\frac{1}{x}\frac{dR}{dx}-\frac{\beta}{x}R-0 \tag{12}$$

with the above generalized Bessel equation we find

$$\alpha=0,\quad m=0,\quad p=\tfrac{1}{2},\quad p^2\nu^2=-\beta,\quad a=2i\sqrt{\beta},\quad \nu=0$$

Hence, the solution of differential equation (12) is in the form

$$R=c_1J_0(2i\sqrt{\beta x})+c_2Y_0(2i\sqrt{\beta x}) \tag{13a}$$

or

$$R=c_1I_0(2\sqrt{\beta x})+c_2K_0(2\sqrt{\beta x}) \tag{13b}$$

which involves Bessel functions.

LIMITING FORM FOR SMALL *Z*

For small values of $z(z\to 0)$, the retention of the leading terms in the series results in the following approximations for the values of Bessel functions [5, p. 360]

$$J_\nu(z)\cong(\tfrac{1}{2}z)^\nu\frac{1}{\Gamma(\nu+1)}\qquad \nu\neq-1,-2,-3\ldots \tag{14a}$$

$$Y_\nu(z)\cong-\frac{1}{\pi}\left(\frac{2}{z}\right)^\nu\Gamma(\nu)\qquad \nu\neq 0\quad\text{and}\quad Y_0(z)\cong\frac{2}{\pi}\ln z \tag{14b}$$

$$I_z(z)\cong(\tfrac{1}{2}z)^\nu\frac{1}{\Gamma(\nu+1)}\qquad \nu\neq-1,-2,-3\ldots \tag{15a}$$

$$K_\nu(z)\cong\frac{1}{2}\left(\frac{2}{z}\right)^\nu\Gamma(\nu)\qquad \nu\neq 0\quad\text{and}\quad K_0(z)\cong-\ln z \tag{15b}$$

LIMITING FORM FOR LARGE Z

For large values of $z(z \to \infty)$ the values of Bessel functions can be approximated as [5, pp. 364, 377]

$$J_v(z) \cong \sqrt{\frac{2}{\pi z}} \cdot \cos\left(z - \frac{\pi}{4} - \frac{v\pi}{2}\right) \tag{16a}$$

$$Y_v(z) \cong \sqrt{\frac{2}{\pi z}} \cdot \sin\left(z - \frac{\pi}{4} - \frac{v\pi}{4}\right) \tag{16b}$$

$$I_v(z) \cong \frac{e^z}{\sqrt{2\pi z}} \quad \text{and} \quad K_v(z) \cong \sqrt{\frac{\pi}{2z}} \cdot e^{-z} \tag{16c}$$

DERIVATIVES OF BESSEL FUNCTIONS [3, pp. 161–163]

$$\frac{d}{dz}[z^v W_v(\beta z)] = \begin{cases} \beta z^v W_{v-1}(\beta z) & \text{for} \quad W \equiv J, Y, I & \text{(17a)} \\ -\beta z^v W_{v-1}(\beta z) & \text{for} \quad W \equiv K & \text{(17b)} \end{cases}$$

$$\frac{d}{dz}[z^{-v} W_v(\beta z)] = \begin{cases} -\beta z^{-v} W_{v+1}(\beta z) & \text{for} \quad W \equiv J, Y, K & \text{(18a)} \\ \beta z^{-v} W_{v+1}(\beta z) & \text{for} \quad W \equiv I & \text{(18b)} \end{cases}$$

For example, by setting $v = 0$, we obtain

$$\frac{d}{dz}[W_0(\beta z)] = \begin{cases} -\beta W_1(\beta z) & \text{for} \quad W \equiv J, Y, K & \text{(19a)} \\ \beta W_1(\beta z) & \text{for} \quad W = I & \text{(19b)} \end{cases}$$

INTEGRATION OF BESSEL FUNCTIONS

$$\int z^v W_{v-1}(\beta z) dz = \frac{1}{\beta} z^v W_v(\beta z) \qquad \text{for} \quad W \equiv J, Y, I \tag{20}$$

$$\int \frac{1}{z^v} W_{v+1}(\beta z) dz = -\frac{1}{\beta z^v} W_v(\beta z) \qquad \text{for} \quad W \equiv J, Y, K \tag{21}$$

For example, by setting $v = 1$ in equation (20), are obtain

$$\int z W_0(\beta z) dz = \frac{1}{\beta} z W_1(\beta z) \qquad \text{for} \quad W \equiv J, Y, I \tag{22}$$

Infinite integrals involving Bessel functions are [1, pp. 394–395]

$$\int_0^\infty e^{-pz^2} z^{\nu+1} J_\nu(az)\,dz = \frac{a^\nu}{(2p)^{\nu+1}} e^{-a^2/4p} \tag{23}$$

$$\int_0^\infty e^{-pz^2} z J_\nu(az) J_\nu(bz)\,dz = \frac{1}{2p} e^{-(a^2+b^2)/4p} I_\nu\left(\frac{ab}{2p}\right) \tag{24}$$

The indefinite integral of the square of Bessel functions is given by [1, p. 135; 2, p. 110]

$$\int r G_\nu^2(\beta r)\,dr = \tfrac{1}{2} r^2 [G_\nu^2(\beta r) - G_{\nu-1}(\beta r) G_{\nu+1}(\beta r)] \tag{25a}$$

$$= \tfrac{1}{2} r^2 \left[G_\nu'^2(\beta r) + \left(1 - \frac{\nu^2}{\beta^2 r^2}\right) G_\nu^2(\beta r) \right] \tag{25b}$$

where $G_\nu(\beta r)$ is any Bessel function of the first or second kind of order ν.

The indefinite integral of the product of two Bessel functions can be expressed in the form [9, equation 9]

$$\int r G_\nu(\beta r) \bar{G}_\nu(\beta r)\,dr = \frac{r^2}{2} \left\{ G_\nu'(\beta r) \bar{G}_\nu'(\beta r) + \left[1 - \left(\frac{\nu}{\beta r}\right)^2\right] G_\nu(\beta r) \bar{G}_\nu(\beta r) \right\} \tag{26a}$$

or in the form [1, p. 134; 2, p. 110]

$$\int r G_\nu(\beta r) \bar{G}^\nu(\beta r)\,dr = \tfrac{1}{4} r^2 [2 G_\nu(\beta r) \bar{G}_\nu(\beta r) - G_{\nu-1}(\beta r) \bar{G}_{\nu+1}(\beta r)$$
$$- G_{\nu+1}(\beta r) \bar{G}_{\nu-1}(\beta r)] \tag{26b}$$

where $G_\nu(\beta r)$ and $\bar{G}_\nu(\beta r)$ can be any Bessel function of the first or second kind. We note that equations (25a,b) are special cases of the integrals (26a,b).

WRONSKIAN RELATIONSHIP

The wronskian relationship for the Bessel functions

$$J_\nu(\beta r) Y_\nu'(\beta r) - Y_\nu(\beta r) J_\nu'(\beta r) = \frac{2}{\pi \beta r} \tag{27}$$

is useful in the simplification of expressions involving Bessel functions.

TABLE IV-1 Numerical Values of Bessel Functions

$J_0(z)$

z	0	0.1	0.2	0.3	0.4	0.5	0.6	0.7	0.8	0.9
0	1.0000	0.9975	0.9900	0.9776	0.9604	0.9385	0.9120	0.8812	0.8463	0.8075
1	0.7652	0.7196	0.6711	0.6201	0.5669	0.5118	0.4554	0.3980	0.3400	0.2818
2	0.2239	0.1666	0.1104	0.0555	0.0025	−0.0484	−0.0968	−0.1424	−0.1850	−0.2243
3	−0.2601	−0.2921	−0.3202	−0.3443	−0.3643	−0.3801	−0.3918	−0.3992	−0.4026	−0.4018
4	−0.3971	−0.3887	−0.3766	−0.3610	−0.3423	−0.3205	−0.2961	−0.2693	−0.2404	−0.2097
5	−0.1776	−0.1443	−0.1103	−0.0758	−0.0412	−0.0068	+0.0270	+0.0599	0.0917	0.1220
6	0.1506	0.1773	0.2017	0.2238	0.2433	0.2601	0.2740	0.2851	0.2931	0.2981
7	0.3001	0.2991	0.2951	0.2882	0.2786	0.2663	0.2516	0.2346	0.2154	0.1944
8	0.1717	0.1475	0.1222	0.0960	0.0692	0.0419	0.0146	−0.0125	−0.0392	−0.0653
9	−0.0903	−0.1142	−0.1367	−0.1577	−0.1768	−0.1939	−0.2090	−0.2218	−0.2323	−0.2403
10	−0.2459	−0.2490	−0.2496	−0.2477	−0.2434	−0.2366	−0.2276	−0.2164	−0.2032	−0.1881
11	−0.1712	−0.1528	−0.1330	−0.1121	−0.0902	−0.0677	−0.0446	−0.0213	+0.0020	0.0250
12	0.0477	0.0697	0.0908	0.1108	0.1296	0.1469	0.1626	0.1766	0.1887	0.1988
13	0.2069	0.2129	0.2167	0.2183	0.2177	0.2150	0.2101	0.2032	0.1943	0.1836
14	0.1711	0.1570	0.1414	0.1245	0.1065	0.0875	0.0679	0.0476	0.0271	0.0064
15	−0.0142	−0.0346	−0.0544	−0.0736	−0.0919	−0.1092	−0.1253	−0.1401	−0.1533	−0.1650

When $z > 15.9$, $J_0(z) \simeq \sqrt{\left(\dfrac{2}{\pi z}\right)}\left\{\sin(z + \tfrac{1}{4}\pi) + \dfrac{1}{8z}\sin(z - \tfrac{1}{4}\pi)\right\}$

$$\simeq \frac{0.7979}{\sqrt{z}}\left\{\sin(57.296z + 45)^\circ + \frac{1}{8z}\sin(57.296z - 45)^\circ\right\}$$

TABLE IV-1 *(Continued)*

$J_1(z)$

z	0	0.1	0.2	0.3	0.4	0.5	0.6	0.7	0.8	0.9
0	0.0000	0.0499	0.0995	0.1483	0.1960	0.2423	0.2867	0.3290	0.3688	0.4059
1	0.4401	0.4709	0.4983	0.5220	0.5419	0.5579	0.5699	0.5778	0.5815	0.5812
2	0.5767	0.5683	0.5560	0.5399	0.5202	0.4971	0.4708	0.4416	0.4097	0.3754
3	0.3391	0.3009	0.2613	0.2207	0.1792	0.1374	0.0955	0.0538	0.0128	−0.0272
4	−0.0660	−0.1033	−0.1386	−0.1719	−0.2028	−0.2311	−0.2566	−0.2791	−0.2985	−0.3147
5	−0.3276	−0.3371	−0.3432	−0.3460	−0.3453	−0.3414	−0.3343	−0.3241	−0.3110	−0.2951
6	−0.2767	−0.2559	−0.2329	−0.2081	−0.1816	−0.1538	−0.1250	−0.0953	−0.0652	−0.0349
7	−0.0047	+0.0252	0.0543	0.0826	0.1096	0.1352	0.1592	0.1813	0.2014	0.2192
8	0.2346	0.2476	0.2580	0.2657	0.2708	0.2731	0.2728	0.2697	0.2641	0.2559
9	0.2453	0.2324	0.2174	0.2004	0.1816	0.1613	0.1395	0.1166	0.0928	0.0684
10	0.0435	0.0184	−0.0066	−0.0313	−0.0555	−0.0789	−0.1012	−0.1224	−0.1422	−0.1603
11	−0.1768	−0.1913	−0.2039	−0.2143	−0.2225	−0.2284	−0.2320	−0.2333	−0.2323	−0.2290
12	−0.2234	−0.2157	−0.2060	−0.1943	−0.1807	−0.1655	−0.1487	−0.1307	−0.1114	−0.0912
13	−0.0703	−0.0489	−0.0271	−0.0052	+0.0166	0.0380	0.0590	0.0791	0.0984	0.1165
14	0.1334	0.1488	0.1626	0.1747	0.1850	0.1934	0.1999	0.2043	0.2066	0.2069
15	0.2051	0.2013	0.1955	0.1879	0.1784	0.1672	0.1544	0.1402	0.1247	0.1080

When $z > 15.9$, $J_1(z) \simeq \sqrt{\left(\frac{2}{\pi z}\right)}\left\{\sin\left(z - \tfrac{1}{4}\pi\right) + \frac{3}{8z}\sin\left(z + \tfrac{1}{4}\pi\right)\right\}$

$$\simeq \frac{0.7979}{\sqrt{z}}\left\{\sin(57.296z - 45)^\circ + \frac{3}{8z}\sin(57.296z + 45)^\circ\right\}$$

$Y_0(z)$

z	0	0.1	0.2	0.3	0.4	0.5	0.6	0.7	0.8	0.9
0	$-\infty$	−1.5342	−1.0811	−0.8073	−0.6060	−0.4445	−0.3085	−0.1907	−0.0868	+0.0056
1	0.0883	0.1622	0.2281	0.2865	0.3379	0.3824	0.4204	0.4520	0.4774	0.4968
2	0.5104	0.5183	0.5208	0.5181	0.5104	0.4981	0.4813	0.4605	0.4359	0.4079
3	0.3769	0.3431	0.3071	0.2691	0.2296	0.1890	0.1477	0.1061	0.0645	0.0234
4	−0.0169	−0.0561	−0.0938	−0.1296	−0.1633	−0.1947	−0.2235	−0.2494	−0.2723	−0.2921
5	−0.3085	−0.3216	−0.3313	−0.3374	−0.3402	−0.3395	−0.3354	−0.3282	−0.3177	−0.3044
6	−0.2882	−0.2694	−0.2483	−0.2251	−0.1999	−0.1732	−0.1452	−0.1162	−0.0864	−0.0563
7	−0.0259	+0.0042	0.0339	0.0628	0.0907	0.1173	0.1424	0.1658	0.1872	0.2065
8	0.2235	0.2381	0.2501	0.2595	0.2662	0.2702	0.2715	0.2700	0.2659	0.2592
9	0.2499	0.2383	0.2245	0.2086	0.1907	0.1712	0.1502	0.1279	0.1045	0.0804
10	0.0557	0.0307	0.0056	−0.0193	−0.0437	−0.0675	−0.0904	−0.1122	−0.1326	−0.1516
11	−0.1688	−0.1843	−0.1977	−0.2091	−0.2183	−0.2252	−0.2299	−0.2322	−0.2322	−0.2298
12	−0.2252	−0.2184	−0.2095	−0.1986	−0.1858	−0.1712	−0.1551	−0.1375	−0.1187	−0.0989
13	−0.0782	−0.0569	−0.0352	−0.0134	+0.0085	+0.0301	+0.0512	0.0717	0.0913	0.1099
14	0.1272	0.1431	0.1575	0.1703	0.1812	0.1903	0.1974	0.2025	0.2056	0.2065
15	0.2055	0.2023	0.1972	0.1902	0.1813	0.1706	0.1584	0.1446	0.1295	0.1132

When $z > 15.9$, $Y_0(z) \simeq \sqrt{\left(\frac{2}{\pi z}\right)}\left\{\sin(z - \frac{1}{4}\pi) - \frac{1}{8z}\sin(z + \frac{1}{4}\pi)\right\}$

$$\simeq \frac{0.7979}{\sqrt{z}}\left\{\sin(57.296z - 45)^\circ - \frac{1}{8z}\sin(57.296z + 45)^\circ\right\}$$

TABLE IV-1 *(Continued)*

$Y_1(z)$

z	0	0.1	0.2	0.3	0.4	0.5	0.6	0.7	0.8	0.9
0	$-\infty$	−6.4590	−3.3238	−2.2931	−1.7809	−1.4715	−1.2604	−1.1032	−0.9781	−0.8731
1	−0.7812	−0.6981	−0.6211	−0.5485	−0.4791	−0.4123	−0.3476	−0.2847	−0.2237	−0.1644
2	−0.1070	−0.0517	+0.0015	+0.0523	0.1005	0.1459	0.1884	0.2276	0.2635	0.2959
3	0.3247	0.3496	0.3707	0.3879	0.4010	0.4102	0.4154	0.4167	0.4141	0.4078
4	0.3979	0.3846	0.3680	0.3484	0.3260	0.3010	0.2737	0.2445	0.2136	0.1812
5	0.1479	0.1137	0.0792	0.0445	0.0101	−0.0238	−0.0568	−0.0887	−0.1192	−0.1481
6	−0.1750	−0.1998	−0.2223	−0.2422	−0.2596	−0.2741	−0.2857	−0.2945	−0.3002	−0.3029
7	−0.3027	−0.2995	−0.2934	−0.2846	−0.2731	−0.2591	−0.2428	−0.2243	−0.2039	−0.1817
8	−0.1581	−0.1331	−0.1072	−0.0806	−0.0535	−0.0262	+0.0011	+0.0280	0.0544	0.0799
9	+0.1043	0.1275	0.1491	0.1691	0.1871	0.2032	0.2171	0.2287	0.2379	0.2447
10	0.2490	0.2508	0.2502	0.2471	0.2416	0.2337	0.2236	0.2114	0.1973	0.1813
11	0.1637	0.1446	0.1243	0.1029	0.0807	0.0579	0.0348	0.0114	−0.0118	−0.0347
12	−0.0571	−0.0787	−0.0994	−0.1189	−0.1371	−0.1538	−0.1589	−0.1821	−0.1935	−0.2028
13	−0.2101	−0.2152	−0.2182	−0.2190	−0.2176	−0.2140	−0.2084	−0.2007	−0.1912	−0.1798
14	−0.1666	−0.1520	−0.1359	−0.1186	−0.1003	−0.0810	−0.0612	−0.0408	−0.0202	+0.0005
15	0.0211	0.0413	0.0609	0.0799	0.0979	0.1148	0.1305	0.1447	0.1575	0.1686

When $z > 15.9$, $$Y_1(z) \simeq \sqrt{\left(\frac{2}{\pi z}\right)}\left\{\sin\left(z - \tfrac{3}{4}\pi\right) + \frac{3}{8z}\sin\left(z - \tfrac{1}{4}\pi\right)\right\}$$

$$\simeq \frac{0.7979}{\sqrt{z}}\left\{\sin(57.296z - 135)^\circ + \frac{3}{8z}\sin(57.296z - 45)^\circ\right\}$$

$I_0(z)$

	z	0	0.1	0.2	0.3	0.4	0.5	0.6	0.7	0.8	0.9
	0	1.0000	1.0025	1.0100	1.0226	1.0404	1.0635	1.0920	1.1263	1.1665	1.2130
	1	1.2661	1.3262	1.3937	1.4693	1.5534	1.6467	1.7500	1.8640	1.9896	2.1277
	2	2.2796	2.4463	2.6291	2.8296	3.0493	3.2898	3.5533	3.8417	4.1573	4.5027
	3	4.8808	5.2945	5.7472	6.2426	6.7848	7.3782	8.0277	8.7386	9.5169	10.369
$10 \times$	4	1.1302	1.2324	1.3442	1.4668	1.6010	1.7481	1.9093	2.0858	2.2794	2.4915
$10 \times$	5	2.7240	2.9789	3.2584	3.5648	3.9009	4.2695	4.6738	5.1173	5.6038	6.1377
$10 \times$	6	6.7234	7.3663	8.0718	8.8462	9.6962	10.629	11.654	12.779	14.014	15.370
$10^2 \times$	7	1.6859	1.8495	2.0292	2.2266	2.4434	2.6816	2.9433	3.2309	3.5468	3.8941
$10^2 \times$	8	4.2756	4.6950	5.1559	5.6626	6.2194	6.8316	7.5046	8.2445	9.0580	9.9524
$10^3 \times$	9	1.0936	1.2017	1.3207	1.4514	1.5953	1.7535	1.9275	2.1189	2.3294	2.5610

When $z \geqslant 10, I_0(z) \simeq \dfrac{0.3989\epsilon^z}{z^{1/2}}\left\{1 + \dfrac{1}{8z} + \dfrac{9}{128z^2} + \dfrac{75}{1024z^3}\right\}$.

$K_0(z)$

	z	0	0.1	0.2	0.3	0.4	0.5	0.6	0.7	0.8	0.9
	0	∞	2.4271	1.7527	1.3725	1.1145	0.9244	0.7775	0.6605	0.5653	0.4867
	1	0.4210	0.3656	0.3185	0.2782	0.2437	0.2138	0.1880	0.1655	0.1459	0.1288
$10^{-1} \times$	2	1.1389	1.0078	0.8926	0.7914	0.7022	0.6235	0.5540	0.4926	0.4382	0.3901
$10^{-1} \times$	3	0.3474	0.3095	0.2759	0.2461	0.2196	0.1960	0.1750	0.1563	0.1397	0.1248
$10^{-2} \times$	4	1.1160	0.9980	0.8927	0.7988	0.7149	0.6400	0.5730	0.5132	0.4597	0.4119
$10^{-2} \times$	5	0.3691	0.3308	0.2966	0.2659	0.2385	0.2139	0.1918	0.1721	0.1544	0.1386
$10^{-3} \times$	6	1.2440	1.1167	1.0025	0.9001	0.8083	0.7259	0.6520	0.5857	0.5262	0.4728
$10^{-3} \times$	7	0.4248	0.3817	0.3431	0.3084	0.2772	0.2492	0.2240	0.2014	0.1811	0.1629
$10^{-4} \times$	8	1.4647	1.3173	1.1849	1.0658	0.9588	0.8626	0.7761	0.6983	0.6283	0.5654
$10^{-4} \times$	9	0.5088	0.4579	0.4121	0.3710	0.3339	0.3006	0.2706	0.2436	0.2193	0.1975

When $z \geqslant 10,\ K_0(z) \simeq \dfrac{1.2533e^{-z}}{z^{1/2}}\left\{1 - \dfrac{1}{8z} + \dfrac{9}{128z^2} - \dfrac{75}{1024z^3}\right\}$.

TABLE IV-1 *(Continued)*

$I_1(z)$

	z	0	0.1	0.2	0.3	0.4	0.5	0.6	0.7	0.8	0.9
	0	0	0.0501	0.1005	0.1517	0.2040	0.2579	0.3137	0.3719	0.4329	0.4971
	1	0.5652	0.6375	0.7147	0.7973	0.8861	0.9817	1.0848	1.1963	1.3172	1.4482
	2	1.5906	1.7455	1.9141	2.0978	2.2981	2.5167	2.7554	3.0161	3.3011	3.6126
	3	3.9534	4.3262	4.7343	5.1810	5.6701	6.2058	6.7927	7.4357	8.1404	8.9128
$10\times$	4	0.9759	1.0688	1.1706	1.2822	1.4046	1.5389	1.6863	1.8479	2.0253	2.2199
$10\times$	5	2.4336	2.6680	2.9254	3.2080	3.5182	3.8588	4.2328	4.6436	5.0946	5.5900
$10\times$	6	6.1342	6.7319	7.3886	8.1100	8.9026	9.7735	10.730	11.782	12.938	14.208
$10^2\times$	7	1.5604	1.7138	1.8825	2.0679	2.2717	2.4958	2.7422	3.0131	3.3110	3.6385
$10^2\times$	8	3.9987	4.3948	4.8305	5.3096	5.8366	6.4162	7.0538	7.7551	8.5266	9.3754
$10^3\times$	9	1.0309	1.1336	1.2467	1.3710	1.5079	1.6585	1.8241	2.0065	2.2071	2.4280

When $z \geqslant 10$, $I_1(z) \simeq \dfrac{0.3989e^z}{z^{1/2}}\left\{1 - \dfrac{3}{8z} - \dfrac{15}{128z^2} - \dfrac{105}{1024z^3}\right\}$.

$K_1(z)$

	z	0	0.1	0.2	0.3	0.4	0.5	0.6	0.7	0.8	0.9
	0	∞	9.8538	4.7760	3.0560	2.1844	1.6564	1.3028	1.0503	0.8618	0.7165
	1	0.6019	0.5098	0.4346	0.3725	0.3208	0.2774	0.2406	0.2094	0.1826	0.1597
$10^{-1}\times$	2	1.3987	1.2275	1.0790	0.9498	0.8372	0.7389	0.6528	0.5774	0.5111	0.4529
$10^{-1}\times$	3	0.4016	0.3563	0.3164	0.2812	0.2500	0.2224	0.1979	0.1763	0.1571	0.1400
$10^{-2}\times$	4	1.2484	1.1136	0.9938	0.8872	0.7923	0.7078	0.6325	0.5654	0.5055	0.4521
$10^{-2}\times$	5	0.4045	0.3619	0.3239	0.2900	0.2597	0.2326	0.2083	0.1866	0.1673	0.1499
$10^{-3}\times$	6	1.3439	1.2050	1.0805	0.9691	0.8693	0.7799	0.6998	0.6280	0.5636	0.5059
$10^{-3}\times$	7	0.4542	0.4078	0.3662	0.3288	0.2953	0.2653	0.2383	0.2141	0.1924	0.1729
$10^{-4}\times$	8	1.5537	1.3964	1.2552	1.1283	1.0143	0.9120	0.8200	0.7374	0.6631	0.5964
$10^{-4}\times$	9	0.5364	0.4825	0.4340	0.3904	0.3512	0.3160	0.2843	0.2559	0.2302	0.2072

When $z \geqslant 10$, $K_1(z) \simeq \dfrac{1.2533e^{-z}}{z^{1/2}}\left\{1 + \dfrac{3}{8z} - \dfrac{15}{128z^2} + \dfrac{105}{1024z^3}\right\}$.

RECURRENCE RELATIONS

The recurrence formulae for the Bessel functions are given as [1, pp. 45 and 66; 5, p. 361]

$$W_{\nu-1}(z) + W_{\nu+1}(z) = \frac{2\nu}{z} W_\nu(z) \tag{28a}$$

$$W_{\nu-1}(z) - W_{\nu+1}(z) = 2W'_\nu(z) \tag{28b}$$

$$W_{\nu-1}(z) - \frac{\nu}{z} W_\nu(z) = W_\nu(z) \tag{28c}$$

$$-W_{\nu+1}(z) + \frac{\nu}{z} W_\nu(z) = W'_\nu(z) \tag{28d}$$

where $W = J$ or Y or any linear combination of these functions, the coefficients in which are independent of z and ν.

A systematic tabulation of various integrals involving Bessel functions is given in references 6 and 7.

In Table IV-1 we present the numerical values of $J_n(z)$, $Y_n(z)$, $I_n(z)$, and $K_n(z)$ functions for $n = 0$ and 1 [2, pp. 215–221], and in Table IV-2 we present the first 10 roots of $J_n(z)$ function for $n = 0, 1, 2, 3, 5$.

Finally, in Tables IV-3 and IV-4 we present the roots of $\beta J_1(\beta) - cJ_0(\beta) = 0$ and $J_0(\beta)Y_0(c\beta) - Y_0(\beta)J_0(c\beta) = 0$, respectively [8, p. 493; 5, pp. 414–415].

TABLE IV-2 First 10 Roots of $J_n(z) = 0$ $n = 0, 1, 2, 3, 4, 5$

	J_0	J_1	J_2	J_3	J_4	J_5
1	2.4048	3.8317	5.1356	6.3802	7.5883	8.7715
2	5.5201	7.0156	8.4172	9.7610	11.0647	12.3386
3	8.6537	10.1735	11.6198	13.0152	14.3725	15.7002
4	11.7915	13.3237	14.7960	16.2235	17.6160	18.9801
5	14.9309	16.4706	17.9598	19.4094	20.8269	22.2178
6	18.0711	19.6159	21.1170	22.5827	24.0190	25.4303
7	21.2116	22.7601	24.2701	25.7482	27.1991	28.6266
8	24.3525	25.9037	27.4206	28.9084	30.3710	31.8117
9	27.4935	29.0468	30.5692	32.0649	33.5371	34.9888
10	30.6346	32.1897	33.7165	35.2187	36.6990	38.1599

TABLE IV-3 First Six Roots of $\beta J_1(\beta) - cJ_0(\beta) = 0$

c	β_1	β_2	β_3	β_4	β_5	β_6
0	0	3.8317	7.0156	10.1735	13.3237	16.4706
0.01	0.1412	3.8343	7.0170	10.1745	13.3244	16.4712
0.02	0.1995	3.8369	7.0184	10.1754	13.3252	16.4718
0.04	0.2814	3.8421	7.0213	10.1774	13.3267	16.4731
0.06	0.3438	3.8473	7.0241	10.1794	13.3282	16.4743
0.08	0.3960	3.8525	7.0270	10.1813	13.3297	16.4755
0.1	0.4417	3.8577	7.0298	10.1833	13.3312	16.4767
0.15	0.5376	3.8706	7.0369	10.1882	13.3349	16.4797
0.2	0.6170	3.8835	7.0440	10.1931	13.3387	16.4828
0.3	0.7465	3.9091	7.0582	10.2029	13.3462	16.4888
0.4	0.8516	3.9344	7.0723	10.2127	13.3537	16.4949
0.5	0.9408	3.9594	7.0864	10.2225	13.3611	16.5010
0.6	1.0184	3.9841	7.1004	10.2322	13.3686	16.5070
0.7	1.0873	4.0085	7.1143	10.2419	13.3761	16.5131
0.8	1.1490	4.0325	7.1282	10.2516	13.3835	16.5191
0.9	1.2048	4.0562	7.1421	10.2613	13.3910	16.5251
1.0	1.2558	4.0795	7.1558	10.2710	13.3984	16.5312
1.5	1.4569	4.1902	7.2233	10.3188	13.4353	16.5612
2.0	1.5994	4.2910	7.2884	10.3658	13.4719	16.5910
3.0	1.7887	4.4634	7.4103	10.4566	13.5434	16.6499
4.0	1.9081	4.6018	7.5201	10.5423	13.6125	16.7073
5.0	1.9898	4.7131	7.6177	10.6223	13.6786	16.7630
6.0	2.0490	4.8033	7.7039	10.6964	13.7414	16.8168
7.0	2.0937	4.8772	7.7797	10.7646	13.8008	16.8684
8.0	2.1286	4.9384	7.8464	10.8271	13.8566	16.9179
9.0	2.1566	4.9897	7.9051	10.8842	13.9090	16.9650
10.0	2.1795	5.0332	7.9569	10.9363	13.9580	17.0099
15.0	2.2509	5.1773	8.1422	11.1367	14.1576	17.2008
20.0	2.2880	5.2568	8.2534	11.2677	14.2983	17.3442
30.0	2.3261	5.3410	8.3771	11.4221	14.4748	17.5348
40.0	2.3455	5.3846	8.4432	11.5081	14.5774	17.6508
50.0	2.3572	5.4112	8.4840	11.5621	14.6433	17.7272
60.0	2.3651	5.4291	8.5116	11.5990	14.6889	17.7807
80.0	2.3750	5.4516	8.5466	11.6461	14.7475	17.8502
100.0	2.3809	5.4652	8.5678	11.6747	14.7834	17.8931
∞	2.4048	5.5201	8.6537	11.7915	14.9309	18.0711

From Carslaw and Jaeger [8].

TABLE IV-4 First Five Roots of $J_0(\beta)Y_0(c\beta) - Y_0(\beta)J_0(c\beta) = 0$

c	β_1	β_2	β_3	β_4	β_5
1.2	15.7014	31.4126	47.1217	62.8302	78.5385
1.5	6.2702	12.5598	18.8451	25.1294	31.4133
2.0	3.1230	6.2734	9.4182	12.5614	15.7040
2.5	2.0732	4.1773	6.2754	8.3717	10.4672
3.0	1.5485	3.1291	4.7038	6.2767	7.8487
3.5	1.2339	2.5002	3.7608	5.0196	6.2776
4.0	1.0244	2.0809	3.1322	4.1816	5.2301

REFERENCES

1. G. N. Watson, *A Treatise on the Theory of Bessel Functions*, 2nd ed., Cambridge at the University Press, London, 1966.
2. N. W. McLachlan, *Bessel Functions for Engineers*, 2nd. ed., Oxford at the Clarendon Press, London, 1961.
3. F. B. Hildebrand, *Advanced Calculus for Engineers* Prentice-Hall, Englewood Cliffs, N.J., 1949.
4. T. K. Sherwood and C. E. Reed, *Applied Mathematics in Chemical Engineering*, McGraw-Hill, New York, 1939.
5. M. Abramowitz and I. A. Stegun, *Handbook of Mathematical Functions*, National Bureau of Standards, Applied Mathematic Series 55, U.S. Government Printing Office, Washington, D.C., 20402, 1964.
6. I. S. Gradshteyn and I. M. Ryzhik, *Table of Integrals, Series, and Products* (trans. from the Russian and ed. by A. Jeffrey), Academic press, New York, 1965.
7. Y. L. Luke, *Integrals of Bessel Functions*, McGraw-Hill, New York, 1962.
8. H. S. Carslaw and J. G. Jaeger, *Conduction of Heat in Solids*, Oxford at the Clarendon Press, London, 1959.
9. G. Cinelli, *Int. J. Eng. Sci.* **3**, 539–559, 1965.
10. E. C. Titchmarsh, *Eigenfunction Expansions*, Part I, Clarendon Press, London, 1962.

APPENDIX V

NUMERICAL VALUES OF LEGENDRE POLYNOMIALS OF THE FIRST KIND

x	$P_1(x)$	$P_2(x)$	$P_3(x)$	$P_4(x)$	$P_5(x)$	$P_6(x)$	$P_7(x)$
0.00	0.0000	−.5000	0.0000	0.3750	0.0000	−.3125	0.0000
.01	.0100	−.4998	−.0150	.3746	.0187	−.3118	−.0219
.02	.0200	−.4994	−.0300	.3735	.0374	−.3099	−.0436
.03	.0300	−.4986	−.0449	.3716	.0560	−.3066	−.0651
.04	.0400	−.4976	−.0598	.3690	.0744	−.3021	−0.862
.05	.0500	−.4962	−.0747	.3657	.0927	−.2962	−.1069
.06	.0600	−.4946	−.0895	.3616	.1106	−.2891	−.1270
.07	.0700	−.4926	−.1041	.3567	.1283	−.2808	−.1464
.08	.0800	−.4904	−.1187	.3512	.1455	−.2713	−.1651
.09	.0900	−.4878	−.1332	.3449	.1624	−.2606	−.1828
.10	.1000	−.4850	−.1475	.3379	.1788	−.2488	−.1995
.11	.1100	−.4818	−.1617	.3303	.1947	−.2360	−.2151
.12	.1200	−.4784	−.1757	.3219	.2101	−.2220	−.2295
.13	.1300	−.4746	−.1895	.3129	.2248	−.2071	−.2427
.14	.1400	−.4706	−.2031	.3032	.2389	−.1913	−.2545
.15	.1500	−.4662	−.2166	.2928	.2523	−.1746	−.2649
.16	.1600	−.4616	−.2298	.2819	.2650	−.1572	−.2738
.17	.1700	−.4566	−.2427	.2703	.2769	−.1389	−.2812
.18	.1800	−.4514	−.2554	.2581	.2880	−.1201	−.2870
.19	.1900	−.4458	−.2679	.2453	.2982	−.1006	−.2911
.20	.2000	−.4400	−.2800	.2320	.3075	−.0806	−.2935
.21	.2100	−.4338	−.2918	.2181	.3159	−.0601	−.2943
.22	.2200	−.4274	−.3034	.2037	.3234	−.0394	−.2933

x	$P_1(x)$	$P_2(x)$	$P_3(x)$	$P_4(x)$	$P_5(x)$	$P_6(x)$	$P_7(x)$
.23	.2300	−.4206	−.3146	.1889	.3299	−.0183	−.2906
.24	.2400	−.4136	−.3254	.1735	.3353	.0029	−.2861
.25	.2500	−.4062	−.3359	.1577	.3397	.0243	−.2799
.26	.2600	−.3986	−.3461	.1415	.3431	.0456	−.2720
.27	.2700	−.3906	−.3558	.1249	.3453	.0669	−.2625
.28	.2800	−.3824	−.3651	.1079	.3465	.0879	−.2512
.29	.2900	−.3738	−.3740	.0906	.3465	.1087	−.2384
.30	.3000	−.3650	−.3825	.0729	.3454	.1292	−.2241
.31	.3100	−.3558	−.3905	.0550	.3431	.1492	−.2082
.32	.3200	−.3464	−.3981	.0369	.3397	.1686	−.1910
.33	.3300	−.3366	−.4052	.0185	.3351	.1873	−.1724
.34	.3400	−.3266	−.4117	−.0000	.3294	.2053	−.1527
.35	.3500	−.3162	−.4178	−.0187	.3225	.2225	−.1318
.36	.3600	−.3056	−.4234	−.0375	.3144	.2388	−.1098
.37	.3700	−.2946	−.4284	−.0564	.3051	.2540	−.0870
.38	.3800	−.2834	−.4328	−.0753	.2948	.2681	−.0635
.39	.3900	−.2718	−.4367	−.0942	.2833	.2810	−.0393
.40	.4000	−.2600	−.4400	−.1130	.2706	.2926	−.0146
.41	.4100	−.2478	−.4427	−.1317	.2569	.3029	.0104
.42	.4200	−.2354	−.4448	−.1504	.2421	.3118	.0356
.43	.4300	−.2226	−.4462	−.1688	.2263	.3191	.0608
.44	.4400	−.2096	−.4470	−.1870	.2095	.3249	.0859
.45	.4500	−.1962	−.4472	−.2050	.1917	.3290	.1106
.46	.4600	−.1826	−.4467	−.2226	.1730	.3314	.1348
.47	.4700	−.1686	−.4454	−.2399	.1534	.3321	.1584
.48	.4800	−.1544	−.4435	−.2568	.1330	.3310	.1811
.49	.4900	−.1398	−.4409	−.2732	.1118	.3280	.2027
.50	.5000	−.1250	−.4375	−.2891	.0898	.3232	.2231
.51	.5100	−.1098	−.4334	−.3044	.0673	.3166	.2422
.52	.5200	−.0944	−.4258	−.3191	.0441	.3080	.2596
.53	.5300	−.0786	−.4228	−.3332	.0204	.2975	.2753
.54	.5400	−.0626	−.4163	−.3465	−.0037	.2851	.2891
.55	.5500	−.0462	−.4091	−.3590	−.0282	.2708	.3007
.56	.5600	−.0296	−.4010	−.3707	−.0529	.2546	.3102
.57	.5700	−.0126	−.3920	−.3815	−.0779	.2366	.3172
.58	.5800	.0046	−.3822	−.3914	−.1028	.2168	.3217
.59	.5900	.0222	−.3716	−.4002	−.1278	.1953	.3235
.60	.6000	.0400	−.3600	−.4080	−.1526	.1721	.3226
.61	.6100	.0582	−.3475	−.4146	−.1772	.1473	.3188
.62	.6200	.0766	−.3342	−.4200	−.2014	.1211	.3121
.63	.6300	.0954	−.3199	−.4242	−.2251	.0935	.3023
.64	.6400	.1144	−.3046	−.4270	−.2482	.0646	.2895

x	$P_1(x)$	$P_2(x)$	$P_3(x)$	$P_4(x)$	$P_5(x)$	$P_6(x)$	$P_7(x)$
.65	.6500	.1338	−.2884	−.4284	−.2705	.0347	.2737
.66	.6600	.1534	−.2713	−.4284	−.2919	.0038	.2548
.67	.6700	.1734	−.2531	−.4268	−.3122	−.0278	.2329
.68	.6800	.1936	−.2339	−.4236	−.3313	−.0601	.2081
.69	.6900	.2142	−.2137	−.4187	−.3490	−.0926	.1805
.70	.7000	.2350	−.1925	−.4121	−.3652	−.1253	.1502
.71	.7100	.2562	−.1702	−.4036	−.3796	−.1578	.1173
.72	.7200	.2776	−.1469	−.3933	−.3922	−.1899	.0822
.73	.7300	.2994	−.1225	−.3810	−.4026	−.2214	.0450
.74	.7400	.3214	−.0969	−.3666	−.4107	−.2518	.0061
.75	.7500	.3438	−.0703	−.3501	−.4164	−.2808	−.0342
.76	.7600	.3664	−.0426	−.3314	−.4193	−.3081	−.0754
.77	.7700	.3894	−.0137	−.3104	−.4193	−.3333	−.1171
.78	.7800	.4126	.0164	−.2871	−.4162	−.3559	−.1588
.79	.7900	.4362	.0476	−.2613	−.4097	−.3756	−.1999
.80	.8000	.4600	.0800	−.2330	−.3995	−.3918	−.2397
.81	.8100	.4842	.1136	−.2021	−.3855	−.4041	−.2774
.82	.8200	.5086	.1484	−.1685	−.3674	−.4119	−.3124
.83	.8300	.5334	.1845	−.1321	−.3449	−.4147	−.3437
.84	.8400	.5584	.2218	−.0928	−.3177	−.4120	−.3703
.85	.8500	.5838	.2603	−.0506	−.2857	−.4030	−.3913
.86	.8600	.6094	.3001	−.0053	−.2484	−.3872	−.4055
.87	.8700	.6354	.3413	.0431	−.2056	−.3638	−.4116
.88	.8800	.6616	.3837	.0947	−.1570	−.3322	−.4083
.89	.8900	.6882	.4274	.1496	−.1023	−.2916	−.3942
.90	.9000	.7150	.4725	.2079	−.0411	−.2412	−.3678
.91	.9100	.7422	.5189	.2698	.0268	−.1802	−.3274
.92	.9200	.7696	.5667	.3352	.1017	−.1077	−.2713
.93	.9300	.7974	.6159	.4044	.1842	−.0229	−.1975
.94	.9400	.8254	.6665	.4773	.2744	.0751	−.1040
.95	.9500	.8538	.7184	.5541	.3727	.1875	.0112
.96	.9600	.8824	.7718	.6349	.4796	.3151	.1506
.97	.9700	.9114	.8267	.7198	.5954	.4590	.3165
.98	.9800	.9406	.8830	.8089	.7204	.6204	.5115
.99	.9900	.9702	.9407	.9022	.8552	.8003	.7384
1.00	1.0000	1.0000	1.0000	1.0000	1.0000	1.0000	1.0000

From W. E. Byerly, *Fourier Series and Spherical, Cylindrical, and Ellipsoidal Harmonics*, Dover Publications, New York, 1959, pp. 280–281.

APPENDIX VI

SUBROUTINE TRISOL

to Solve Tridiagonal Systems by Thomas Algorithm

```
C ** MAIN PROGRAM THAT READS INPUT DATA AND CALLS
C     TRISOL TO SOLVE THE SYSTEM OF EQUATIONS BY
C     THOMAS ALGORITHM
C
C   M = Dimension of the matrix
C   A = Off-diagonal term (lower) NOTE:A(1)=0
C   B = Diagonal term
C   C = Off-Diagonal term (upper) NOTE:C(M)=0
C   D = on input - right hand-side
C         on output- solution
C
        IMPLICIT REAL*8(A-H,O-Z)
        DIMENSION A(10),B(10),C(10),D(10)
        READ(1,*) M
        READ(1,*) (A(I),I=1,M)
        READ(1,*) (B(I),I=1,M)
        READ(1,*) (C(I),I=1,M)
        READ(1,*) (D(I),I=1,M)
        CALL TRISOL(M,A,B,C,D)
        DO 10 I=1,M
     10 WRITE(3,100)D(I)
    100 FORMAT(F10.6)
        STOP
        END
C
C ** SUBROUTINE TRISOL uses the Thomas algorithm
C     to solve a tri-diagonal matrix equation
C
C   M = Dimension of the matrix
C   A = Off-diagonal term (lower)
C   B = Diagonal term
C   C = Off-Diagonal term (upper)
C   D = on input - right-hand-side
C         on output- solution
C
        SUBROUTINE TRISOL(M,A,B,C,D)
        IMPLICIT REAL*8(A-H,O-Z)
        DIMENSION A(1),B(1),C(1),D(1)
```

```
C
C ESTABLISH UPPER TRIANGULAR MATRIX
C
      DO 10 I=2,M
      R=A(I)/B(I-1)
      B(I)=B(I)-R*C(I-1)
   10 D(I)=D(I)-R*D(I-1)
C
C BACK SUBSTITUTION
C
      D(M)=D(M)/B(M)
      DO 20 I=2,M
      J=M-I+1
   20 D(J)=(D(J)-C(J)*D(J+1))/B(J)
C
C SOLUTION STORED IN D
C
      RETURN
      END
```

APPENDIX VII

PROPERTIES OF DELTA FUNCTIONS

The symbol $\delta(x)$, known as Dirac's delta function, is zero for every value of x except the origin $x = 0$ where it is infinite in such a way that

$$\int_{-\infty}^{\infty} \delta(x)\, dx = 1 \tag{1}$$

Such a definition is not meaningful in the true mathematical sense, but the theory of distributions justifies its use as well as its derivatives. Then $\delta(x)$ has the following properties

$$\delta(x - b) = 0 \qquad \text{everywhere } x \neq b \tag{2}$$

For every continuous function $F(x)$ we write

$$\int_{-\infty}^{\infty} F(x)\delta(x - b)dx = F(b) \tag{3}$$

$$\int_{-\infty}^{\infty} F(x)\delta(x - 0)\, dx = F(0) \tag{4}$$

$$F(x)\delta(x - b) = F(b)\delta(x - b) \tag{5}$$

Derivatives of Delta Function

$$\int_{-\infty}^{\infty} F(x)\delta'(x)dx = -\int_{-\infty}^{\infty} F'(x)\delta(x)dx = -F'(0) \tag{6}$$

$$\int_{-\infty}^{\infty} F(x)\delta''(x)dx = -\int_{-\infty}^{\infty} F'(x)\delta'(x)\,dx = -F''(0) \tag{7}$$

The delta function itself is the derivative of the step function $H(x)$, that is

$$H'(x) = \delta(x) \tag{8}$$

where

$$\begin{aligned} H(x) &= 1 \quad \text{for} \quad x > 0 \\ &= 0 \quad \text{for} \quad x < 0 \end{aligned} \tag{9}$$

Note that the derivative of $H(x)$ is zero for $x < 0$, zero for $x > 0$, and undefined for $x = 0$.

INDEX